Carbohydrate Chemistry

Volume 33

A Specialist Periodical Report

Carbohydrate Chemistry

Monosaccharides, Disaccharides and Specific Oligosaccharides

Volume 33

A Review of the Literature Published during 1999

Senior Reporter

R.J. Ferrier, *Industrial Research Limited, Lower Hutt, New Zealand*

Reporters

R. Blattner, *Industrial Research Limited, Lower Hutt, New Zealand*

R.A. Field, *University of St.Andrews, St.Andrews, UK*

R.H. Furneaux, *Industrial Research Limited, Lower Hutt, New Zealand*

J.M. Gardiner, *UMIST, Manchester, UK*

J.O. Hoberg, *Victoria University of Wellington, Wellington, New Zealand*

K.P.R. Kartha, *University of St.Andrews, St.Andrews, UK*

D.M.G. Tilbrook, *University of Western Australia, Nedlands, Australia*

P.C. Tyler, *Industrial Research Limited, Lower Hutt, New Zealand*

R.H. Wightman, *Heriot-Watt University, Edinburgh, UK*

ISBN 0-85404-233-4
ISSN 0951-8428

A catalogue record of this book is available from the British Library

Published by The Royal Society of Chemistry,
Thomas Graham House, Science Park, Milton Road, Cambridge CB4 0WF, UK
Registered Charity Number 207890

For further information see our web site at www.rsc.org

Typeset by Computape (Pickering) Ltd, Pickering, North Yorkshire, UK
Printed by Athenaeum Press Ltd, Gateshead, Tyne and Wear, UK

Preface

Yet again our team has completed this volume well behind schedule, and is all too aware of the inappropriate delay in reporting on the literature of our field published in the last (or second last?) year of the millennium. Half of our members are under undue work-related stress, and it is unreasonable to expect them to continue in these circumstances – especially with an ever-expanding and increasingly complex job. Volume 34, covering the literature for the year 2000, may therefore be the last of the current series – unless of course less pressurized volunteers can be found to allow it to continue. It was interesting to note at a recent meeting of Senior Reporters of the Specialist Periodical Reports that I was alone in recording our need to call a halt, and I had to wonder whether activity in carbohydrate science has recently been increasing in a selective manner. Given the birth of glycobiology and the overdue wide recognition of the sugars as substances with 'real' organic chemistry and also with immense potential significance in pharmaceutical science, perhaps this is not unexpected. Electronic searching methods are seemingly at the point of largely replacing traditional abstracting/reviewing hard copy procedures in the field.

This volume describes much that follows logically from what has gone before with heavy emphasis on the reporting of developing aspects of the organic chemistry of the carbohydrates. However, as has been the case for several years, the eye catching new work is specifically focused in the field between chemistry and biology and medicine. Nowhere is this more apparent than in the area of the synthesis of oligosaccharides and glycoconjugates of importance in medicine where Danishefsky's syntheses of cancer antigens based on oligosaccharides and glycopeptides may have brought the future to hand. With rapidly developing methods based, for example, on the use of computer-selected reactants in one pot procedures, and intramolecular glycosidic and specific enzymic couplings, we are surely at the threshold of the age of the application of complex carbohydrates to specific medical problems. How appropriate that the signs may also be pointing to the use of molecular biological methods in the field of oligosaccharide synthesis.

John Gardiner, who has been responsible for writing the key and demanding chapter on the use of carbohydrates in the synthesis of chiral non-carbohydrate compounds, must hand over his responsibilities with this volume. His contributions to the last six volumes are acknowledged with gratitude, and John ˇoberg is welcomed to the reporting team.

`net Freshwater and Alan Cubitt have provided excellent liaison with the

Royal Society of Chemistry over several years, and their administrative and practical help has been very much appreciated.

R. J. Ferrier
November 2001

Contents

·bohydrate Chemistry, Volume 33

Abbreviations

The following abbreviations have been used:

Ac	acetyl
Ade	adenin-9-yl
AIBN	2,2-azobisisobutyronitrile
All	allyl
Ar	aryl
Ara	arabinose
Asp	aspartic acid
BBN	9-borabicyclo[3.3.3]nonane
Bn	benzyl
Boc	*t*-butoxycarbonyl
Bu	butyl
Bz	benzoyl
CAN	ceric ammonium nitrate
Cbz	benzyloxycarbonyl
CD	circular dichroism
Cer	ceramide
CI	chemical ionization
Cp	cyclopentadienyl
Cyt	cytosin-1-yl
Dahp	3-deoxy-D-*arabino*-2-heptulosonic acid 7-phosphate
DAST	diethylaminosulfur trifluoride
DBU	1,8-diazabicyclo[5.5.0]undec-5-ene
DCC	dicyclohexylcarbodi-imide
DDQ	2,3-dichloro-5,6-dicyano-1,4-benzoquinone
DEAD	diethyl azodicarboxylate
DIBALH	di-isobutylaluminium hydride
DMAD	dimethyl acetylenedicarboxylate
DMAP	4-(dimethylamino)pyridine
DMF	*N,N*-dimethylformamide
DMSO	dimethyl sulfoxide
Dmtr	dimethoxytrityl
e.e.	enantiomeric excess
Ee	1-ethoxyethyl
ESR	electron spin resonance
Et	ethyl
FAB	fast-atom bombardment
Fmoc	9-fluorenylmethylcarbonyl

Fru	fructose
FTIR	Fourier transform infrared
Fuc	fucose
Gal	galactose
GalNAc	2-acetamido-2-deoxy-D-galactose
GLC	gas–liquid chromatography
Glc	glucose
GlcNAc	2-acetamido-2-deoxy-D-glucose
Gly	glycine
Gua	guanin-9-yl
Hep	L-*glycero*-D-*manno*-heptose
HMPA	hexamethylophosphoric triamide
HMPT	hexamethylphosphorous triamide
HPLC	high performance liquid chromatography
IDCP	iodonium dicollidine perchlorate
Ido	idose
Im	imidazolyl
IR	infrared
Kdo	3-deoxy-D-*manno*-2-octulosonic acid
LAH	lithium aluminium hydride
LDA	lithium di-isopropylamide
Leu	leucine
LTBH	lithium triethylborohydride
Lyx	lyxose
Man	mannose
*m*CPBA	*m*-chloroperbenzoic acid
Me	methyl
Mem	(2-methoxyethoxy)methyl
Mmtr	monomethoxytrityl
Mom	methoxymethyl
Ms	methanesulfonyl (mesyl)
MS	mass spectrometry
NAD	nicotinamide adenine dinucleotide
NBS	*N*-bromosuccinimide
NeuNAc	*N*-acetylneuraminic acid
NIS	*N*-iodosuccinimide
NMNO	*N*-methylmorpholine *N*-oxide
NMR	nuclear magnetic resonance
NOE	nuclear Overhauser effect
ORD	optical rotatory dispersion
PCC	pyridinium chlorochromate
PDC	pyridinium dichromate
Ph	phenyl
Phe	phenylalanine
Piv	pivaloyl
Pmb	*p*-methoxybenzyl

Pr	propyl
Pro	proline
p.t.c.	phase transfer catalysis
Py	pyridine
Rha	rhamnose
Rib	ribose
Ser	serine
SIMS	secondary-ion mass spectrometry
TASF	tris(dimethylamino)sulfonium(trimethylsilyl)difluoride
Tbdms	*t*-butyldimethylsilyl
Tbdps	*t*-butyldiphenylsilyl
Tipds	tetraisopropyldisilox-1,3-diyl
Tips	triisopropylsilyl
Tf	trifluoromethanesulfonyl (triflyl)
Tfa	trifluoroacetyl
TFA	trifluoroacetic acid
THF	tetrahydrofuran
Thp	tetrahydropyranyl
Thr	threonine
Thy	thymin-1-yl
Tips	1,1,3,3-tetraisopropyldisilox-1,3-diyl
TLC	thin layer chromatography
Tms	trimethylsilyl
TPP	triphenylphosphine
Tps	tri-isopropylbenzenesulfonyl
Tr	triphenylmethyl (trityl)
Ts	toluene-*p*-sulfonyl (tosyl)
Ura	uracil-1-yl
UDP	uridine diphosphate
UDPG	uridine diphosphate glucose
UV	ultraviolet
Xyl	xylose

1
Introduction and General Aspects

Review material published this year includes an essay 'the unexpected and the unpredictable in organic synthesis' which is a valuable account by Mukaiyama of the extraordinary history of his research, sections on stereoselective routes to sugars and glycosides being of special relevance.[1]

As the subject continues to be driven increasingly by biological challenges a larger proportion of the text/review literature reflects this trend. The third of a series of major texts on Bioorganic Chemistry entitled 'Carbohydrates' follows two on nucleic acids (1996) and proteins (1997). Thirteen chapters by distinguished authors are introducted by Professor Ray Lemieux whose death in 2000 marked the end of the most notable of modern carbohydrate chemical careers. The topics covered in the book lay appropriately heavy emphasis on glycobiological aspects of the subject.[2]

Two volumes of *Advances in Carbohydrate Chemistry and Biochemistry* appeared in 1999, one taking the form of useful Tables of Contents, and Subject and Author indices for volumes 1–53, the first of which appeared in 1945.[3] The other contains obituary notices on Lord Todd, Professor Melvin Calvin and Dr Margaret A. Clarke, the first two winning Nobel Prizes for their work which illuminated issues associated with the roles played by the carbohydrates in natural processes, and the third making major contributions to sugar chemistry within the industrial/commercial scene. The volume also contains reviews on *N*-thiocarbonyl derivatives of sugars – isothiocyanates, thioamides, thioureas and thiocarbamates (J.M. García Fernández and C. Ortiz Mellet), the synthesis of chiral polyamides from carbohydrate-derived monomers (O. Varela and H.A. Orgueira) and hydrazine derivatives of carbohydrates (H. El Khadem and A.J. Fatiadi).[4]

An extensive range of other reviews has been published, many being referred to at the beginning of relevant chapters. Of general significance are assessments of recent developments in polymer-supported syntheses of oligosaccharides,[5] of recent advances in solid- and solution-phase methods relevant to the generation of carbohydrate and glycoconjugate libraries and of the use of carbohydrates as molecular scaffolds for library synthesis,[6] of the preparation of biologically active carbocyclic oligosaccharides and their structure-activity relationships as glycosidase inhibitors,[7] and of the synthesis of 'glycopolymers' made from sugars with natural and unnatural linkages and from copolymers comprising carbohydrates in part.[8]

Carbohydrate Chemistry, Volume 33

Further reviews of relevance in glycobiology have dealt with the synthesis and function of glycoconjugates,[9] with recent developments in this field[10] and with the use of carbohydrate mimetics in the study of carbohydrate-mediated biomolecular recognition.[11,12] Bertozzi has reported on her new approaches to the preparation of stable *C*-glycosidic glycopeptide mimetics.[13]

Of general relevance in glycobiology is a review of the impact of evolutionary considerations in respect of oligosaccharide diversity and biological function.[14] A paper on single yeast cell studies of the enzymic hydrolysis of a tetramethylrhodamine-labelled triglucoside could revolutionize *in vitro* whole cell analysis of oligosaccharide processing,[15] and the finding that low molecular weight dextrans enter the nucleus and are better inhibitors of breast cancer cell growth than are larger oligomers is of obvious potential value.[16]

References

1 T. Mukaiyama, *Tetrahedron*, 1999, **55**, 8609.
2 *Bioorganic Chemistry: Carbohydrates*, 1999, ed. S.M. Hecht, Oxford University Press, Oxford.
3 *Adv. Carbohydr. Chem. Biochem.*, 1999, Vol. 54.
4 *Adv. Carbohydr. Chem. Biochem.*, 1999, Vol. 55.
5 H.M.I. Osborn and T.H. Khan, *Tetrahedron*, 1999, **55**, 1807.
6 M.J. Sofia and D.J. Silva, *Curr. Opin. Drug Discovery Deliv.*, 1999, **2**, 365 (*Chem. Abstr.* 1999, **131**, 257 761).
7 S. Ogawa, *Carbohydr. Mimics*, 1998, 87 (*Chem. Abstr.*, 1999, **130**, 153 862).
8 J. Kadokawa, H. Tagaya and K. Chiba, *Synlett*, 1999, 1845.
9 T. Ogawa, *Kichin, Kitosan Kenkyu*, 1998, **4**, 275 (*Chem. Abs.*, 1999, **130**, 25 233).
10 B.G. Davis, *J. Chem. Soc., Perkin Trans. 1*, 1999, 3215.
11 C.-H. Wong, *Acc. Chem. Res.*, 1999, **32**, 376.
12 P. Sears and C.-H. Wong, *Angew. Chem., Int. Ed. Engl.*, 1999, **38**, 2301.
13 L.A. Marcaurelle and C.R. Bertozzi, *Chem. Eur. J.*, 1999, **5**, 1384.
14 P. Gagneux and A. Varki, *Glycobiology*, 1999, **9**, 747.
15 X.C. Le, W. Tan, C.H. Scaman, A. Szpacenko, E. Arriaga, Y. Zhang, N.J. Dovichi, O. Hindsgaul and M.M. Palcic, *Glycobiology*, 1999, **9**, 219.
16 P. Bittoun, T. Avramoglou, J. Vassy, M. Crépin, F. Chaubet, S. Fermandjian, *Carbohydr. Res.*, 1999, **322**, 247.

2
Free Sugars

1 Theoretical Aspects

AM1 and *ab initio* calculations performed on D-glucopyranose and its mono-alkoxy anions indicated that the C-1- and C-4-hydroxyl groups are the most acidic for the α- and β-anomers, respectively. In all cases, the primary hydroxyl group was the least acidic.[1]

In a molecular dynamics study on the hydration properties of an 85 w/w aq. solution of glucose, the radial distribution function, H-bond residence times, hydration number, and the mean life time and size of glucose and water clusters have been computed,[2] and in a theoretical investigation of the mutarotation of glucose assisted by up to five H_2O molecules the participation of water dimers and trimers, which provide a strain-free hydrogen bond network for ready H-transfer, has been considered.[3]

2 Synthesis

A paper entitled 'The unexpected and unpredictable in organic synthesis' covering Mukaiyama's work in the aldol condensation field contained a section on the synthesis of carbohydrates from D-glyceraldehyde and L-erythrose.[4]

2.1 Tetroses and Pentoses. – All D-tetroses and D-pentoses have been synthesized from 2,3-*O*-isopropylidene-D-glyceraldehyde by consecutive one-carbon chain-elongation using the Li salt of ethyl ethylthiomethyl sulfoxide. The additions proceeded with high *anti*-diastereoselectivity and the formations of the major products, D-erythrose and D-ribose, are outlined in Scheme 1.[5]

D-Xylose has been converted into D-lyxose *via* mesylate **1**, which on treatment with one molar equivalent of NaOH in aqueous DMF gave **2**, possibly by way of the 1,2-β-D-*lyxo*-epoxide.[6] Conversion of D-xylose into L-xylose involved resolution of the racemic xylitol derivative **3** by lipase-catalysed hydrolysis to give a mixture of the D-alcohol and unreacted L-acetate. The latter was hydrolysed chemically, then oxidized (Pfitzer Moffat) and deacetalized.[7] L-Xylose was the starting compound for the synthesis of L-*ribo*-nucleosides, the required inversion of configuration at the 3-position

Carbohydrate Chemistry, Volume 33

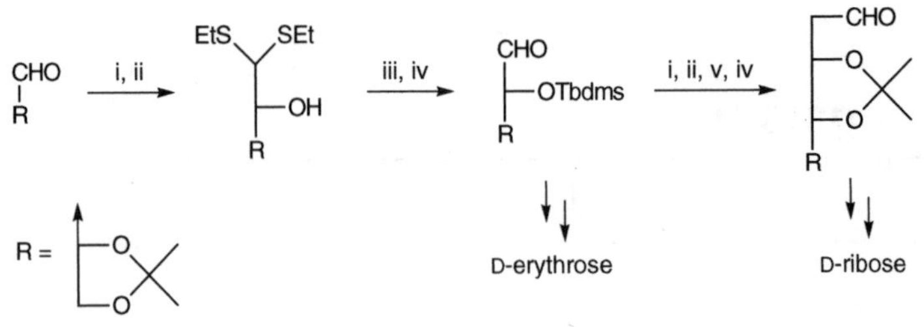

Reagents: i, $Li^+ \bar{C}H(SEt)S(O)Et$; ii, LAH; iii, TbdmsCl, Im; iv, $HgCl_2$, HgO, aq. acetone; v, $Me_2C(OMe)_2$, H^+

Scheme 1

being achieved by oxidation–reduction (CrO_3/pyridine–LAH).[8] In the synthesis of [3′,4′,5′,5″-^{2}H]-ribonucleosides from diacetone-D-glucose, the deuterium atoms were introduced consecutively by reduction of a keto group with $NaBD_4$, hydrogen-deuterium exchange α to an aldehyde using D_2O and reduction of an ester group with $LiAlD_4$, as shown in Scheme 2.[9]

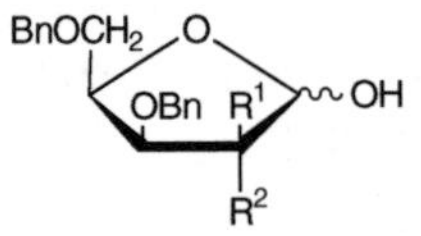

1 R^1 = H, R^2 = OMs
2 R^1 = OH, R^2 = H

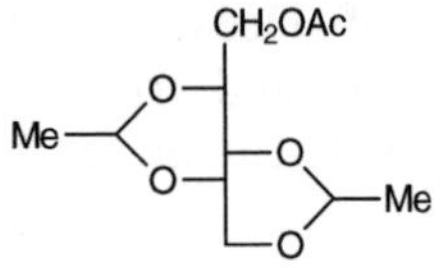

3

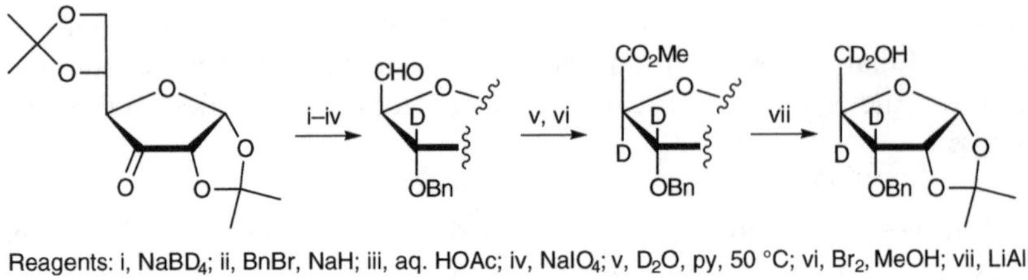

Reagents: i, $NaBD_4$; ii, BnBr, NaH; iii, aq. HOAc; iv, $NaIO_4$; v, D_2O, py, 50 °C; vi, Br_2, MeOH; vii, $LiAlD_4$

Scheme 2

The synthesis of racemic, fluorinated xylulose derivatives based on a Wittig approach and two syntheses of labelled 1-deoxy-D-xylulose are covered in Chapters 8 and 15, respectively, and the preparation of 3-deoxy-3-*C*-methylene-pentofuranosides (as precursors of novel nucleoside analogues) from commercial 3-methyL-2-butenal is referred to in Chapter 14.

2.2 Hexoses. – In a new protocol for the synthesis of hexopyranoses from furfuraldehyde, resolution was achieved at an early stage by asymmetric dihydroxylation (*e.g.* **4**→**5**, Scheme 3). Careful choice of conditions for the reduction at the 4-position and for the second dihydroxylation gave preferential access to either enantiomer of manno-, gulo- and talo-pyranose, as the 6-silylated 1-esters, such as benzoate **6** in the example chosen.[10]

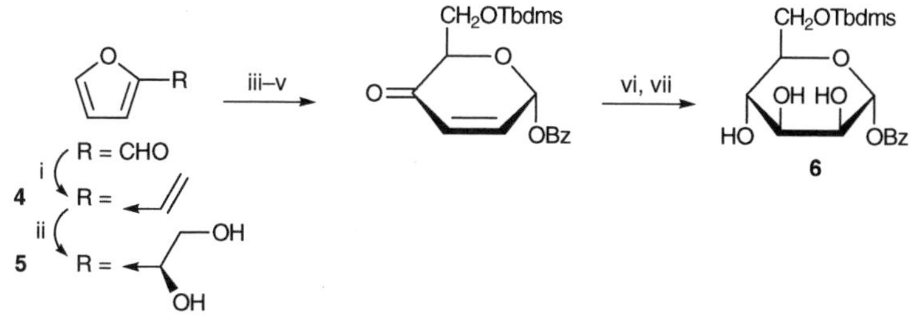

Reagents: i, $TmsCH_2MgCl$; ii, Sharpless, AD-mix-α; iii, TbdmsCl, Et_3N, DMAP; iv, NBS, H_2O; v, BzCl, Et_3N; vi, $NaBH_4$, $CeCl_3$; vii, OsO_4, NMO

Scheme 3

Construction of 6,6,6-trideoxy-trifluoro-hexopyranose ring systems by inverse-demand Diels Alder reaction is covered in Chapter 8. The syntheses of 4-deoxyfructose 6-phosphate and of a branched hexulose phosphate from achiral starting materials by routes involving biocatalytic reactions are referred to in Chapters 12 and 14, respectively, and the synthesis of L-ascorbic acid from chlorobenzene *via* L-gulono-1,4-lactone is dealt with in Chapter 16.

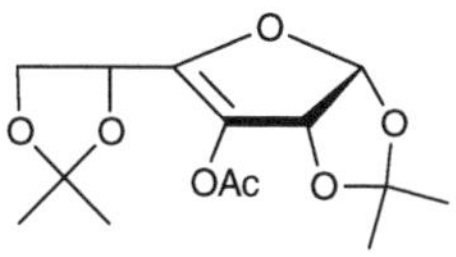

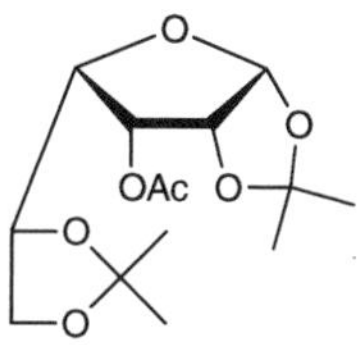

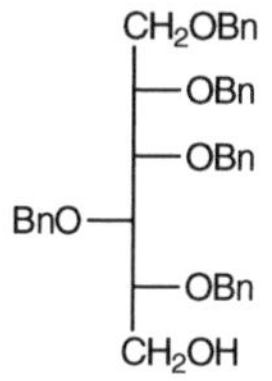

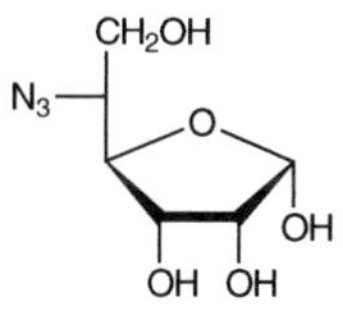

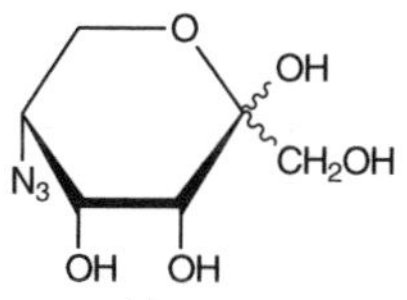

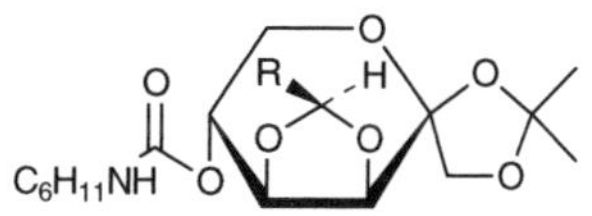

Stereospecific *cis*-hydrogenation of the known, diacetone-D-glucose-derived enol acetate **7** over a Rh/Al catalyst offers easy access to D-gulose derivative **8**. The preparation of imidazolo-sugars from **8** is referred to in Chapter 18.[11] L-Glucose has been obtained from D-gulono-1,4-lactone *via* 2,3,4,5,6-penta-*O*-benzyl-L-glucitol (**9**), which was oxidized with CrO_3/pyridine complex to give 2,3,4,5,6-penta-*O*-benzyl-*aldehydo*-L-glucose in 59% yield.[12] Immobilized glucose isomerase, which accepts D-*allo*- and D-*talo*-configured substrates, has been applied to the synthesis of 5-functionalized 2-ketoses from the corresponding hexofuranose derivatives, for example **10**→**11**. The 2-ketoses thus obtained were converted into powerful glycosidase inhibitors (see Chapter 18).[13] Treatment of 1,2-*O*-isopropylidene-D-fructopyranose with chloral/DCC caused the concomitant formation of a trichloroethylidene acetal and introduction of a carbamoyl function at C-5. Configurational inversion occurred at the central hydroxyl group of the *cis-trans*-triol (see Vol. 32, Chapter 6, ref. 8; Vol. 30, p. 99, refs. 17–19) furnishing D-tagatose *via* intermediate **12**.[14]

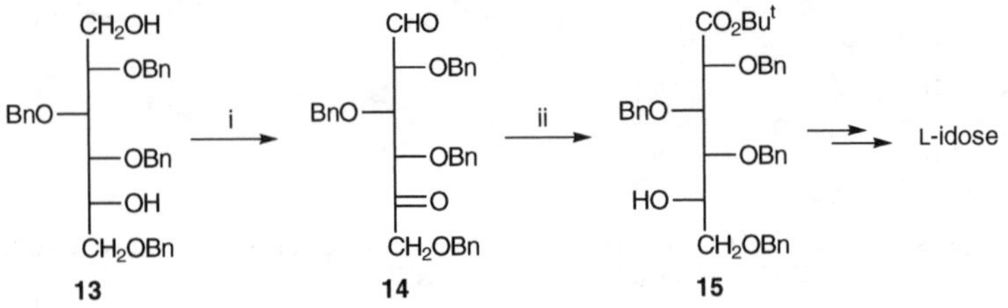

13 **14** **15** → L-idose

Reagents: i, Swern; ii, Bu^tOSmI_2

Scheme 4

Intramolecular Tishchenko oxidoreduction of a protected hexos-5-ulose intermediate to give an aldonic acid ester (**14**→**15**, Scheme 4) was the key-step in the conversion of tetra-*O*-benzyl-D-glucitol (**13**) to L-idose in 65% overall yield. A D-galactitol derivative was similarly converted to L-altrose.[15]

The synthesis of the previously unknown L-*ribo*-hexos-5-ulose from an L-*arabino*-configured precursor, involving a stereocontrolled oxidation–reduction as the key-step, is covered in Chapter 15, and the isomerization of D-glucose- to D-allose-derivatives by use of Mitsunobu reactions is referred to in Chapter 7.

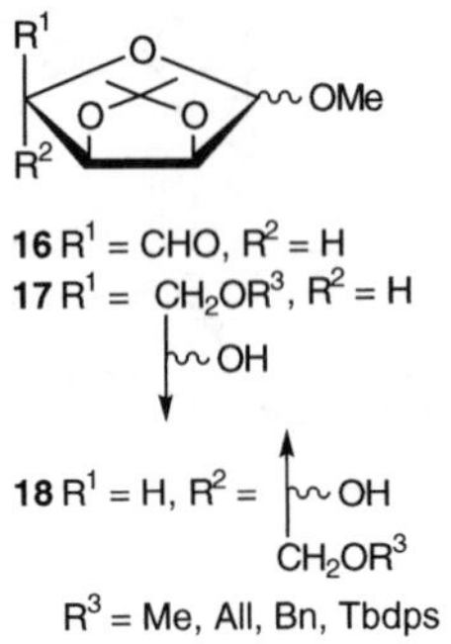

16 R^1 = CHO, R^2 = H
17 R^1 = CH_2OR^3, R^2 = H
18 R^1 = H, R^2 = $CH(OH)CH_2OR^3$
R^3 = Me, All, Bn, Tbdps

Chain-extensions of pentodialdo-1,4-furanoses with C_1 Grignard reagents ($ROCH_2MgCl$, R = Me, All, Bn or Tbdps) gave the expected stereoisomeric hexoses (*e.g.* **16**→**17**), accompanied in some cases by C-4-inverted products (in the illustrated example **18**). The selectivity at both C-4 and C-5 varied with the protecting group of the reagent.[16]

2.3 Chain-extended Sugars. – A review (27 pp., 79 refs.) on the synthetic applications of indium-mediated reactions in aqueous media contained several examples of carbohydrate homologation at the ‘reducing’ as well as at the ‘non-reducing’ end.[17]

An asymmetric hetero-Diels Alder reaction has been used to make optically active 5-*C*-aryl pentopyranoses, as outlined in Scheme 5.[18]

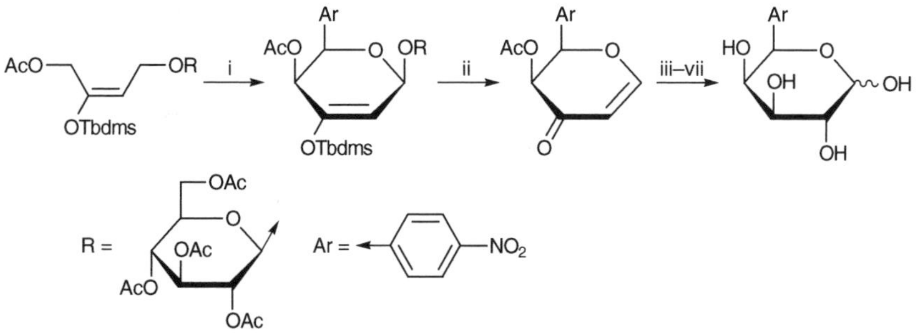

Reagents: i, ArCHO, Eu(fod)$_3$; ii,TFA; iii, NaBH$_4$, CeCl$_3$; iv, Et$_3$N, CH$_2$Cl$_2$; v, OsO$_4$, Ba(ClO$_3$)$_2$; vi, Ac$_2$O, TfOH, vii, Et$_3$N, MeOH

Scheme 5

A new, one-pot procedure involving a cascade of four enzymatic steps (phosphorylation, oxidation, aldol condensation with butanal and dephosphorylation), leading from glycerol to trideoxy-D-*xylo*-hept-2-ulose, is covered in Chapter 12

2.3.1 Chain-extension at the 'Non-reducing End'. – Further use has been made of protected dialdoses for this purpose: *t*-Butyl 7-deoxyocturonic esters **20** have been synthesized in from aldehyde **19** by exposure to MeCO$_2$Bu^t-LDA. Conversion of **20** and related compounds into castanospermine analogues is referred to in Chapter 18.[19] On exposure of aldehyde **21** to LDA at −30 °C, a complex mixture was obtained which contained, in addition to the reduction product **22**, the chain-extended compounds **23** and **24**, formed by deacetonation of some of the sample and subsequent aldol condensation between the acetone thus liberated and the starting aldehyde.[20] Reaction of aldehyde **25** with but-3-enylmagnesium bromide, followed by iodolactonization, gave a 4:1 mixture of the bis-THF compound **26** and its 2′,5′-*syn*-isomer.[21]

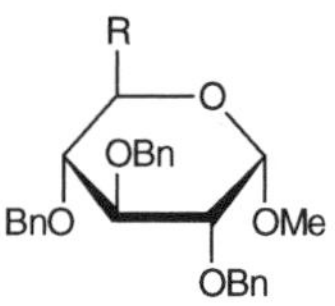

19 R = CHO
20 R = CH$_2$CO$_2$Bu^t

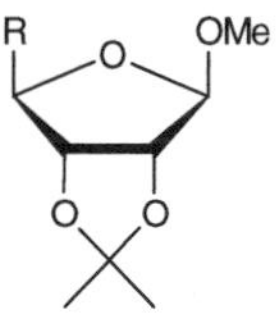

21 R = CHO
22 R = CH$_2$OH
20 R = CH$_2$COMe

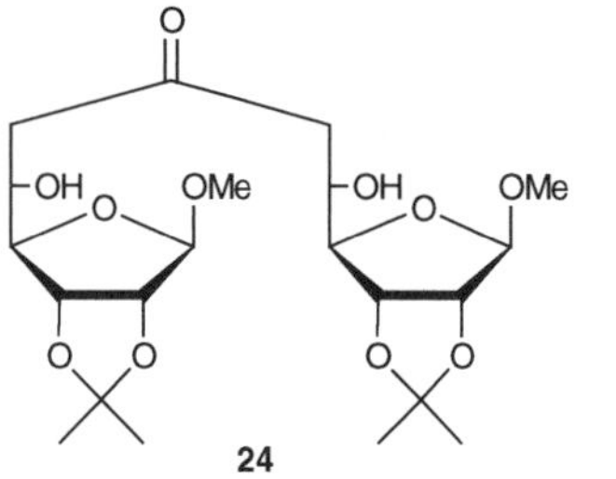

24

The phosphonate isostere **29** of methyl-α-D-mannoside 6-phosphate resulted from the reaction of aldehyde **27** with the carbanion of tetraethyl methylenebisphosphonate to furnish the 6,7-unsaturated intermediate **28**, followed by concomitant reduction of the double bond and debenzylation on exposure to

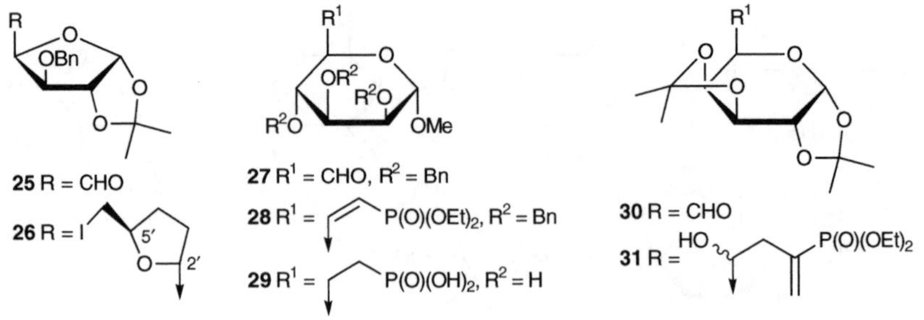

25 R = CHO
26 R = I
27 R[1] = CHO, R[2] = Bn
28 R[1] = P(O)(OEt)2, R[2] = Bn
29 R[1] = P(O)(OH)2, R[2] = H
30 R = CHO
31 R = HO P(O)(OEt)2

H_2–Pd/C.[22] Allyl phosphonate **31** was prepared from dialdose derivative **30** on exposure to diethyl 1-bromo-2-propenylphosphonate in the presence of Zn/Ag on graphite.[23] Extending the chain of the known aldehyde **32** by treatment with $(EtO)_2POCH_2CN$/NaH gave the unsaturated tetradeoxy-nonurononitrile **33**, which underwent *in situ* a highly stereoselective intramolecular Michael addition to afford the annulated furanose **34**.[24]

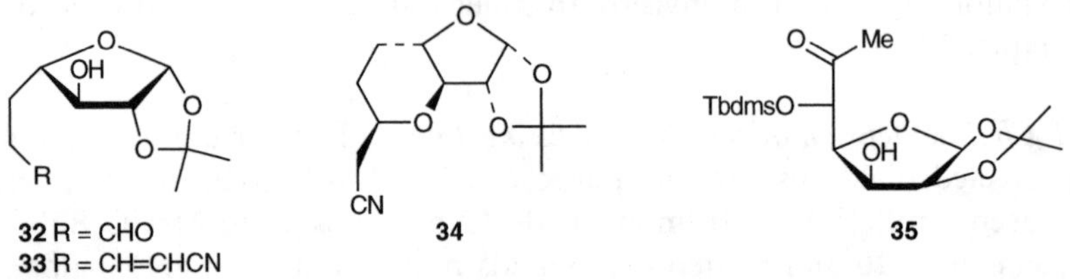

32 R = CHO
33 R = CH=CHCN
34
35

The 7-deoxy-β-D-*gluco*-heptos-6-ulose deivative **35**, available by reaction of 1,2-*O*-isopropylidene-5-*O*-Tbdms-D-mannono-3,6-lactone with methyl lithium, gave (6*S*)-6-*C*-methyl-D-mannose (**36**) on reduction with $NaBH_4$, followed by desilylation and acetal hydrolysis. When the reaction sequence was reversed, *i.e.* with desilylation prior to reduction, the (6*R*)-isomer **37** was obtained as the main product.[25]

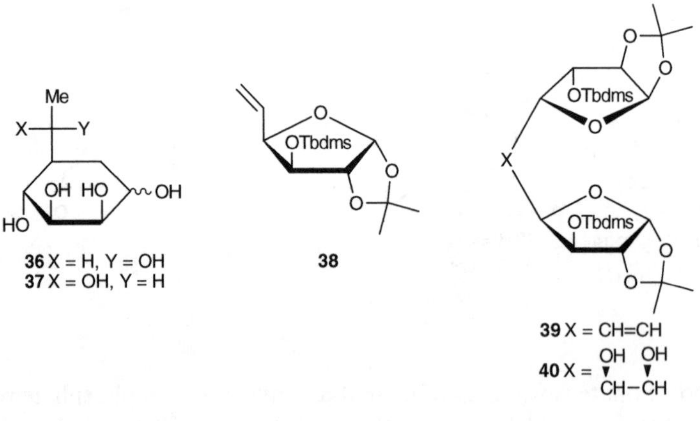

36 X = H, Y = OH
37 X = OH, Y = H
38
39 X = CH=CH
40 X = CH(OH)–CH(OH)

Metathetic dimerization of ω-unsaturated hexofuranoses over Grubbs' catalyst furnished unsaturated products, in several cases with ≥95% *Z*-

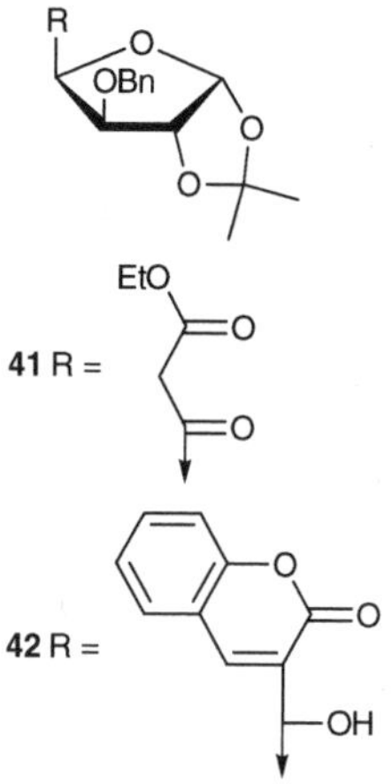

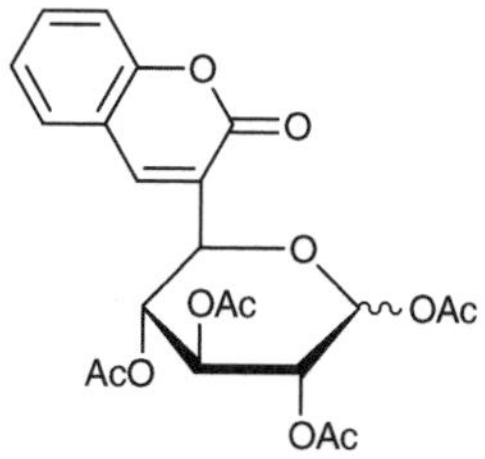

selectivity (*e.g.* **38**→**39**). Dihydroxylation (OsO_4, NMO) of the new double bond gave decadialdoses, such as **40**.[26]

For the synthesis of compound **43**, incorrectly termed a 'coumarin *C*-glycoside', a one-pot Knoevenagel condensation of the known β-keto sugar ester **41** with salicylaldehyde was used, followed by reduction of the C-5 carbonyl group to furnish intermediate alcohol **42**. Rearrangement to the chain-extended target-pyranose, isolated as the peracetate **43**, took place on deprotection.[27] A number of chromenes with C-2 linked to L-arabinose, *e.g.* **45**, resulted from the condensation of protected *E*-7-nitrohept-6-enes with substituted salicylaldehydes in the presence of basic alumina and subsequent replacement of the nitro groups of the initial products, *e.g.* **44**, by cyanide (Scheme 6). In this particular case, only the (6*S*)-isomer **45** was formed, due to stereoelectronic factors.[28]

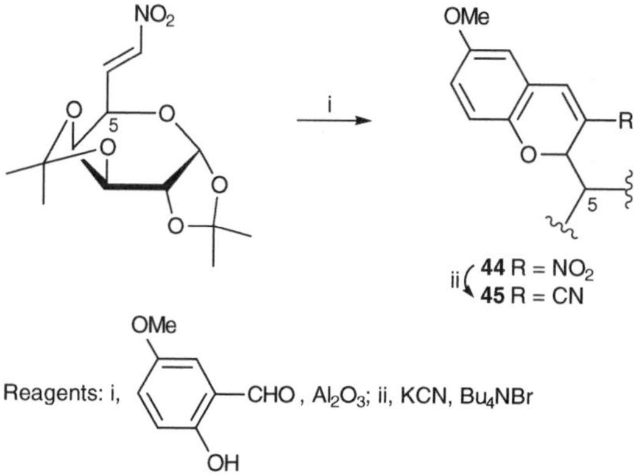

Reagents: i, 5-methoxysalicylaldehyde (OMe, OH, CHO), Al_2O_3; ii, KCN, Bu_4NBr

Scheme 6

New examples of the 'nitrile oxide/isoxazoline route' (see Vol. 31, p.7, ref. 21) and of the 'phosphonate route' (see Vol. 32, Chapter 2, refs. 28, 29) to higher sugars have been published. The former approach led to C_{11}-monosaccharides by cycloaddition of di-*O*-isopropylidene-D- or -L-arabinononitrile oxide and the 3-*O*-benzyl analogue of the glucose-derived alkene **38** or an analogous, mannose-derived alkene. The initial addition products were functionalized isoxazolines, such as **46**.[29,30] In the latter approach C_{13}- and C_{15}-monosaccharides were produced from C_6-aldehydes and C_7- or C_9-phosphonates, respectively (*e.g.* **47** + **48** → **49**).[31,32] Reaction of phosphonate **48** with Tbdps-protected glycolaldehyde gave a silyloxy-enone which cyclized on desilylation, then aromatized affording 2-furyl sugar **50**.[32]

46

47 R = CHO

48 R =

49 R =

50 R =

2.3.2 Chain-extension at C-1 of Ald-2-uloses – Extended-chain uloses have been prepared by highly selective boron aldol additions. On treatment with dicyclohexylboron chloride, L-erythrulose derivative **51**, for example, formed a *Z*-enolate, which reacted with benzaldehyde to furnish the *syn*/*syn*-product **53**, whereas under similar conditions the dibenzoate **52** formed an *E*-enolate and hence the *syn*/*anti* adduct **54**.[33,34]

51 R = Bn
52 R = Bz

53 R = Bn, X = H, Y = OH
54 R = Bz, X = OH, Y = H

2.3.3 Chain-extension at the 'Reducing End' – The molybdic acid-mediated skeletal rearrangement of 2-*C*-hydroxymethyl-D-allose to D-*altro*-heptulose (sedoheptulose), which results in a 2:12 equilibrium mixture, has been exploited in a facile synthesis of the latter compound from D-allose (see Vol. 32, Chapter 2, ref. 41).[35]

Several papers have been published on reactions of phosphoranes with protected aldehydo sugars, osuloses and sugar lactones. On condensation of 2,3:4,5-di-*O*-isopropylidene-D-xylose with benzoylmethylenetriphenylphosphorane, enone derivative **55** was formed. Conversion to ketoxime **56** and oxidative cyclization with I_2/KI/Na_2CO_3 furnished, after deprotection, 3-phenyl-5-(tetrahydroxybutyl)isoxazole **57**.[36] 2,3:4,6-Di-*O*-isopropylidene-β-D-*arabino*-hexos-2-ulopyranose (**58**) reacted with (cyanomethylene)triphenylphosphorane to give, after catalytic hydrogenation of the double bond, 4-octulosononitrile **59**.[37] Reaction of osulose **58** with phosphorane **61**, synthesized from 2-ethyl-1,3-propanediol in four steps, including enzyme-catalysed, desymmetrizing acetylation, furnished **60**, after saturation of the double bond. This was further processed to give spiroketal **62**.[38]

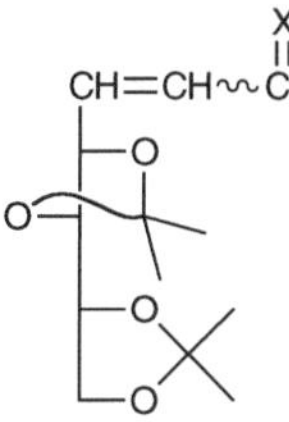

55 X = O
56 X = NOH

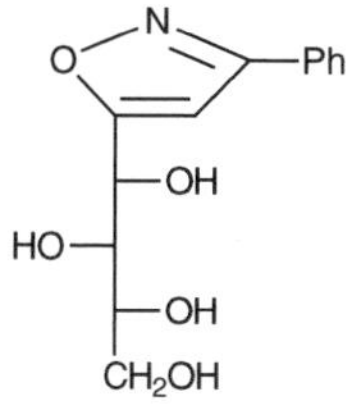

57

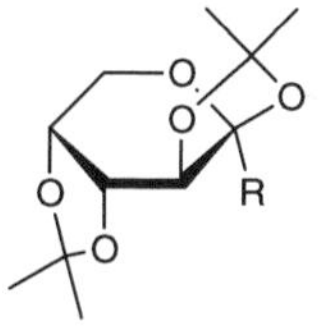

58 R = CHO
59 R = CH_2CH_2CN
60 R = $CH_2CH_2CH(Et)CH_2OAc$

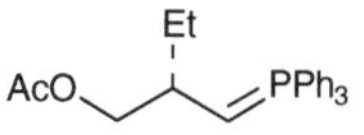

61

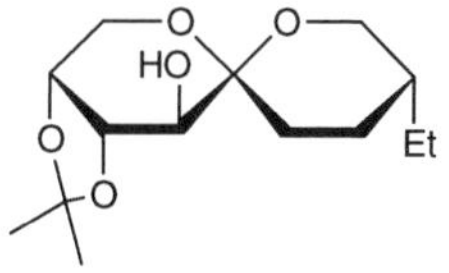

62

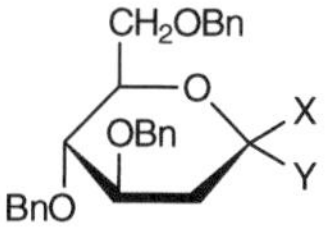

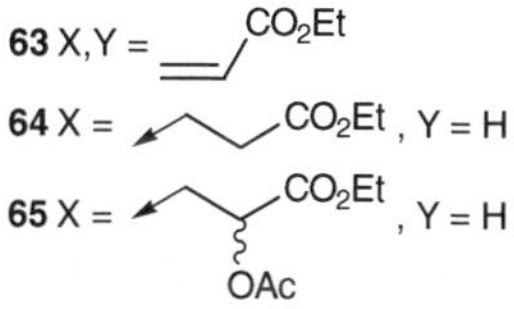

63 X,Y = $=CHCO_2Et$
64 X = $CH_2CH_2CO_2Et$, Y = H
65 X = $CH_2CH(OAc)CO_2Et$, Y = H

Compound **63** and similar *exo*-glycals were prepared by conventional Wittig alkylidenation of protected hexono-1,5-lactones. They were either hydrogenated, or dihydroxylated then deoxygenated, to furnish β-*C*-glycosides, such as **64** and **65**, respectively, after acetylation.[39] An efficient approach to the construction of the spiroketal moiety **67** of papulacandins was based on the condensation of the D-arabinono-1,4-lactone derivative **66** with 2-lithiated 3-phenylsulfonyl-4,5-dihydrofuran, as shown in Scheme 7.[40]

Acetylenic *C*-glycosides, for example compound **68**, have been obtained by reaction of a sugar lactone with a carbohydrate-derived lithium acetylide (see Vol. 31, p. 48, ref. 303). They were deoxygenated at C-1′ and further transformed to *C*-linked disaccharides, such as **69**. An iterative process has been developed on the basis of these reactions to furnish *C*-linked-β-(1→6)-D-galactooligosaccharides **70**.[41] The synthesis of aza-sugar-containing, *C*-linked

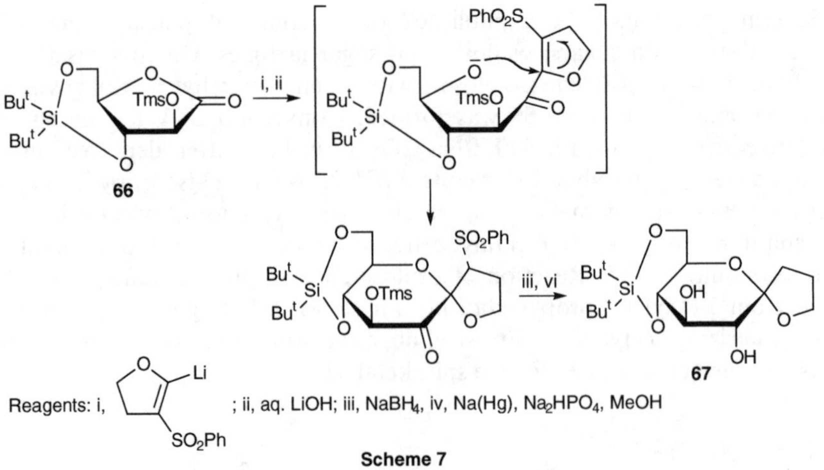

Reagents: i, [2-lithio-3-(phenylsulfonyl)-4,5-dihydrofuran]; ii, aq. LiOH; iii, $NaBH_4$, iv, Na(Hg), Na_2HPO_4, MeOH

Scheme 7

disaccharides from acetylenic *C*-glycosides is covered in Chapter 18, and related *C*-linked disaccharides are discussed in Chapter 3.

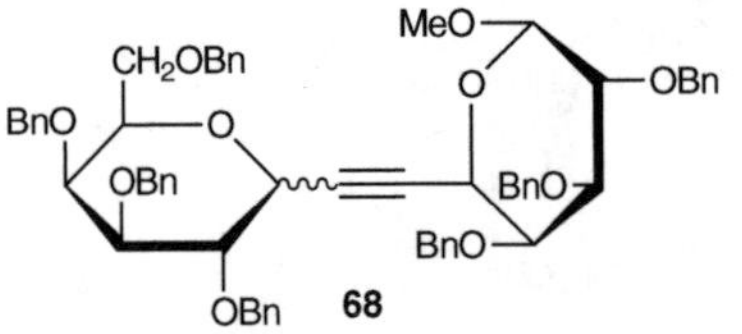

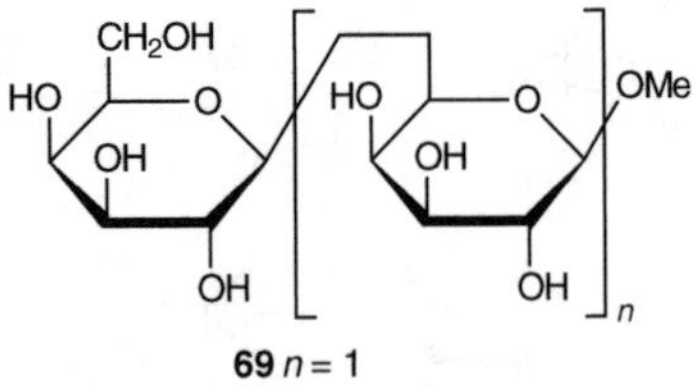

69 $n = 1$
70 $n = 2$ or 3

3 Natural Products

1-Deoxy-1-(4-hydroxyphenyl)-L-sorbose and -L-tagatose (**71**) and (**72**) were isolated from the roots of the Nepalese medicinal plant *Dactylorhizia hatagires* and named Dactylose A and B, repectively,[42] and sarioside (**73**), found in the shrub *Picramnia antidesma*, has been shown by X-ray crystallography to be a sugar derivative with the carbon-chain extended from the 'non-reducing end'.[43]

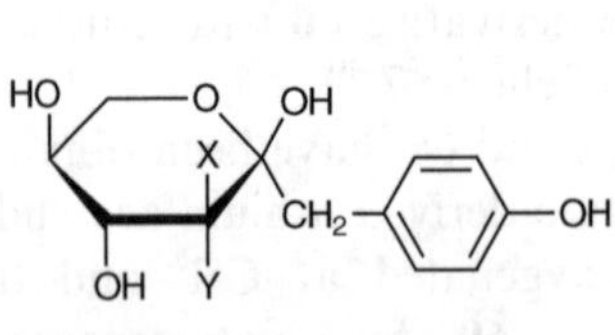

71 X = H, Y = OH
72 X = OH, Y = H

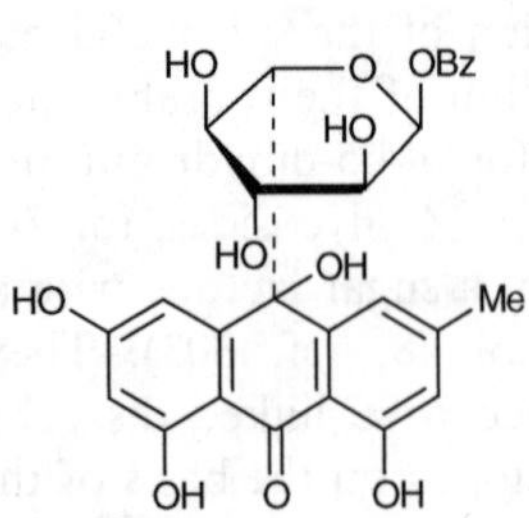

4 Other Aspects

The thermodynamic parameters of the interaction of HCl with D-arabinose in water have been determined from electromotive force measurements between 278.15 and 318.15 K.[44]

The molybdic acid-catalysed stereospecific intramolecular rearrangement of 2-ketohexoses to the corresponding 2-*C*-(hydroxymethyl)aldoses has been studied in some detail (see also ref. 18 above). The C-3–C-4 bond cleavage has been confirmed by the formation [1-^{13}C]D-hamamelose from [3-^{13}C]D-fructose. The equilibria, which ranged from 1:14 (D-fructose) to 1:32 (L-sorbose) in favour of the branched aldoses, were shifted towards the ketoses by the addition of boric acid [3:1 (D-fructose) to 7:1 (L-sorbose)].[45]

The kinetics and mechanism of the Ru(VIII)-catalysed oxidation of L-sorbose and maltose with alkaline sodium metaperiodate have been investigated,[46] and in a spectrophotometric study, the kinetics of oxidation of aldoses, amino sugars and methyl glycosides by tris(pyridine-2-carboxylato)-manganese(III) have been examined.[47]

The ability of three bis-boronic acids, *e.g.* **74**, and of the quaternary ammonium-phenylboronic acid **75** to transport fructose selectively through a lipid membrane as its boronate esters has been investigated.[48]

References

1 B. Mulroney, J.B. Peel and J.C. Traeger, *J. Mass. Spectrom.*, 1999, **34**, 544 (*Chem. Abstr.*, 1999, **131**, 32 101).
2 E.R. Caffarena and J.R. Grigera, *Carbohydr. Res.*, 1999, **315**, 63.
3 S. Yamabe and T. Ishikawa, *J. Org. Chem.*, 1999, **64**, 4519.
4 T. Mukaiyama, *Tetrahedron*, 1999, **55**, 8609.
5 Y. Arroyo-Gómez, J.A. López-Sastre, J.F. Rodríguez-Amo, M. Santos-García and M.A. Sanz-Tejedor, *Tetrahedron: Asymmetry*, 1999, **10**, 973.
6 V. Popsavin, S. Grabez, B. Stojanovic, M. Popsavin and D. Miljkovic, *Carbohydr. Res.*, 1999, **321**, 110.
7 G.D. Gamalevich, B.N. Morozov, A.L. Vlasyuk and E.P. Serebryakov, *Tetrahedron*, 1999, **55**, 3665.
8 E. Moyroud and P. Strazewski, *Tetrahedron*, 1999, **55**, 1277.
9 A. Trifonova, A. Földesi, Z. Dinya and J. Chattopadhyaya, *Tetrahedron*, 1999, **55**, 4747.
10 J.M. Harris, M.D. Keranen and G.A. O'Doherty, *J. Org. Chem.*, 1999, **64**, 2982.

11 H. Siendt, T. Tschamber and J. Streith, *Tetrahedron Lett.*, 1999, **40**, 5191.
12 H. Hajkó, A. Lipták and V. Pozsgay, *Carbohydr. Res.*, 1999, **321**, 116.
13 M.H. Fechter and A.E. Stütz, *Carbohydr. Res.*, 1999, **319**, 55.
14 M. Frank, R. Miethchen aand D. Degenring, *Carbohydr.Res.*, 1999, **318**, 167.
15 M. Adinolfi, G. Barone, F. DeLorenzo and A. Iadonisi, *Synlett*, 1999, 336.
16 H. Stepowska and A. Zamojski, *Tetrahedron*, 1999, **55**, 5519.
17 C.-J. Li and T.-H. Chan, *Tetrahedron*, 1999, **55**, 11149.
18 M. Helliwell, I.M. Phillips, R.G. Pritchard and R.J. Stoodley, *Tetrahedron Lett.*, 1999, **40**, 8651.
19 E. Bartnicka and A. Zamojski, *Tetrahedron*, 1999, **55**, 2061.
20 H. Stepowska and A. Zamojski, *Carbohydr. Res.*, 1999, **321**, 105.
21 P. Bertraud, H. El Sukkari, J.-P. Gesson and B. Renoux, *Synthesis,* 1999, 330.
22 C. Vidil, A. Morère, M. García, V. Barragan, B. Hamadoui, H. Rochefort and J.L. Montero, *Eur. J. Org. Chem.*, 1999, 447.
23 R. Csuk and C. Schröder, *J. Carbohydr. Chem.*, 1999, **18**, 285.
24 J.L. Marco and S. Fernandez, *J. Chem. Res. (S)*, 1999, 544.
25 A. Martin, M.P. Watterson, A.R. Brown, F. Imitaz, B.G. Winchester, D.J. Watkin and G.W.J. Fleet, *Tetrahedron: Asymmetry*, 1999, **10**, 355.
26 P. Hadwiger and A.E. Stütz, *Synlett*, 1999, 1787.
27 N.N. Saha, V.N. Desai and D.D. Dhavale, *J. Org. Chem.*,1999, **64**, 1715.
28 J.M.J. Tronchet, S. Zerelli and G. Bernardinelli, *J. Carbohydr. Chem.*, 1999, **18**, 343.
29 K.E. McGhie and R.M. Paton, *Carbohydr. Res.*, 1999, **321**, 24.
30 R.O. Gould, K.E. McGhie and R.M. Paton, *Carbohydr. Res.*, 1999, **322**, 1.
31 S. Jarosz, S. Skora, A. Stefanowicz, M. Mach and J. Frelek, *J. Carbohydr. Chem.*, 1999, **18**, 961.
32 S. Jarosz, M. Mach and S. Skora, *Synlett*, 1999, 313.
33 J.A. Marco, M. Carda, E. Falomir, C. Palomo, M. Diarbide, J.A. Ortiz and A. Linden, *Tetrahedron Lett.*, 1999, **40**, 1065.
34 M. Carda, E. Falomir, J. Murga, F. Gonzalez and J.A. Marco, *Tetrahedron Lett.*, 1999, **40**, 6845.
35 Z. Hricovíniová-Bíliková and L. Petrus, *Carbohydr. Res.*, 1999, **320**, 31.
36 J.M. Bañez-Sanz, J.A. López-Sastre, M. R. Patiño Molina, T. Santacana Gómez and C. Romero-Ávila García, *J. Carbohydr. Chem.*, 1999, **18**, 403.
37 I. Izquierdo, M.T. Plaza, R. Robles, C. Rodriguez, A. Ramirez and A.J. Mota, *Eur. J. Org. Chem.*, 1999, 1269.
38 I. Izquierdo, M.T. Plaza, M. Rodriguez and J. Tamayo, *Tetrahedron: Asymmetry*, 1999, **10**, 449.
39 J. Xie, A Molina and S. Czernecki, *J. Carbohydr. Chem.*, 1999, **18**, 481.
40 J.C. Carretero, J.E. de Diego and C. Hamdouchi, *Tetrahedron*, 1999, **55**, 15159.
41 Y.-C. Xin, Y.-M. Zhang, J.-M. Mallet, C.P.J. Glaudemans and P. Sinaÿ, *Eur. J. Org. Chem.*, 1999, 471.
42 H. Kizu, E. Kaneko and T. Tomimori, *Chem. Pharm. Bull.*, 1199, **47**, 1618.
43 M.R. Hernandez-Medel, C.O. Ramirez-Corzas, M.N. Rivera-Dominguez, J. Ramirez-Mendez, R. Santillan and S. Rojas-Lima, *Phytochemistry*, 1999, **50**, 1379.
44 K. Zhuo, J. Wang, Q. Zhang, Z. Yan and J. Lu, *Carbohydr. Res.*, 1999, **316**, 26.
45 Z. Hricovíniová-Bíliková, M. Hricovíni, M. Petrusová, A.S. Serianni and L. Petrus, *Carbohydr. Res.*, 1999, **319**, 38.

46 N. Gupta, S. Rahmani and A.K. Singh, *Oxid. Commun.*, 1999, **22**, 237 (*Chem. Abstr.*, 1999, **131**, 199 896).
47 K.K.S. Gupta and B.A. Begum, *Carbohydr. Res.*, 1999, **315**, 70.
48 S.J. Gardiner, B.D. Smith, P.J. Duggan, M.J. Karpa and G.J. Griffin, *Tetrahedron*, 1999, **55**, 2857.

3
Glycosides and Disaccharides

1 *O*-Glycosides

1.1 Synthesis of Monosaccharides Glycosides. – In an important review of the history of his research group's work on glycoside synthesis over the years Mukaiyama has concentrated on the stereoselectivity of glycosylation processes,[1] and Schmidt and colleagues have reported further on newer aspects of glycoside bond formation – particularly those based on the use of glycosyl trichloroacetimidates and phosphites.[2] In more specific work Dondoni and Marra have reviewed the use of 'thiazolylketoses' in the preparation of ketosides.[3]

1.1.1 Methods of synthesis of glycosides. – Solid, essentially neutral and environmentally benign catalysts are being used increasingly for catalysing glycoside formation from free sugars. Glucose has been converted to long chain alkyl glycosides by use of H-Beta zeolites, and the same products can be derived similarly from butyl glucosides by transglycosylation.[4] These latter pyranosides have also been made from the free sugar by use of dealuminated HY zeolites[5] or Al-MCM-41 mesoporous materials.[6] Montmorillonite K-10 catalysed the condensation of 3,4-di-*O*-protected olivose (2,6-dideoxy-D-*arabino*-hexose) with (hydroxymethyl)cyclohexane with 96% efficiency, and a 6-unsubstituted glucoside to give 1,6-linked disaccharide derivatives with 77% efficiency.[7]

Fructose in suspension in THF with long chain alcohols and iron(III) chloride gives the β-pyranosides in modest yields. Novel liquid crystal properties were claimed for the products.[8]

A novel method for activating the C-1 hydroxyl group of a sugar for glycoside formation involves the *in situ* formation of a glycosyl imidate by treatment with an oximoxy-trisdimethylaminophosphonium salt. The first-formed *O*-glycosyloxime undergoes the Beckmann rearrangement in solution at room temperature (see Scheme 1).[9] The method requires development before it be claimed as an efficient new glycosylation process. The related approach involving application of the Mitsunobu reaction to 1-*O*-unprotected sugars has been used to give perfluoroalkyl and pentafluorophenyl glycosides.[10]

In more standard work C_7–C_{16} alkyl glycosides of several common aldose

Carbohydrate Chemistry, Volume 33

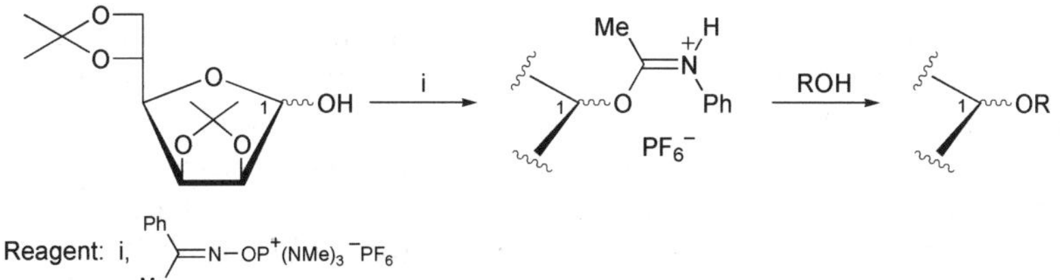

Reagent: i, Ph(Me)C=N–OP⁺(NMe)$_3$ $^-$PF$_6$

Scheme 1

monosaccharides and disaccharides have been made from the peracetylated β-sugars with tin(IV) chloride as catalyst.[11] The use of β-D-mannopyranose 1,2,6-orthoacetates in the preparation of α-mannopyranosides is illustrated in Scheme 2, alkyl glycosides and disaccharides being prepared in 30–70% yield in this way.[12] Mixed glycosyl carbonates, made in good yield from 1-*O*-unprotected sugars and succinimidyl carbonates, decarboxylate on treatment with trimethylsilyl triflate to give glycosides and disaccharides. With 2,3,4,6-tetra-*O*-benzyl-D-glucose β-carbonates were formed with high selectivity and these collapsed again with good β-selectivity. Yields were in the 60–90% range so the method may have practicability for making β-glucosides and related compounds.[13]

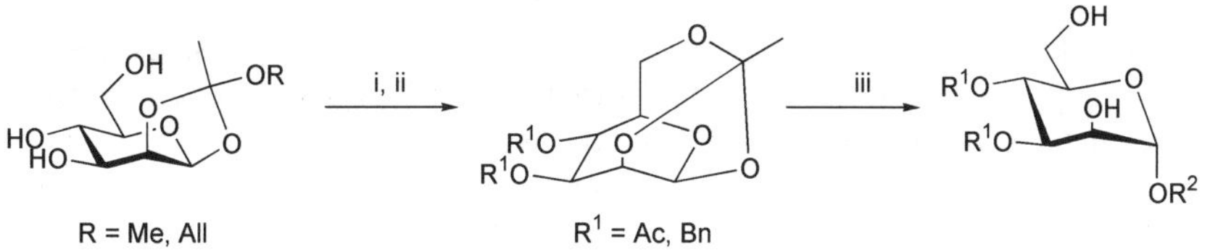

R = Me, All

R^1 = Ac, Bn

Reagents: i, PyH$^+$Tf$^-$, Py; ii, standard *O*-substitution; iii, R^2OH, TmsOTf

Scheme 2

Glycosyl halides remain important glycosylating agents. Tetra-*O*-pivaloyl-α-D-mannopyranosyl fluoride with $BF_3.OEt_2$ in dichloromethane affords stereospecific access to α-mannosides of primary, secondary and benzylic alcohols as well as phenols.[14]

A study of the effect of the nature of the O-4 substituent on the stereoselectivity of glycosylation by 2,3-di-*O*-benzyl-fucosyl bromides has been reported.[15]

O-Protected 2-acetamido-2-deoxy-α-D-glucosyl chloride in the presence of sodium iodide and a crown ether gives access to β-glucosides at room temperature and mostly α-isomers at 90 °C.[16] Presumably glycosyl iodide intermediates are involved, as they are when mercury(II) is used as activator.[17,18]

Activation of the α-and β-anomers of *O*-protected xylopyranosyl sulfoxides, glycosyl bromides and thioglycosides with Tf_2O, AgOTf and PhSOTf, respectively, leads to reactions with nucleophiles that are independent of the anomeric configurations and give β-pyranosides. The use of hindered bases diverts the reactions in favour of the 1,2-orthoesters.[19] On the other hand, the same group has reported that phenyl 2,3-di-*O*-benzyl-4,6-*O*-benzylidene-1-

thio-α- or β-D-glucopyranosides (or the corresponding sulfoxides), activated with triflic anhydride in the presence of a highly hindered base, give α-glucosides with high selectivity (α:β > 19:1) and in good yield. From the mannosyl analogues β-glycosides are obtained. These selectivities are discussed, glycosyl triflates being deemed to be reaction intermediates.[20]

Differently protected glycosyl phosphates activated by TmsOTf are powerful glycosyl donors,[21] and effective mild 1,2-*cis*-glycosylation can be conducted with *O*-benzyl protected diethyl glycosyl phosphites activated by 2,6-di-*tert*-butylpyridinium iodide. Yields are typically >80% and for glucosides α,β-ratios are > 9:1.[22] Compound **1**, promoted by TmsOTf in nitromethane, gives mainly α-glycosides with some deacetylation occurring at O-2 perhaps because 1,2-orthoester intermediates were involved.[23]

Increasing use of enzymes for the preparation of glycosides is being reported. A cellulase from *Aspergillus niger* and a β-glucosidase from almonds have been used to produce octyl β-D-glucopyranoside from cellulose.[24] In related work the same group has produced alkyl β-D-glucosaminides from chitosan by use of the *exo*-β-D-glucosaminidase from a *Penicillium*.[25,26] β-Glucuronides of cholesterol absorption inhibitors (*e.g.* **2**) have been made using a transferase from bovine and dog liver microsomes and UDPglucuronic acid as glycosyl source.[27]

Allyl, (2-trimethylsilyl)ethyl and benzyl β-D-mannopyranosides were prepared from the free sugar by use of a thermostable glycosidase which was also effective with D-glucose,[28] and in related work the β-hexyl, -heptyl and -octyl glycosides of D-fucose were made using almond β-glucosidase which also was active with D-galactose.[29] Galactosylation of *N*-protected serine methyl esters was effected by β-galactosidases from various sources in the presence of moderate proportions of organic solvents. Best results were with lactose as donor and an enzyme from *E. coli*. Yields were up to 28%.[30]

Transfer from sucrose, catalysed by levansucrase from *Bacillus subtilis*, gave access to 2-*O*-β-D-fructofuranosylglycerol.[31]

Epoxypropyl β-D-glucopyranoside and -cellobioside acted as irreversible inhibitors of sweet almond β-glucosidase.[32]

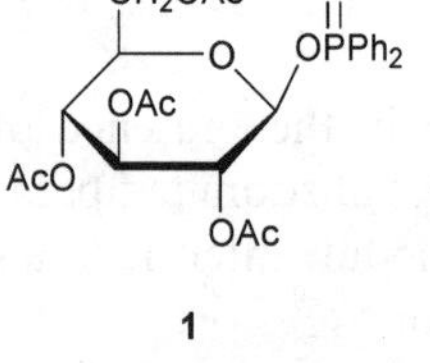

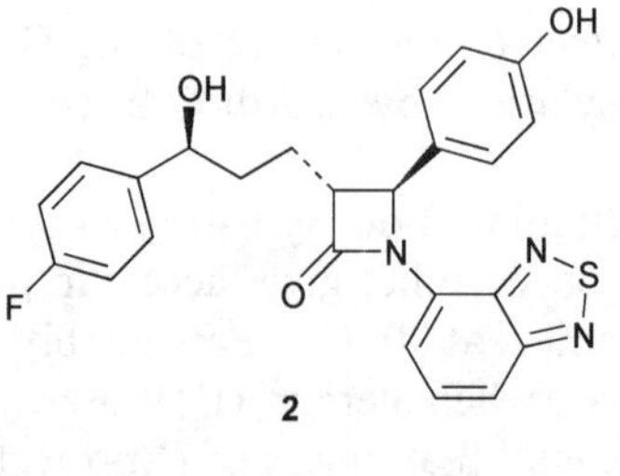

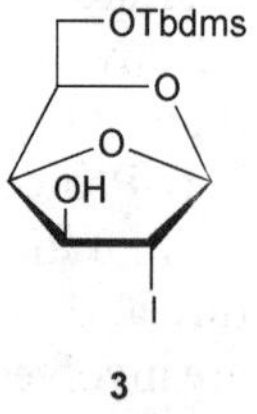

The methyl glycoside of 5-deoxy-5-hydroxymethyl-β-D-fructopyranose, made by enzymic methods, is noted in Chapter 14.

1.1.2 Classes of glycosides. – In this section different groups of glycosides which have received particular attention are treated. These are furanosides, β-mannopyranosides, 2-deoxyglycosides, amino-sugar glycosides, sugar acid glycosides, glycosides with aromatic ring-containing aglycons, glycosides with bridging aglycons and compounds having more than one glycosidically linked sugar.

A mesoporous molecular sieve has been used as a mild acid catalyst to produce alkyl fructofuranosides. The alcohols up to C_4 gave quantitative yields, but the processes were less efficient with higher alcohols.[33] 1,6-Anhydro-α-D-galactofuranose tribenzoate was ring opened with acetic anhydride to give the 6-*O*-acetylglycofuranosyl acetate from which the ethyl thioglycoside was made. Both of these compounds were used as glycosylating agents to produce galactofuranosyl di- and tri-saccharides.[34] Intramolecular transfer from a substituent on O-2 of *S*-ethyl 1-thio-α-D-arabinofuranoside, after de-*O*-protection, was used to make methyl 3,5-di-*O*-acetyl-β-D-arabinofuranoside which was coupled *via* O-5 to *myo*-inositol *via* a phosphate bridge.[35] The 1,4-anhydro-D-galactopyranose compound **3**, made from D-galactal, has been used to make octyl 2-deoxygalactofuranoside derivatives as well as *C*-glycosides.[36]

The preparation of β-mannopyranosides continues to attract attention. A review has been prepared on the use of glycosyl triflates as intermediates in their preparation from thioglycosides.[37] A specific example of this chemistry has been referred to above.[20] Roush's group has developed specific routes to α- and β-glycosides of 2-deoxy-sugars based on reductive dehalogenation of 2-deoxy-2-halogeno-compounds. 2-Deoxy-2-iodo-α-D-manno- and talo-pyranosyl acetates afford only α-glycosides,[38] whereas 2-deoxy-2-iodoglucopyranosyl acetates, activated with TmsOTf, give β-products.[39] Parallel work has shown that 2-deoxy-2-iodo- and 2-bromo-2-deoxy glucopyranosyl trichloroacetimidates also afford these latter anomers with selectivities >20:1.[40] Continuing their work in this area Franck's group has used compound **4**, which reacts with alcohols in the presence of acid to give **5** which could be rearranged to **6** and hence transformed to 2-deoxy β-glycosides (Scheme 3). The β-*manno*-

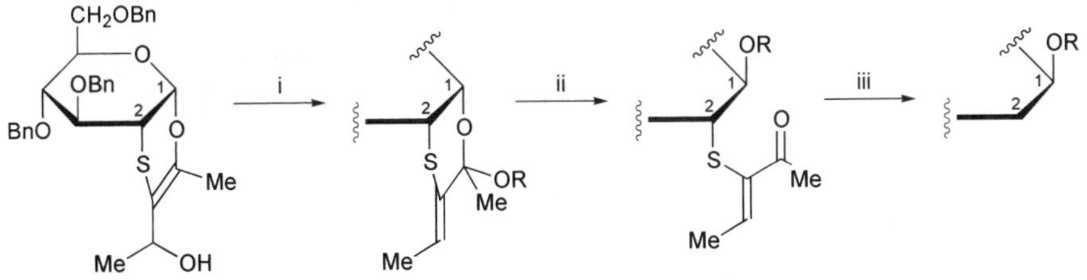

Reagents: i, ROH, CF_3CO_2H; ii, TsOH; iii, Raney Ni

Scheme 3

analogue of **4** also led to these products.[41] Very similar work has opened a route to aryl 2-deoxy-α-glucosides (Scheme 4). Other glycals and *ortho*-thioquinones were also investigated.[42]

A library of glycosides of β-*N*-acetylglucosamine, several containing het-

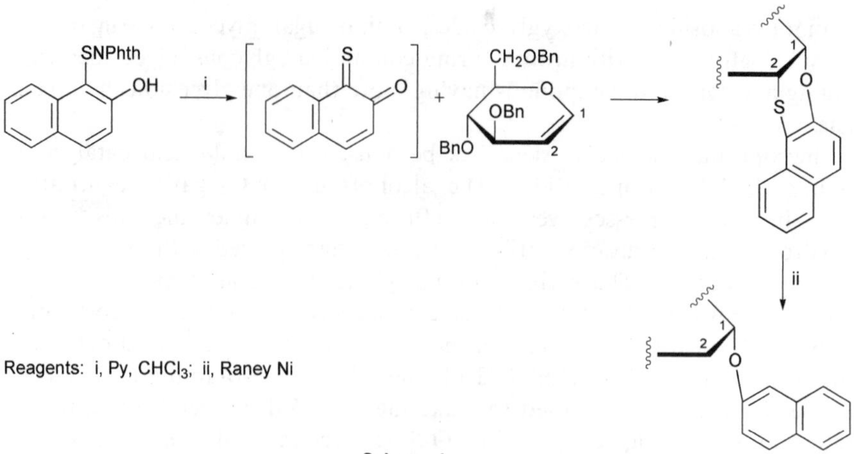

Reagents: i, Py, $CHCl_3$; ii, Raney Ni

Scheme 4

eroatoms in the aglycons, have been made by standard reactions and tested as *in vivo* glycosyl transferase acceptors.[43] Reference has already been made to the preparation of glycosides of glucosamine derivatives by chemical[16–18] and enzymic[25,26] methods. See Chapter 9 for other amino-sugar glycosides.

Both anomers of the glycosyl trichloroacetimidates of *O*-protected methyl glucuronate have been used in the synthesis of the β-glucuronide of alcohol **7**, the glycoside being a metabolite of a multi-drug resistance reversing compound.[44] A novel route to α-glycosides of 3-deoxy-ulosonic acids involves TmsOTf-promoted addition of alcohols to ketene dithioacetals, *e.g.* **8**, followed by hydrolysis of the dithianyl residue and oxidation.[45] The 8-hydrazidyloctyl α-glycoside of *N*-acetylneuraminic acid has been coupled to interleukin-1α (2.9 sugar units per molecule) by the acyl azide method,[46] and the further NeuNAc derivative **9** was made as a colorimetric substrate for assay of the neuraminidase of influenza A and B. It was not a substrate for other viral or bacterial enzymes.[47]

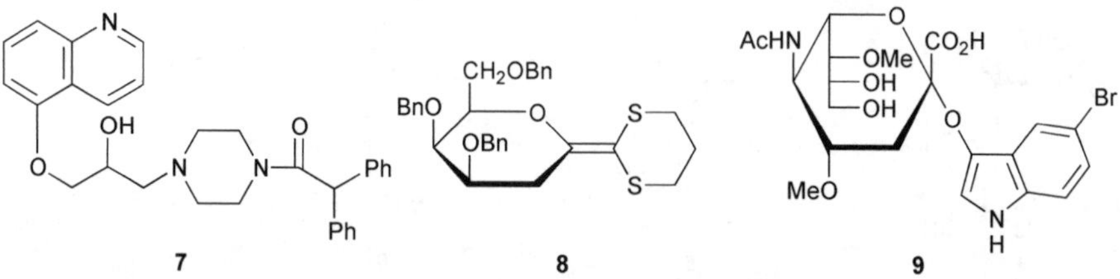

7 **8** **9**

Compounds **10–12** have been made for biological studies: they are respectively an antithrombotic agent,[48] a near UV filter compounds in human eye lenses[49] and a mustard glycoside required in connection with studies of gene-directed enzyme-prodrug therapy.[50] The aryl glucuronide **13** was made for cancer prodrug monotherapy and antibody-directed enzyme prodrug therapy. Enzymolysis releases 9-aminocamptothecin which is considerably more toxic

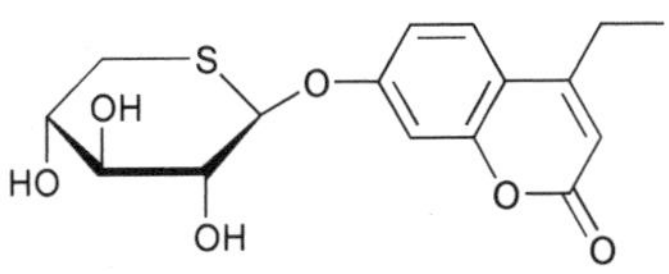

10

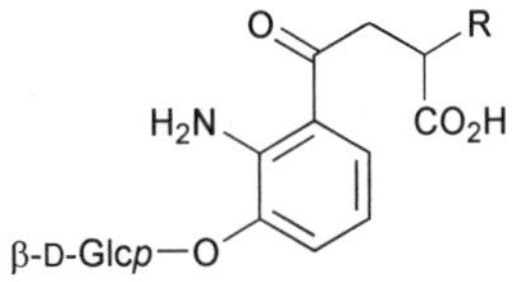

11 R = H, NH_2

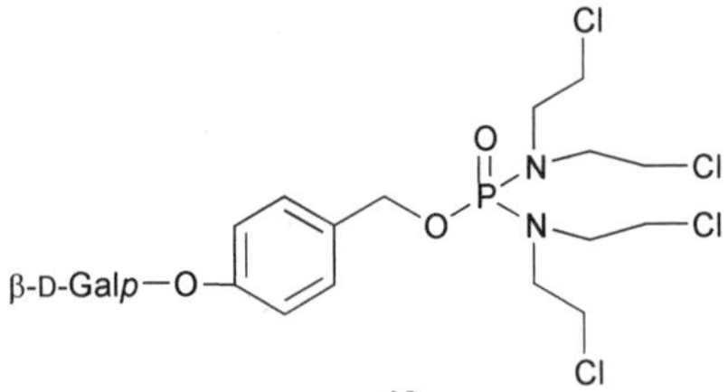

12

than the glycoside and equal in cytotoxicity.[51] The benzylic porphyrin glycoside **14** was made for studies of its metal bonding and aggregating properties.[52] β-D-Glucopyranosides of two complex quinoline and imidazo[1,2-*a*]pyridine derivatives have been made as good reversible proton pump inhibitors,[53] and a β-D-glucosyloxymethyl-2*H*-benzofuro[3,2-*g*]-1-benzopyran-2-one has been synthesized.[54]

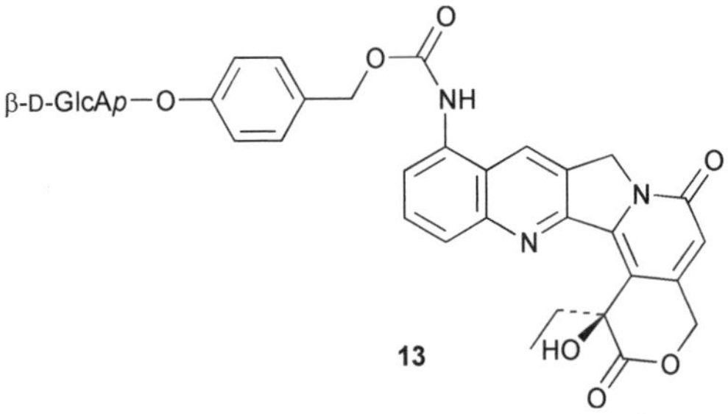

13

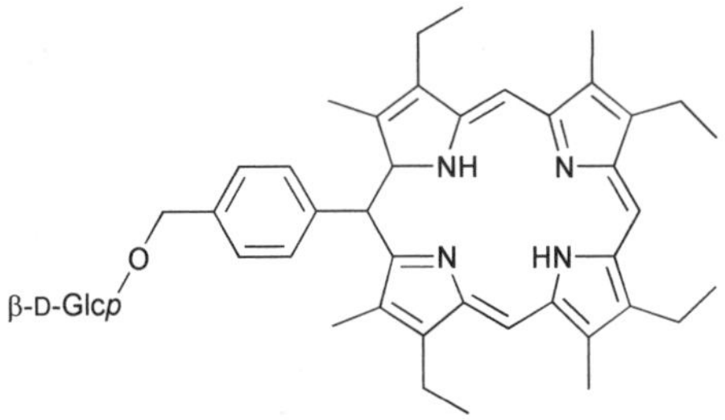

14

Many *O*-glycosides have been made for the purposes of linking the carbohydrate to another type of compound by way of a bridging unit. Ethylene glycol commonly forms the link and has been used to bond the 3-amino-2,3,6-trideoxyhexose angolosamine to acid **15** to give an ester with DNA-cleaving properties.[55] β-D-Glucopyranose has been similarly linked to block-copolymers of styrene and acrylic acid to give glycopolymers which formed micelle-like spheres, vesicles and tubules.[56] Mannose, lactose and α-D-Gal-(1→3)-D-Gal glycosides of 8-*N*-acryloylamino-3,6-dioxaoctanol were copolymerized with a view to using the mannose to bind to bacterial mannose receptors and the other sugars as antigens.[57] β-D-Galactopyranose linked by chemoenzymic methods to a polyethyleneglycol polymer carrying distearoylphosphatidic acids gave functionally active glyco-lipo polymers.[58]

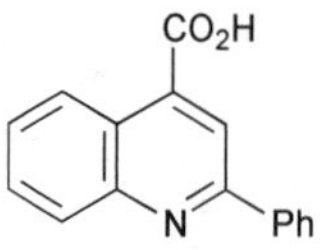

15 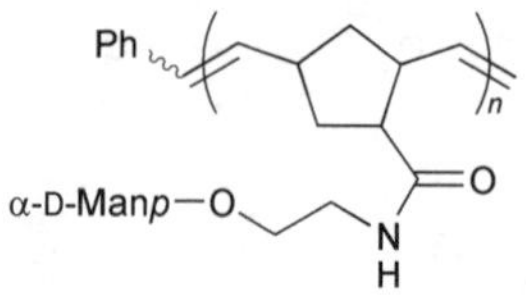 **16** 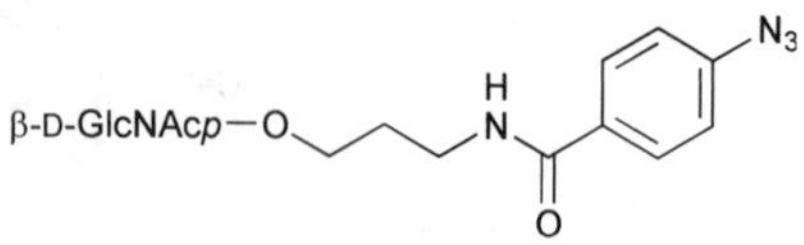 **17**

Aminoalcohols can also be used to act as linking agents. Compounds with structure **16** bind to cell surface receptors to induce a variety of biological responses,[59] and the putative photoaffinity label **17** and 14 analogues were tested as acceptor substrates for Nod C glycosyltransferase from a rhizobium.[60]

ω-Bromoalkanol glucosides and galactosides have been used to make active transport systems for antimicrobials, for example **18**, which are a set of glycosyl-linked derivatives of cipralfloxacin.[61]

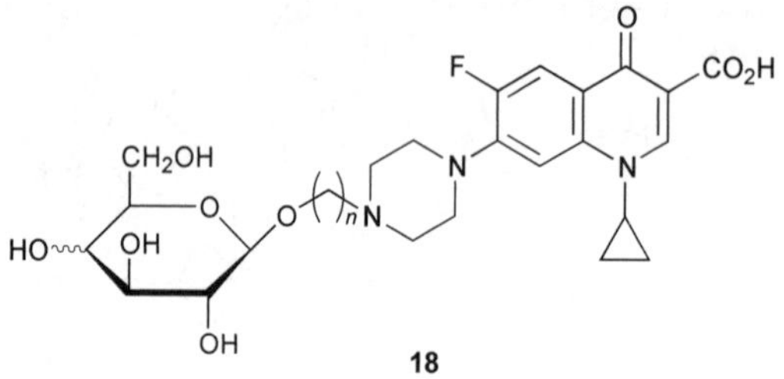

18

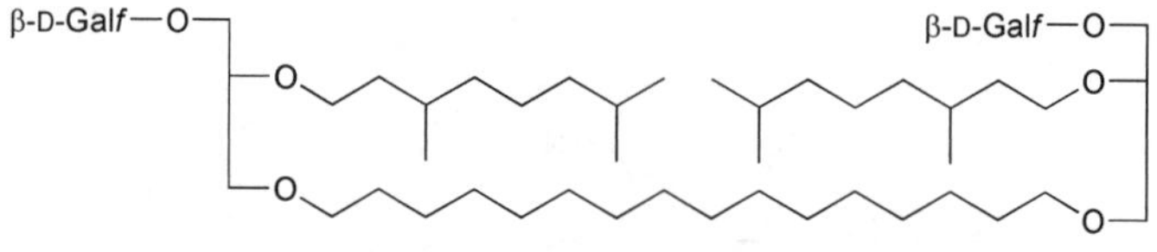

19

Advances in interest in complex glycosidic compounds is increasing the number of synthetic compounds bearing more than one sugar moiety. Compound **19** has been made (together with symmetrical and unsymmetrical analogues) by use of pent-4-enyl glycoside technology.[62] Tetra-*O*-acetyl-β-D-glucopyranosyl 'stopper' end substituents have been used in a polyether rotaxane having a central azobenzene unit the *trans*- to *cis*- orientation isomerization of which was effected photochemically.[63]

The feature **20** formed the central part of an β-galactose-based neoglycopeptide made to inhibit the binding of globotriosylceramide with verotoxin. The amido group was peptide-linked, while the amino groups were spacer-bonded to 3,7-anhydro-2-deoxy-D-*glycero*-L-*gluco*-octonic acid (an α-D-galactopyranose *C*-glycoside).[64] In a related study 12-β-D-glucopyranosyloxydodecyl groups (and β-D-galacto- and α-D-manno-analogues) were bonded to the oxygen atoms of **21** to give compounds with cationic centres that bind to DNA

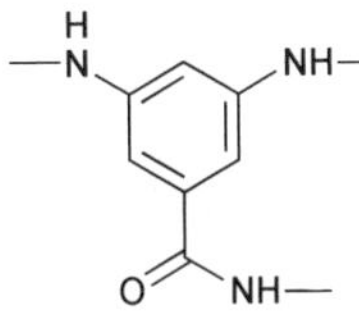

20 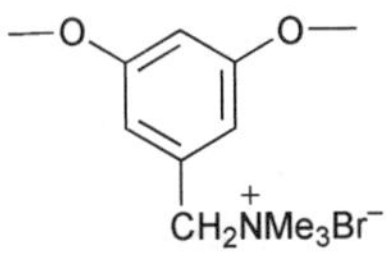 **21** 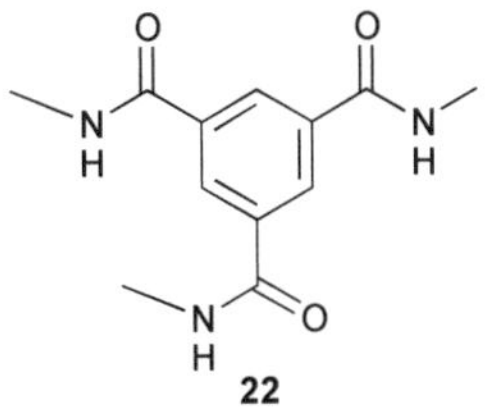 **22** 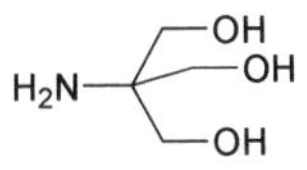 **23**

and sugar end groups to act as cell-targeting ligands.[65] Unit **22** formed the centres of multivalent dendritic saccharides having α-D-mannopyranose linked *via* spacers to the amido groups or to trivalent benzenoid units which were bonded to **22**. Some simpler, related spacer-linked mannosides were also reported, and X-ray analyses were conducted on di- and tri-valent ligands bound to concanavalin A. The multivalent arrays did not increase binding but led to aggregation.[66] Compound **23** formed the core part of a synthetic glycolipid amphiphilic nitrone, a new spin trap able to trap free radicals in aqueous media. The nitrone was bonded to the amino group by a polyamide spacer and two of the hydroxyl groups bore β-D-galactopyranosyl groups while the other was carbamate-linked to a hydrophobic chain.[67] The same aminotriol was used to build amphiphilic dendritic galactosides (an alkyloxy steroid amide spacer group linked to NH_2 and the sugars similarly linked to OH) required for targeting liposomes to the hepatic asialoglycoprotein receptor.[68] (A digalactosylated ceramide is noted in ref. 76). Fluorescein-labelled lysinyl 'tree' compounds containing two, four or eight mannose residues were made to probe the interactions with mannose receptors. D-Galactosyl analogues were made as control compounds.[69]

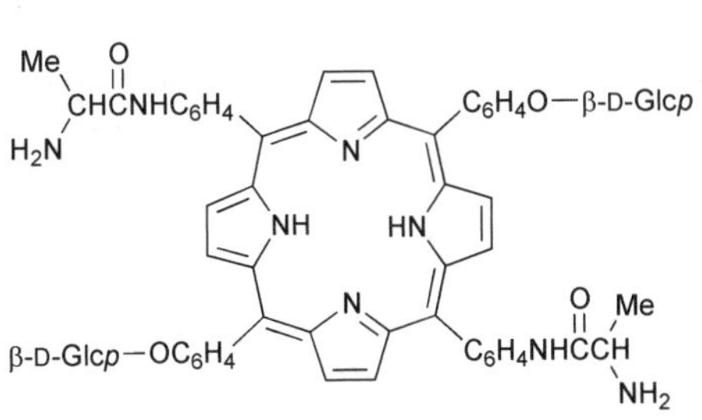

24 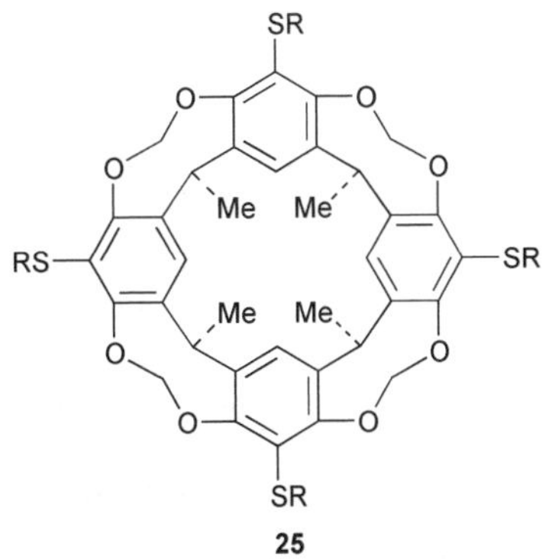 **25**

Compound **24** and analogues were made as glycosylated porphyrins bearing amino acids for use in photodynamic therapy against malignant cells.[70] The cluster compounds **25** (R = β-D-Glc(*p*)OCH_2CH_2- and maltose and maltotriose analogues) were produced for binding studies.[71]

Polygalactosylation of serum albumin is noted in ref. 82, and a silyl linker has been used to develop solid phase syntheses of oligomers of threonine and serine having sugars (especially GalNAc) substituted on each hydroxyl group. By this approach the glycophorin A_M fragment was made.[71a]

1.2 Synthesis of Glycosylated Natural Products and Their Analogues. – Compounds with acyclic aglycons continue to attract attention and glycerides remain of significant interest. The synthesis of the novel aminoglycoglycerolipid **26**, a species-specific antigen of *Mycoplasma fermentans*, has been

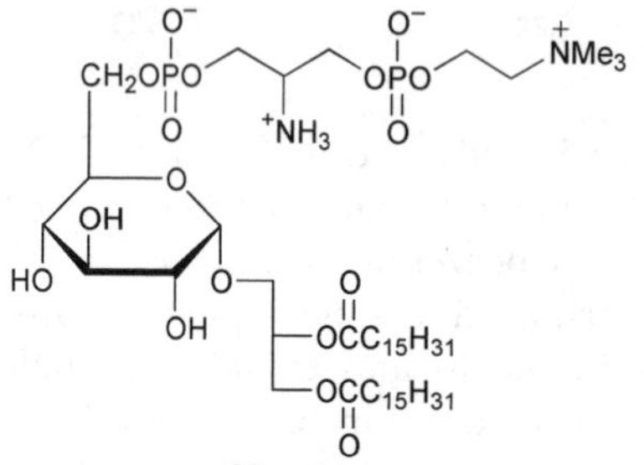

completed, the work establishing the absolute configuration at the asymmetric centre of the phosphate side chain.[72] The related glucoconjugate **27**, having retinoic acid ester functions, and its diastereomer at the secondary glycerol centre have been made as substrate models for epidermal β-glucocerebrosidase. Both isomers bound to the enzyme, the illustrated one hydrolysing four times faster than the other.[73] The analogue of **27**, which is the 1′-β-D-glucopyranoside of D-*glycero*-butane-1′,3′,4′-triol having an acetal derived from the symmetrical $(C_{16}H_{33})_2CO$ ketone at O-3′, O-4′, undergoes self-assembly. Conformational analysis by detailed NMR experiments was carried out.[74] D-Ribitol-based compound **28** and the analogue with an L-ribitol linking unit were made to establish the D-absolute configuration of the ribitol moiety in the C-polysaccharide of *Streptococcus pneumoniae*.[75] A ceramide based on 2-acylaminoctadecan-1,3-diol having α-D-galactopyranosyl substituents at both the hydroxyl groups (and the isomer with the sugars α- and β-linked) were made to study their effects on the proliferation of murine spleen cells.[76] The set of glycoside derivatives **29** with $n = 1$, 2, 3, 5 or 7 was made and screened for immunostimulant activity and several members were found to enhance the production of tetanus antibodies in mice and to augment cytotoxic T cells.[77]

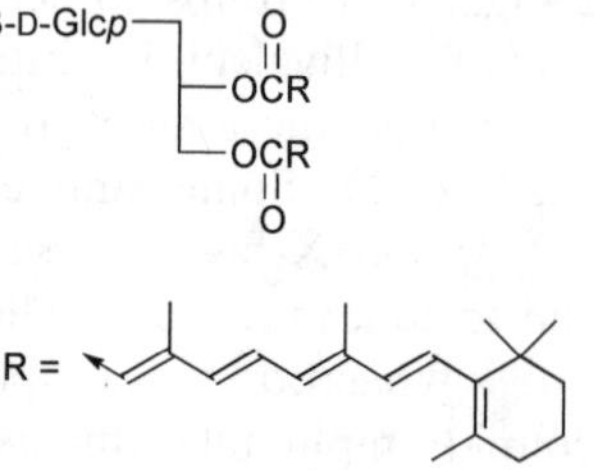

Compound **30**, a derivative of galactosylhydroxylysine (GHL), which is released into the serum during bone resorption, has been made and trans-

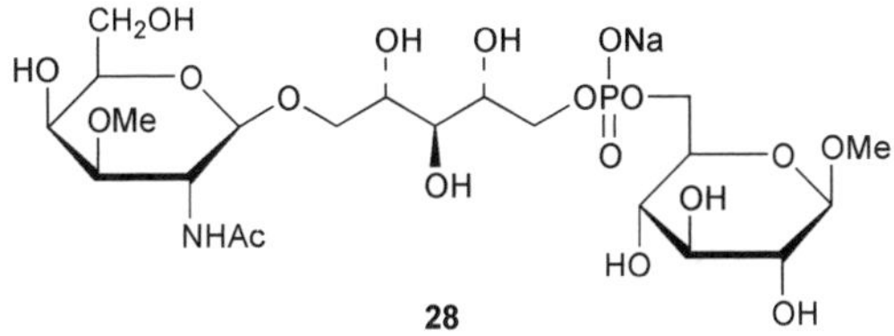

28

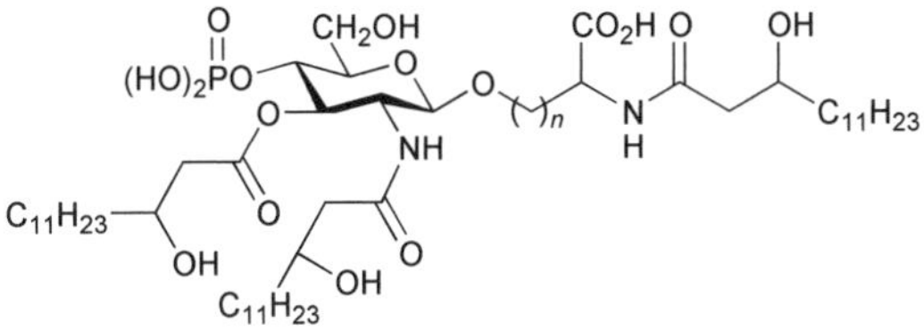

29

formed into immunogens for generation of anti-GHL antibodies. It was also tagged with fluorescent and chemiluminescent labels.[78]

In the area of *O*-glycosidic aminoacids and peptides *O*-β-D-xylopyranosyl-serine 2-phosphate and its derivative carrying a β-D-galactopyranosyl group at O-4 were made as well as the 6-sulfate of the latter compound.[79] The analogous α-D-GalNAc-*O*-Ser and -*O*-Thr building blocks for glycopeptide synthesis have been made by the unusual means of Michael addition of the suitably carboxy- and *N*-protected amino acids to C-1 of 3,4,6-tri-*O*-benzyl-2-nitro-D-galactal in the key step.[80] Stepwise peptide synthesis was used to produce β-D-Glc*p*O-Thr-(Ala)-Phe and the analogue with serine in place of threonine.[81] Human serum albumin has been galactosylated (*ca.* 30 units/protein molecule) by use of a 1-thiogalactosamine derivative. The product was labelled with technetium and used with good results as a hepatic receptor imaging agent.[82]

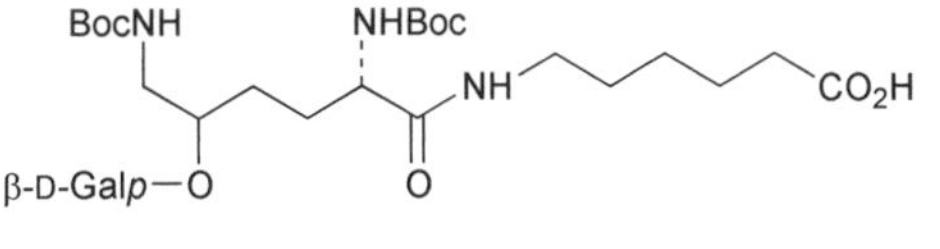

30 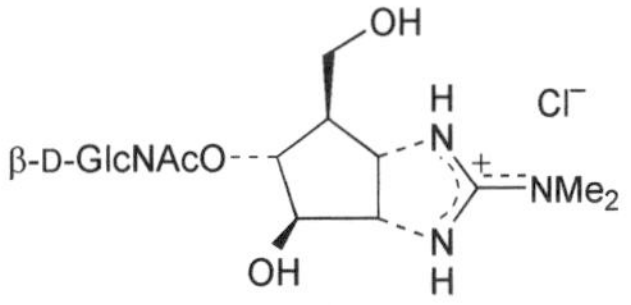 **31**

In the field of glycosylated carbocycles two glucosaminyl *myo*-inositol monophosphoglycerides, which are analogues of an early intermediate in the biosynthesis of GPI membrane anchors, have been made.[83] A related, extensive study has described the syntheses, structures and biological activities of several related compounds, some having α-D-Man*p*-(1→4)-linked to the glucosamine, some acylglycerol phosphate esters and some cyclic phosphate ester groups.[84] In the field of potential chitinase inhibitors compound **31** was made from a chitobiose oxime derivative which was radical cyclized to give a main product having the required vicinal *cis*-related amino groups suitable for production of the fused ring system of the product.[85]

In work connected with the synthesis of C1027, an enediyne anti-tumour antibiotic, the trichloroacetimidate **32** was condensed with the tertiary, allylic alcohol **33** to give 56% of the β-glycoside in what must be a very unusual coupling.[86] An allied synthesis has given the analogue **34** of the neocarzinostatin chromophore.[87] From carbomycin A a new semisynthetic route to forocidin (**35**, R = 3,6-dideoxy-3-dimethylamino-D-glucosyl, mycaminosyl) has been developed.[88]

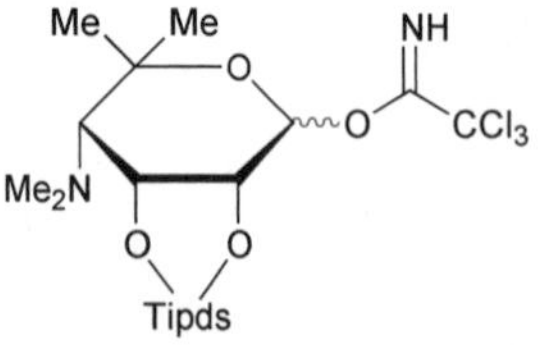

32 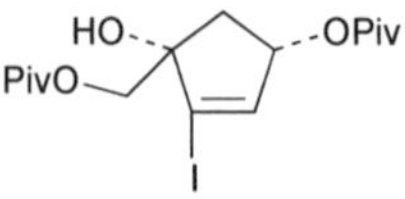 **33** 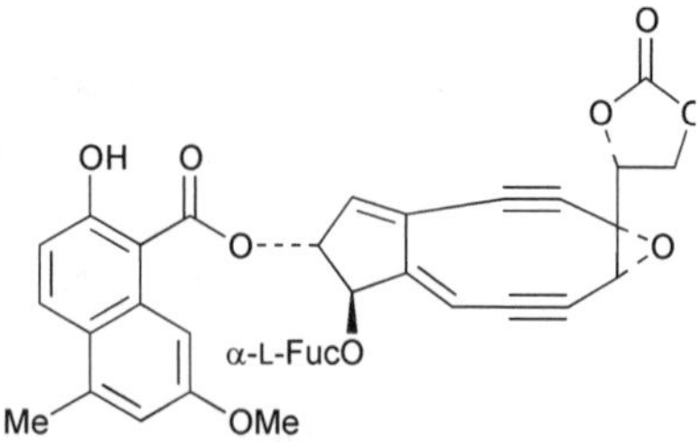 **34**

Helferich β-glucosylation has been used for the total synthesis of methyl β-D-glucopyranlosyloxyjasmonate and an epimer.[89] Glycosylations using *O*-benzoylated trichloroacetimidates have been applied to sapogenins on the 50 g scale with yields approaching theoretical.[90] Improved syntheses of estrone and related aromatic steroidal 3-β-glucuronides have been effected using glycosyl trichloroacetimidates.[91] Other aryl glycosides to have been reported are several flavone derivatives[92] and compound **36** which is aceroside IV present in a Japanese tree.[93] Both reports describe the use of standard glycosyl bromide reactions. Two Chinese groups have also used this approach in the synthesis of salidroside [2-(*p*-allyloxyphenyl)ethyl β-D-glucopyranoside],[94,95] and the related disaccharide arylethyl glucoside acteoside (**37**) has also been prepared.[96]

The β-D-*galacto*-and α-D-*manno*-analogues of potassium lespedezate[38] (**38**)

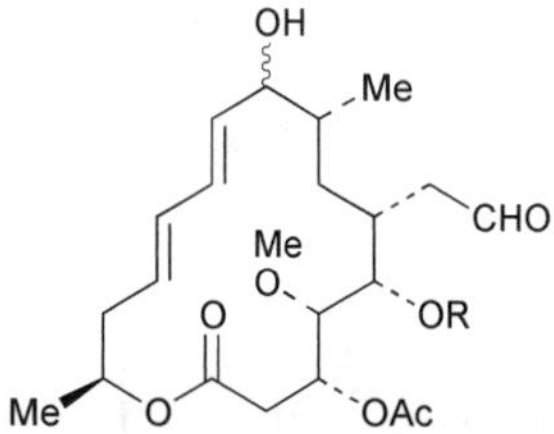

35 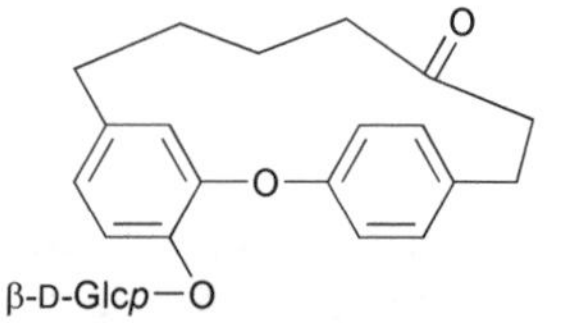 **36**

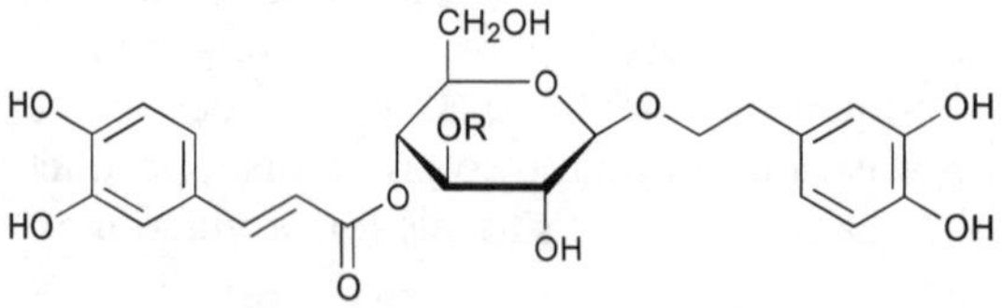

37 R = α-L-Rha*p* 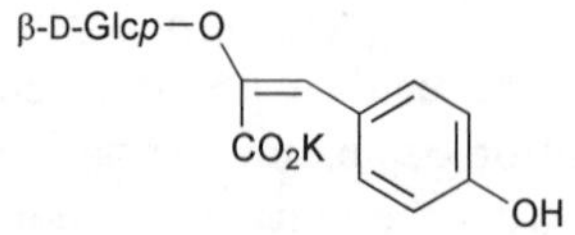 **38**

are as effective leaf-opening compounds as the natural glucoside, and as they may not be hydrolysed in plants they may be useful as control compounds.[97,98]

The recently discovered modified nucleoside 5-(β-D-glucopyranosyloxymethyl)-2β-deoxyuridine (**39**) and its α-anomer have been made by trichloroacetimidate-based methods.[99] Likewise the β-D-$GlcNH_2$ analogue was made and incorporated into oligodeoxynucleotides. Binding affinity, especially to complementary RNA, was reduced.[100] In closely related work 8-(glycosyloxy)purine nucleosides were produced as analogues of the chitin synthase inhibitor guanofosfocin, α-D-mannofuranose and -pyranose being the sugars used.[101] A 3-α-L-fucopyranosyl ribonucleoside is noted in the disaccharide section.

1.3 *O*-Glycosides Isolated from Natural Products. – As always, only a selection on papers published on this aspect of glycoside chemistry is dealt with: either those reporting interesting structural features within the carbohydrate moieties or those describing notable biological activities. The α-D-galactopyranoside and β-D-glucopyranoside of lactones **40** and **41**, respectively, have been isolated from a fungal phytopathogen[102] and a *Phyllanthus* plant.[103] The latter has leaf-closing properties as does the β-D-glucopyranoside

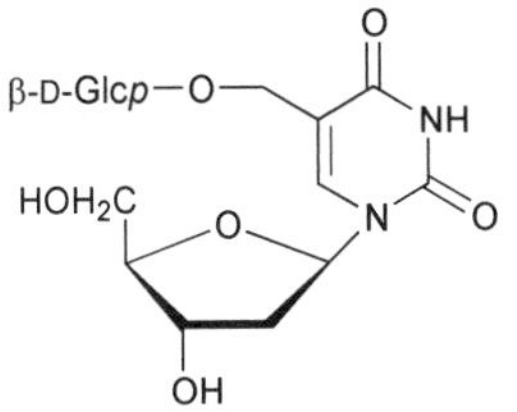

β-D-Glc*p*—O
HOH₂C
NH
N
O
OH
39

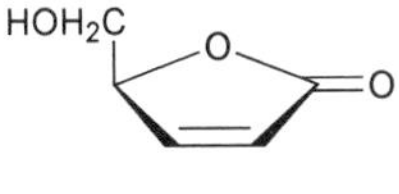

HOH₂C
O
40

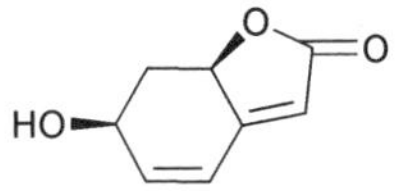

HO
O
41

of 11-hydroxyjasmonate isolated from an *Albizzia* plant,[104] while the slow periodical movements of *Mimosa* are caused by the 5-β-D-glucopyranoside of potassium 2,5-dihydroxybenzoate.[105,106]

Marine sources continue to yield glycosphingolipids, one being the terminal α-D-glucopyranoside of compound **42**.[107] Another is a related α-D-galactoside having a singly unsaturated aglycon and choline phosphonate linked to C-6 of the sugar unit.[108]

New antiinflammatory macrolides with kijanose [2,3,4,6-tetradeoxy-3-*C*-methyl-4-(methoxycarbonyl)amino-3-*C*-nitro-D-*xylo*-hexose], or the corresponding reduced 3-amino analogue, and a hetero-2,6-dideoxyhexose trisaccharide substituent, have been isolated from fermentation broths of a marine bacterium growing on a brown alga.[109]

1.4 Synthesis of Disaccharides and Their Derivatives. – *1.4.1 General.* A novel method for condensing *O*-benzylated -OH sugars with alcohols involves the use of the acid $H_4SiW_{12}O_{40}$ in acetonitrile.[110] An example is given under 'Deoxy-sugar disaccharides' (Section 1.4.7). Several hexopyranosyl-(1→4)-α-D-mannopyranosyl phosphates including β-D-talose and β-D-gulose derivatives have been reported, the disaccharides being prepared by trichloroacetimidate methods and the phosphate bonds by the glycosyl hydrogen phosphonate procedure. They were required for studies of the acceptor substrate specificity of biosynthetic enzymes of *Leishmania*.[111]

Disaccharides having 2′-deoxy-2′-fluoroglycosyl non-reducing moieties have been produced from glycals by additions involving the use of Selectfluor to introduce F^+ at C-2 and selectively protected sugars to provide the aglycons. In this way 2-deoxy-2-fluoro-α-D-mannopyranosyl-(1→6)-di-*O*-isopropylidene-D-galactose has been made.[112]

3-Nitro-2-pyridyl and 5-nitro-2-pyridyl glycosides have been found to be better than corresponding *p*-nitrophenyl compounds as glycosyl donors in enzymic transglycosylations catalysed by glycosidases such as β-galactosidases, β-glucosidases and *N*-acetylhexosaminidases. Their high solubility in water and high reactivity permit reactions at high concentrations of donors and hence rapid product formation.[113] Several β-glycosidases from *Thermus thermophilus*, using nitrophenyl β-glycosidic donors cause *o*-nitrophenyl β-D-gluco- and -galactopyranoside to self-condense faster than they react with other acceptors, the products being the nitrophenyl 1,3-linked homodisaccharide glycosides. In the case of *p*-nitrophenyl β-L-fucopyranoside in the presence of methyl α-D-galactopyranoside β-L-Fuc-(1→6)-α-D-Gal-OMe is the main product.[114] Of potential importance is the use of phenylboronic acid derivatives as promoters to activate specific groups as acceptors for glycosylation of polydroxy compounds. See ref. 129 and Chapter 7.

1.4.2 Non-reducing disaccharides. A review on organic transformations catalysed by zeolite-like materials briefly covers their effects on the TmsOTf-promoted dimerization of sugar derivatives unprotected at O-1. 2,3,4,6-Tetra-*O*-benzyl-D-glucose gives the α,α- α,β- and β,β-trehaloses in 76% yield and in the ratios 2.8:10:0.1. In the case of 2,3,5-tri-*O*-benzyl-D-arabinose only the α,α-product was obtained (62%).[115] In related work reaction of 2,3,4,6-tetra-*O*-acetyl-β-D-glucose with acetobromoglucose in the presence of silver triflate and 2,4-lutidine gave an orthoester-linked disaccharide product in 92% yield. With TmsOTf this rearranged to β,β-trehalose octaacetate in 81% yield. In like manner non-reducing β,β-tetramers were made from lactose and maltose esters.[116]

Chiral disaccharide phosphines **43**, made from α,α-trehalose-derived 4,6-*O*-protected-2,3-anhydro-D-*allo*-derivatives and corresponding phosphine oxides, were prepared.[117] See Section 1.4.7 for 3,3′-dideoxytrehalose.

1.4.3 Glucosyl disaccharides. Sophorose [β-D-Glc-(1→2)-D-Glc] has been identified within the structures of mono-*O*-alkyldiglycosylglycerols isolated from natural sources,[118] while *p*-nitrophenyl laminaribioside [based on β-D-Glc-(1→3)-D-Glc] was produced in 19% yield from laminaran and *p*-nitrophenyl β-D-glucopyranoside by cleavage with an enzyme from a marine mollusc.[119] A laminarin glycoside was made by intramolecular methods from compound **44** which is stereochemically constrained by the novel use of a rigid spacer. In this

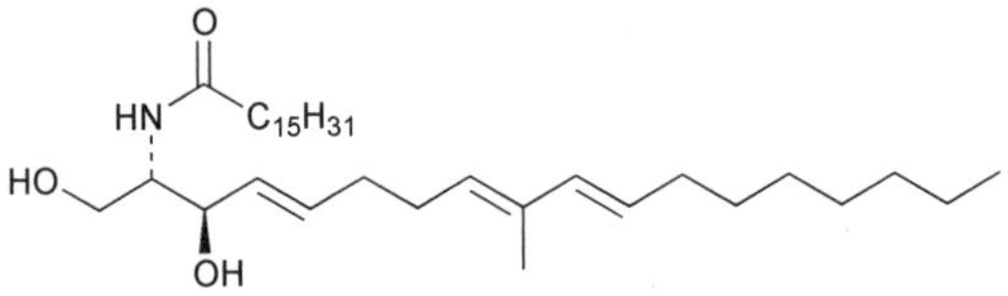

42

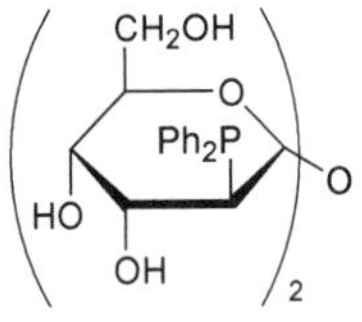

43

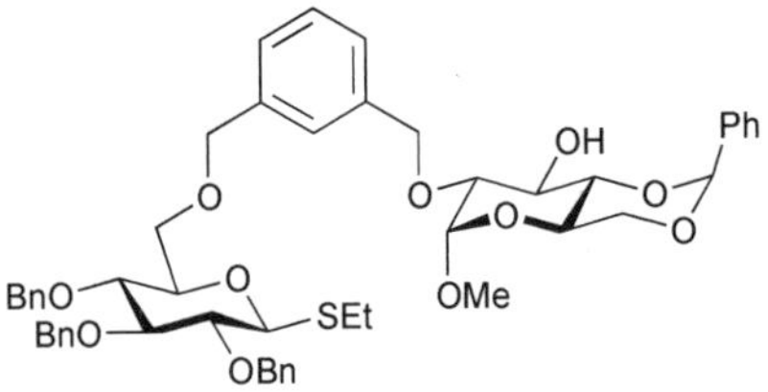

44

way the stereochemistry of coupling was cleverly controlled. The α-(1→3)-linked isomer was also made as well as the 1,4- and 1,6-linked compounds by this approach.[120] Schmidt and colleagues, who developed this approach, have studied the conformational mobility of tricyclic compounds comprising cellobiose 6,6- and 3,6-linked by *m*-xylylene bridges.[121] A related approach to 1,4-glucobioses based on the use of various 6,6′-diacyl and silyl bridges has also been reported. The work was extended to produce tetrasaccharides by this procedure.[122]

A less sophisticated approach to maltosides involved coupling of 2,3,4,6-tetra-*O*-benzyl-D-glucose with a 4-*O*-unprotected glucoside by use of $TmsClO_4$, $(CCl_3CO)_2O$ in the presence of a specific sieve. Good yields and α-selectivities were recorded and isomaltose analogues [α-(1→6) linked] were also made.[123]

Methyl α-(4-^{2}H)-cellobioside was made from methyl 2,3-di-*O*-benzyl-α-D-*xylo*-hex-4-uloside which was reduced by use of $NaB^2H(OAc)_3$ prior to 6-substitution and β-glucosylation.[124]

A calorimetric study of structural relaxation in a maltose glass has been reported.[125]

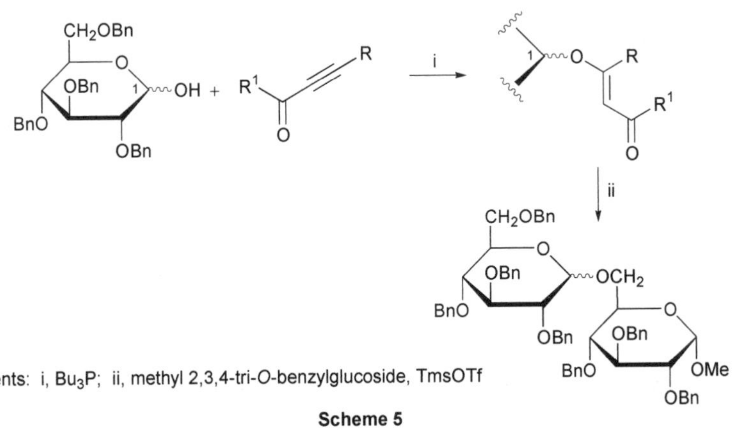

Reagents: i, Bu_3P; ii, methyl 2,3,4-tri-*O*-benzylglucoside, TmsOTf

Scheme 5

A novel glycosylation method illustrated in Scheme 5 utilizes a novel enol ether and gave the 1,6-link glucobiosides in 82% yield with α,β ratio 4.5:1. The advantage of the approach is the short reaction times required and the low temperatures (−50 °C).[126]

Methyl 3,4-di-*O*-benzyl-α-D-glucopyranoside was glycosylated selectively at O-6 with a 6′-specifically substituted trichloroacetimidate to give access to a 2,6′-diunprotected compound with α,β selectivity 2:1.[127] Several glucosyl derivatives (of which SPh compounds were best) bonded *via* O-6 by silyl tethers to resins were studied as donors in disaccharide synthesis procedures.[128]

Interestingly, phenylboronic acid derivatives, rather than acting as diol protecting agents, can be used to activate specific hydroxyl groups, and acetobromoglucose as a glycosylating agent, coupled with methyl α-L-fucoside in the presence of a diarylborinate derivative, leads for example to 3-D-glucopyranosyl-L-fucose with good selectivity. Other examples of selective substitution were given[129] (see also Chapter 7).

Compound **45** was α-glucosylated at O-6 to give an isomaltose-bearing monoterpene indole alkaloid found in a Thai medicinal plant.[130]

Isomaltulose (6-*O*-α-D-glucopyranosyl-D-fructose) gave the glycosylated acetylated furfural derivative **46** together with acetylated allyl glucosides amongst the products of its treatment with allyl alcohol and HCl followed by acetylation.[131]

α-D-Glc-(1→2)-β-D-Gal has been *O*-bonded to hydroxylysine in work aimed at type II collagen glycopeptide synthesis,[132] and α-D-Glc-(1→4)-β-D-Gal linked to a hydroxyl group on the central carbon atom of a C_{33} alkane (and some modifications thereof) has been isolated from sponges and found to inhibit the proliferation of activated T-cells.[133]

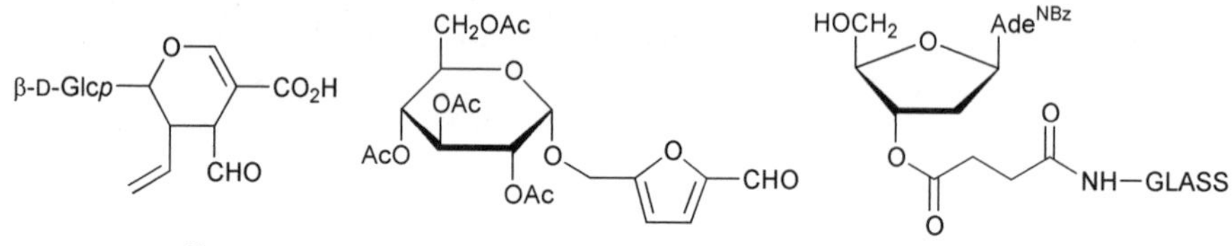

45 **46** **47**

β-D-Glc-(1→6)- and -(1→3)-D-Man glycosides have been made by glucosylations using acetobromoglucose and partially protected α-mannosides. They proceeded by way of orthoester-bonded disaccharides that were rearranged to β-glucosyl compounds by treatment with TmsOTf.[134]

Glass bound nucleoside **47** was β-D-glucosylated at O-5′ and the glycosylated nucleoside was incorporated into one end of oligonucleotides that were phosphate-bonded to O-6 of methyl α-D-glucopyranoside thus giving oligomers with glucose residues at both ends.[135]

1.4.4 Mannosyl disaccharides. α-D-Man-(1→6)-α-D-Man benzyl glycoside derivatives have been made by use of a hyperbranched polyester soluble support and a photolabile *o*-nitrobenzyl aglycon.[135a] Methyl α-D-mannopyranoside and ethyl 2,3,4-tri-*O*-benzyl-1-thio-α-D-mannopyranoside each 6-*O*-linked to acid groups of a peptide, on activation of the thioglycosidic centre, gave variously linked cyclic methyl mannosyl-mannosides, the regio- and stereo-selectivities of the observed couplings being rationalized by molecular modelling.[136] ^{13}C-Labelled methyl 2-*O*-α-D-mannopyranosyl-α-D-mannopyranoside has been prepared (from U-^{13}C-D-glucose),[137] and 3-*O*-α-D-mannopyranosyl-D-mannose was made *via* an orthoester-linked intermediate.[134]

3,4,6-Tri-*O*-benzyl-2-*O*-(*p*-methoxybenzyl)-α-D-mannosyl fluoride was linked *via* a *p*-methoxybenzylidene bridge to O-4 of benzyl 2,3,6-tri-*O*-benzyl-β-D-glucopyranoside to give compound **48** with the (S)-configuration at the acetal centre. On the other hand, the configuration at this centre was *R* when the acetal was made in the reverse way with the *p*-methoxybenzyl group at O-4 of the benzyl glycoside and the hydroxyl group at C-2 of the fluoride. The acetal products can act as precursors of 4-*O*-mannosyl-D-glucosides, the stereochemical significance of the acetal centre on the glycosylation now being open to examination.[138] A different approach to making mannosyl disaccharides intramolecularly involves such linked systems as in **49**, made by the Tebbe reaction applied to the corresponding benzoate, which, on treatment with phenylselenyl chloride and silver triflate, gave α-D-Man-(1→6)-D-Glc in 80% yield as the only product. The α-(1→4)- isomer was also made by the approach, and when analogous α- and β-glucosyl donors were used, yields were similar and α,β-ratios were 1:2–1:20.[139]

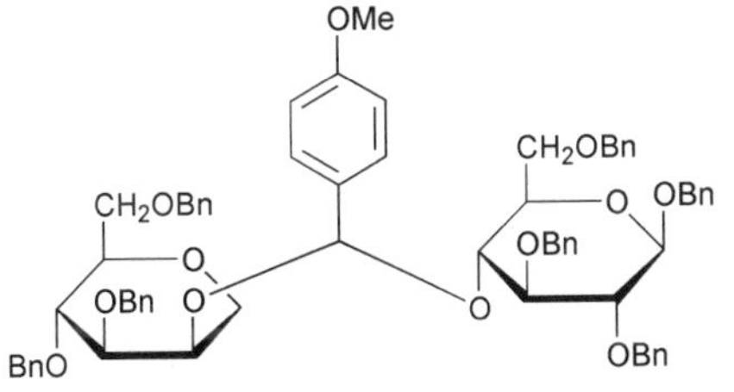

48 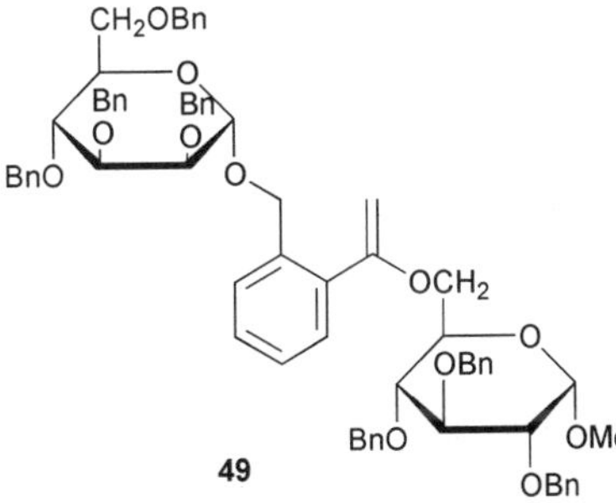 **49**

α-D-Man-(1→4)-α-D-GlcA-(1→2)-*myo*-inositol is a new signalling oligosaccharin found in roses and thought to be derived from membrane-bound glycophosphosphingolipids.[140] Phenyl 3,4,6-tri-*O*-benzyl-2-*O*-isopropenyl-1-thio-α-D-mannopyranoside has been used to prepare a derivative of β-D-mannopyranosyl-(1→6)-D-galactose with a free hydroxyl group at C-2′, and the in the course of the work an analogously protected α-D-Glc-(1→6)-D-Gal compound was described.[141]

α-D-Man-(1→5)-β-D-Ara*f*-1-octyl has been isolated from *Mycobacterium tuberculosis* together with (1→2)-linked mannobiosyl- and mannotriosyl-arabinosyl analogues, all of which have been synthesized from octyl 2,3-di-*O*-benzyl-β-D-arabinofuranoside.[142]

In the area of D-mannofuranosyl disaccharides 6-*O*-allyl-1,2-anhydro-3,5-di-

O-benzyl-β-D-mannose has proved to be a useful glycosylating agent from which derivatives of α-D-Man*f*-(1→6)-D-Glc, α-D-Man*f*-(1→6)-D-Gal and α-D-Man*f*-(1→5)-D-Xyl have been made in good yield.[143]

1.4.5 Galactosyl disaccharides. Ethyl 2,3,6-tri-*O*-benzyl-1-thio-β-D-galactopyranosides bearing different substituted benzoates at O-4 have been assessed as iodonium-promoted galactosylating agents. The *p*-methoxybenzoate gave best α-selectivity and various α-galactosyl disaccharides were made with its assistance.[144]

Ethyl 2,3,4,6-tetra-*O*-benzyl-1-thio-β-D-galactopyranoside is a better glycosylating agent than *p*-tolyl 4-*O*-acetyl-2,6-di-*O*-benzyl-1-thio-β-D-galactopyranoside so that thiophilic activators applied to a mixture give α-D-Gal-(1→3)-β-D-Gal-*S*-Tol in good yield. When the ethylthio glycoside has a disarming benzoate at C-2 it becomes a poorer glycosylating agent than 2-*O*-benzyl tolylthio compounds.[145] The 2-(tetradecyl)hexadecyl glycoside of β-D-Gal-(1→6)-β-D-Gal and related compounds have been made during work on sphingolipid analogues.[146]

Most synthetic work on galactobioses has been conducted enzymically. A thermophilic bacterial enzyme made β-D-Gal-(1→3)-D-Gal-*o*-$C_6H_4NO_2$ as the main product from *o*-nitrophenyl β-D-galactoside, and the corresponding α-(1→3)-linked product was derived stereoselectively from *o*-nitrophenyl α-D-galactoside by use of a coffee bean enzyme.[147] In related work using a bacterial β-galactosidase, β-D-galactose has been transferred to O-4 of several glycosides of Gal and GlcNAc.[148] *p*-Nitrophenyl β-glycosides of various D-galactose analogues carrying CH_3, CH_2F, $C{\equiv}CH$, $CH{=}CH_2$, CH_2CH_3, or $CH(OH)CH_3$ groups at C-5 have acted as glycosyl donors in the presence of snail or barley enzymes to give 1,4- and 1,6-linked modified galactobiosides based on methyl α-D-galactopyranoside as acceptor.[149]

In the field of galactosylglucose compounds an α-D-Gal-(1→4)-D-Glc glycoside has been made in high yield by the intramolecular approach involving a phenyl thiogalactoside linked 2′→3 by a succinoyl bridge to an (glucoside with free hydroxyl group at C-4. Reactions of similar 6′→3 tethered compounds were also investigated.[150] Galactosylation with tetra-*O*-benzyl-α-D-galactopyranosyl iodide of diacetoneglucose in benzene afforded D-Gal-(1→3)-D-Glc disaccharides in 91% yield with α,β-ratio 9:1. With acetonitrile as solvent, however, β-products predominated. In the course of the work fucose analogues of trehalose were made with the α,α- and β,β-linkages.[151] Peracetylated 3-nitro-2-pyridyl-β-D-galactoside with methyl 2,3,6-tri-*O*-benzyl-α-D-glucopyranoside reacted in the presence of TmsOTf to give the lactoside product in 78% yield.[152] Allyl lactoside has been made by glycosylation on a resin to which the glucosyl residue was held by an alkene tether which was cleaved by use of Grubbs' reagent.[153] β-D-Galactose has been linked 1→3, 1→4 and 1→6 to Glc-1-NHAc by enzymic methods using a bacterial β-galactosidase and *p*-nitrophenyl β-D-galactopyranoside as donor. Yields were up to 41% with 30% acetone as solvent, but the reaction showed poor regioselectivity (35:47:18).[154]

β-D-Gal-(1→4)-β-D-Man-1-OPO_3H_2, the phosphodisaccharide repeating unit of the antigenic lipophosphoglycan of *Leishmania donovani*, has been made from lactal peracetate *via* the 1,2-anhydro-β-manno-derivative,[155] and α-D-Gal-(1→2)-D-Man was produced as a thioglycoside by glycosylation of a suitably protected thiomannoside with a β-1-thiogalactoside derivative.[156]

Chemical and enzymic methods continue to be used to produce *N*-acetyllactosamine and its derivatives. A novel approach starts with lactulose which, with benzylamine followed by acetic acid/methanol, gives *N*-benzyllactosamine which was converted to LacNAc and other *N*-substituted derivatives.[157] Relative performances of variously *O*-substituted phenyl 1-thiogalactosides and tetra-*O*-acetyl-α-D-galactosyl trichloroacetimidate were compared as glycosylating agents of 6-*O*-Tbdps-GlcNAcOAll. Best yields (82%) were obtained with the 2,3,4-tri-*O*-benzoyl-6-*O*-Tbdps thioglycoside, high β-(1→4) selectivity being obtained.[158] Roy's group also reported on the use in oligosaccharide synthesis of a product formed by selective β-galactosylation at O-4 of a 6-protected glucosamine derivative.[159] The synthesis of an LacNAc-peptide derivative from a resin-bound glycal compound has also been reported.[160]

p-Nitrophenyl β-D-galactopyranoside as donor and D-GlcNAc give 75% yield of LacNAc exclusively with a bacterial β-galactosidase, whereas with the *o*-substituted compound the yield was 40% and 2% of the β-(1→6) linked isomer was also formed. With the phenyl glycosidic donor only 14% lactosaminide was obtained together with traces of the (1→6)-linked by-product.[161] An enzymic method for making LacNAc from orotic acid (a pyrimidine carboxylic acid to provide UTP), galactose and GlcNAc yielded the product at a concentration of 107 g L^{-2} on a 2–5 L scale.[162]

A panel of 13 substituted LacNAc derivatives was synthesized to allow the assessment of enzyme-substrate interactions for a transferase.[163] Derivatives of 5′-carba-β-lactosaminide are referred to in Chapter 18.

An enzymic method was used to make a glycoside of β-D-Gal-(1→3)-α-GalNAc which was subsequently sialylated at O-3′,[164] and the same disaccharide was made chemically and coupled to L-serine *en route* to T-antigen.[165] Related work provided the disaccharide linked to a peptide from which glycopeptide antifreeze protein was produced.[166]

Preparation of a derivative of β-D-Gal-(1→4)-α-D-ManNAc led to compounds based on dimeric or trimeric ethylenoxy-α,ω-ols bearing the disaccharide on each alcohol group.[167]

D-Galactofuranose compounds remain of interest for biological reasons and β-D-Gal*f*-(1→3)-D-GlcNAc, β-D-Gal*f*-(1→6)-D-GlcNAc[168] and β-D-Gal*f*-(1→4)-α-L-Rha*p*-OC_8H_{17}[169] have been made, the last as a probe for potential inhibitors of mycobacterial galactosyl transferases.

1.4.6 Amino-sugar disaccharides. 2-Amino-2-deoxyhexoses continue as key components of important disaccharides, compound **50** being a useful glycosylating agent, activated by Tf_2O, in solid and solution phase glycosylations when several other *N*-substituted glucosaminyl sulfoxides were not. In consequence, a library of solid phase β-linked disaccharides was made. The trifluoroacetyl group is readily removed with LiOH in MeOH.[170]

Work in connection with analogues of Lipid A has produced β-D-GlcNH$_2$-(1→6)-(α-D-GlcNH$_2$ having phosphate groups at O-1 and O-4′ and short chain acyl groups on all the amino and hydroxy functions.[171] Other reports have described divergent syntheses of Lipid A and analogues,[172] and related compounds lacking substituents at O-1 and O-3 whose immunostimulant activities were determined.[173] A further analogue contained a fluorescent label bonded by a bridging unit to O-6′ and an ethylene glycol spacer at O-1, the compound exhibiting similar bioactivity to that of natural Lipid A.[174]

An enzymic procedure, using a β-*N*-acetylhexosaminidase, has given *p*-nitrophenyl β-chitobioside (22%),[175] and the 1,3-linked isomer has been made similarly.[176] In the course of this work GlcNAc was β-(1→3)-linked to ethyl 6-*O*-benzyl-1-thio-β-D-GlcNH$_2$ and to the analogous thiogalactoside compound.

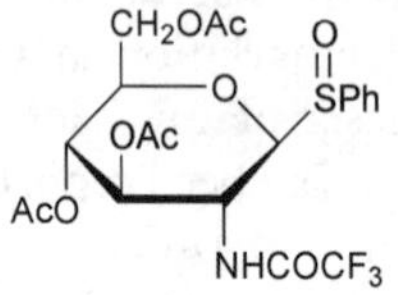

CH_2OAc O SPh OAc AcO $NHCOCF_3$

50

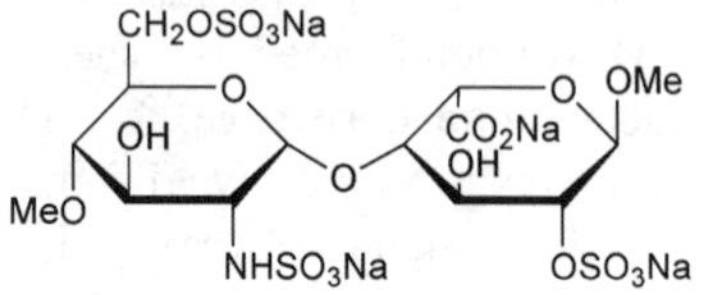

CH_2OSO_3Na O OMe OH O CO_2Na OH MeO $NHSO_3Na$ OSO_3Na

51

In the area of heparin chemistry, disaccharide **51**, closely related to the compound responsible for the binding of heparin to platelets, was converted into cluster compounds containing 2–3 units which were more active than the disaccharide.[177] In related studies a propyl glycosidic analogue of compound **51** without the methyl ether group, and its analogue without the O-6′ sulfate were described.[178]

Work in the area of moenomycin A has produced disaccharide **52**,[179] and in a combinatorial approach 1300 disaccharide derivatives closely related to **52**, but linked through O-1 *via* a phosphate to lipids, were made on a solid support involving a photolabile linkage. Several compounds with antibiotic activity (including against vancomycin-resistant bacteria) were obtained.[180]

The *N*,*N*-dibenzyl glycosylating agent **53** affords high β-selectivity, and by its use GalNH$_2$ has been β-linked separately to each of the hydroxyl groups of octyl β-D-galactopyranoside.[181] GalNAc has been β-(1→3)-coupled by biochemical methods to the ethylthio glycosides of 6-*O*-benzyl-D-Gal and -D-GlcNH$_2$[176] and β-(1→4) coupled to methylumbelliferyl β-D-glucuronide.[182]

1.4.7 Deoxy-sugar disaccharides. There has been considerable activity in the area of disaccharides containing non-reducing deoxy-sugar moieties. Tri-*O*-benzoyl-2-deoxy-glucosyl trichloroacetimidates, α-fluoride and α-diethyl phosphite have been coupled using lithium perchlorate as activator in various solvents with several partially protected hexose derivatives. The fluoride donor gave the highest α-selectivity.[183] In closely related work with tri-*O*-benzyl-2-deoxy-α-D-gluclopyranosyl fluoride conditions were found for coupling which gave either good α- or β-selectivity.[184] In the synthesis of the namenamicin A–C disaccharide derivatives **54** glycosyl fluoride coupling was again used.[185]

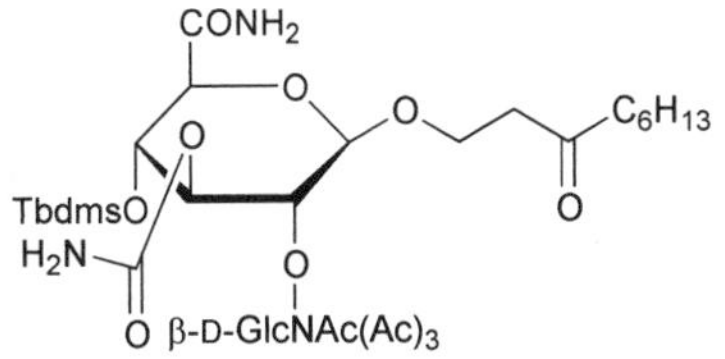

52

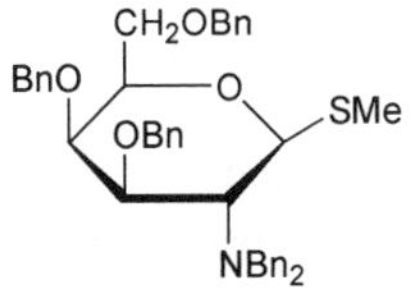

53

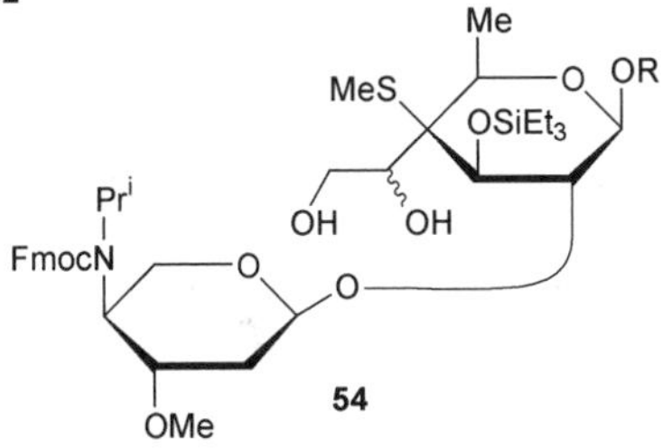

54

Treatment of tetra-*O*-acetyl-2-hydroxyglucal with iodine in aqueous acetone gives a non-reducing bis-2,3-unsaturated dissacharide derivative, which on hydrogenation and deacetylation, affords the 3,3′-dideoxytrehalose **55**.[186] The unusual disaccharide α-L-Fuc-(1→3)-β-D-Rib*f* having sulfate ester groups at O-2 and O-5 of the ribose moiety is the carbohydrate part of a guanosine derivative which is a neuroactive compound produced by the funnel-web spider. It and isomers with fucose bonded separately to O-2′ and O-5′ have been made (see also Chapter 20).[186a]

2,3,4-Tri-*O*-acetyl-α-L-fucopyranosyl propan-1,3-diyl phosphate was used as donor in the synthesis of α-L-Fuc-(1→2)-β-D-Gal-O-4-methylumbelliferone,[187] whereas the thioglycoside method was favoured for the synthesis of the four diastereomers of L-Fuc-(1→2)-D-Gal-OMe.[188] Polycavernoside A, a red algae toxin containing 2,3-di-*O*-methyl-α-L-Fuc-(1→3)-2,4-di-*O*-methyl-β-D-Xyl has been synthesized.[189,190]

α-L-Rha-(1→2)-β-D-Glc occurs with its 4′-aroyl esters in anti-human leukemia plant saponins,[191,192] and the α-(1→3) linked analogue is the carbohydrate of acteoside (verbascoside), a plant phenylethanoid glycoside, which has potential for wound healing.[193] Its synthesis by the thioglycoside coupling method has been reported.[194] Condensation of 2,3,4-tri-*O*-benzyl-L-rhamnose with methyl 2,3,4-tri-*O*-benzyl-α-D-glucopyranoside with heteropoly acid ($H_4SiW_{12}O_{40}$) as catalyst in acetonitrile gives the α-L-Rha-(1→6)-Glc glycoside with >99% α-selectivity in 76% yield.[110] The three α-L-rhamnosyl derivatives of methyl α-D-glucopyranosiduronic acid were made by trichloroacetimidate coupling, the acid function being introduced in the last step.[195]

6-Deoxyglucosyl disaccharides are available from compound **56** by opening the sulfur-containing ring with the hydroxy nucleophile of an acceptor sugar derivative using NIS/TfOH as promotor. Reduction with Raney nickel of the 6′,6′-disulfide products gives the 6′-deoxydisaccharides. Similar results were obtained by use of the seleno analogue of **56**.[196]

In the field of dideoxy disaccharides 4-deoxy-α-L-Rha-(1→4)-GlcNAc-OMe and the corresponding 4-deoxy-4-fluoro analogue were made by thioglycoside coupling as analogues of the mycobacterial arabinogalactan linkage disaccharide.[197] Reaction of di-*O*-acetyl-L-fucal with the appropriate *C*-glycoside in the presence of montmorillonite gave a 2,3-unsaturated *O*-glycoside from which the pentadeoxydisaccharide *C*-glycoside **57**, related to angucyclines, was made.[198]

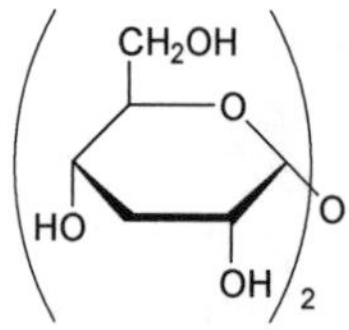

55 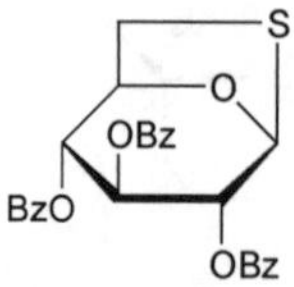 **56** 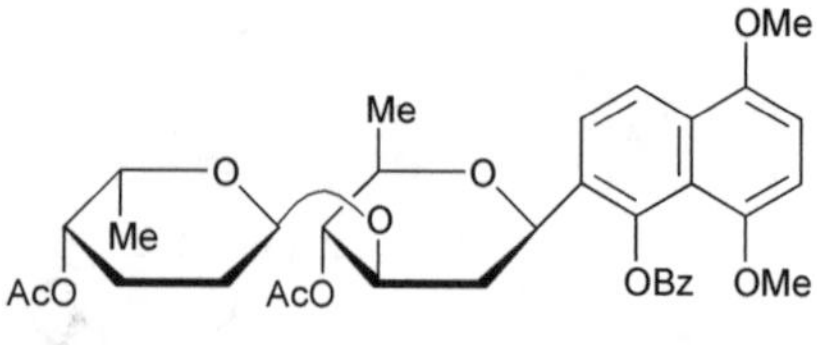 **57**

The apiosylglucosides **58**, **59** (kelampayosides A,B plant products with antiulcerogenic activity), were made using the thioglycoside and trichloroacetimidate methods for building the disaccharide and the aryl glycosidic bonds, respectively.[199]

1.4.8 Sugar acid disaccharides. β-D-GlcA-(1→3)-β-D-Gal has been made as its allyl[200] and 2-aminoethyl[201] glycosides, the former also as its 3′-sulfate. Also in connection with GAG disaccharides β-D-GlcA-(1→3)-β-D-GalNAc was prepared as its 4,6,2′-trisulfate for use in methods that involved the combinatorial approach.[202]

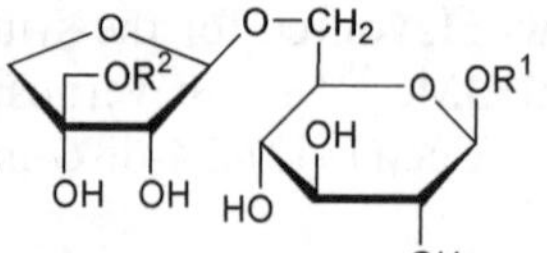

58 R^2 = H

59 R^2 =

R^1 =

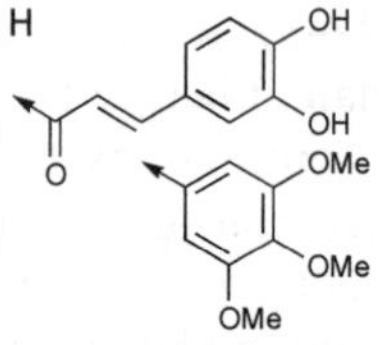

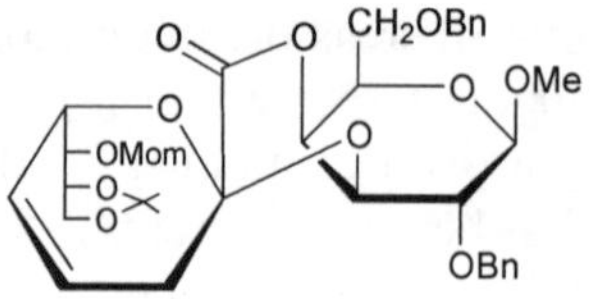

60

In a paper covering neuraminic acid synthesis the methyl *O,N*-acetylated 2-(methylthio)ethyl ester-2-thioglycoside was made and found to be highly selective as an α-glycosylating agent, and in the course of this work the unsaturated disaccharide lactone **60** was made.[203] α- and β-NeuNAc-(2→6)-β-D-GlcO$(CH_2)_{21}$Me, sea cucumber ganglioside analogues, have been prepared[204] as well as a ceramide glycoside of the same α-linked disaccharide.[205] The biantennary compound **61** was made as a Sia-Le^x mimic, but the analogue with the disaccharide replaced by a second fucosyl substituent was a more potent binder of E-selectin.[206] A 3-thio-thioglycoside was used to make α-NeuNAc-(2→8)-α-NeuNAc *via* α-NeuNAc-(2→3)-D-GalNH$_2$.[207] A second paper on the synthesis of this 2→8 linked compound described a direct glycosylation approach and also the preparation of NeuNAc-(2→3)-D-Gal.[208]

1.4.9 Other disaccharides. Enzymic cleavage of a xylan in supercritical CO_2 in the presence of octanol gave β-(1→4) linked octyl β-xylobioside and the analogous trioside.[209] β-D-Xyl-(1→3)-L-Ara was made as its 2-acetate-2′-*p*-methoxybenzoate and coupled to a cholestane derivative to give an exceptionally potent antititumour saponin found in a member of the lily family.[210]

7-*O*-Carbamoyl-α-Hep-(1→3)-Hep has been synthesized as its α- and β-allyl glycosides.[211]

1.5 Disaccharides with Anomalous Linking. – α-D-Glc-1-OCH$_2$CH$_2$O-1-β-δ-Gal having the 2,3-diol of the α-linked sugar involved as part of an 18-crown-6 system is bound by a lectin from *Kluyveromyces bulgaricus*.[212] Tetra-*O*-acetyl-β-D-glucopyranosyl isothiocyanate and methyl 6-amino-6-deoxy-α-D-glucopyranoside were used to make a disaccharide derivative linked by a thiourea bridge. The corresponding α- and β-mannosyl, galactosyl, cellobiosyl and lactosyl isothiocyanates were also used to make similarly coupled compounds.[213]

1.6 Reactions, Complexation and Other Features of *O*-Glycosides. – The alkaline hydrolysis of *p*-nitrophenyl α- and β-glycosides can be selectively influenced by cyclodextrins α-CD, for example, accelerating the cleavage of the α-D-mannoside, β-D-glucoside and β-D-galactoside (*i.e.* the 1,2-*trans*-isomers) relative to their anomers.[214]

The acid-catalysed methanolysis of methyl 2-deoxy-4-*O*-methyl-α-D-*arabino*-hexoside in the presence of the reducing agent 4-methylmorpholine borane led to anomerization and the formation of small amounts of the corresponding 1,5-anhydrohexitol and 2-deoxy-1,4-di-*O*-methyl-D-*arabino*-hexitol which were taken to represent the products of trapping of the cyclic and acyclic oxocarbenium ion intermediates. From these studies it was concluded that aldofuranosides react by *endo* C–O cleavage, keto-furanosides and -pyranosides cleave by *exo* bond fission and aldopyranosides react by both pathways.[215] Acid-catalysed transglycosylation reactions involving methyl

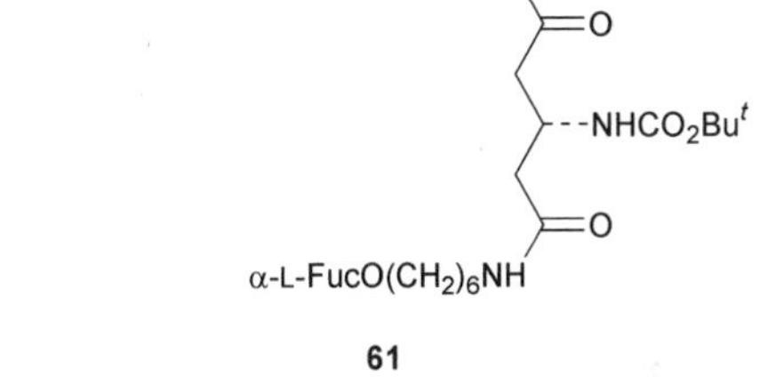

α-NeuAc-(2→3)-β-D-GalO(CH$_2$)$_6$NH
=O
--NHCO$_2$Bu^t
=O
α-L-FucO(CH$_2$)$_6$NH

61

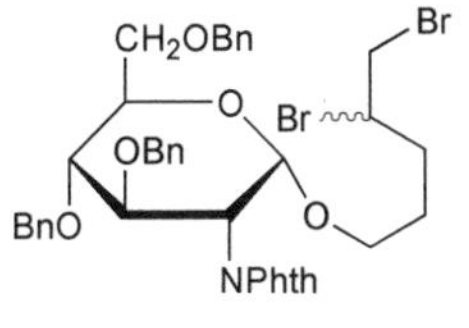

CH$_2$OBn O Br OBn Br BnO O NPhth

62

tetra-*O*-methyl-glycopyranosides and the butyl alcohols and cyclohexanol have been described.[216]

$CuBr_2$, LiBr allow the quantitative bromination of glycosides having alkene functionality within the aglycons whereas bromine leads to complex products. For example, dibromide **62** was obtained in 99% yield by use of the former reagent while only in 10% yield when bromine was used. The mechanisms of the various processes were considered in detail.[217]

Reductive ozonolysis of allyl β-D-fructopyranoside gave the hemiacetals **63** which were characterized by use of X-ray diffraction methods. Several derivatives were reported.[218]

O-Substituted propargyl β-D-glucoside and the dimeric analogue having two glucosyl substituents on 2-butyne-1,4-diol have been treated with $B_{10}H_{12}$ to give products, *e.g.* **64**, with a carborane cage in the aglycons.[219]

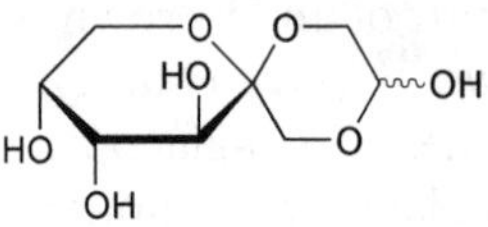

63 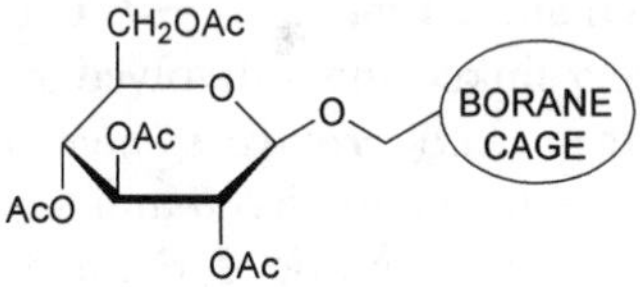 **64** 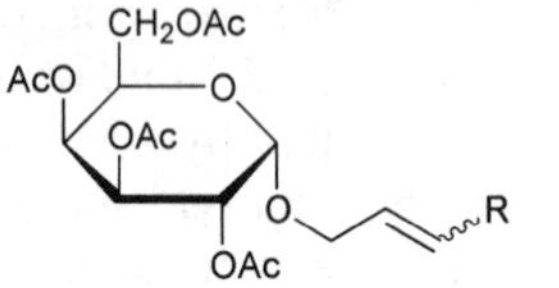 **65** R = CH_2Tms, Ph, $CH_2CH_2CO_2Me$, *etc*

Grubbs' catalyst is attracting considerable attention in carbohydrate chemistry, allyl tetra-*O*-acetyl-α-D-galactoside having been cross coupled in metathesis processes with various alkenes to give compounds **65** in 30–94% yield and

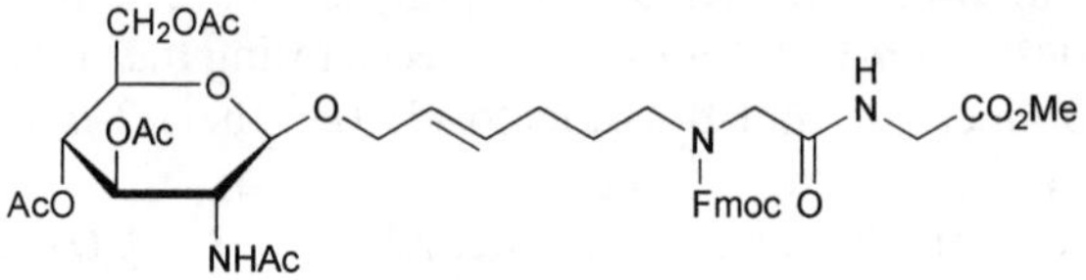

66

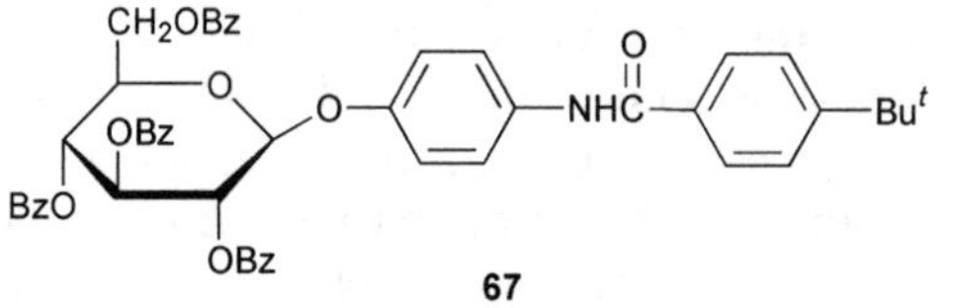

67

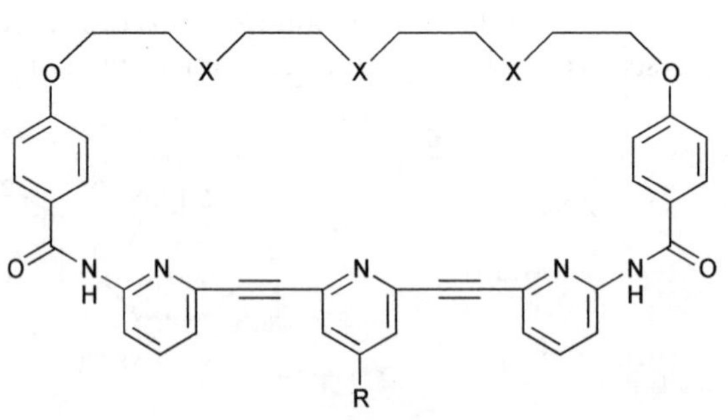

68

R	X
H	O
H	CH_2
OBu	O

E/*Z* ratios 2:1–19:1.[220] *N*-Alkenylpeptides, *e.g.* **66**, were also made by this approach; in this case the yield was 49% and the sugar dimer and peptoid dimer were also obtained in 11% and 27% respectively.[221]

The aryl glycloside **67** is bound within a macrocyclic tetraamide to give a rotaxane complex,[222] and related work has revealed that compounds of the set **68** bind glycosides with appreciable selectivity. For example, octyl β-D-glucopyranoside and -galactoside and methyl β-D-ribofuranoside were bound. Epimers can be distinguished by the macrocycles.[223] (2,2-Dipyrid-4-yl)methyl 2-acetamido-2-deoxy-α-D-galactopyranoside self-associates in the presence of Fe(II) to give a trimer that binds to lectins.[224]

The gelation properties of amino-containing glycosides, *e.g.* *p*-aminophenyl 4,6-*O*-benzylidene-α-D-glucopyranoside which has strong gelling characteristics in eight solvents, have been examined, and the effects on them of metal ions have been studied.[225]

2 *S*-, *Se*- and *Te*-Glycosides

Alkyl and aryl thioglycosides and *S*-linked disaccharides have been made from glycosylthioiminium salts in two phase systems with a phase-transfer catalyst.[226] *o*-Nitrophenyl 1-thiofuranosides were obtained in good yield from the 2,3,5-tri-*O*-benzyl ethers of D-arabinose, D-ribose and D-xylose by treatment with *o*-nitrophenylsulfenyl chloride followed by Et_3P on $(EtO)_3P$.[227] Thioglycosides (mainly β-anomers) of *N*-acetylneuraminic acid were obtained using the thiols, the *O*- and *N*-acetylated methyl ester and tin(IV) chloride as catalyst. Trimethylsilyl bromide produced the (glycosyl bromide from which the aryl α-thioglycosides were made. Benzylthiol, however, again led to the β-anomer.[228] New 2,3-unsaturated thioglycosides were produced from tri-*O*-acetyl-D-glucal and thiols with $LiBF_4$ as catalyst.[229] The branched chain compounds **69** were synthesized for use in the total synthesis of altohyrtin C, a potent anti-tumour macrolide isolated from sponges.[230]

Many uses of thioglycosides as glycosylating agents for the synthesis of *O*-glycosides and disaccharides have been referred to in the papers already cited in this chapter. Further uses are referred to in Chapter 4. Several derivatives of ethyl 1-thio-α-D-mannopyranoside were compared as glycosyl donors used with various promoters. Iodonium dicollidine perchlorate was more effective with the *O*-benzylated donor and gave α-products but, surprisingly, was inactive with the 4,6-*O*-benzylidene derivative. For this, MeOTf and $MeSS^+(Me)_2$ OTf were activators and led to β-thioglycosides.[231]

Particular thioglycosides to have been made are the β-galactosyl compounds linked to thioserine and a serine-like derivative (by a Mitsunobu-like process involving the *O*-acetylated 1-thio-sugar and appropriate alcohols).[232] The multiply-substituted **70** was amide-linked to a solid support to provide a '5-dimensional diversity scaffold' for the preparation of many differently substituted galactosides. Variable substituents such as urethanes gave access to many possibilities.[233]

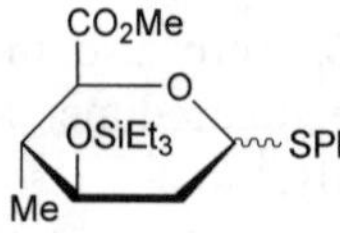

69 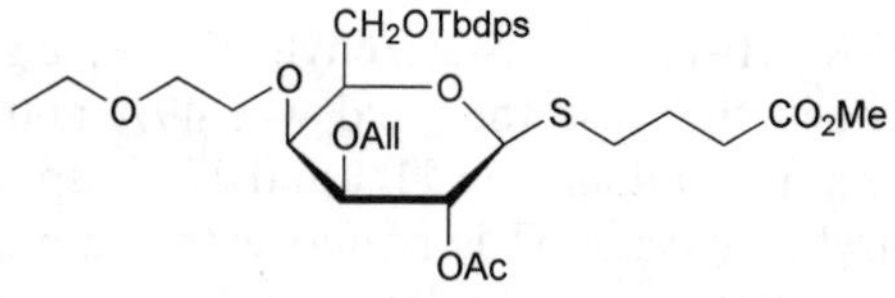 70

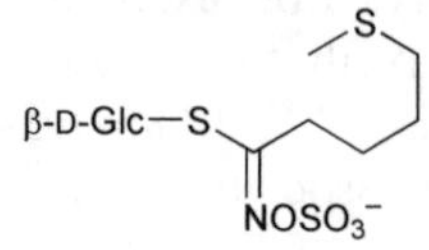

71 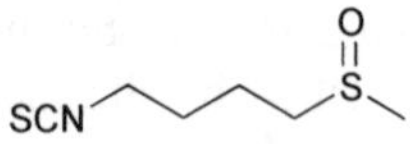 72

Peroxide oxidation of glucoerucin (**71**), a glucosinolate obtained from *Eruca sativa,* gave a sulfoxide that, on cleavage with myrosinase, afforded sulforaphane (**72**) a potential anticarcinogen.[234] The deuterium-labelled desulfoglucosinolates **73–75** were made from tetra-*O*-acetyl-1-thio-D-glucose to improve the sensitivity of quantitative LC–MS analysis of glucosinolates.[235]

3 *C*-Glycosides

3.1 General. – There are clear indications of increased interest in compounds of this series – especially those with sugars *C*-linked to other natural products, notably amino acids and other sugars.

Several reviews have appeared on the synthesis of *C*-glycoside carbohydrate mimetics[236,237] including two on *C*-linked disaccharides[238,239] and one from Vogel's group on the use of 'naked sugars' in *C*-glycoside preparation.[240] The conversion of glycosyl stannanes into *C*-glycosides by radical substitution methods has also been reviewed.[241]

3.2 Pyranoid Compounds. – Such is the activity with compounds of this category that they can be treated in sub-sets.

3.2.1 Compounds with short-chain 'aglycons'. Interesting chemistry continues to be demonstrated by relatively simple compounds. Thus, for example, the

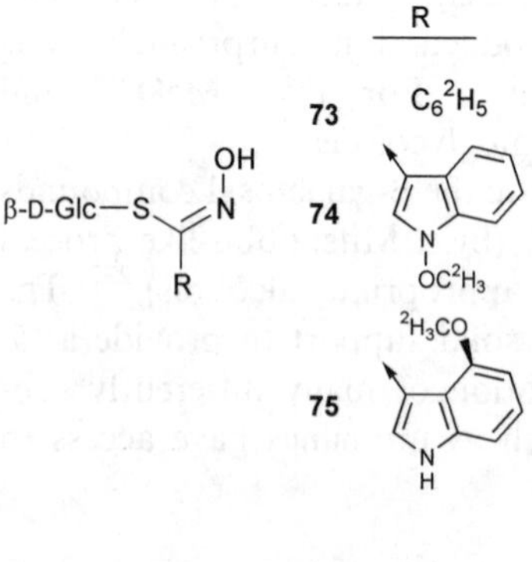

73 74 75

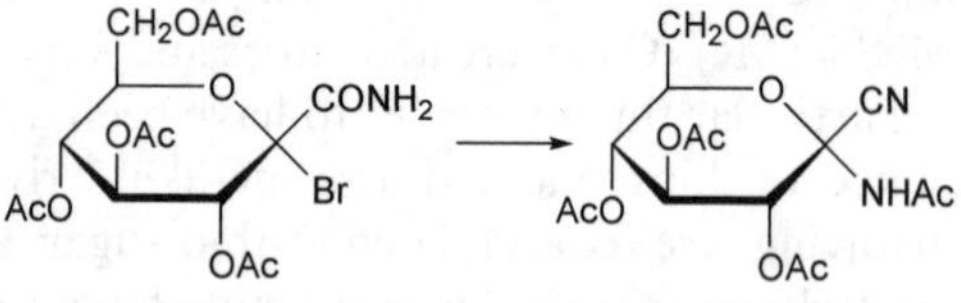

76 77

bromo derivative **76**, made by photobromination of a simple *C*-glycoside, on treatment with silver carbonate in acetonitrile, gave the α-amino acid derivative **77** in 76% yield in a very novel Ritter-like rearrangement involving the solvent. The process was assumed to proceed by way of a *spiro*-oxazole.[242] In a further unusual process the tellurium glycosides **78** reacted with 2,6-dimethylbenzoisonitrile under light to give the Te-substituted imino *C*-glycosides **79**.[243]

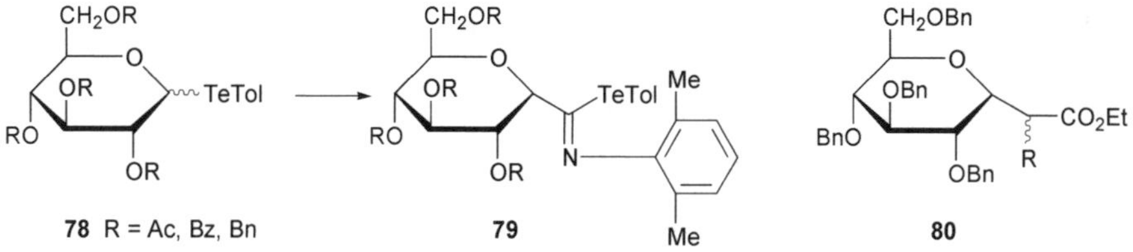

78 R = Ac, Bz, Bn **79** **80**

Straightforward Wittig chemistry applied to *O*-substituted aldonolactones gives C-1-*exo*-alkenes from which compounds such as **80** (R = H, OAc) are obtainable.[244] Scheme 6 illustrates an unusual way of making related glycosyl acetaldehydes.[245]

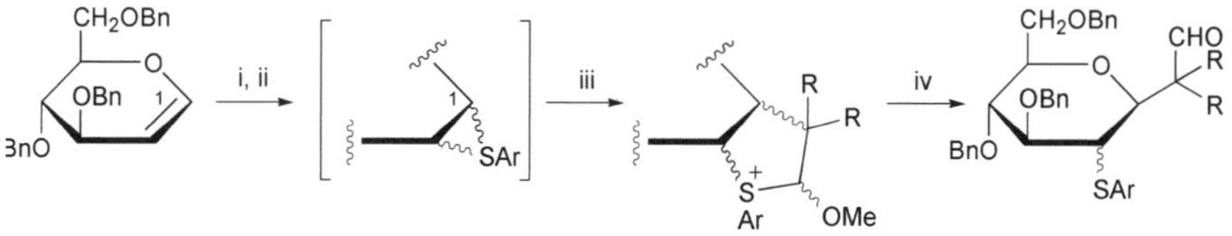

Reagents: i, ArSCl; ii, $SnCl_4$; iii, $R_2C{=}CHOMe$; iv, H_2O

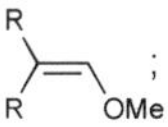

Scheme 6

A stereoselective method for introducing a C_2 unit at C-1 of glycosyl moieties depends on vinyl transfer from a dimethylvinylsilyl group at O-2. The reaction, however, lacks regioselectivity, with considerable radical transfer from C-1 to C-5 occurring (Scheme 7).[246] In related work heterolytic activation led to the formation of bicyclic *C*-glycosides see Section 3.2.4). For glycosyl radical addition to acrylonitrile to give the well known 3-glycosylpropionitriles, complex azobisnitriles (*e.g.* **81**) acting as radical initiators with Bu_3SnH,

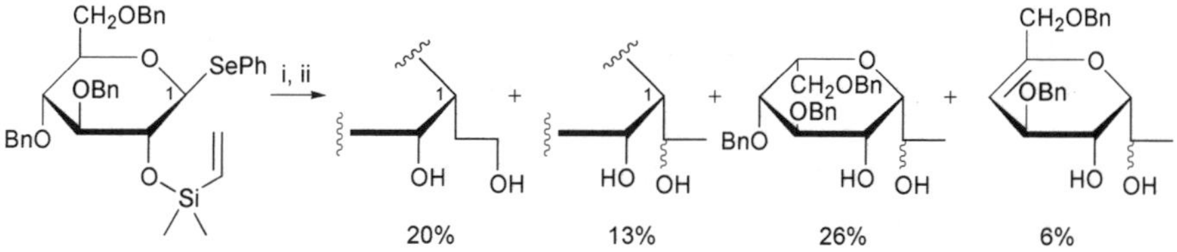

20% 13% 26% 6%

Reagents: i, Bu_3SnH, AIBN; ii, H_2O_2, KF, $KHCO_3$, MeOH/THF

Scheme 7

give better stereocontrol than the more commonly used AIBN. Acetobromoglucose gave 68% of the α-*C*-glycoside with no β-anomer.[247]

Opening of heptopyranuronic-1,7-lactones with allyltrimethylsilane in the presence of TmsOTf followed by methanol gives allyl-*C*-glycosides of the methyl uronates.[248] The well known reaction of the same silane with tri-*O*-acetyl-D-glucal and -galactal to give 2,3-unsaturated allyl *C*glycosides can be promoted with indium trichloride. In the former case the yield was 95% and the α,β ratio 9:1, whereas the galactal ester reacted exclusively to give the α-anomer (78% yield).[249] β-D-Glucopyranosyl *C*-glycosides having allyl, 1-buten-4-yl and 1-penten-5-yl aglycons were lactosylated at O-4 enzymically and then, with β-galactosidase, converted to the cellobioside unsaturated C-glycosides which were then epoxidized.[249a]

3.2.2 Compounds wityh alkynyl 'aglycons'. An allylic rearrangement reaction of tri-*O*-acetyl-D-glucal occurs on treatment with 3-TbdpsO-1-Tms-propyne and $SnCl_4$ as promoter, and the product **(82)** was obtained in 84% yield. The reaction was not successful when the Tbdps group in the reagent was replaced by the more electron-withdrawing acetyl group.[250] (See *Chem. Commun.*, 1998, 2665 for earlier work.)

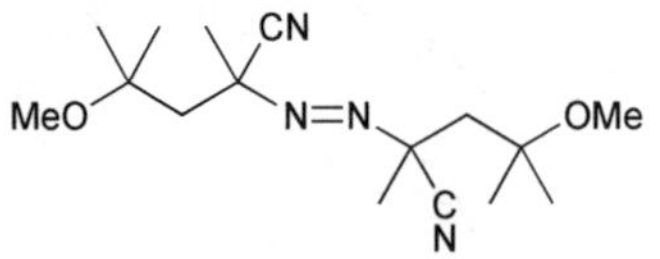

MeO CN N=N OMe C N

81

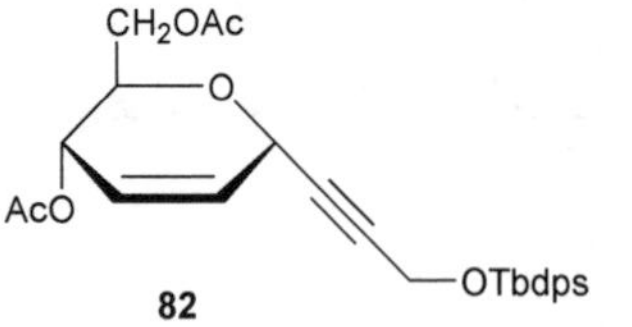

CH_2OAc O AcO OTbdps

82

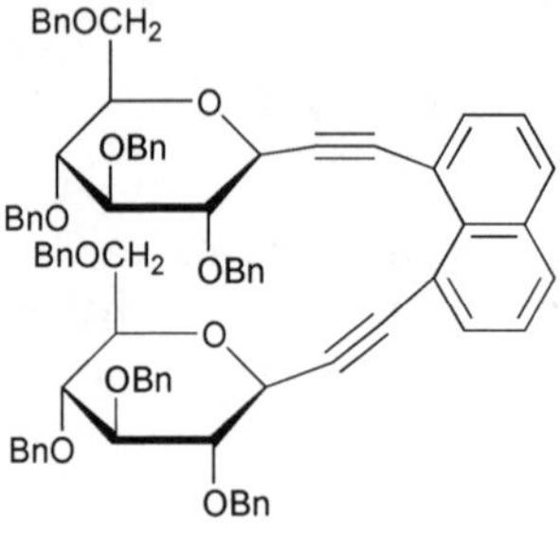

$BnOCH_2$ O OBn BnO $BnOCH_2$ OBn O OBn BnO OBn

83

Vasella's group have continued their work with ethynyl *C*-glycosides and reported the naphthalene derivative **83** in the course of studies related to aspects of cellulose chemistry.[251] Further work on disaccharide analogues *C*-linked 1→4 by but-1,3-diynyl bridges has been reported,[252] and following from this the cyclic tetrayne **84** was made by forming the bisalkyne bridge in the last step. It and the acyclic precursor bind a range of metal ions.[253]

Other work has reported the preparation of aryl *C*-glycosides from glycosyl alkynes (Scheme 8).[254] The synthesis of a naphthalenyl *C*-glycoside from an ethynyl analogue is noted in ref. 259 below.

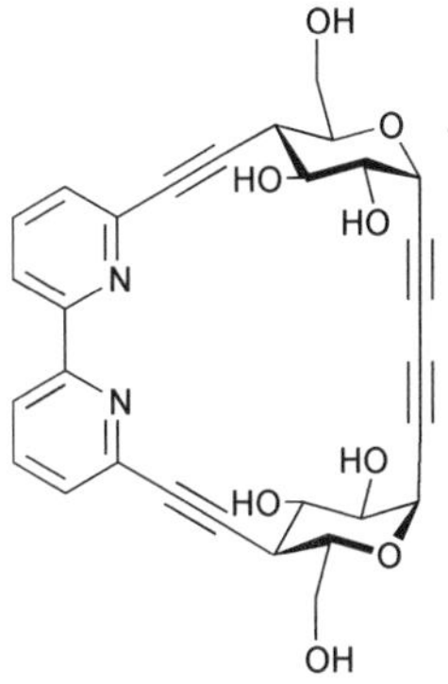

84

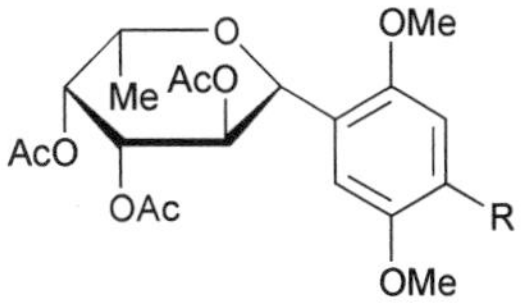

85 R = $C_6H_4CO_2Me$-*p*, $H_2CC_6H_4CO_2Me$-*p*, $COC_6H_4CO_2Me$-*m*

3.2.3 Compounds with aromatic 'aglycons'. The construction of the aromatic ring of a benzenoid *C*-glycoside is illustrated in Scheme 8.[254] Satoh's group has reported the synthesis of the 2,5-dimethoxyphenyl *C*-glycosides of several sugars made by Lewis acid-catalysed coupling of the substituted benzenes with the sugar peresters. This was followed by stannylation at the 4-position of the aglycons and hence more complex 4-substituted compounds, *e.g.* **85**, were made.[255–257] In the course of the work related compounds having sugars *C*bonded to each of the rings of diarylmethanes were described.[258]

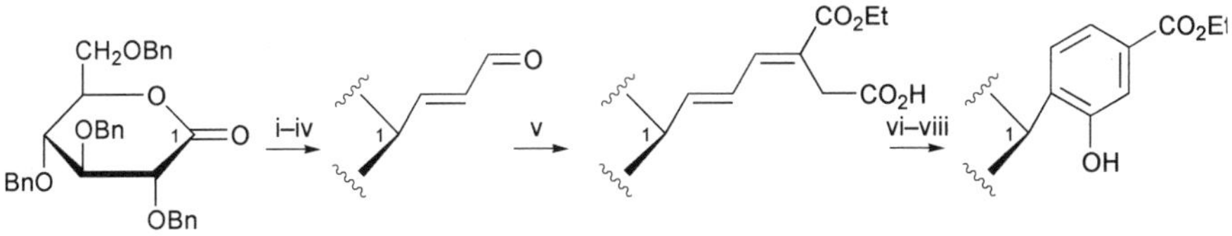

Reagents: i, Li—≡—CH₂OThp; ii, Et_3SiH, BF_3•OEt_2; iii, REDAL; iv, MnO_2; v, $Ph_3P{=}C(CO_2Et)CH_2CO_2H$; vi, $ClCO_2Et$, Et_3N; vii, NaOH, EtOH; viii, HCl

Scheme 8

An ingeneous preparation of a naphthalenyl *C*-glycoside from a *C*-alkynyl compound is illustrated in Scheme 9.[259] Condensation of unprotected olivose (2,6-dideoxy-D-*arabino*-hexose) with 1,5-dihydroxynaphthalene in the presence of TmsOTf led to the dihydroxynaphthyl β-*C*-glycoside which on oxidation

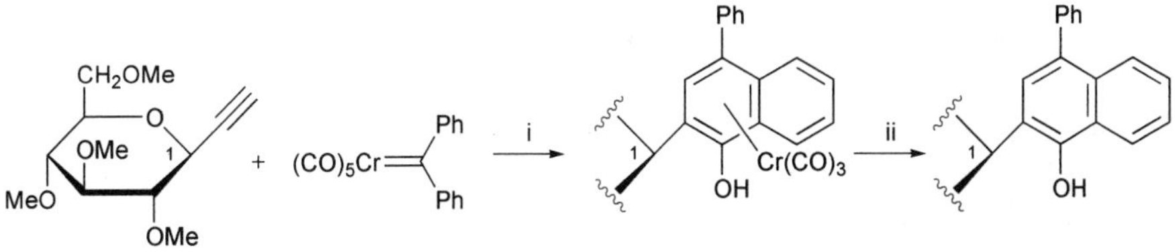

Reagents: i, THF; ii, O_2

Scheme 9

with oxygen under light gave compound **86** required for the synthesis of urdamycinone, a *C*-glycosylangucycline antibiotic.[260] (See ref. 198 for closely related work.)

An extension of the Dondoni group work on thiazolylketoses describes this approach to *C*-glycosylthiazoles.[261] 2-Lithioindole reagents used with 1,2-anhydro-3,4,6-tri-*O*-benzyl-β-D-mannose gave access to peptide-containing *C*-linked glycosylindole derivatives, *e.g.* **87**.[262]

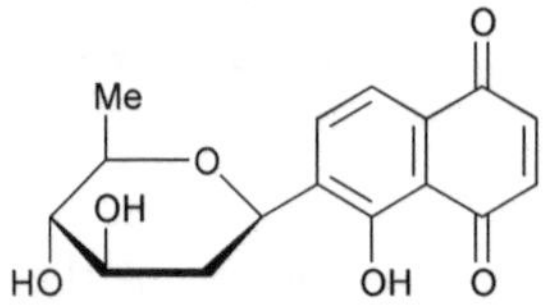

Me O OH HO OH O O

86

CH_2OH O OH OH HO H N NH-Phe-Thr Trp-Glu-Ala O

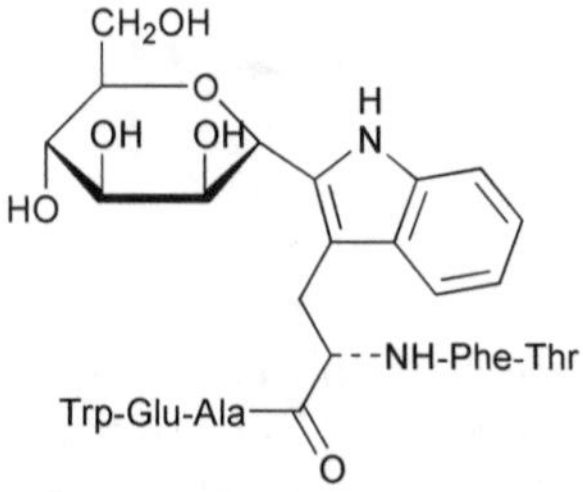

87

3.2.4 1,2-Fused bicyclic systems. Reports have been made on the cycloaddition of various *O*-substituted glucals and dichloroketene. The cyclobutanone products, with the carbonyl group bonded to C-2, were treated with various reagents.[263] Reactions of tri-*O*-acetyl-2-*C*-vinylglucal with maleic anhydride and maleimide give the expected tricyclic Diels Alder products.[264] Activation of compound **88** with silver triflate gives the cyclized **89** and reactions of this nature were employed in the total synthesis of 2,3-dideoxy-D-*manno*-2-octulosonic acid.[265] A similar approach affords means of preparing related bicyclic products containing four-membered rings, compound **90** behaving in this manner when activated with $EtAlCl_2$. This is referred to in a comprehensive review of desulfonylation reactions which contains several carbohydrate examples.[266]

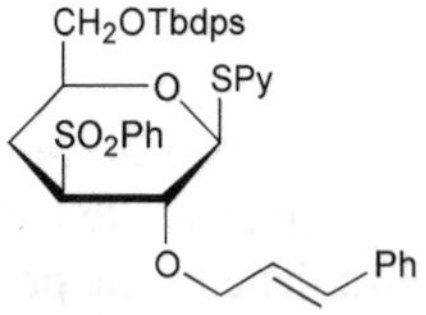

$CH_2OTbdps$ O SPy SO_2Ph O Ph

88

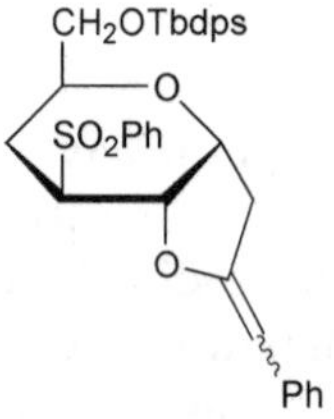

$CH_2OTbdps$ O SO_2Ph O Ph

89

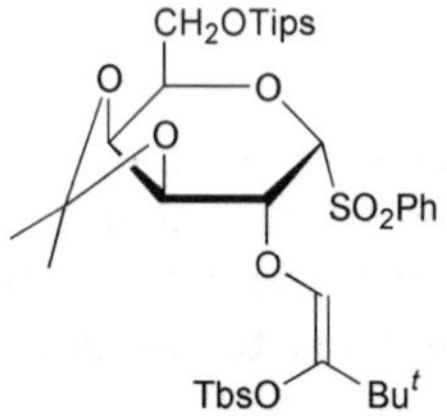

CH_2OTips O O O SO_2Ph O TbsO Bu^t

90

Grubbs' ruthenium-based catalysed methathesis reaction has enabled compound **91** to be converted to the bicyclic **92** as indicated in Scheme 10, the overall yield being about 70%. The (anomer cyclized similarly to give the *trans*-fused product in slightly lower yield.[267] In closely related work the 2-*O*-allyl-β-*C*-allyl glycoside reacted similarly to give a dioxabicyclo[5.4.0] system which, like its hydroxylated product, represents the A/B ring features of the cigua-

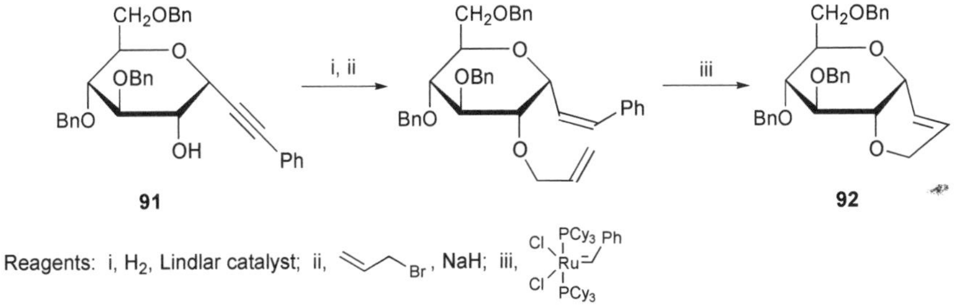

Reagents: i, H_2, Lindlar catalyst; ii, allyl bromide, NaH; iii, $Cl_2(PCy_3)_2Ru{=}CHPh$

Scheme 10

toxins. From the ene, compound **93** was made.[268] A third application of the method converted 1-methylene-2-alkenyl ethers into the set of compounds **94**.[269]

Quite a different approach to bicyclic systems gives access to compounds with a fused aromatic ring. Reaction of a glycal with SMe and formyl substituents at C-1 and C-2, respectively, and a carbonyl group at C-3, with a 2-oxomaleic acid diester, gave the product **95**.[270]

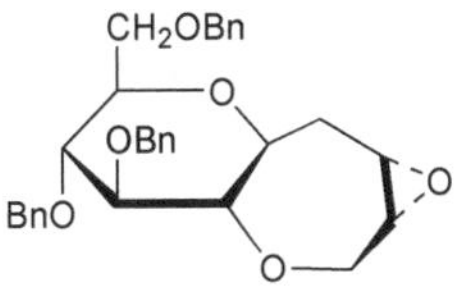

93 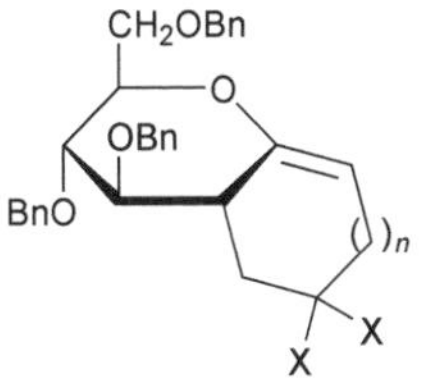 **94** X = H, H, *n* = 0, 2, 3; X = O, *n* = 1, 2 **95**

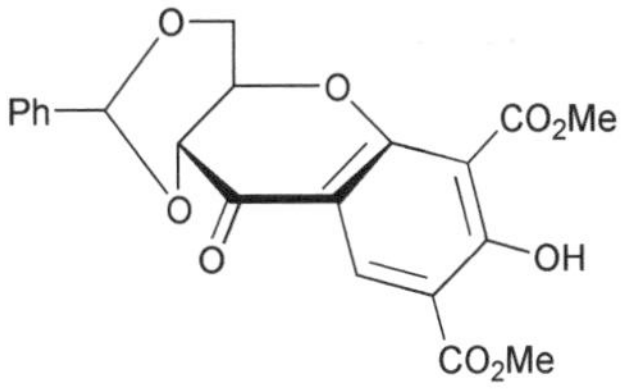

3.2.5 C-Linked disaccharides. *C*-Linked disaccharides continue to attract interest, D-mannose-containing compounds receiving particular attention this year. As previously, they will be described in the abbreviated form used for *O*-linked disaccharides with the group that replaces the linking oxygen atom being specified. The following compounds have been made in derivatized form: β-D-Glc-(1-CH_2-1)-β-D-Glc (by application of the Ramburg-Bäcklund desulfonylation process);[271] β-D-Gal-(1-CH_2-1)-α-D-Man, β-D-Gal-(1-CH_2-6)-α-D-Gal and β-D-Gal-(1-CH_2-4)-α-D-Glc (all by the process illustrated in Scheme 11),[272] α-D-Man-[1-CH(OH)-2]-α-D-Man (by SmI_2-catalysed coupling of a 2-deoxy-2-*C*-formyl mannoside with a mannosyl sulfoxide);[273] α-D-Man-(1-CH_2-3)-α-D-Man (by a similar process followed by deoxygenation at the linking centre);[274] β-D-Man-(1-CH_2-6)-α-D-Glc (by linking a 7,7-dibromo-7,7-dideoxy-hept-6-enose derivative with a mannonolactone in the presence of BuLi to give an ethynyl-bridged compound which was reduced);[275] β-D-Man-[1CH(NO_2)-2]-D-Glc (from a nitromethyl-*C*-mannoside added, in the presence of base, to a 1,2-dideoxy-1-nitro-hex-1-enitol derivative);[276] α-D-Man-(1-CH_2-3)-D-GalNAc (by mannosyl radical addition to polyfunctional methylene cyclohexane derivative; the product is a new galactosidase inhibitor);[277] β-D-Man-[1-CH(OAc)-3]-α-L-GulA (made from a common cyclohexenone deriva-

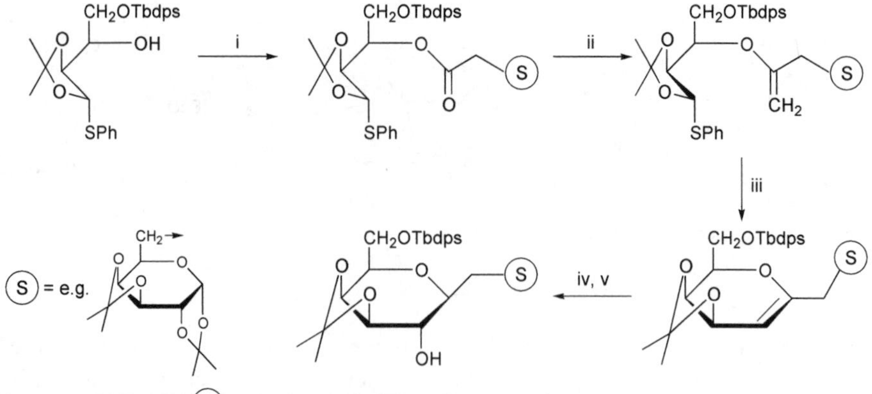

Reagents: i, HO_2CCH_2(S); ii, Tebbe; iii, MeOTf; iv, BH_3; v, Na_2O_2

Scheme 11

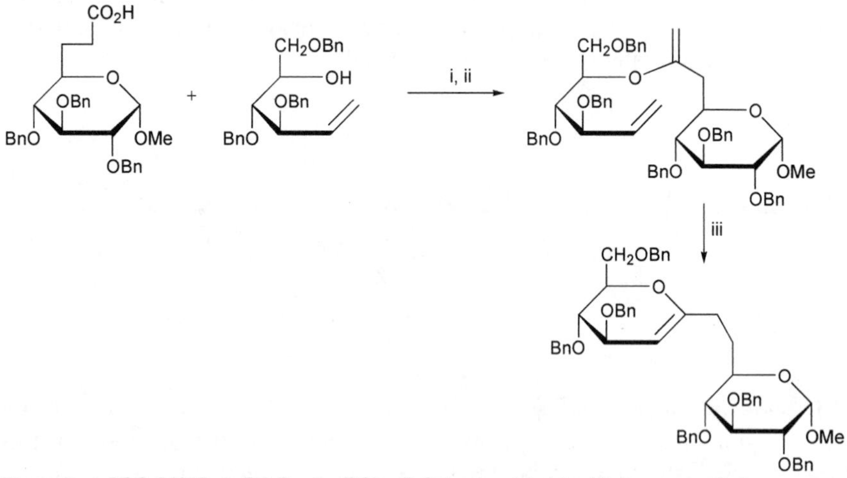

Reagents: i, DCC, DMAP; ii, CH_2Br_2, Zn, $TiCl_4$; iii, Schrock catalyst (molybdenum complex)

Scheme 12

tive and related to the disaccharide of bleomycins);[278] and NeuAc-[2-CH(OAc)-3]-D-Gal (by SmI_2-catalysed coupling of a 2-sulfonyl NeuAc derivative to a 3-deoxy-3-*C*-formyl galactoside).[279]

A method related to that illustrated in Scheme 11 was used to make a further *C*-linked disaccharide with a glycal function (Scheme 12).[280] In Scheme 13 a phosphonate/aldehyde coupling was employed to give the complex *C*-linked disaccharide derivative **96**.[281]

3.2.6 C-Glycosylated amino-acids and related compounds. The 2-deoxy-β-D-galactosylalanine derivative **97** was made using the glycosyl triphenylphosphonium salt added to a 2-amino-2-deoxy-D-glyceraldehyde derivative to give an alkene that was hydrogenated.[282] The related compounds **98** were synthe-

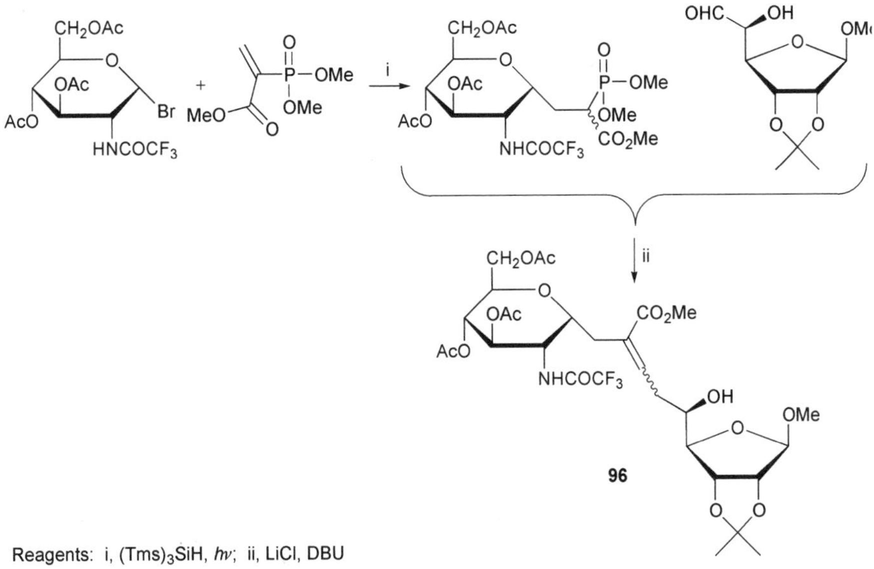

96

Reagents: i, $(Tms)_3SiH$, $h\nu$; ii, LiCl, DBU

Scheme 13

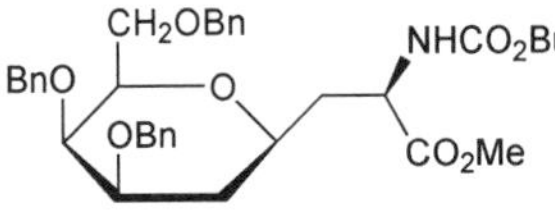

97

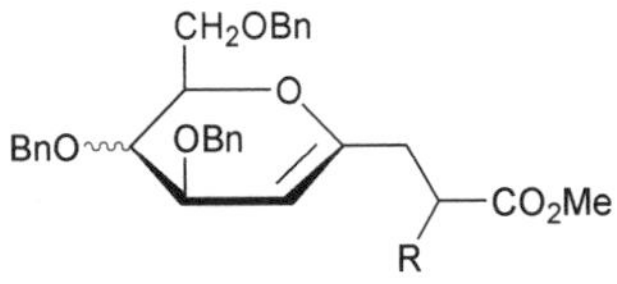

98 R = NHBoc, NPhth

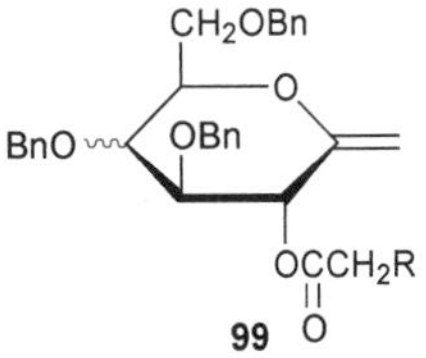

99

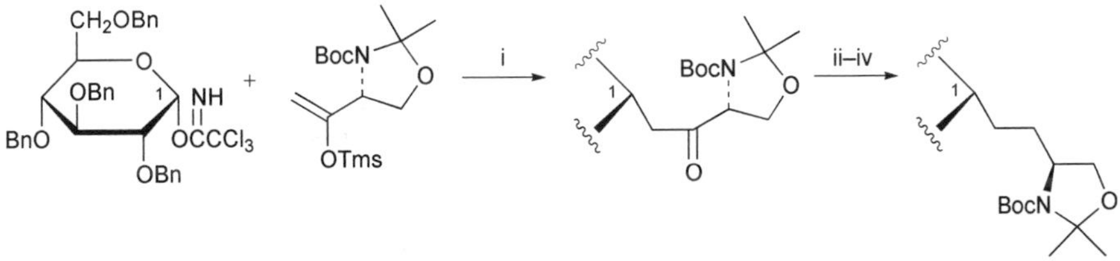

Reagents: i, $BF_3{\cdot}Et_2O$; ii, $NaBH_4$; iii, 1,1'-thiocarbonyldiimidazole; iv, Bu_3SnH, AIBN

Scheme 14

sized by entirely different procedures involving Claisen-Ireland rearrangements of isomers **99**.[283]

The reactions shown in Scheme 14 illustrate a route to a *C*-linked isostere of *O*-glycosylserine. The β-anomer and related derivatives of asparagine were also described.[284] A different approach to compounds of this type used the Romberg-Bäcklund desulfonation method to produce *C*-glycoside **101** from **100**. In the same paper the synthesis of compound **102** was described. It involved the use of the 1-methylene sugar which was hydroborated and coupled with a vinyl iodide (Suzuki coupling).[285] Compounds **103**, converted

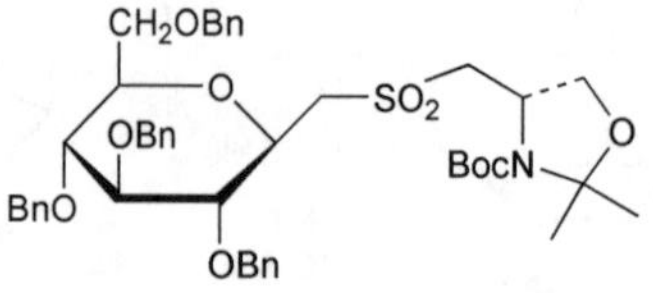

CH_2OBn O OBn SO_2 BocN O BnO OBn

100

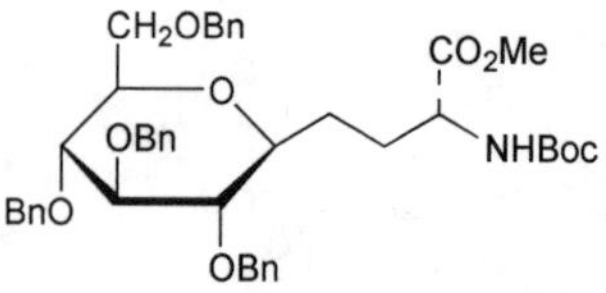

CH_2OBn O OBn CO_2Me NHBoc BnO OBn

101

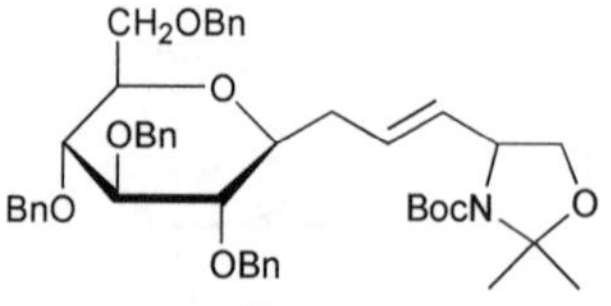

CH_2OBn O OBn BnO OBn BocN O

102

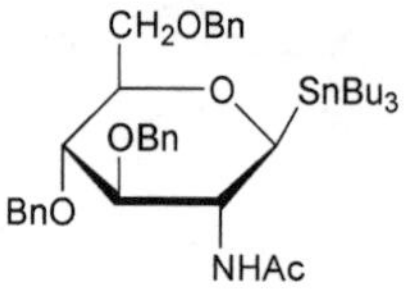

CH_2OBn O $SnBu_3$ OBn BnO NHAc

103

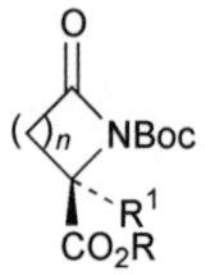

O $(\ )_n$ NBoc R^1 CO_2R

104

CH_2OBn O OBn O R^1 NHBoc $(\ )_n$ CO_2R BnO OBn

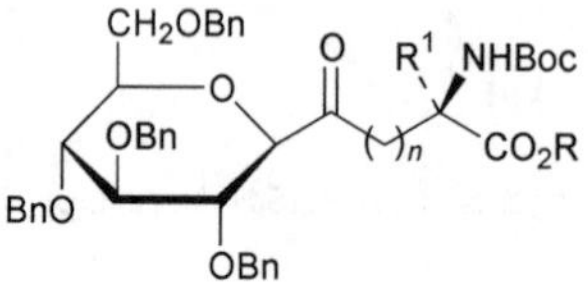

105 $n = 1$, $R^1 = H$, $R = Bn$
$n = 2$, $R^1 = H$, $R = Et$
$n = 3$, $R^1 = Et$, $R = Et$

to the 1-lithio compounds and treated with the lactams **104**, gave access to the *C*-glycosyl amino-acids **105**.[286]

Scheme 15 illustrates the preparation of an α-*C*-mannoside of an isostere of tyrosine which was incorporated into *C*-glycopeptides.[287]

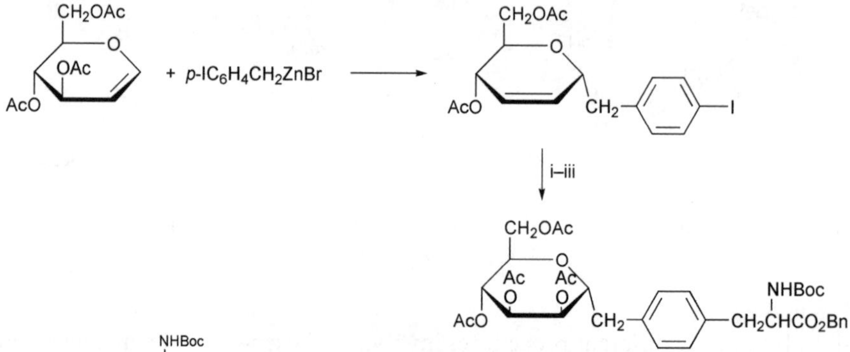

CH_2OAc O OAc AcO + p-$IC_6H_4CH_2ZnBr$ ⟶ CH_2OAc O AcO CH_2–C_6H_4–I

i–iii

CH_2OAc O Ac O Ac O AcO CH_2–C_6H_4–$CH_2CH(NHBoc)CO_2Bn$

Reagents: i, $IZnCH_2CH(NHBoc)CO_2Bn$; ii, NMO, OsO_4; iii, Ac_2O, Py

Scheme 15

3.2.7 Other C-glycosides. Compounds **106** (R = Me, Bn *etc.*), which exist in equilibrium with their furanose isomers, were made by Amadori rearrangement of glucosylamines formed by Mitsunobu condensation of tetra-*O*-acetyl-D-glucose and *N*-2-nitrobenzenesulfonyl amino acids.[288]

The oximo-linked *C*-glycopeptide **107** was made by imine coupling between *C*-acetonyl tetra-*O*-benzyl-α-D-glucopyranoside and a peptide bearing an oxamino group.[289] In related fashion, compound **108,** *C*-glucotropaeolin, was made by oximation of the 3-phenylacetonyl *C*-glycoside.[290]

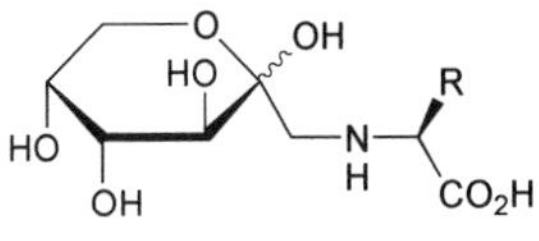

106

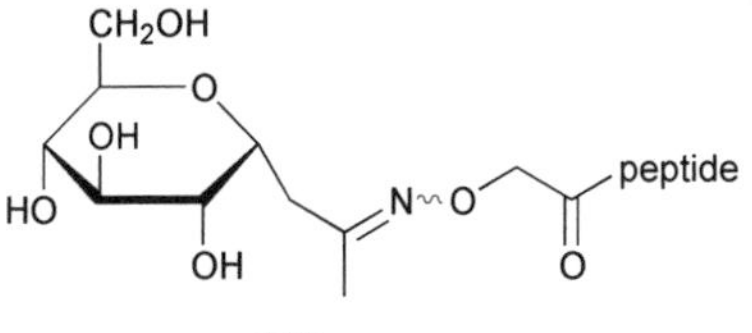

107

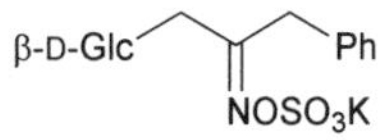

108

Coupling of a phosphinylmethyl *C*-galactoside with a 3-carbon aminoalde-hyde followed by further elaboration gave the *C*-β-D-galactosylceramide **109**.[291] The long chain *C*-glycoside **110** is the first member of a new class of polyketides from a marine dinoflagellate to contain gentiobiose and a sulfate ester.[292]

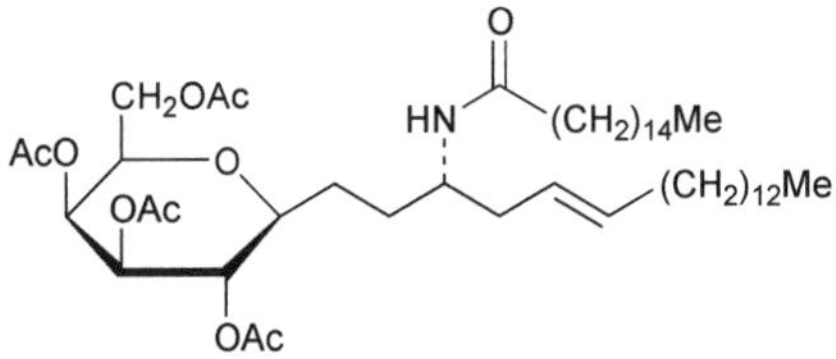

109

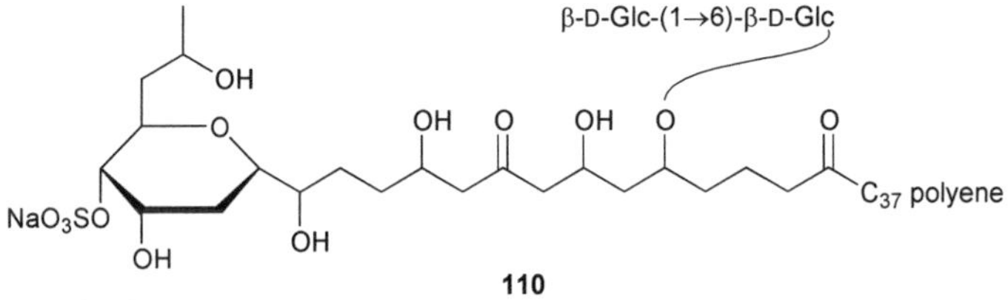

110

3.3 Furanoid Compounds. – Two reports have appeared on the preparation of allyl *C*-glycosides, *e.g.* **111**, from a range of branched-chain furanosyl acetates using allyltrimethylsilane and a Lewis acid.[293] The second was based on 5-*O*-benzyl-2,3-dideoxy-3-*C*-methyl-D-*erythro*- and D-*threo*-pentosyl acetate.[294] Spontaneous and specific cyclization to give *C*-glycoside **112** occurred during the Wittig preparation of the iodoalkene **113**.[295]

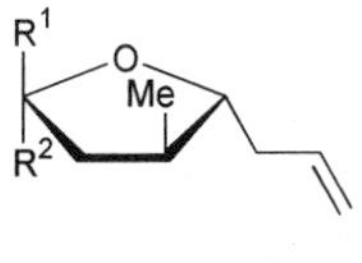

R^1	R^2
Pr^i	H
H	Pr^i
Pr^i	Pr^i
Me	Me

111

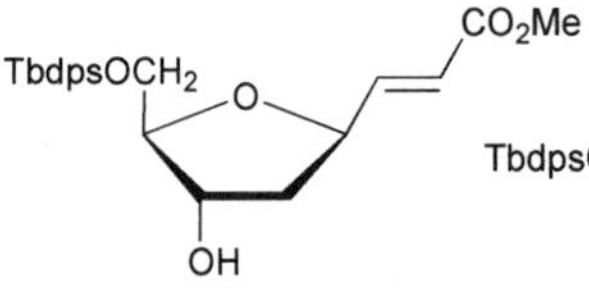

112

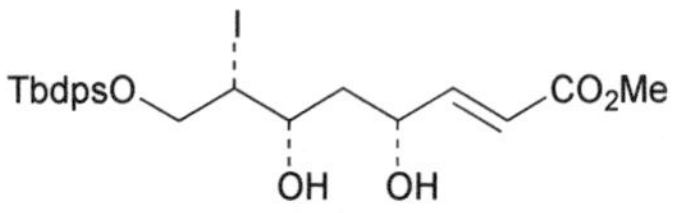

113

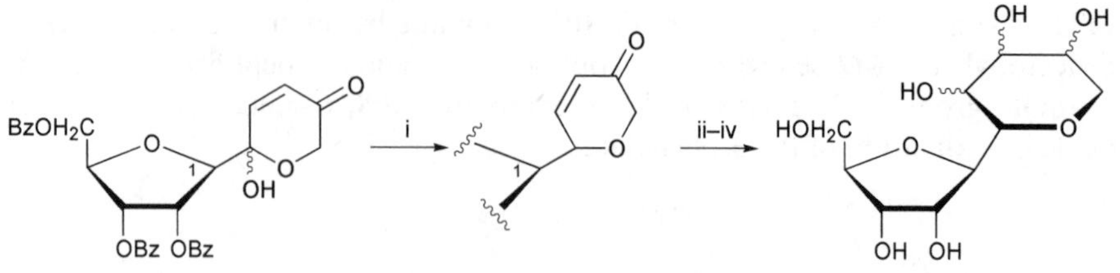

Reagents: i, Et_3SiH, $BF_3 \cdot Et_2O$; ii, $NaBH_4$; iii, OsO_4; iv, Na_2CO_3, MeOH

Scheme 16

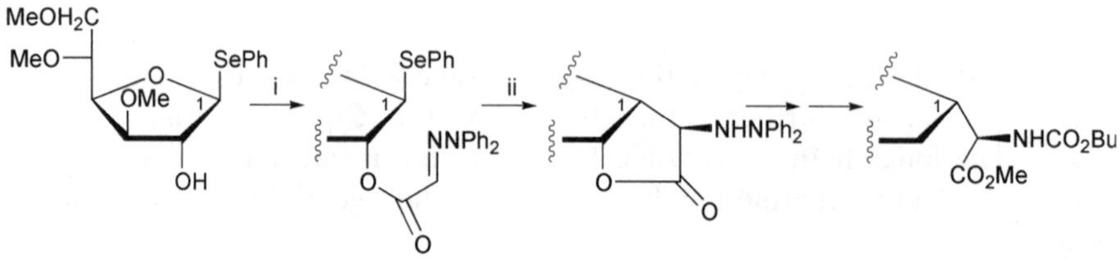

Reagents: i, $HO_2CCH{=}NNPh_2$, DCC, DMAP; ii, Bu_3SnH, AIBN

Scheme 17

The hemiacetal **114** was used to prepare six different, directly linked *C*-ribosyl-pentopyranosides **115** (Scheme 16).[296]

Intramolecular cyclization has been used in a further method of making *C*-glycosyl amino-acids (Scheme 17).[297] Epimers **116** [(+)-goniofufurone] have been made by different routes, the key step being the reaction of free sugars with Meldrum's acid which gives the required bicyclic *C*-glycosides (*Carbohydr. Res.*, 1992, **225**, 159).[298] An extensive series of nucleoside analogues with fluorophores instead of the usual bases have been described, *e.g.* **117** (R = stilbenzyl, pyrenyl, [*p*-terphenyl]yl, (benzoterthiophen)yl, (terthiophen)yl.[299] Further in the area of aryl *C*-furanosides, compound **118** has been produced from ribonolactone as a ligand for binding to Pd^{2+}.[300] *O*-Substituted arabinals with Grignard reagents gave aryl 2,3-unsaturated furanosides; however, with pyranoid analogues the products were acyclic pentadienyl aromatics.[300a] Pyrolysis of hydrochlorides of compounds **119** *etc.* gives access to the *C*-glycosylimidazoles **120**.[301]

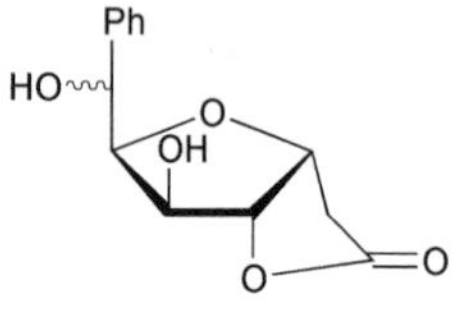

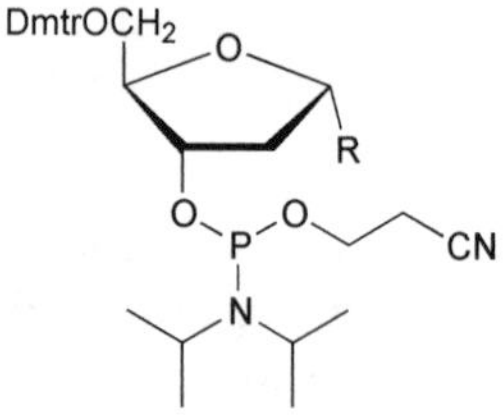

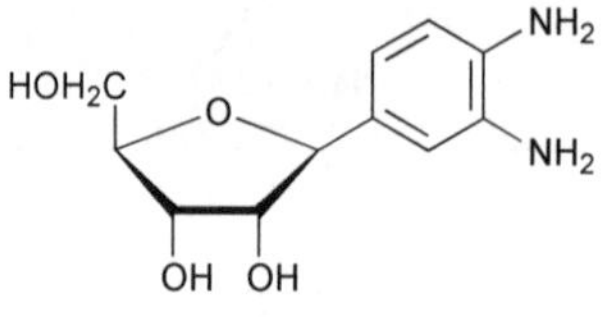

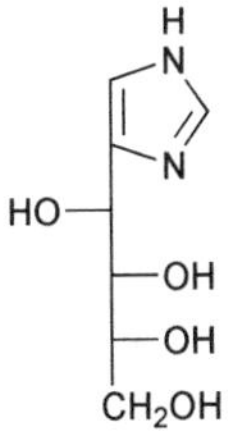

119

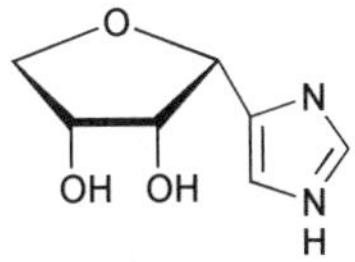

120

When anomers **121** underwent the [1,2]-Wittig rearrangement (Scheme 18) they gave product **122** almost specifically (62% yield), which means that the β-anomer underwent rearrangement with retention of configuration at the anomeric centre while the α-compound did not react. When the work was repeated in the D-xylofuranose series the α-anomer reacted with retention and the β with inversion. The results were rationalized on the basis of steric hindrance at the anomeric centres of radical intermediates. The findings represented in Scheme 18 are applicable in zaragozic acid A synthesis.[302]

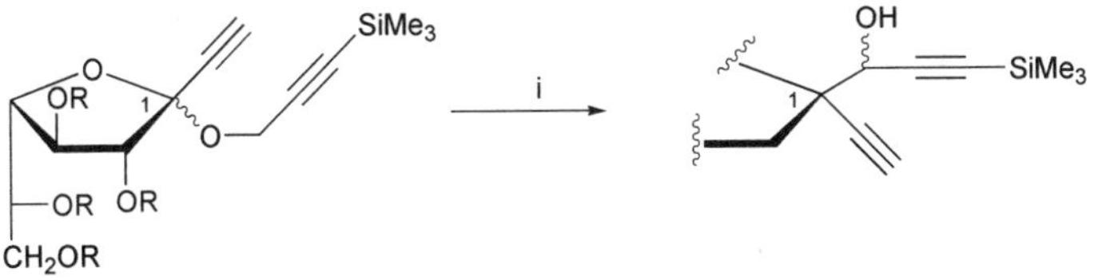

121 α:β = 15:85

122 62% (>95% this anomer)

Reagents: i, Bu^nLi

Scheme 18

The displacement with participation of the dioxolane ring oxygen atom illustrated in Scheme 19 represents an unusual process.[303]

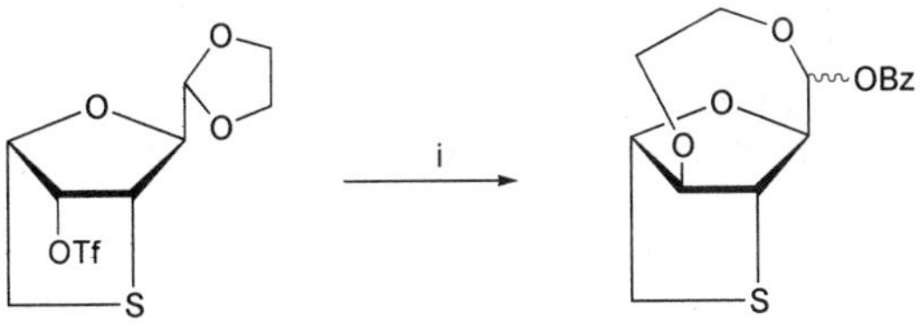

Reagents: i, LiOBz, DMF

Scheme 19

References

1 T. Mukaiyama, *Tetrahedron*, 1999, **55**, 8609.
2 R.R, Schmidt, J.C. Castro-Palomino and O. Retz, *Pure Appl. Chem.*, 1999, **71**, 729.

3 A. Dondoni and A. Marra, *Chem. Commun.*, 1999, 2133.
4 A. Corma, S. Iborra, S. Miquel and J. Primo, *J. Catal.*, 1998, **180**, 218 (*Chem. Abstr.*, 1999, **130**, 139 526).
5 J.F. Chapat, A. Finiels, J. Joffre and C. Moreau, *J. Catal.*, 1999, **185**, 445 (*Chem. Abstr.*, 1999, **131**, 214 469).
6 M.J. Climent, A. Corma, S. Iborra, S. Miquel, J. Primo and F. Rey, *J. Catal.*, 1999, **183**, 76 (*Chem. Abstr.*, 1999, **130**, 296 901).
7 T. Jyojima, N. Miyamoto, Y. Ogawa, S. Matsumura and K. Toshima, *Tetrahedron Lett.*, 1999, **40**, 5023.
8 V. Ferrières, T. Benvegnu, M. Lefeuvre, D. Plusquellec, G. MacKenzie, M.J. Watson, T.A. Haley, J.W. Goodby, R. Pindak and M.K. Durbin, *J. Chem. Soc., Perkin Trans.*, 2, 1999, 951.
9 S. Thiebaut, C. Gerardin-Charbonnier and C. Selve, *Tetrahedron*, 1999, **55**, 1329.
10 D. Gueyrard, P. Rollin, T.T.T. Nga, M. Ourévitch, J.-P. Bégué and D. Bonnet-Delpon, *Carbohydr. Res.*, 1999, **318**, 171.
11 S. Konstantinovic, J. Predojevic, Z. Petrovic, A. Spasojevic, B. Dimitrijevic and G. Milosevic, *J. Serb. Chem. Soc.*, 1999, **64**, 169 (*Chem. Abstr.*, 1999, **131**, 88 091).
12 S. Hiranuma, O. Kanie and C.-H. Wong, *Tetrahedron Lett.*, 1999, **40,** 6423.
13 I. Azumaya, T. Niwa, M. Kotani, T. Iimori and S. Ikegami, *Tetrahedron Lett.*, 1999, **40**, 4683.
14 I.L. Scott, R.V. Market, R.J. DeOrazio, H. Meckler and T.P. Kogan, *Carbohydr. Res.*, 1999, **137**, 210.
15 A.G. Gerbst, N.E. Ustuzhanina, A.A. Gracher, D.E. Tsvetkov, E.A. Khatuntseva and N.E. Nifant'ev, *Mendeleev Commun.*, 1999, 114 (*Chem. Abstr.*, 1999, **131**, 185 146).
16 V.O. Kur'yanov, A.E. Zemlyakov, T.A. Chupakhina and V.Ya. Chirva, *Bioorg. Khim.*, 1999, **25**, 319 (*Chem. Abstr.*, 1999, **131**, 5 453).
17 A.E. Zemelyakov, V.O. Kur'yanov, E.A. Sidorova and V. Ya. Chirva, *Bioorg. Khim*, 1998, **24**, 623.
18 A.E. Zemelyakov, V.O. Kur'yanov, E.A. Aksenova and V. Ya. Chirva, *Ukr. Khim, ZH. (Russ. Ed.)*, 1997, **63,** 46 (*Chem. Abstr.*, 1999, **130**, 182 696).
19 D. Crich, Z. Dai and S. Gastaldi, *J. Org. Chem.*, 1999, **64**, 5224.
20 D. Crich and W. Cai, *J. Org. Chem.*, 1999, **64**, 4926.
21 O.J. Plante, R.B. Andrade and P.H. Seeberger, *Org. Lett.*, 1999, **1**, 211 (*Chem. Abstr.*, 1999, **131**, 73 867).
22 H. Tanaka, H. Sakomoto, A. Sano, S. Nakamura, M. Nakajima and S. Hashimoto, *Chem. Commun.*, 1999, 1259.
23 J.-I. Kadokawa, J. Ebana, T. Nagaoka, M. Karasu, H. Tagaya and K. Chiba, *Nippon Kagaku Kaishi*, 1999, 625 (*Chem. Abstr.*, 1999, **131**, 310 773).
24 S. Matsumura, H. Tsuruta and K. Toshima, *Nihon Yukagakkaishi*, 1999, **48**, 15, (*Chem. Abstr.*, 1999, **130**, 182 683).
25 S. Matsumura, E. Yao and K. Toshima, *Biotechnol. Lett.*, 1999, **21**, 451 (*Chem. Abstr.*, 1999, **131**, 310 783).
26 S. Matsumura, E. Yao, K. Sakiyama and K. Toshima, *Chem. Lett.*, 1999, 373.
27 P. Reiss, D.A. Burnett and A. Zaks, *Bioorg. Med. Chem.*, 1999, **7**, 2199.
28 N. Kildemark and K.G.I. Nilsson, *Carbohydr. Lett.*, 1998, **3**, 211 (*Chem. Abstr.*, 1999, **130**, 81 712).
29 T. Kobayasi, S. Adachi and R. Matsuno, *Biotechnol. Lett.*, 1999, **21**, 105 (*Chem. Abstr.*, 1999, **130**, 311 986).

30 K.-C. Becker and P. Kuhl, *J. Carbohydr. Chem.*, 1999, **18**, 121.
31 F. Gonzalez-Muñoz, A. Pérez-Oseguera, J. Cassani, M. Jiménez-Estrada, R. Vazquez-Duhalt and A. López-Munguía, *J. Carbohydr. Chem.*, 1999, **18**, 275.
32 T. Stach, M. Matulova, V. Farkas, Z. Sulova, V. Patoprsty and K. Linek, *Chem. Pap.*, 1999, **53**, 218 (*Chem. Abstr.*, 1999, **131**, 322 839).
33 A.M. van der Heijden, F. van Rantwijk and H. van Bekkum, *J. Carbohydr. Chem.*, 1999, **18**, 131.
34 S.K. Sarkar, A.K. Choudhury, B. Mukhapadhyay and N. Roy, *J. Carbohydr. Chem.*, 1999, **18**, 1121.
35 J. Désiré and J. Prandi, *Carbohydr. Res.*, 1999, **317**, 110.
36 V. Jaouen, A. Jagou, L. Lemee and A. Veyerières, *Tetrahedron*, 1999, **55**, 9245.
37 B. Yu and L.-X. Dai, *Chemtracts*, 1999, **12**, 629 (*Chem. Abstr.*, 1999, **131**, 286 705).
38 W.R. Roush and S. Narayan, *Org. Lett.*, 1999, **1**, 899 (*Chem. Abstr.*, 1999, **131**, 272 092).
39 W.R. Roush and C.E. Bennett, *J. Am. Chem. Soc.*, 1999, **121**, 3541.
40 W.R. Roush, B.W. Gung and C.E. Bennett, *Org. Lett.*, 1999, **1**, 891 (*Chem. Abstr.*, 1999, **131**, 272 093).
41 A. Dios, C. Nativi, G. Capozzi and R.W. Franck, *Eur. J. Org. Chem.*, 1999, 1869.
42 G. Capozzi, C. Falciani, S. Menichetti, C. Nativi and B. Raffaelli, *Chem. Eur. J.*, 1999, **5**, 1748.
43 Y. Ding, Y. Miura, J.R. Etchison, H.H. Freeze and O. Hindsgaul, *J. Carbohydr. Chem.*, 1999, **18**, 471.
44 T. Suzuki, K. Mabuchi and N. Fukazawa, *Bioorg. Med. Chem. Lett.*, 1999, **9**, 659.
45 J. Mlynarski and A. Banaszek, *Pol. J. Chem.*, 1999, **73**, 973 (*Chem. Abstr.*, 1999, **131**, 73 901).
46 T. Chiba, K. Moriya, S. Nabeshima, H. Hayashi, Y. Kobayashi, S. Sasayama and K. Onozaki, *Glycoconjugate J.*, 1999, **16**, 499.
47 A. Liav, J.A. Hansjergen, K.E. Achyathan and C.D. Shimasaki, *Carbohydr. Res*, 1999, **317**, 198.
48 Y. Collette, K. Ou, J. Pires, M. Baudry, G. Descotes, J.-P. Praly and V. Barberousse, *Carbohydr. Res.*, 1999, **318**, 162.
49 M.K. Manthey, J.F. Jamie and R.J.W. Truscott, *J. Org. Chem.*, 1999, **64**, 3930.
50 A.K. Ghosh, S. Khan and D. Farquhar, *Chem. Commun.*, 1999, 2527.
51 Y.-L. Leu, S.R. Roffler and J.-W. Chern, *J. Med. Chem.*, 1999, **42**, 3623.
52 C. Schell and H.K. Hombrecher, *Chem. Eur. J.*, 1999, **5**, 587.
53 B.B. Lohray, V. Bhushan, A. Reddy, L.N. Sekar, V.V. Rao, V. Saibaba, N.J. Reddy, P. Kumar and K.N. Reddy, *Indian J. Chem., Sect. B: Org. Chem. Incl. Med. Chem.*, 1999, **388**, 635 (*Chem. Abstr.*, 1999, **131**, 299 604).
54 B. Zorc, M.J.-M. Takac and P. Rodighiero, *Acta Pharm.(Zagreb)*, 1999, **49**, 11 (*Chem. Abstr.*, 1999, **130**, 296 906).
55 K. Toshima, R. Takano, Y. Maeda, M. Suzuki, A. Asai and S. Matsumura, *Angew. Chem., Int. Ed. Engl*, 1999, **38**, 3733.
56 Z.-C. Li, Y.-Z,. Liang and F.-M. Li, *Chem. Commun.*, 1999, 1557.
57 J. Li, S. Zacharek, X. Chen, J. Wang, W. Zhang, A. Janczuk and P.G. Wang, *Biorg. Med. Chem.*, 1999, **7**, 1549.
58 S. Zalipsky, N. Mullah, A. Dibble and T. Flaherty, *Chem. Commun.*, 1999, 653.
59 L.E. Strong and L.L. Kiessling, *J. Am. Chem. Soc.*, 1999, **121**, 6193.
60 E. Kamst, K. Zegelaar-Jaarsveld, G.A. van der Marel, J.H. van Boom, B.J.J. Lugtenberg and H.P. Spaink, *Carbohydr. Res.*, 1999, **321**, 176.

61 M.E. Jung, E.C. Yang, B.T. Vu, M. Kiankarimi, E. Spyrou and J. Kaunitz, *J. Med. Chem.*, 1999, **42**, 3899.

62 G. Lecollinet, R. Auzély-Velty, M. Danel, T. Benvegnu, G. MacKenzie, J.W. Goodby and D. Plusquellec, *J. Org. Chem.*, 1999, **64**, 3139.

63 C. Kauffmann, W.M. Müller, F. Vögtte, S. Weinman, S. Abramson and B. Fuchs, *Synthesis*, 1999, 849.

64 P. Arya, K.M.K. Kutterer, H. Qin, J. Roby, M.L. Barnes, S. Lin, C.A. Lingwood and M.G. Peter, *Bioorg. Med. Chem.*, 1999, **7**, 2823.

65 T. Ren and D. Liu, *Tetrahedron Lett.*, 1999, **40**, 7621.

66 S.M. Dimick, S.C. Powell, S.A. McMahon, D.N. Moothoo, J.H. Naismith and E.J.Toone, *J. Am. Chem. Soc.*, 1999, **121**, 10286.

67 O. Ouari, A. Polidori, B. Pucci, P. Tordo and F. Chalier, *J. Org. Chem.*, 1999, **64**, 3554.

68 L.A.J.M. Sliedregt, P.C.N. Rensen, E.T. Rump, P.J. van Santbrink, M.K. Bijsterbosch, A.R.P.M. Valentijn, G.A. van der Marel, J.H. van Boom, T.J.C. van Berkel and E.A.L. Biessen, *J. Med. Chem.*, 1999, **42**, 609.

69 C. Grandjean, C. Rommens, H. Gras-Masse and O. Melnyk, *Tetrahedron Lett.*, 1999, **40**, 7235.

70 V. Sol, J.C. Blais, V. Carré, R. Granet, M. Guilloton, M. Spiro and P. Krausz, *J. Org. Chem.*, 1999, **64**, 4431.

71 O. Hayashida, K. Nishiyama, Y. Matsuda and Y. Aoyama, *Tetrahedron Lett.*, 1999, **40**, 3407.

71a K. Nakamura, A. Ishii, Y. Ito and Y. Nakahara, *Tetrahedron*, 1999, **55**, 11253.

72 Y. Nishida, Y. Takamori, H. Ohrui, I. Ishizuka, K. Matsuda and K. Kobayashi, *Tetrahedron Lett.*, 1999, **40**, 2371.

73 D. Redoulés and J. Perié, *Tetrahedron Lett.*, 1999, **40**, 4811.

74 J. Song and R.I. Hollingsworth, *J. Am. Chem. Soc.*, 1999, **121**, 1851.

75 H. Qin and T.B. Grindley, *Can. J. Chem.*, 1999, **77**, 481.

76 T. Sakai, M. Morita, N. Matsunaga, A. Akimoto, T. Yokoyama, H. Iijima and Y. Koezuka, *Bioorg. Med. Chem. Lett.*, 1999, **9**, 697.

77 D.A. Johnson, C.G. Sowell, C.L. Johnson, M.T. Livesay, D.S. Keegan, M.J. Rhodes, J.T. Ulrich, J.R. Ward, J.L. Cantrell and V.G. Brookshire, *Bioorg. Med. Chem. Lett.*, 1999, **9**, 2273.

78 M. Adanczyk and R.E. Roddy, *Tetrahedron: Asymm.*, 1999, **10**, 3157.

79 J.-I. Tamura and J. Nishihara, *Bioorg. Med. Chem. Lett.*, 1999, **9**, 1911.

80 G.A. Winterfeld, Y. Ito, T. Ogawa and R.R. Schmidt, *Eur. J. Org. Chem.*, 1999, 1167.

81 K. Burger, M. Kluge, S. Fehn, B. Kokseh, L. Hennig and G. Müller, *Angew. Chem., Int. Ed. Engl.*, 1999, **38**, 1414.

82 R.-J. Zhang, J. Jian, G.-L. Liang, W.-X. Wan, Y.-H. Tao and B.-C. Wang, *Nuc. Sci. Tech.*, 1998, **9**, 189 (*Chem. Abstr.*, 1999, **130**, 25240).

83 A. Crossman Jr., J.S. Brimacombe, M.A.J. Ferguson and T.K. Smith, *Carbohydr. Res.*, 1999, **321**, 42.

84 H. Dietrich, J.-F. Espinosa, J.L. Chiara, J. Jimenez-Barberu, Y. Leon, I. Varela-Nieto, J.-M. Mato, F.H. Cano, C. Foces-Foces and M. Martin-Lomas, *Chem. Eur. J.*, 1999, **5**, 320.

85 S. Takahashi, H. Terayama, H. Koshino and H. Kuzuhara, *Tetrahedron*, 1999, **55**, 14871.

86 I. Sato, Y. Akahori, T. Sasaki, T. Kikuchi and M. Hirama, *Chem. Lett.*, 1999, 867.

87 A.G. Myers, J. Liang and M. Hammond, *Tetrahedron Lett.*, 1999, **40**, 5129.
88 D.C. Limburg, M.J. Gourley and J.S. Williamson, *Nat. Prod. Lett.*, 1998, **12**, 97 (*Chem. Abstr.*, 1999, **130**, 25 253)..
89 M. Inoue and T. Kitahara, *Tetrahedron*, 1999, **55**, 4621.
90 S. Deng, B. Yu, J. Xie and Y. Hui, *J. Org. Chem.*, 1999, **64**, 7265.
91 B. Werschkun, K. Gorziza and J. Thiem, *J. Carbohydr. Chem.*, 1999, **18**, 629.
92 B. Alluis and O. Dangles, *Helv. Chem. Acta*, 1999, **82**, 2201.
93 G.I. Gonzalez and J. Zhu, *J. Org. Chem.*, 1999, **64**, 914.
94 D.-Q. Song and J.-S. Zhen, *Zhongguo Yiyao Gongye Zazhi*, 1998, **29**, 446 (*Chem. Abstr.*, 1999, **130**, 252 476).
95 S. Zhang, G. Shang, Z. Li, A. Wang and M. Cai, *Zhongguo Yaowu Huaxue Zazhi*, 1997, **7**, 256 (*Chem. Abstr.*, 1999, **130**, 25 245).
96 T. Kawada, R. Asano, S. Hayashida and T. Sakuno, *J. Org. Chem.*, 1999, **64**, 9268.
97 M. Ueda, Y. Sawai and S. Yamamura, *Tetrahedron*, 1999, **55,** 10925.
98 M. Ueda, Y. Sawai and S. Yamamura, *Tetrahedron Lett.*, 1999, **40**, 3757.
99 M. de Kort, E. Ebrahimi, E.R. Wijsman, G.A. van der Marel and J.H. van Boom, *Eur. J. Org. Chem.*, 1999, 2337.
100 J. Hunziker, *Biorg. Med. Chem. Lett.*, 1999, **9**, 201.
101 H. Sugimara, H. Kamamori and K. Stansfield, *Tennen Yuki Kagobatsu Toronkai Koen Yoshishu*, 1998, 655 (*Chem. Abstr.*, 1999, **131**, 73 906).
102 S.E. Keates, F.A. Loewus, G.L. Helms and D.L. Zink, *Phytochemistry*, 1998, **49**, 2397 (*Chem. Abstr.*, 1999, **130**, 186 572).
103 M. Ueda, M. Asano, Y. Sawai and S. Yamamura, *Tetrahedron*, 1999, **55**, 5781.
104 M. Ueda and S. Yamamura, *Tetrahedron Lett.*, 1999, **40**, 7823.
105 M. Ueda and S. Yamamura, *Tetrahedron Lett.*, 1999, **40**, 2981.
106 M. Ueda and S. Yamamura, *Tetrahedron*, 1999, **55,** 10937.
107 K.M. Jenkins, P.R. Jensen and W. Fenical, *Tetrahedron Lett.*, 1999, **40**, 7637.
108 S.C. Fields, *Tetrahedron*, 1999, **55**, 12237.
109 Z.-D. Jiang, P.R. Jensen and W. Fenical, *Bioorg. Med. Chem. Ltd.*, 1999, **9**, 2003.
110 K. Toshima, H. Nagai and J. Matsumura, *Synlett*, 1999, 1420.
111 I.A. Ivanova, A.J. Ross, M.A.J. Ferguson and A.V. Nickolev, *J. Chem. Soc. Perkin Trans 1*, 1999, 1743.
112 S.P. Vincent, M.D. Burkhart, C.-Y. Tsai, Z. Zhang and C.-H. Wong, *J. Org. Chem.*, 1999, **64**, 5254.
113 T. Yasukochi, C. Inaba, K. Fukase and S. Kusumoto, *Tetrahedron Lett.*, 1999, **40**, 6585.
114 V. Chiffoleau-Giraud, P. Spangenberg, M. Dion and C. Rabiller, *Eur. J. Org. Chem.*, 1999, 757.
115 S.E. Sen, S.M. Smith and K.A. Sullivan, *Tetrahedron*, 1999, **55**, 12657.
116 W. Wang and F. Kong, *Tetrahedron Lett.*, 1999, **40**, 1361.
117 K. Konehara, T. Hashizume, K. Ohe and S. Uemura, *Tetrahedron: Asymmetry,* 1999, **10**, 4029.'
118 S. Aoki, K. Higuchi, A. Kato, N. Murakami and M. Kobayashi, *Tetrahedron*, 1999, **55**, 14865.
119 T.N. Zvyagintseva, T.N. Makar'eva, S.P. Ermakova and L.A. Elyakova, *Bioorg. Khim*, 1998, **24**, 219 (*Chem. Abstr.,* 1999, **131**, 102 446).
120 M. Müller, U. Huchel, A. Geyer and R.R. Schmidt, *J. Org. Chem.*, 1999, **64**, 6190.

121 A. Geyer, U. Huchel and R.R. Schmidt, *Magn. Reson. Chem.*, 1999, **37**, 145 (*Chem. Abstr.,* 1999, **130**, 209 879).
122 M. Wakao, K. Fukase and S. Kusumoto, *Synlett.*, 1999, 1911.
123 M. Wakao, Y. Nakai, K. Fukase and S. Kusumoto, *Chem. Lett.*, 1999, **40**, 27.
124 P. Söderman and G. Widmalm, *J. Org. Chem.*, 1999, **64**, 4199.
125 T.R. Noel, R. Parker, S.M. Ring and S.G. Ring, *Carbohydr. Res.*, 1999, **319**, 166.
126 Y. Osa, K. Takeda, T. Sato, E. Kaji, Y. Mizuno and H. Takayanagi, *Tetrahedron Lett.*, 1999, **40**, 1531.
127 K. Fukase, Y. Nakai, K. Egusa, J.A. Porco, Jr. and S. Kusumoto, *Synlett*, 1999, 1911.
128 T. Doi, M. Sugiki, H. Yamada, T. Takahashi and J.A. Porco, *Tetrahedron Lett.*, 1999, **40**, 2141.
129 K. Oshima and Y. Aoyama, *J. Am. Chem. Soc.*, 1999, **121**, 2315.
130 O. Ohmori, H. Takayama and N. Aimi, *Tetrahedron Lett.*, 1999, **40**, 5039.
131 Z. Fang, *Jingxi Huagong*, 1999, **16**, 49 (*Chem. Abstr.,* 1999, **130**, 182 680).
132 J. Broddefalk, M. Forsgren, I. Sethson and J. Kihlberg, *J. Org. Chem.*, 1999, **64**, 8948.
133 V. Constantino, E. Fattorusso, A. Mangoni, M. DiRosa and A. Ianaro, *Bioorg. Med. Chem. Lett.*, 1999, **9**, 271.
134 W. Wang and F. Kong, *Angew. Chem., Int. Ed. Engl.,* 1999, **38**, 1247.
135 M. Adinolfi, G. Barone, L. Denapoli, L. Guariniello, A. Iadonisi and G. Piccialli, *Tetrahedron Lett.*, 1999, **40**, 2607.
135a A.B. Kantchev and J.R. Parquette, *Tetrahedron Lett.,* 1999, **40**, 8049.
136 R.J. Tennant-Eyles, B.G. Davis and A.J. Fairbanks, *Chem. Commun.*, 1999, 1037.
137 A.K. Misra, J.M. Brown, S.W. Homans and R.A. Field, *Carbohydr. Lett.*, 1998, **3**, 217 (*Chem. Abstr.*, 1999, **130**, 81 713).
138 M. Lergenmüller, T. Nukada, K. Kuramochi, A. Dan, T. Ogawa and Y. Ito, *Eur. J. Org. Chem.*, 1999, 1367.
139 G. Scheffler and R.R. Schmidt, *J. Org. Chem.*, 1999, **64**, 1319.
140 C.K. Smith, C.M. Hewage, S.C. Fry and I.H. Sadler, *Phytochemistry*, 1999, **52**, 387.
141 S.C. Ennis, A.J. Fairbanks, R.J. Tennant-Eyles and H.S. Yeates, *Synlett*, 1999, 1387.
142 V. Subramanian and T.L. Lowary, *Tetrahedron*, 1999, **55**, 5965.
143 X. Ding and F. Kong, *J. Carbohydr. Chem.*, 1999, **18**, 775.
144 A.V. Demchenko, E. Rousson and G.-J. Boons, *Tetrahedron Lett.*, 1999, **40**, 6523.
145 A.K. Choudhury, I. Mukherjee, B. Mukhopadhyay and N. Roy, *J. Carbohydr. Chem.*, 1999, **18**, 361.
146 N. Hada, E. Hayashi and T. Takeda, *Carbohydr. Res.*, 1999, **316**, 58.
147 P. Spangenberger, V. Chiffoleau-Girard, C. André, M. Dion and C. Rabiller, *Tetrahedron: Asymm.*, 1999, **10**, 2905.
148 E. Farkas and J. Thiem, *Eur. J. Org. Chem.*, 1999, 3073.
149 D.A. MacManus, V. Grabowska, K. Biggadike, M.I. Bird, S. Davies, E.N. Vulfson and T. Gallacher, *J. Chem. Soc., Perkin Trans. 1*, 1999, 295.
150 T. Ziegler, R. Dettmann, Ariffadhillah and U. Zettl, *J. Carbohydr. Chem.*, 1999, **18**, 1079.
151 M.J. Hadd and J. Gervay, *Carbohydr. Res.*, 1999, **220**, 61.
152 T. Yasukochi, K. Fukase and S. Kusumoto, *Tetrahedron Lett.*, 1999, **40**, 6591.

153 L. Knerr and R.R. Schmidt, *Synlett*, 1999, 1802.
154 K. Priya and D. Loganathan, *Tetrahedron*, 1999, **55**, 1119.
155 M. Upreti and R.A. Vishwakarma, *Tetrahedron Lett.*, 1999, **40**, 2619.
156 T. Ziegler, R. Dettmann and J. Grabowski, *Synthesis*, 1999, 1661.
157 T. Wrodnigg and A.E. Stütz, *Angew. Chem., Int. Ed. Engl.*, 1999, **38**, 827.
158 Z. Gan, S. Cao, Q. Wu and R. Roy, *J. Carbohydr. Chem.*, 1999, **18**, 755.
159 S. Cao, Z. Gan and R. Roy, *Carbohydr. Res.*, 1999, **318**, 75.
160 K.A. Savin, J.C.G. Woo and S.J. Danishefsky, *J. Org. Chem.*, 1999, **64**, 4183.
161 A. Vetere, L. Novelli and S. Paoletti, *J. Carbohydr. Chem.*, 1999, **18**, 515.
162 T. Endo, S. Koizumi, K. Tabata, S. Kakita and A. Ozaki, *Carbohydr. Res.*, 1999, **316**, 179.
163 C.L.M. Stutts, B.A. Macher, R. Bhatti, O.P. Srivastava and O. Hindsgaul, *Glycobiology*, 1999, **9**, 661.
164 R. Gonzalez and J. Thiem, *Carbohydr. Res.*, 1999, **317**, 180.
165 P.M. St Hilaire, L. Cipolla, A. Franco, U. Tedebark, D.A. Tilly and M. Meldal, *J. Chem. Soc., Perkin Trans.1*, 1999, 3559.
166 W.-T. Jiaang, K.-F. Hsiao, S.-T. Chen and K.-T. Wang, *Synthesis*, 1999, 1687.
167 Q. Li, H. Li, M.-S. Cai, Z.-J. Li and R.-L. Zhou, *Tetrahedron:Asymm.*, 1999, **10**, 2675.
168 C. Gallo-Rodriguez, L. Gondolfi and R.M. de Lederkremer, *Org. Lett.*, 1999, **1**, 245 (*Chem. Abstr.*, 1999, **131**, 73 857).
169 A.K. Pathak, G.B. Besra, D. Crick, J.A. Maddry, C.B. Morehouse, W.J. Suling and R.C. Reynolds, *Biorg. Med. Chem.*, 1999, **7**, 2407.
170 D.J. Silva, H. Wang, N.M. Allanson, R.K. Jain and M.J. Sofia, *J. Org. Chem*, 1999, **64**, 5926.
171 M. Oikawa, T. Shintaku, H. Sekljic, K. Fukase and S. Kusumoto, *Bull. Chem. Soc. Jpn.*, 1999, **72**, 1857.
172 W.-C. Liu, M. Oikawa, K. Fuxase, Y. Suda and S. Kusumoto, *Bull. Chem. Soc. Jpn.*, 1999, **72**, 1377.
173 D.A. Johnson, D.S. Keegan, C.G. Sowell, M.T. Livesay, C.L. Johnson, L.M. Taubner, A. Harris, K.R. Myers, J.D. Thompson, G.L. Gustafson, M.J. Rhodes, J.T. Ulrich, J.R. Ward, Y.M. Yorgensen, J.L. Cantrell and V.G. Brookshire, *J. Med. Chem.*, 1999, **42**, 4640.
174 M. Oikawa, H. Furuta, Y. Suda and S. Kusumoto, *Tetrahedron Lett.*, 1999, **40**, 5199.
175 J. Kubish, L. Weignerova, S. Kötter, T.K. Lindhorst, P. Sedmera and V. Křen, *J. Carbohydr. Chem.*, 1999, **18**, 975.
176 K.G.I. Nilsson, A. Eliasson, H. Pan and M. Rohman, *Biotechnol. Lett.*, 1999, **21**, 11 (*Chem. Abstr.*, 1999, **130**, 312 003)..
177 S. Koshida, Y. Suda, Y. Fukui, J. Ormsby, M. Sobel and S. Kusumoto, *Tetrahedron Lett.*, 1999, **40**, 5725.
178 B.L. Ferla, L. Lay, M. Guerrini, L. Poletti, L. Panza and G. Russo, *Tetrahedron*, 1999, **55**, 9867.
179 D. Weigelt, R. Krähmer, K. Brüschke, L. Hennig, M. Findeisen, D. Müller and P. Welzel, *Tetrahedron*, 1999, **55**, 687.
180 M.J. Sofia, N. Allanson, N.T. Hatzenbuhler, R. Jain, R. Kakarla, N. Kogan, R. Liang, D. Liu, D.J. Silva, H. Wang, D. Gange, J. Anderson, A. Chen, F. Chi, R. Dulina, B. Huang, M. Kamau, C. Wang, E. Baizman, A. Branstrom, N. Pristol, R. Goldman, K. Han, C. Longley, S. Midha and H.R. Axelrod, *J. Med. Chem.*, 1999, **42**, 3193.

181 H. Jiao and O. Hindsgaul, *Angew. Chem., Int. Ed. Engl.*, 1999, **38**, 346.
182 Y. Miura, Y. Ding, A. Manzi, O. Hindsgaul and H.H. Freeze, *Glycobiology*, 1999, **7**, 1053.
183 H. Schene and H. Waldmann, *Synthesis*, 1999, 1411.
184 K. Toshima, K.-I. Kasumi and S. Matsumura, *Synlett*, 1999, 813.
185 D.S. Weinstein and K.C. Nicolaou, *J. Chem. Soc. Perkin Trans. 1*, 1999, 545.
186 F.W. Lichtenthaler and B. Wender, *Carbohydr. Res.*, 1999, **319**, 47.
186a J. McCormick, Y. Li, K. McComick, H.J. Duynstee, A.K. van Engen, G.A. van der Marel, B. Ganem, J.H. van Boom and J. Meinwald, *J. Am. Chem. Soc.* 1999, **121**, 5661.
187 H. Vankayalapati and G. Singh, *Tetrahedron Lett.*, 1999, **40**, 3925.
188 P. Sengupta, S. Sarbajna and S. Basu, *J. Carbohydr. Chem.*, 1999, **18**, 87.
189 L.A. Paquette, L. Barriault and D. Pissarnitski, *J. Am. Chem. Soc.*, 1999, **121**, 4542.
190 L. Barriault, S.L. Boulet, K. Fujiwara, A. Murai, L.A. Paquette and M. Yotsu-Yamashita, *Biorg. Med. Chem. Lett.*, 1999, **9**, 2069.
191 M. Kuroda, Y. Mimaki and Y. Sashida, *Phytochemistry*, 1999, **52**, 435.
192 Y. Mimaki, M. Kuroda, A. Ide, A. Kameyama, A. Yokosuka and Y. Sashida, *Phytochemistry*, 1999, **50**, 805.
193 H. Sudo, K. Kijima, H. Otsuka, T. Ide, E. Hirata, Y. Takeda, M. Isaji and Y. Kurashina, *Chem. Pharm. Bull.*, 1999, **47**, 1341.
194 H.I. Duynstee, M.C. de Koning, H. Ovaa, G.A. van der Marel and J.H. van Boom, *Tetrahedron: Asymmetry*, 1999, **10**, 2623.
195 C.L. Battistelli, C. DeCastro, A. Iadonisi, R. Lanzetta, L. Mangoni and M. Parilli, *J. Carbohydr. Chem.*, 1998, **18**, 69.
196 R.V. Stick, D.M.G. Tilbrook and S.J. Williams, *Aust. J. Chem.*, 1999, **52**, 685.
197 P.G. Hultin and R.M. Buffie, *Carbohydr. Res.*, 1999, **322**, 14.
198 K. Krohn and C. Bäuerlein, *J. Carbohydr. Chem.*, 1999, **18**, 807.
199 H.I. Duynstee, M.C. de Koning, G.A. Van der Marel and J.H. van Boom, *Tetrahedron*, 1999, **55**, 9881.
200 L.O. Kononov, A.V. Kornilov, A.A. Sherman, E.V. Zyryanov, G.V. Zatonsky, A.S. Shashkov and N.E. Nifant'ev, *Bioorg. Khim.*, 1998, **24**, 608 (*Chem. Abstr.*, 1999, **131**, 45 016).
201 J. Tamura, Y. Miura and H.H. Freeze, *J. Carbohydr. Chem.*, 1999, **18**, 1.
202 A. Lubineau and D. Bonnaffé, *Eur. J. Org. Chem.*, 1999, 2523.
203 T. Takahashi, H. Tsukamoto, M. Kurosaki and H. Yamada, *Tennen Yuki Kagobutsu Toronkai Koen Yoshishu*, 1997, **39th**, 49 (*Chem. Abstr.*, 1999, **130**, 352 485).
204 R. Higuchi, T. Mori, T. Sugata, K. Yamada and T. Miyamoto, *Eur. J. Org. Chem.*, 1999, 3175.
205 R. Higuchi, T. Mori, T. Sugata, K. Yamada and T. Miyamoto, *Eur. J. Org. Chem.*, 1999, 145.
206 M. Sakagami, K. Horie, K. Higashi, H. Yamada and H. Hamana, *Chem. Pharm. Bull.*, 1999, **47**, 1237.
207 N. Hossain and G. Magnusson, *Tetrahedron Lett.*, 1999, **40**, 2217.
208 A.V. Demchenko and G.-J. Boons, *Chem. Eur. J.*, 1999, **5**, 1278.
209 S. Matsumura, T. Nakamura, E. Yao and K. Toshima, *Chem. Lett.*, 1999, 581.
210 S. Deng, B. Yu, Y. Lou and Y. Hui, *J. Org. Chem.*, 1999, **64**, 202.
211 A. Reiter, A. Zamyatina, H. Schindl, A. Hofinger and P. Kosma, *Carbohydr. Res.*, 1999, **317**, 39.

212 B. Dumont-Hornebeck, J.-P. Joly, J. Coulon and Y. Chapleur, *Carbohydr. Res.*, 1999, **321**, 314.
213 J.M. Benito, C.O. Mellet, K. Sadalapure, T.K. Lindhorst, J. Defaye and J.M. García Fernández, *Carbohydr. Res.*, 1999, **320**, 37.
214 T. Ohe, Y. Kajiwara, T. Kida, W. Zhang, Y. Nakatsuji and P. Ikeda, *Chem. Lett.*, 1999, 921.
215 P.J. Garegg, K.-J. Johannsson, P. Konradsson and B. Lindberg, *J. Carbohydr. Chem.*, 1999, **18**, 31.
216 C.K. Lee, J.H. Jun and Y.H. Seo, *Bull. Korean Chem. Soc.*, 1998, **19**, 1233 (*Chem. Abstr.*, 1999, **130**, 81 716).
217 R. Rodebaugh, J.S. Debenham, B. Fraser-Reid and J.P. Snyder, *J. Org. Chem.*, 1999, **64**, 1758.
218 H. Regeling, F. Sunghwa, B. Zwanenburg, R. de Gelder and G.J.F. Chittenden, *Carbohydr. Polym.*, 1998, **37**, 323.
219 G.B. Giovenzana, L. Lay, D. Monti, G. Pamisano and L. Panza, *Tetrahedron*, 1999, **55**, 14123.
220 R. Roy, R. Dominique and S.K. Das, *J. Org. Chem.*, 1999, **64**, 5408.
221 Y. J. Hu and R. Roy, *Tetrahedron Lett.*, 1999, **40**, 3305.
222 C. Seel, A.H. Parham, O. Safarowsky, G..M. Hüber and F. Vögtle, *J. Org. Chem.*, 1999, **64**, 7236.
223 M. Inouye, J. Chiba and H. Nakazumi, *J. Org. Chem.*, 1999, **64**, 8170.
224 S. Sakai, Y. Shigemasa and T. Sasaki, *Bull. Chem. Soc. Jpn.*, 1999, **72**, 1313.
225 N. Amanokura, Y. Kanekiyo, S. Shinkai and D.N. Reinhoudt, *J. Chem. Soc., Perkin Trans.*, 2, 1999, 1995.
226 T. Fujihira, T. Takido and M. Seno, *J. Mol. Catal. A: Chem.*, 1999, **137**, 65 (*Chem. Abstr.*, 1999, **130**, 125 276).
227 I. Fokt, J. Bogusiak and W. Szeja, *Carbohydr.Lett.*, 1998, **3**, 191 (*Chem. Abstr.*, 1999, **130**, 81 711).
228 C. Marino, O. Varela and R.M. de Lederkremer, *An. Asoc. Quim. Argent.*, 1998, **86**, 237 (*Chem. Abstr.*, 1999, **130**, 153 905).
229 B.S. Babu and K.K. Balasubramanian, *Tetrahedron Lett.*, 1999, **40**, 5777.
230 D.A. Evans, B.W. Trotter, P.J. Coleman, B. Cote, L.C. Dias, H.A. Rajapakse and A.N. Tyler, *Tetrahedron*, 1999, **55**, 8671.
231 M. Yun, Y. Shin, S. Yoon, I.H. Chun, J.E. Shin, E.N. Jeong, *Bull. Korean Chem. Soc.*, 1998, **19**, 1239 (*Chem. Abstr.*, 1999, **130**, 81 731).
232 R.A. Falconer, I. Jablonkai and I. Toth, *Tetrahedron Lett.*, 1999, **40**, 8663.
233 C. Kallus, T. Opatz, T. Wunberg, W. Schmidt, S. Henke and H. Kunz, *Tetrahedron Lett.*, 1999, **40**, 7783.
234 R. Iori, R. Bernardi, D. Gueyrard, P. Rollin and S. Palmieri, *Biorg. Med. Chem. Lett.*, 1999, **9**, 1047.
235 A.A.B. Robertson and N.P. Botting, *Tetrahedron* 1999, **55**, 13269.
236 L. Cipolla, B.L. Ferla and F. Nicotra, *Carbohydr. Polym.*, 1999, **37**, 291 (*Chem. Abstr.*, 1999, **130**, 153 868).
237 P.H. Gross, *Carbohydr. Polym.*, 1998, **37**, 215 (*Chem. Abstr.*, 1999, **130**, 153 867).
238 P. Sinaÿ, *Pure Appl. Chem.*, 1998, **70**, 1495 (*Chem. Abstr.*, 1999, **130**, 182 658).
239 R.M. Paton, *Carbohydr. Mimics*, 1998, 49 (*Chem. Abstr.*, 1999, **130**, 153 861).
240 P. Vogel, R. Ferritto, K. Kraehenbuehl and A. Baudat, *Carbohydr. Mimics*, 1998, 19 (*Chem. Abstr.*, 1999, **130**, 153 860).
241 D.H. Braithwaite, C.W. Holzapfel, D.B.G. Williams, *S. Afr. J. Chem.*, 1998, **51**, 162 (*Chem. Abstr.*, 1999, **130**, 223 501).

242 V. Gyóllai, L. Somsák and L. Szilagyi, *Tetrahedron Lett.*, 1999, **40**, 3969.
243 S. Yamago, H. Miyazoe, R. Goto and J.-i. Yoshida, *Tetrahedron Lett.*, 1999, **40**, 2347.
244 J. Xie, A. Molina and S. Czernecki, *J. Carbohydr. Chem.*, 1999, **18**, 481.
245 I.P. Smoliakova, M. Han, J. Gong, R. Caple and W.A. Smit, *Tetrahedron*, 1999, **55**, 4559.
246 Y. Yahiro, S. Ichikawa, S. Shuto and A. Matsuda, *Tetrahedron Lett.*, 1999, **40**, 5527.
247 K. Gotanda, M. Matsugi, M. Suemura, C. Ohira, A. Sano, M. Oka and Y. Kita, *Tetrahedron*, 1999, **55**, 10315.
248 O. Gaertzen, A.M. Misske, P. Wolbers and H.M.R. Hoffmann, *Tetrahedron Lett.*, 1999, **40**, 6359.
249 R. Gosh, D. De, B. Shown and S.B. Maiti, *Carbohydr. Res.*, 1999, **321**, 1.
249a J.K. Fairweather, R.V. Stick, D.M.G. Tilbrook and H. Driguez, *Tetrahedron*, 1999, **55**, 3695.
250 M. Isobe, R. Saeeng, R. Nishizawa, M. Konobe and T. Nishikawa, *Chem. Lett.*, 1999, 467.
251 J. Xu and A. Vasella, *Helv. Chim. Acta*, 1999, **82**, 1728.
252 T.V. Bohner, R. Beaudegnies and A. Vasella, *Helv. Chim. Acta*, 1999, **82**, 143.
253 R. Burli and A. Vasella, *Helv. Chim. Acta*, 1999, **82**, 485.
254 C. Fuganti and S. Serra, *Synlett*, 1999, **82**, 1241.
255 T. Kuribayashi, Y. Mizuno, S. Gohya and S. Satoh, *J. Carbohydr. Chem.*, 1999, **18**, 371.
256 T. Kuribayashi, S. Gohya, Y. Mizuno and S. Satoh, *J. Carbohydr. Chem.*, 1999, **18**, 383.
257 T. Kuribayashi, S. Gohya, Y. Mizuno and S. Satoh, *J. Carbohydr. Chem.*, 1999, **18**, 393.
258 T. Kuribayashi, S. Gohya, Y. Mizuno, M. Shimojima, K. Ito and S. Satoh, *Synlett*, 1999, 737.
259 D. Paetsch and K.H. Dötz, *Tetrahedron Lett.*, 1999, **40**, 487.
260 G. Matsuo, Y. Miki, M. Nakata, S. Matsumura and K. Toshima, *J. Org. Chem.*, 1999, **64**, 7101.
261 A. Dondoni and A. Marra, *Chem. Commun.*, 1999, 2133.
262 S. Manabe and Y. Ito, *J. Am. Chem. Soc.*, 1999, **121**, 9754.
263 K.N. Cho, J. Oh, T. Yoon, K.H. Chun and J.E.N. Shin, *J. Korean Chem. Soc.*, 1999, **43**, 375 (*Chem. Abstr.*, 1999, **131**, 286 707).
264 M. Hayashi, K. Tsukada, H. Kawabata and C. Lamberth, *Tetrahedron*, 1999, **55**, 12287.
265 D. Craig, M.W. Pennington and P. Warner, *Tetrahedron*, 1999, **55**, 13495.
266 C. Najera and M. Yus, *Tetrahedron*, 1999, **55**, 10547.
267 M.A. Leeuwenburgh, C. Kulker, H.I. Duynstee, H.S. Overkleeft, G.A. van der Marel and J.H. van Boom, *Tetrahedron*, 1999, **55**, 8253.
268 L. Eriksson, S. Guy, P. Perlmutter and R. Lewis, *J. Org. Chem.*, 1999, **64**, 8396.
269 O. Dirat, T. Vidal and Y. Langlois, *Tetrahedron Lett.*, 1999, **40**, 4801.
270 K. Methling, S. Aldinger, K. Pesake and M. Michalik, *J. Carbohydr. Chem.*, 1999, **18**, 429.
271 F.K. Griffin, D.E. Paterson and R.J.K. Taylor, *Angew Chem. Int. Ed. Engl.*, 1999, **38**, 2939.
272 N. Khan, X. Cheng and D.R. Mootoo, *J. Am. Chem. Soc.*, 1999, **121**, 4918.

273 O. Jarreton, T. Skrydstrup, J.-F. Espinosa, J. Jiménez-Bardero and J.-M. Beau, *Chem. Eur. J.*, 1999, **5**, 430.
274 S.L. Krintel, J. Jiménez-Barbero and T. Skrydstrup, *Tetrahedron Lett.*, 1999, **40**, 7565.
275 M.A. Leeuwenburgh, S. Picasso, H.S. Overkleeft, G.A. van der Marel, P. Vogel and J.H. van Boom, *Eur. J. Org. Chem.*, 1999, 1185.
276 D.-P. Pham-Huu, M. Petrušová, J.N. BeMiller and L. Petruš, *Tetrahedron Lett.*, 1999, **40**, 3053.
277 C. Pasquarello, R. Demange and P. Vogel, *Bioorg. Med. Chem. Lett.*, 1999, **9**, 793.
278 P. Gerber and P. Vogel, *Terahedron Lett.*, 1999, **40**, 3165.
279 H.G. Bazin, Y. Du, T. Polat and R.J. Linhardt, *J. Org. Chem.*, 1999, **64**, 7254.
280 M.H.D. Postema and D. Calimente, *Tetrahedron Lett.*, 1999, **40**, 4755.
281 H.-D. Junker, N. Phung and W.-D. Fessner, *Tetrahedron Lett.*, 1999, **40**, 7063.
282 A. Lieberknecht, H. Griesser, B. Krämer, R.D. Bravo, P.A. Colinas and R.J. Grigera, *Tetrahedron*, 1999, **55**, 6475.
283 T. Vidal, A. Haudrechy and Y. Langlois, *Tetrahedron Lett.*, 1999, **40**, 5677.
284 A. Dondoni, A. Marra and A. Massi, *J. Org. Chem.*, 1999, **64**, 933.
285 A.D. Campbell, D.E. Paterson, T.M. Rayaham and R.J.K. Taylor, *Chem. Commun.*, 1999, 1599.
286 B. Westermann, A. Walter and N. Diedrichs, *Angew. Chem., Int. Ed. Engl.*, 1995, **38**, 3384.
287 A.J. Pearce, S. Ramaya, S.N. Thorn, G.B. Bloomberg, D.S. Walter and T. Gallagher, *J. Org. Chem.*, 1999, **64**, 5453.
288 J.J. Turner, N. Wilschut, H.S. Overkleeft, W. Klaffke, G.A. van der Marel and J.H. van Boom, *Tetrahedron Lett.*, 1999, **40**, 7039.
289 F. Peri, L. Cipolla, B. La Ferla, P. Dumy and F. Nicotra, *Glycoconjugate J.*, 1999, **16**, 399.
290 V. Aucagne, D. Gueyrard, A. Tatibouët, S. Cottaz, H. Driguez, M. Lafosse and P. Rollin, *Tetrahedron Lett.*, 1999, **40**, 7319.
291 A. Dondoni, D. Perrone and E. Turturici, *J. Org. Chem.*, 1999, **64**, 5557.
292 J. Kobayashi, T. Kubota, M. Takahashi, M. Ishibashi, M. Tsuda and H. Naoki, *J. Org Chem.*, 1999, **64**, 1478.
293 J.T. Shaw and K.A. Woerpel, *Tetrahedron*, 1999, **55**, 8747.
294 C.H. Larsen, B.H. Ridgway, J.T. Shaw and K.A. Woerpel, *J. Am. Chem. Soc.*, 1999, **121**, 12208.
295 D. Egron, T. Durand, A. Roland, J.-P. Vidal and J.-C. Rossi, *Synlett*, 1999, 435.
296 M. Yuasa, S. Kanazouva, N. Nishimura, T. Higuchi and I. Maeba, *Carbohydr. Res.*, 1999, **315**, 98.
297 J. Zhang and D.L.J. Clive, *J. Org. Chem.*, 1999, **64**, 770.
298 R. Bruns, A. Wernicke and P. Koll, *Tetrahedron*, 1999, **55**, 9793.
299 C. Strässler, N.E. Davis and E.T. Kool, *Helv. Chim. Acta*, 1999, **82**, 2160.
300 K. Tanaka and M. Shionoya, *J. Org. Chem.*, 1999, **64**, 5002.
300a M. Tingoli, B. Panunzi and F. Santacroce, *Tetrahedron Lett.*, 1999, **40**, 9329.
301 T. Tschamber, H. Rudyk and D. LeNouën, *Helv. Chim. Acta*, 1999, 2015.
302 K. Tomooka, M. Kikuchi, K. Igawa, P.-H. Keong and T. Nakai, *Tetrahedron Lett.*, 1999, **40**, 1917.
303 V. Popsavin, O. Berić, M. Popsavin, J. Csanádi, D. Vujić and R. Hrabal, *Tetrahedron Lett.*, 1999, **40**, 3629.

4
Oligosaccharides

1 General

As previously, this chapter deals with specific tri- and higher saccharides; disaccharides are dealt with in Chapter 3. Most references relate to chemical, enzymic or chemico-enzymic syntheses. Chemical features of the cyclodextrins are noted separately but their properties as complexing agents are not treated. With the increasing use of enzymic synthetic methods in the field, the rising importance of complex structures (as for example in glycoproteins and dendrimers) and rapid developments in the application of combinatorial procedures, more examples are appearing in the literature of oligosaccharide mixtures that are difficult to classify by the approach used in these reports. Work in glycobiology is revealing more and more the importance of the oligosaccharides in biology and medicine, and progress in synthetic work is advancing rapidly to meet the challenges offered.

A glimpse into the future is provided by Wong's group in their seminal paper 'Programmable 1-Pot Oligosaccharide Synthesis' in which they report the quantification of factors that affect the reactivities of glycosyl donors and acceptors. In consequence a computer programme 'OptiMer' has been developed for selecting the optimal buildings blocks for one-pot syntheses of linear and branched-chain oligosaccharides. To illustrate the power of the approach the synthesis of the branched tetrasaccharide β-D-Gal*p*-(1→3)-[α-L-Fuc*p*-(1→4)]-β-D-Gal*p*-(1→4)-D-GlcNH$_2$ was reported.[1] Wong has also published with Ley's group on the synthesis of 15 complex oligosaccharides up to undecamers found in the glycodelins and related to the *N*-linked glycoprotein oligosaccharides. The simplest are linear trisaccharides and the most complex are decamers and undecamers displaying Le^x, Le^y, SiaLe^x and T antigen epitopes on a 3,6-branched mannotriose core. In general two lactosamine units were attached to the core to give a heptasaccharide to which NeuNAc and L-Fuc were attached enzymically. An example is given in the undecasaccharide section.[2]

A Russian and a Japanese review have been published on the synthesis of derivatives and analogues of oligosaccharides[3] and on solid phase syntheses,[4] respectively, and Hindsgaul has reviewed the generation of oligosaccharide libraries by use of combinatorial principles.[5] He has also highlighted the use of the *p*-acetoxybenzyl and *p*-[2-(trimethylsilyl)ethoxymethoxy]benzyl groups for

Carbohydrate Chemistry, Volume 33

protecting purposes in oligosaccharide synthesis. The former can be removed either with hot methoxide or with methoxide and mild oxidation with iron(III) chloride, while the latter is cleavable by fluoride treatment.[6]

Synthesis of heparin oligosaccharide mimics and their activity as anti-coagulants have been covered in a specific review with a historical perspective.[7]

Considerable attention has been paid to reviewing aspects of the use of enzymes in oligosaccharide chemistry. 'Glycanase-catalysed synthesis of non-natural oligosaccharides' deals with oligomers made by use of glycosyl fluorides as donors,[8] and the use of an *E. coli* transferase using lactose as donor and oligosaccharides as acceptors has been surveyed.[9]

Further reviews on the use of recombinant α-(1→3)-galactotransferase for making α-galactosated oligosaccharides[10] and on the specificity of glycosyl-transferases from unnatural sugar nucleotide donors in making unnatural oligosaccharides[11] have appeared.

In new enzymic studies galacto-oligosaccharides have been made using lactose as donor and an enzyme from *Aspergillus oryzae*, sometimes immobilized on chitosan,[12,13] and α-glucosidases from *Bacillus stearothermophilus* and brewers' yeast have been compared with respect to their ability to glucosylate maltose. The bacterial enzyme was more active and gave tri- and tetra-saccharides, while the latter gave only trisaccharides in lower yield.[14] Chitosan with 24% degree of acetylation has been depolymerized by a mixture of cellulase, α-amylase and proteinase, and oligomers with DP 3-10 were isolated by membrane separation.[15]

2 Trisaccharides

2.1 General. – Compounds in Sections 2.2–2.4 are treated according to their non-reducing end sugars. Unless otherwise specified, the sugar name abbreviations (Glc *etc.*) imply the pyranosyl ring forms.

2.2 Linear Homotrisaccharides. – β-1,6-Linked glucotriose has been made by condensing a tetra-*O*-protected β-glucosyl phenyl carbonate with a 2,3,4-tri-*O*-protected β-thioglycoside with trityl tetrakis(pentafluorophenyl) borate [$Tr^+B^-(C_6F_5)_4$] as specific activator of the glycosyl carbonate. The resulting disaccharide thioglycoside was then coupled with methyl 2,3,4-tri-*O*-benzyl-α-D-glucoside using the same activator but now in the presence of sodium periodate. Both glycosidic bonds were formed with high β-selectivity.[16] The isomeric α-1,6-linked glucotriose has been made as main product of a trimer mixture using the solid phase approach with a *p*-acylaminobenzyl linker which was cleaved by use of DDQ and trichloroacetimidate glycosylation.[17]

A novel approach was used in the synthesis of the α-1,3-linked L-rhamnose-trimer which involved the controlled use of 1,2-orthoesters. Compound **1** was made by coupling tri-*O*-acetyl-α-L-rhamnosyl bromide with allyl α-L-rhamno-pyranoside with silver triflate as promoter, and converted by treatment with trimethylsilyl triflate to the disaccharide triol **2**. Further application of the triacetylglycosyl bromide with silver triflate gave the trisaccharide orthoester equivalent of compound **1** which was converted to the α-1,3-linked trimer. The glycosylation steps occurred with almost 80% regioselectivity. Application of this procedure to the preparation of a 2-*O*-galactosyl-[4-*O*-xylosyl]-rhamnoside is reported in Section 2.5.[18] The α-1,3-linked Rha trimer has been found as a tetraacyl ester of its octyl glycoside together with several related disaccharide esters in the stems of *Mezzettia leptopoda* and to have weakly cytotoxic properties.[19]

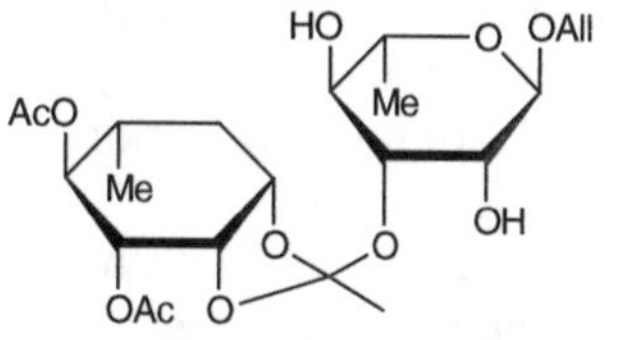

2.3 Linear Heterotrisaccharides. – α-D-Gal-(1→3)-β-D-Gal-(1→4)-D-Glc has received considerable attention, notably because of its relevance as an epitope involved in rejection during xenotransplantation. It has been synthesized by chemical methods on a 50 g scale as its β-glycosyl azide[20] which was used, following reduction of the azido group, for binding to polyacrylamide for testing for inhibition of binding of human serum to mouse laminin.[21] Use of a penicillin α-galactosidase allowed access to the trisaccharide by galactosylation of lactose, and analogous use of LacNAc gave the *N*-acetylamino trisaccharide.[22] Parallel enzymic work involving galactosylation of 6′-substituted lactoses led to the 6′-acetyl, -propanoyl and -butanoyl derivatives of the trisaccharide.[23]

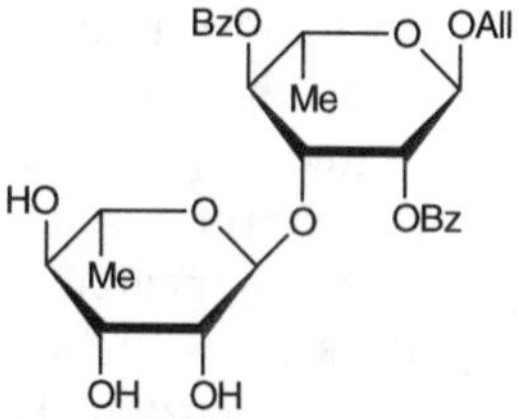

Interest is also appreciable in α-D-Gal-(1→4)-β-D-Gal-(1→4)-D-Glc (globotriose, see refs. 52 and 53 for mention of monoamino derivatives and their ceramide aglycon analogues) which has been incorporated into a dendrimer as a receptor of verotoxins produced by pathogenic *E. coli*.[25] Carrying a fluorescent label its *p*-acrylaminophenyl glycoside has been copolymerized with acrylamide for studies of the binding of lectins.[26] It has also been incorporated into *C*-linked serine analogue derivatives by way of the allyl *C*-glycoside which was ozonolysed prior to a Wittig-like aglycon chain extension.[27] A yeast β-D-galactosidase was used in the preparation of β-D-Gal-(1→4)-β-D-Gal-(1→4)-D-Glc with lactose as the source of galactose as well as the acceptor,[28] and this trimer has been produced as part of a crown ether having the macrocycle incorporating O-2 and O-3 of the glucose moiety.[29]

Acetobromolactose coupled with methyl 4,6-*O*-benzylidene-α-D-glucopyranoside with silver triflate as promoter gave a mixed orthoacetate involving the oxygen atoms at C-1 and C-2 of the lactose and C-3 of the acceptor. Treatment with trimethylsilyl triflate resulted in a derivative of β-D-Gal-(1→4)-β-D-Glc-(1→3)-D-Glc.[30]

Other trisaccharides which have D-galactose at the non-reducing end to have been reported are β-D-Gal-(1→3)-β-D-Gal-(1→4)-β-D-Xyl-*O*-serine (chemicoenzymic methods),[31] α-D-Gal-(1→3)-3-*C*-Me-β-D-Gal-(1→4)-β-D-Glc-*O*-octyl and the 1→4,1→4 linked analogue (the branch-point introduced from the corresponding 3-ulose derived from a lactoside derivative, and the galactosyl unit by use of an enzyme),[32] 3-*O*-SO_3^- – β-D-Gal-(1→3)-β-D-GlcNAc-(1→3)-β-D-Gal-OAll (and the compound with a sulfate group at C-6′),[33] β-D-Gal-(1→3)-2,6-di-NH_2-β-D-Xyl-(1→6) or -(1→2)-D-Man (for study of the effects of M^{2+} on the conformations of the central unit)[34] and α-D-Gal-(1→2)-α-LD-Hep-(1→4)-Kdo (isolated from a bacterial lipopolysaccharide).[35]

Enzymic methods were used to produce β-D-Man-(1→4)-β-D-GlcNAc-(1→4)-D-GlcNAc[36] and the bacterial O-antigen β-D-Man-(1→4)-α-L-Rha-(1→3)-β-D-GalO$(CH_2)_4C_6H_4(NO_2)$*p*.[37]

A saponin with hepatoprotective properties, isolated from flowers of *Pueraria* sp., contains the trisaccharide α-L-Rha-(1→2)-β-D-Gal-(1→2)-D-GlcA,[38] and parallel work has found a further saponin with anti-ulcerogenic properties in the flowers of *Spartium junceum* (Spanish broom); its carbohydrate component is the epimer α-L-Rha-(1→2)-β-D-Glc-(1→2)-β-D-GlcA.[39] Synthetic studies have produced tricolorin G which is α-L-Rha-(1→2)-β-D-Glc-(1→2)-D-Fuc with a long chain alkyl aglycon terminating in a carboxylic acid group that lactonizes with O-2 of the rhamnose unit. In the course of the work tricolorin A, which has a further L-Rha unit linked to O-3 of the Rha, was also made.[40]

α-L-Fuc-(1→2)-β-D-Gal-(1→3)-2-OAc-D-Gal has been made as a mimic of the terminal trisaccharide of a tumour antigen; conformational and antibody binding studies of it and the natural antigen were reported.[41] Related studies have led to the synthesis of α-L-Fuc-(1→2)-β-D-Gal-(1→4)-D-GlcNAc with a prop-1,3-diyl tether between O-6 of the Gal and O-3 of the GlcNAc, which is a conformationally constrained derivative of human blood group H substance. Interestingly it was made by first introducing the tether which then facilitated intramolecular glycosylation between the Gal and GlcNAc units.[42]

2,3-Di-*O*-Me-α-L-Fuc-(1→3)-2,4-di-*O*-Me-β-D-Xyl-(1→3)-2,4-dideoxy-2-*C*-Me-β-D-Glc-CH_2CO_2Me, part of a red algae toxin has been made,[43] as has an incompletely described 2,3-di-*O*-Me-α-L-Fuc-α-L-Rha--2-Tal peptide glycoside which occurs as a mycobacterial hapten.[44]

In the area of dideoxysugar trisaccharides β-D-Ole-(1→4)-β-D-Cym-(1→4)-D-Ole (Ole and Cym = 2,6-dideoxy-3-*O*-methyl-*arabino*-and-*ribo*-hexose, respectively) and the same compound without the *O*-methyl group on the central sugar have been isolated from the plant *Marsdenia roylei*.[45] The trisaccharides **3** and **4** are carbohydrate components of the antibiotics cororubicin and chromomycin A_3, respectively. The first has been synthesized by chemical means,[46] and the latter has been isolated from the natural product by use of samarium iodide at low temperatures. It was then converted to an *O*-substituted glycal derivative for reattachment to synthetic aglycons for study of their binding to DNA.[47] In a fine example of developing approaches to oligosaccharide synthesis compound **5** was made in 25% yield as its *O*-

benzylated phenyl thioglycoside in one step using tunable glycosyl sulfoxides as glycosylating agents.[48]

Compounds with aminosugars at the non-reducing end to have been reported are β-D-GlcNAc-(1→4)-α-D-Man-(1→4)-D-Man, the repeating unit of the O-specific cell wall polysaccharide of *E.coli*, which has been synthesized using 2-azido-*O*-benzylated sugars activated with *p*-nitrobenzenesulfonyl chloride, silver triflate and triethylamine,[49] and the β-1,2;α-1,3 linked isomer which was made in constrained conformation by a methylene acetal bridge between O-4 and O-6″ which mimics an intramolecular H-bond in the natural trisaccharide recognized by α-D-Man-specific lectins. Complexing of the cyclic product and its conformational analysis were reported.[50] The trisaccharide **6** was concluded to be the site of binding of heparins to blood platelets following binding studies on partial structures of the pentasaccharide corresponding to the antithrombin-III-binding region of heparin.[51] Monoamino analogues of globotriose [α-D-Gal-(1→4)-β-D-Gal-(1→4)-D-Glc] have been synthesized with the modified function at C-6′, C-2″, C-4″ and C-6″[52] and converted into aminodeoxy analogues of globotriosylceramide.[53]

Considerable attention continues to be given to gangliosides and related

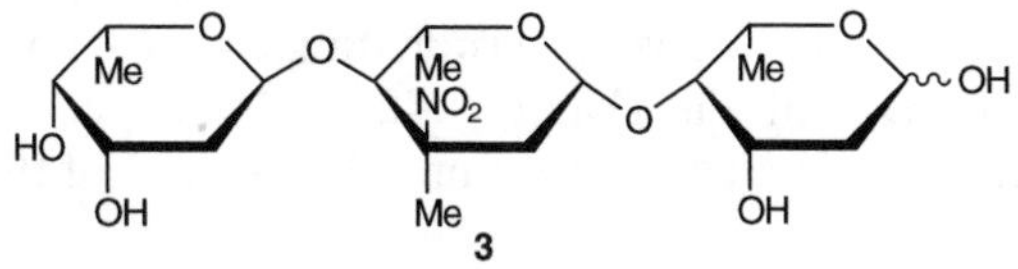

3

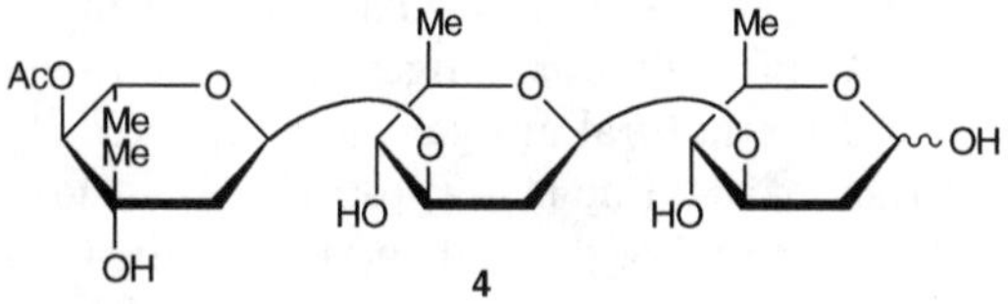

4

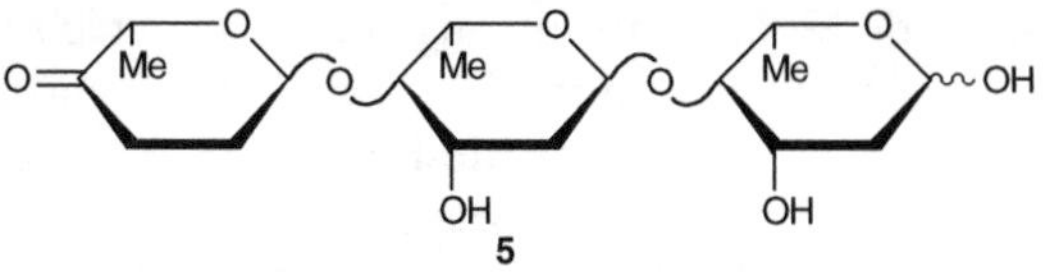

5

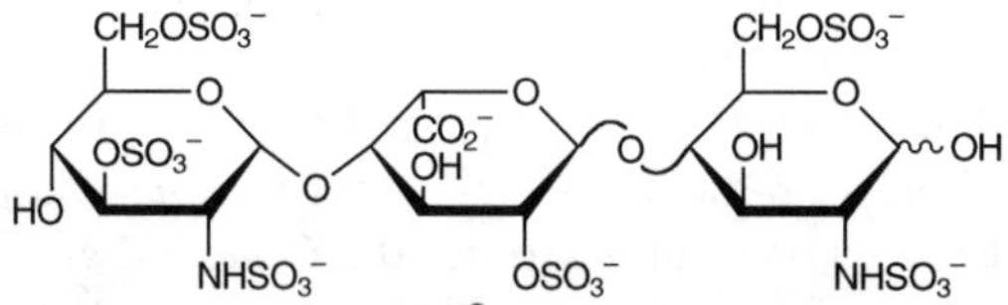

6

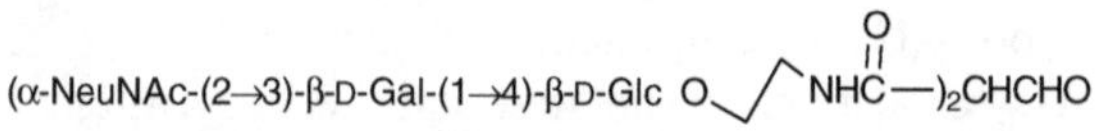

(α-NeuNAc-(2→3)-β-D-Gal-(1→4)-β-D-Glc O–CH2CH2–NHC(=O)—)$_2$CHCHO

7

compounds. The synthesis of the GM$_3$ trisaccharide α-NeuNAc-(2→3)-β-D-Gal-(1→4)-D-Glc has been conducted on a water soluble polymer,[54] and it has been converted into the bivalent linker **7** for coupling to thiolated proteins for the study of the immune response in mice.[55] Isolation of its ceramide glycoside (GM_3 ganglioside) can be effected from the natural branched pentasaccharide glycoside, GM_1, by conversion to the lactone involving the NeuNAc carboxylic acid group followed by partial acid hydrolysis.[56] Otherwise it has been synthesized with a fluorescent dansyl group in the aglycon by chemicoenzymic methods on a polymer support.[57]

The last paper also reports the synthesis of the unnatural α-(2→6)-β-(1→4)-linked analogous 'pseudo-GM_3'[57] and other workers have produced the related α-NeuNAc-(2→6)-β-D-Gal-(1→4)-β-D-GlcNAc bonded to an aminoalcohol linker aglycon[58] and the same trisaccharide with a 9-deoxy-9-fluoro substituent in the NeuNAc moiety and with ^{13}C labels at C-3 of this unit and throughout the galactose has been made chemoenzymically. In the course of this work modified NeuNAc was made with an azidodeoxy group at C-9 and separately with a deoxy group at C-8.[59]

The tumour 2,3-sialyl-T antigen α-NeuNAc-(2→3)-β-D-Gal-(1→3)-GalNAc has been prepared as a derivative for use in automated solid phase glycopeptide synthesis by chemical methods,[60] and also by a chemoenzymic route with a spacer aglycon.[61] In a massive piece of 'state of the art' work Danishefsky's group have made it linked to serine and threonine by their 'cassette' approach whereby the non-reducing end disaccharide was bonded to the GalNAc-amino acid glycosides. In the course of the work many related and more elaborate oligosaccharides, including the Le^y blood group antigen, were made.[62]

Other trisaccharides with acidic non-reducing terminals to have been reported are α-Kdn-(2→6)-β-D-Gal-(1→4)-D-GlcNAc (Kdn = 3-deoxy-D-*glycero*-D-*galacto*-2-nonulosonic acid), which was synthesized by chemoenzymic methods,[63] and α-D-GlcA-(1→3)-α-L-Rha-(1→2)-L-Rha which was isolated from the green alga *Chlorella vulgaris*.[64]

β-D-Xyl-(1→2)-β-D-Man-(1→4)-D-Glc, a component of *Hyriopsis schlegelii* glycophospholipid, has been made by chemical methods, the β-mannosyl linkage being effected by use of an α-glycosyl sulfoxide.[65] For the α-rhamnosyl link in β-L-Xyl-(1→2)-α-L-Rha-(1→4)-D-Man an orthoester involving O-1 and O-2 of L-Rha and O-4 of D-Man was employed.[66]

From bioactive tetramic acid glycosides of a sponge a polyene carrying 5-deoxy-2-*O*-Me-β-D-Ara*f*-(1→4)-β-D-Ara*p*-(1→2)-D-Xyl was isolated.[67]

2.4 Branched Homotrisaccharides. – The beautiful one-pot procedure illustrated in Scheme 1, which depends on initial activation of the bromide and subsequent activation of the thioglycoside, gave 76% of the *O*-substituted 3,6-di-*O*-β-D-glucopyranosyl-α-D-glucoside, and the corresponding 2,6- and 4,6-disubstituted analogues were obtained similarly in 72 and 64%, respectively.[68] Selective substitution with tetra-*O*-benzoyl-α-D-mannopyranosyl trichloroacetimidate of the α-D-mannoside has given the 3,6-di-*O*-α-D-mannopyranosyl mannoside of 8-(methoxycarbonyl)-octan-1-ol which was extended *via* the aglycon to give a glycophospholipid.[69]

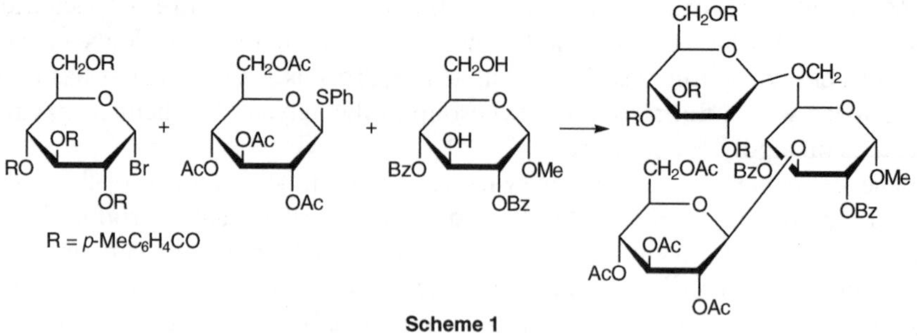

R = *p*-MeC_6H_4CO

Scheme 1

2.5 Branched Heterotrisaccharides. – Compounds in this section are categorized according to their reducing end sugars.

Selective substitutions on diosgenin glucoside with glycosyl trichloroacetimidates have been used to make β-D-Xyl-(and α-L-Ara*f*)-(1→3)-[α-L-Rha-(1→2)]-β-D-GlcO-diosgenin,[70] and 4,6-di-*O*-(β-D-Gal*f*)-D-Glc has been synthesized by use of 2,3,5,6-tetra-*O*-benzoyl-D-galactofuranosyl acetate and β-ethyl thioglycoside.[71]

β-D-GlcNAc-(1→3)-[β-D-Glc-(1→4)]-D-Gal, the repeating trisaccharide of group B type 1A *Streptococcus* capsular polysaccharide has been made on a polymer support. The issue of transfer of acyl protecting groups during glycosylations was addressed, and large pivaloates in the acceptors and 2-*O*-pivaloates in the donors were used to suppress these side reactions.[72]

β-D-Gal-(1→2)-[β-D-Xyl-(1→4)]-L-Rha has been made using the orthoester rearrangement technique,[18] and also in the area of deoxysugar trisaccharides compound **8** has been made, using tri-*O*-acetyl-D-glucal as primary sugar derivative, as a Sia Le^x analogue. It binds 30 times more strongly to E-selectin than does Sia Le^x.[73] A related trisaccharide, 5-thio-α-L-Fuc-(1→3)-[β-D-Gal-(1→4)]-D-GlcNAc, is a thio analogue of the Le^x trisaccharide,[74] and compounds also related to selectin binders to have been made are α-NeuNAc-(2→6)-[α- and β-Gal-(1→3)]-GalNAc linked to a spacer group.[75]

β-D-Glc-(and β-D-Xyl)-(1→2)-[β-D-Glc-(1→4)]-D-GlcA, which enhance the absorption of magnesium ions in mice, have been found in saponins of European horse chestnut.[76]

2.6 Analogues of Trisaccharides and Compounds with Anomalous Linking. – Ene-yne coupling over a ruthenium catalyst of a propargyl and an allyl glycoside gave a linked, conjugated diene which, on Diels Alder addition with but-1-en-3-one and deprotection, afforded the pseudo-trisaccharide **9**. An example of a hetero Diels Alder reaction involving the use of an imine was also given which resulted in a product with a dihydropyridine as the central ring.[77] Galactosylation of allyl 2,3-di-*O*-benzyl-4,6-di-*O*-(3-hydroxypropyl)-β-D-glucopyranoside with a galactosyl difluorophosphate resulted in a glucoside with two β-galactosyl substituents linked 1,4- and 1,6-through propyl groups.[78] β-(1→4)-Linked glucotriose with sulfur as the linking atoms has been made by use of solid phase methods,[79] and the trimannoses **10** and **11**, each with one C–C inter-unit link, and the latter with sulfur as the ring heteroatom of the CH_2-linked unit, have been made. The C–C linkage was introduced by radical addition of glycosyl moieties to the double bond of methyl 6,7-dideoxy-2,3-*O*-isopropylidene-α-D-*lyxo*-hex-6-enofuranosid-5-ulose.[80]

See Chapter 19 for trisaccharide analogues having a carbocyclic central moiety.

3 Tetrasaccharides

Compounds of this set and higher oligosaccharides are classified according to whether they have linear or branched structures and then by the nature of the sugars at the reducing ends.

3.1 Linear Homotetrasaccharides. – Heptaacetyl maltosyl bromide coupled with its hydrolysis product heptaacetyl maltose by silver triflate/2,4-lutidine promotion affords an orthoester-linked product which, with trimethylsilyl triflate, gives access to the α-(1→4)-β,β-(1↔1)-α-(4←1) non-reducing gluco-tetraose. Related compounds based on lactose and maltose were also made.[81] On the other hand, the reducing α-(1→4) linked glucotetraose has been made by intramolecular coupling of a phenyl thiomaltoside joined O-6–O-6′ by a phthalate bridging diester to a maltoside with a free hydroxyl group at C-4′.[82]

A cellulase from *Trichoderma viride* transfers β-D-Glc to O-6 of different glucose di- and tri-saccharides, and by its use various trisaccharides as well as β-D-Glc-(1→6)-β-D-Glc-(1→6)-β-D-Glc-(1→4)-D-Glc were made in low yield.[83,84] β-D-Glc-(1→6)-β-D-Glc-(1→3)-β-D-Glc-(1→6)-D-Glc is the sugar component of an anthraquinone glycoside found in *Cassia torosa* and with anti-allergic properties.[85]

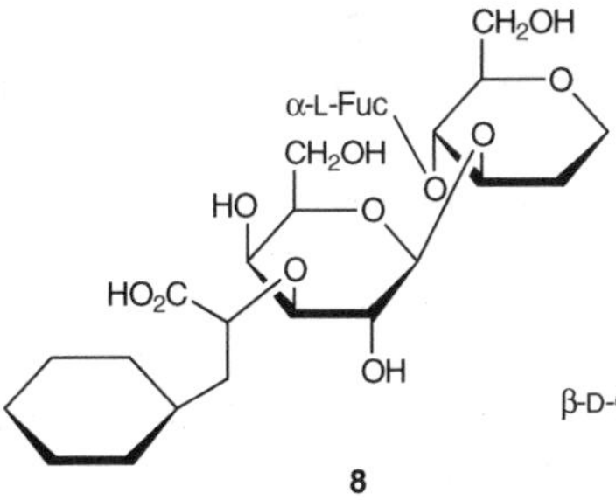

8

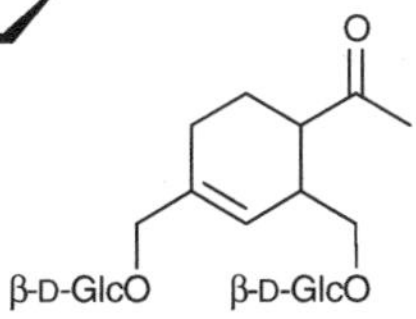

9

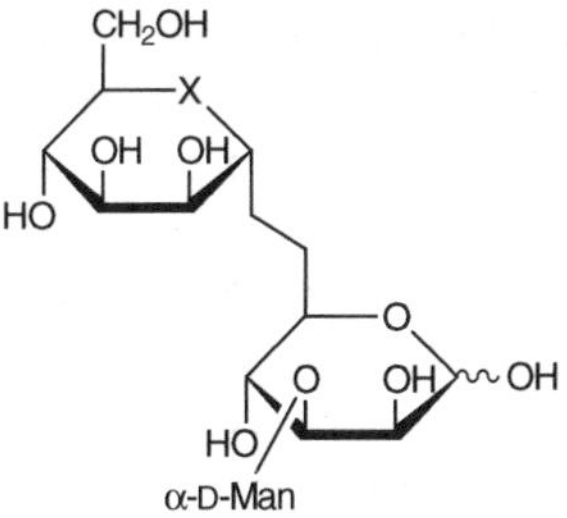

10 X = O
11 X = S

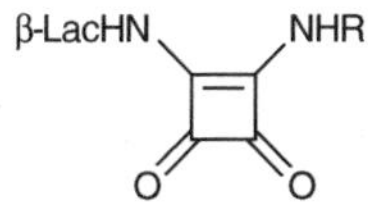

12 R = C_7H_{15} or Sepharose

3.2 Linear Heterotetrasaccharides. – β-D-Gal-(1→4)-β-D-Glc-(1→4)-β-D-Gal-(1→4)-D-Glc (a lactose dimer) was made on a polymer by use of an unsaturated tether which, on cleavage with Grubbs' catalyst, gave the tetramer as its allyl glycoside.[86] Two reports on the enzymic synthesis of the human milk oligosaccharide β-D-Gal-(1→4)-β-D-GlcNAc-(1→3)-β-D-Gal-(1→4)-D-Glc (lacto-*N*-neotetraose) have been reported, the first[87] also recording the related preparation of the 1→3,1→3,1→4 linked isomer (lacto-*N*-tetraose), and the second[88] describing the use of the novel squaric acid linking units illustrated by **12**, as the starting materials for the synthesis. This is established by use of the diethoxy compound 'diethyl squarate' and is cleavable with ammonia. The procedures were applied in solution and on a polymer support, and the same two tetramers have been made by chemical methods that featured the use of the 2,3-dimethylmaleoyl *N*-protecting group.[89]

The previously mentioned orthoester approach has also been applied to making the nonreducing lactose dimer β-D-Gal-(1→4)-β-D-Glc-(1↔1)-β-D-Glc(4(←1)-β-D-Gal.[81] A fully and selectively *O*-substituted derivative of α-L-Rha-(1→4)-α-D-Gal-(1→2)-α-L-Rha-(1→4)-β-D-Gal was made as a key to synthesizing extended- or branched-chain higher saccharides with further sugars attached selectively to the rhamnose moieties. This work was conducted during immunological studies of the detailed function of rhamnogalacturonan I, which occurs in plants.[90]

The mannose-terminating α-NeuNAc-(2→3)-β-D-Gal-(1→4)-β-D-GlcNAc-(1→2)-α-D-Man linked to serine, which is found in the laminin-binding protein α-dystroglycan, has been made by two Japanese groups, one using chemical procedures and introducing the amino acid after the tetrasaccharide had been formed.[91] The other group started with β-D-GlcNAc-(1→2)-D-Man from which they made the trisaccharide enzymically, thence the serinyl glycoside and finally, with a recombinant sialyltransferase, the target compound.[92] The tetrasaccharide fragment of cobra venom α-L-Fuc-(1→3)-β-D-GlcNAc-(1→2)-α-D-Man-(1→6)-D-Man has been produced by a chemical 2+2 approach.[93]

Selective GlcNAc and then Gal transferases have been used to make the LacNAc-terminating α-D-Gal-(1→4)-β-D-GlcNAc-(1→3)-β-D-Gal-(1→4)-D-GlcNAc,[94] and a tetrasaccharide and a hexasaccharide corresponding to the 'regular' sequence of heparin have been synthesized, the former having the structure α-L-IdoA-(1→4)-α-D-$GlcNH_2$-(1→4)-α-L-IdoA-(1→4)-α-D-$GlcNH_2$ with sulfate groups at N-2, O-6 and O-2′ of each disaccharide unit.[95]

From the leaves of *Duranta repens* saponins with plant growth inhibiting properties and containing α-L-Rha-(1→3)-β-D-Api-(1→4)-α-L-Rha-(1→2)-L-Ara have been isolated,[96] and the closely related compound α-L-Rha-(1→3)-β-D-Glc-(1→3)-α-L-Rha-(1→2)-L-Ara has been identified in a saponin derived from the roots of *Dipsacus asperoides*.[97] α-D-Man-(1→2)-α-D-Man(1→2)-α-D-Man-(1→5)-β-D-Ara*f*-OC_8H_{17}, found in a *Mycobacterium tuberculosis* lipopolysaccharide, has been synthesized by use of glycosylation steps based on mannose ethyl thioglycosides.[98] Fraser-Reid has employed his *n*-pentenyl glycosylation procedure to make β-D-GlcA-(1→3)-β-D-Gal-(1→3)-β-D-Gal-

(1→4)-D-Xyl glycosidically linked to serine, which represents the linking region of *O*-bonded proteoglycans.[99]

3.3 Branched Heterotetrasaccharides. – Compound **13**, which comprises GM_2 linked through a spacer group to a B-cell stimulating glycolipid, has been synthesized as an immunostimulant for use in cancer vaccines.[100] The diosgenyl saponin containing β-D-Xyl-(1→3)-β-D-Glc-(1→4)-[α-L-Rha-(1→3)]-β-D-Glc has been synthesized together with trisaccharide analogues.[101]

The D-Gal-terminating α-D-Man-(1→4)-[α-D-Gal-(1→6)]-β-D-Gal-(1→6)-β-D-Gal glycosidically linked to a C_{30} alkan-16-ol, which is found in the glycosphingolipids of an earthworm, has been made by use of stepwise thioglycoside linking,[102] and a similar approach has given β-D-Gal-(1→6)-[α-L-Fuc-(1→3)]-β-D-Gal-(1→6)-β-D-Gal linked to a branched C_{30} alkyl group to mimic a glycolipid of the parasite *Echinococcus mululocularis.*[103] α-L-Ara*f*-(1→2)-[α-L-Ara*f*-(1→6)]-β-D-Gal-(1→6)-D-Gal has also been synthesized by chemical methods that featured the use of the (methoxydimethyl)methyl protecting group. It is of interest in relation to the medicinal value of extracts of *Echinacea purpurea.*[104]

In the area of 2-amino-sugar tetrasaccharides Danishefsky has applied the 'cassette' approach to the synthesis of glycophorin α-NeuNAc-(2→3)-β-D-Gal-(1→3)-[α-NeuNAc-(1→6)]-α-D-GalNAc-O-serine,[62] and chemoenzymic methods have been applied to make the related $SiaLe^x$ bound by a spacer group to an integrin peptide ligand. The product binds concurrently to selectins and to integrins.[105]

β-Xyl-(1→2)-α-L-Ara*p*-(1→3)-[β-D-Gal-(1→2)]-β-D-GlcA is part of a saponin isolated from the seeds of the tea plant and which has potent gastroprotective properties,[106] and the *Shigella flexneri* O-antigen β-D-GlcNAc-(1→2)-α-L-Rha-(1→2)-[α-D-Glc-(1→3)]-L-Rha has been synthesized.[107]

3.4 Analogues of Tetrasaccharides and Compounds with Anomalous Linking. – Condensation of amine **14** with a fully substituted 4-O-Tf-β-D-Fuc-(1→4)-β-D-Glc-(1→4)-β-D-GlcOMe derivative gave the target pseudo-tetrasaccharide methyl acarboside, but only in 4% yield. It did not inhibit a cellulase but was active against β-glucosidases.[108]

Compound **15** has been made as a mimic of ganglioside GM_1, the carbocycle being obtained by Diels Alder chemistry.[109] The final product is a high affinity binder for cholera toxin and binds in the same manner as does GM_1.[110] Related studies have yielded the $SiaLe^x$ mimic **8**[73] (which was arbitrarily cartegorized as a trisaccharide), and the closely related compound having cyclohexane-*trans*-1,2-diol in place of the 1,2-dideoxy sugar was identified as the only one in a mixture of ten related substances to bind strongly to selectins. Transfer NOE NMR spectroscopy was used to identify this selectivity.[111] In the same area of study the Sia Le^x mimic **16** has been made with the ulosonic acid moiety being derived from D-glucono-δ-lactone.[112] An α-NeuNac-α-

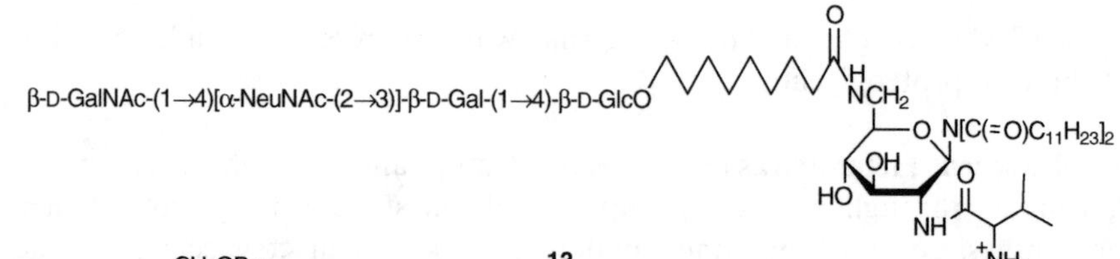

NeuNGc-(2→3)-β-D-Gal-(1→4)-β-D-Glc-O-Cer has been isolated from a starfish, the terminal sialic acid being α-2-ester-linked to the glycolyl group of *N*-glycolylneuraminic acid.[113]

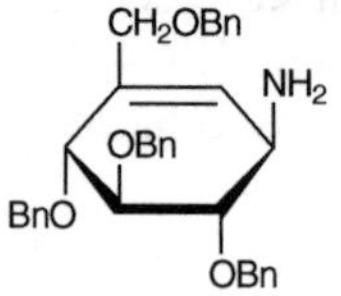

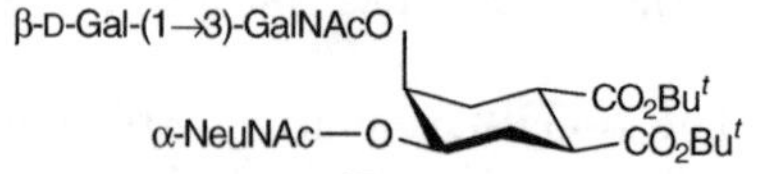

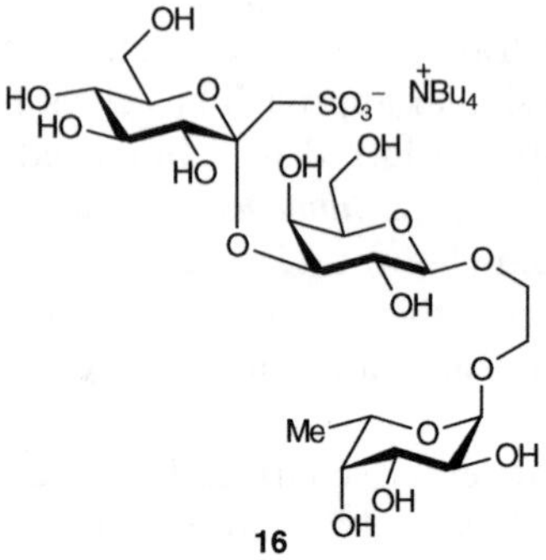

In the area of anomalously linked compounds the di-, tri- and tetramers of the *C*-linked β-(1→6)-galactose oligomers were made by the addition of lithio hept-6-yne derivatives to tetra-*O*-benzyl-D-galactono-δ-lactone.[114]

Compound **17** gives NMR spectra that indicate it has no secondary structure, unlike the related amide-linked oligomers which are derived from the carbohydrate-based amino acid which is epimeric at the carbon atom to which the carbonyl group is bonded.[115] Extended studies from Fleet's group have led to amide-linked tetramers derived from precursors **18**–**20**.[116]

4 Pentasaccharides

4.1 Linear Heteropentasaccharides. – A set of glucopentaoses with a modification at the reducing end (*e.g.* **21**) were made by enzymic transfer of α-cyclodextrin to O-4 of 6-azido-6-deoxy-D-glucose with cyclodextrin glycosyltransferase followed by degradation with β-amylase. This gave a set of modified 6-azido-6-deoxy-maltosaccharides with the pentamer the largest. Reduction gave **21**, and various other modifications were the corresponding saccharides terminating in the analogous amino acid and seven-membered 1,6-iminoglucitol.[117] One-pot sequential reactions (Vol. 28, p. 65, ref. 10) involving

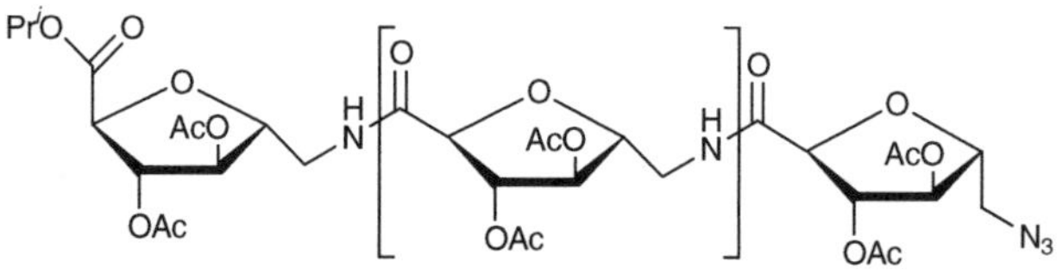

17

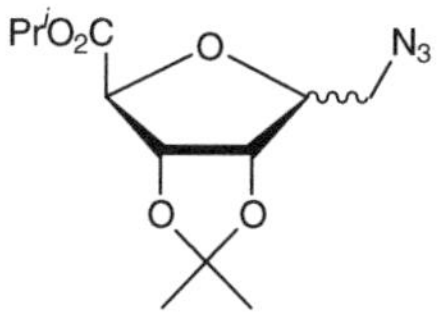

18, **19**

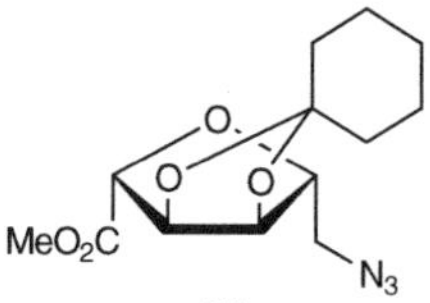

20

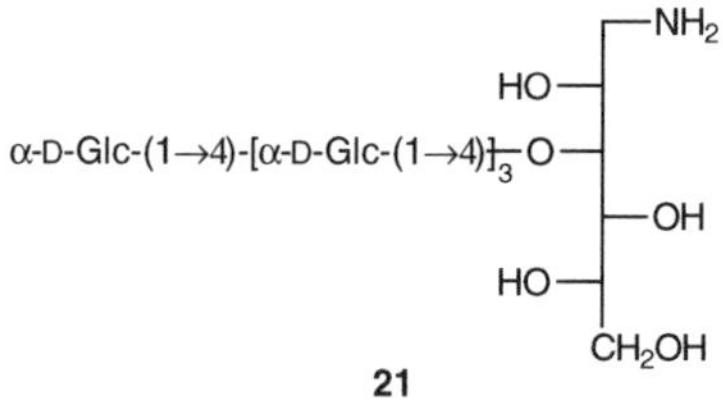

21

glycosylation of a partially protected glucosyl diosgenin with a disaccharide monohydroxy ethyl thioglycoside followed by a fully substituted disaccharide trichloroacetimidate gave the α-L-Rha-(1→3)-linked tetrasaccharide α-(1→4)-linked to the glucoside.[118]

The antithrombotic **22** has the heparin pentasaccharide conjugated to the active site inhibitor NAPAP and shows anti-thrombin and AT III-mediated anti-Xa activities. It also shows a prolonged *in vivo* half life relative to NAPAP.[119] Other heparan sulfate pentasaccharide analogues to have been made are **23**–**25**.[120]

Several reports have appeared on pentasaccharides **26** with different sugars attached as X to 'lacto-*N*-neotetraose' which commonly occurs in bacteria on mammals. Two with X = α-2-Fuc-(1→2) and α-D-Gal-(1→3) have been found in the milk of coati.[121] Chemoenzymic methods have been employed to give **26** [X = β-D-Gal-(1→3)] but with the glucosamine moiety unacetylated,[122] and **26** [X = α-NeuNAc-(2→3)] was assembled from lactose by three sequential enzymic glycosylations.[123] The related β-D-Gal-(1→3)-β-D-GalNAc-(1→3)- and (1→4)-α-D-Gal-(1→4)-β-D-Gal-(1→4)-D-Glc isomeric pentasaccharides have been made as 3-aminopropyl glycosides by chemical methods from the globotrioside.[124]

β-D-GlcA-(1→3)-β-D-GlcNAc-(1→4)-β-D-GlcA-(1→3)-β-D-GlcNAc-(1→4)-β-D-GlcA-O-aryl, which is structurally related to a segment of hyaluronic acid, has been prepared by chemical methods.[125]

The total synthesis of a GPI anchor of yeast has been reported. It comprises in outline α-D-Man-(1→2)-6-O-P-α-d Man-(1→2)-α-D-Man-(1→2)-α-D-Man-(1→4)-α-D-GlcNH$_2$ linked to an inositol phosphate carrying a ceramide group.[126]

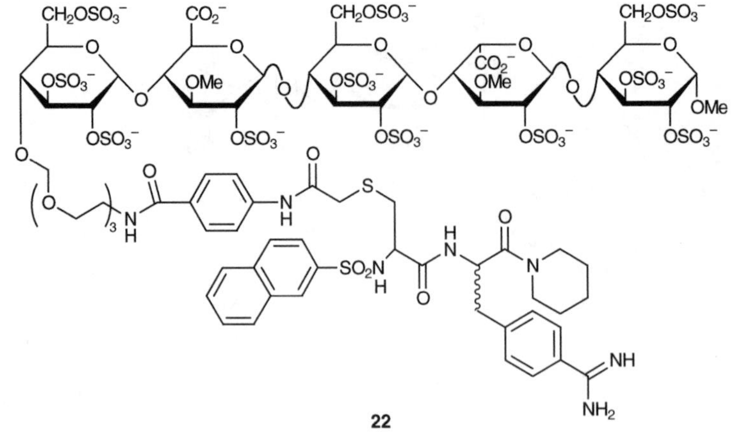

22

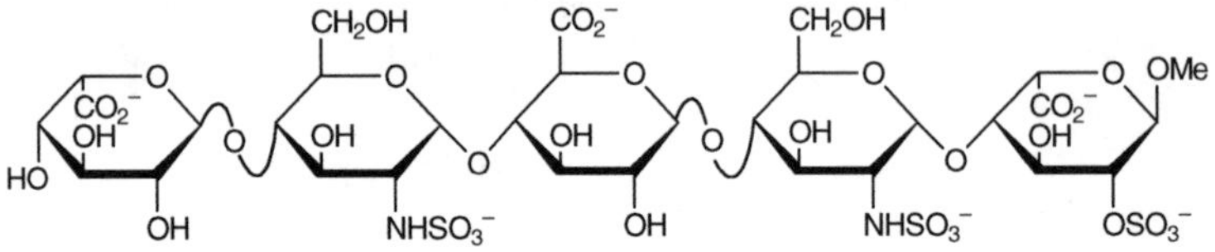

23

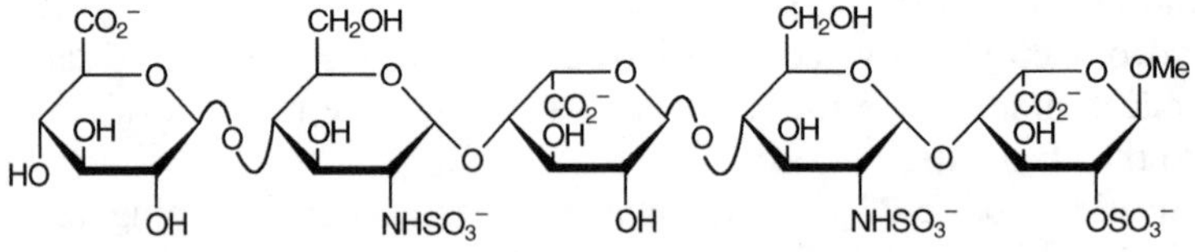

24

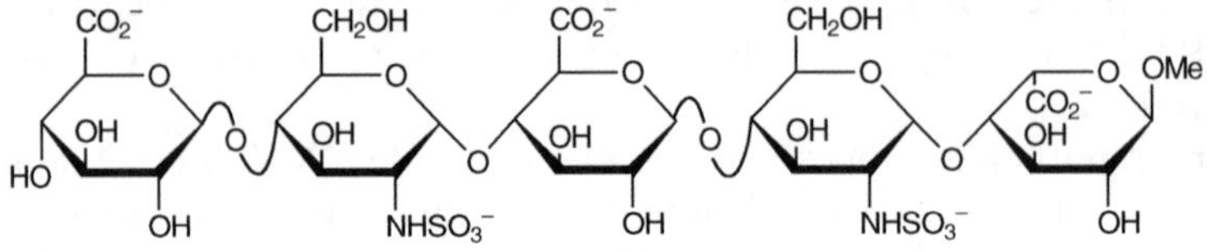

25

X—β-D-Gal-(1→4)-β-D-GlcNAc-(1→3)-β-D-Gal-(1→4)-D-Glc

26

4.2 Branched Heteropentasaccharides. – Two papers have reported the chemical synthesis of lacto-*N*-fucopentaose, β-D-Gal-(1→4)-[α-L-Fuc-(1→3)]-β-D-GlcNAc-(1→3)-β-D-Gal-(1→4)-D-Glc, which is the Le^x tetrasaccharide with an additional glucose unit at the reducing end, by chemical means.[127,128] Related work has given 3-fucosyllactotetraose, β-D-Gal-(1→3)-β-D-GlcNAc-(1→3)-β-D-Gal-(1→4)-[α-L-Fuc-(1→3)]-D-Glc.[129]

3,6-Di-*O*-β-D-lactosyl-D-mannose has been made in good yield by use of acetobromolactose and *via* 1,2-orthoesters also linked to a mannose acceptor. Rearrangement by use of timethylsilyl triflate gave the product.[130]

α-L-Fuc-(1→2)-β-D-Gal-(1→4)-[α-L-Fuc-(1→3)]-β-D-GlcNAc-(1→3)-α-D-

GalNAc which, relative to the Le^y oligosaccharide has GalNAc in place of Gal, has been made by the cassette method and coupled to three contiguous serine groups of a peptidolipid. The produce is a good immunogenic mimic of a tumour-associated cell surface Le^y mucin.[131]

An unusual (but not novel) method was used to obtain the branching β-mannose residue in the *N*-linked glycopeptide unit α-D-Man-(1→6)-[α-D-Man-(1→3)]-β-D-Man-(1→4)-β-D-GlcNAc-(1→4)-β-D-GlcNAc-asparagine. A trisaccharide with β-D-Gal linked (1→4) to a fully substituted chitobiosyl azide was selectively mannosylated at O-3 and O-6 of the galactose moiety which was then stereochemically inverted at C-2 and C-4 by way of the ditriflate.[132]

β-D-Glc-(1→6)-[β-D-Xyl-(1→2)]-β-D-Glc-(1→3)-[α-L-Rha-(1→2)]-α-L-Ara is the pentasaccharide of a saponin from the seeds of *Zizyhus jujuba* (a plant used in traditional Chinese medicine) which, from a set of such compounds, displayed the best immunoadjuvent properties.[133] β-D-Glc-(1→3)-α-L-Rha-(1→2)-[β-D-Glc-(1→3)]-β-D-Glc-(1→2)-6-deoxy-β-D-Glc was found carrying several hydrophobic acyl substitutes and with a C-15 alkyl aglycon with a carboxyl group ester linked to the branching glucose unit in a Convolvulaceae plant used in Portuguese traditional medicine.[134]

5 Hexasaccharides

As has become customary in these volumes, an abbreviated method is now used for representing higher saccharides. Sugars will be numbered as follows, and linkages will be indicated in the usual way:

1 D-Glc*p*	**2 D-Man*p***	**3 D-Gal*p***
4 D-Glc*p*NAc	**5 D-Gal*p*NAc**	**6 Neu*pN*Ac**
7 L-Rha*p*	**8 L-Fuc*p***	**9 D-Xyl*p***
10 D-Glc*p*NH$_2$	**11 D-Glc*p*A**	**12 D-Qui (6-deoxy-D-glucopyranose)**
13 L-*Glycero*-D-*manno*-heptose	**14 L-Ara*f***	**15 Kdn**

5.1 Linear Homohexasaccharides. – Laminarasaccharides (β-1,3-linked glucoses) up to the hexaose as their glycosides of long chain alcohols have been partially sulfated and tested for anti-HIV properties. The highly sulfated products with hydrophobic aglycons were most active.[135] The antigenic capsular polysaccharide of *Salmonella typhi* is a polymer of α-1,4-linked *N*-acetyl-D-galactosaminuronic acid. Oligomers up to the hexamer have been synthesized and the tetramer inhibited antibody binding to the polysaccharide.[136] A linear hexamer comprising α-(1→2) linked 4-amino-4,6-dideoxy-D-mannose with 2,4-dihydroxybutanoic acid amide-bonded at each amino group, which is the terminal *O*-antigen of *Vibrio cholerae* O:1, has been prepared by chemical methods and attached to a cluster polyamide and then to albumin to give a product for use in studies of the development of vaccines against cholera.[137]

5.2 Linear Heterohexasaccharides. – The known linear 6-*O*-galactosyl *p*-nitrophenyl maltopentaoside, made by use of enzymic galactosyl transfer from lactose, and which was synthesized as a substrate for human α-amylase in serum, was selectively hydrolysed to the linear 6-*O*-galactosyl maltotriose. This, on oxidation to the aldonolactone inhibits human salivary α-amylase, but not as strongly as does the trisaccharide glycoside having a benzyl group in place of the galactosyl residue.[138]

Compound **27**, 'globo-H', a cancer antigen has been made by Danishefsky's

(α8)1→2(β3)1→3(β5)1→3(α3)1→4(β3)1→4(1)

27

group by the 'glycal assembly' approach and found to induce strong humoural immune response in (particularly prostate cancer) patients.[139] The same important target has been attained by Zhu and Boons using a highly convergent approach.[140] The lactosamine/lactose hexamers **28**, **29** carrying a ^{13}C label in residue 4 have been made enzymically and used in the study of galectin-1/carbohydrate interactions by NMR methods.[141]

(β3)1→4(β4)1→3(β3)1→4(β4)1→3(β3)1→4(1)

28

(β3)1→4(β4)1→3(β3)1→4(β4)1→3(β3)1→4(4)

29

In the course of work on the preparation of polysialylglycoconjugates the α-(2→8) linked tetramer of NeuNAc has been isolated from the mild acid hydrolysate of colominic acid. After peracetylation, during which 2,9-lactonization occurred, the tetramer was 1,3-coupled with a lactose derivative *via* its phenyl thioglycoside to give the hexamer **30**.[142]

(α6) $\frac{2\to8}{1\to9}$ (α6) $\frac{2\to8}{1\to9}$ (α6) $\frac{2\to8}{1\to9}$ (α6) 1→3(β3)1→4(1)

30

(β11) 1→4(α1) 1→4 (β11) 1→4(α1) 1→4 (β11) 1→4(α1) OMe

31

The heparan related 31 having sulfate esters at all free hydroxyl groups of the glucose moieties and at O-4 of the non-reducing end group and methyl groups at O-2, O-3 of the uronic acid groups has been made and isomers having L-iduronic acid in place of the glucuronic acid at position 2 and at positions 2, 4 and 6 have been synthesized. The last of these is suitable for making simple heparin mimetics able to inhibit the coagulation of thrombin and factor Xa.[143] In related work two hexasaccharides isolated from porcine intestinal dermatan sulfate had double bonds conjugated with the hexuronic acid moiety at the non-reducing end, but otherwise were trimers of α-L-IdoA-(1→3)-β-D-GalNAc-(1→4) with sulfate groups at O-4 of the amino-sugar units and the occasional sulfate at O-6 of the same unit.[144]

Recombinant *E.coli* containing an incorporated chitin pentaose synthase and a β-1,4-galactosyltransferase produced β-D-Gal-(1→4)-[β-D-GlcNAc-(1→4)]$_4$-GlcNAc in a nice exemplification of a novel and potentially powerful approach to oligosaccharide synthesis.[145]

5.3 Branched Homohexasaccharides. – Three reports have been produced on the chemical synthesis of the phytoalexin elicitor hexasaccharide **32**;[146–148] in the last case the product was made with a photoreactive aglycon and thereby the position of binding in soyabean root was located. A relevant branched oligomannose has been made in the course of work reported under heptasaccharides.

(β1) 1→6 (β1) 1→6 (β1) 1→6 (β1)
3 ↑ 1 (β1) on the second unit; 3 ↑ 1 (β1) on the fourth unit

32

5.4 Branched Heterohexasaccharides. – In the area of the ganglioside type of compounds **33** has been synthesized together with its analogues having 7-, 8- and 9-deoxy NeuNAc as the branching sugar unit,[149] as well as the similar **34**

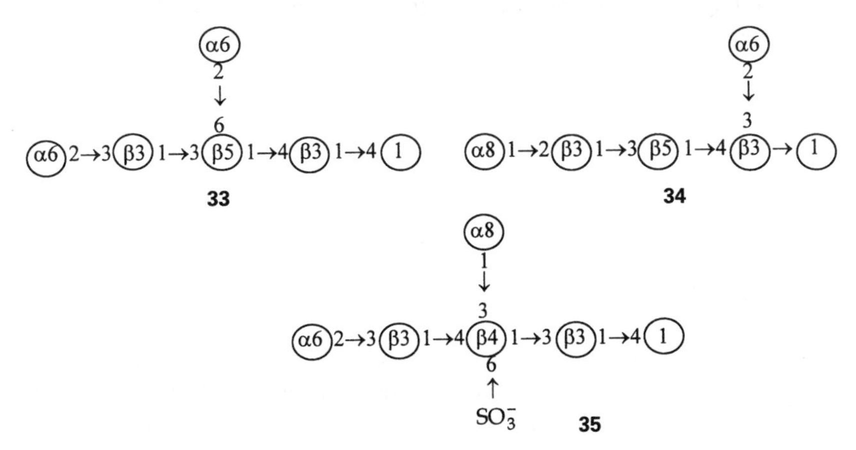

33 **34** **35**

made in connection with cancer vaccine work[150] and the sulfated Le^x ganglioside **35**.[146]

Chemo-enzymic methods were used to make 36 which is the disialylated hexamer of Type 8 Group B *Streptococcus* capsular polysaccharide.[152] Likewise the dimer **37** of the repeating unit of the antigen O2 polysaccharide of *Stentrophomonas maltophilia* has been made.[153]

(α6)
2
↓
3
(α6)2→3(β3)1→4(β7)1→4(β1)1→4(β3)O$(CH_2)_3N_3$

36

(β9)
1
↓
2
(β9)1→2(α7)1→4(α2)1→3(α7)1→4(2)

37

Branched hexasaccharide **38** (with $GlcNH_2$ instead of GlcNAc at the reducing end) is the hexasaccharide of rat brain GPI anchor. It is glycosidically linked to the inositol phosphate moiety and has phosphate esters at O-2 and O-6 of the first and third mannose units. The synthesis required 11 high yielding steps.[154] (See ref. 126 for a related synthesis.) The first synthesis of the 'core class II' disialyl hexasaccharide **39** with the sialic acid lactone- as well as glycosidically-linked has been described.[155]

(α5)
1
↓
4
(α2)1→2(α2)1→2(α2)1→6(α2)1→4(α6)

38

(α6) 1→4/2→3 (β3)1→4(β4)
1
↓
6
(α6) 1→4/2→3 (β3)1→3(α5) O-Thr

39

6 Heptasaccharides

The behaviour of 1,3-di-*O*-dodecyl-2-(β-D-maltoheptaosyl)glycerol in forming micelles has been examined.[156] In the area of branched mannoheptaoses compounds **40**[157] and **41**[158] have been made by chemical methods. In the

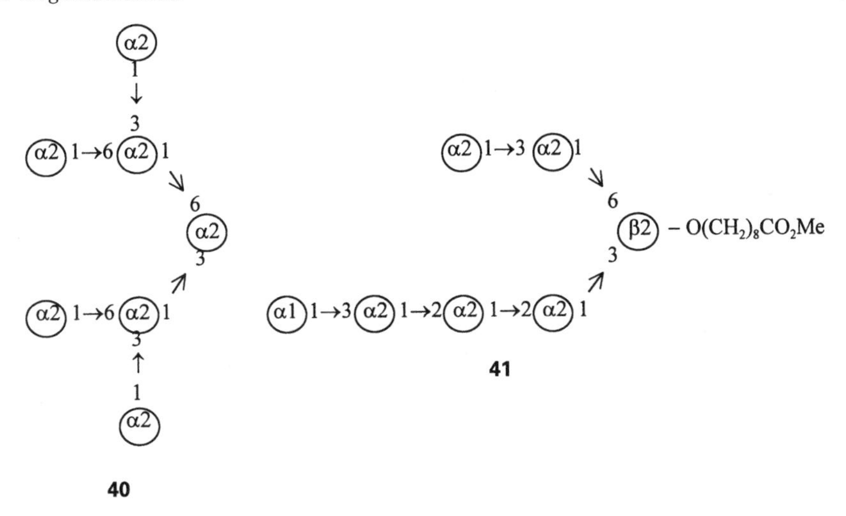

former paper the trityl/cyanoethylidene coupling method was used, and in the latter glycosyl fluorides and thio- and seleno-glycosides were employed. In the course of this work several related lower saccharides were also prepared.

Several branched oligomers of *S*-linked glucoses up to heptamers (*cf.* **32**) have been made, some being found to be active in eliciting phytoalexin accumulation in soybean. The structures were based on the β-(1→6)-*S*-linked glucopentaose with β-(1→3)-*S*-linked branches at the 2 and 3, and 3 and 4 glucose units.[159]

7 Octasaccharides

The α-(1→5)-linked L-arabinofuranosyl octamer has been produced by a 4+4 approach which involved coupling of *O*-acylated α-furanosyl trichloroacetimidates and acceptor α-L-arabinofuranosides without hydroxyl protection.[160]

Compound **42**, containing three LacNAc units, has been synthesized by chemical methods for the kinetic characterization of cloned glycosyltransferases.[161]

(β3)1→4(β4)1→3(β3)1→4(β4)1→3(β3)1→4(β4)1→6(α2)1→6(2)

42

In the area of anomalously linked octamers a compound with an amide-linked structure akin to that of **17**, but based on the amine monomer derived from **18**, **19** with the α-aminomethyl-*C*-uronoside configuration, has been found to adopt a left-handed α-helical conformation.[162]

Several fragments of the phosphoglycan of *Leishmania major* have been prepared, the most complex having structure **43**. Glycosyl H-phosphonates were used for the construction of the phosphodiester links.[163]

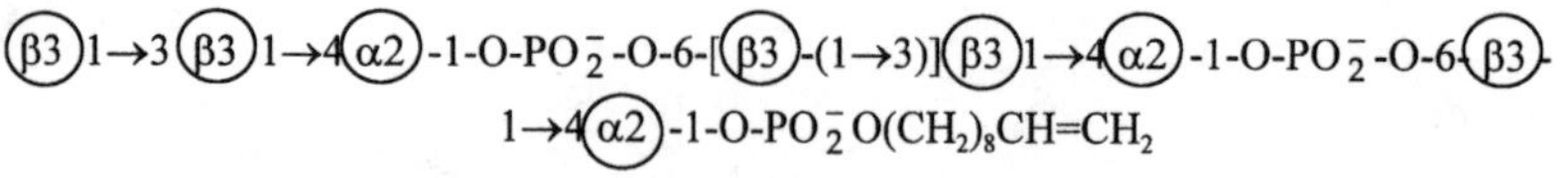

(β3)1→3(β3)1→4(α2)-1-O-PO_2^--O-6-[(β3)-(1→3)](β3)1→4(α2)-1-O-PO_2^--O-6(β3)-
1→4(α2)-1-O-PO_2^-O$(CH_2)_8CH{=}CH_2$

43

8 Higher Saccharides

The branched maltononaose with structure **44** has been made in 40 steps from glucose and maltose using trichloroacetimidate glycosylations.[164] In similar work, but in the D-mannose series, the monomer 45, related to the cell wall mannan of *Candida albicans*, has also been synthesized.[165]

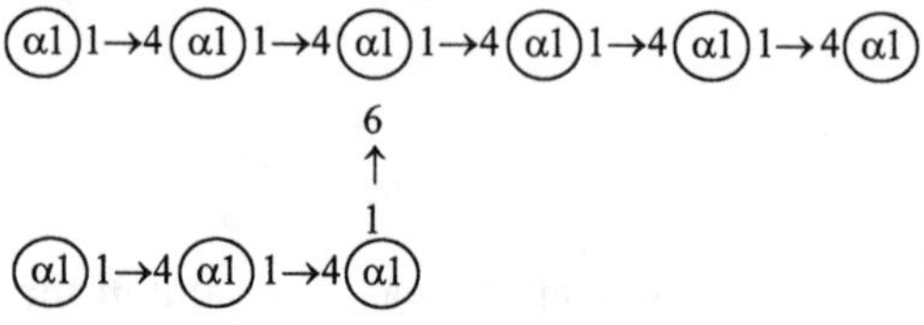

(α1)1→4(α1)1→4(α1)1→4(α1)1→4(α1)1→4(α1)
6
↑
1
(α1)1→4(α1)1→4(α1)

44

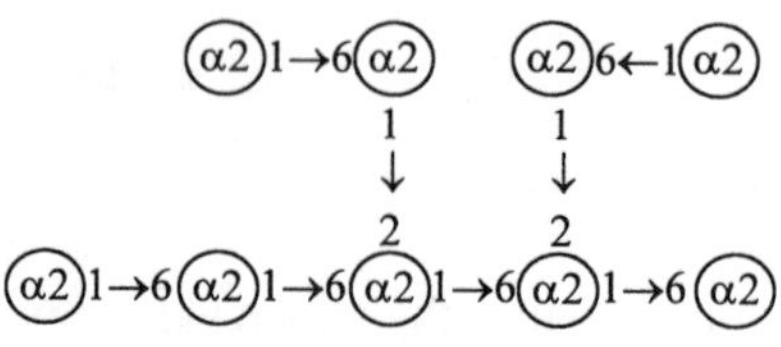

(α2)1→6(α2)   (α2)6←1(α2)
1   1
↓   ↓
2   2
(α2)1→6(α2)1→6(α2)1→6(α2)1→6(α2)

45

Also in the oligomannose series compound **46**, the parent oligomer of oligomannoses used during glycoprotein *N*-glycan processing, has been

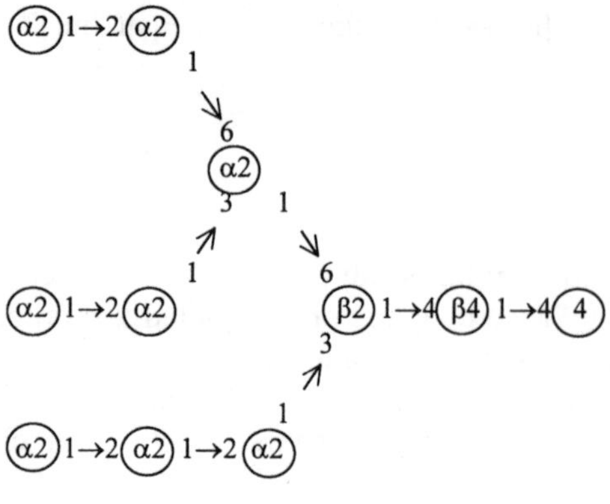

(α2)1→2(α2)
1
↘
6
(α2)
3   1
↗   ↘
1   6
(α2)1→2(α2)   (β2)1→4(β4)1→4(4)
3
↗
1
(α2)1→2(α2)1→2(α2)

46

subjected to detailed dynamic NMR and molecular dynamics simulation in water. It was concluded from the agreeing results that the molecule is subject to a high degree of internal flexibility. Nevertheless the overall topology, including its hydrogen bonding relationships could be determined.[166]

Two duodecamers, one based on a 5-sulfated α-D-arabinofuranose and terminating in a *C*-glycoside, and the other based on D-glucosamine carrying sulfates at N-2 and O-6, have been linked to generate complex biologically active glycosaminoglycan-like conjugates.[167]

Vasella's work on β-(1→4) acetylene linked oligoglucoses has produced a linear compound with 16 4-deoxyglucosyl units each joined by a buta-1,3-diyne linkage.[168]

Oligomers with structures **47** up to the '18-mer' have been made to assess the size of the heparin sequence involved in thrombin inhibition.[169] Related work by the same group has yielded '15, 17 and 19-mers' comprising the anti-thrombin binding pentasaccharide extended at the non-reducing end by a thrombin-binding domain,[170] and also oligomers up to a 20-mer of low sulfate heparin mimetics of structure **48**.[171]

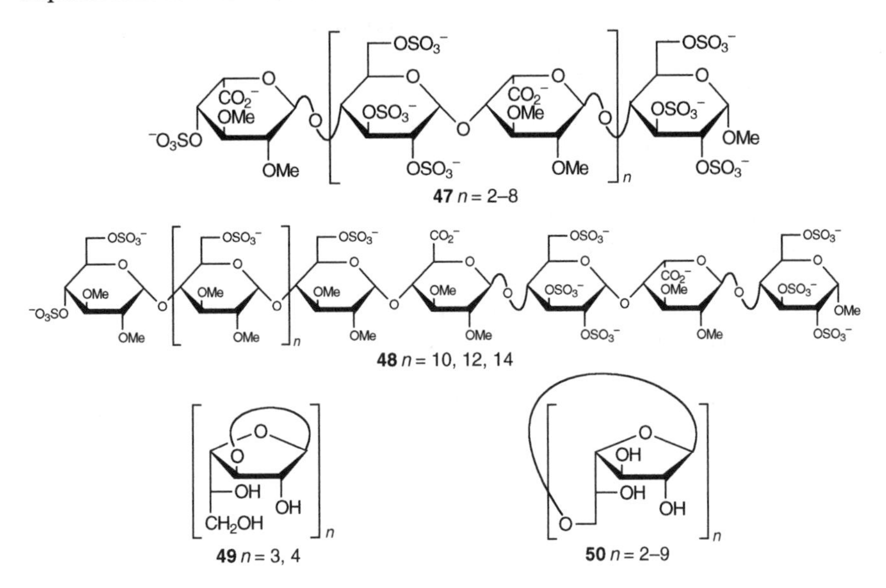

9 Cyclodextrins

9.1 General Matters. – β-(1→3) and β-(1→6)-linked cyclogalactofuranoses **49** and **50** have been investigated by Monte Carlo simulations and shown to favour unsymmetrical conformations. The natures of the molecular surfaces were described.[172] An NMR procedure, involving a selective long range INEPT experiment followed by a 2-D heteronuclear multiple bond correlation experiment, enables the position of substitution of monosubstituted β-cyclodextrins to be identified.[173] The same authors have written a further paper on a closely related topic.[174]

9.2 Branched Cyclodextrins. – $6^A,6^X$-Dithio-β-CD (where X = C, D or E) treated with 6,6′-dideoxy-6,6′-diiodo-α,α-trehalose gave branched derivatives with the disaccharide doubly linked through two sulfur atoms across the base of the annulus. In this way 'sugar bowls' were prepared.[175]

9.3 Cyclodextrin Ethers. – A report on the microwave-promoted methylation of β-CD with dimethyl sulfate recorded a 76% yield.[176] By treatment with allyl bromide or 1-allyloxy-2,3-epoxypropane β-CD has been converted to the mono-2-*O*-allyl, -2-*O*-(3-allyloxy-2-hydroxypropyl)- and -6-*O*-(3-allylloxy-2-hydroxypropyl) ethers. These, on addition of bisulfite, were converted to highly water soluble adducts with the allyl group converted to 3-propylsulfonyl. They solubilize such compounds as naphthalene in aqueous conditions.[177]

The benzyl bromide **51** coupled with 6-monohydroxy permethyl β-CD gave the monoether which, with a tolyl-terpyridyl ruthenium complex, gave the luminescent CD derivative with the benzyl substituent **52**.[178] In related work permethyl α-CD having free hydroxyl groups at 6^A and 6^D was converted to the dimesylate and then with *m*- and *p*-iodophenol to the bis(iodophenyl) ether, and finally with diphenylphosphine and a palladium catalyst to a derivative with the bidentate ligand **53** feature. This coordinated with Pd and Pt species.[179]

In the area of silyl ethers further silylation of β-CD heptakis Tbdms 6-ether with TbdmsCl gave a product with an eighth substituent at O-2 (39%). Similar reaction with PivCl or TsCl gave the respective monoesters at C-2 (36%, 26%, respectively). This provides a way of making fully protected β-CD derivatives but for single free hydroxyl groups at O-2 or O-3.[180]

Mono-6-silyl ethers give access to permethylated compounds having one triflate ester at the 6-position from which, for example, monoamino derivatives are obtainable.[181]

9.4 Cyclodextrin Esters. – The phosphorylation of per-(6-bromo-6-deoxy)-β-CD with alkylene phosphochloridites has been examined.[182] The three possible isomers of penta 6-*O*-mesitylenesulfonyl β-CD have been characterized by conversion to the corresponding 3,6-anhydro-derivatives which were identified by ^1H NMR methods,[183] and the mono-6-(2-naphthoate) of the β-CD has been studied as a complexing host.[184]

9.5 Amino Derivatives. – Considerable interest has been maintained in mono-(6-amino-6-deoxy)-CDs and their *N*-substituted derivatives which can be made by *N*-substitution, or by introduction of substituted amino functions or from 6-azido-6-deoxy compounds. Substituted amino groups to have been introduced include: ω-aminoalkylamino,[185] *m*-(aminomethyl)benzylamino,[186] silicated acylamino (by use of silicate-bonded *N*-propylurea; the products can be used as chiral stationary phases for enantiomer resolution),[187] cholester-3-yloxycarbonylamino,[188] N^{α}-dansyl-L-lysinylamino[189] and *N*-tryptophanyl.[190]

Several mono-6-amino-6-deoxy derivatives have been linked through nitrogen to give dimers. Linking units to have been used were made by coupling 2-hydroxylethyl-*N*-compounds with α,ω-alkyl dibromides,[191] or by amide-bonding α,ω-dicarboxylic acids with long alkyl chains having thio or disulfide internal groups.[192] Related compounds have been derived by treatment of 6-azido-6-deoxy CDs with α,ω-diamines in the presence of triphenylphosphine and CS_2. The derived bridging chains consist of two thiourea molecules *N,N*-joined by the α,ω-dicarboxylic acids alkdiyl groups.[193] Of special note is the 6-

amino-6-deoxy-β-CD compound having the *N*-substituent **54** which complexes metal ions that can be used for catalytic purposes.[194]

Opening of the epoxide ring of a β-CD derivative containing a 2,3-anhydro-D-allose unit with sodium azide has led to the 3-amino-3-deoxy-D-glucose analogue from which the 3-*N*-(imidazo)acetyl derivative was obtained. This has better catalytic properties for the hydrolysis of guest esters then does the D-*altro*-isomer.[195] Related amino-CDs have been made with l-tryptophan linked to the amino group *via* aminoethyl or 1-aminopropyl chains.[196] Other workers have described 3-[2-(aminoethyl)amino]-3-deoxy derivatives with amido- and imino- substituents on the free amino groups.[197]

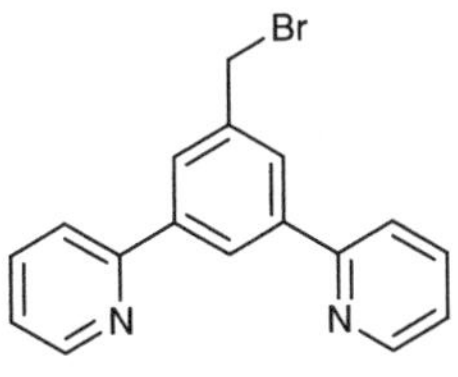

51 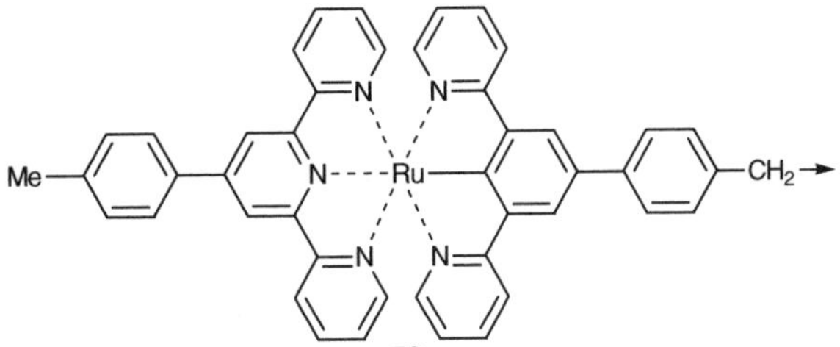 **52** 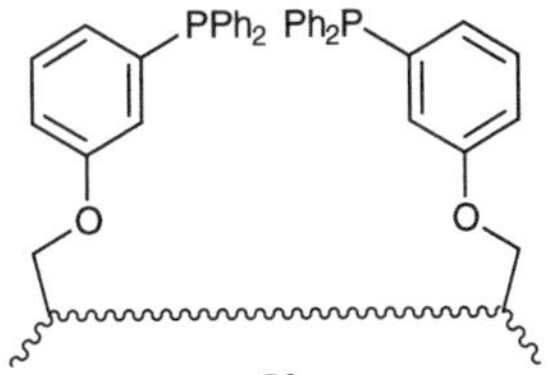 **53** 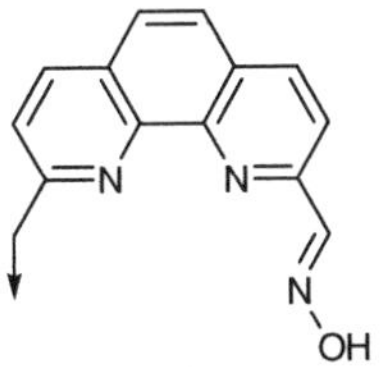 **54** 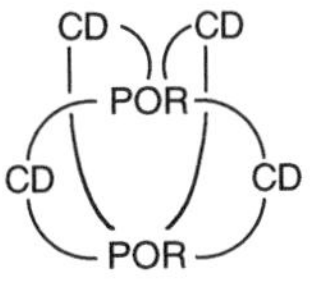 **55**

Several derivatives of β-CD with substituted amino groups at each of the primary positions have been made from the heptakis iodide or azide. From the former per 6-alkylamino-6-deoxy-[198] derivatives and compounds with *N*-(glycosylthio)acetyl substituents[199] have been reported. Treatment of the heptakis azide with 4-aminomethyl-2,2′-bipyridyl and Ph_3P, CO_2 gave the product having *N*-(2′,2′-bipyridyl-4-yl)methylurea substituents.[200]

Attention has been paid to compounds with bridges between amino groups on two different anhydroglucose rings. Thus amino groups at C-6 of the A and D units of β-CD have been bridged by linked alanine chains,[201] and by linked thioureas which were converted to guanidinium groups.[202] 6-Amino groups of the A and E glucoses of γ-CD were converted into *N*-*p*-formylbenzoyl derivatives. With pyrrole and BF_3, followed by *p*-chloranil, porphyrins were produced to give complex products containing four bifunctional CDs and two cofacial porphyrin rings (see **55**).[203]

9.6 Thio and Seleno Derivatives. – Heptakis 6-deoxy-6-iodo β-CD treated with 1-thiohexoses or glycosyl thiouronium salts has given products with glycosyl substituents S-bonded to each glucose unit,[199] and γ-CD having trehalose doubly S-linked across the base of the anulus to give 'sugar bowls' has already been mentioned.[175]

Treatment of the per 2,3-*manno*-anhydride derived from β-CD with sodium sulfide in DMF gave dimeric products with S-linking through C-3 of two 3-thio-D-*altro*-units (30%) and through C-2 of 2-thio-D-*gluco* units (13%).[204] In related work the mono-*manno*-epoxide treated with α-thiotoluene and Cs_2CO_3 gave the 2-benzylthio-glucosyl and 3-benzylthio-altrosyl products which were debenzylated to give the thiols. Of these the 3-thio isomer was the more effective in promoting acyl transfers of guest esters. In the course of the work the thiols were oxidized to give CD dimers linked by disulfide bridges.[205]

Heptakisthio-β-CD has been attached *via* the sulfur atoms by amide-containing links to O-1 of β-D-Gal and β-D-GlcNAc to give cluster-like products.[206]

In the area of seleno compounds mono-6-*p*-anisylseleno-,[207] phenylseleno-, benzylseleno- and *m*-tolylseleno-[208] derivatives have been reported, and these give complexes with aliphatic alcohols. Treatment of mono-6-*O*-tosyl-β-CD with disodium propane-1,3-diselenate gave a dimeric product with the CDs linked by the propane 1,3-diselenyl bridge. Related organoseleno-bridged products, including 2,2′-linked compounds, were described. Guest binding affinities were increased by the addition of Pt(IV) ions.[209]

9.7 Halogenated Derivatives. – The well known per-6-bromo- and 6-iodo-compounds have been used for several of the synthesis described above. High yields of the bromo- and chloro- derivatives of α-, β- and γ-CD have been obtained by use of (bromomethylene)morpholinium bromide and the corresponding dichloro compound, respectively.[210] DAST was used in the preparation of the known mono-6-deoxy-6-fluoro-β-CD, and the heptakis analogue, and the mono-2,2,2-trifluoroethylthio compound was also described. The mono-substituted products showed 'reasonable' water solubility.[211]

9.8 Deoxy Derivatives. – Reduction of the heptakis *manno*-2,3-anhydride derived from β-CD with lithium triethylborohydride gave the all-3-deoxy-D-*arabino*-derivative which, with several O-2-substituted derivatives, was evaluated for capillary GC enantio-discriminations.[212]

9.9 Oxidized Derivatives. – Reaction of β-CD 2-iodosobenzoic acid in DMSO gave the monoaldehyde.[213] Compounds containing ketonic groups (^{13}C NMR evidence) were produced on bromine oxidation of α- and β-CD partially *O*-substituted with 2-hydroxypropyl ethers. The products had a catalytic effect on peroxomonosulfate-oxidation of aryl alkyl sulfoxides, pyridine, aniline and several amino acids.[214]

References

1 Z. Zhang, I.R. Ollmann, X.-S. Ye, R. Wischant, T. Baasov and C.-H. Wong, *J. Am. Chem. Soc.*, 1999, **121**, 734.
2 D. Depré, A. Düffels, L.G. Green, R. Lenz, S.V. Ley and C.-H. Wong, *Chem. Eur. J.*, 1999, **5**, 3326.
3 N.E. Nifant'ev, *Ross. Khim. Zh.*, 1998, **42**, 134 (*Chem. Abstr.*, 1999, **130**, 153 869).
4 S. Manabe and Y. Ito, *Kobonshi*, 1998, **47**, 766 (*Chem. Abstr.*, 1999, **130**, 14 125).
5 O. Hindsgaul, *Curr. Opin. Chem. Biol.*, 1999, **3**, 291 (*Chem. Abstr.*, 1999, **131**, 144 755).
6 L. Jobron and O. Hindsgaul, *J. Am. Chem. Soc.*, 1999, **121**, 5835.
7 M. Petitou, J.P. Herault, A. Bernat, P.A. Driguez, P. Duchaussoy, J.C. Lormeau and J.M. Herbert, *Ann. Pharm. Fr.*, 1999, **57**, 232 (*Chem. Abstr.*, 1999, **131**, 170 533).
8 S. Shoda, M. Fujita and S. Kobayashi, *Trends Glycosci. Glycotechnol.*, 1998, **10**, 279 (*Chem. Abstr.*. 1999, **133**, 66 667).
9 Q.-M. Gu, *J. Environ. Polym. Degrad.*, 1999, **7**, 1 (*Chem. Abstr.*, 1999, **131**, 59 025).
10 W. Zhang, W. Xie, J. Wang, X. Chen, J. Fang, Y. Chen, J. Li, L. Yu, D. Chen and P.G. Wang, *Curr. Org. Chem.*, 1999, **3**, 241 (*Chem. Abstr.*, 1999, **131**, 102 431).
11 J.M. Elhalbi and K.G. Rice, *Curr. Med. Chem.*, 1999, **6**, 93 (*Chem. Abstr.*, 1999, **130**, 237 740).
12 L. Gui, D. Wei, Y. Cui and J. Yu, *Huadong Ligong Daxue Xuebao*, 1998, **24**, 422 (*Chem. Abstr.*, 1999, **130**, 66 698).
13 X. Wang, D. Wei, Y. Cui and J. Yu, *Huadong Ligong Daxue Xuebao*, 1999, **25**, 144 (*Chem. Abstr.*, 1999, **131,** 199 913).
14 S. Mala, H. Dvorakova, R. Hrabal and B. Kralova, *Carbohydr. Res.*, 1999, **322**, 209.
15 H. Zhang, Y. Du, X. Yu, M. Mitsutomi and S.-I Aiba, *Carbohydr. Res.*, 1999, **320**, 257.
16 T. Mukaiyama, Y. Wakiyama, K. Miyazaki and K. Takeuchi, *Chem. Lett.*, 1999, 933.
17 K. Fukase, Y. Nakai, K. Egusa, J.A. Porco, Jr. and S. Kusumoto, *Synlett*, 1999, 1074.
18 Y. Du and F. Kong, *J. Carbohydr. Chem.*, 1999, **18**, 655.
19 B. Cui, H. Chai, T. Santisuk, V. Reutrakul, N.R. Fanworth, G.A. Cordell, J.M. Pezzuto and A.D. Kinghorn, *J. Nat. Prod.*, 1998, **61**, 1535 (*Chem. Abstr.*, 1999, **130**, 14 132).
20 W. Zhangi, J. Wang, J. Li, L. Yu and P.G. Wang, *J. Carbohydr. Chem.*, 1999, **18**, 1009.
21 J.-Q. Wang, X. Chen, W. Zhang, S. Zacharek, Y. Chen and P.G. Wang, *J. Am. Chem. Soc.*, 1999, **121**, 8174.
22 S. Singh, M. Scigelova and D.H.G. Crout, *Chem. Commun.*, 1999, 206.
23 L. Weignerová, P. Sedmera, Z. Huňková, P. Halada, V. Křen, M. Casali and S. Riva, *Tetrahedron Lett.*, 1999, **40**, 9297.
24 H.C. Hansen and G. Magnusson, *Carbohydr. Res.*, 1999, **322**, 181.
25 K. Matsuoka, M. Terabatake, Y. Esumi, D. Terunuma and H. Kuzuhara, *Tetrahedron Lett.*, 1999, **40**, 7839.

26 H. Dohi, Y. Nishida, M. Mizumo, M. Shinkai, T. Kobayashi, T. Takeda, H. Uzawa and K. Kobayashi, *Bioorg. Med. Chem.*, 1999, **7**, 2053.
27 S.D. Debenham, J. Cossrow and E.J. Toone, *J. Org. Chem.*, 1999, **64**, 9153.
28 H.-Y. Shin and J.-W. Yang, *J. Microbiol. Biotechnol.*, 1998, **8**, 484 (*Chem. Abstr.*, 1999, **130**, 168 575).
29 B. Dumont-Hornebeck, J.-P. Joly, J. Coulon and Y. Chapleur, *Carbohydr. Res.*, 1999, **320**, 147.
30 W. Wang and F. Kong, *J. Carbohydr. Chem.*, 1999, **18**, 451.
31 T. Yasukochi, K. Fukase and S. Kusumoto, *Tetrahedron Lett.*, 1999, **40**, 6591.
32 X. Qian, K. Sujino, A. Otter, M.M. Palcic and O. Hindsgaul, *J. Am. Chem. Soc.*, 1999, **121**, 12063.
33 J. Xia, C.F. Piskorz, R.D. Locke, E.V. Chandrasekaran, J.L. Alderfer and K.L. Matta, *Bioorg. Med. Chem. Lett.*, 1999, **9**, 2941.
34 H. Yuasa and H. Hashimoto, *J. Am. Chem. Soc.*, 1999, **121**, 5089.
35 E. Katzenellenbogen, A. Gamian and E. Romanowska, *Carbohydr. Lett.*, 1998, **3**, 223 (*Chem. Abstr.*, 1999, **130**, 81 732).
36 M. Scigelova, S. Singhand and D.H.G. Crout, *J. Chem. Soc., Perkin Trans. 1*, 1999, 777.
37 Y. Zhao and J.S. Thorson, *Carbohydr. Res.*, 1999, **319**, 184.
38 J. Kinjo, K. Aoki, M. Okawa, Y. Shii, T. Hirakawa, T. Nohara, Y. Nakajima, T. Yamazaki, T. Hosono, M. Soneya, Y. Niho and T. Kurashige, *Chem. Pharm. Bull.*, 1999, **47**, 708.
39 E. Yesilada and Y. Takaishi, *Phytochemisry*, 1999, **51**, 903.
40 A. Fürstner and T. Müller, *J. Am. Chem. Soc.*, 1999, **121**, 7814.
41 S. Canevari, D. Colombo, F. Compostella, L. Panza, F. Ronchetti, G. Russo and L. Toma, *Tetrahedron*, 1999, **55**, 1469.
42 S.A. Wacowich-Sgarbi and D.R. Bundle, *J. Org.Chem.*, 1999, **64,** 9080.
43 K. Fujiwara and A. Murai, *Tennen Yuki Kagobutsu Toronkai Koen Yoshishu*, 1997, 85 (*Chem. Abstr.*, 1999, **131**, 5431).
44 B. Jayalakshmi, *Indian J. Chem., Sect. B: Org. Chem. Incl. Med. Chem.*, 1998, **37B**, 1087 (*Chem. Abstr.*, 1999, **130**, 196 871).
45 A. Kumar, A. Khare and N.K. Khare, *Phytochemistry*, 1999, **52**, 675.
46 L. Noecker, F. Duarte, S.A. Bolton, W.G. McMahon, M.T. Diaz and R.M. Giuliano, *J. Org. Chem.*, 1999, 64, 6275.
47 K.M. Specht, C.R. Harris, G.A. Molander and D. Kahne, *Tetrahedron Lett.*, 1999, **40**, 1855.
48 J. Gildersleeve, A Smith, K. Sakurai, S. Raghavan and D. Kahne, *J. Am. Chem. Soc.*, 1999, **121**, 6176.
49 S. Koto, K. Asami, M. Hirooka, K. Nagura, M. Takizawa, S. Yamamoto, N. Okamoto, M. Sato, H. Tajima, T. Yoshida, N. Nonaka, T. Sato, S. Zen, K. Yago and F. Tomonaga, *Bull. Chem. Soc. Jpn.*, 1999, **72**, 765.
50 N. Navarre, N. Amiot, A. van Oijen, A. Imberty, A. Poveda, J. Jiménez-Barbero, A. Cooper, M.A. Nutley and G.-J. Boons, *Chem. Eur. J.*, 1999, **5**, 2281.
51 S. Koshida, Y. Suda, M. Sobel, J. Ormsby and S. Kusumoto, *Bioorg. Med. Chem. Lett.*, 1999, **9**, 3127.
52 H.C. Hansen and G. Magnusson, *Carbohydr. Res.*, 1999, **322,** 166.
53 H.C. Hansen and G. Magnusson, *Carbohydr. Res.*, 1999, **322**, 190.
54 A. Tuchinsky and U. Zehavi, *React. Funct. Polym.*, 1999, **39**, 147 (*Chem. Abstr.*, 1999, **130**, 196 893).

55 W. Zou, M. Abraham, M. Gilbert, W.W. Wakarchuk and H.J. Jennings, *Glycoconjugate, J.*, 1999, **16**, 507.
56 L. Mauri, R. Casellato, G. Kirschner and S. Sonnino, *Glycoconjugate J.*, 1999, **16**, 197.
57 K. Yamada, S. Matsumoto and S.-I. Nishimura, *Chem. Commun.*, 1999, 507.
58 S. Figueroa-Pérez and V. Vérez-Bencomo, *Carbohydr. Res.*, 1999, **317**, 29.
59 T. Miyazaki, T. Sakakibara, H. Sato and Y. Kajihara, *J. Am. Chem. Soc.*, 1999, **121**, 1411.
60 S. Komba, M. Meldal, O. Werdelin, T. Jensen and K. Bock, *J. Chem. Soc., Perkin Trans. 1*, 1999, 415.
61 R. Gonzalez and J. Thiem, *Carbohydr. Res.*, 1999, **317**, 180.
62 J.B. Schwarz, S.D. Kuduk, X.-T. Chen, D. Sames, P.W. Glunz and S.J. Danishefsky, *J. Am. Chem. Soc.*, 1999, **121**, 2662.
63 Y. Kajihara, S. Akai, T. Nakagawa, R. Sato, T. Ebata, H. Kodama and K. Sato, *Carbohydr. Res.*, 1999, **315**, 137.
64 K. Ogawa, Y. Ikeda and S. Kondo, *Carbohydr. Res.*, 1999, **321**, 128.
65 D. Crich and Z. Dai, *Tetrahedron*, 1999, **55**, 1569.
66 W. Wang and F. Kong, *J. Carbohydr. Chem.*, 1999, **18**, 263.
67 N.U. Sata, S.-i. Wada, S. Matsunaga, S. Watabe, R.W.M. van Soest and N. Fusetani, *J. Org. Chem.*, 1999, **64**, 2331.
68 H. Yamada, T. Kato and T. Takahashi, *Tetrahedro Lett.*, 1999, **40**, 4581.
69 B. Becker, R.H. Furneaux, F. Reck and O.A. Zubkov, *Carbohydr. Res.*, 1999, **315**, 148.
70 C. Li, B. Yu and Y. Hui, *J. Carbohydr. Chem.*, 1999, **18**, 1107.
71 S.K. Sarkar, A.K. Choudhury, B. Mukhapadhyay and N. Roy, *J. Carbohydr. Chem.*, 1999, **18**, 1121.
72 T. Nukada, A. Berces and D.M. Whitfield, *J. Org. Chem.*, 1999, **64**, 9030.
73 G. Thoma, W. Kinzy, C. Bruns, J.T. Patton, J.L. Magnani and R. Bänteli, *J. Med. Chem.*, 1999, **42**, 4909.
74 O. Tsuruta, H. Yuasa, H. Hashimoto, S. Kurono and S. Yazawa, *Bioorg. Med. Chem. Lett.*, 1999, **9**, 1019.
75 L.A. Simeoni, N.E. Byranmoi and N.V. Bovin, *Bioorg. Khim.*, 1999, **25**, 62 (*Chem. Abstr.*, 1999, **131**, 214 478).
76 Y.Li, H. Matsuda, S. Wen, J. Yamahara and M. Yoshikawa, *Bioorg. Med. Chem. Lett.*, 1999, **9**, 2473.
77 S.C. Schürer and S. Blechert, *Chem. Commun.*, 1999, 1203.
78 I. Neda, P. Sakhaii, A. Wabmaan, U. Niemeyer, F. Günther and J. Engel, *Synthesis*, 1999, 1625.
79 G. Hummel and O. Hindsgaul, *Angew. Chem., Int. Ed. Engl.*, 1999, **38**, 1782.
80 O. Tsuruta, H. Yuasa, S. Kurono and H. Hashimoto, *Bioorg. Med. Chem. Lett.*, 1999, **9**, 807.
81 W. Wang and F. Kong, *Tetrahedron Lett.*, 1999, **40**, 1361.
82 M. Wakao, K. Fukase and S. Kusumoto, *Synlett*, 1999, 1911.
83 H. Kono, M.R. Waelchili, M. Fujiwara, T. Erata and M. Takai, *Carbohydr. Res.*, 1999, **319**, 29.
84 H. Kono, M.R. Waelchili, M. Fujiwara, T. Erata and M. Takai, *Carbohydr. Res.*, 1999, **321**, 67.
85 M. Kanno, T. Shibano, M. Takido and S. Kitanaka, *Chem. Pharm. Bull.*, 1999, **47**, 915.
86 L. Knerr and R.R. Schmidt, *Synlett*, 1999, 1802.

87 T. Murata, T. Inukai, M. Suzuki, M. Yamagishi and T. Usui, *Glycoconjugate J.*, 1999, **16**, 189.
88 O. Blixt and T. Norborg, *Carbohydr. Res.*, 1999, **139**, 80.
89 M.R.E. Aly, E.I. Ibrahim, E.H. El Ashry and R.R. Schmidt, *Carbohydr. Res.*, 1999, **316**, 121.
90 J.R. Rich, R.S. McGavin, R. Gardner and K.B. Reimer, *Tetrahedron Asymm.*, 1999, **10**, 17.
91 J. Seifert, T. Ogawa and Y. Ito, *Tetrahedron Lett.,* 1999, **40**, 6803.
92 I. Matsuo, M. Isomura and K. Ajisaka, *Tetrahedron Lett.*, 1999, **40**, 5047.
93 Z.-J. Li, H. Li and M.-S. Cai, *Carbohydr. Res.*, 1999, **320**, 1.
94 O. Blixt, I. van Die, T Norbog and D.M. van den Eijnden, *Glycobiology*, 1999, **9**, 1061.
95 C. Tabeur, J.-M. Mallet, F. Bono, J.-M. Herbert, M. Petitou and P. Sinaÿ, *Bioorg. Med. Chem.*, 1999, **7**, 2003.
96 S. Hiradate, H. Yada, T. Ishii, N. Nakajima, M. Ohnishi-Kameyama, H. Sugie, S. Zungsontiporn and Y. Fujii, *Phytochemistry*, 1999, **52**, 1223.
97 Z.-C. Miao and R. Feng, *Huaxue Xuebao*, 1999, **57**, 801 (*Chem. Abstr.*, 1999, **131**, 272 099).
98 V. Subramanian and T.L. Lowary, *Tetrahedron*, 1999, **55**, 5965.
99 J.G. Allen and B. Fraser-Reid, *J. Am. Chem. Soc.*, 1999, **121**, 468.
100 W. Dullenkopf, G. Ritter, S.R. Fortunato, L.J. Old and R.R. Schmidt, *Chem. Eur. J.*, 1999, **5**, 2432.
101 S. Deng, B. Yu, Y. Hui, H. Yu and X. Han, *Carbohydr. Res.*, 1999, **317**, 53.
102 N. Hada, A. Matsusaki, M. Sugita and T. Takeda, *Chem. Pharm. Bull.*, 1999, **47**, 1265.
103 N. Hada, E. Hayashi and T. Takeda, *Carbohydr. Res.*, 1999, **316**, 58.
104 A. Borbas, L. Janossy and A. Liptak, *Carbohydr. Res.*, 1999, **318**, 98.
105 S.-I. Nishimura, M. Matsuda, H. Kitamura and T. Nishimura, *Chem. Commun.*, 1999, 1435.
106 T. Murakami, J. Nakamura, H. Matsuda and M. Yoshikawa, *Chem. Pharm. Bull.*, 1999, **47**, 1759.
107 L.A. Mulard and J. Ughetto-Monfrin, *J. Carbohydr. Chem.*, 1999, **18**, 721.
108 R.V. Stick, D.M.G. Tilbrook and S.J. Williams, *Aust. J. Chem.*, 1999, **52**, 895.
109 A. Bernardi, G. Boschin, A. Checchia, M. Lattanzio, L. Manzoni, D. Potenza and C. Scolastiro, *Eur. J. Org. Chem.*, 1999, 1311.
110 A. Bernardi, A. Checchia, P. Brocca, S. Sonnino and F. Zuccotto, *J. Am. Chem. Soc.*, 1999, **121**, 2032.
111 D. Henrichsen, B. Ernst, J.L. Magnani, W.-T. Wang, B. Meyer and T. Peters, *Angew Chem., Int. Ed. Engl.*, 1999, **38**, 98.
112 A. Borbás, G. Szabovik, Z. Antal, P. Herczegh, A. Agócs and A. Lipták, *Tetrahedron Lett.*, 1999, **40**, 3639.
113 M. Inogaki, R. Isobe and R. Higuchi, *Eur. J. Org. Chem.*, 1999, 771.
114 Y.-C. Xin, Y.-M. Zhang, J.-M. Mallet, C.P.J. Glaudemans and P. Sinaÿ, *Eur. J. Org. Chem.*, 1999, 471.
115 M.D. Smith, D.D. Long, A. Martín, D.G. Marquess, T.D.W. Claridge and G.W.J. Fleet, *Tetrahedron Lett.*, 1999, **40**, 2191.
116 D.D. Long, N.L. Hungerford, M.D. Smith, D.E.A. Brittain, D.G. Marquess, T.D.W. Claridge and G.W.J. Fleet, *Tetrahedron Lett.*, 1999, **40**, 2195.
117 R. Uchida, A. Nasu, S. Tokutake, K. Kasai, K. Tobe and N. Yamaji, *Chem. Pharm. Bull.*, 1999, **47**, 187.

118 B.Yu, H. Yu, Y. Hui and X. Han, *Tetrahedron Lett.*, 1999, **40**, 8591.
119 R.C. Buijsman, J.E.M. Basten, T.G. van Dinther, G.A. van der Marel, C.A.A. van Boeckel and J.H. van Boom, *Bioorg. Med. Chem. Lett.*, 1999, **9**, 2013.
120 J. Kovensky, P. Duchaussoy, F. Bono, M. Salmivirta, P. Sizun, J.-M. Herbert, M. Petitou and P. Sinaÿ, *Bioorg. Med. Chem.*, 1999, **7**, 1567.
121 T. Urashima, M. Yamamoto, T. Nakamura, I. Arai, T. Saito, M. Namikie, K.-I. Yamaaka and K. Kawahara, *Comp. Biochem. Physiol., Part A: Mol. Integr. Physiol.*, 1999, **123A**, 187 (*Chem. Abstr.*, 1999, **131**, 228 890).
122 J. Fang, X. Chen, W. Zhang, J. Wang, P.A. Andreana and P.G. Wang, *J. Org. Chem.*, 1999, **64**, 4089.
123 K.F. Johnson, *Glycoconjugate J.*, 1999, **16**, 141.
124 W. Zou, J.-R. Brisson, S. Larocque, R.L. Gardner and H.J. Jennings, *Carbohydr. Res.*, 1999, **315**, 251.
125 K.M. Halkes, T.M. Slaghek, T.K. Hypponen, J.P. Kamerling and J.F.G. Vliegenthart, *Pol. J. Chem.*, 1999, **73**, 1123 (*Chem. Abstr.*, 1999, **131**, 144 778).
126 T.G. Mayer and R.R. Schmidt, *Eur. J. Org. Chem.*, 1999, 1153.
127 S. Cao, Z. Gan and R. Roy, *Carbohydr. Res.*, 1999, **318**, 75.
128 Y.-M. Zhang, J. Esnault, J.-M. Mallet and P. Sinaÿ, *J. Carbohydr. Chem.*, 1999, **18**, 419.
129 Y. Ishizuka, T. Nemato, M. Fujiwara, K.-i. Fujita and H. Nakanishi, *J. Carbohydr. Chem.*, 1999, **18**, 523.
130 W. Wang and F. Kong, *Angew. Chem., Int. Edn. Engl.*, 1999, **38**, 1247.
131 P.W. Glunz, S. Hinterman, J.B. Schwarz, S.D. Kuduk, X.-T. Chen, L.J. Williams, D. Sames, S.J. Danishefsky, V. Kudryashov and K.O. Lloyd, *J. Am. Chem. Soc.*, 1991, **121**, 10636.
132 I. Matsuo, M. Isomura and K. Ajisaka, *J. Carbohydr. Chem.*, 1999, **18,** 841.
133 H. Matsuda, T. Murakami, A. Ikebata, J. Yamahara and M. Yoshikawa, *Chem. Pharm. Bull*, 1999, **47**, 1744.
134 E.M.M. Gasper, *Tetrahedron Lett.,* 1999, **40**, 6861.
135 K. Katsuraya, H. Nakashima, N. Yamamoto and T. Uryu, *Carbohydr. Res.*, 1999, **315**, 234.
136 L.K. Shi-Shun, J.-M. Mallet, M. Moreau and P. Sinaÿ, *Tetrahedron*, 1999, **55**, 14043.
137 J. Zhang and P. Kovac, *Carbohydr. Res.*,1999, **321**, 157.
138 M. Takada, K. Ogawa, T. Murata and T. Usui, *J. Carbohydr. Chem.*, 1999, **18**, 149.
139 G. Ragupathi, S.F. Slovin, S. Adluri, D. Sames, I.J. Kim, H.M. Kim, M. Spassova, W.G. Bornmann, K.O. Lloyd, H.I. Scher, P.O. Livingston and S.J. Danishefsky, *Angew. Chem., Int. Ed. Engl.*, 1999, **38**, 563.
140 T. Zhu and G.-J. Boons, *Angew. Chem., Int. Ed. Engl.*, 1999, **38,** 3495.
141 S. Di Virgilio, J. Glushka, K. Moremen and M. Pierce, *Glycobiology*, 1999, **9**, 353.
142 H. Ando, H. Ishida and M. Kiso, *J. Carbohydr. Chem.*, 1999, **18**, 603.
143 P. Duchaussoy, G. Jaurand, P.-A. Driguez, I. Lederman, F. Gourvenec, J.-M. Strassel, P. Sizun, M. Petitou and J.-M. Herbert, *Carbohydr. Res.*, 1999, **317**, 63.
144 S. Yamada, Y. Yamane, K. Sakamoto, H. Tsuda and K. Sugahara, *Eur. J. Biochem.*, 1998, **258**, 775.
145 E. Bettler, E. Samain, V. Chazalet, C. Bosso, A. Heyraud, D.H. Joziasse, W.W. Wakarchuk, A. Imberty and R.A. Geremia, *Glycoconjugate J.*, 1999, **16**, 205.
146 W. Wang and F. Kong, *Carbohydr. Res.*, 1999, **315**, 117.

147 W. Wang and F. Kong, *J. Org. Chem.*, 1999, **64**, 5091.
148 R. Geurtsen, F. Côté, M.G. Hahn and G.-J. Boons, *J. Org. Chem.*, 1996, **64**, 7828.
149 N. Sawada, H. Ishida, B.E. Collins, R.L. Schnaar and M. Kiso, *Carbohydr. Res.*, 1999, **316**, 1.
150 J.R. Allen and S.J. Danishefsky, *J. Am. Chem. Soc.*, 1999, **121**, 10875.
151 S. Komba, C. Galustian, H. Ishida, T. Feizi, R. Kannagi and M. Kiso, *Angew. Chem., Int. Ed. Engl.*, 1999, **38**, 1131.
152 E. Eichler, H.J. Jennings, M. Gilbert and D.M. Whitfield, *Carbohydr. Res.*, 1999, **319**, 1.
153 W. Wang and F. Kong, *Carbohydr. Res.*, 1999, **315**, 128.
154 D. Tailler, V. Ferrière, K. Pekari and R.R. Schmidt, *Tetrahedron Lett.*, 1990, **40**, 679.
155 L. Singh, Y. Nakahara, Y. Ito and Y. Nakahara, *Tetrahedron Lett.*, 1999, **40**, 3769.
156 M. Hato, H. Minamikawa and J.B. Seguer, *J. Phys. Chem.*, 1998, **102**, 11035 (*Chem. Abstr.*, 1999, **130**, 125 284).
157 L.V. Backinowsky, P.I. Abronina, N.K. Kochetkov and A.A. Grachev, *Bioorg. Khim*, 1998, **24**, 715 (*Chem. Abstr.*, 1999, **131**, 45 017).
158 A. Düffels and S.V. Ley, *J. Chem. Soc., Perkin Trans. 1*, 1999, 375.
159 V. Ding, M.-O. Contour-Galcera, J. Ebel, C. Ortiz-Mellet and J. Defaye, *Eur. J. Org. Chem.*, 1999, 1143.
160 Y. Du, Q. Pan and F. Kong, *Synlett*, 1999, 1648.
161 J.C. McAuliffe, M. Fukuda and O. Hindsgaul, *Biorg. Med. Chem. Lett.*, 1999, **9**, 2855.
162 T.D.W. Claridge, D.D. Long, N.N. Hungerford, R.T. Aplin, M.D. Smith, D.G. Marquess and G.W.J. Fleet, *Tetrahedron Lett.*, 1999, **40**, 2199.
163 A.P. Higson, Y.E. Tsvetkov, M.A.J. Ferguson and A.V. Nicolaev, *Tetrahedron Lett.*, 1999, **40**, 9281.
164 I. Damager, C.E. Olson, B.L. Moller and M.S. Motawia, *Carbohydr. Res.*, 1999, **230**, 19.
165 J. Ning and F. Kong, *Tetrahedron Lett.*, 1999, **40**, 1357.
166 R.J. Woods, A. Pathiaseril, M.R. Wormald, C.J. Edge and R.A. Dwek, *Eur. J. Biochem.*, 1998, **258**, 372.
167 J. Rong, K. Nordling, I. Björk and U. Lindahl, *Glycobiology*, 1999, **9**, 1331.
168 T.V. Bohner, O.-S. Becker and A. Vasella, *Helv. Chim. Acta,* 1999, **82**, 198.
169 P. Duchaussoy, G. Jaurand, P.-A. Driguez, I. Lederman, M.-C. Ceccato, F. Gourvenec, J.-M. Strassel, P. Sizun, M. Petitou and J.-M. Herbert, *Carbohydr. Res.*, 1999, **317**, 85.
170 M. Petitou, P. Duchaussoy, P.-A. Driguez, J.-P.Hérault, J.-C. Lormeau and J.-M. Herbert, *Bioorg. Med. Chem. Lett.*, 1999, **9**, 1155, 1161.
171 P.-A. Driguez, I. Lederman, J.-M. Strassel, J.-M. Herbert and M. Petitou, *J. Org. Chem.*, 1999, **64**, 9512.
172 H. Gohlke, S. Immel and F.W. Lichtenthaler, *Carbohydr. Res.*, 1999, **321**, 96.
173 P. Forgo and V.T. D'Souza, *J. Org. Chem.*, 1999, **64**, 306.
174 P. Forgo and V.T. D'Souza, *Tetrahedron Lett.*, 1999, **40**, 8533.
175 K. Koga, K. Ishida, T. Yamada, D.-Q. Yuan and K. Fujita, *Tetrahedron Lett.*, 1999, **40**, 923.
176 G. Zou and Z. Tan, *Guangzhou Huagong*, 1998, **26**, 17 (*Chem. Abstr.*, 1999, **130**, 81 737).

177 G. Wenz, T. Höfler, *Carbohydr. Res.*, 1999, **322**, 153.
178 M. Chavarot and Z. Pikramenou, *Tetrahedron Lett.*, 1999, **40**, 6865.
179 D. Armspach and D. Matt, *Chem. Commun.*, 1999, 1073.
180 S.-H. Chiu and D.C. Myles, *J. Org. Chem.*, 1999, **64**, 332.
181 N. Lupesch, C.K.Y. Ho, G. Jia and J.J. Krepinski, *J. Carbohydr. Chem.*, 1998, **18**, 99.
182 M.K. Grachev, I.G. Mustafin and E.E. Nifant'ev, *Russ. J. Gen. Chem.*, 1998, **68**, 1451 (*Chem. Abstr.*, 1999, **130**, 267 646).
183 H. Yamamura, D. Iida, S. Araki, K. Kobayashi, R. Katakai, K. Kano and M. Kawai, *J. Chem. Soc., Perkin Trans. 1*, 1999, 3111.
184 X.-M. Gao, L.-H. Tong, Y.-L. Zhang, A.-Y. Hao and Y. Inoue, *Tetrahedron Lett.*, 1999, **40**, 969.
185 S.D. Kean, B.L. May, P. Clements, S.F. Lincoln and C.J. Easton, *J. Chem. Soc., Perkin Trans. 2*, 1999, 1257, 1711.
186 K.K. Park, H.S. Lim and J.W. Park, *Bull. Korean Chem. Soc.*, 1999, **20**, 211 (*Chem. Abstr.*, 1999, **130**, 267 647).
187 L.-f. Zhang, Y.-C. Wong, L. Chen, C.B. Ching and S.-C. Ng, *Tetrahedron Lett.*, 1999, **40**, 1815.
188 R. Auzély-Velty, B. Perly, O. Taché, T. Zemb, P. Jéhan, P. Guenot, J.-P. Dalbiez and F. Djedaini-Pilard, *Carbohydr. Res.*, 1999, **318**, 82.
189 A. Ueno, A. Ikeda, H. Ikeda, T. Ikeda and F. Toda, *J. Org. Chem.*, 1999, **64**, 382.
190 Y. Liu, B.-H. Han, S.-X. Sun, T. Wada and Y. Inoue, *J. Org. Chem.*, 1999, **64**, 1487.
191 M.M. Luo, R.G. Xie, W. Lu, P.F. Xia and H.M. Zhao, *Chin. Chem. Lett.*, 1998, **9**, 135 (*Chem. Abstr.*, 1999, **131**, 144 771).
192 H. Yamamura, S. Yamada, K. Kohno, N. Okuda, S. Araki, K. Kobayashi, R. Katakai, K. Kano and M. Kawai, *J. Chem. Soc., Perkin Trans. 1*, 1999, 2943.
193 F. Charbonnier, A. Marsura and I. Pinter, *Terahedron Lett.,* 1999, **40**, 6581.
194 R. Breslow and N. Nesnas, *Tetrahedron Lett.*, 1999, **40**, 3335.
195 W.-H. Chen, S. Hayashi, T. Tahara, Y. Nogami, T. Koga, M. Yamaguchi and K. Fujita, *Chem. Parm. Bull.*, 1999, **47**, 588.
196 C. Donze, E. Rizzarelli and G. Vecchio, *Supramol. Chem.*, 1998, **10**, 33 (*Chem. Abstr.*, 1999, **130**, 196 873).
197 A.Y. Hao, L.H. Tong and M. Zhu, *Chin. Chem. Lett.*, 1998, **9**, 13 (*Chem.. Abstr.*, 1999, **131**, 144 763).
198 J. Yu, Y. Zhao, M.J. Holterman and D.L. Venton, *Bioorg. Med. Chem. Lett.*, 1999, **9**, 2705.
199 J.J. García-López, F. Hernández-Mateo, J. Isac-García, J.M. Kim, R. Roy, F. Santoyo-González and A. Vargas-Berenguel, *J. Org. Chem.*, 1999, **64**, 522.
200 F. Charbonnier, T. Humbert and A. Marsura, *Tetrahedron Lett.*, 1999, **40**, 4047.
201 R. Corradini, G. Buccella, G. Galaverna, A. Dossena and R. Marchelli, *Tetrahedron Lett*, 1999, **40**, 3025.
202 S.L. Hauser, E.S. Cotner and P.J. Smith, *Tetrahedron Lett.*, 1999, **40**, 2865.
203 W.-H. Chen, T.-M. Yan, Y. Tagashira, M. Yamaguchi and K. Fujita, *Tetrahedron Lett.*, 1999, **40**, 891.
204 J. Yan, R. Watanabe, M. Yamaguchi, D.-Q. Yuan and K. Fujita, *Tetrahedron Lett.*, 1999, **40**, 1513.
205 M. Fukudome, Y. Okabe, D-Q. Yuan and K. Fujita, *Chem. Commun.*, 1999, 1045.
206 T. Furuike and S. Aiba, *Chem. Lett.*, 1999, 69.

207 Y. Liu, C.-C. You, T. Wada and Y. Inoue, *J. Org. Chem.*, 1999, **64**, 3630.
208 Y. Liu, B. Li, B.-H. Han, T. Wada and Y. Inoue, *J. Chem. Soc. Perkin Trans. 2*, 1999, 563.
209 Y. Liu, C.-C. You, Y. Chen, T. Wada and Y. Inoue, *J. Org. Chem.*, 1999, **64**, 7781.
210 K. Chmurski and J. Defaye, *Pol. J. Chem.*, 1999, **73**, 967 (*Chem. Abstr.*, 1999, **131**, 73 871).
211 J. Diakur, Z. Zuo and L.I. Wiebe, *J. Carbohydr. Chem.*, 1999, **18**, 209.
212 D.R. Kelly and A.K. Mish'al, *Tetrahedron: Asymmetry*, 1999, **10**, 3627.
213 J. Hu, C.F. Ye, Y.D. Zhao, J.B. Chang and R.Y. Guo, *Chin. Chem. Lett.*, 1999, **10**, 273 (*Chem. Abstr.*, 1999, **131**, 157 878).
214 M.E. Deary and D.M. Davies, *Carbohydr. Res.*, 1999, **317**, 10.

5
Ethers and Anhydro-sugars

1 Ethers

1.1 Methyl Ethers. – The partial methylation of a number of pentosides and hexosides with diazomethane in the presence of certain transition metal chlorides has been studied. For methyl pyranosides of pentoses and hexoses, tin(II) and antimony(III) salts promoted substitution mainly at O-3. However, methyl β-L-rhamnopyranoside demonstrated higher reactivity at O-2 in all cases, and cerium(III) and zinc(II) salts promoted substitution at O-2 of pentosides.[1] The synthesis of partially methylated D-galactose derivatives from 1,6-anhydro-D-galactose has been reported,[2] and the isolation and characterization of per-*O*-methyl α- and β-D-galactofuranosides and α-L-fucofuranoside from the reaction mixtures of per-*O*-methylation of D-galactose and L-fucose have been described.[3]

1.2 Other Alkyl and Aryl Ethers. – The use of 'Wacker'reaction conditions ($PdCl_2$–CuCl/THF–H_2O) to effect de-*O*-allylation has been found again to lead to ketone products rather than deallylation.[4] Tritylation of 1,2-*O*-isopropylidene-α-D-glucofuranose under forcing conditions has afforded significant quantities of the 3,6-diether by way of pyridine hydrochloride-catalysed detritylation–retritylation of the initially formed 5,6-ditrityl ether.[5] The deetherification of 5′-dimethoxytrityl nucleosides has been achieved under aprotic neutral conditions using stannous chloride,[6] and *O*-(benzotriazol-1-yl)-*N,N,N′,N′*-tetramethyluronium tetrafluoroborate has been used to effect de-*O*-dimethoxytritylation in the presence of Tbdms ethers.[7]

The *p*-dodecyloxybenzyl ether protecting group has been used to provide sufficient lipophilicity for selective adsorption onto C_{18}-silica gel of partially protected disaccharides.[8] The phase-transfer-catalysed benzylation of various 4,6-*O*-benzylidene-D-glucopyranosides has afforded preferentially the 2-*O*-benzylated products,[9] while the acid-catalysed *O*-benzylation of some carbohydrate derivatives using complex quinonemethide substrates has been studied as model reactions for the formation of lignin-carbohydrate materials (Scheme 1).[10]

A two-phase system employing aqueous $NaBrO_3$–$Na_2S_2O_4$ has been used to achieve *O*-debenzylation. Bromine radicals are produced which effect benzylic bromination and the products undergo spontaneous hydrolysis.[11] Triisobutylaluminium has been described as a reagent of choice for the regioselective

Carbohydrate Chemistry, Volume 33

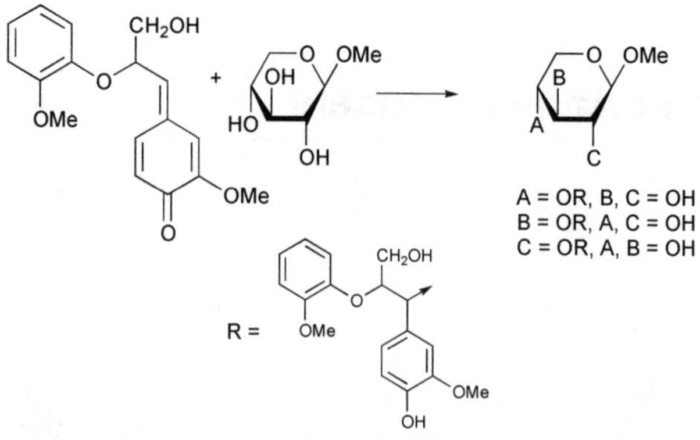

A = OR, B, C = OH
B = OR, A, C = OH
C = OR, A, B = OH

Scheme 1

mono-*O*-debenzylation of a variety of perbenzylated mono- and di-saccharides.[12] The selective 2-*O*-debenzylation of a perbenzylated allyl-*C*-glycoside has been described (Scheme 2). Only the α-*C*-glycoside reacted so that separation of the anomeric mixture was achieved.[13]

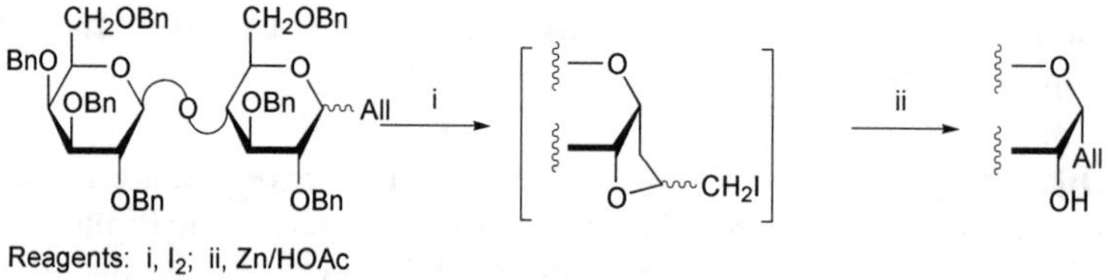

Reagents: i, I_2; ii, Zn/HOAc

Scheme 2

New mild conditions [$MeOC_6H_4CH_2OH$, $Yb(OTf)_3$] have been employed for the preparation of Pmb ethers from alcohols,[14] and $CeCl_3$–NaI conditions have been utilized for the cleavage of Pmb ethers.[15] The selective release of the 5-OH group from a 5,6-di-*O*-Pmb ether of a glucofuranose derivative has been achieved using EtSH, $SnCl_2.2H_2O$.[16]

The direct transformation of unprotected sucrose into various ethers has been reviewed in the context of the preparation of derivatives with industrial interest.[17] The base-promoted reaction of sucrose with *tert*-butyl chloromethyl ketone has afforded a moderate yield of the mono-ether substituted at O-2 in the pyranose moiety,[18] and ampiphilic hydroxyalkyl sucrose ethers have been prepared from unprotected sucrose with epoxydodecane and a tertiary amine in water.[19]

2-Chloropropionic acid undergoes reaction with sugar alcohols and sodium hydride with inversion of configuration to give the carboxylic acid ethers directly,[20] and 1-bromo-3-tetrahydropyranyloxypropane has afforded carbohydrate ethers under similar conditions.[21] A range of 3-*O*-(3-*O*-alkylglyceryl)-D-glucopyranoses have been prepared and their ampiphilic characteristics compared to those of some 3-*O*-alkyl-D-glucopyranoses.[22]

The galactose derivative **1** has been used as a scaffold for solid-phase combinatorial synthesis whereby different ethers were incorporated at O-2 and O-6.[23] A 6-*O*-aryl-D-galactose substituted porphyrin has been prepared and (with the aryl moiety bonded to the porphyrin) incorporated into liposomes and lipoproteins.[24]

1 R = solid phase

2
R = $C_{12}H_{25}$, $C_{18}H_{37}$
$R^1 = R^2$ = Me, R^3 = X
R = R^1; R^2, R^3 = X
$R^1 = R^2 = R^3$ = X

1.3 Silyl Ethers. – 1,3-Dichlorotetraisopropyldisiloxane has been generated *in situ* from the corresponding silane using $PdCl_2/CCl_4$ in a new and cheaper procedure for synthesizing 3′,5′-*O*-Tips protected nucleosides. The 3′,5′-di-*O*-Tbdms derivatives were also prepared using *tert*-butyldimethylsilane in a similar procedure.[25] The dehydrogenative silylation of alcohols has been achieved using a trialkylsilane and tri(pentafluorophenyl)borane. The reaction conditions are compatible with the presence of ethers, esters, ketones, alkenes and halogens.[26] Some glycosyl donors have been tethered to solid-phase resins *via* a trialkylsilyl ether linker before activation and disaccharide formation,[27] and the long chain alkylsilyl glycosides **2** have been reported.[28] The introduction of vicinal bulky silyl ethers onto equatorial hydroxy groups in a pyranose ring can effect a conformational inversion affording conformers with the substituents axial.[29]

A mild selective means of cleaving Tbdms ethers employs $Zn(BF_4)_2$ in water. The conditions are compatible with the presence of Tbdps, Bn and allyl ethers as well as Thp, ester, aldehyde and ketone groups.[30] Alternatively, IBr has been used for the same selective cleavage in high yields, in the presence of Tbdps, Bn, MBn, ester and acetal functionality.[31]

2 Intramolecular Ethers (Anhydro-Sugars)

2.1 Oxirans. – The use of epoxy-sugars in synthesis has been reviewed.[32] The reduction of epoxides **3** and **4** with $LiAlH_4$ during the synthesis of muscarine analogues has afforded predominantly the regioisomers **5** and **6**.[33]

2.2 Other Anhydrides. – An improvement on the diazotization process for the conversion of 2-amino-2-deoxy-D-glucose into 2,5-anhydro-D-mannose has been described.[34] Treatment of the galactofuranoside **7** with $SnCl_4$ has resulted in formation of the 1,6-anhydrofuranose **8**. Acetolysis of this material has afforded a glycosyl acetate, which was converted into galactofuranosyl donors.[35]

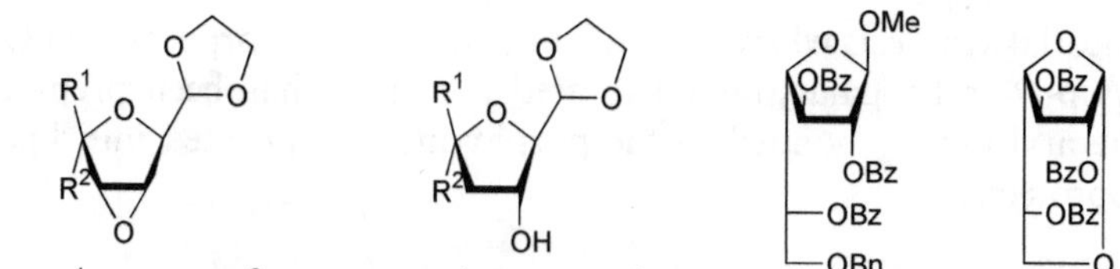

3 $R^1 = CH_2OH$, $R^2 = H$
4 $R^1 = H$, $R^2 = CH_2OH$

5 $R^1 = CH_2OH$, $R^2 = H$
6 $R^1 = H$, $R^2 = CH_2OH$

7

8

The 3,6-anhydro-sugars **9** and **10** have been prepared from 1,2-*O*-isopropylidene-α-D-glucofuranose-5,6-cyclic sulfate and -5,6-cyclic sulfite respectively.[36]

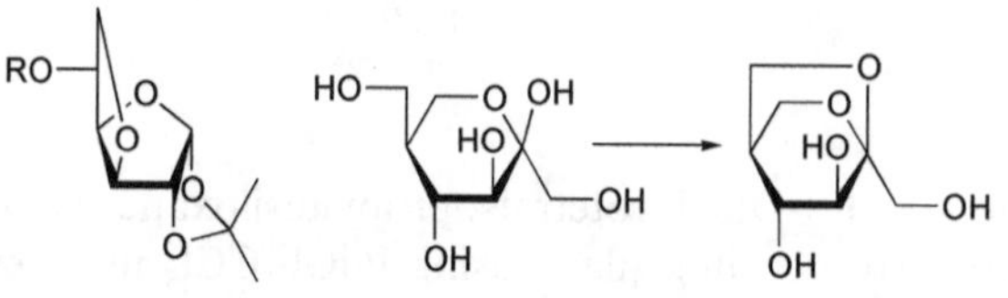

9 R = SO_3Na
10 R = H

11

12

Base treatment of 1-*O*-acetyl-2,3-*O*-isopropylidene-5-*O*-methanesulfonyl-α,β-D-ribofuranose has afforded 1,5-anhydro-2,3-*O*-isopropylidene-β-D-ribofuranose,[37] and acid treatment of the branched hexulose **11** has given rise to anhydro-sugar **12**.[38] Similar treatment of **14**, derived from the aldonolactone **13** (Scheme 3) has produced the 2,6-anhydroketose **15**.[39] In a study of rigid L-fucose mimics the 2,7-anhydro-heptulose derivative **16** has been functionalized to **17** and **18** as potential fucosyl transferase inhibitors.[40]

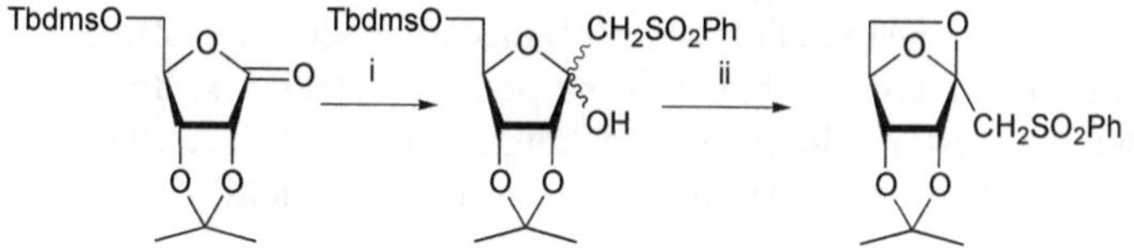

13 **14** **15**

Reagents: i, $LiCH_2SO_2Ph$; ii, $BF_3{\cdot}OEt_2$

Scheme 3

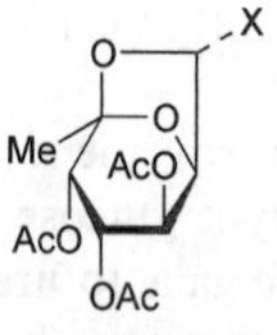

16 X = H
17 X = Br
18 X = NH_2

The ring-opening polymerization of 1,5-anhydro-2,3-di-*O*-(*p*-azidobenzyl)-β-D-ribofuranose with the use of Lewis acid catalysts has been studied. All catalysts afforded the (1,5)-α-D-ribofuranan polymer.[41]

References

1 E.V. Evtushenko, *Carbohydr. Res.*, 1999, **316**, 187.
2 S. Sarbajna and N. Roy, *Indian J. Chem., Sect. B: Org. Chem. Incl. Med. Chem.*, 1999, **38B**, 361 (*Chem. Abstr.*, 1999, **131**, 185 138).
3 D.D. Asres and H. Perreault, *Can. J. Chem.*, 1999, **77**, 319.
4 Z.-J. Li, H. Li, Y.-P. Lu, X.-L. Shi and M.-S. Cai, *Carbohydr. Res.*, 1999, **317**, 191.
5 N. Morishima and Y. Mori, *Chem. Pharm. Bull.*, 1999, **47**, 1481.
6 A. Khalafi-Nezhad and R. Fareghi Alamdari, *Iran. J. Chem. Eng.*, 1998, **17**, 58 (*Chem. Abstr.*, 1999, **131**, 45 034).
7 K.S. Ramasamy and D. Averett, *Synlett*, 1999, 709.
8 V. Pozsgay, *Org. Lett.*, 1999, **1**, 477 (*Chem. Abstr.*, 1999, **131**, 102 450).
9 K. Khanbabaee, K. Loetzerich, M. Borges and M. Grosser, *J. Prakt. Chem.*, 1999, **341**, 159 (*Chem. Abstr.*, 1999, **130**, 223 502).
10 M. Toikka and G. Brunow, *J. Chem. Soc., Perkin Trans. 1*, 1999, 1877.
11 M. Adinolfi, G. Barone, L. Guariniello and A. Iadonisi, *Tetrahedron Lett.*, 1999, **40**, 8439.
12 M. Sollogoulo, S.K. Das, J.-M. Mallet and P. Sinaÿ, *C.R. Acad. Sci., Ser. IIC: Chim.*, 1999, **2**, 441 (*Chem. Abstr.* 1999, **131**, 322 836).
13 L. Lay, L. Cipolla, B. La Ferla, F. Peri and F. Nicotra, *Eur. J. Org. Chem.*, 1999, 3437.
14 G.V.M. Sharma and A.K. Mahalingam, *J. Org. Chem.*, 1999, **64**, 8943.
15 A. Cappa, E. Marcantoni, E. Torregiani, G. Bartoli, M.C. Bellucci, M. Bosco and L. Sambri, *J. Org. Chem.*, 1999, **64**, 5696.
16 A. Bouzide and G. Sauvé, *Tetrahedron Lett.*, 1999, **40**, 2883.
17 G. Descotes, J. Gagnaire, A. Bouchu, S. Thevent, N. Giry-Panaud, P. Salanski, S. Belniak, A. Wernicke, S. Porwanski and Y. Queneau, *Pol. J. Chem.*, 1999, **73**, 1069 (*Chem. Abstr.*, 1999, **131**, 144 750).
18 N. Giry-Panaud, G. Descotes, A. Bouchu and Y. Queneau, *Eur. J. Org. Chem.*, 1999, 3393.
19 J. Gagnaire, G. Toraman, G. Descotes, A. Bouchu and Y. Queneau, *Tetrahedron Lett.*, 1999, **40**, 2757.
20 T. Staroske, J. Görlitzer, R.M.H. Entress, M.A. Cooper and D.H. Williams, *J. Chem. Soc., Perkin Trans. 1*, 1999, 1105.
21 I. Neda, P. Sakhau, A. Waßmaan, U. Niemeyer, F. Günther and J. Engel, *Synthesis*, 1999, 1625.
22 Ph. Bault, P. Gode, G. Goethals, P. Martin and P. Villa, *Commun. Jorn. Com. Esp. Deterg.*, 1999, **29**, 419 (*Chem. Abstr.*, 1999, **131**, 19 198).
23 C. Kallus, T. Opatz, T. Wunberg, W. Schmidt, S. Henke and H. Kunz, *Tetrahedron Lett.*, 1999, **40**, 7783.
24 C. Schell and H.K. Hombrecher, *Bioorg. Med. Chem.*, 1999, **7**, 1857.
25 C. Ferreri, C. Costantino, R. Romeo and C. Chatgilialoglu, *Tetrahedron Lett.*, 1999, **40**, 1197.

26 J.M. Blackwell, K.L. Foster, V.H. Beck and W.E. Piers, *J. Org. Chem.*, 1999, **64**, 4887.
27 T. Doi, M. Sugiki, H. Yamada, T. Takahashi and J.A. Porco, *Tetrahedron Lett.*, 1999, **40**, 2141.
28 A.M.A. Aisa and H. Richler, *Carbohydr. Res.*, 1999, **321**, 168.
29 H. Yamada, M. Nakatani, T. Ikeda and Y. Marumoto, *Tetrahedron Lett.*, 1999, **40**, 5573.
30 B.C. Ranu, U. Jana and A. Majee, *Tetrahedron Lett.*, 1999, **40**, 1985.
31 K.P.R. Kartha and R.A. Field, *Synlett*, 1999, 311.
32 W. Voelter, R. Thurmer, R.A. Al-Qawasmeh, T.H. Al-Tel, R. Abdel-Jalil and Y. Al-Abed, *Pol. J. Chem.*, 1999, **73**, 55 (*Chem. Abstr.*, 1999, **130**, 196 834).
33 V. Popsavin, M. Popsavin, L. Radić, O. Perić and V. Ćirin-Novta, *Tetrahedron Lett.*, 1999, **40**, 9305.
34 S. Claustre, F. Bringaud, L. Azéma, R. Baron, J. Périé and M. Wilson, *Carbohydr. Res.*, 1999, **315**, 339.
35 S.K. Sarkar, A.K. Choudhury, B. Mukhopadhyay and N. Roy, *J. Carbohydr. Chem.*, 1999, **18**, 1121.
36 M.P. Molas, M.I. Matheu, S. Castillon, J. Isac-Garcia, F. Hernandez-Mateo, F.G. Calvo-Flores and F. Santoyo-Gonzalez, *Tetrahedron*, 1999, **55**, 14649.
37 A. Fleetwood and N.A. Hughes, *Carbohydr. Res.*, 1999, **317**, 204.
38 S. David, *Eur. J. Org. Chem.*, 1999, 1415.
39 A. Alzérreca, E. Hernández, E. Mangual and J.A. Prieto, *J. Heterocycl. Chem.*, 1999, **36**, 555.
40 K.H. Smelt, Y. Blériot, K. Biggadike, S. Lynn, A.L. Lane, D.J. Watkin and G.W.J. Fleet, *Tetrahedron Lett.*, 1999, **40**, 3255.
41 C.-P. Wu, S.-J. Zheng, C.-Y. Pan and T. Uryu, *Chin. J. Polym. Sci.*, 1999, **17**, 123 (*Chem. Abstr.*, 1999, **131**, 5 439).

6
Acetals

The direct transformation of unprotected sucrose into various acetals has been covered in a review on the preparation of sucrose derivatives of industrial interest.[1]

1 Isopropylidene, Benzylidene and Methylidene Acetals

The 8,9-isopropylidene acetal **1** was an important intermediate in the synthesis of KDN-derivatives with a single unprotected hydroxyl group (see Chapters 5 and 7).[2]

Selective radical-chain epimerization at C-5 of 3-deoxy-1,2:5,6-di-*O*-isopropylidene-α-D-*ribo*-hexofuranose (**2**), on exposure to tri-*t*-butoxysilanethiol as protic polarity-reversal catalyst and 2,2-di-(*t*-butylperoxy)butane as radical initiator, gave a 68:32 D-*ribo*/L-*lyxo*-equilibrium mixture from which the L-*lyxo*-compound **3** was available in 25% isolated yield. There were no signs of epimerization at other centres.[3]

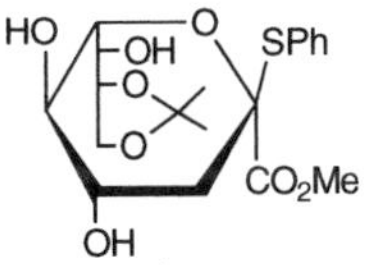

HO OH O SPh O O CO_2Me OH **1**

R O O O **2** R = **3** R =

O O OR^1 R^2O O OAll **4** R^1, R^2 = Me Me **5** R^1 = H, R^2 = Bu^t

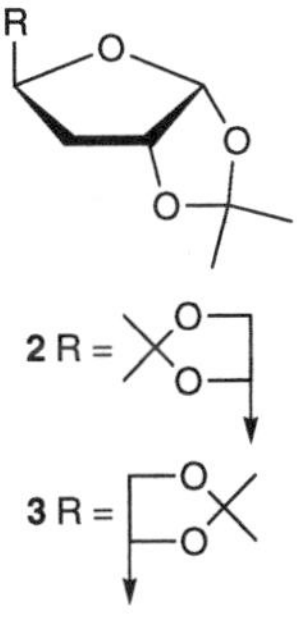

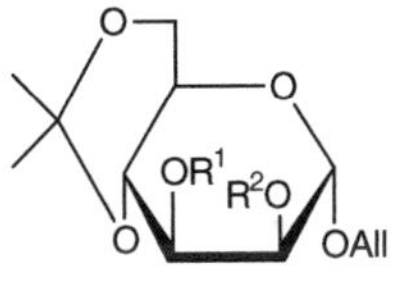

Chelation-controlled acetal opening by methylmagnesium iodide in benzene took place at the 1,3-dioxolane ring of diacetal **4** to afford 3-*t*-butyl ether **5** exclusively in high yield. Many other examples of acetal-opening of this kind are given.[4] Efficient, selective removal of the terminal 1,3-dioxolane ring of 3-*O*-protected diacetoneglucose derivatives (**6**→**7**) has been achieved under non-acidic conditions by heating with thiourea in refluxing aqueous ethanol,[5] whereas methanolic iodine removed the 1,2-*O*-isopropylidene group of **6**

(R = Bn), furnishing methyl glycoside **8** with a free 2-OH group.[6] Deacetonation under the influence of LDA is referred to in Chapter 2, ref. 24.

Use of a Dean-Stark apparatus loaded with molecular sieves in the 4,6-*O*-benzylidenation of methyl α-D-glucopyranoside with $PhCH(OMe)_2$/CSA in chloroform has led to a significantly improved product yield.[7]

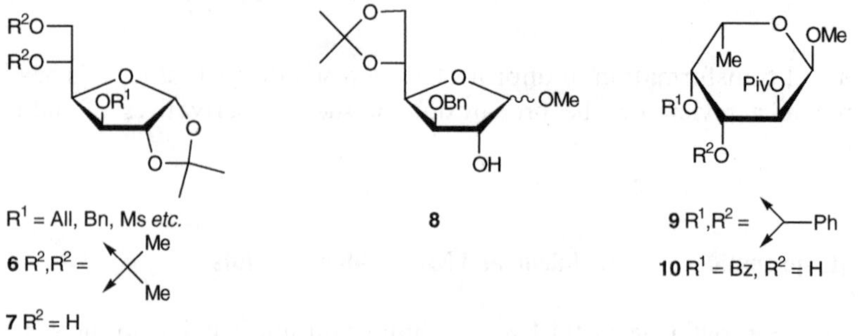

R¹ = All, Bn, Ms *etc.*
6 R²,R² = CMe₂
7 R² = H
8
9 R¹,R² = CHPh
10 R¹ = Bz, R² = H

Benzylidene acetal **9** has been cleaved to give the 4-*O*-benzoate **10** in 92% yield by use of $NaBrO_3/Na_2S_2O_4$ in a two-phase system. (The reagents produce bromine atoms which diffuse to the organic phase and effect benzal bromination.) Methyl 2,3-di-*O*-acetyl-4,6-*O*-benzylidene-α-D-glucopyranoside reacted with high yield but low regioselectivity.[8]

The 4,6-*O*-benzylidene derivatives of several methyl D-hexopyranosides acted as gelators of various organic solvents.[9]

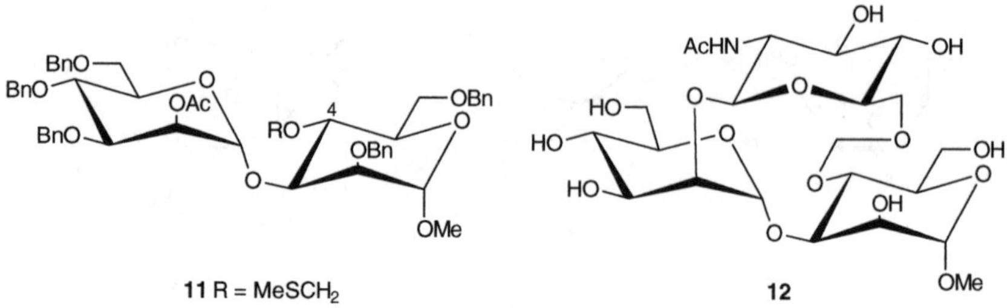

11 R = $MeSCH_2$
12

The methylene acetal-bridge in trisaccharide **12**, which mimics an intramolecular hydrogen bond in the β-D-Glc*N*Ac-(1→2)-α-D-Man-(1→3)-D-Man fragment of a biantennary glycan, was introduced prior to formation of the macrocycle by NIS/TfOH-promoted coupling of the methyl thiomethyl ether group at O-4 of the disaccharide moiety **11** to the 6-OH group of a Glc*N*Ac acceptor. The synthesis was then completed by intramolecular glycosylation to form the trisaccharide macrocycle, followed by deprotection.[10]

2 Other Acetals

The use of 1,2:5,6-di-*O*-cyclohexylidene-D-mannitol as chiral inducer in certain asymmetric homologation reactions is covered in Chapter 24.

The concomitant formation of a trichloroethylidene acetal and introduction of a carbamoyl function with configurational inversion at the central hydroxyl group of a *cis-trans*-contiguous triol (see Vol. 32, Chapter 6, ref. 8; Vol. 30, p. 99, refs. 17–19) has now been applied to the preparation of D-tagatose derivatives from D-fructose (see Chapter 2, ref. 14), and to the epimerization of a cyclohexanepentaol derivative (see Chapter 18, ref. 126).

Treatment of methyl α-D-glucopyranoside with ethyl 4,4-diethoxypentanoate (3 mol equiv.) in DMF in the presence of catalytic CSA gave the ethoxycarbonylbutylidene acetals **13** in 70% yield. With 6 mol equiv. of reagent, methyl α-D-mannopyranoside afforded an isomeric mixture of the four diacetals **14** (68%); sucrose and trehalose furnished as the main-products the 4,6-mono- and the 4,6:4′6′-di-acetals, respectively. The ethyl esters were hydrolysed to the corresponding acid salts by treatment with aqueous NaOH.[11]

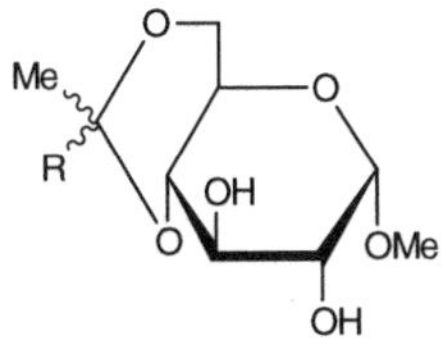

13 R = $EtO_2CCH_2CH_2$

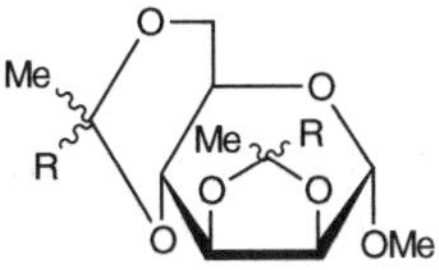

14 R = $EtO_2CCH_2CH_2$

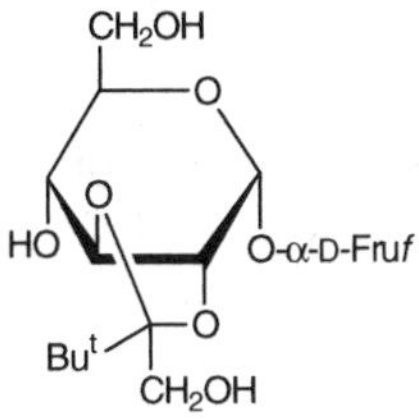

15

Acetal **15** was formed as the main-product in 45% yield on exposure of sucrose to *t*-butyl chloromethyl ketone in DMF under base catalysis. The main by-product (22%) was the 2-ether **16** (see Chapter 5), indicating that under these conditions 2-OH is the most reactive hydroxyl group of sucrose.[12]

Under conventional conditions [$Ph_2C(OMe)_2$/TfOH], the expected 2,3:4,6-di-*O*-acetal **17** of methyl α-D-glucopyranoside was obtained in 40% yield, accompanied by equal amounts of the 2-*O*-(methoxydiphenylmethyl)-4,6-*O*-diphenylmethylidene derivative **18** with a free 3-OH group. Compound **19**, unprotected at the 2-position, was available by selective cleavage of the dioxolane ring of **17** with phenylmagnesium bromide.[13]

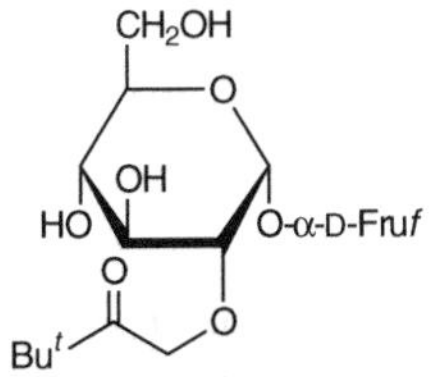

16

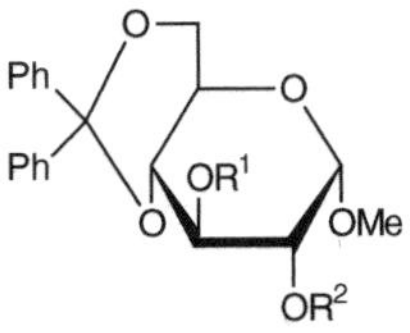

17 R^1,R^2 = CPh_2

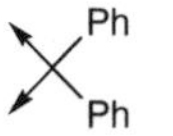

18 R^1 = H, R^2 = C(OMe)Ph_2

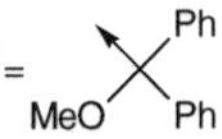

19 R^1 = Tr, R^2 = H

From the core part of a *Proteus* sp. lipopolisaccharide the tetrasaccharide moiety **20** has been isolated in which an open-chain Gal*N*Ac residue is linked as a cyclic acetal to positions 4 and 6 of D-Gal*p*NH_2.[14]

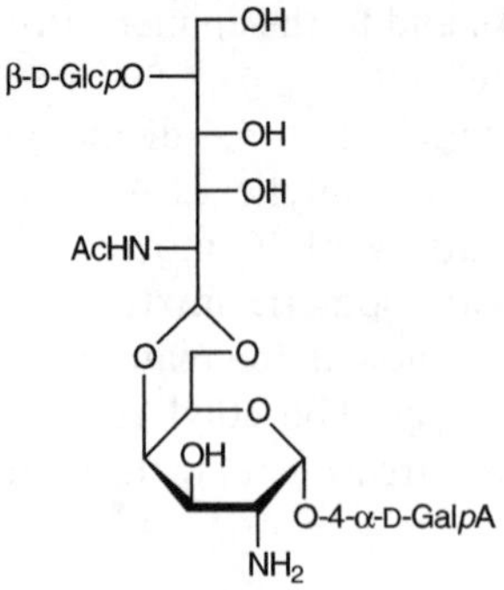

The structure of the 1,2-*O*-substituted glucosyl-acetal derivative **21**, isolated from twigs and thorns of *Castela polyandra*, has been determined by X-ray crystallography.[15]

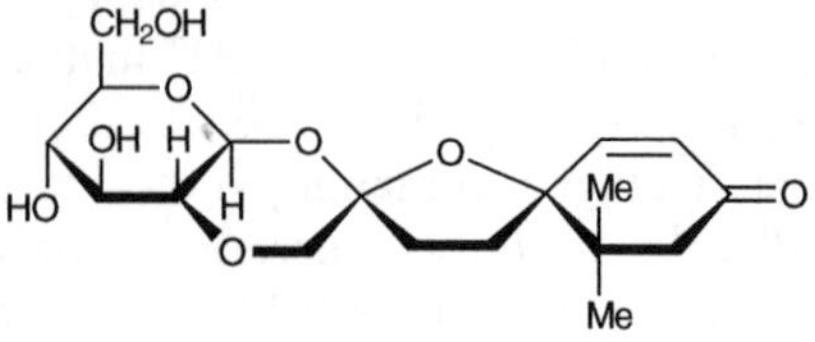

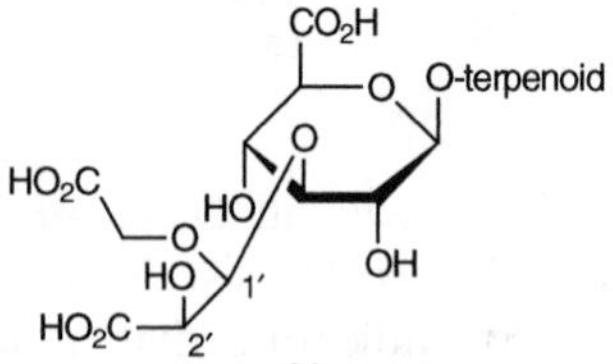

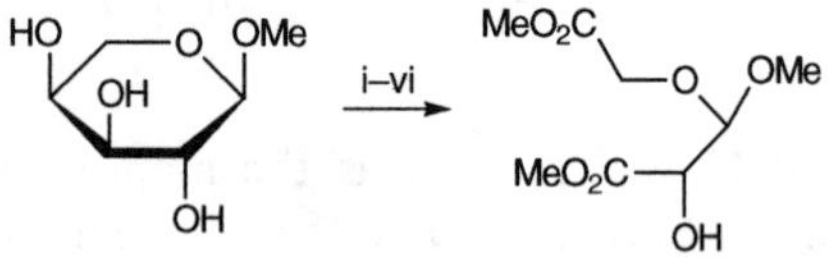

Reagents: i, $Me_2C(OMe)_2$, H^+; ii, TbdmsCl, Im; iii, aq. HOAc; iv, $Pb(OAc)_4$; v, $NaClO_2$, H_2NSO_3H; vi, CH_2N_2

Scheme 1

The absolute stereochemistry at positions 1′ and 2′ in the acetal substituent of the Glc*p*A residue of the saponin betavulgaroside IV (**22**) has been determined by synthesis of all four isomers of this moiety from the four isomeric methyl arabinopyranosides. An example is shown in Scheme 1.[16]

References

1 G. Descotes, J. Gagnaire, A. Bouchu, S. Thevenet, N. Giry-Panaud, P. Salanski, S. Belniak, A. Wernicke, S. Porwanski and Y. Queneau, *Pol. J. Chem.*, 1999, **73**, 1069 (*Chem. Abstr.*, 1999, **131**, 144 750).

2 S. Akai, T. Nakagaa, Y. Kajihara and K.-i. Sato, *J. Carbohydr. Chem.*, 1999, **18**, 639.

3 H.S. Dang and B.P. Roberts, *Tetrahedron Lett.*, 1999, **40**, 4271.

4 W.-L. Cheng, Y.-J. Shaw, S.-M. Yeh, P.P. Kanakamma, Y.-H. Chen, C. Chen,

J.-C. Shieu, S.-J. Yiin, G.-H. Lee, Y. Wang and T.-Y. Luh, *J. Org. Chem.*, 1999, **64**, 532.
5 S. Majumdar and A. Bhattacharjya, *J. Org. Chem.*, 1999, **64**, 5682.
6 J. Molina Arévalo and C. Simone, *J. Carbohydr. Chem.*, 1999, **18**, 535.
7 R.A. Spanerello and D.D. Saavedra, *Org. Prep. Proced. Int.*, 1999, **31**, 460 (*Chem. Abstr.*, 1999, **131**, 351 548).
8 M. Adinolfi, G. Barone, L. Guariniello and A. Iadonisi, *Tetrahedron Lett.*, 1999, **40**, 8439.
9 K. Yuza, N. Amanokura, Y. Ono, T. Akao, H. Shinmori, M. Takeuchi, S. Shinkai and D.N. Reinhardt, *Chem. Eur. J.*, 1999, **5**, 2722.
10 N. Navarre, N. Amiot, A von Oijen, A. Imberty, A. Poveda, J. Jiménez-Barbero, A. Cooper, M.A. Nutley and G.-J. Boons, *Chem. Eur. J.*, 1999, **5**, 2281.
11 S. Carbonel, C. Fayet and J. Gelas, *Carbohydr.Res.*, 1999, **319**, 63.
12 N. Giry-Panaud, G. Descotes, A. Bouchu and Y. Queneau, *Eur. J. Org. Chem.*, 1999, 3393.
13 L. Di Donna, A. Napoli, S. Siciliano and G. Sindona, *Tetrahedron Lett.*, 1999, **40**, 1013.
14 E. Vinogradov and K. Bock, *Angew. Chem., Int. Ed. Engl.*, 1999, **38**, 671.
15 P.A. Grieco, J. Haddad, M.M. Pineiro-Nunez and J.C. Hufmann, *Phytochemistry*, 1999, **51**, 575.
16 T. Murakami, H. Matsuda, M. Inadzuki, K. Hirano and M. Yoshikawa, *Chem. Pharm. Bull.*, 1999, **47**, 1717.

7
Esters

1 Carboxylic Esters

1.1 Synthesis. – *1.1.1 Chemical Acylation and Deacylation.* Reaction between tetra-*O*-acetyl-α-D-glucopyranosyl fluoride and carboxylic acids with $BF_3.OEt_2$/di-*tert*-butylmethylpyridine as promoter provided glucosyl carboxylates in 62–73% yield with high β selectivity.[1] Acetylation of a series of methyl hexosides with acetic anhydride and a Zeolite gave high yields of the tetraacetates,[2] while benzoylation, under Mitsunobu conditions, provided complete benzoylation at O-6 and regioselective protection at the remaining three alcohol groups.[3] A regioselective acylation of 1,5-anhydro-D-fructose was accomplished with dodecanoyl chloride and pyridine at the primary O-6 centre.[4] Regiospecific syntheses of twelve novel primary butanoyl and phenylalkyloyl monoesters of D-mannose and xylitol have also been reported.[5] Esterification of benzyl and naphthyl 6-*O*-acetyl-α- or 6-deoxy-α- or β-D-gluco- or manno-pyranosides with trifluoroethyl butanoate led predominantly to 3-*O*-butanoyl derivatives.[6] Other esterifications to have been reported include the formation of penta-*O*-galloyl-glucose and hexa-*O*-galloyl-*myo*-inositol,[7] a phase-transfer catalysed transesterification of di-*O*-isopropylidene-glucose with methyl esters under microwave assistance (see Vol. 32, p. 97, ref. 20),[8] and an O-3 acylation of 4,6-*O*-benzylidene-β-D-glucopyranosides by use of a copper chelate soluble in THF.[9] This last example gives only moderate yields, but good regioselectivities.

Fukase *et al.* reported the *O*-substitution of alcohol **1** (R = TrOCO) with propargyloxycarbonyl chloride to give 100% yield of the carbonate (**2**) which is

Ph OH O OAll 3 NHR **1** → O OCOCH$_2$C≡CH 3 NHR **2**

stable in TFA but is readily cleaved at room temperature with $Co_2(CO)_8$ and TFA, thus providing an excellent protecting group against acids.[10] The relative reactivities of the hydroxyl groups of octyl D-gluco-, D-manno- and D-galacto-

Carbohydrate Chemistry, Volume 33

pyranosides were investigated using DMAP-catalysed monoacetylation in chloroform. Product distributions depended on the concentration of the DMAP, the structure of the substrate (in particular the natures of the intramolecular hydrogen bonding networks) and on reaction time.[11] The synthesis of anomerically mixed *O*-glycosyl carbamates by treatment of 2,3,4,6-tetra-*O*-acetyl-D-hexoses with alkyl isocyanates has been reported.[12]

Surfactants containing two butyl α-D-glucoside moieties per molecule have been synthesized by use of glutaryl, succinyl and terephthaloyl links through O-2 or O-6 [Scheme 1, O-6 linking depicted, X = $(CH_2)_2$, $(CH_2)_3$, C_6H_4-*o*]. The critical micellar concentrations (CMC) for the new compounds were ten fold less than those for butyl glucoside.[13]

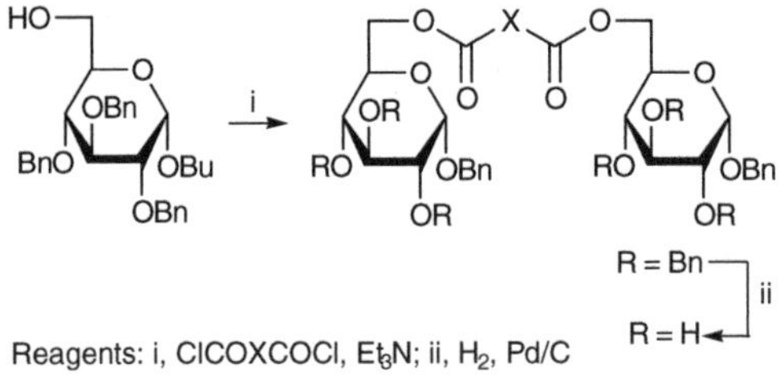

Reagents: i, ClCOXCOCl, Et_3N; ii, H_2, Pd/C

Scheme 1

A combinatorial solid-phase synthesis using galactose as the scaffold gave libraries containing urethanes and ethers at each available hydroxyl group of the sugar. The strategy involved selective protection and deprotection of the galactose anomerically linked to an aminomethyl poylstyrene.[14] The synthesis of diacetate **4** from thio-furanoside **3** was accomplished in 51% using the two-step strategy indicated in Scheme 2.[15] The intermediate acetal was isolated by chromatography then treated with IDCP and TMSOTf. Furanoside **4** was subsequently used in a synthesis of methyl β-D-arabinofuranoside 5-(*myo*-inositol phosphates) (see Chapter 3).

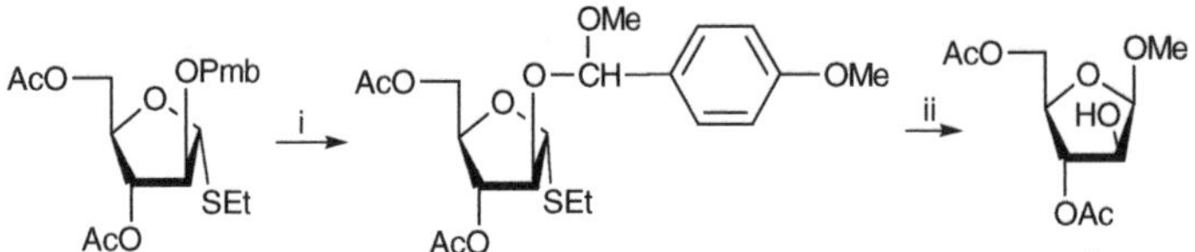

Reagents: i, DDQ, MeOH; ii, IDCP, TMSOTf

Scheme 2

A review on the direct transformation of unprotected sucrose into various esters, ethers and acetals in the context of derivatives of industrial interest has been written.[16] Other reports of the esterification of sucrose include descriptions of gallic esters as new antioxidants,[17] and the preparation of polyesters by transesterification from methyl oleate induced by tetrabutylammonium bromide and tetrabutyl titanate.[18] Three other reports include a review on the use of sucrose fatty esters in microemulsions,[19] a study on acylations with vinyl laurate in DMSO with simple catalysts such as Celite or Eupergit C[20] and a study on acylations using acid chlorides in basic aqueous media.[21] In this final report, it was found that the C-2 OH group is the most active, but there is fast subsequent acyl migration that enriches the amount of substitution at the primary positions.

Further reports on the use of dibutyltin oxide in selective substitutions of various carbohydrate diols and triols have appeared. Sites of reactions (which were mainly esterifications) were: 3-*O*-benzyl-1,2-*O*-isopropylidene-D-glucofuranose, O-6; methyl α-L-rhamnopyranoside, O-3; methyl 4,6-di-*O*-benzylidene-α-D-glucopyranoside, O-2.[21a] This methodology was also used in the selective benzoylation of several alkyl and phenylthio 4,6-*O*-Tipds-hexopyranosides. In each case the proportion of 3-esters was increased – sometimes markedly – by initial use of Bu_2SnO.[22] The 2-deoxy-glucopyranosyl esters **5** and **6**, which were used as donors in transesterifications with plant-derived enzymes, were synthesized from tri-*O*-benzyl-D-glucal in good yields (Scheme 3).[23]

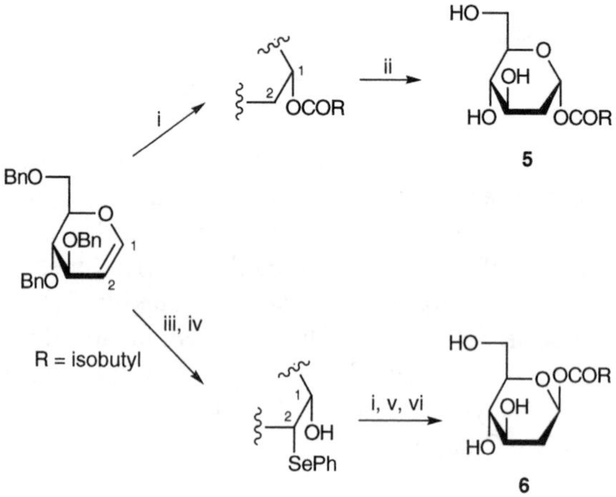

Reagents; i, RCO_2H, Ph_3P HBr; ii, H_2, Pd/C; iii, PhSeCl; iv, Na_2CO_3; v, Bn_3SnH; vi, H_2, Pd/C

Scheme 3

The arylamido trisaccharide glycoside **7** was copolymerized with acrylamide and the allyl glycoside **8**,[24] and the binding of the products with lectins was studied. Additionally, the preparation of carbonate **9** and its copolymerization with L-lactide has been reported.[25] Similarly, new bolaamphiphiles [sugar-$OOCHN(CH_2)_nNHCOO$-sugar], in which the sugars are D-galactose or lactose were prepared using conventional methods from per-*O*-acetyl sugars and alkyl-diisocyanates, with the alkyl group containing 6, 10 or 12 carbon atoms.[26]

A number of specific ester derivatives of note have been prepared, for example, ortho-pivaloate **10** from D-mannitol and the 1,2,4-orthoester 3-

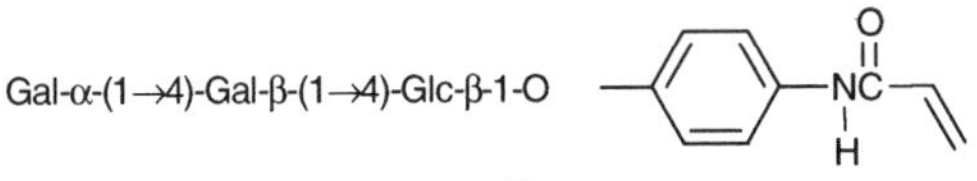

Gal-α-(1→4)-Gal-β-(1→4)-Glc-β-1-O

7

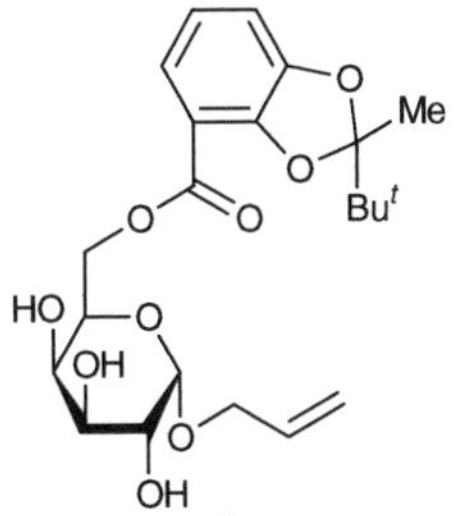

8

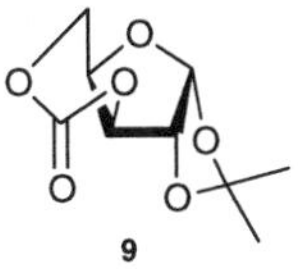

9

pivaloate from erythritol using HF-supported catalyst.[27] Additionally, the ferrocene-containing compounds **11** were synthesized from the corresponding acids and tetra-*O*-acetylglucosyl bromide.[28] The synthesis of thiocyanate **13**, which exists in equilibrium with cyclic thiocarbamate **12**, to an extent

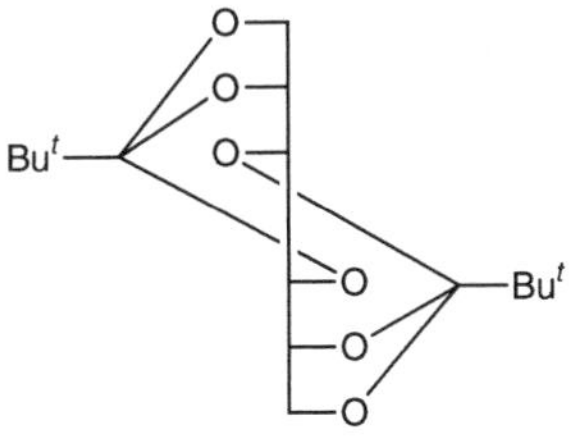

10

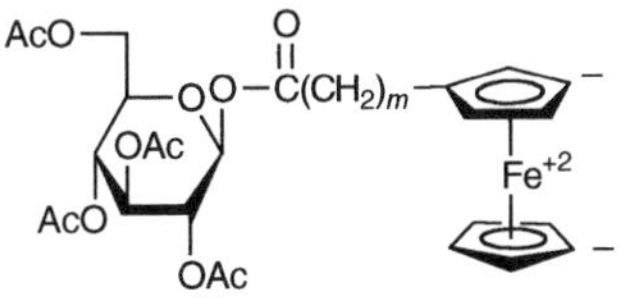

11 $m = 1, 3$

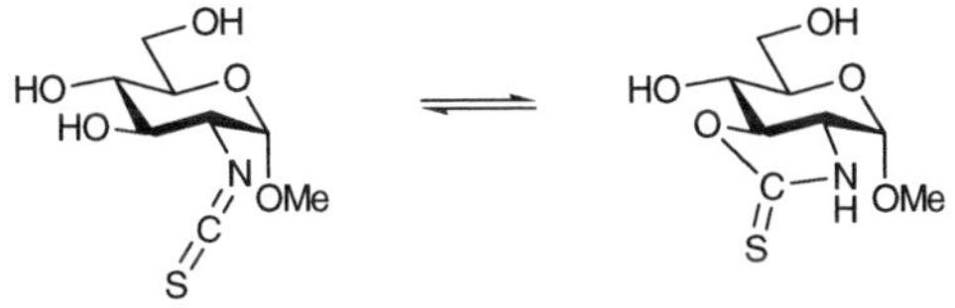

13 **12**

dependent on the temperature, has also been accomplished.[29] Two interesting conversions involve the formation of the trehalose-spanning porphyrin derivative **14** (Scheme 4) which is formed with its isomer in which the sugar-bound aromatic rings are substituted opposite to each other on the porphyrin.[30] An unusual rearrangement involving participation of a dioxolane ring oxygen atom in the displacement of a triflate ester to produce the unusual 1-benzoates **15** of a 2,5-anhydro-3-thio-L-mannose derivative is indicated in Scheme 5.[31]

Peracetylated sugars, on treatment with a mixture of ethylenediamine and acetic acid (1:1.2) in THF, gave products formed by selective deacetylation at the anomeric position. Yields ranged from 86 to 100% for nine sugars.[32] The issue of transfer of acyl protecting groups to the acceptor alcohols and others derived from the donors, which can lead to by-products during glycosylation

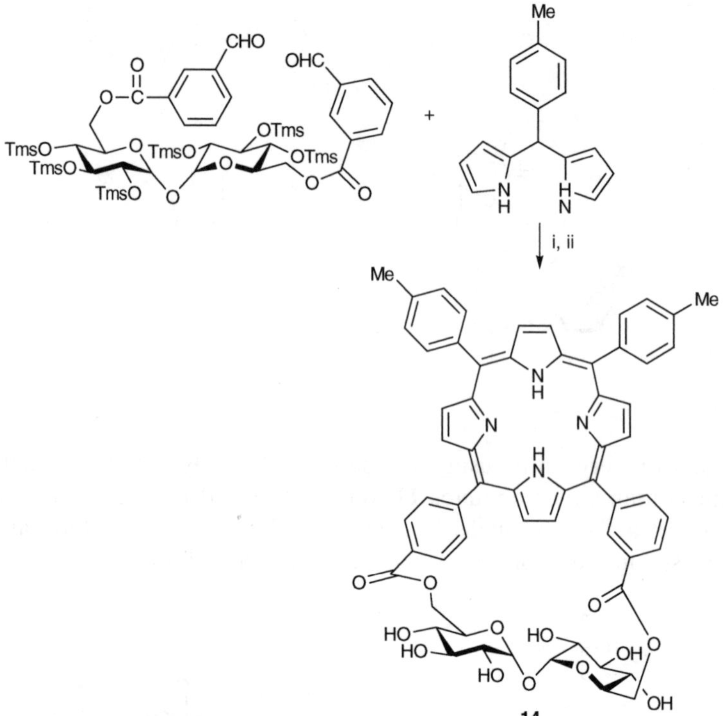

Reagents: i, BF_3, Et_2O; ii, Bu_4NF

Scheme 4

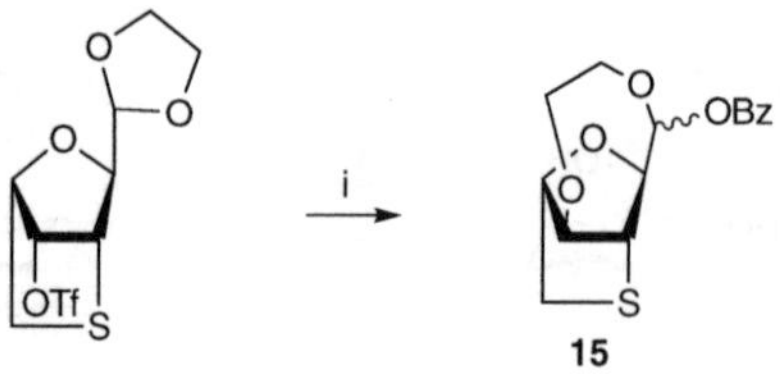

Reagents: i, LiOBz, DMF

Scheme 5

reactions with acylated thioglycosides, has been addressed. The side-reactions can be suppressed by use of pivaloyl-protected donors.[33] During the synthesis of verbascoside, the removal of phenoxyacetyl groups in the presence of a caffeoyl moiety was accomplished with 0.001M K_2CO_3 in $MeOH/CH_2Cl_2$.[34]

1.1.2 Enzymic Acylation and Deacylation. Numerous papers covering enzyme-catalysed acylations have appeared this year and a review on recent developments was published.[35] A variety of sugars, including D-glucose, -galactose, -mannose and -fructose, were 6-*O*-acylated with methyl or vinyl esters of various acids using a lipase from *Candida antarctica B*. The acids used included acetic, caprylic, capric, lauric, myristic, palmitic, stearic, 12-*R*-hydroxystearic, oleic and evucic.[36] Similarly, the lipase-catalysed synthesis of glucose 6-

palmitate was performed, also using the same enzyme,[37] and this same lipase, along with that of *Mucor milhei*, was used in the synthesis of 6-*O*-esters of glycopyranoses from the free sugar and carboxylic acids.[38] These same two lipases were also used in the regioselective acylation of D-fructose with various fatty acids.[39] Acetylation of 4-*O*-isobutyryl-D-galactal occurred selectively at O-6 in 57% yield by use of lipase PS and vinyl acetate, and in the same work selective deesterification at O-6 of methyl 2-azido-2-deoxy-3,4,6-tri-*O*-isobutyryl-D-galactopyranoside (70%) and ethyl 2-azido-2-deoxy-3,4,6-tri-*O*-isobutyryl-1-thio-β-D-glucopyranoside (59%) were reported.[40]

Immobilized lipase from *Candida* sp.1619 was used to catalyse the esterification of saccharides and alditols with fatty acids in *tert*-butanol,[41] acylation of mono- and di-saccharide derivatives with 2-bromomyristic acid was accomplished with *Candida antarctica* and *Mucor milhei* enzymes,[42] and an additional report on the regioselective enzymic introduction of fatty acids into saccharides to give esters for use as surfactants has also been reported.[43] In the esterification of xylitol with oleic acid, mostly tri- and tetra-esters were obtained using lipase from *Candida cylindracea.*[44] Ester **16** was prepared by enzymatic transesterification from an alkyl lactate to butyl α-D-glucopyranoside.[45]

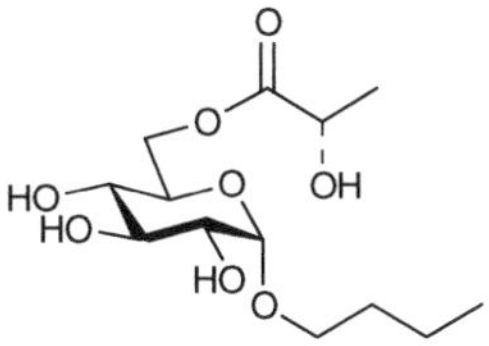

16

Enzymatic acylation of 2-amino-9-β-D-arabinofuranosyl-6-methoxy-9*H*-purine (506U78) to give the 5′-acetate was accomplished using Norozyme-435, which is an immobilized preparation derived from *Candida antarctica.* This provided a more soluble and bioavailable version of the anti-leukemic agent.[46] Norozyme-435 was also used in the selective acylation at the primary positions of sugars of the sophorolipid microbial glycolipids using vinyl acetate or acrylate.[47] Other enzymic esterifications of sugars to have been reported include the synthesis of 6,6′-diesters of sucrose,[48] of 6′- and 3-esters of 2-*O*-β-D-glucopyranosylglycerol,[49] of ^{14}C- and ^{3}H- labelled 1-*O*-valproyl-β-D-glucopyranuronic acid,[50] and the butanoates of swainsonine as shown in Scheme 6. In

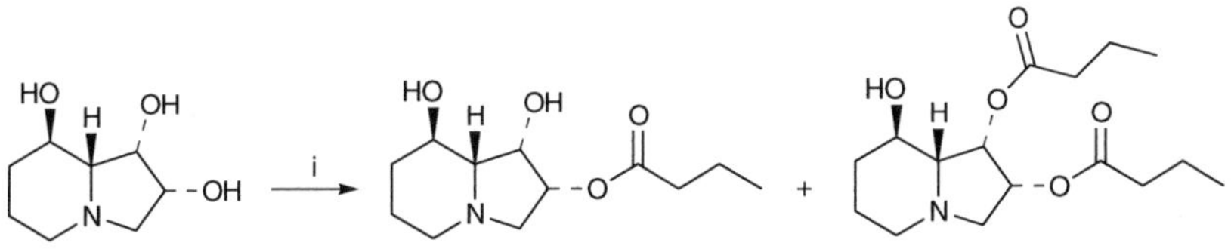

17 **18**

Reagents: i, enzyme, pyridine, trichloroethyl butanoate

Scheme 6

the last report three different enzymes were used with subtilisin Carlsberg and trichloroethyl butanoate giving **17** as the only product in 25% yield. A porcine pancreatic lipase produced derivative **18** in 31% yield with 6% of **17**.[51]

Several deacetylations have been reported with varying degrees of success (see also ref. 40). *Candida* lipase deacetylation of α-glucose pentaacetate gave 1,2,3,6-tetra-*O*-acetyl-α-D-glucose,[52] and Nicolosi *et al.* published an account of the selective deacetylation of pentaacetyl-β-D-glucosamine (**19**). By use of Novozyme **19** was converted to triol **20** in 95% yield over 10 days, and this was then converted to the diol **21** by use of vinyl acetate and Novozyme in 95% yield.[53] The resolution of racemic conduritol B has been accomplished *via* its tetraacetate as shown in Scheme 7.[54]

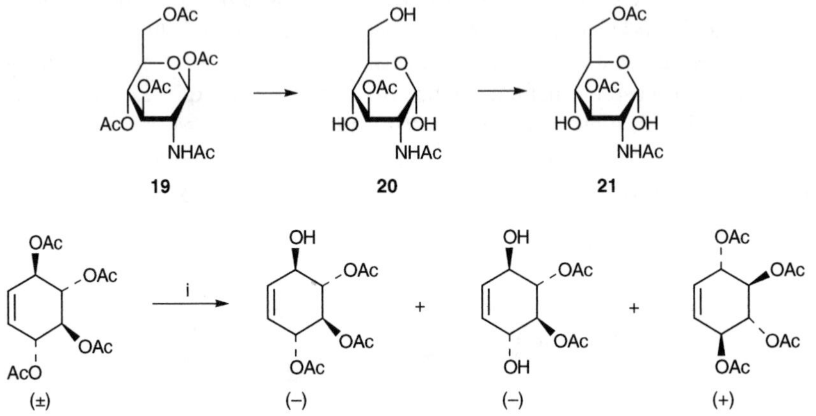

Reagents: i, Lipozyme, Bu^{t}OMe, Bu^{n}OH

Scheme 7

1.2 Natural Products. – Five cyclolanostanol glycosides were isolated from the underground components of *Cimicifuga simplex* which are used in traditional Chinese medicine. The glycosides all have the 2-*O*-malonyl-β-D-xylopyranosyl carbohydrate substituent **22**, with the malonate ester group being cleaved when the natural product is allowed to stand in methanol solutions at room temperature.[55] Potent immunosuppressive activity ($IC_{50} \sim 4–10 \times 10^{-5}$M) was observed with four acylated flavonol glycosides **23** isolated from *Persicaria lapathifolia*.[56]

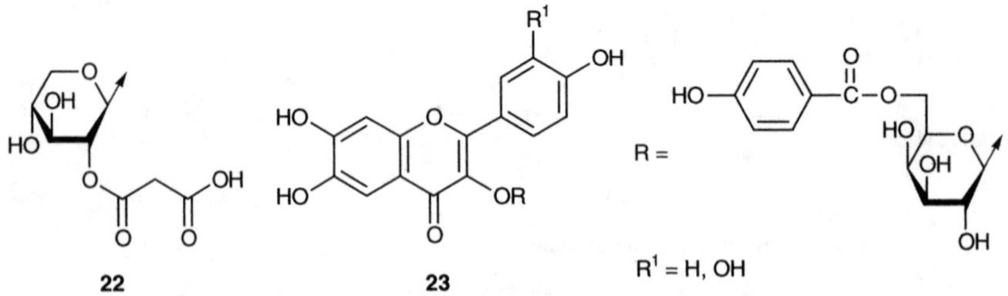

R^1 = H, OH

Two reports in the area of ellagitannin chemistry were published The derivative **24** was synthesized by (*S*)-atropodiastereoselective esterification of a glucose derivative with hexabenzyloxy diphenic acid (Scheme 8),[57] while the first synthesis of tellimagandin II was accomplished.[58]

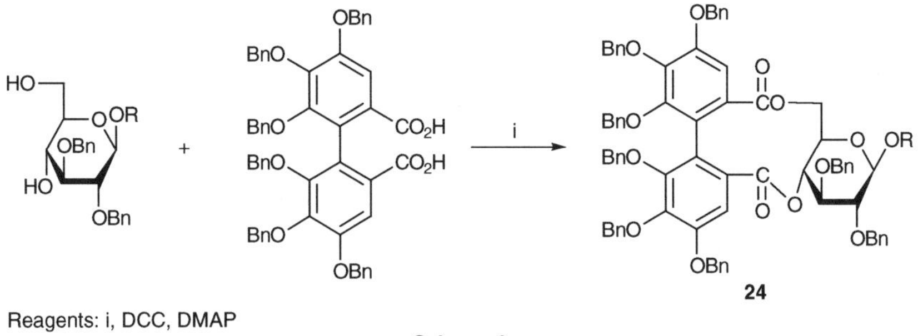

Reagents: i, DCC, DMAP

Scheme 8

2 Phosphates and Related Esters

A kinetic study on the hydrolysis of adenosine 2′,3′-cyclic monophosphate (cAMP) catalysed by Cu(II) terpyridyl concluded that the copper acts as a nucleophilic catalyst. This contrasts with its behaviour as a basic catalyst in the transesterification of RNA.[59] A mini-review of potent sialyl transferase inhibitors was presented which included six examples, the most potent being the neuraminic acid/uridine-5′-phosphate-based **25**.[60]

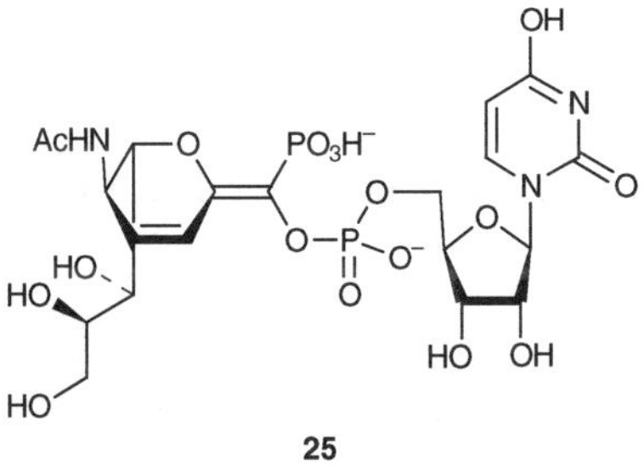

25

A synthesis of per-phosphorylated cellobiose and maltotriose glycosides linked *via* a spacer to the heparin pentasaccharide was achieved, the disaccharide components corresponding to the thrombin binding domain.[61] Phosphate and phosphonate analogues **26** of SiaLe^x were also prepared *via* the aldehyde derivatives indicated in Scheme 9. These condensations were accomplished by use of dihydroxyacetone phosphate or the corresponding phosphonate and various aldolases such as RAMA, RhaA or FucA.[62]

Synthesis of methyl 4,6-*O*-benzylidene-α-D-glucopyranoside 2,3-cyclic phosphite ethyl ester was performed by reaction of ethyl dichlorophosphite with the diol.[63] Preparation of the furanoside derivative **27**, has also been reported.[64]

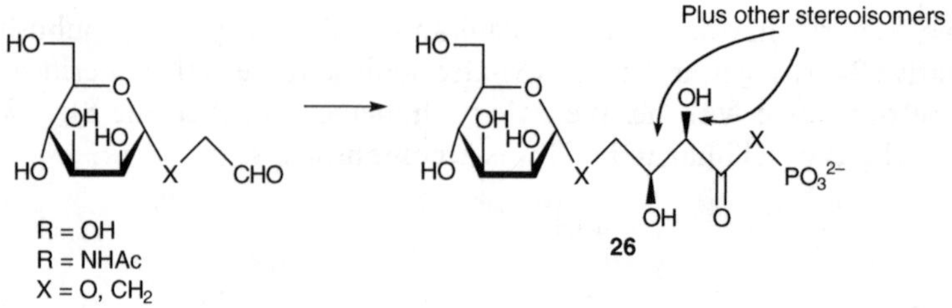

HO– OH HO HO O X CHO

R = OH
R = NHAc
X = O, CH2

HO– OH HO HO O X OH OH O X PO_3^{2-}

Plus other stereoisomers

26

Scheme 9

The glycosyl phosphate **28** was prepared by substitution of the glycosyl nitrate group (Scheme 10) and an L-fucopyranosyl and other analogues were also reported.[65]

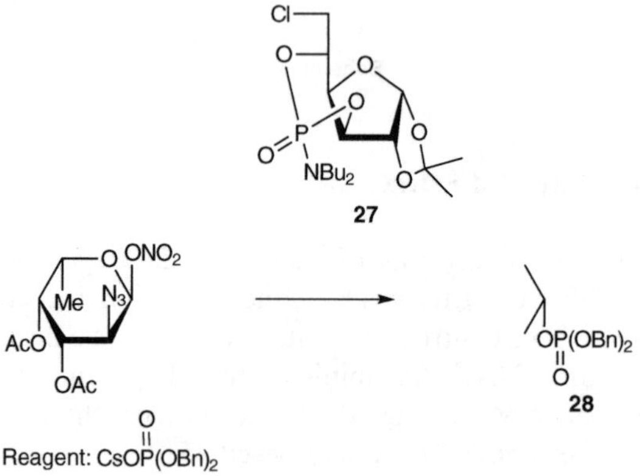

Cl O O O O O=P NBu2

27

O ONO2 Me N3 AcO OAc

OP(OBn)2 O

28

Reagent: $CsOP(O)(OBn)_2$

Scheme 10

A new method for the preparation of glycosyl phosphodiester links *via* hydrogen phosphonate esters is illustrated in Scheme 11.[66] Glycosyl donors with EtS, Cl, and Br leaving groups and other activators were also examined in the preparation of **29**.

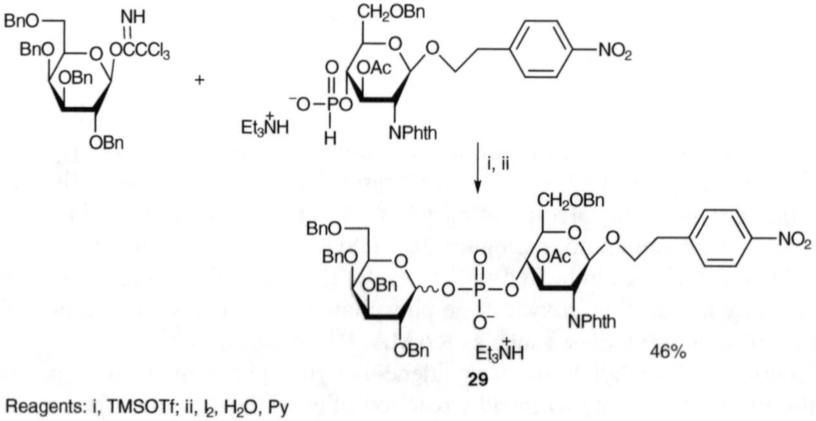

BnO– BnO O OCCCl3 NH OBn OBn

\+

CH2OBn O O OAc NPhth –NO2 $^-$O–PO O H $Et_3\overset{+}{N}H$

i, ii

BnO– BnO O OBn OBn O–P–O O O$^-$ CH2OBn O O OAc NPhth –NO2 $Et_3\overset{+}{N}H$ 46%

29

Reagents: i, TMSOTf; ii, I_2, H_2O, Py

Scheme 11

D-Ribose 2-phosphate, 3-phosphate, 4-phosphate, 2,3-bis-phosphate, 3,4-bis-phosphate, 2,4-bis-phosphate, cyclic 2,3-phosphate, cyclic 3,4-phosphate, and cyclic 2,4-phosphate were synthesized and characterized.[67] The synthesis of the phosphodisaccharide repeating unit [β-D-Gal-(1→4)-α-D-Man-1-P] of the antigenic lipophosphoglycan of *Leishmania denovani* parasite was also accomplished.[68] In this example, treatment of the galactosylmannose hepta-acetate with BuLi and $(PhO)_2POCl$ followed by hydrogenation with Adam's catalyst provided the target compound. The phosphonomethyl isoester **31** of D-*glycero*-tetrulose 1-phosphate **32** was synthesized for use in the investigation of the slow-binding inhibition of rabbit muscle aldolase (Scheme 12). Oxidation of **30** followed by hydrogenation and treatment with $Ba(OH)_2$ provided **31** in good yields.[69]

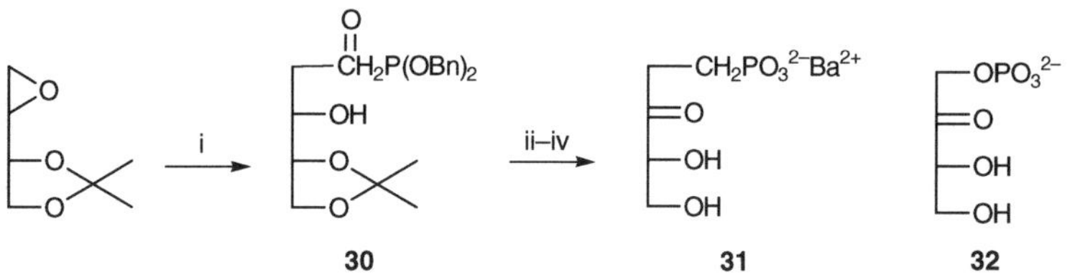

Reagents: i, Me(O)P(OBn)$_2$, BuLi, BF$_3$OEt$_2$; ii, DCC,DMSO, iii, H$_2$, Pd/C; iv, Ba(OH)$_2$;

Scheme 12

S-Glycosyl *O,O*-diethyl phosphorodithioate derivatives of L-arabino-, D-ribo- and D-xylo-furanose were prepared by treatment of the reducing monosaccharides with diphenyl phosphorochloridate and diethyl phosphorodithioate under phase transfer conditions.[70] By use of phosphorodiamidite methodology, derivative **33** was made from the 6-hydroxy glycoside, converted to the methacrylamido compound **34** and polymerized to give the poly-methacrylamido derivative of the 6-phosphocholine-α-D-glucopyranoside.[71] Similar methodology was used in the formation of 2-*O*-hydroxyethyl 2-deoxy-α-D-*threo*-pentopyranoside trisphosphate, which is an analogue of *myo*-inositol 1,4,5-trisphophate.[72]

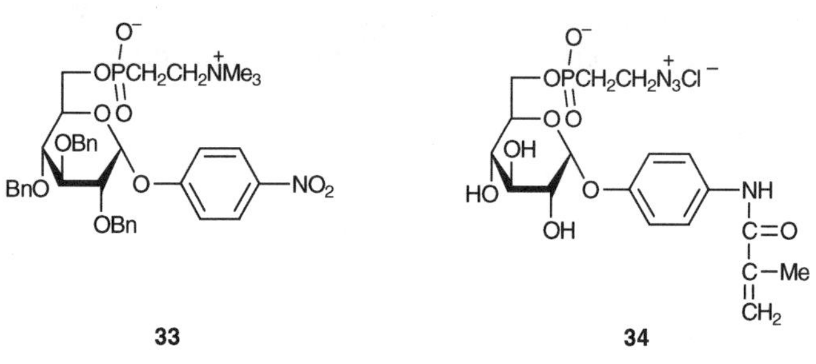

Ab initio calculations have been used to model the conformation of glycosyl phosphates about the C-1–O–P–O bond.[73] These led to better understanding of the interactions with metal ions and the catalytic mechanism of glycosyl transferase.

3 Sulfates and Related Esters

Lactose 3-*O*-sulfate, which has not previously been isolated from natural sources, was obtained from the milk of beagle dogs.[74] Recent reviews on sulfate and sulfonate esters describe the synthesis and biological activities of sulfated oligosaccharides as mimics of $SiaLe^A$ and $SiALe^X$,[75] and the $S_NP(V)$ reaction of dialkyl chlorophosphates with lithiated methyl sulfones and sulfonates.[76] An example of the latter involves conversion of the 3-*O*-mesylated allose derivative **35** to the Horner-Wadsworth-Emmons reagent followed by reaction with a carbohydrate-derived aldehyde to give **36**, thus linking the two monosaccharide units *via* a sulfonate bridge (Scheme 13).

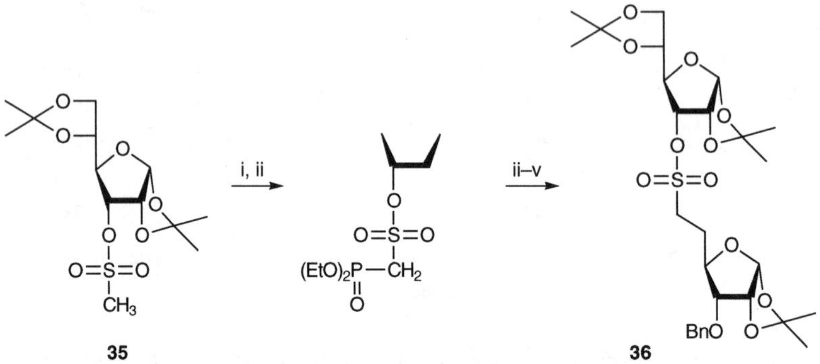

Reagents: i, $KN(SiMe_3)_2$, –78 °C; ii, $(EtO)_2P(=O)Cl$; iii, *n*-BuLi, –78 °C; iv, sugar aldehyde; v, $NaBH_4$

Scheme 13

Sulfated laminaro oligosaccharides, ranging from the tetra- to hexa-ose, were prepared with the degree of sulfation ranging from 0.7 to 2.8. The products were tested as anti-HIV agents, and highly sulfated compounds and those with hydrophobic aglycons were the most active.[77] With the known propensity of heparin to bind to platelets and with a disaccharide unit being responsible for this binding activity, compound **37** was synthesized and subjected to binding studies.[78] Oligomeric clusters of **37**, containing 2–3 units, were also tested and shown to be more active then **37**; however, heparin itself was found to be even more active. The first total synthesis of 6-sulfo-de-*N*-acetyl-$SiaLe^X$ ganglioside **38** was accomplished,[79] and a combinatorial approach towards the synthesis of sulfated GAG disaccharides using **39** as a key intermediate was described.[80] During this work it was found that sulfate groups are more stable than previously assumed and can be introduced prior to the final steps. Several examples of other sulfated oligosaccharides are noted in Chapter 4. The synthesis of sulfates **40** as new surfactants was achieved by treatment of methyl α-D-glucopyranoside with $SOCl_2$, then oxidation of the cyclic sulfite with $RuCl_3$/$NaIO_4$ and ester ring-opening with $NH_2(CH_2)_nMe$.[81] Finally, a regioselective enzymic method for the sulfation of chitotriose has been developed by

C.-H. Wong, by use of 3-phosphoadenosine-5-phosphosulfate as source of sulfate (Scheme 14).[82]

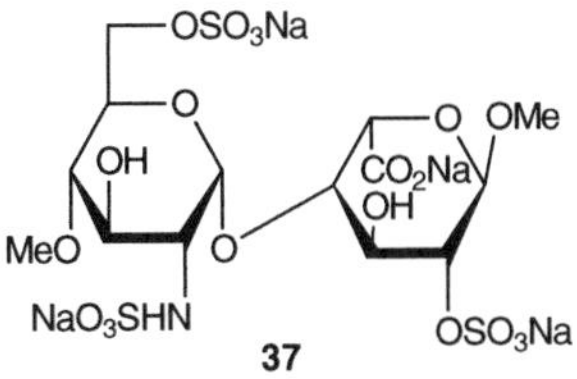

37

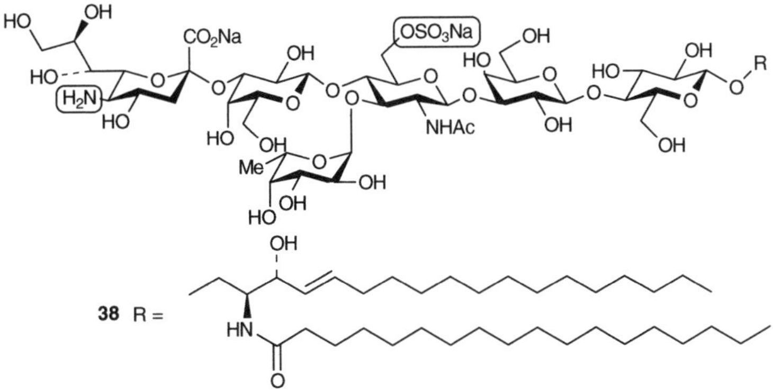

38 R =

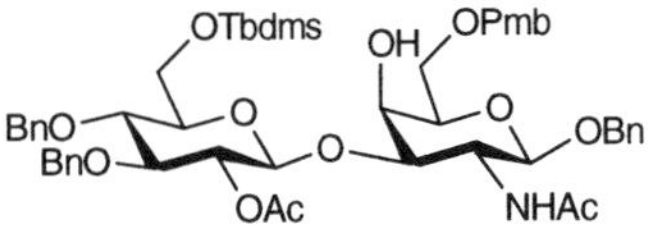

39

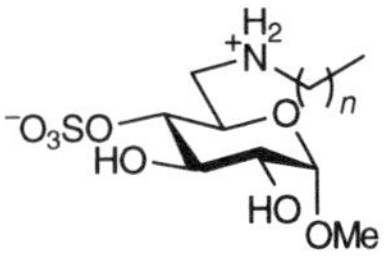

40 n = 5, 9, 13

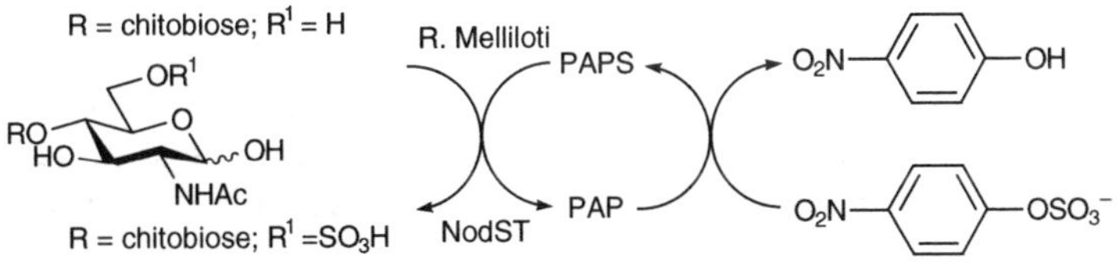

R = chitobiose; R^1 = H

R. Melliloti

PAPS

PAP

NodST

R = chitobiose; R^1 = SO_3H

Scheme 14

Reduction of 1,6:2,5-dianhydro-3,4-di-*O*-methanesulfonyl-1-thio-D-glucitol with LAH gave the corresponding 3,4-diol in 51% yield with six by-products which result from reactions involving a sulfonium ion intermediate the mechanism of the reactions of which was discussed.[83]

4 Other Esters

Several boronic acids have been studied for their ability to complex with sugars. The bis-boronic acids **41** and **42** complex with glucose to give 1,2:3,5-bisesters. Interestingly, **41** is selective for glucose relative to fructose or

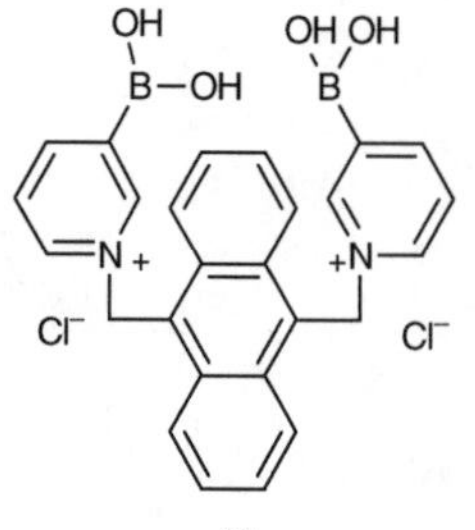

41 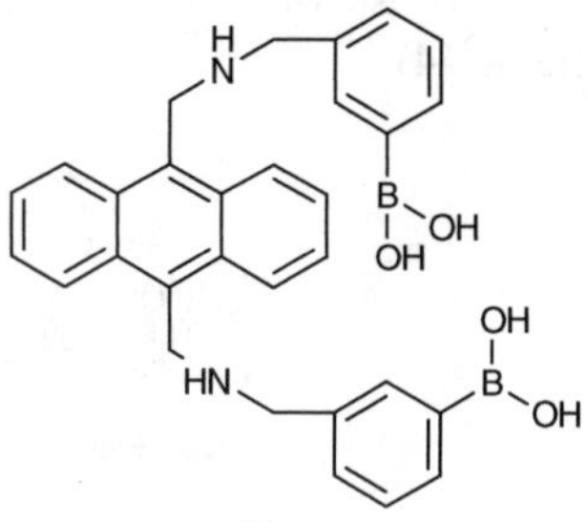 **42**

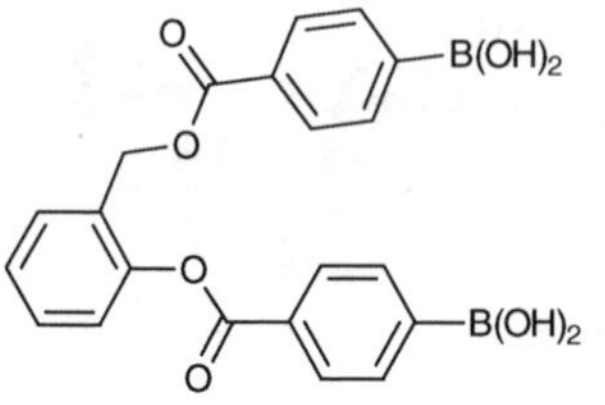

43 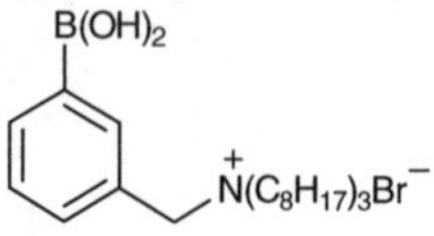 **44**

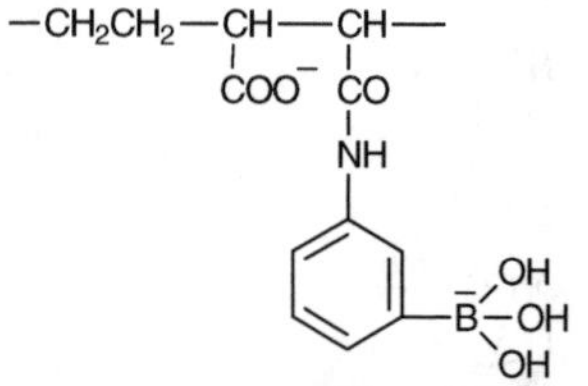

45

galactose and shows increased fluorescence when complexed.[84,85] The ability of boronic acids **43** and **44** to transport fructose selectively through a lipid membrane as the boronate complexes has been studied.[86] Transport with **43** was slower than with mono-boronic acids, but the selectivity for fructose over glucose increased. Transport with **44** was very fast, but the loss of **44** into the aqueous phase made the membrane unstable over time. The polymeric

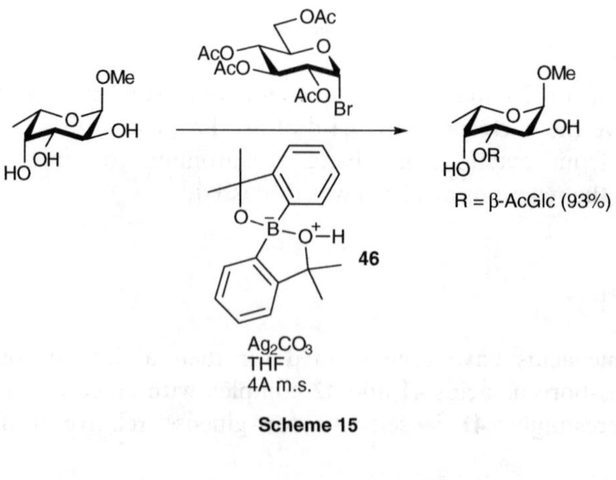

R = β-AcGlc (93%)

46

Ag_2CO_3
THF
4A m.s.

Scheme 15

polyamide **45** complexed adenosine 5′-phosphate which, upon removal of the AMP, created a cleft that recognizes and binds AMP with high affinity.[87]

Complex phenylborinic acids, such as **46**, react with diols of trihydroxy-glycosides in a manner that causes selective O-activation of one of the esterified alcohol groups. Thus glycosylation of the complex derived from methyl α-L-fucopyranoside gives 3-linked disaccharides, whereas simple boronic acids protect the *cis*-3,4-diol and lead to the formation of 2-linked products as shown in Scheme 15.[86]

References

1 K.-I. Oyama and T. Kondo, *Synlett*, 1999, 1627.
2 P.M. Bhaskar and D. Loganathan, *Synlett*, 1999, 129.
3 M. Kim, B. Grzeszezyk and A. Zamojski, *Carbohydr. Res.*, 1999, **320**, 244.
4 S.M. Andersen, I. Lundt, J. Marcussen and S. Yu, *Carbohydr. Res.*, 1999, **320**, 250.
5 P. Pouillart, O. Douillet, B. Scappini, A. Gozzini, V. Santini, A. Grossi, G. Pagliai, P. Stripoli, L. Rigacci and G. Ronco, *Eur. J. Pharm. Sci.*, 1999, **7**, 93 (*Chem. Abstr.*, 1999, **130**, 252 586).
6 B. Danieli, F. Peri, G. Roda, G. Carrea and S. Riva, *Tetrahedron*, 1999, **55**, 2045.
7 K.S. Feldman, A. Sambandam, S.T. Lemon, R.B. Nicewonger, G.S. Long, D.F. Battaglia, S.M. Ensel and M.A. Laci, *Phytochemistry*, 1999, **51**, 867.
8 S. Deshays, M. Liagre, A. Loupy, J.-L. Luche and A. Petit, *Tetrahedron*, 1999, **55**, 10851.
9 J.J. Gridley, H.M.I. Osborn and W.G. Suthers, *Tetrahedron Lett.*, 1999, **40**, 6991.
10 Y. Fukase, K. Fukase and S. Kusumoto, *Tetrahedron Lett.*, 1999, **40**, 1169.
11 T. Kurahashi, T. Mizutani and J. Yoshida, *J. Chem. Soc., Perkin Trans. 1*, 1999, 465.
12 C. Prata, N. Mora, J.-M. Lacombe, J.-C. Maurizis and B. Pucci, *Carbohydr. Res.*, 1999, **321**, 4.
13 M.J.L. Castro, J. Koensky and A.F. Cirelli, *Tetrahedron*, 1999, **55**, 12711.
14 C. Kallus, T. Opatz, T. Wunberg, W. Schmidt, S. Henke and H. Kunz, *Tetrahedron Lett.*, 1999, **40**, 7783.
15 J. Désiré and J. Prandi, *Carbohydr. Res.*, 1999, **317**, 110.
16 G. Descotes, J. Gagnaire, A. Beuchu, S. Thevenet, N. Giry-Panaud, P. Salanski, S. Belniak, A. Wernicke, S. Porwanski and Y. Queneau, *Pol. J. Chem.*, 1999, **73**, 1069 (*Chem. Abstr.*, 1999, **131**, 144 750).
17 P. Potier, V. Maccario, M.-B. Giudicelli, Y. Queneau and O. Dangles, *Tetrahedron Lett.*, 1999, **40**, 3387.
18 D. Xie, D. Gao and J. Zheng, *Zhongguo Youzhi*, 1998, **23**, 60 (in Chinese) (*Chem. Abstr.*, 1999, **130**, 66 702).
19 N. Garti, V. Clement, M. Leser, A. Aserin and M. Fanun, *J. Mol. Liq.*, 1999, **80**, 253 (*Chem. Abstr.*, 1999, **131**, 185 136).
20 F.J. Plou, M.A. Cruces, E. Pastor, M. Ferrer, M. Bernabe and A. Ballesterose, *Biotechnol. Lett.*, 1999, **21**, 635 (*Chem. Abstr.*, 1999, **131**, 286 715).
21 S. Thevenet, A. Wernicke, S. Belniak, G. Descotes, A. Beuchu and Y. Queneau, *Carbohydr. Res.*, 1999, **318**, 52.
21a (a)A.K.M.S. Kabir, M.M. Matin and S. Majumder, *Chittagong Univ. Stud., Part*

II, 1997, **21**, 65 (*Chem. Abstr.*, 1999, **130**, 196 854; (b) A.K.M.S. Kabir, Md. Alauddin, M.M. Matin and S.C. Battacharjee, *Chittagong Univ. Stud., Part II*, 1997, **21**, 59 (*Chem. Abstr.*, 1999, **130**, 196 853; (c) A.K.M.S. Kabir, M.M. Matin and M.J.U. Bhuiyan, *Chittagong Univ. Stud., Part II*, 1997, **21**, 33 (*Chem. Abstr.*, 1999, **130**, 196 852.)

22 T. Ziegler, R. Dettmann and J. Grabowski, *Synthesis*, 1999, 1661.

23 G.S. Ghangas, *Phytochemistry*, 1999, **52**, 785.

24 H. Dehi, Y. Nishida, M. Mizuno, M. Shinkai, T. Kobayashi, T. Takeda, H. Uzawa and K. Kobayashi, *Bioorg. Med. Chem.*, 1999, **7**, 2053.

25 X. Chen and R.A. Gross, *Macromolecules*, 1999, **32**, 308.

26 C. Prata, N. Mora, A. Polidori, J.-M. Lacombe and B. Pucci, *Carbohydr. Res.*, 1999, **321**, 15.

27 H. Klein, R. Mietchen, H. Reinke and M. Michalik, *J. Prakt, Chem.*, 1999, **341**, 41 (*Chem. Abstr.*, 1999, **130**, 209 857.)

28 S. Fu, H. Liu and H. Zhang, *Huaxue Shiji*, 1998, **20**, 361 (*Chem. Abstr.*, 1999, **130**, 267 639).

29 J.G. Fernandez-Bolanos, E. Zafra, I. Robina and J. Fuentes, *Carbohyd. Lett.*, 1999, **3**, 239 (*Chem. Abstr.*, 1999, **130**, 311 999.)

30 E. Davoust, R. Granat, P. Krausz, V. Carré and M. Guilloton, *Tetrahedron Lett.*, 1999, **40**, 2513.

31 V. Popsavin, O. Berič, M. Popsavin, J. Csanádi, D. Vuji and R. Hrabal, *Tetrahedron Lett.*, 1999, **40**, 3629.

32 J. Zhang and P. Kováč, *J. Carbohydr. Chem.*, 1999, **18**, 461.

33 T Nukada, A. Berces and D.M. Whitfield, *J. Org. Chem.*, 1999, **64**, 9030.

34 H.J. Duynstee, M.C. de Koning, H. Ovaa, G.A. van der Marel and J.H. van Boom, *Tetrahedron: Asymmetry*, 1999, **10**, 2623.

35 Z.Y. Li and J.J. Liu, *Youji Huaxue*, 1999, **19**, 121 (*Chem. Abstr.*, 1999, **131**, 45 005).

36 B. Haase, G. Machmuller and M.P. Schneider, *Schriftenr. "Nachwachsende Rohst."*, 1998, **10**, 218 (*Chem. Abstr.*, 1999, **130**, 110 465).

37 L. Cao, U.T. Bornscheuer and R.D. Schmid, *J. Mol. Catal. B: Enzym.*, 1999, **6**, 279 (*Chem. Abstr.*, 1999, **130**, 237 748).

38 P. Degn, L.H. Pedersen, J.O. Duus and W. Zimmermann, *Biotechnol. Lett.*, 1999, **21**, 275 (*Chem. Abstr.*, 1999, **131**, 73 851).

39 V. Sereti, H. Stamatis and F.N. Kolisis, *Prog. Biotechnol.*, 1998, **15**, 725 (*Chem. Abstr.*, 1999, **130**, 237 747).

40 P. Pfau and H. Kunz, *Synlett*, 1999, 1817.

41 X. Kou and J. Xu, *Ann. N.Y. Acad. Sci.*, 1998, **864** (Enzyme Engineering XIV), 352 (*Chem. Abstr.*, 1999, **130**, 209 892).

42 C. Gao, M.J. Whitcombe and E.N. Vulfson, *Enzyme Microb. Technol.*, 1999, **25**, 264 (*Chem. Abstr.*, 1999, **131**, 299 606).

43 C. Gao, A. Millquist-Fureby, M.J. Whitcombe and E.N. Vulfson, *J. Surfactants Deterg.*, 1999, **2**, 293 (*Chem. Abstr.*, 1999, **131**, 199911).

44 C. Hatanaka, T. Haraguchi, S. Ide, M. Goto and K. Kumada, *Kitakyushu Kogyo Koto Senmon Gakko Kenkyu Hokoku* 1999, **32**, 127 (in Japanese) (*Chem. Abstr.*, 1999, **130**, 282263).

45 M.-P. Bousquet, R.-M. Willemot, P. Monsan and E. Boures, *Biotechnol. Bioeng.*, 1999, **62**, 225 (*Chem. Abstr.*, 1999, **130**, 139 516).

46 M. Mahmoudian, J. Eaddy and M. Dawson, *Biotechnol. Appl. Biochem.*, 1999, **29**, 229 (*Chem. Abstr.*, 1999, **131**, 185 188).

47 K.S. Bisht, R.A. Gross and D.L. Kaplan, *J. Org. Chem.*, 1999, **64**, 780.
48 A.F. Artamonov, M.T. Aldabergenova, F.S. Nigmatullina and B.Zh. Dzhiembaev, *Chem. Nat. Compd.*, 1999, **34**, 431 (*Chem. Abstr.*, 1999, **131**, 102 451).
49 D. Colombo, F. Compostella, F. Ronchetti, A. Scala, L. Toma, H. Tokuda and H. Nishino, *Bioorg. Med. Chem.*, 1999, **7**, 1867.
50 N. Yamamwa, S. Muramatsu, K. Suzuki, M. Uchiyama and E. Nakajima, *Radioisotopes*, 1999, **48**, 383 (*Chem. Abstr.*, 1999, **131**, 130 176).
51 G.G. Perrone, K.D. Barrow and I.J. McFarlane, *Bioorg. Med. Chem.*, 1999, **7**, 831.
52 A. Bastida, R. Fernández-Lafuente, G. Fernández-Lorente, J.M. Guisán, G. Pagani and M. Terreni, *Bioorg. Med. Chem. Lett.*, 1999, **9**, 633.
53 G. Nicolosi, C. Spatafora and C. Tringali, *Tetrahedron: Asymmetry*, 1999, **10**, 2891.
54 C. Sanfilippo, A. Patti and G. Nicolosi, *Tetrahedron: Asymmetry*, 1999, **10**, 3273.
55 A. Kusano, M. Shibano and G. Kusano, *Chem. Pharm. Bull.*, 1999, **47**, 1175.
56 S.-H. Park, S.R. Oh, K.Y. Jung, I.S. Lee, K.S. Ahn, J.H. Kim, Y.S. Kim, J.J. Lee and H.-K. Lee, *Chem. Pharm. Bull.*, 1999, **47**, 1484.
57 K. Khanbabaee and K. Lötzerich, *Eur. J. Org. Chem.*, 1999, 3079.
58 K.S. Feldman and K. Sahasrabudhe, *J. Org. Chem.*, 1999, **64**, 209.
59 L.A. Jenkins, J.K. Bashkin, J.D. Pennock, J. Florian and A. Warshel, *Inorg. Chem.*, 1999, **38**, 3215.
60 P.N. Schroder and A. Giannis, *Angew. Chem., Int. Ed. Engl.*, 1999, **38**, 1379.
61 R.C. Buijsman, J.E.M. Basten, C.M. Dreef-Tromp, G.A. van der Marel, C.A.A. van Boeckel and J.H. van Boom, *Bioorg. Med. Chem.*, 1999, **7**, 1881.
62 C.-C. Li, F. Moris-Varas, G. Weitz-Schmidt and C.-H. Wong, *Bioorg. Med. Chem.*, 1999, **7**, 425.
63 J.J. Hu, Y. Ju and Y.F. Zhao, *Chin. Chem. Lett.*, 1999, **10**, 457 (*Chem. Abstr.*, 1999, **131**, 286 708).
64 M.P. Koroteev, N.M. Pugashova, S.B. Khrebtova, E.E. Nifantyev, O.S. Zhuzova, T.P. Ivanova, N.M. Peretolchina and G.K. Gerasimova, *Bioorg. Khim.*, 1998, **24**, 58 (*Chem. Abstr.*, 1999, **130**, 352 471).
65 P.A. Illarionov, V.I. Torgov, I.C. Hancock and V.N. Shibaev, *Tetrahedron Lett.*, 1999, **40**, 4247.
66 P.J. Garegg, J. Hansson, A.-C. Helland and S. Oscarson, *Tetrahedron Lett.*, 1999, **40**, 3049.
67 S. Pitsch, C. Spinner, K. Atsumi and P. Ermert, *Chimia*, 1999, **53**, 291 (*Chem. Abstr.*, 1999, **131**, 130 177).
68 M. Upreti and R.A. Vishwakarma, *Tetrahedron Lett.*, 1999, **40**, 2619.
69 P. Page, C. Blanski and J. Périé, *Eur. J. Org. Chem.*, 1999, 2853.
70 J. Bogusiak, *Pol. J. Chem.*, 1999, **73**, 619 (*Chem. Abstr.*, 1999, **130**, 282 245).
71 Y. Nishida, Y. Takamori, K. Matsuda, H. Ohrui, T. Yamada and K. Kobayashi, *J. Carbohydr. Chem.*, 1999, **18**, 985.
72 F. Roussel, M. Hilly, F. Chrétien, J.-P. Mauger and Y. Chapleur, *J. Carbohydr. Chem.*, 1999, **18**, 697.
73 I. Ivaroska, I. Andre and J.P. Carver, *J. Phys, Chem. B*, 1999, **103**, 2560 (*Chem. Abstr.*, 1999, **131**, 19 192).
74 W.A. Bubb, T. Urashima, K. Kohso, T. Nakamura, I. Arai and T. Saito, *Carbohydr. Res.*, 1999, **318**, 123.
75 C. Auge, F. Dragon, R. Lemoine, Ch. Le Narvor and A. Lubineau, *Ann. Pharm. Fr.*, 1999, **57**, 216 (*Chem. Abstr.*, 1999, **131**, 170 532).

76 F. Eymery, B. Iorga and P. Savignac, *Tetrahedron*, 1999, **55**, 13109.
77 K. Katsuraya, H. Nakashima, N. Yamamoto and T. Uryu, *Carbohydr. Res.*, 1999, **315**, 234.
78 S. Koshida, Y. Suda, Y. Fukui, J. Ormsby, M. Sobel and S. Kusumoto, *Tetrahedron Lett.*, 1999, **40**, 5725.
79 S. Komba, C. Galustian, H. Ishida, T. Feizi, R. Kannagi and M. Kiso, *Angew. Chem., Int. Ed. Engl.*, 1999, **38**, 1131.
80 A. Lubineau and D. Bonnaffé, *Eur. J. Org. Chem.*, 1999, 2523.
81 H.G. Bazin and R.J. Linhardt, *Synthesis*, 1999, 621.
83 E. Bozó, S. Boros, J. Kuszmann and E. Gács-Baitz, *Tetrahedron*, 1999, **55**, 8095.
82 M.D. Burkart, M. Izumi and C.-H. Wong, *Angew. Chem., Int. Ed. Engl.*, 1999, **38**, 2747.
84 H. Eggert, J. Frederiksen, C. Morin and J.C. Norrild, *J. Org. Chem.*, 1999, **64**, 3846.
85 M. Bielecki, H. Eggert and J.C. Norrild, *J. Chem. Soc., Perkin Trans. 2*, 1999, 449.
86 S.J. Gardiner, B.D. Smith, P.J. Duggan, M.J. Karpa and G.J. Griffin, *Tetrahedron*, 1999, **55**, 2857.
87 Y. Kanekiyo, Y. Ono, K. Inoue, M. Sano and S. Shinkai, *J. Chem. Soc., Perkin Trans. 2*, 1999, 557.
88 K. Oshima and Y. Aoyama, *J. Am. Chem. Soc.*, 1999, **121**, 2315.

8

Halogeno-sugars

1 Fluoro-sugars

New aspects of the synthesis of fluorinated carbohydrates have been reviewed with particular emphasis on applications of glycosyl fluorides as donors in di- and oligosaccharide synthesis.[1] Phenyl 1-selenoglycosides have been converted into glycosyl fluorides using halonium ion activation,[2] whereas thioglycoside **1** on exposure to DAST has afforded glycosyl fluoride **2**, presumably *via* a 1,2-sulfonium intermediate.[3] Treatment of glycals with SiF_4/1,3-dibromo-5,5-dimethylhydantoin gave 2-bromo-glycosyl fluorides which were readily converted into the corresponding 2-deoxy-glycosyl fluorides. Similarly, the combination of SiF_4/phenyliodonium triacetate has produced the 2-hydroxy-1-fluorination products, *i.e.* glycosyl fluorides from glycals.[4]

A review on recent advances in electrophilic fluorination has many examples of the synthesis of fluorinated derivatives of carbohydrates.[5] The electrophilic fluorination of glycals using 'Selectfluor' has been further studied,[6] and mechanistic studies have allowed optimization of conditions for the coupling of secondary alcohols of carbohydrates, amino acids, phosphates and phosphonates at C-1 with fluorination at C-2 during selectfluor activation of glycals.[7] Similar activation of glycal derivatives has featured in the synthesis of 2-deoxy-2-fluoroglycosides of hydroxylated amino acids.[8]

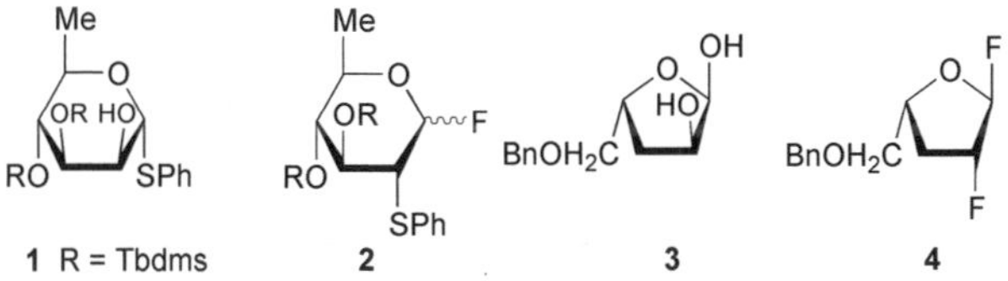

Me O OR HO RO SPh
1 R = Tbdms

Me O OR F RO SPh
2

O OH HO BnOH$_2$C
3

O F BnOH$_2$C F
4

The use of DAST and CF_3SiMe_3 in the synthesis of fluorinated carbohydrate moieties has been reviewed,[9] and bis(2-methoxyethyl)aminosulfur trifluoride has been reported to be more thermally stable than its analogue DAST and to have similar fluorinating reactivity.[10] DAST treatment of diol **3** has produced the 1,2-difluoro compound **4** which was converted into 2′,3′-dideoxy-2′-fluoro-L-*threo*-pentofuranosyl nucleosides.[11] 2′,3′-Dideoxy-3′,3′-difluoro- and 2′,3′-dideoxy-2′,2′-difluoropyranosyl analogues of gemcitabine have been synthesized. Key steps were formation of the difluoromethylene

Carbohydrate Chemistry, Volume 33

moieties by reaction of appropriately protected uloses with DAST.[12] Some 3-deoxy-3-*C*-methyl-3-nitro-hexopyranosides on exposure to DAST have afforded only the products of replacement of the primary hydroxy groups with fluorine.[13] Similar treatment of pentofuranosides **5** (either anomer) has produced the 3-deoxy-3-fluoro compounds **6** as the major products[14] and UDP-6-deoxy-6-fluoro-galactose has been prepared by way of DAST treatment of a suitable galactopyranoside derivative.[15]

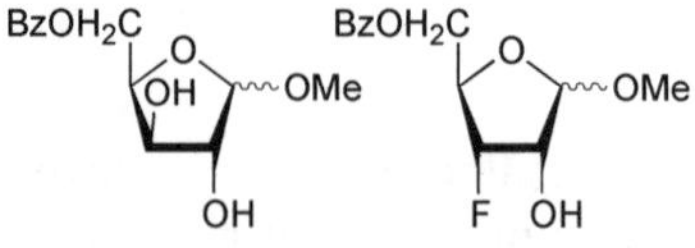

BzOH$_2$C, OH, OMe, OH, **5**; BzOH$_2$C, OMe, F, OH, **6**

The first 'per-fluorinated' carbohydrate has been synthesized in 20% yield (Scheme 1). The compound was characterized by ^{19}F NMR, MS and elemental analysis.[16] A synthesis of 2,6-dideoxy-2,6,6,6-tetrafluoro-L-talopyranose has been described and its daunomycinone glycoside prepared.[17] A mixture of 5-*C*-trifluoromethyl-D-glucuronolactone and -L-iduronolactone has been synthesized. A key step was the radical addition of the trifluoromethyl moiety to a ketene dithioacetal derived from 1,2-*O*-isopropylidene-α-D-*xylo*-1,5-pentodialdofuranose.[18] An efficient synthesis of 2,6-dideoxy-6-fluoro-D-*arabino*-hexopyranose from D-glucose has provided ready access to gram quantities.[19] Fluoride ion displacement of 5,6-*O*-cyclic sulfates of some hexofuranose derivatives has generated the corresponding 6-deoxy-6-fluoro derivatives,[20] and a similar displacement of a 2-*O*-imidazolylsulfonate derivative of L-ribose was utilized in a practical synthesis of the nucleoside L-FMAU.[21] The synthesis of an ethyl 4-deoxy-4-fluoro-1-thio-α-L-rhamnopyranose derivative from L-rhamnose or L-fucose as a glycosyl donor was achieved, but it proved non-trivial and a number of routes were investigated.[22]

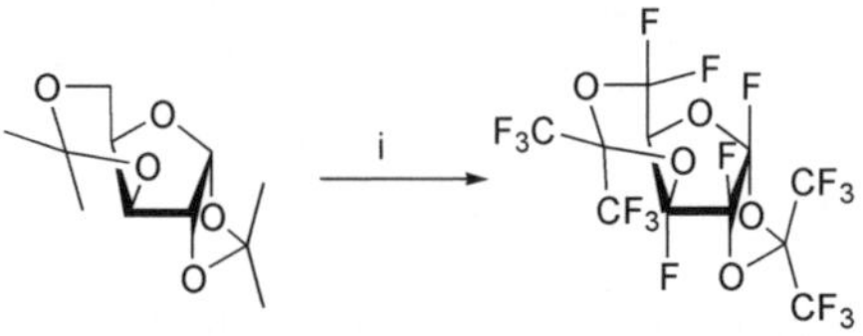

Reagents: i, F$_2$/He, –90 °C to RT

Scheme 1

A total synthesis of ethyl 2,6-dideoxy-6,6,6-trifluoro-β-D- and -L-*arabino*-hexopyranosides provided each enantiomer separately[23] while 1-fluoro and 1,1-difluoro analogues of 1-deoxyxylulose have also been prepared by total synthesis.[24] The synthesis of some fluorinated inositols is covered in Chapter 18 and several deoxyfluoro analogues of ulosonic acids are mentioned in Chapter 16.

The optimum conditions for alkaline hydrolysis of 1,3,4,6-tetra-*O*-acetyl-2-

deoxy-2-[^{18}F]fluoro-D-glucose that minimizes the amount of C-2 epimerization have been found,[25] while an attempt to prepare 2-deoxy-2-[^{18}F]-D-glucose by fluoride ion displacement of 2-*O*-methanesulfonyl-D-mannose afforded only D-glucopyranosyl fluoride, presumably *via* the 1,2-epoxide.[26] 1,3,4,6-Tetra-*O*-acetyl-2-deoxy-2-[^{19}F]fluoro-β-D-glucopyranose has been prepared in order to test the epimeric purity at C-2 (by ^{19}F NMR) of 2-deoxy-2-[^{18}F]-D-glucose prepared using $^{18}F^-$ in a 'no-carrier-added' synthesis.[27] 2-Deoxy-2,2-[^{18}F]di-fluoro-D-glucose has been prepared by way of reaction of acetyl hypofluorite with tri-*O*-acetyl-2-fluoro-D-glucal.[28]

2 Chloro-, Bromo- and Iodo-sugars

The use of $(Me_3Si)_2/I_2$ for the transformation of peracetylated mono- and di-saccharides into their respective acetylated glycosyl iodides has been described[29] and perbenzylated glycosyl phosphites have been converted into their respective glycosyl iodides with 2,6-di-*tert*-butylpyridinium iodide.[30] New electrophilic reagents [R_4NBr (or R_4NI)/$PhI(OAc)_2$/$TMSN_3$] for the 1,2-functionalization of glycals have allowed the synthesis of 2-bromo- or 2-iodo-glycosyl azides or acetates with generally good selectivity,[31,32] and the combination of CAN/NaI/HOAc provides 2-deoxy-2-iodoglycosyl acetates from glycals with very good selectivity.[33]

A three-step synthesis of 6-deoxy-6-iodo-D-fructose (**9**) from D-fructose features the iodination of pyranose **7** to give the acyclic hexos-2-ulose derivative **8** (Scheme 2).[34] A reexamination of the reaction of 6-*O*-acetyl-sucrose with sulfuryl chloride has established that the main product after dechlorosulfation and acetylation was **10** and not **11** as previously reported. A mechanism for the chlorination at C-4′ was proposed.[35] 1,1,2,2-Tetraphenyldisilane in ethanol has proved effective for the radical reductive dehalogenation of some carbohydrate bromides.[36] The conversion of a nucleoside 5′-aldehyde into a 6′-bromo-5′,6′-unsaturated derivative is discussed in Chapter 20.

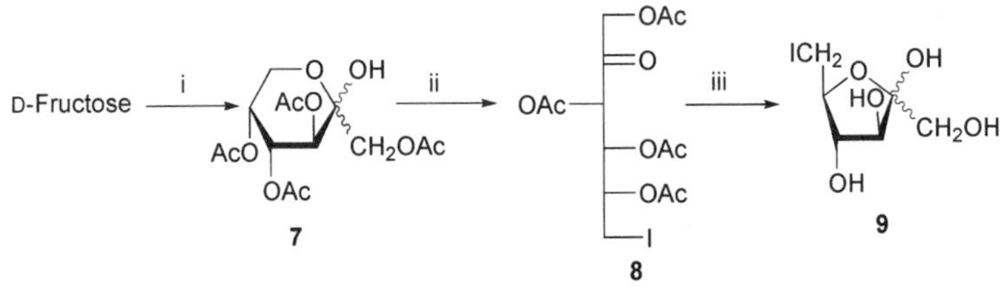

Reagents: i, Ac_2O, $ZnCl_2$; ii, $ClPPh_2$, imidazole, I_2; iii, $Ba(OH)_2$

Scheme 2

The radical bromination of some 2,5-anhydroaldonic acid derivatives such as **12** and **15** has afforded α-bromo-adducts **13** and **16** from which the corresponding azides **14** and **17** were prepared (Scheme 3). These served as precursors to α-amino acids with a rigid carbohydrate scaffold.[37]

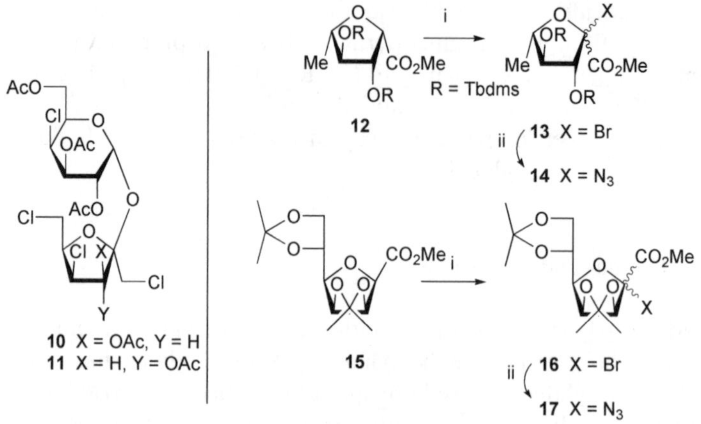

Reagents: i, NBS/Bz_2O_2/CCl_4 or CH_3CCl_3, 75 °C; ii, NaN_3, DMF

Scheme 3

References

1 K. Dax, M. Albert, J. Ortner and B.J. Paul, *Curr. Org. Chem.*, 1999, **3**, 287 (*Chem. Abstr.*, 1999, **131**, 73 846).
2 G. Horne and W. Mackie, *Tetrahedron Lett.*, 1999, **40**, 8697.
3 K.C. Nicolaou, H.J. Mitchell, H. Suzuki, R.M. Rodriguez, O. Baudoin and K.C. Fylaktakidou, *Angew. Chem., Int. Ed. Engl.*, 1999, **38**, 3334.
4 M. Shimizu, Y. Nakahara and H. Yoshioka, *J. Fluorine Chem.*, 1999, **97**, 57 (*Chem. Abstr.*, 1999, **131**, 185 148).
5 S.D Taylor, C. Kotoris and G. Hum, *Tetrahedron*, 1999, **55**, 12431.
6 J. Ortner, M. Albert, H. Weber and K. Dax, *J. Carbohydr. Chem.*, 1999, **18**, 297.
7 S.P. Vincent, M.D. Burkart, C.-Y. Tsai, Z. Zhang and C.-H. Wong, *J. Org. Chem.*, 1999, **64**, 5264.
8 M. Albert, B.J. Paul and K. Dax, *Synlett*, 1999, 1483.
9 C. Lamberth, *Recent Res. Dev. Synth. Org. Chem.*, 1998, 1 (*Chem. Abstr.*, 1999, **131**, 199 893).
10 G.S. Lal, G.P. Pez, R.J. Pesaresi, F.M. Prozonic and H. Cheng, *J. Org. Chem.*, 1999, **64**, 7048.
11 S.C.H. Cavalcanti, Y. Xiang, M.G. Newton, R.F. Schinazi, Y.-C. Cheng and C.K. Chu, *Nucleosides Nucleotides*, 1999, **18**, 2233.
12 R. Fernandez and S. Castillon, *Tetrahedron*, 1999, **55**, 8497.
13 A.T. Carmona, P. Borrachero, F. Cabrera-Escribano, M. Jesus Dianez, M. Dolores Estrada, A. Lopez Castro, R. Ojeda, M. Gomez-Guillén and S. Pérez-Garrido, *Tetrahedron: Asymm.*, 1999, **10**, 1751.
14 I.A. Mikhailopulo and G.E. Sivets, *Helv. Chim. Acta*, 1999, **82**, 2052.
15 C.-L. Schengrund and P. Kováč, *Carbohydr. Res.*, 1999, **319**, 24.
16 T.-Y. Lin, H.-C. Chang and R.J. Lagow, *J. Org. Chem.*, 1999, **64**, 8127.

17 K. Nakai, Y. Takagi and T. Tsuchiya, *Carbohydr. Res.*, 1999, **316**, 47.
18 G. Foulard, T. Brigaud and C. Portella, *J. Fluorine Chem.*, 1998, **91**, 179 (*Chem. Abstr.*, 1999, **130**, 81 710).
19 J. Nieschalk and D. O'Hagan, *J. Fluorine Chem.*, 1998, **91**, 159 (*Chem. Abstr.*, 1999, **130**, 81 709).
20 J. Fuentes, M. Angulo and M. Angeles Pradera, *Carbohydr. Res.*, 1999, **319**, 192.
21 J. Du, Y. Choi, K. Lee, B.K. Chun, J.H. Hong and C.K. Chu, *Nucleosides Nucleotides*, 1999, **18**, 187.
22 P.G. Hultin and R.M. Buffie, *Carbohydr. Res.*, 1999, **322**, 14.
23 C.M. Hayman, L.R. Hanton, D.S. Larsen and J.M. Guthrie, *Aust. J. Chem.*, 1999, **52**, 921.
24 D. Bouvet and D. O'Hagan, *Tetrahedron*, 1999, **55**, 10481.
25 G.-J. Meyer, K.H. Matzke, K. Hamacher, F. Fuchtner, J. Steinbach, G. Notohamiprodje and S. Zijlstra, *Appl. Radiat. Isot.*, 1999, **51**, 37 (*Chem. Abstr.*, 1999, **131**, 102 436).
26 T. De Groot, G. Bormans, R. Busson, L. Mortelmans and A. Verbruggen, *J. Labelled Compd. Radiopharm.*, 1999, **42**, 147 (*Chem. Abstr.*, 1999, **130**, 237 752).
27 G. Zhao, X. Long, D. Li and Y. Wang, *Tongweisu*, 1997, **10**, 129 (*Chem. Abstr.*, 1999, **130**, 95 721).
28 M.J. Adam, *J. Labelled Compd. Radiopharm.*, 1999, **42**, 809 (*Chem. Abstr.*, 1999, **131**, 272 084).
29 K.K.P. Ravindranathan and R.A. Field, *Carbohydr. Lett.*, 1998, **3**, 179 (*Chem. Abstr.* 1999, **130**, 81 758).
30 H. Tanaka, H. Sakomoto, A. Saro, S. Nakamura, M. Nakajima and S. Hashimoto, *Chem. Commun.*, 1999, 1259.
31 A. Kirschning, M. Jesberger and H. Monenschein, *Tetrahedron Lett.*, 1999, **40**, 8999.
32 A. Kirschning, M.A. Hashem, H. Monenschein, L. Rose and K.-U. Schöning, *J. Org. Chem.*, 1999, **64**, 6522.
33 W.R. Roush, S. Narayan, C.E. Bennett and K. Briner, *Org. Lett.*, 1999, **1**, 895 (*Chem. Abstr.*, 1999, **131**, 272 094).
34 M. Fellahi and C. Morin, *Carbohydr. Res.*, 1999, **322**, 142.
35 C.K. Lee, H.C. Kang and A. Linden, *J. Carbohydr. Chem.*, 1999, **18**, 241.
36 O. Yamazaki, H. Togo, S. Matsubayashi and M. Yokoyama, *Tetrahedron*, 1999, **55**, 3735.
37 M.D. Smith, D.D. Long, A. Martin, N. Campbell, Y. Blériot and G.W.J. Fleet, *Synlett*, 1999, 1151.

9
Amino-sugars

1 Natural Products

The branched-chain diamino-sugars **1** have been reported as components of anti-inflammatory macrolide glycosides, lobophorins A and B, isolated from the fermentation of a marine bacterium growing on a brown alga.[1] The Amadori-type compound 2′-*N*-(1-deoxy-β-D-fructopyranos-2-yl)cephaeline (**2**) and an isomer were isolated from dried roots of *Cephaelis ipecacaunha*. The former was not considered an artifact of extraction, because while **2** could be obtained in 24% yield by heating cepaeline with D-glucose in HOAc–Et_3N at 95–100 °C, no equivalent Amadori product was observed for another alkaloid present in the extract that readily undergoes Amadori reaction.[2]

2 Syntheses

Syntheses covered in this section are grouped according to the method used for introducing the amino-functionality.

2.1 Reviews. – Syntheses of 4- and 5-aminodeoxy-aldoses and 5- and 6-aminodeoxy-ketoses by chemical and enzymatic methods and their conversion to iminoalditols by intramolecular reductive amination have been reviewed,[3] as have preparations of 6-deoxygenated amino-sugars.[4]

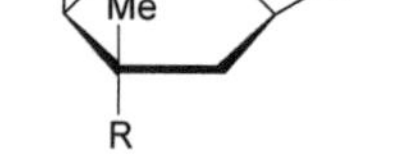

1 R = NH_2 or NO_2 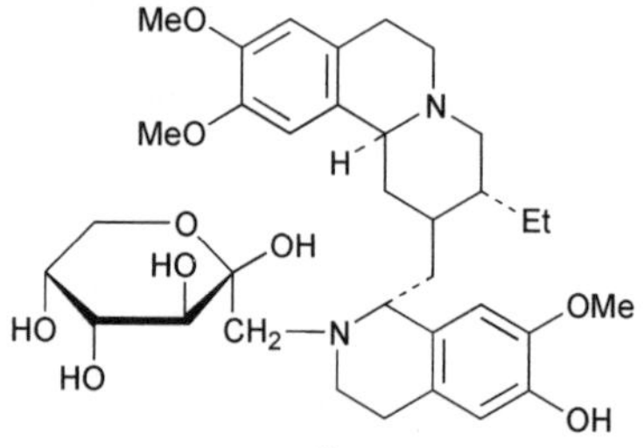 **2** 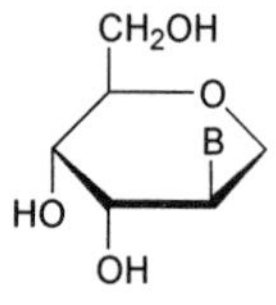 **3** B = U, $Ad^{N\text{-}Bz}$, *etc.*

2.2 By Epoxide Ring Opening. – Nucleophilic opening of the epoxide ring of 1,5;2,3-dianhydro-4,6-*O*-benzylidene-D-allitol with sodium or lithium salts of adenine, uracil or guanine derivatives led to the D-*alto*-configured isonucleoside analogues **3** and derivatives of them that are suitable for incorporation in oligonucleotides.[5] (See Chapter 20 for related nucleoside analogues.) In the synthesis of intermediates towards the construction of ecteinascidin 743, the D-glucose-derived 5,6-epoxide **4** has been converted into the 5,6-epimine **5** and then in 14 steps into the tetracyclic alkaloid (not shown) *via* the 5-amino-2-azido-derivative **6** (Scheme 1).[6]

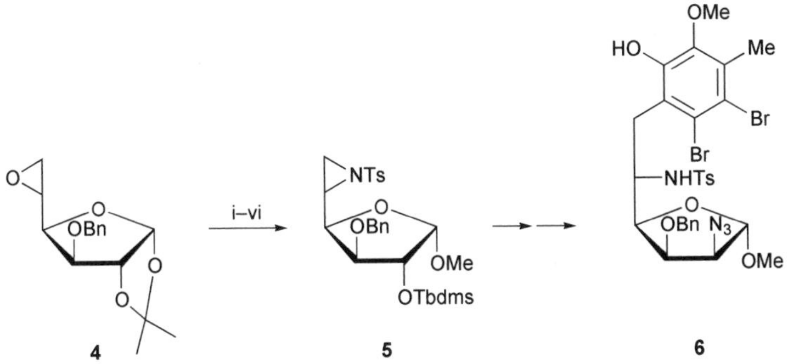

Reagents: i, $TsNH_2$, Cs_2CO_3; ii, MsCl, Et_3N; iii, MeOH, HCl; iv, $SnCl_4$; v, NaOH, MeOH; vi, TbdmsCl, imidazole

Scheme 1

2.3 By Nucleophilic Displacement. – (*E*)-(Pyrrolidin-2-ylidene)glycinates such as **9** were synthesized as exemplified in Scheme 2, involving intramolecular 1,3-dipolar cycloaddition reaction of azido-alkene **8**. Compound **8** was prepared from 2,3,5-tri-*O*-benzyl-D-arabinose by Horner-Emmons olefination of the 5-*O*-Tbdms protected derivative **7**, followed by introduction of the azido-group by sulfonate displacement reaction with inversion.[7] 6,6′-Dideoxy-6,6′-di-(1-hydroxy-2-butylamino)-trehalose and the corresponding compound with additional methylene groups (C-7, 7′) have been synthesized by amine and cyanide displacement reactions on a 6,6′-ditriflate, as analogues of the anti-tuberculosis agent ethambutol, but did not display significant activity.[8] The 1-oxacepham derivative **13** has been prepared by conversion of resin-bound D-xylofuranose diacetal derivative **10** to the 5-triflate **11** and then the 5-amino-derivative **12**, followed by concomitant release and cyclization on treatment with Lewis acid (Scheme 3).[9]

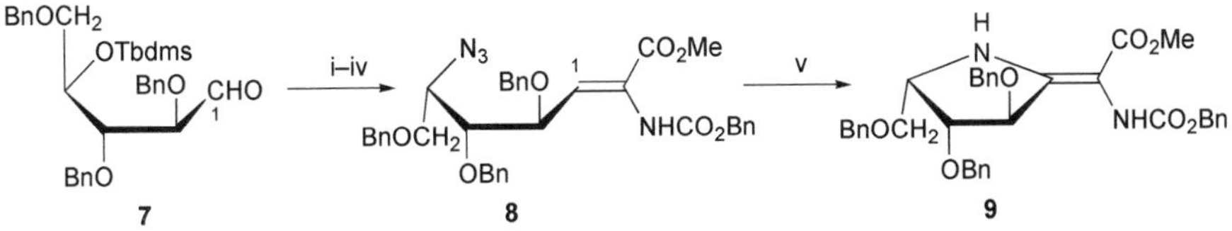

Reagents: i, $(MeO)_2P(O)CH(NHCO_2Bn)CO_2Me$, DBU; ii, HF, Py; iii, Tf_2O, 2,6-lutidine; iv, NaN_3, $BnEt_3\overset{+}{N}\cdot\overset{-}{Br}$; v, 60 °C, THF

Scheme 2

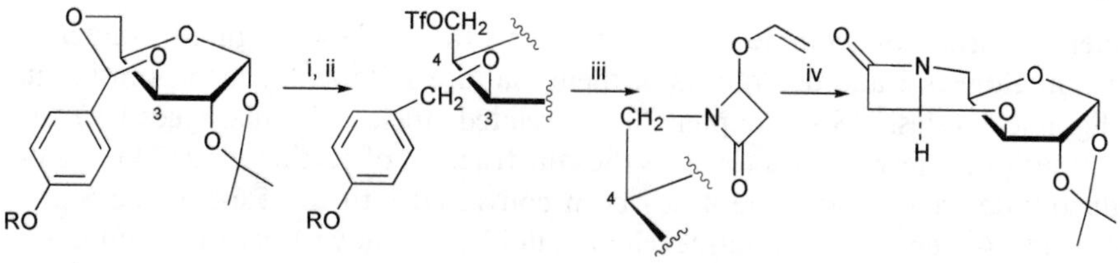

R = resin

Reagents: i, $Bu^{i}_{2}AlH$; ii, Tf_2O, lutidine; iii, BuLi, 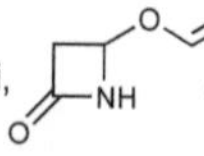 , Bu_4NHSO_4; iv, $BF_3{\cdot}OEt_2$, CH_2Cl_2

Scheme 3

Phenyl 6-deoxy-6-(morpholin-4-yl)-β-D-glucopyranoside has been prepared directly from the corresponding 6-*O*-tosylate and found to be an inhibitor of mouse α-glucosidase (K_i <16 μM).[10] The 6-amino-6-deoxy-hexose derivatives **14** were obtained by reaction of the corresponding 5,6-cyclic sulfate with sodium azide in DMF, followed by catalytic hydrogenation.[11] New surfactants **15** were obtained from methyl α-D-glucopyranoside by conversion to the 4,6-cyclic sulfate (i, $SOCl_2$; ii, $NaIO_4$) and reaction with the appropriate amine in DMF, followed by *in situ* hydrolytic removal of the sulfate group. Compound **15** ($n = 5$) was a good surfactant in combination with low molecular weight carboxymethylcellulose.[12] Spectroscopic studies (including NMR) on amino-sugars synthesized by triflate displacement reactions with suitably protected amines have been reported.[13] The preparation of 6-*O*-alkylamino-substituted permethylated cyclodextrins is covered in Chapter 4, while analogues of the spiramycins (4-amino-4-deoxy-glycosylamines) are covered in Chapter 20.

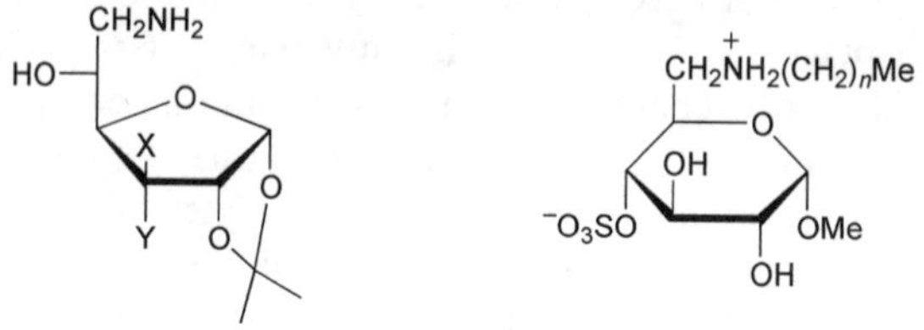

14 X = OAc, OBn or OMs; Y = H or X = H, Y = N_3

15

2.4 By Amadori Reaction and Heyns Rearrangement. – The Heyns rearrangement (step i) has been employed for an efficient conversions of lactulose **16** into various lactosamine derivatives **17**, as exemplified in Scheme 4.[14] Deprotection of the glycosylamine derivatives **18**, produced by Mitsunobu condensation of amino acid-derived 2-nitrobenzenesulfonamides with 2,3,4,6-tetra-*O*-acetyl-D-glucose, led to Amadori products **19** (Scheme 5);[15] see also Vol. 30, pp. 128–129.

2.5 From Azido-sugars. – 2-Azido-3,4,6-tri-*O*-benzyl-2-deoxy-D-galactono-1,5-lactone was smoothly converted into the corresponding 2-acetamido-

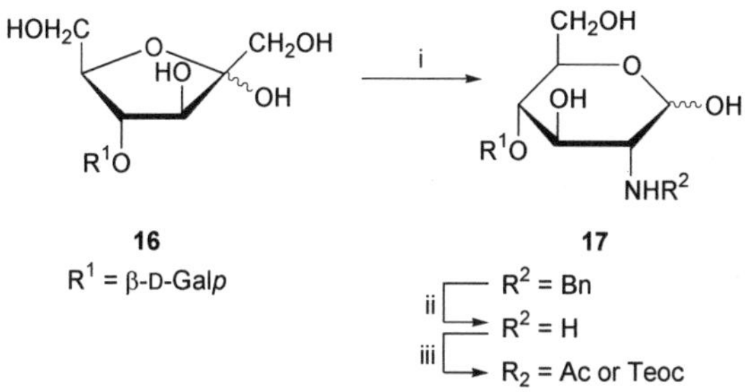

Reagents: i, $BnNH_2$ then AcOH, MeOH; ii, H_2Pd/C; iii, Ac_2O, MeOH, $NaHCO_3$; or Cl_3CCH_2OCOCl, $NaHCO_3$, H_2O

Scheme 4

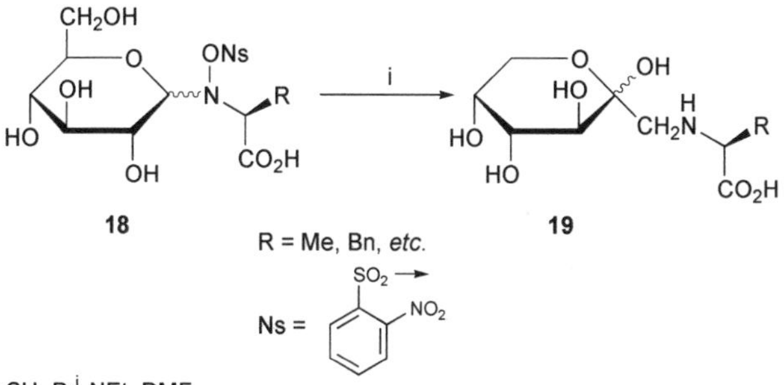

Reagents: i, PhSH, Pr^i_2NEt, DMF

Scheme 5

derivative (i, Raney Ni; ii, Ac_2O, Py), but the D-*gluco*-isomer gave the 2-acetamido-2-deoxy-D-mannonolactone **20** instead in 50% yield as a result of epimerization at C-2.[16] The 2-*tert*-butoxycarbonylamino-2,3,6-trideoxy-3-dimethylamino-D-*arabino*-hexoside **21** has been prepared from the corresponding phenyl 3-azido-1-thio-glycoside and its DNA-cleaving ability was established.[17] 5-Deoxy-5-imidazolyl-D-xylose **22** (X = CH) was obtained from 5-azido-5-deoxy-1,2-*O*-isopropylidene-α-D-xylofuranose by reduction, cyclization and hydrolysis [i, Ph_3P; ii, $NCCH(NH_2)CONH_2$, KOH, DMF; iii, CF_3CO_2H, H_2O, iv, H_2SO_4, H_2O]. Synthesis of the 5-triazolyl analogue **22** (X = N) involved direct condensation of the azido-sugar with $NCCH_2CONH_2$. Both routes involved formation and hydrolysis of bicyclic glycosylamines intermediates **23**.[18] Isomeric methyl 5-deoxy-5-(1-hydroxy-2-butylamino)-D-arabinofuranosides have been prepared from the corresponding 5-azidoglycosides utilizing reductive amination, as analogues of the anti-tuberculosis agent ethambutol, but did not display significant activity.[8] Analogues of the aminoglycoside antibiotic neamine with a 2-, 3-, 4- or 6-aminodeoxy-α-D-glucopyranosyl moiety replacing the 2,6-diamino-2,6-dideoxy-α-D-glucopyranosyl moiety, have been synthesized by glycosylation of a 2-deoxystreptamine derivative with the corresponding phenyl azido-tri-*O*-benzyl-deoxy-1-thio-β-D-glucopyranosides, but they did not bind RNA as well as did neamine.[19] Syntheses of 5′-deoxy-5′-trifluoromethyl-daunomycinone and -doxorubicin are

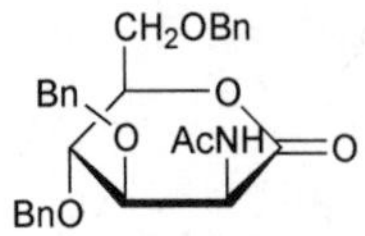

20 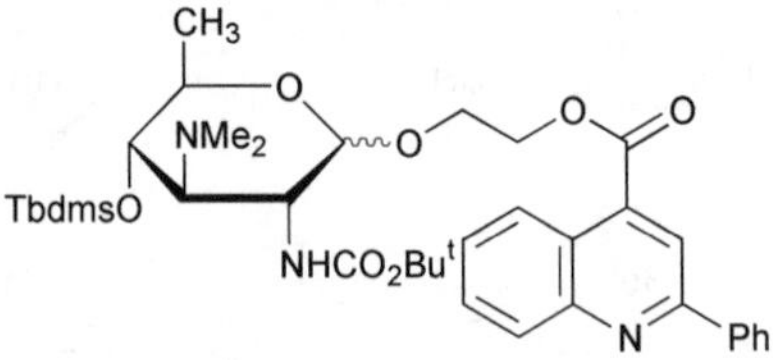 **21** 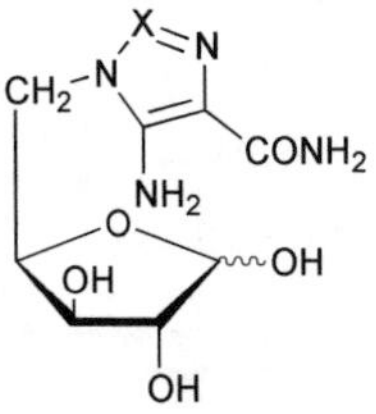 **22** X = CH or N

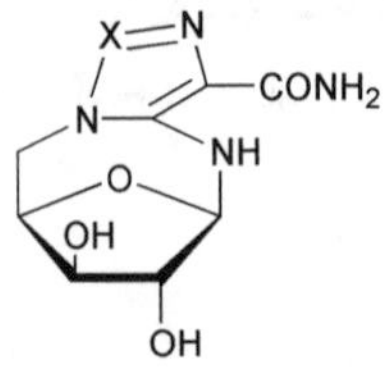

23 X = CH or N 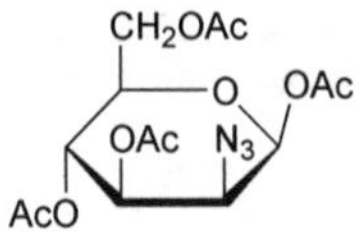 **24**

covered in Chapters 8 and 19, while the synthesis of 3*S*,4*R*-3,4-dihydroxyglutamic acid from D-ribose is reported in Chapter 24.

2.6 From Unsaturated Sugars. – A mixture of the 2-azido-2-deoxy-β-D-mannose derivative **24** and its α-D-*gluco*-isomer was prepared by azidonitration of tri-*O*-acetyl-D-glucal. When used for the $BF_3.Et_2O$-catalysed glycosidation of di- and tri-ethylene glycols, only **24** reacted as a donor, leading eventually to the production of β-D-Gal*p*-(1→4)-α-D-ManNAc-1→*O*$(CH_2CH_2)_n$OH for use in anti-metastasis studies.[20] The 2-amino-2-deoxy-D-talose derivative **26** was obtained with high stereoselectivity and in high yield and by hydrogen-bond directed *cis*-dihydroxylation of the 3,4-ene **25**, prepared from tri-*O*-acetyl-D-glucal (Vol. 26, p.113), using stoichiometric quantities of osmium tetroxide at low temperature (Scheme 6). In contrast, use of catalytic quantities of osmium tetroxide in aqueous solvent led to a mixture of the D-*altro*- and D-*talo*-isomers in the ratio 2.6:1. The 2-amino-2-deoxy-D-allose derivative **28** was similarly obtained in high stereoselectivity from 3,4-ene **27**, also derived from tri-*O*-acetyl-D-glucal, using catalytic osmium tetroxide and quinuclidine *N*-oxide as oxidant (Scheme 7).[21] Similar studies on the corresponding 2-amino-1,5-anhydro-hex-3-enitols is reported in Chapter 18.

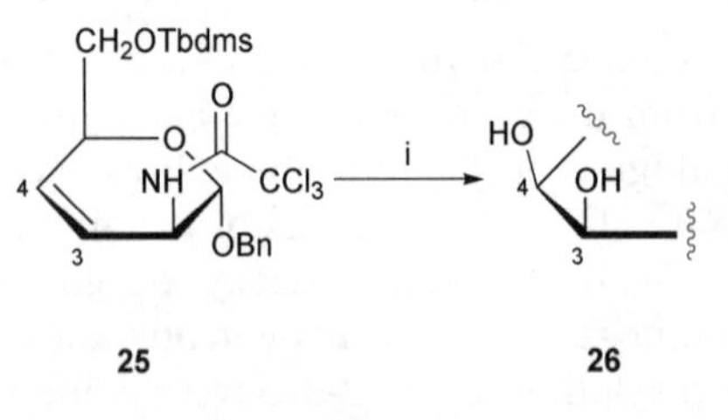

Reagents: i, OsO_4, TMEDA, CH_2Cl_2, –78 °C, then MeOH, HCl

Scheme 6

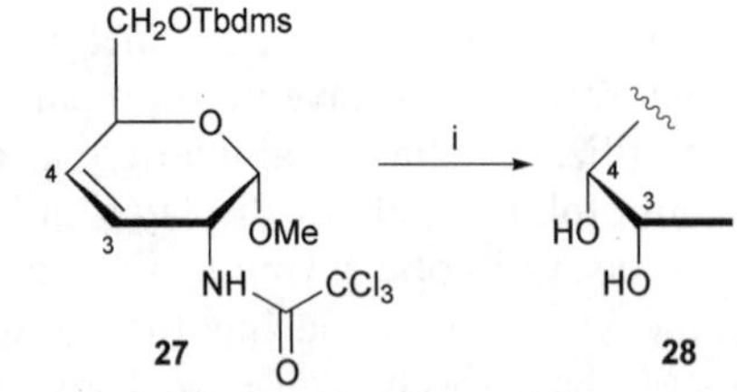

Reagents: i, OsO_4, quinuclidene *N*-oxide, CH_2Cl_2, 4 equiv. H_2O

Scheme 7

The C-1 substituted glycals **29** have been synthesized by an intramolecular Claisen-Ireland rearrangement of C-1 exo-methylene glucoside derivatives (see Chapters 3 and 13). Compound **29** (R = Me) was transformed into the bicyclic amino-sugar derivative **30** by way of an azidonitration reaction (Scheme 8).[22] The 4-deoxy-4-dialkylamino-hex-2-enopyranosides **31** were synthesized by $Pd(PPh_3)$-catalysed allylic displacement of a 4-acetoxy group with retention of configuration.[23] Construction of sialic acid analogues, involving conjugate addition of benzylamine as the means to introduce the nitrogen atom, is covered in Chapter 16.

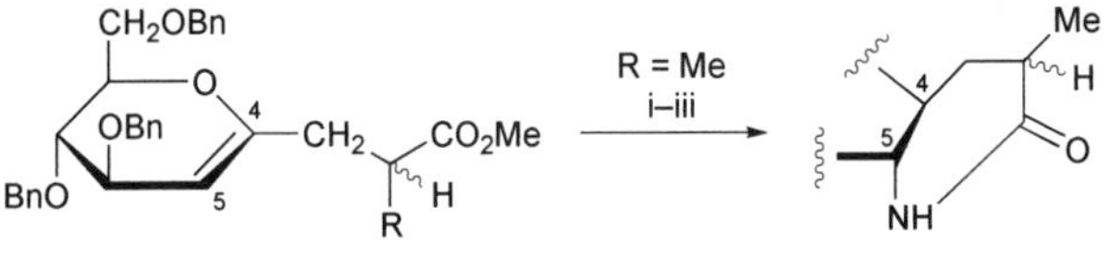

29 R = Me, $NHCO_2Bu^t$ or NPhth **30**

Reagents: i, $(NH_4)_2Ce(NO_3)_6$, NaN_3; ii, Et_3SiH, $BF_3{\cdot}OEt_2$; iii, $NaBH_4$, $NiCl_2$

Scheme 8

2.7 From Dicarbonyl Compounds, Aldosuloses, Dialdoses and Ulosonic Acids. The diester derivative **32**, a partial structure analogue of the microbial phytotoxin tagetitoxin, has been synthesized from the known 1,6-anhydro-3-amino-3-deoxy-D-gulose, produced from 1,6-anhydro-D-glucopyranose by periodate cleavage, condensation with nitromethane, and reduction. Selective esterification was achieved by way of an *N*-benzyl-2-*O*,3-*N*-carbonyl-derivative.[24] Reduction of an oxime features as the key step in the construction of *C*-

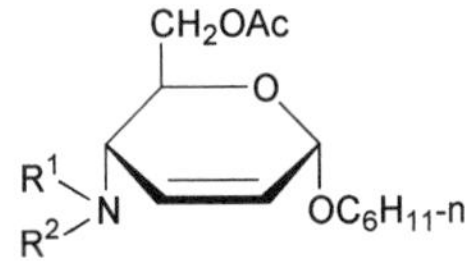

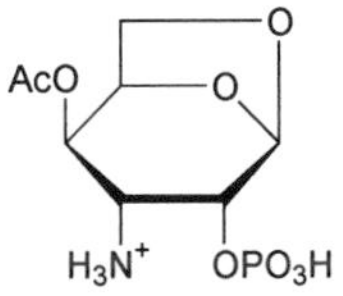

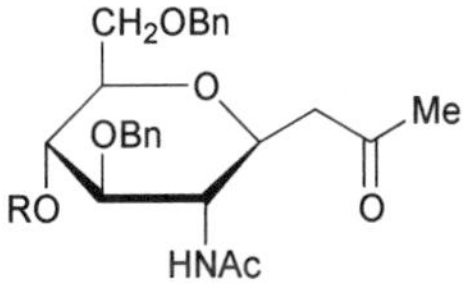

31 $R^1 = R^2$ = Et, or R^1,R^2 =

32

33 R = β-D-Gal*p*-$(Bn)_4$

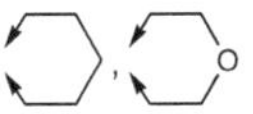

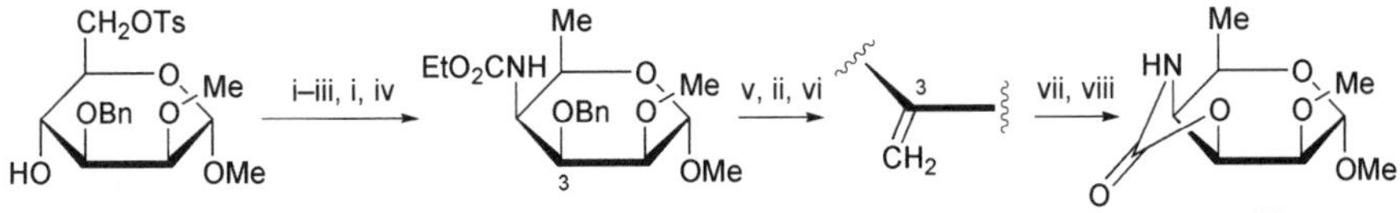

34 **35**

Reagents: i, $LiAlH_4$; ii, DMSO, $(CF_3CO)_2O$, Et_3N; iii, NH_2OH; iv, $ClCO_2Et$, $NaHCO_3$; v, H_2, Pd/C; vi, $Ph_3P{=}CH_2$; vii, $I^+(collidine)_2{\cdot}ClO_4^-$; viii, Bu_3SnH

Scheme 9

glycosidic lactosamine derivative **33** from the corresponding 1-*C*-allyl-lactose derivative,[25] and in the synthesis of callipeltose **35**, a component sugar of the natural product callipeltoside A, from the mannoside **34** (Scheme 9).[26]

Intramolecular reductive amination of the D-*xylo*-hex-5-ulosonamide derivatives **36**, prepared from tetra-*O*-benzyl-D-glucono-1,5-lactone, with $NaBH_3CN$ or $NaBH_4$ under acidic conditions has been studied. With $NaBH_3CN$, HCO_2H in boiling acetonitrile, **36** (R = H) gave the expected D-glucono-1,5-lactam **37** in 83% yield, but the *N*-benzyl analogue **36** (R = Bn) gave D-talono-1,5-lactam **38** as the major product (49%), as a result of epimerization at both C-2 and C-4 prior to reduction (Scheme 10). Reaction of **36** (R = Bn) with $NaBH_4$, CF_3CO_2H in acetonitrile at room temperature gave mostly the L-idulonamide **39** by reduction of the keto-group.[27]

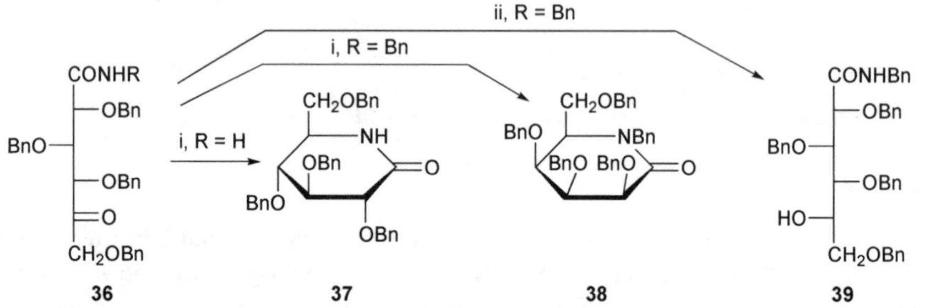

Reagents: i, $NaBH_3CN$, HCO_2H, MeCN; ii, $NaBH_4$, CF_3CO_2H

Scheme 10

The poly(sugar-amine) **40** was obtained by reductive amination ($NaBH_3CN$) of D-galactodialdose (from galactose oxidase-catalysed oxidation of D-galactose) with ethylenediamine. Similarly, polymer **41** was obtained by self condensation of the galactose oxidase-catalysed oxidation product of 2-amino-2-deoxy-D-galactose upon reductive amination.[28]

Syntheses of the ascorbic acid-derived lactams **42** have been described.[29]

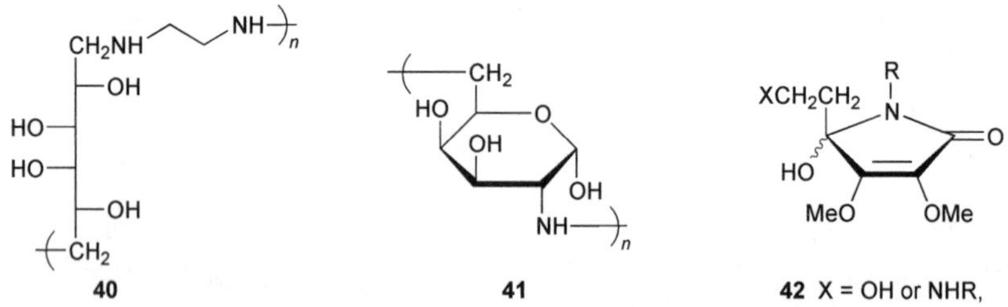

42 X = OH or NHR,
R = Bu^n, Bn or n-C_6H_{11}

2.8 From Chiral Non-carbohydrates. – The 2-amino-2-deoxy-L-mannose derivative **45** was obtained by Julia olefination of 2,3-*O*-isopropylidene-L-glyceraldehde **43** with the L-serine-derived chiral sulfone **44**, followed by OsO_4-catalysed dihydroxylation (Scheme 11).[30] The methyl α-glycosides of L-dauno-

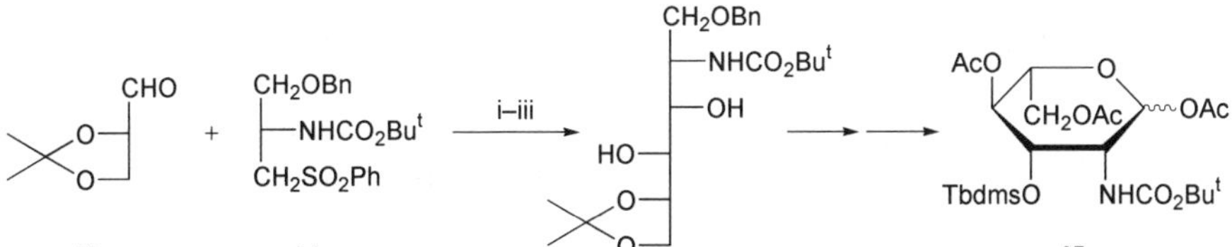

Reagents: i, BuLi; ii, Na–Hg, Na_2HPO_4, MeOH; iii, OsO_4, NMO, Bu^tOH, H_2O

Scheme 11

samine and L-ristosamine, **47** and **48**, respectively, have been synthesized from (*S*)-1-(2-furyl)ethanol (**46**), prepared by kinetic resolution of racemic material by lipase catalysed esterification (Scheme 12). Introduction of the nitrogen

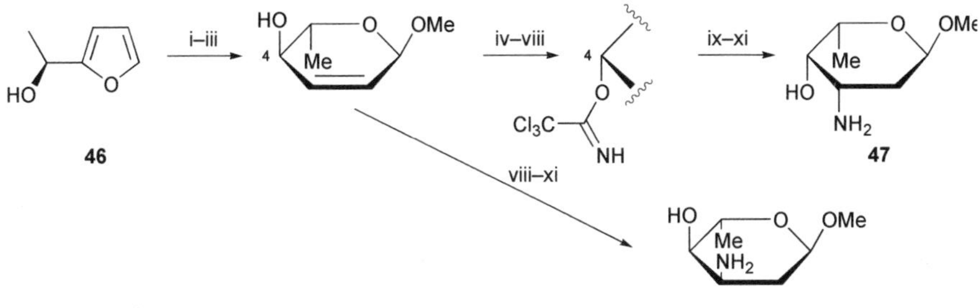

Reagents: i, electrochemical methoxylation; ii, H_3O^+, Me_2CO; iii, MeI, Ag_2O; iv, $ZnCl_2$; v, $NaBH_4$, $CeCl_3$; vi, DEAD, Ph_3P, BzOH; vii, NaOMe, MeOH; viii, Cl_3CCN, NaH; ix, $Hg(OCOCF_3)_2$; x, $NaBH_4$; xi, $Ba(OH)_2$, MeOH

Scheme 12

atom utilized a mercuration–demercuration sequence (steps ix and x). The enantiomers were obtained in the same way starting with (*R*)-1-(2-furyl)-ethanol.[31] Methyl 3-*epi*-D-daunosaminide **52** and methyl L-daunosaminide **47** have been synthesized from the adduct **49**, obtained from the highly stereo-selective asymmetric conjugate addition of lithium (*R*)-*N*-benzyl-(-methylben-zylamide to methyl (*E*,*E*)-hexa-2,4-dienoate. The key step, OsO_4-cataylsed

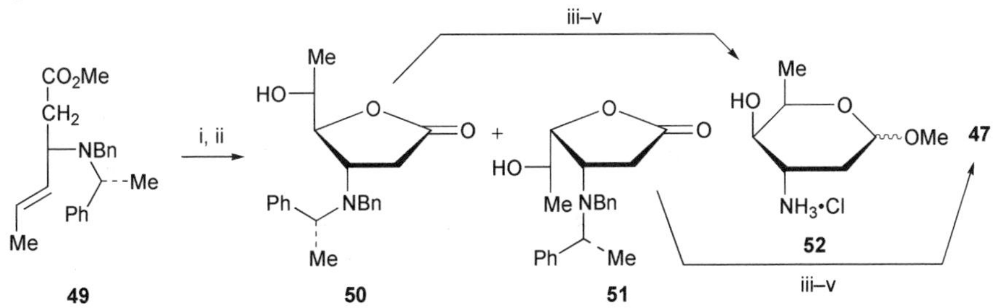

Reagents: i, OsO_4, $K_3Fe(CN)_6$, K_2CO_3, Bu^tOH; ii, Na_2SO_3; iii, H_2, Pd/C, $(Bu^tO_2C)_2O$; iv, $Bu^i{}_2AlH$; v, MeOH, HCl

Scheme 13

dihydroxylation, yielded separable isomers **50** (48%) and **51** (32%) (Scheme 13).[32]

The methyl α-glycoside of L-kedarosamine **54** has been synthesized from the D-lactaldehyde derivative **53**. The nitrogen atom was introduced by intramolecular opening of an epoxide by an allylic *O*-(*N*-benzoylcarbamate) group with concomitant N→O-benzoyl migration (Scheme 14).[33] Methyl vicenisaminide **57** has been synthesized from the chiral epoxy-alcohol **55**, obtained by Sharpless asymmetric epoxidation (Scheme 15). Chain elongation of intermediate **56** utilized a selective allylation with allylmagnesium bromide and a chiral boron reagent [(−)-$(IPC)_2BOMe$].[34]

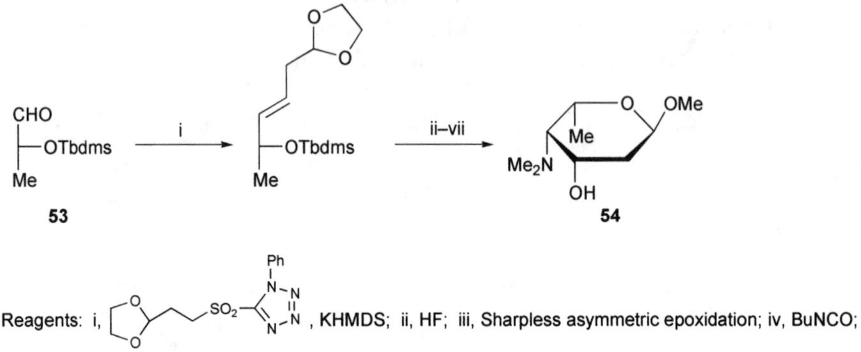

Reagents: i, [dioxolanyl-ethyl phenyltetrazolyl sulfone], KHMDS; ii, HF; iii, Sharpless asymmetric epoxidation; iv, BuNCO; v, NaHMDS then MeI; vi, $LiAlH_4$; vii, Resin(H^+-form), MeOH

Scheme 14

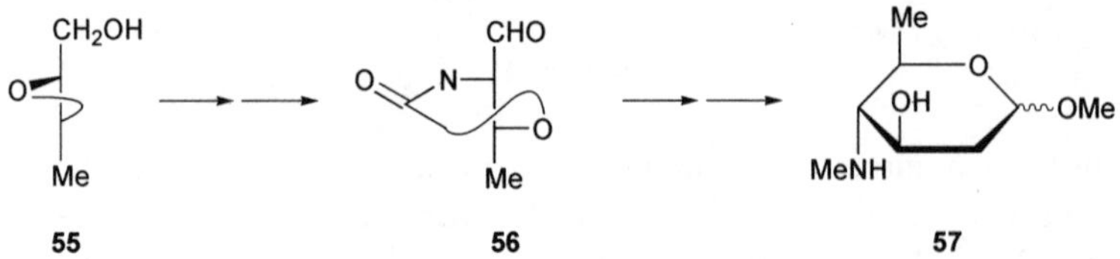

Scheme 15

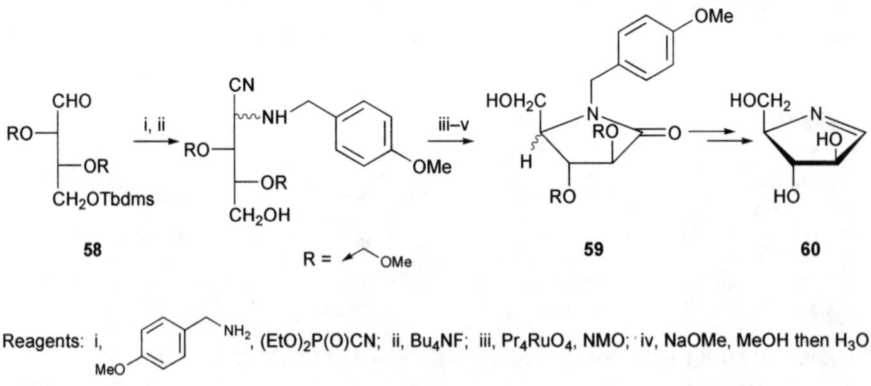

Reagents: i, [4-methoxybenzylamine], $(EtO)_2P(O)CN$; ii, Bu_4NF; iii, Pr_4RuO_4, NMO; iv, NaOMe, MeOH then H_3O v, $LiBH_4$

Scheme 16

Necristine **60**, and its 4-epi- and 5-deoxy-analogues, have been synthesized from diethyl D-tartrate *via* the known D-threose derivative **58** and the separable mixture of isomeric lactams **59** (Scheme 16), the nitrogen atom being introduced by aminonitrile formation (step i).[35]

A review on the synthesis and applications in synthesis of optically active α-furfurylamine derivatives included syntheses of aza-sugars and 1-deoxy-aza-sugars.[36]

3 Reactions and Derivatives

3.1 Interconversion and Degradation Reactions. – Butyl 2-amino-2-deoxy-β-D-glucopyranoside was obtained by treatment of chitosan with *Penicillium funicolosum* (as a source of exo-hexosaminidase) in buffered aqueous butanol, the conversion being 210 mg g^{-1} in 2 days.[37] *N*-Acetyl-lactosamine has been produced enzymatically from orotic acid (the precursor for UTP), D-galactose and 2-acetamido-2-deoxy-D-glucose, with accumulation of product to a concentration of 107 g L^{-1} after 38 hours.[38]

Siastatin B (**61**) has been converted in a multi-step procedure into the galactosidase inhibitor **62**.[39] Various *N*-acylated 2-amino-2-deoxy-D-glucoses have been converted to the furanose derivatives **63** (i, $FeCl_3$, Me_2CO; ii, AcOH, MeOH, H_2O) and then into the corresponding hex-2-enofuranose derivatives **64** (Scheme 17).[40]

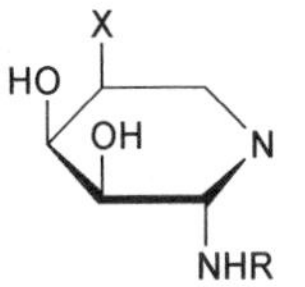

61 X = CO_2H, R = Ac
62 X = CH_2OH, R = $COCF_3$

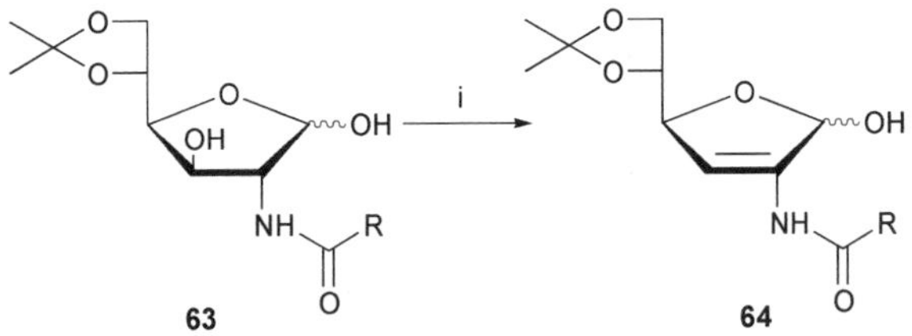

R = $(CH_2)_nMe$, *n* = 4, 6, 8, 10

Reagents: i, Resin(OH^-), MeOH

Scheme 17

From a study of the reaction of hyaluronic acid and its monomers (GlcA and GalNAc) with reactive oxygen species, it has been concluded that most of the hyaluronic acid degradation under such conditions occurs at GlcA units, which are converted to *meso*-tartaric acid.[41] The conversion of a 2-amino-2-deoxy-1-*C*-tributyltin-D-glucose derivative into *C*-glycosylated amino acids is covered in Chapters 3 and 17.

3.2 *N*-Acyl and *N*-Carbamoyl Derivatives. – The influence on the ^1H and ^{13}C NMR chemical shifts of allyl 2-amino-3,4-di-*O*-benzoyl-6-*O*-*tert*-butyldimethylsilyl-2-deoxy-β-D-glucopyranoside of protonation or replacement of the 2-amino group with an acetamido-, phthalimido-, tertrachlorophthalimido- or azido-group has been determined.[42] Chloambucil adducts such as **65** and its D-*allo*-isomer,[43] and *S*-nitrosoamides such as **66**,[44] have been synthesized as cytotoxic agents. Application of the Ugi and Passerini reactions to the 2-deoxy-2-isocyano-D-glucose derivative **67** provided various glycopeptide analogues such as **68** and **69**, respectively (Scheme 18).[45] Phenyl 3,4,6-tri-*O*-acetyl-

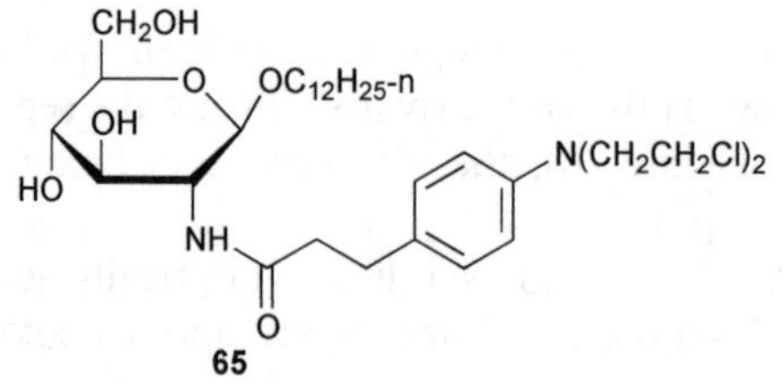

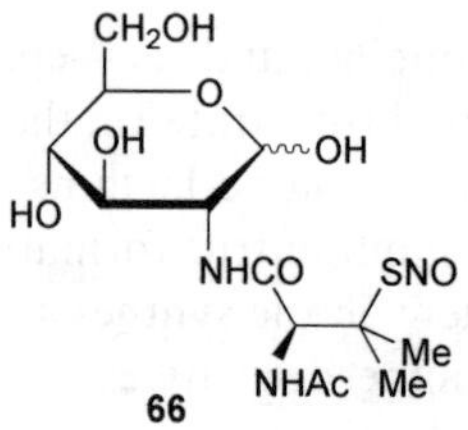

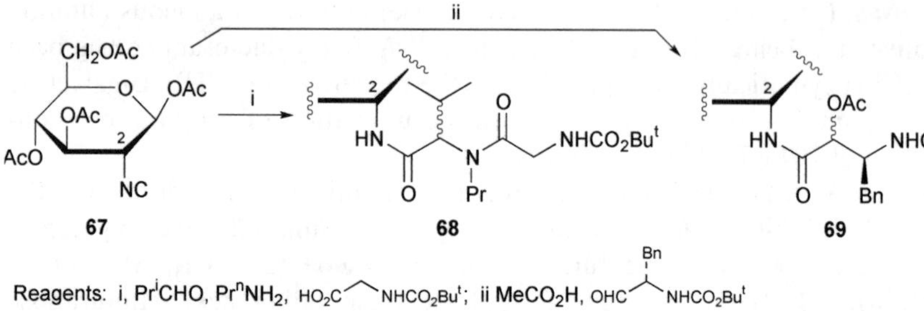

Reagents: i, Pr^iCHO, Pr^nNH_2, $HO_2CCH_2NHCO_2Bu^t$; ii $MeCO_2H$, $OHCCH(Bn)NHCO_2Bu^t$

Scheme 18

2-deoxy-2-trifluoroacetamido-β-D-glucopyranosyl sulfoxide has been shown to be a good donor in both solution and solid phase glycosylations, when several related analogues with other *N*-protecting groups were not. The *N*-$COCF_3$ group was easily removed (LiOH, MeOH).[46] Chemists at Intercardia, Inc, have prepared more than 1300 disaccharide-phospholipid partial structure analogues of moenomycin A, mainly varying in the acyl substituent on a 2-amino-sugar and in the carbamoyl substituents on a 3-amino-sugar, by automated synthesis on a solid support.[47]

Simple large scale preparations of the 2-amino-2-deoxy-D-glucoside derivatives **72–74** from the *N*-acetyl derivative **70** *via* **71**, have been published (Scheme 19).[48] Selective de-*N*-acetylation of 4-nitrophenyl β-chitobioside with chitin deacetylase from *Colletotrichium lindemuthianum* gave mono-amine **75**, useful as a substrate that can distinguish otherwise apparently identical

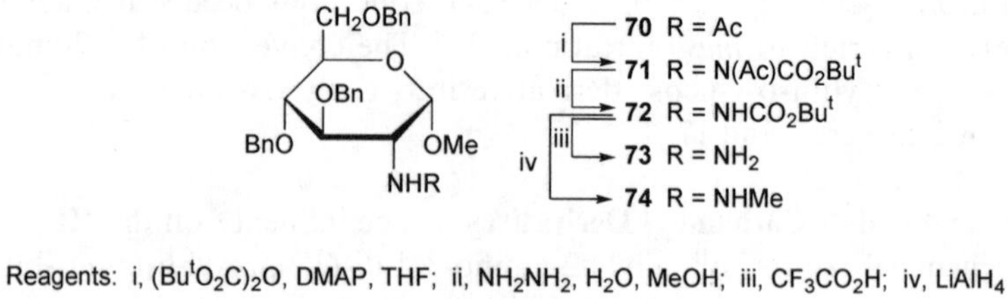

70 R = Ac
71 R = $N(Ac)CO_2Bu^t$
72 R = $NHCO_2Bu^t$
73 R = NH_2
74 R = NHMe

Reagents: i, $(Bu^tO_2C)_2O$, DMAP, THF; ii, NH_2NH_2, H_2O, MeOH; iii, CF_3CO_2H; iv, $LiAlH_4$

Scheme 19

β-D-GlcNH$_2$-(1→4)-β-D-GlcNAc-1→OC$_6$H$_4$NO$_2$-*p*

75

β-D-GlcNAc-(1→4)-β-D-GlcNH$_2$

76

chitinases.[49] In contrast, derivative **76** with a single *N*-acetyl-group on the non-reducing terminal unit has been obtained by action of the same enzyme in reverse on chitobiose.[50]

Spirothiazolidinone **77** was obtained by reaction of 2-amino-2-deoxy-D-glucose with reagent **78**, and was converted into the spirothiazoloxazole derivative **79** (Scheme 20).[51]

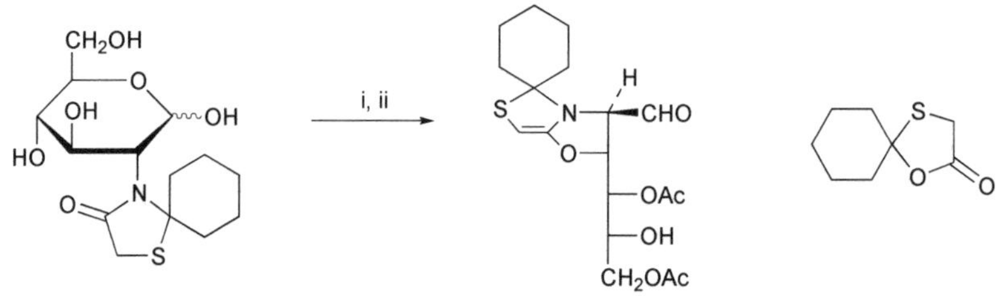

CH$_2$OH, OH, OH, HO, O, N, S

77

i, ii

S, N, H, CHO, O, OAc, OH, CH$_2$OAc

79

S, O, O

78

Reagents: i, AcOH, NaOAc, Ac$_2$O; ii, H$_2$O, HOAc

Scheme 20

3.3 Urea and Thiourea Derivatives. – Urea derivative **80** was prepared by reaction of the corresponding free-amino-sugar derivative with di(4-nitrophenyl)carbamate, and converted into a mono-*N*-nitroso derivative (with N_2O_3, CH_2Cl_2).[52] Various thiourea-linked oligosaccharides **81** have been synthesized by condensation of methyl 6-amino-6-deoxy-α-D-glucopyranoside with glycosyl isothiocyanates in pyridine, followed by Zemplen deacetylation which had to be conducted at less than 0 °C to avoid anomerization.[53]

3.4 *N*-Alkyl and *N*-Alkenyl Derivatives. – Use of methyl 3,4,6-tri-*O*-benzyl-2-*N*,*N*-dibenzylamino-2-deoxy-1-thio-β-D-galactopyranoside as a glycosyl donor provides glycosides with high β-selectivity.[54] *N*-Demethylation of the 3-dimethylamino-group on the desosamine unit of erythromycin to give a 3-methylamino-group has been achieved by reaction with 1-chloroethyl chloroformate then methanol.[55] Full details on the synthesis of an *N*-linked pseudodisaccharide and its ability to inhibit glucosidase II have been published.[56]

3.5 Lipid A Analogues. – Conjugates **82** containing tumour-associated octa- and nona-peptide antigens covalently linked to a lipid A analogue have been synthesized and shown to amplify peptide-specific immune responses in mice.[57] Analogues of lipid A **83**, lacking the 1-phosphate and 3-*O*-hydroxytetradecanoate group, have been synthesized and evaluated as immunostimulants.[58] See Chapter 3 for other references to Lipid A-based work.

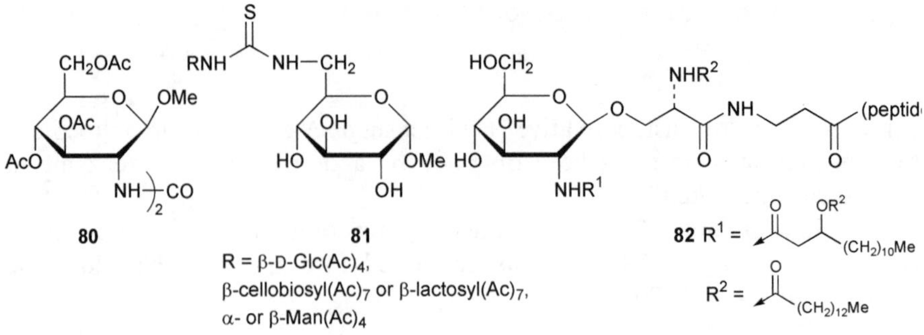

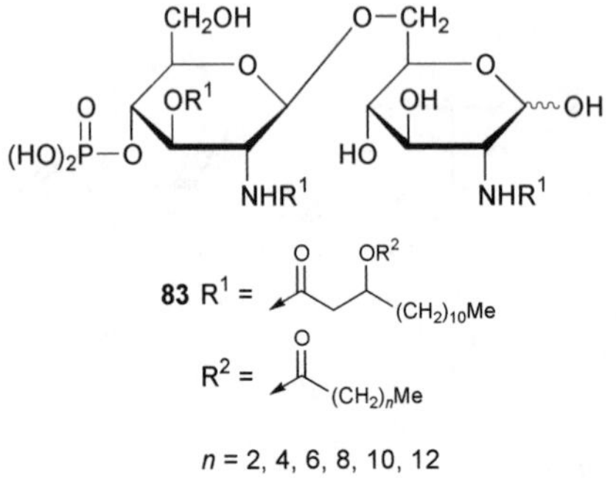

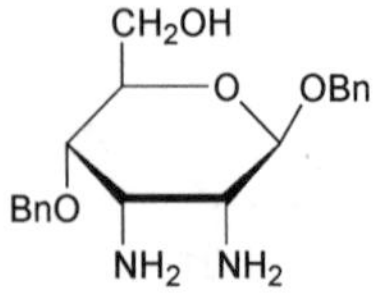

80

81
R = β-D-Glc(Ac)$_4$,
β-cellobiosyl(Ac)$_7$ or β-lactosyl(Ac)$_7$,
α- or β-Man(Ac)$_4$

82

83

n = 2, 4, 6, 8, 10, 12

84

4 Diamino-sugars

The 2,3-diamino-glycoside **84**, designed as a HIV protease transition state mimic, was synthesized from a 2-deoxy-2-phthalimido-β-D-glucopyranoside, the 3-amino-group being introduced by displacement with inversion of a 3-triflate by azide ion.[59] The 2,3-diamino-sugar **86**, bearing a chlorambucil moiety, has been synthesized from the 2-nitro-sugar **85** by an elimination, amine addition sequence (Scheme 21).[60] The 2,3-diamino-hex-4-uronoside **91**

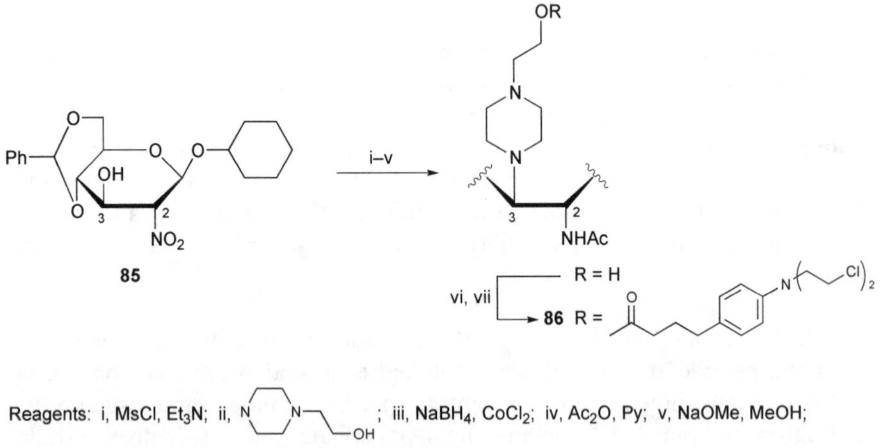

85 → **86**

Reagents: i, MsCl, Et_3N; ii, [1-(2-hydroxyethyl)piperazine]; iii, $NaBH_4$, $CoCl_2$; iv, Ac_2O, Py; v, NaOMe, MeOH; vi, chlorambucil, DMAP, DCC; vii, HCl, MeOH

Scheme 21

was synthesized from 2-acetamido-2-deoxy-D-glucose *via* the glycoside **87** and the oxazoline **88**, which opened with $TmsN_3$ to give epimeric azides **89** and **90** in a ratio of 1:3 (Scheme 22). The minor epimer was converted to the desired product. The 2,6-anhydro-4,5-diaminohex-2-enonic derivative **93**, synthesized from the known diamino-derivative **92** (Scheme 23), was a potent, selective inhibitor of influenza A sialidase (IC_{50} 2 nM).[61]

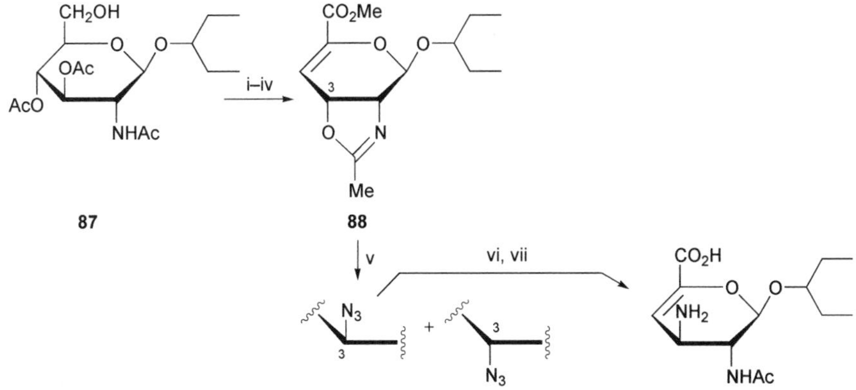

Reagents: i, SO_3•Py, DMSO, CF_3CO_2H; ii, NH_2SO_3, $NaClO_2$; iii, MeOH, TBTU; iv, TmsOTf; v, $TmsN_3$, Bu^tOH; vi, $SnCl_2$, MeOH; vii, H_2O, Et_3N

Scheme 22

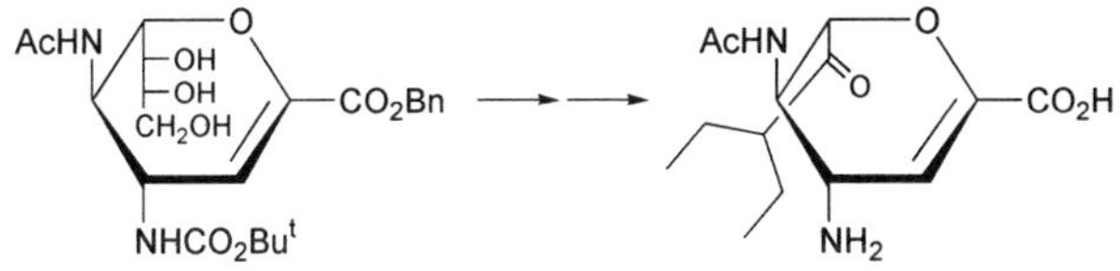

Scheme 23

2,4-Diamino-2,4-dideoxy-D-*threo*-tetronolactam **94** was obtained from the known calcium D-erythronate by sequential reaction with HBr in HOAc, K_2CO_3 in Me_2CO then liquid NH_3. The C-2 epimer was obtained in the same way from the D-erythronate.[62]

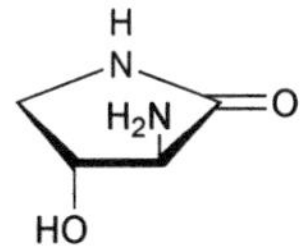

Trisaccharides **95** containing a central 2,4-diamino-2,4-dideoxy-D-xylosyl residue have been synthesized. In the presence of Zn^{2+} or Hg^{2+}, 20–40% of this central residue is complexed *via* the amino-goups in the $^1C_4$ conformation.[63]

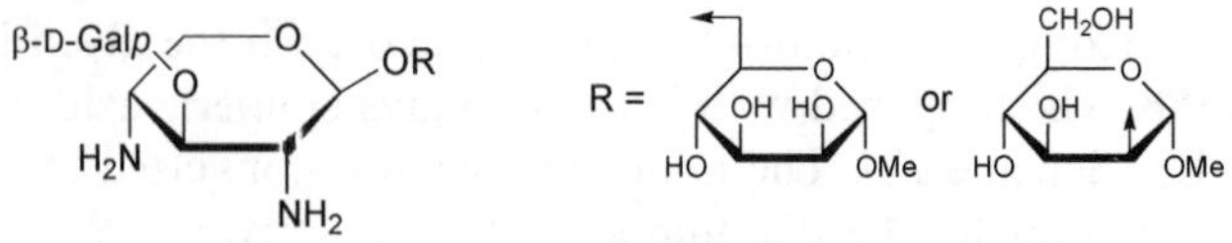

95

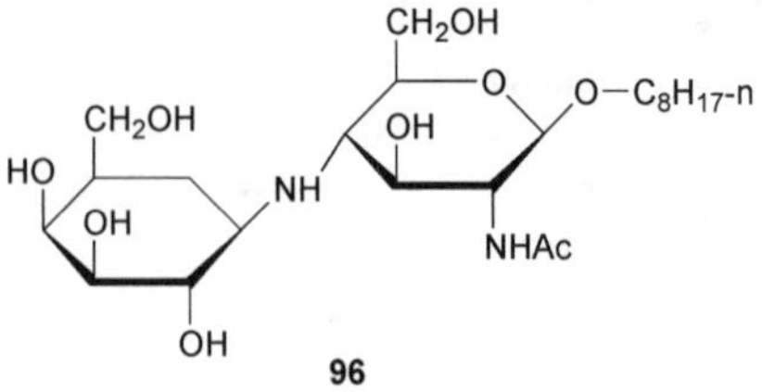

96

The construction of the *N*-linked pseudodisaccharide analogue **96** of lactosamine, and its 3-linked 3-amino-sugar analogue, involved reaction of amino-sugar and cyclitol epoxide derivatives. Compound **96** was an acceptor for α-(1→3)-fucosyltransferase, but its linkage isomer was not.[64] Glycosyl phosphatidyl inositol anchors with photo- and radio-labels attached by acylation of a 6-amino-group on the $GlcNH_2$ residue, such as **97**, have been synthesized.[65] The universal solid support **98** for use in oligonucleotide synthesis was prepared from 2,3;5,6-di-*O*-isopropylidene-α-D-mannofuranose. The 5,6-diamino-functionality was introduced by azide displacement applied to a 5,6-dimesylate, the product being described (without comment) as having retained stereochemistry at C-5. Compound **98** can be used as a primer for DNA or RNA synthesis *via* a 2-phosphate ester linkage. Cleavage of this linkage can be affected under mild basic conditions, utilizing intramolecular 2,3-cyclic phosphate formation, assisted by Zn^{2+} ion complexation of the deprotected 5,6-diamino-function.[66]

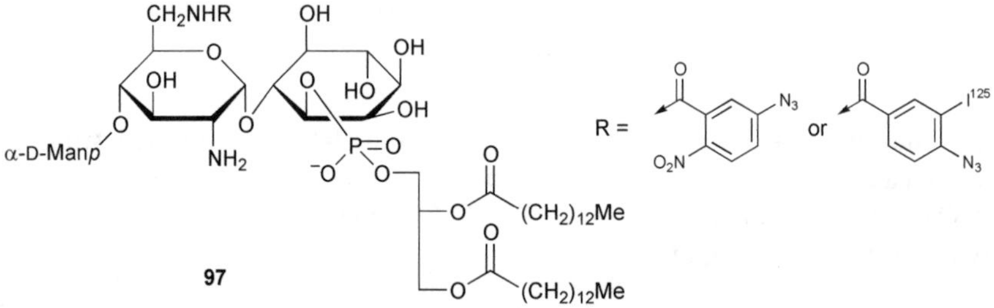

97

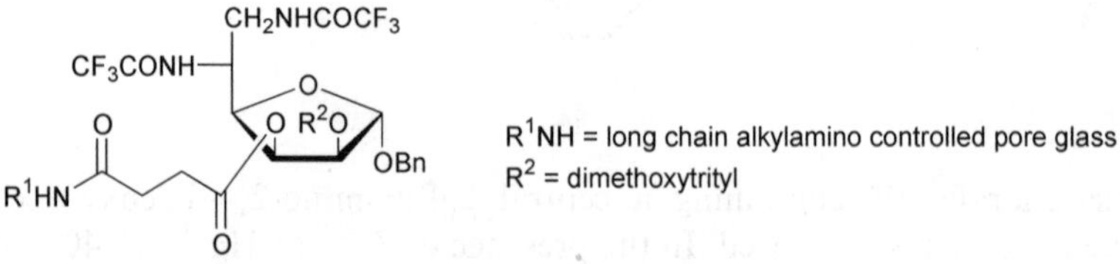

98

R^1NH = long chain alkylamino controlled pore glass
R^2 = dimethoxytrityl

References

1 Z.-D. Jiang, P.R. Jensen and W. Fenical, *Bioorg. Med. Chem. Lett.*, 1999, **9**, 2003
2 A. Itoh, Y. Ikuta, Y. Baba, T. Tanahashi and N. Nagakura, *Phytochemistry*, 1999, **52**, 1169.
3 M.H. Fechter, A.E. Stütz and A. Tauss, *Curr. Org. Chem.*, 1999, **3**, 269 (*Chem. Abstr*, 1999, **131**, 73 845).
4 L.A. Otsomaa and A.M.P. Kosikinen, *Prog. Chem. Org. Nat. Prod.*, 1998, **74**, 197 (*Chem. Abstr.*, 1999, **130**, 267 628).
5 B. Allart, R. Busson, J. Rozenski, A. Van Aerschot and P. Herdewijn, *Tetrahedron*, 1999, **55**, 6527.
6 A. Endo, T. Kann and T. Fukuyama, *Synlett*, 1999, 1103.
7 Y. Konda, T. Sato, K. Tsushima, M. Dodo, A. Kusunoki, M. Sakayanagi, N. Sato, K. Takeda and Y. Harigaya, *Tetrahedron*, 1999, **55**, 12723.
8 R.C. Reynolds, N. Bansal, J. Rose, J. Friederich, W.J. Suling and J.A. Maddry, *Carbohydr. Res.*, 1999, **317**, 164.
9 B. Furman, R. Thurmer, Z. Kaluza, W. Voelter and M. Chmielewski, *Tetrahedron Lett.*, 1999, **40**, 5909.
10 M. Balbaa, N. Abdel-Hady, F. El-Rashidy, L. Awad, E.H. El-Ashry and R.R. Schmidt, *Carbohydr. Res.*, 1999, **317**, 100.
11 J. Fuentes, M. Angulo and M. Angeles Pradera, *Carbohydr. Res.*, 1999, **319**, 192.
12 H.G. Bazin and R.J. Linhardt, *Synthesis*, 1999, 621.
13 Z. Ahmed, S.N.-U.-H. Kazmi, A.Q. Khan and A. Malik, *J. Chem. Soc. Pak.*, 1998, **20**, 54 (*Chem. Abstr.*, 1999, **130**, 52 653).
14 T. Wrodnigg and A.E. Stütz, *Angew. Chem., Int. Ed. Engl.*, 1999, **38**, 827.
15 J.J. Turner, N. Wilschut, H.S. Overkleeft, W. Klaffke, G.A. Van der Marel and J.H. van Boom, *Tetrahedron Lett.*, 1999, **40**, 7039.
16 J. Xie, A. Molina and S. Czernecki, *J. Carbohydr. Chem.*, 1999, **18**, 481.
17 K. Toshima, R. Takano, Y. Maeda, M. Suzuki, A. Asai and S. Matsumura, *Angew. Chem., Int. Ed. Engl.*, 1999, **38**, 3733.
18 D.F. Ewing, G. Goethals, G. Mackenzie, P. Martin, G. Ronco, I. Vanbaelinghem and P. Villa, *J. Carbohydr. Chem.*, 1999, **18**, 441.
19 W.A. Greenberg, E.S. Priestley, P.S. Sears, P.B. Alper, C. Rosenbohm, M. Hendrix, S.-C. Hung and C.-H. Wong, *J. Am. Chem. Soc.*, 1999, **121**, 6527.
20 Q. Li, H. Li, M.-S. Cai, Z.-J. Li and R.-L. Zhou, *Tetrahedron: Asymmetry*, 1999, **10**, 2675.
21 T.J. Donohoe, K. Blades and M. Helliwell, *Chem. Commun.*, 1999, 1733.
22 T. Vidal, A. Haudrechy and Y. Langlois, *Tetrahedron Lett.*, 1999, **40**, 5677.
23 T.M.B. de Brito, L.P. da Silva, V.L. Siqueira and R.M. Srivastava, *J. Carbohydr. Chem.*, 1999, **18**, 609.
24 B.R. Dent, R.H. Furneaux, G.J. Gainsford and G.P. Lynch, *Tetrahedron*, 1999, **55**, 6977.
25 L. Lay, L. Cipolla, B. La Ferla, F. Peri and F. Nicotra, *Eur. J. Org. Chem.*, 1999, 3437.
26 M.K. Gurjar and R. Reddy, *Carbohydr. Lett.*, 1998, **3**, 169 (*Chem Abstr.*, 1999, **130**, 81 748).
27 R. Kovarikova, M. Ledvina and D. Saman, *Collect. Czech. Chem. Commun.*, 1999, **64**, 673 (*Chem. Abstr.*, 1999, **131**, 5 452).
28 X.-C. Liu and J.S. Dordick, *J. Am. Chem. Soc.*, 1999, **121**, 466.
29 M.A. Khan and H. Adams, *Carbohydr. Res.*, 1999, **322**, 279.

30 L. Ermolenko, N.A. Sasaki and P. Potier, *Tetrahedron Lett.*, 1999, **40**, 5187.
31 B. Szechner, O. Achmatowicz and K. Badowska-Roslonek, *Pol. J. Chem.*, 1999, **73**, 1133 (*Chem. Abstr.*, 1999, **131**, 157 893).
32 S.G. Davies, G.D. Smyth and A.M. Chippindale, *J. Chem. Soc., Perkin Trans.1*, 1999, 3089.
33 M.J. Lear and M. Hirama, *Tetrahedron Lett.*, 1999, **40**, 4897.
34 Y. Matsushima, H. Arai, H. Itoh, T. Eguchi, K. Kakinuma and K. Shindo, *Tennen Yuki Kagobatsu Toronkai Koen Yoshishu,* 1997, 75 (*Chem. Abstr.*, 1999, **131**, 5 446).
35 Y.J. Kim, A. Takatsuki, N. Kogoshi and T. Kitahara, *Tetrahedron*, 1999, **55**, 8353.
36 W.-S. Zhou, Z.-H. Lu, Y.-M. Xu, L.-X. Liao and Z.-M. Wang, *Tetrahedron*, 1999, **55**, 11959.
37 S. Matsumura, E. Yao, K. Sakiyama and K. Toshima, *Chem. Lett.*, 1999, 373.
38 T. Endo, S. Koizumi, K. Tabata, S. Kakita and A. Ozaki, *Carbohydr. Res.*, 1999, **316**, 179.
39 E. Shitara, Y. Nishimura, F. Kojima and T. Takeuchi, *J. Antibiotics*, 1999, **53**, 348.
40 J.G. Fernandez-Bolanos, V. Ulgar, I. Robina and J. Fuentes, *Carbohydr. Res.*, 1999, **322**, 284.
41 M. Jahn, J.W. Baynes and A. Spiteller, *Carbohydr. Res.*, 1999, **321**, 228.
42 L. Olsson, Z.J. Jia and B. Fraser-Reid, *Pol. J. Chem.*, 1999, **73**, 1091 (*Chem. Abstr.*, 1999, **131**, 144 773).
43 F. Iglesias-Guerra, J.I. Candela, J. Bautista, F. Alcudia and J.M. Vega-Pérez, *Carbohydr. Res.*, 1999, **316**, 71.
44 Y. Hou, J. Wang, P.R. Andreana, G. Cantauria, S. Tarasia, L. Sharp, P.G. Braunschweiger and P.G. Wang, *Bioorg. Med. Chem. Lett.*, 1999, **9**, 2255.
45 T. Ziegler, H.-J. Kaisers, R. Scholmer and C. Koch, *Tetrahedron,* 1999, **55**, 8397.
46 D.J. Silva, H. Wang, N.M. Allanson, R.K. Jain and M.J. Sofia, *J. Org. Chem.*, 1999, **64**, 5926.
47 M.J. Sofia, N. Allanson, N.T. Hatzenbuhler, R. Jain, R. Kakarla, N. Kogan, R. Liang, D. Liu, D.J. Silva, H. Wang, D. Gange, J. Anderson, A. Chen, F. Chi, R. Dulina, B. Huang, M. Kamau, C. Wang, E. Baizman, A. Branstrom, N. Pristol, R. Goldman, K. Han, C. Longley, S. Midha and H.R. Axelrod, *J. Med. Chem.*, 1999, **42**, 3193.
48 C. Henry, J.-P. Joly and Y. Chapleur, *J. Carbohydr. Chem.*, 1999, **18**, 689.
49 K. Tokuyasu, H. Ono, Y. Kitagawa, M. Ohnishi-Kameyama, K. Hayashi and Y. Mori, *Carbohydr. Res.*, 1999, **316**, 173.
50 K. Tokuyasu, H. Ohno, K. Hayashi and Y. Mori, *Carbohydr. Res.*, 1999, **322**, 26.
51 M.S. Al-Thebeiti, *J. Carbohydr. Chem.*, 1999, **18**, 667.
52 A. Temeriusz, B. Piekorska-Bartoszewicz, M. Weychert and I. Wawer, *Pol. J. Chem.*, 1999, **73**, 1011 (*Chem. Abstr.*, 1999, **131**, 73 892).
53 J.M. Benito, C.O. Mellet, K. Sadalapure, T.K. Lindhorst, J. Defaye and J.M. Garcia Fernandez, *Carbohydr. Res.*, 1999, **320**, 37.
54 H. Jiao and O. Hindsgaul, *Angew. Chem., Int. Ed. Engl.*, 1999, **38**, 346.
55 J.E. Hengeveld, A.K. Gupta, A.H. Kemp and A.V. Thomas, *Tetrahedron Lett.*, 1999, **40**, 2497.
56 I. Carvalho and A.H. Haines, *J. Chem. Soc., Perkin Trans. 1,* 1999, 1795.
57 K. Ikeda, K. Miyajima, Y. Maruyama and K. Achiwa, *Chem. Pharm. Bull.*, 1999, **47**, 563.

58 D.A. Johnson, D.S. Keegan, C.G. Sowell, M.T. Livesay, C.L. Johnson, L.M. Taubner, A. Harris, K.R. Myers, J.D. Thompson, G.L. Gustafson, M.J. Rhodes, J.T. Ulrich, J.R. Ward, Y.M. Yorgensen, J.L. Cantrell and V.G. Brookshire, *J. Med. Chem.*, 1999, **42**, 4640.
59 S. Kurihara, T. Tsumuraya and I. Fujii, *Bioorg. Med. Chem. Lett.*, 1999, **9**, 1179.
60 J.M. Vega-Perez, J.I. Candela, E. Blanco and F. Iglesias-Guerra, *Tetrahedron*, 1999, **55**, 9641.
61 P.W. Smith, J.E. Robinson, D.N. Evans, S.L. Sollis, P.D. Howes, N. Trivedi and R.C. Bethell, *Bioorg. Med. Chem. Lett.*, 1999, **9**, 601.
62 G. Limberg, I. Lundt and J. Zavilla, *Synthesis*, 1999, 178.
63 H. Yuasa and H. Hashimoto, *J. Am. Chem. Soc.*, 1999, **121**, 5089.
64 S. Ogawa, N. Matsunaga, H. Li and M. Palcic, *Eur. J. Org. Chem.*, 1999, 631.
65 T.G. Mayer, R. Weingart, F. Münstermann, T. Kawada, T. Kurzchalia and R.R. Schmidt, *Eur. J. Org. Chem.*, 1999, 2563.
66 A.V. Azhayev, *Tetrahedron*, 1999, **55**, 787.

10
Miscellaneous Nitrogen-containing Derivatives

1 Glycosylamines and Related Glycosyl-*N*-bonded Compounds

1.1 Glycosylamines and Maillard Reaction Products. – N^1-(β-D-Ribofuranosyl)-4-oxonicotinamide (**1**) has been isolated from the West African plant *Rothmannia longifolia*; it had previously been reported only in human urine.[1] The mutarotation on *N*-(4-chlorophenyl)-β-D-glucopyranosylamine in MeOH, HCl has been studied by polarimetry.[2]

Direct condensation of the free sugars with the amines has been used for the synthesis of bis-*N*-glycosyl-diethylenetriamines **2**,[3] *N*-glycosylated derivatives (**3**) of the antibiotic coprafloxacin in a search for improved transport proper-

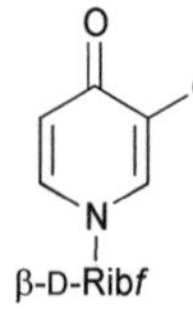

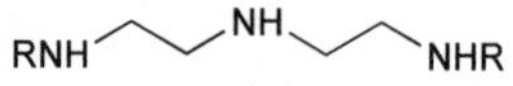

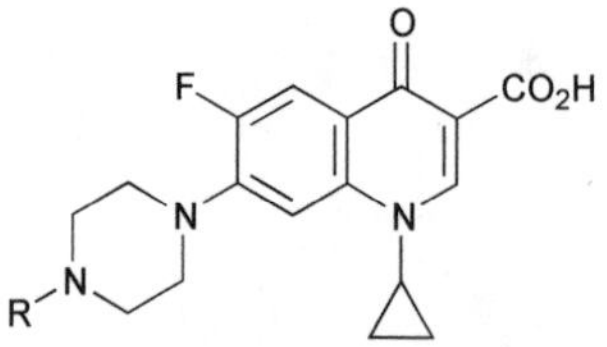

ties[4] and a variety of indolocarbazole *N*-glycosides such as **4**, related to rebeccamycin, which feature an intramolecular NH..sugar ring O hydrogen bond, especially in non-polar solvents.[5] Syntheses of the cytotoxic *S*-nitroso-fructosylamine derivative **5**,[6] and various *N*-(hexopyranosyl)amino-triazolothiadiazoles,[7] have been reported. NMR and other spectroscopic studies of alkaloidal-*N*-glycosides have been discussed.[8]

The reaction of penta-*O*-benzoyl-D-glucopyranose with piperidine has been studied with the aid of ^{14}C-labelled substrates. The first products formed are

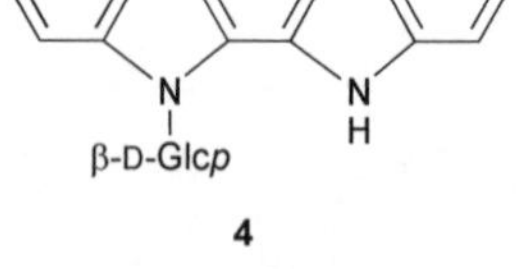

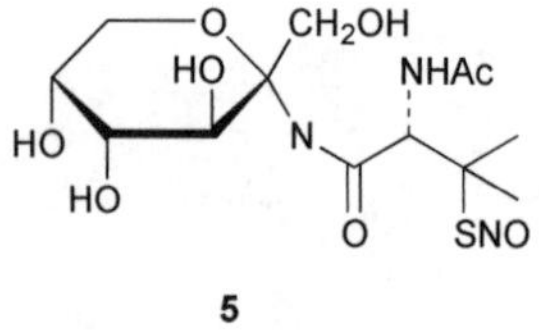

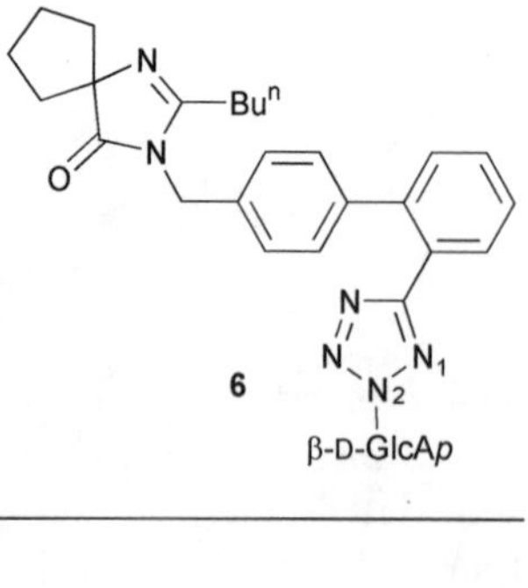

Carbohydrate Chemistry, Volume 33

N-(3,4,6-tri-*O*-benzoyl-β-D-glucopyranosyl)piperidine, which can be isolated in 44% yield, along with *N*-benzoylpiperidine (from the C-1 benzoate) and piperidinium benzoate (from the C-2 benzoate). Benzoyl migration then leads to a small amount of *N*-(2,4,6-tri-*O*-benzoyl-β-D-glucopyranosyl)piperidine.[9]

The antibiotic pyralomicin 2C was synthesized by Mitsunobu coupling of 2,3,4,6-tetra-*O*-(methoxymethyl)-D-glucose to the pyrrole nitrogen in the aglycon followed by methanolysis.[10] The *N*-2-glucuronosyl derivative **6**, a natural metabolite of the angiotensin agonist drug irbesartan, and its *N*-1-glucuronosyl isomer, have been synthesized by the Koenigs-Knorr method followed by deprotection (LiOOH, THF, H_2O).[11] A set of twelve unprotected and *O*-acetylated *N*-(D-glycopyranosyl)-imidazoles and -2-methylimidazoles were prepared by reaction of acetylated D-glucosyl, 2-deoxy-D-*arabino*-hexosyl and D-xylosyl bromides with excess of the corresponding imidazole. An NMR study showed that on protonation the anomeric equilibrium shifted to increase the proportion of axial C-1 substituent, opposite of that predicted for a reverse anomeric effect.[12]

A variety of rebeccamycin analogues have been synthesized. Rebeccamycin (**7**) and its 11-dechloro-analogue were synthesized by NaH-induced coupling of a 1,2-anhydro-α-D-glucopyranose derivative with a chloroindole-3-carboxamide then elaborating the aglycon.[13] Similarly, four symmetrical bisphenolic isomers (*e.g.* **8**) with potent activity against human topoisomerase were synthesized by treating tetra-*O*-benzyl-D-glucose with a 3-(benzyloxyindol-3-yl)succinimide derivative then elaborating the aglycon.[14] A set of such analogues varying in the substituent on the phthalimido nitrogen atom have been tested for topoisomerase inhibition, cytotoxicity and anti-cancer activity. Compound **9** was significantly active against human stomach cancer cells.[15,16]

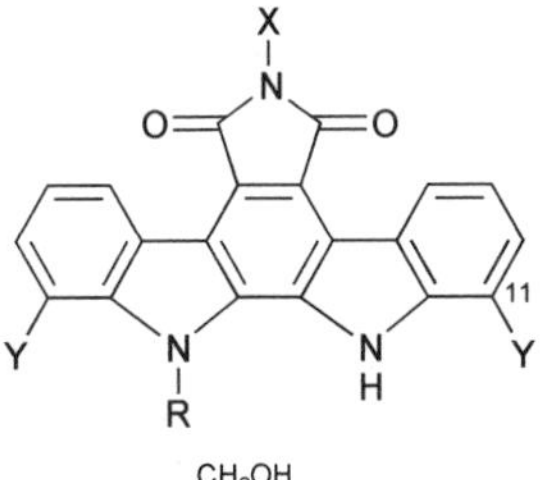

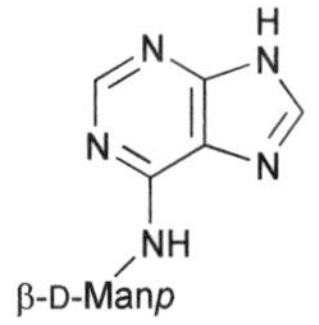

7 R = (4-*O*-methyl-β-D-glucopyranosyl), Y = Cl, X = H

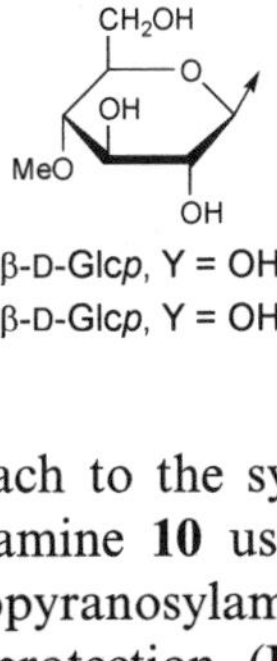

8 R = β-D-Glc*p*, Y = OH, X = H

9 R = β-D-Glc*p*, Y = OH, X = $NHCH(CH_2OH)_2$

10 (β-D-Man*p*-NH-purine)

A new approach to the synthesis of *N*-(heteroaryl)glycosylamines such as the D-mannosylamine **10** used palladium(0)-catalysed coupling of a per-*O*-benzylated-glycopyranosylamine with an *N*-protected chloro-heteroaromatic, followed by deprotection (BBr_3). These compounds are analogues of the antitumour microbial metabolites spicamicin and septacidin, varying only in the sugar residue.[17] Coupling of lactosylamine to squaric acid then the product either to a fatty amine or a solid support gave primers **11** that were used for enzymic oligosaccharide synthesis.[18] A review on the synthesis of trehazolin, which is a 2-β-D-glucosylamino-oxazolinocyclopentitol, and a number of analogues has appeared.[19]

A preparation of the oxazoline **12** involving the reaction of starch with benzonitrile and anhydrous hydrogen fluoride, followed by acetyl chloride and triethylamine, has been reported.[20] Siastatin B (**13**) has been converted into the hydrochloride salt of the aza-sugar glycosylamine **14**, a galactosidase inhibitor, by a multi-step procedure.[21] The 5-thio-D-glucosylamines **15** were synthesized by reaction of 5-thio-D-glucose pentacetate and the corresponding arylamine in the presence of $HgCl_2$. The α-anomers were shown to be modest inhibitors of glucoamylase (K_i values of 0.27–0.87 mM).[22]

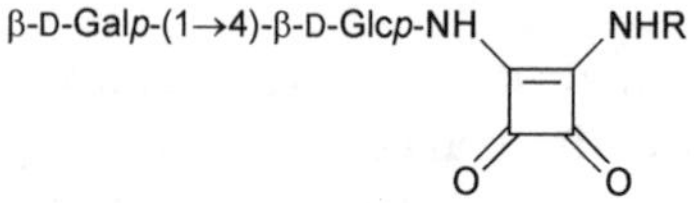

β-D-Gal*p*-(1→4)-β-D-Glc*p*-NH ... NHR

11 R = C_7H_{11}-n or agarose beads

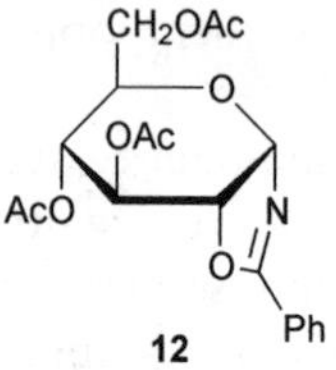

12

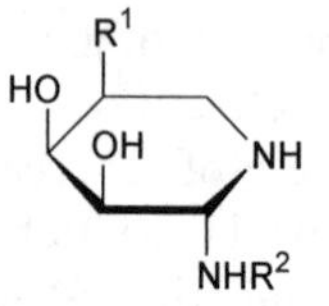

13 $R^1 = CO_2H$, $R^2 = Ac$
14 $R^1 = CH_2OH$, $R^2 = COCF_3$

The reaction of D-glucose with α-*N*-(*tert*-butoxycarbonyl)-protected L-lysine and L-arginine was studied as a model for the crosslinking of proteins under Maillard reaction conditions. The coupled products **16** were isolated from the reaction in phosphate buffer at 70 °C for 17 hours, following *N*-deprotection.[23] The blue pigment **17**, formed on Maillard-type reaction of D-xylose and glycine in aqueous bicarbonate, has been identified by NMR and mass spectrometry.[24]

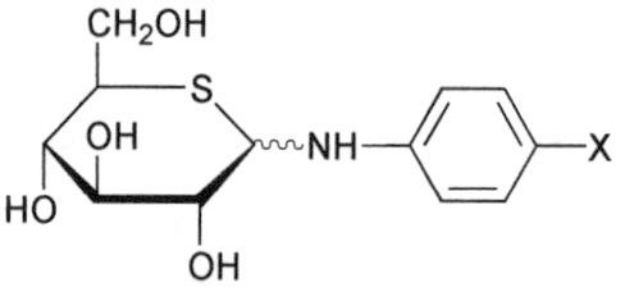

15 X = H, OMe, NO_2 or CF_3

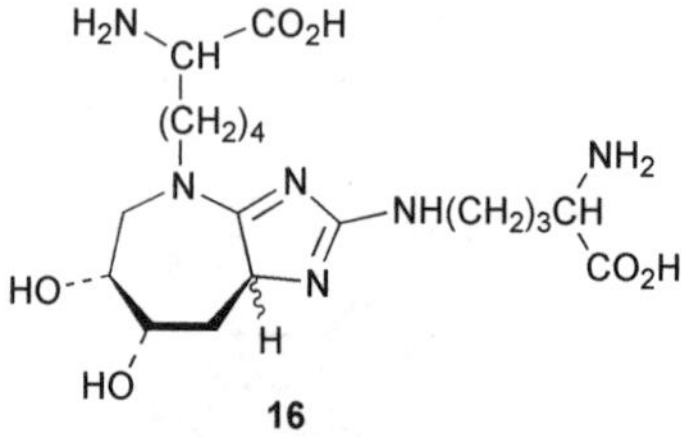

16

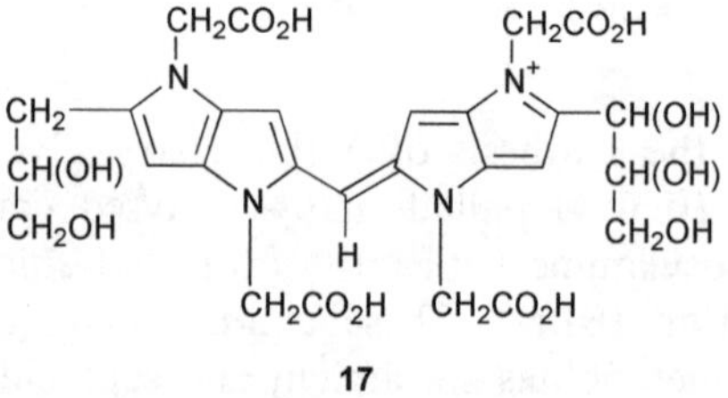

17

1.2 Glycosylamides Including *N*-Glycopeptides. – The *N*,*N'*-diacetylchitobiosylamide derivatives **18** and their monosaccharide analogues were synthesized by acylation of the corresponding *O*-acetyl protected glycosylamines followed by deprotection (NH_3). The disaccharide derivatives showed chitinase inhibitory properties.[25] A conjugate of DNA with enhanced stability (greater T_m, nuclease resistance) and lectin recognition ability was synthesized from β-lactosylamine *via* amide **19**, which was bonded to the 8-position of guanine in salmon testis DNA by diazo-coupling.[26]

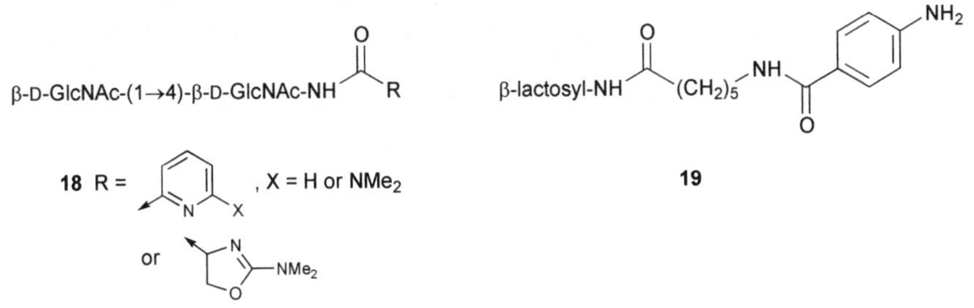

β-D-GlcNAc-(1→4)-β-D-GlcNAc-NH–C(O)–R

18 R = 2-pyridyl (X = H or NMe_2) or 2-(NMe_2)-oxazolinyl

β-lactosyl-NH–C(O)–$(CH_2)_5$–NH–C(O)–C_6H_4–NH_2

19

The β-D-glucosylamine clusters with a cyclodextrin core (**21** and **23**) were obtained as shown in Scheme 1 by condensation of iodide **20** and chloroacetamide **22**, respectively, with isothiouronium salt **24**, prepared from the corresponding glycosyl azide (with i, Bu_3P then chloroacetic anhydride; ii, thiourea). Analogous β-D-galactosyl-, α-D-mannosyl- and 2-acetamido-2-deoxy-D-glucosyl-amine clusters were also prepared.[27] The enzyme-catalysed 3- and 4-*O*-galactosylation of *N*-acetyl-β-D-glucopyranosylamine is covered in Chapter 3, and the synthesis of a ganglioside tetrasaccharide coupled *via* a spacer arm to a 2,6-dideoxy-β-D-glucopyranosylamide derivative in Chapter 4.

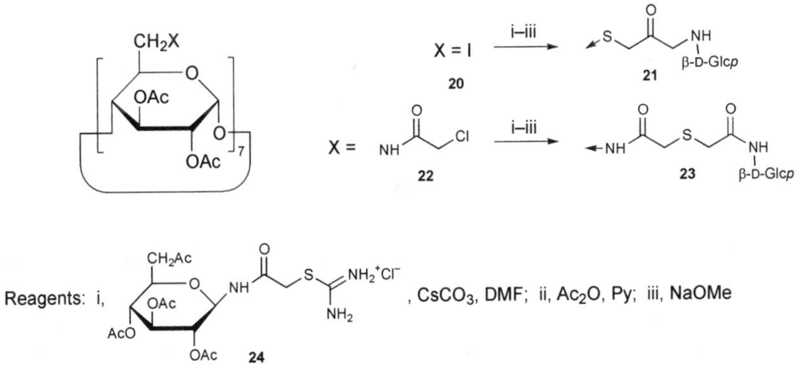

Reagents: i, **24**, $CsCO_3$, DMF; ii, Ac_2O, Py; iii, NaOMe

Scheme 1

Glycopeptides have been included in a review of recent developments in glycoconjugates.[28] Side-chain amide-linked *N*-(β-D-glucopyranosyl)- and *N*-(2-acetamido-2-deoxy-D-glucopyranosyl)-amino acids have been synthesized by condensation of the corresponding amino acid derivatives with a free carboxylic acid in the side-chain with *O*-acylated or benzylated β-glycosyl azides, activated by a trialkylphosphine.[29,30] An *O,N*-protected N^4-(β-cellobiosyl)-asparagine has been synthesized from octa-*O*-acetyl-β-cellobiosylamine and incorporated into a panel of peptides. The products were shown to be more resistant towards enzymic hydrolysis than the unglycosylated analogues.[31]

(2-Deoxy-2-fluoro-glycosylamido)-amino acid derivatives, such as **25**, have been prepared by electrophilic fluorination of acetylated D-galactal (Scheme 2), D-glucal and L-rhamnal, conversion of the product to a 1-azide, followed by reduction and coupling to an amino acid derivative.[32] The liposaccharide derivative **26** of the peptide somatostatin had the best bioavailability characteristics among a number of analogues synthesized with glycosylamide residues at the *N*- or *C*-termini.[33] Full details have been published on the of Ugi and Passerini reactions applied to tetra-*O*-acetyl-β-D-glucopyranosyl isocyanide to make various *N*-linked glycosyl α-hydroxyamides and peptides (*cf.* Vol. 32, Chapter 10, ref. 27).[34]

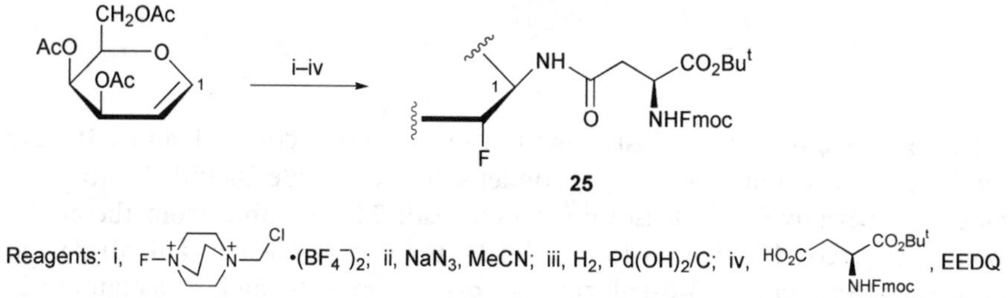

Reagents: i, F–N⁺(CH₂CH₂)₃N⁺–CH₂Cl •$(BF_4^-)_2$; ii, NaN_3, MeCN; iii, H_2, $Pd(OH)_2/C$; iv, HO_2C–CH₂–CH(NHFmoc)–CO_2Bu^t, EEDQ

Scheme 2

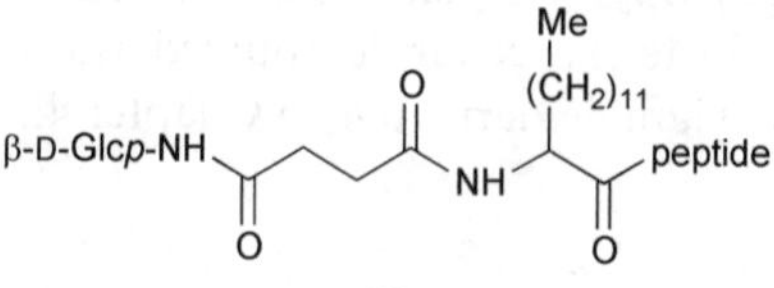

26

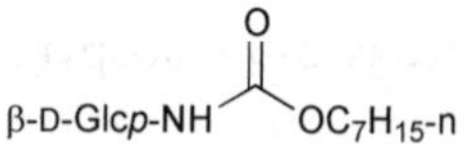

27

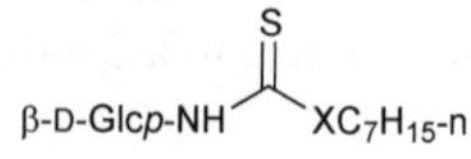

28 X = O or NH

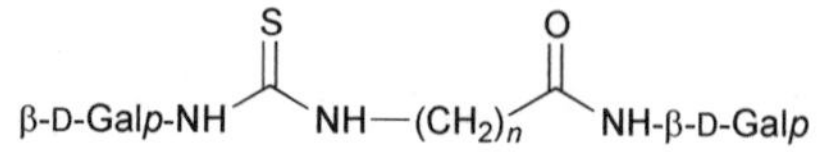

29 *n* = 6, 10 or 12

1.3 *N*-Glycosyl-carbamates, -ureas, -isothiocyanates, -thioureas and Related Compounds. – Amphiphiles with sugar and lipid head groups, such as *N*-glycosylcarbamate **27** or *N*-glycosyl-thiocarbamate or -thiourea **28**, and bolamphiphiles with two sugar groups, such as the amide- and thiourea-linked **29**, have been synthesized by conventional means and their detergent properties have been evaluated.[35,36] Condensation of 2-amino-2-deoxy-D-glucose with 2-halophenyl isothiocyanates gave the (4*R*,5*R*)-imidazolidin-2-thiones **30**, whereas 2-methoxyphenyl isothiocyanate gave the (4*R*,5*S*)-isomer **31**, although it epimerized to the more stable (4*R*,5*R*)-isomer in DMSO solution at room temperature. Acetylation at −15 °C (Ac_2O, Py) gave the corresponding penta-*O*-acetates. At 80 °C, *N*-acetylation also occurs. In DMSO solution there was rapid loss of acetic acid to give *e.g.* **32**.[37] Fused gluco-imidazolidin-2-thiones such as **32** were obtained by heating an adducts such as **30** (X = Cl) in hot aqueous acetic acid. The corresponding imidazolidin-2-ones such as **33** (Y = O) were obtained by heating 2-amino-2-deoxy-D-glucose with an aryl isocyanate. Atropisomerism (restricted rotation about the *N*-aryl bond) in these bicyclic compounds and their *O*-acetylated derivatives was evident from NMR studies, particularly for the thiones, but the barriers were insufficient to allow the atropisomers to be isolated separately.[38] Rotationally constricted analogues, isolable by chromatography, were obtained by condensation of 3,4,6-tri-*O*-acetyl-2-amino-2-deoxy-D-glucose with 2,6-disubstituted-aryl iso(thio)cyanates, for use as atropisomerically selective auxiliaries.[39]

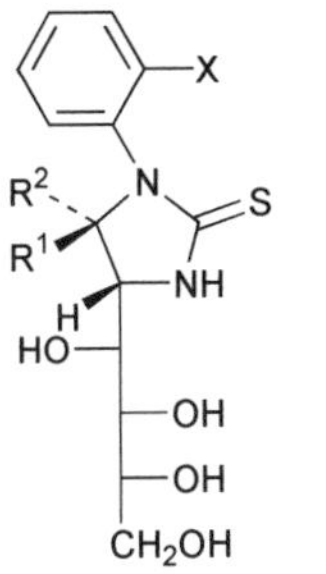

30 X = F, Cl or Br
R^1 = OH, R^2 = H
31 X = OMe
R^1 = H, R^2 = OH

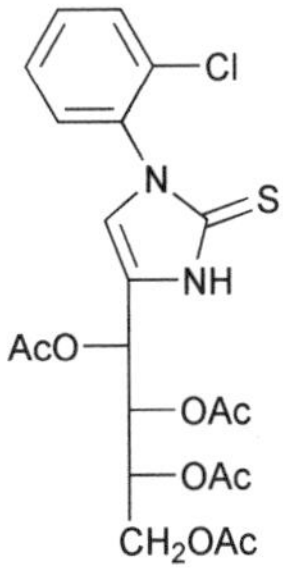

32

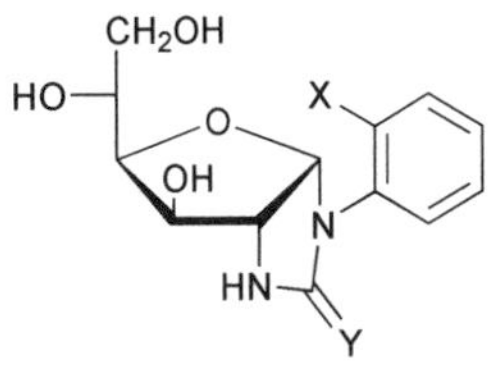

33 Y = O or S

N-(2-Deoxy-glycopyranosyl)thioureas have been synthesized from tri-*O*-acetyl-D-glucal and -D-galactal by sequential reaction with PhSeCl, KSCN, NH_3 then Bu_3SnH.[40] Various thiourea-linked oligosaccharides such as **34** have been synthesized by coupling a per-*O*-acetylated glycosyl isothiocyanate with an unprotected amino-sugar glycoside, followed by Zemplen deacetylation at 0 °C to avoid the unexpected anomerization seen at higher temperatures.[41] *N*-Glycosyl-thiosemicarbazides such as **35** have been prepared from the corresponding β-D-glucopyranosyl and β-D-xylopyranosyl isothiocyanates.[42]

2 Azido-sugars

2.1 Glycosyl Azides and Triazoles.

– Glycosyl azides such as **36** have been obtained by reaction of 1,2-anhydro-α-D-gluco-, α-D-galacto- and β-D-altro-pyranoses bearing a range of *O*-protecting groups (obtained by epoxidation of glycals) with the reagent formed from LiN_3 and Bu^i_2AlH.[43] Two groups have reported an improved, mild, high yielding synthesis of β-glycopyranosyl azides by treatment of per-*O*-acetylated glycosyl chlorides or bromides with a hypervalent azido-silicate reagents made by reaction of $TmsN_3$ with Bu_4NF or, better, TBAT [Bu_4N.(triphenyldifluorosilicate)].[44,45] A new reagent for effecting halo-azidation of glycals was prepared from the combination of $PhI(OAc)_2$, $Et_4N^+ X^-$ and $TmsN_3$. Iodo-azidation of tri-*O*-benzyl-D-galactal with this reagent led to the 2-deoxy-2-iodo-β-D-galactosyl azide **37** with high stereoselectivity, whereas the corresponding bromo-azidation gave a mixture containing three isomeric 2-bromo-2-deoxy-D-galactosyl azides.[46] A polymer-supported version of this reagent was obtained using a trimethylammonium form resin in place of the Et_4N^+ salt.[47]

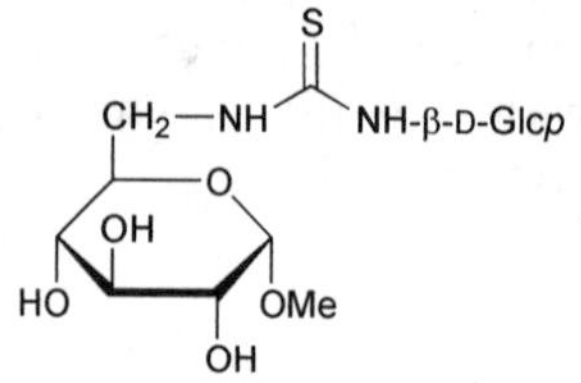

CH₂—NH—C(=S)—NH-β-D-Glc*p*; HO, OH, OMe, OH

34

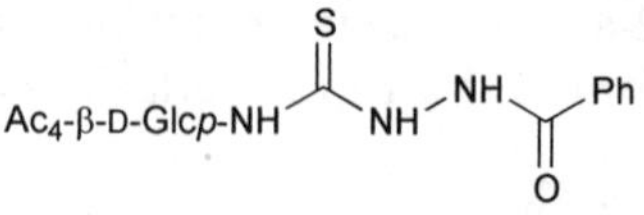

Ac₄-β-D-Glc*p*-NH—C(=S)—NH—NH—C(=O)—Ph

35

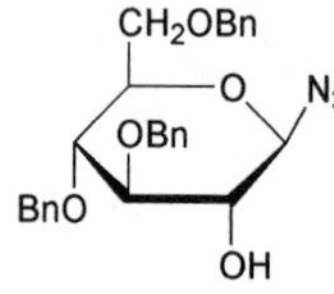

CH₂OBn, OBn, BnO, N₃, OH

36

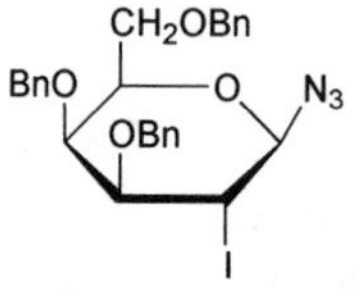

CH₂OBn, BnO, OBn, N₃, I

37

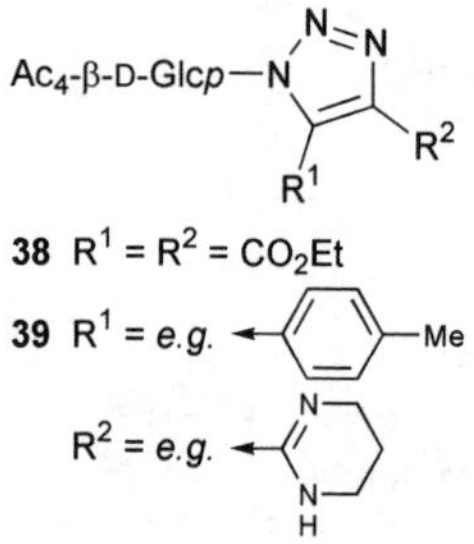

Ac₄-β-D-Glc*p*—N (triazole), R¹, R²

38 R¹ = R² = CO₂Et

39 R¹ = *e.g.* ← C₆H₄—Me

R² = *e.g.* ← (tetrahydropyrimidin-2-yl)

The triazole **38** was obtained by heating the corresponding glycosyl azide with diethyl acetylenedicarboxylate,[48] and alternatively substituted triazoles such as **39** by reaction of the azide with 2-aroylmethylene-1,3-diazacycloalkanes.[49]

The fused triazolo-derivative **40** was obtained on Staudinger reaction (with Ph_3P) of tetra-*O*-acetyl-D-glucopyranosylidene 1,1-diazide, which also involved β-elimination of HOAc and cycloaddition of azide ion. It gave the crystalline internal salt **41** on ammonolysis in methanol (Scheme 3). In contrast, the reaction of *O*-benzylated 1,1-diazide **42** with Ph_3P gave a mixture of many products, from which the lactone **43** and the corresponding ring-opened gluconamide were isolated in low yield.[50]

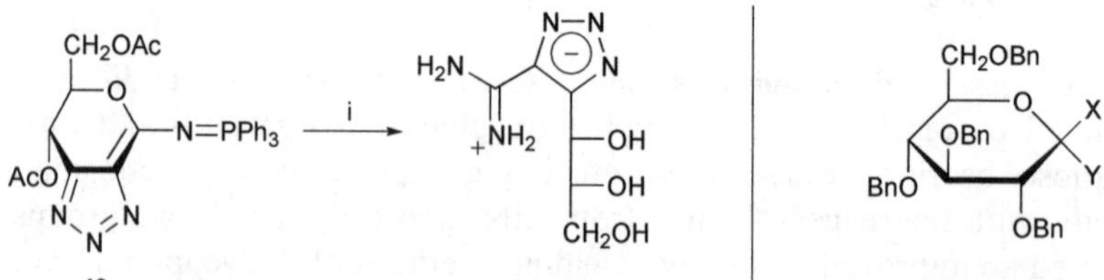

CH₂OAc, AcO, N=PPh₃, N, N, N

40

i →

H₂N, ₊NH₂, N—N, N, OH, OH, CH₂OH

41

CH₂OBn, OBn, BnO, OBn, X, Y

42 X = Y = N₃

43 X,Y = O

Reagent: i, NH₃, MeOH

Scheme 3

2.2 Other Azides and Triazoles. – Several 2-azido-2-deoxyglycosyl donors such as phenyl 4,6-di-*O*-acetyl-2-azido-2-deoxy-3-(4-methoxybenzyl)-1-thio-β-D-glucopyranoside have been made from the corresponding mannoside-2-triflates by displacement with azide.[51] The hypervalent azido-silicate reagent made using $TmsN_3$ and Bu_4NF proved quicker and more efficient for nucleophilic displacement of primary tosylate groups, *e.g.* 1,2,3,4-tetra-*O*-acetyl-6-azido-6-deoxy-β-D-mannopyranose was obtained in 78% yield on reaction of the corresponding 6-tosylate with this reagent for 4 hours, whereas the yield from reaction with NaN_3 in DMF required 48 hours, reacetylation of the crude product, and gave yields in the 30–75% range.[45] A variety of 6-azido-6-deoxy-sugar derivatives have been prepared in yields >80% by reaction of the corresponding 4,6-diols or 5,6-diols, presumably *via* intermediate cyclic phosphoranes as exemplified in Scheme 4, although in the case of methyl 2,3-*O*-benzyl-α-D-galactopyranoside, the major product was methyl 4-azido-2,3-*O*-benzyl-4-deoxy-α-D-galactopyranoside (66%).[52] Similarly, reactions of cyclic 5,6-sulfates with NaN_3 in DMF gave the corresponding 6-azido-6-deoxy-sugars in high yields.[53]

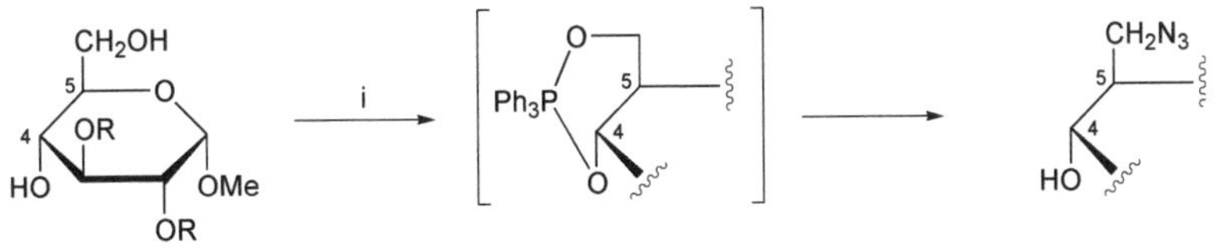

R = Bn or Ac

Reagents: i, Ph_3P, $Pr^iO_2CN{=}NCO_2Pr^i$, then $TmsN_3$

Scheme 4

Reaction of the azido, nitrate ester, and disulfide derivatives **44**–**46** of 5-homoribose with tributyltin deuteride has been studied to probe the mechanism of ribonucleotide reductase action, which involves abstraction of H-3 by a cysteinethiyl radical. While the first two suffered intramolecular abstrac-

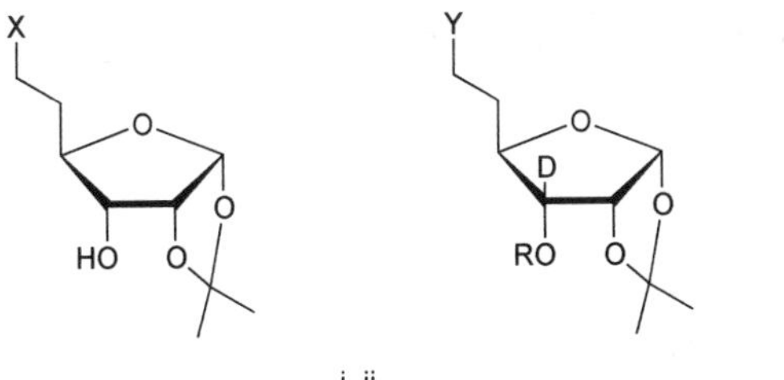

44 X = N_3 —(i, ii)→ **47** Y = NHAc, R = Ac

45 X = ONO_2 —(i)→ **48** Y = OH, R = H

46 X = $-S\frac{1}{2}$

Reagents: i, Bu_3SnD, AIBN; ii, Ac_2O, Py

Scheme 5

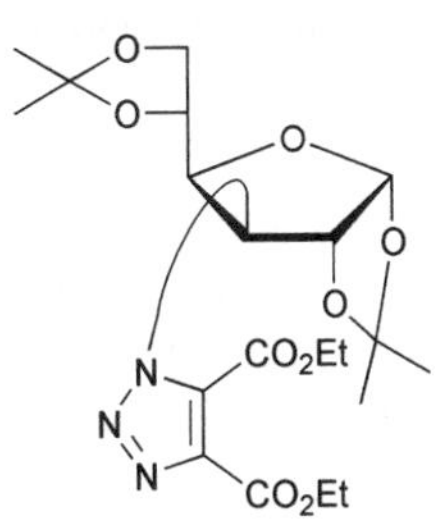

49

tion of H-3 to give products **47** and **48** with a C-3 deuterium atom (Scheme 5), the last gave the 6-thiol but without deuterium incorporation.[54]

The synthesis of triazole **49** by reaction of the corresponding 3-azide with diethyl acetylenedicarboxylate, and some of its reactions have been reported.[48]

3 Nitro-sugars

2-Nitro-sugar derivative **50** has been converted to a variety of 2-nitro-3-amino- and -3-thio derivatives such as **51** by elimination (MsCl, Et_3N) and addition sequences, and thence to 2,3-diamino-sugar derivatives.[55] The *O*-α-D-GalNAc*p*-serine and -threonine derivatives have been synthesized by way of base-catalysed Michael addition of serine and threonine derivatives to 3,4,6-tri-*O*-benzyl-2-nitro-D-galactal.[56] Baer reaction of dialdehyde **52** with nitromethane is known to give a crystalline 1:1 mixture of 3-*C*-methyl-3-nitro-α-D-glucoside **53** and its β-L-glucoside isomer. These have now been separated as their 4,6-*O*-benzylidene derivatives and converted into fluoro-sugar derivatives (see Chapter 8).[57]

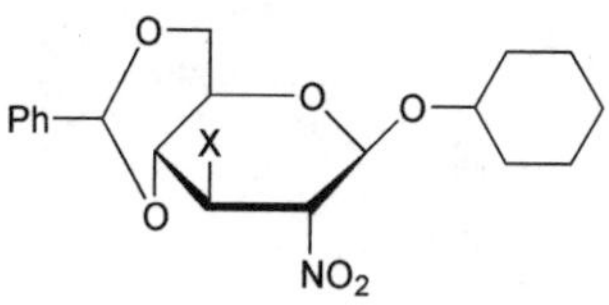

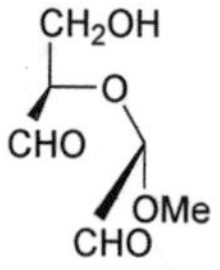

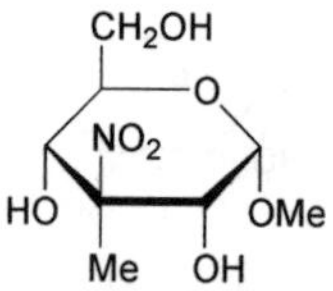

50 X = OH
51 X = NHPr^i, ... , SBu^t

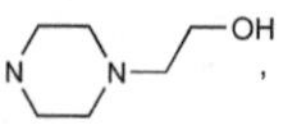

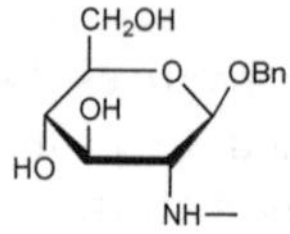

52

53

4 Oximes, Hydroxylamines, Nitrones and Isonitriles

Nitrosylsulfuric acid in ether, acetic acid or formic acid, has been used to effect the de-oximation of the 2-oxime derivatives of 2-keto-aldonic acids, a key final step in a known route to these compounds.[58]

The cyclic nitrone **54** has been synthesized in 11 steps from a glucoside derivative, the keys final steps being shown in Scheme 6. Addition of cyanide to **54** gave mainly the 1-*C*-cyano derivative **55**, which was converted to a range of *N*-bridged bicyclic glycosidase inhibitors (see Chapter 18).[59] Similarly, the D-*ribo*-configured cyclic nitrone **57** has been synthesized from 2,3,5-tri-*O*-benzyl-D-ribose by synthesis of its *O*-(*tert*-butyldiphenylsilyl)oxime derivative **56**, inversion at C-5 then cyclization (Scheme 7). It underwent facile addition of Grignard reagents and lithiated carbanions to give 4-aza-1-*C*-β-glycosides

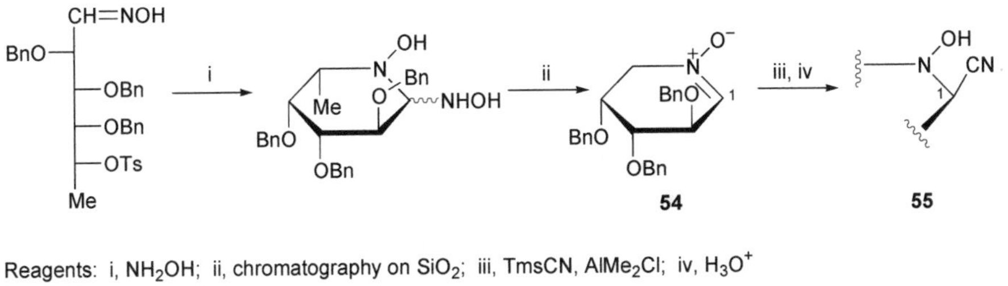

Reagents: i, NH_2OH; ii, chromatography on SiO_2; iii, TmsCN, $AlMe_2Cl$; iv, H_3O^+

Scheme 6

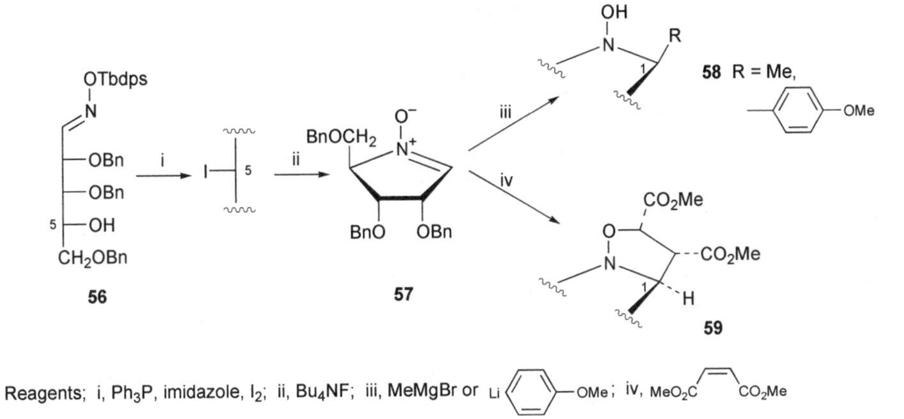

Reagents; i, Ph_3P, imidazole, I_2; ii, Bu_4NF; iii, MeMgBr or Li-C_6H_4-OMe; iv, MeO_2C-CH=CH-CO_2Me

Scheme 7

and -nucleosides, *e.g.* **58**, and dipolar cycloadditions with alkenes to give isoxazolidine derivatives, *e.g.* **59**.[60] The syntheses of 46 new sugar nitrone derivatives and their antibacterial activities have been reported.[61]

Reductive radical cyclization of a chitobiose oxime 5-xanthate derivative to form an *O*-glycosylated diaminocyclitol derivative, a key step in the synthesis of potential chitinase inhibitors, and the SmI_2-mediated reductive cyclization of a 5-hexulose 1-(*O*-benzyloxime), a key step in the synthesis of an epimer of the cyclitolamine trehalosamine, are covered in Chapter 18.

Glycodendrimers such as **60** have been obtained by direct condensation of glucose, lactose or sialic acid with a polyamidoamine (PAMAM) dendrimer with eight *O*-acylhydroxylamine substituents.[62]

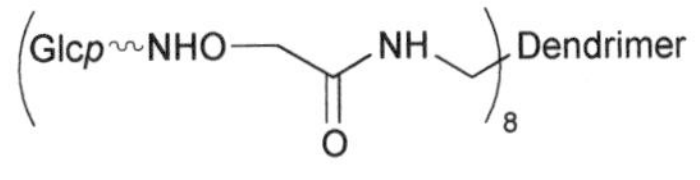

60

3-Hydroxylamino-sugar lactone **62** has been obtained by Michael addition of hydroxylamine to unsaturated lactone **61**. Similar reaction with hydrazine gave the cyclic derivative **63** which could be transformed into the furanolactone **64** (Scheme 8).[63] Strategies employed previously to make hydroxylamino-

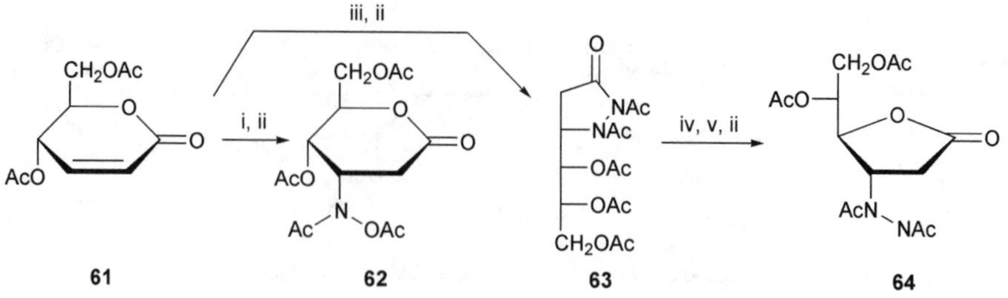

Reagents: i, NH_2OH; ii, Ac_2O, Py; iii, NH_2NH_2; iv, MeOH; v, MeOH, NH_3

Scheme 8

linked disaccharide analogues (*cf.* Vol. 24, p. 133) have been used to synthesize the trisaccharide analogue **65**. Attempts to remove the methoxymethyl protecting group in **65** under acidic conditions caused degradation.[64]

A set of eight cellobiose and -triose analogues containing known glycosidase inhibiting motifs were synthesized by standard or known methods and evaluated as inhibitors of cellobiohydrolases 6A and 7A. While the oximino-lactam derivatives **66** and others were micromolar inhibitors of 6A, none were as potent against 7A.[65] The isoxazoline derivative **67** was a selective β-galactosidase inhibitor (K_i 18 μM), but a number of related compounds that had been obtained as intermediates in the synthesis of iminoalditols showed little inhibition against a panel of 25 glycohydrolases.[66]

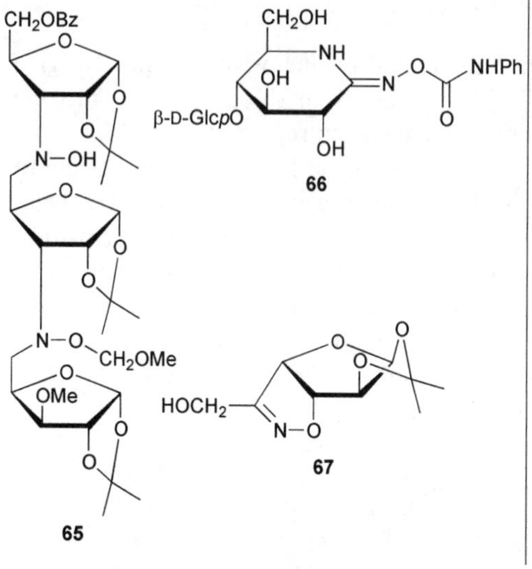

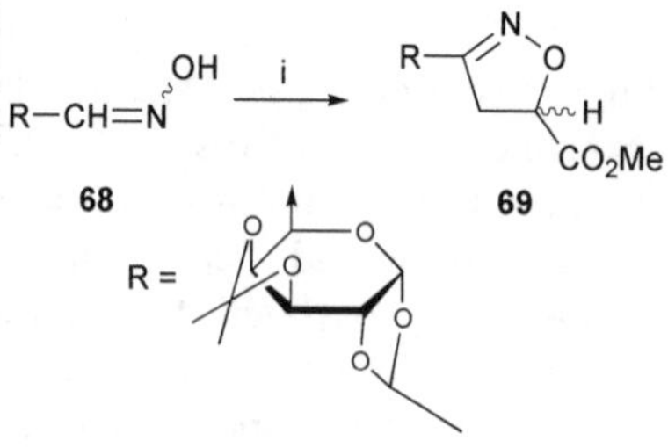

Reagents: i, $CH_2{=}CHCO_2Me$, MnO_2

Scheme 9

The sugar oxime **68** could be oxidized to a nitrile oxide and trapped *in situ* to form the isoxazolines **69** (Scheme 9).[67] The homologous nitrile oxides **70**, derived from diacetoneglucose, underwent intramolecular dipolar cycloadditions to the tetracyclic derivatives **71** and, in each case, an isomer (Scheme 10). Compound **71** ($n = 1$) was further transformed into an oxepanocyclohexane

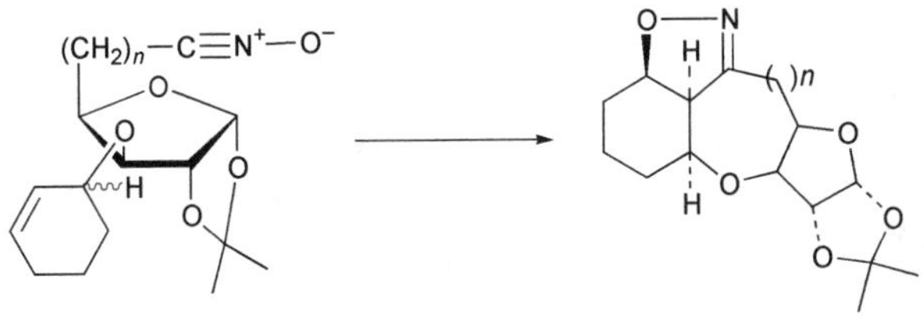

70 $n = 0$ or 1

71 $n = 0$ or 1

Scheme 10

derivative having the skeleton of the natural products forskolin and lasalocid.[68] The intramolecular cycloaddition reactions of *N*-allyl-carbohydrate nitrones such as **72** have been reported. The regioselectivity of the cycloaddition, leading either to six- or seven-membered nitrogen heterocycles (*e.g.* **73** and **74**, respectively), could be controlled by changing the substituent on the nitrogen atom of the *N*-allyl group (Scheme 11).[69]

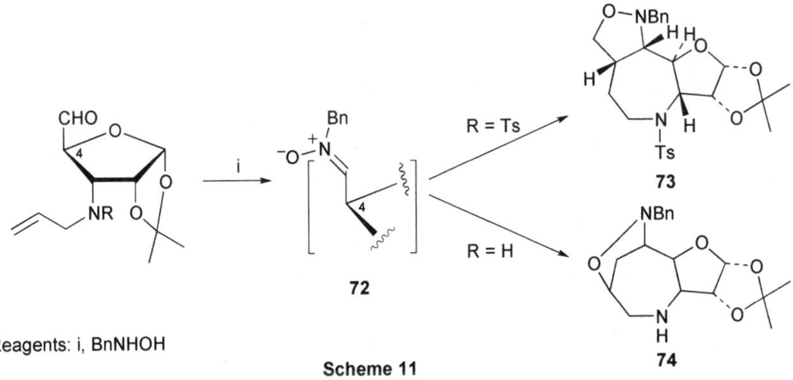

Reagents: i, BnNHOH

Scheme 11

Oxidative intramolecular cycloaddition of the oxime **75** (R = Bn) derived from D-glucose led to isoxazolines **76** that could be further transformed into aminocyclitol derivative **77** (R^2 = Ac, Scheme 12). Similar reactions of oximes and nitrones derived from 5,6-dideoxy-D-*ribo*-hept-5-enose were also re-

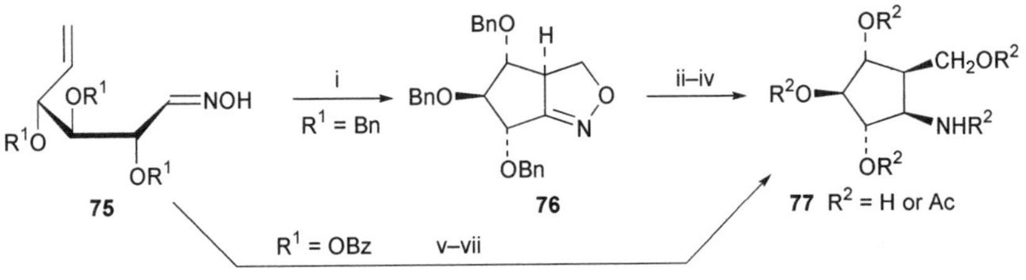

Reagents: i, NaOCl, H_2O, Et_3N, CH_2Cl_2; ii, $NaBH_3CN$, AcOH; iii, H_2, Raney Ni; iv, Ac_2O, Py, DMAP; v, C_6H_6, Δ; vi, NaOMe; vii, H_2, Pd/C

Scheme 12

ported.[70] The same aminocyclopentitol **77** ($R^2 = H$) was obtained directly by thermal cyclization of oxime **75** (R = Bz).[71]

The spirocyclic isoxazole acetals such as **78** have been prepared by nitrile cycloaddition reactions applied to furanoside and pyranoside exocyclic alkenes. Cyclopentanone *exo*-alkenes such as **79** were obtained by intramolecular aldol-like condensations upon reductive opening of the heterocyclic ring (Scheme 13).[72] Syntheses of pyranosido-fused pyridazines and oxazines from 2,3-anhydropentopyranosides are covered in Chapter 14.

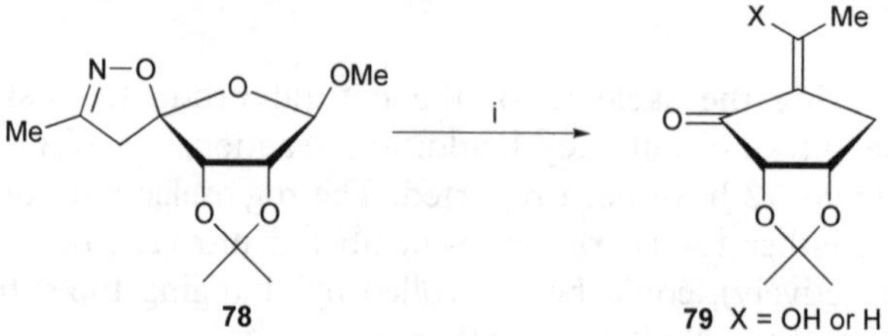

78 → **79** X = OH or H

Reagents: i, H_2, Raney Ni

Scheme 13

5 Nitriles, Tetrazoles and Related Compounds

The synthesis of α,β-glycosyl cyanides by Lewis acid-catalysed reaction of Me_3SiCN with per-*O*-benzylated *S*-α-D-glycopyranosyl phosphorothioates has been described.[73] The novel transformation of the glycosyl bromides such as **80** into *N*-(1-cyano-D-glycopyranosyl)amides such as **81** involved a Ritter-like reaction of nitriles with glycosylium ions (Scheme 14).[74] The epimeric 1-cyano-1-thiocyanato-derivative **83** was synthesized from the bromide **82** (Scheme 15). They could not be thermally isomerized to the corresponding 1-isothiocyanato-derivatives, but gave a variety of spirocyclic products such as **84** on reaction with H_2S.[75]

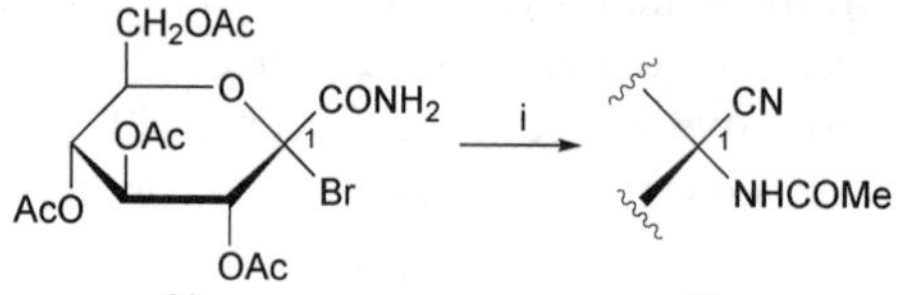

80 → **81**

Reagents: i, Ag_2CO_3, MeCN

Scheme 14

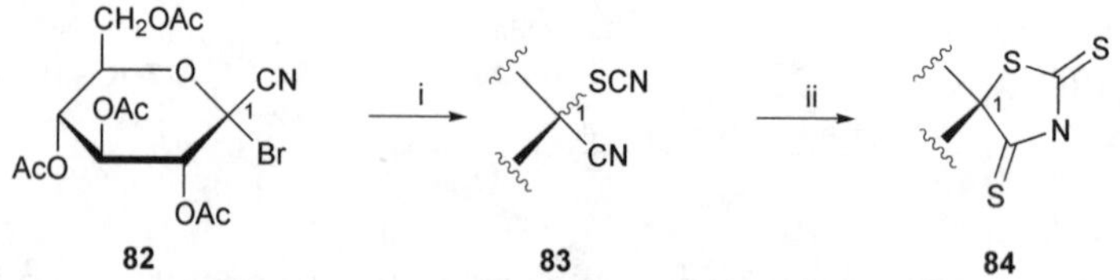

82 → **83** → **84**

Reagents: i, AgSCN or KSCN, $MeNO_2$; ii, H_2S, Et_3N, EtOH, $CHCl_3$

Scheme 15

The 6-cyanide **85**, obtained from the corresponding 6-oximino-derivative, and the known aminonitrile **86** and its C-7 epimer were converted to tetrazoles (by addition of HN_3) and then 1,3,4-oxadiazole derivatives (by acylation).[76,77] The 1,2,4-oxadiazole derivative **88** has been obtained from the cyanohydrin derivative **87** (Scheme 16).[78] The epimeric cyanohydrin derivative **89** has been converted into the spirocyclic derivatives **90** (Scheme 17).[79] Pentitol-1-yl-triazoles such as **91** have been obtained by reaction of perbenzoylated D-mannononitrile with diaza-allene salts such as **92**.[80]

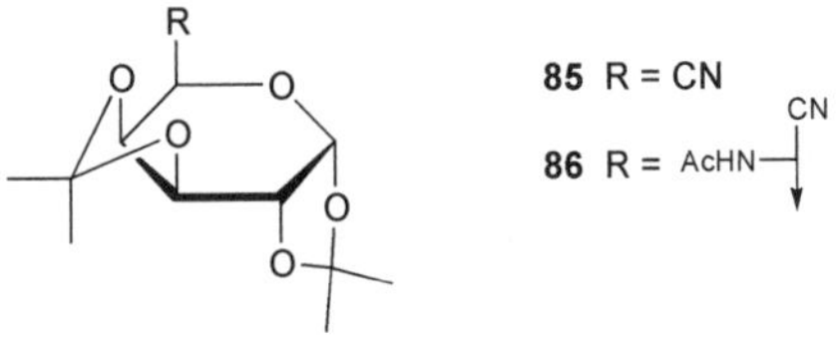

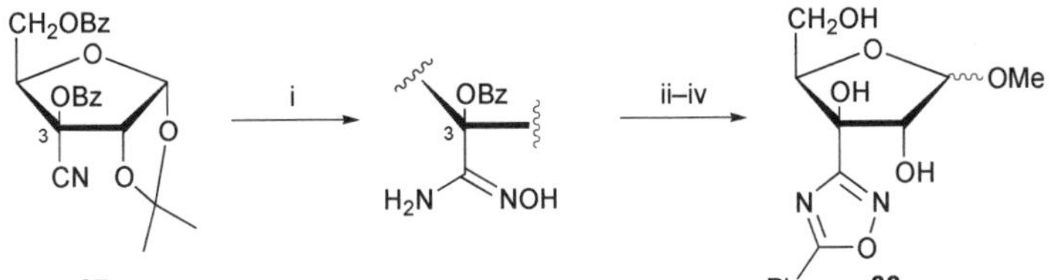

85 R = CN

86 R = AcHN–C(CN)

Reagents: i, NH_2OH; ii, PhCOCl; iii, NaOMe, MeOH; iv, HCl, MeOH

Scheme 16

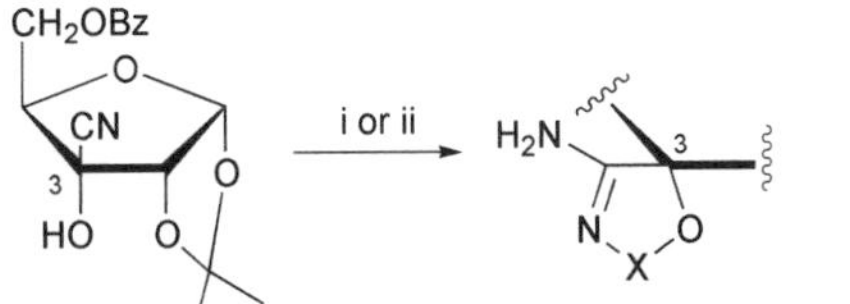

90 R = C=O or SO_2

Reagents: i, $ClSO_2NCS$ then aq. $NaHCO_3$; ii, $ClSO_2NH_2$

Scheme 17

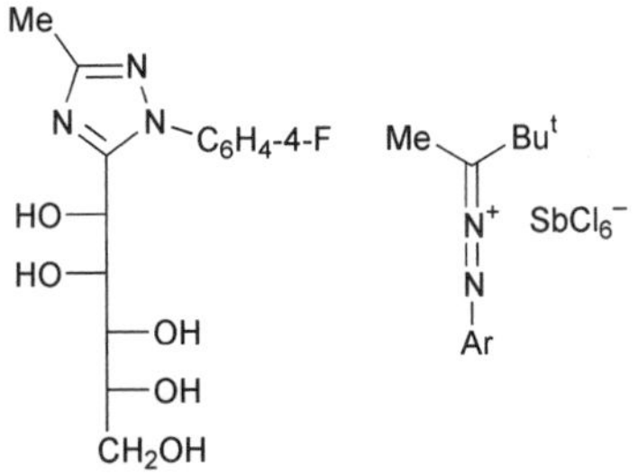

6 Hydrazines, Hydrazones and Related Compounds

Racemic [3-^{13}C]azafagomine **93** has been synthesized utilizing a route reported earlier (Bols *et al.*, *Chem. Eur. J.*, 1997, **3**, 940) (Scheme 18).[81] Azafagomine has been converted to the derivative **94** and used to make a combinatorial library of 125 variable peptides bonded to the NH_2-group, the first such library of azasugar inhibitors.[82]

By X-ray crystallography, the tosylhydrazone derivatives of D-galactose, D-glucose and L-arabinose have been shown to adopt the β-pyranose rather than the acyclic form in the solid state.[83] Tetrazines **96** were formed from **95** by

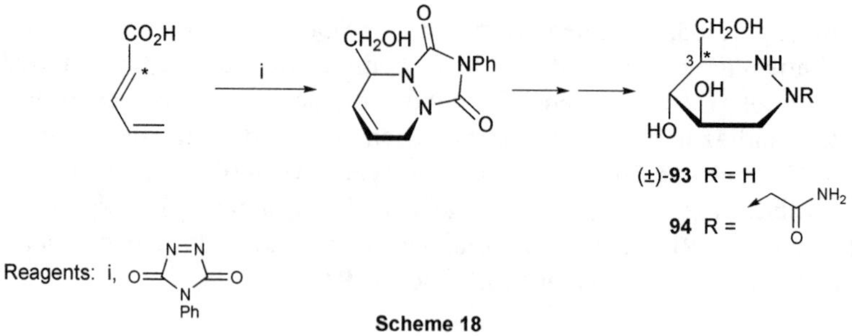

Reagents: i, [4-phenyl-1,2,4-triazoline-3,5-dione]

(±)-93 R = H

94 R = $CH_2C(O)NH_2$

Scheme 18

normal demand Diels-Alder reactions in benzene under microwave irradiation (Scheme 19).[84] Alditol-1-yl substituted 1,2,4-benzotriazines **97** and benzamidazoles **98** were obtained by reduction (H_2, Pd/C) then cyclization (O_2, OH^- or H^+, respectively) of 2-nitrophenylhydrazone derivatives of D-arabinose, D-galactose and D-galacturonic acid.[85] The oxidative cyclization (Br_2, HOAc) of the acetylated hydrazone derivatives formed from monosaccharides and *N*-(5-methyl-1,2,4-triazino[5,6-*b*]indol-3-yl)hydrazine has been reported.[85]

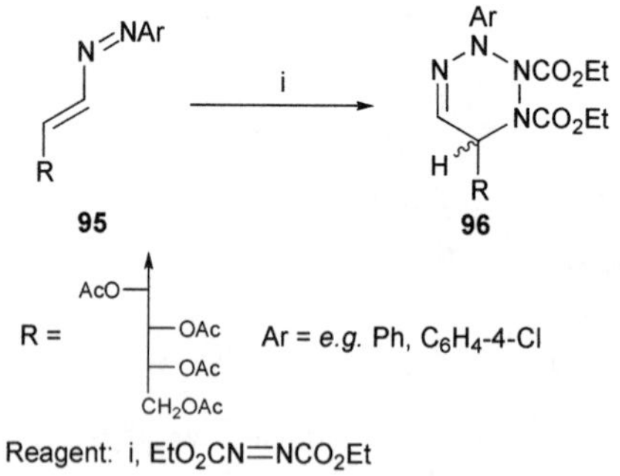

Ar = *e.g.* Ph, C_6H_4-4-Cl

Reagent: i, $EtO_2CN{=}NCO_2Et$

Scheme 19

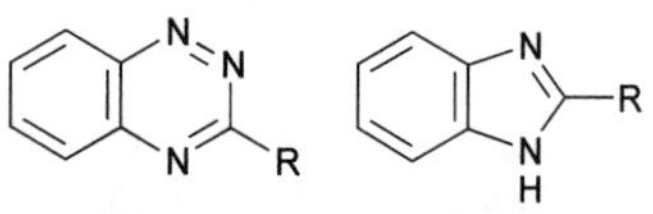

R = alditol-1-yl

7 Other Heterocycles

1-Amino-1-deoxy-D-glucitol has been converted into the 1,2,4-triazine derivative **99**.[87] Various 4-*C*-(tetrafuranosyl)-imidazoles have been obtained by dehydration of the corresponding 4-*C*-(tetritol-1-yl)-imidazoles through pyrolysis of their hydrochloride salts. The same products are seen following reaction of hexoses or hexuloses with formamidine.[88]

Rigid azabicyclo-α-L-fucose mimic **100** has been converted to various bridgehead substituted analogues such as **101** following photobromination at this site.[89]

Further studies (*cf.* Vol. 32, Chapter 9, ref. 13) have been reported on the conversion of 6-*O*-leucyl-enkephalyl-D-mannose either to an Amadori product on reaction in Py and HOAc, or to cyclic derivatives on heating in methanol and thence to the imidazolone **102** on ester hydrolysis.[90] The spirocyclic derivative **103** was obtained by *N*-alkylation of a hydantocidin derivative and

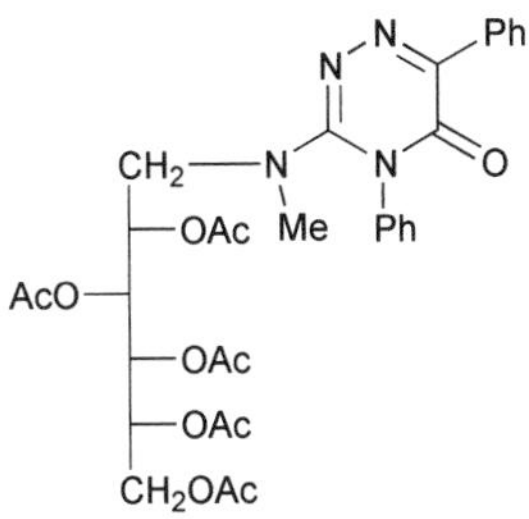

99

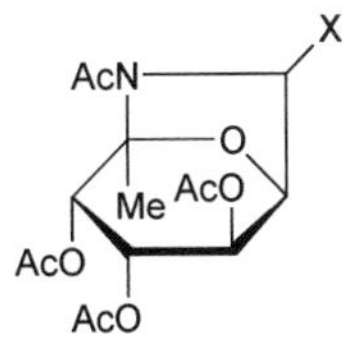

100 X = H
101 X = *exo*-OMe, *exo*-Br or *endo*-N=NNH$_2$

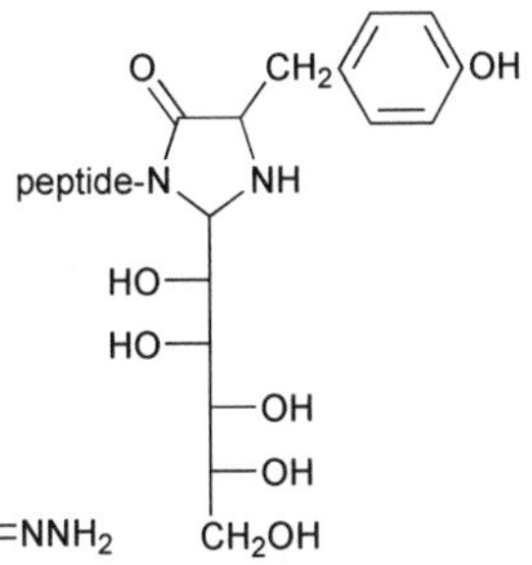

102

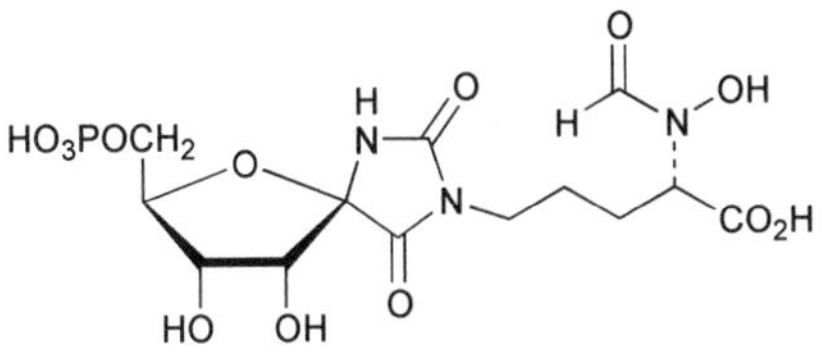

103

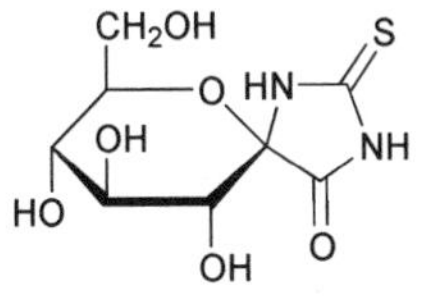

104

shown to be a bisubstrate hybrid inhibitor of adenylsuccinate synthetase (IC_{50} 0.2 μM).[91] The spiro-thiohydantocidin **104**, a potent inhibitor of muscle and liver glycogen phosphorylases, was synthesized by reaction of 2,3,5,7-tetra-*O*-acetyl-α-D-*gluco*-heptonamido-2-ulos-2-yl bromide with AgSCN followed by deacetylation. Hydantocidins were made similarly, but had the opposite configuration at the anomeric centre.[92] Spiro-thiazolidinone derivatives of 2-amino-2-deoxy-D-glucose are covered in Chapter 9, Scheme 20.

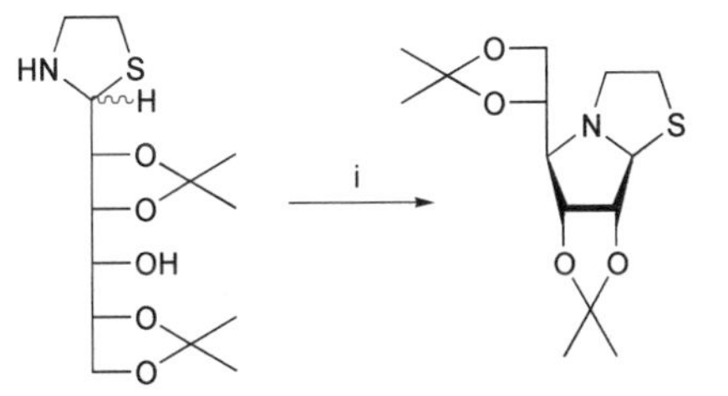

105 106 107

Reagent: i, BuLi then TsCl

Scheme 20

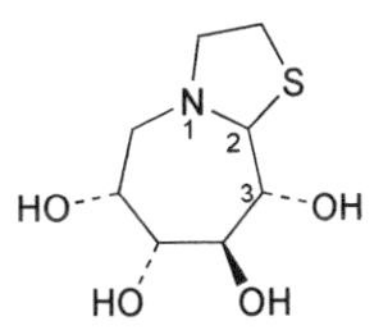

Condensation of unprotected or partially protected aldoses with 2-aminoethanethiol gave adducts with a 1,3-thiazolidine structure, such as **105**, whereas condensation with 1,2- or 1,3-aminopropanols gave mixtures of products with α- and β-pyranosylamine structures.[93,94] Tosylation of **105** gave the cyclized product **106** (Scheme 20). Reaction of 6-*O*-mesyl-D-glucose or -D-mannose with 2-aminoethanethiol gave the 1,3-thiazolo[3,2-*a*]azepine **107** or the expected 3-epimer, respectively.[94] Similar condensations involving 5-bromo-5-deoxy-D-xylose provided a simplified route to bicyclic derivatives **108**

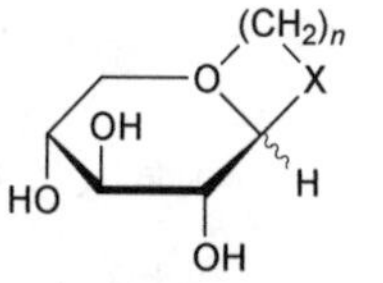

108 X = O, NH, S; *n* = 2
109 X = O, NH; *n* = 3

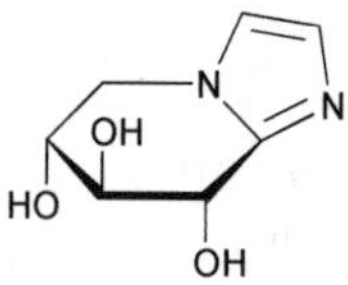

110

and **109**, which were exclusively β-anomers where X = N, but α-anomers in part when X = O or S at equilibrium in solution.[95] Equivalent results were obtained with 5-*O*-tosyl-D-lyxose.[96]

The D-xylopyranoimidazole **110** has been synthesized from diacetoneglucose.[97] The regioisomeric D-*gluco*- and D-*manno*-analogues **112** and **113** have been synthesized from the thionolactam derivative **111** (Scheme 21), the first step in each case giving rise to a pair of separable C-2 epimers, They were evaluated as glycosidase inhibitors.[98] Full details have been published (*cf.* Vol. 29, p. 164; Vol. 30, p. 154) for the synthesis of D-manno-, D- and L-rhamno-furano- and pyrano-tetrazoles.[99,100] Dihydropyridine derivatives such as **115** were obtained from nitroenone **114** (Scheme 22) and could be oxidized with $Ce(NH_4)_2(NO_3)_6$ to the corresponding quinoline derivatives.[101] Annelated pyranose derivatives derived from levoglucosenone are covered in Chapter 14.

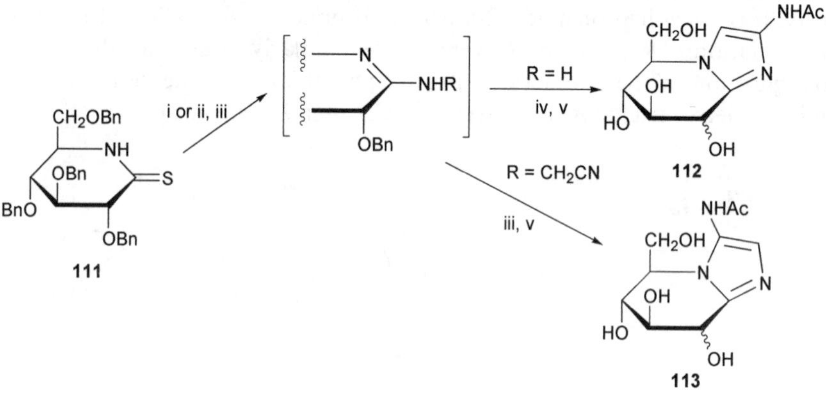

111 **112** **113**

Reagents: i, NH_3, $Hg(OAc)_2$; ii, NH_2CH_2CN, $Hg(OAc)_2$; iii, Ac_2O, Py; iv, ICH_2CN, K_2CO_3; v, H_2, Pd/C

Scheme 21

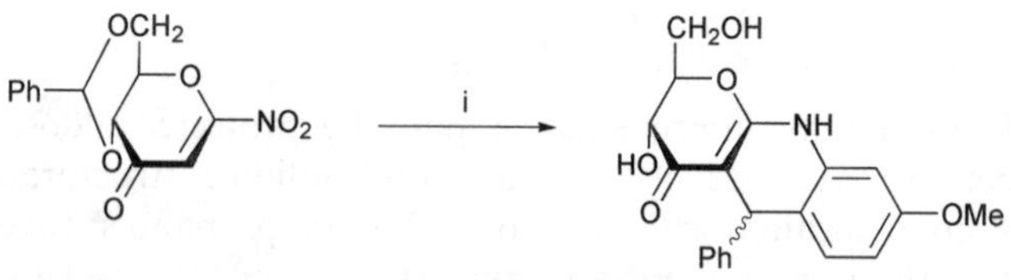

114 **115**

Reagents: i, H_2N–C$_6$H$_4$–OMe; ii, PhCHO

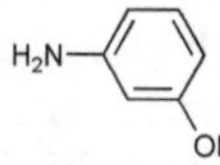

Scheme 22

References

1 G. Bringmann, M. Ochse, K. Wolf, J. Kraus, K. Peters, E.-M. Peters, M. Herderich, L.A. Assi and F.S.K. Tayman, *Phytochemistry*, 1999, **51**, 271.
2 K. Smiataczowa, *Pol. J. Chem.*, 1999, **73**, 1513 (*Chem. Abstr.*, 1999, **131**, 299 630).
3 S.P. Gaucher, S.F. Pedersen and J.A. Leary, *J. Org. Chem.*, 1999, **64**, 4012.
4 M.E. Jung, E.C. Yang, B.T. Vu, M. Kiankarimi, E. Spyrou and J. Kaunitz, *J. Med. Chem.*, 1999, **42**, 3899.
5 E.J. Gilbert, J.D. Chisholm and D.L. Van Vranken, *J. Org. Chem.*, 1999, **64**, 5670.
6 Y. Hou, J. Wang, P.R. Andreana, G. Cantauria, S. Tarasia, L. Sharp, P.G. Braunschweiger and P.G. Wang, *Bioorg. Med. Chem. Lett.*, 1999, **9**, 2255.
7 J.-X. Yu, F.-M. Liu, W.-J. Lu, Y.-P. Li, L.-L. Tian, Y.-T. Liu, C. Liu and M.-S. Cai, *Gaodang Xuexiao Huaxue Xuebao,* 1999, **20**, 1233 (*Chem. Abstr.*, 1999, **131**, 322 835).
8 Z. Ahmed, S.N.-U.-H. Kazmi, A.Q. Khan and A. Malik, *J. Chem. Soc. Pak.*, 1998, **20**, 54 (*Chem. Abstr.*, 1999, **130**, 52 653).
9 A.E. Salinas, *Carbohydr. Res.*, 1999, **316**, 34.
10 K. Tatsuta, M. Takahashi and N. Tanaka, *Tetrahedron Lett.*, 1999, **40**, 1929.
11 S.J. Byard and J.M. Herbert, *Tetrahedron*, 1999, **55**, 5931.
12 C.L. Perrin, M.A. Fabian, J. Brunckova and B.K. Ohta, *J. Am. Chem. Soc.*, 1999, **121**, 6911.
13 M.M. Faul, L.L. Winneroski and C.A. Krumrich, *J. Org. Chem.*, 1999, **64**, 2465.
14 D.E. Zembower, H. Zhang, J.P. Lineswala, M.J. Kuffel, S.A. Aytes and M.M. Ames, *Bioorg. Med. Chem. Lett.*, 1999, **9**, 145.
15 M. Ohkubo, K. Kojiri, H. Kondo, S. Tanaka, H. Kawamoto, T. Nishimura, I. Nishimura, T. Yoshinari, H. Arakawa, H. Suda, H. Morishima and S. Nishimura, *Bioorg. Med. Chem. Lett.*, 1999, **9**, 1219.
16 M. Ohkubo, T. Nishimura, T. Honma, I. Nishimura, S. Ito, T. Yoshinari, H. Arawaka, H. Suda, H. Morishima and S. Nishimura, *Bioorg. Med. Chem. Lett.*, 1999, **9**, 3307.
17 N. Chida, T. Suzuki, S. Tanaka and I. Yamada, *Tetrahedron Lett.*, 1999, **40**, 2573.
18 O. Blixt and T. Norberg, *Carbohydr. Res.*, 1999, **319**, 80.
19 Y. Kobayashi, *Carbohydr. Res.*, 1999, **315**, 3.
20 H. Klein, R. Mietchen, H. Reinke and M. Michalik, *J. Prakt. Chem.*, 1999, **341**, 41 (*Chem. Abstr.*, 1999, **130**, 209 857).
21 E. Shitara, Y. Nishimura, F. Kojima and T. Takeuchi, *J. Antibiotics*, 1999, **52**, 348.
22 K.D. Randell, T.P. Frandsen, B. Stoffer, M.A. Johnson, B. Svensson and B.M. Pinto, *Carbohydr. Res.*, 1999, **321**, 143.
23 M.O. Lederer and H.P. Buhler, *Bioorg. Med. Chem.*, 1999, **7**, 1081.
24 F. Hayase, Y. Takahashi, S. Tominaga, M. Miura, T. Gomyo and H. Kato, *Biosci. Biotechnol. Biochem.*, 1999, **63**, 1512.
25 A. Rottmann, B. Synstad, V. Eijsink and M.G. Peter, *Eur. J. Org. Chem.*, 1999, 2293.
26 K. Matsuura, T. Akasaka, M. Hibino and K. Kokayashi, *Chem. Lett.*, 1999, 247.
27 J.J. García-López, F. Santoyo-González, A. Vargas-Berenguel and J.J. Giménez-Martínez, *Chem. Eur. J.*, 1999, **5**, 1775.

28 B.G. Davis, *J. Chem. Soc., Perkin Trans. 1,* 1999, 3215.
29 M. Mizuno, I. Muramoto, K. Kobayashi, H. Yaginuma and T. Inazu, *Synthesis*, 1999, 162.
30 M. Mizuno, K. Haneda, R. Iguchi, I. Muramoto, T. Kawakami, S. Aimoto, K. Yamamoto and T. Inazi, *J. Am. Chem. Soc.*, 1999, **121**, 284.
31 S. Mehta, M. Meldal, J.Ø. Duus and K. Bock, *J. Chem. Soc., Perkin Trans. 1,* 1999, 1445.
32 M. Albert, B.J. Paul and K. Dax, *Synlett,* 1999, 1483.
33 I. Tóth, J.P. Malkinson, N.S. Flinn, B. Drouillat, A. Horváth, J.Érchegyi, M. Idei, A. Venetianer, P. Artursson, L. Lazorova, B. Szende and G. Kéri, *J. Med. Chem.*, 1999, **42**, 4010.
34 T. Ziegler, H.-J. Kaisers, R. Schlomer and C. Koch, *Tetrahedron*, 1999, **55**, 8397.
35 C. Prata, N. Mora, J.-M. Lacombe, J.-C. Maurizis and B. Pucci, *Carbohydr. Res.*, 1999, **321**, 4.
36 C. Prata, N. Mora, A. Polidori, J.-M. Lacombe and B. Pucci, *Carbohydr. Res.*, 1999, **321**, 15.
37 M. Avalos, R. Babiano, P. Cintas, J.L. Jiménez, J.C. Palacios, G. Silvero and C. Valencia, *Tetrahedron*, 1999, **55**, 4377.
38 M. Avalos, R. Babiano, P. Cintas, F.J. Higes, J.L. Jiménez, J.C. Palacios, G. Silvero and C. Valencia, *Tetrahedron*, 1999, **55**, 4401.
39 M. Avalos, R. Babiano, P. Cintas, F.J. Higes, J.L. Jiménez, J.C. Palacios and G. Silvero, *Tetrahedron: Asymmetry*, 1999, **10**, 4071.
40 C. Uriel and F. Santoyo-González, *Synthesis*, 1999, 2049.
41 J.M. Benito, C.O. Mellet, K. Sadalapure, T.K. Lindhorst, J. Defaye and J.M. García Fernández, *Carbohydr. Res.*, 1999, **320**, 37.
42 J. Yu, F. Liu, Y. Li, L. Cheng, X. Fan and Y. Liu, *Yingyong Huaxue,* 1999, **16**, 41 (*Chem. Abstr.*, 1999, **131**, 337 255).
43 G.S. Lee, H.K. Min and B.Y. Chung, *Tetrahedron Lett.*, 1999, **40**, 543.
44 E.D. Soli and P. DeShong, *J. Org. Chem.*, 1999, **64**, 9724.
45 E.D. Soli, A.S. Manoso, M.C. Patterson, P. DeShong, D.A. Favor, R. Hirschmann and A.B. Smith III, *J. Org. Chem.*, 1999, **64**, 3171.
46 A. Kirschning, M.A. Hashem, H. Monenschein, L. Rose and K.-U. Schöning, *J. Org. Chem.*, 1999, **64**, 6522.
47 A. Kirschning, M. Jesberger and H. Monenschein, *Tetrahedron Lett.*, 1999, **40**, 8999.
48 J. Marco-Contelles and C.A. Jimenez, *Tetrahedron*, 1999, **55**, 10511.
49 X.-M. Chen, Z.-J. Li, Z.-X. Ren and Z.-T. Huang, *Carbohydr. Res.*, 1999, **315**, 262.
50 J. Kovacs, I. Pinter, M. Kajtar-Peredy, G. Argay, A. Kalman, G. Descotes and J.-P. Praly, *Carbohydr. Res.*, 1999, **316**, 112.
51 V. Pozsgay, *J. Org. Chem.*, 1999, **64**, 7277.
52 D. Lafont and P. Boullanger, *J. Carbohydr. Chem.*, 1999, **18**, 675.
53 J. Fuentes, M. Angulo and M. Angeles Pradera, *Carbohydr. Res.*, 1999, **319**, 192.
54 Z. Guo, M.C. Samano, J.W. Krzykawski, S.F. Wnuk, G.J. Ewing and M.J. Robins, *Tetrahedron*, 1999, **55**, 5705.
55 J.M. Vega-Perez, J.I. Candela, E. Blanco and F. Iglesias-Guerra, *Tetrahedron*, 1999, **55**, 9641.
56 G.A. Winterfeld, Y. Ito, T. Ogawa and R.R. Schmidt, *Eur. J. Org. Chem.*, 1999, 1167.
57 A.T. Carmona, P. Borrachero, F. Cabrera-Escribano, M. Jesús Diánez, M.

Dolores Estrada, A. López-Castro, R. Ojeda, M. Gómez-Guillén and S. Pérez-Garrido, *Tetrahedron: Asymmetry*, 1999, **10**, 1751.
58 Yu.M. Mikshiev, V.I. Kornilov, B.B. Paidak and Yu.A. Zhdanov, *Russ. J. Gen. Chem.*, 1999, **69**, 476 (*Chem. Abstr.*, 1999, **131**, 257 776).
59 A. Peer and A. Vasella, *Helv. Chim. Acta,* 1999, **82**, 1044.
60 C.W. Holzapfel and R. Crous, *Heterocycles*, 1998, **48**, 1337 (*Chem. Abstr.*, 1998, **129**, 260 703).
61 G. Zosimo-Landolfo, J.M.J. Tronchet, N. Bizzozero, F. Habashi and A. Kamatari, *Farmaco*, 1998, **53**, 623 (*Chem. Abstr.*, 1999, **130**, 267 670).
62 J.P. Mitchell, K.D. Roberts, J. Langley, F. Koentgen and J.N. Lambert, *Biog. Med. Chem. Lett.*, 1999, **9**, 2785.
63 I. Panfil, D. Mostowicz and M. Chmielewski, *Pol. J. Chem.*, 1999, **73**, 1099 (*Chem. Abstr.*, 1999, **131**, 144 751).
64 J.M.J. Tronchet, M. Koufaki, F. Barbalatrey and M. Geoffroy, *Carbohydr. Lett.*, 1999, **3**, 255 (*Chem. Abstr.*, 1999, **130**, 296 922).
65 S. Vonhoff, K. Piens, M. Pipelier, C. Braet, M. Claeyssens and A. Vasella, *Helv. Chim. Acta,* 1999, **82**, 963.
66 C. Schaller, R. Demange, S. Picasso and P. Vogel, *Bioorg. Med. Chem. Lett.*, 1999, **9**, 277.
67 J. Kiegiel, M. Poplawska, J. Jozwik, M. Kosior and J. Jurczak, *Tetrahedron Lett.*, 1999, **40**, 5605.
68 A. Pal, A. Bhattacharjee, A. Bhattacharjya and A. Patra, *Tetrahedron*, 1999, **55**, 4123.
69 S. Majundar, A. Bhattacharjya and A. Patra, *Tetrahedron*, 1999, **55**, 12157.
70 J.K. Gallos, A.E. Koumbis, V.P. Xiraphaki, C.C. Dellios and E. Coutouli-Argyropoulou, *Tetrahedron*, 1999, **55**, 15167.
71 P.J. Dransfield, S. Moutel, M. Shipman and V. Sik, *J. Chem. Soc., Perkin Trans.1,* 1999, 3349.
72 J.K. Gallos, T.V. Koftis, A.E. Koumbis and V.I. Moutsos, *Synlett,* 1999, 1289.
73 W. Kudelska, *Z. Naturforsch., B: Chem. Sci.*, 1998, **53**, 1277 (*Chem. Abstr.*, 1999, **130**, 110 480).
74 V. Gyóllai, L. Somsák and L. Szilágyi, *Tetrahedron Lett.*, 1999, **40**, 3969.
75 E. Ösz, L. Szilágyi, L. Somsák and A. Bényi, *Tetrahedron*, 1999, **55**, 2419.
76 M.A. Martins-Albo and N.B. D'Accorso, *J. Heterocycl. Chem.*, 1999, **36**, 177.
77 A.A.G. Faraco, M.A. Fontes Prado, N.B. D'Accorso, R.J. Alves, J.D. de Souza Filho and R.F. Prado, *J. Heterocycl. Chem.*, 1999, **36**, 1129.
78 M. Zhang, H. Zhang, Z. Yang, L. Ma, J. Min and L. Zhang, *Carbohydr. Res.*, 1999, **318**, 157.
79 M.-J. Camarasa, M.-L. Jimeno, M.-J. Perez-Perez, R. Alvarez and S. Velazquez, *Tetrahedron*, 1999, **55**, 12187.
80 N.A. Al-Masoudi, Y.A. Al-Soud and I.M. Lagoja, *Carbohydr. Res.*, 1999, **318**, 67.
81 S.U. Hansen and M. Bols, *J. Chem. Soc., Perkin Trans. 1,* 1999, 3323.
82 A. Lohse, K.B. Jensen and M. Bols, *Tetrahedron Lett.*, 1999, **40**, 3033.
83 W.H. Ojala, C.R. Ojala and W.B. Gleason, *J. Chem. Crystallogr.*, 1999, **29**, 19 (*Chem Abstr.*, 1999, **131**, 59 067).
84 M. Avalos, R. Babiano, P. Cintas, F.R. Clemente, J.L. Jiménez, J.C. Palacios and J.B. Sánchez, *J. Org. Chem.*, 1999, **64**, 6297.
85 J. Andersch and D. Sicker, *J. Heterocycl. Chem.*, 1999, **36**, 589.

86 M.A.E. Shaban, A.Z. Nasr and A.E.A. Morgaan, *Pharmazie*, 1999, **54**, 580 (*Chem. Abstr.*, 1999, **131**, 228 935).
87 M.J. Arévalo, M. Avalos, R. Babiano, P. Cintas, M.B. Hursthouse, J.L. Jiménez, M.E. Light, I. López and J.C. Palacios, *Tetrahedron Lett.*, 1999, **40**, 8675.
88 T. Tschamber, H. Rudyk and D. LeNouen, *Helv. Chim. Acta,* 1999, 2015.
89 K.H. Smelt, A.J. Harrison, K. Biggadike, M. Müller, K. Prout, D.J. Watkin and G.W.J. Fleet, *Tetrahedron Lett.*, 1999, **40**, 3259.
90 L.-Varga-Defterdarović, D. Vikić-Topić and S. Horvat, *J. Chem. Soc., Perkin Trans. 1*, 1999, 2829.
91 S. Hanessian, P.-P. Lu, J.-Y. Sanceau, P. Chemla, K. Gohda, R. Fonne-Pfister, L. Prade and S.W. Cowan-Jacob, *Angew. Chem., Int. Ed. Engl.*, 1999, **38**, 3160.
92 E. Osz, L. Somsak, L. Szilagyi, L. Kovacs, T. Docsa, B. Toth and P. Gergely, *Bioorg. Med. Chem. Lett.*, 1999, **9**, 1385.
93 V.V. Alekseev and K.N. Zelenin, *Chem. Heterocycl. Compd.*, 1998, **34**, 919 (*Chem. Abstr.*, 1999, **130**, 282 270).
94 D. Marek, A. Wadouachi and D. Beaupère, *Synthesis*, 1999, 839.
95 D.A. Berges, J. Fan, S. Devinck, N. Liu and N.K. Dalley, *Tetrahedron*, 1999, **55**, 6759.
96 D.A. Berges, N. Zhang and L. Hong, *Tetrahedron*, 1999, **55**, 14251.
97 H. Siendt, T. Tschamber and J. Streith, *Tetrahedron Lett.*, 1999, **40**, 5191.
98 N. Panday and A. Vasella, *Synthesis*, 1999, 1459.
99 B.G. Davis, R.J. Nash, A.A. Watson, C. Smith and G.W.J. Fleet, *Tetrahedron*, 1999, **55**, 4501.
100 B.G. Davis, T.W. Brandstetter, L. Hackett, B.G. Winchester, R.J. Nash, A.A. Watson, R.C. Griffiths, C. Smith and G.W.J. Fleet, *Tetrahedron,* 1999, **55**, 4489.
101 G. Scheffler, M. Justus, A. Vasella and H.P. Wessel, *Tetrahedron Lett.*, 1999, **40**, 5845.

11
Thio-, Seleno- and Telluro-sugars

1 Thiosugars

1.1 Monosaccharides. – *1.1.1 Acyclic compounds.* The 1,3-dithiolane and diphenyl dithioacetals of all D-aldopentoses and of three aldohexoses were conveniently made by reaction of the free sugars with 1,3-propanedithiol and thiophenol, respectively, in 90% TFA. Yields were 78–94%, *i.e.* generally higher than those obtained by traditional methods.[1]

Degradation of agar by use of 0.5 M HCl in EtSH/MeOH (2:1, v/v) furnished a mixture of the constituent sugars as their diethyl dithioacetals from which the main component, the 3,6-anhydro-L-galactose derivative **1**, was isolated in 86% yield.[2]

2,3:5,6-Di-*O*-isopropylidene-D-mannofuranose reacted with cysteamine to give thiazolidine **2** as a single epimer. In contrast, similar condensations of several other partially protected D-mannose- as well as D-glucose-derivatives yielded the corresponding thiazolidines as 1:1 (2*R*)/(2*S*) mixtures. The cyclization of these compounds to [1,3]thiazolo[3,2a]azepanes is referred to in Chapter 24.[3]

1.1.2 Compounds with a ring sulfur atom. A review with 98 refs. on the

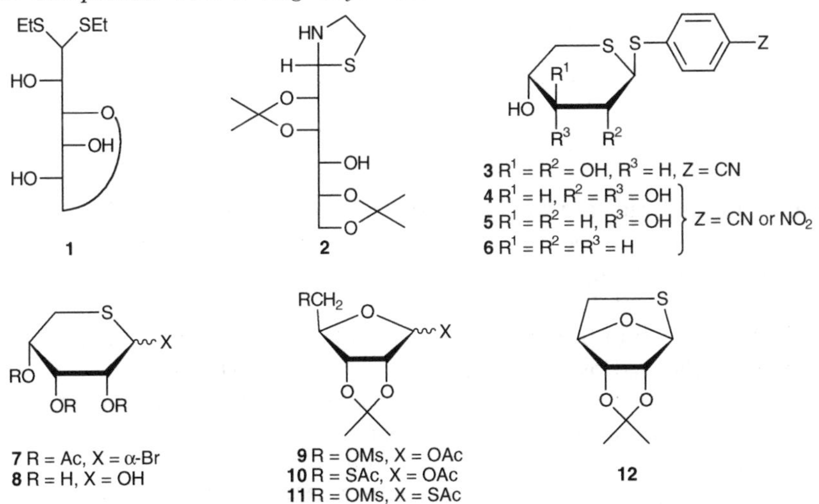

Carbohydrate Chemistry, Volume 33

consequences of replacing the ring oxygen atoms of pyranose sugars with sulfur, with special attention to enzyme inhibition, has been published.[4]

A series of new analogues of the antithrombotic agent beciparcil (**3**) have been reported. The 1,5-dithio-D-ribopyranosides **4** were readily obtained from methyl β-D-ribofuranoside *via* the peracetylated bromide **7**, following a well established protocol (see Vol. 32, Chapter 11, ref. 23); the 2-deoxy- and 2,3-dideoxy analogues **5** and **6**, respectively, were similarly available from the corresponding bromides.[5] Free 5-thio-D-ribose (**8**) has also been prepared independently from 2,3-*O*-isopropylidene-D-ribofuranose by way of the 1-*O*-acetate-5-*O*-mesylate **9** and 1-*O*-acetate-5-*S*-acetate **10**. The 1-*S*-acetate-5-*O*-mesylate **11**, formed as a by-product of this synthesis, was cyclized to the 1,4-anhydro-5-thio compound **12** by treatment with NaOMe.[6] The 3-*O*-methyl- and 4-deoxy-analogues **13** and **14**, respectively, of lead compound **3** were prepared as shown in Scheme 1. A non-radical deoxygenation was necessary

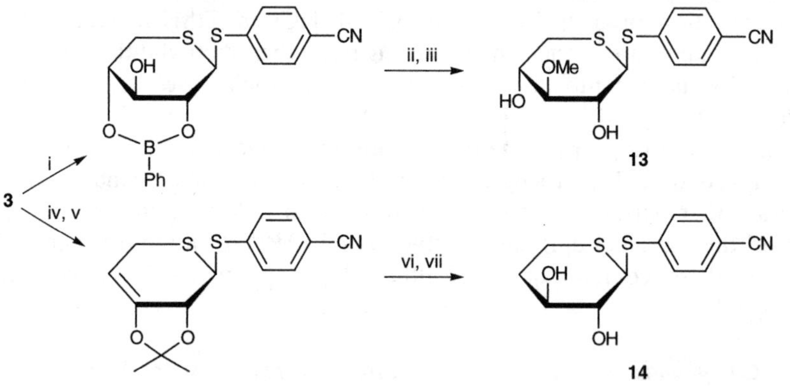

Reagents: i, $PhB(OH)_2$; ii, Ag_2O, MeI; iii, Amberlite IRA(OH^-); iv, $Me_2C(OMe)_2$, H^+; v, (2,4,5-triiodoimidazole), PPh_3; vi, nucleophile (*e.g.* NaN_3); vii, H^+, H_2O; viii, $NaBH_4$

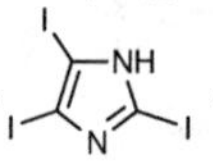

Scheme 1

because of the radical-trapping properties of the aglycon.[7] The bicyclic analogues **17** were formed on exposure of the tri-*O*-acetate **16**, derived from the known methyl glycoside **15**, to 4-cyano- or 4-nitro-benzenethiol and TmsOTf, then deacetylation.[8]

Umbelliferyl 5-thio-β-D-xylopyranoside (**18**), also an antithrombotic agent, has been synthesized by conventional glycosylation.[9] The aromatic 5-thioglucopyranosylamines **19**, formed by condensation of 5-thioglucopyranose pentaacetate with the appropriate arylamines in the presence of $HgCl_2$, were weak glycosidase inhibitors.[10] A 1:1 α/β-mixture of azides **20** was obtained on treatment of acetobromo-5-thioglucose with NaN_3 in HMPA,[11] and the azides **21**, similarly prepared from acetobromo-5-thioxylose with NaN_3 in DMSO, were subjected to thermolysis furnishing tetrahydrothiazepine **22** in 33% isolated yield.[12]

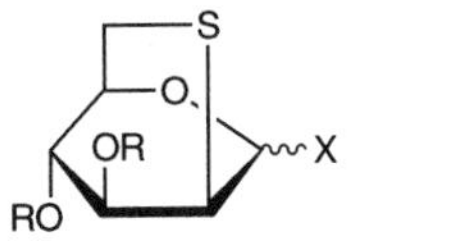

15 R = H, X = α-OMe
16 R = Ac, X = OAc
17 R = H, X = β-S-C_6H_4-Z
Z = CN or NO_2

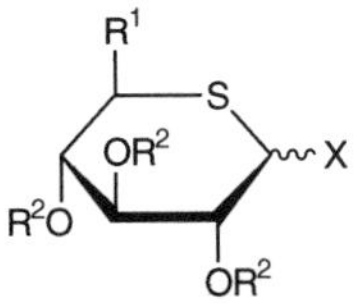

18 R^1 = R^2 = H, X = β-O-(4-ethylcoumarin-7-yl)
19 R^1 = CH_2OH, R^2 = H, X = HN-C_6H_4-Z
Z = H, OMe, NO_2, CF_3
20 R^1 = CH_2OAc, R^2 = Ac, X = N_3
21 R^1 = H, R^2 = Ac, X = N_3

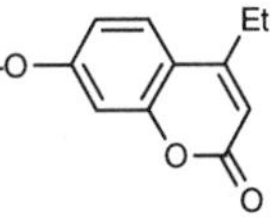

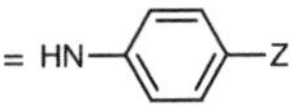

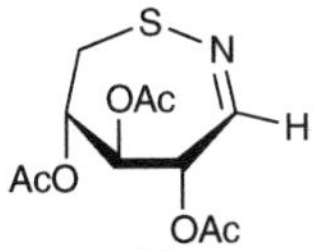

22

The intermediate 1,6-sulfonium ion **23** has been invoked to explain the facile displacement of the 6-*O*-triflate in compound **24** by CN^- to give **25**. (In the corresponding *O*-glycoside **26**, the C-6 leaving group is displaced intramolecularly by O-3).[13]

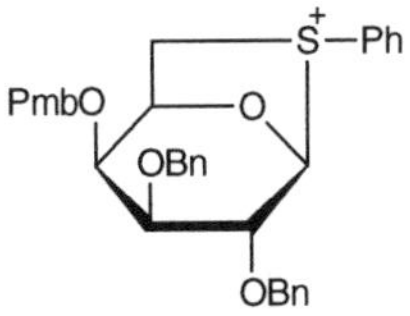

23

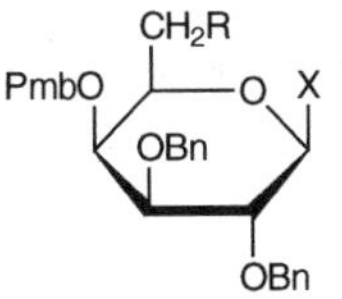

24 R = OTf, X = SPh
25 R = CN, X = SPh
26 R = OTf, X = OBn

The high yielding conjugate addition of HS^- to an α,β-unsaturated sugar ester carrying a good leaving group in the ε-position was the key-step in the synthesis of 'homothiosugars' **27**, as shown in Scheme 2.[14]

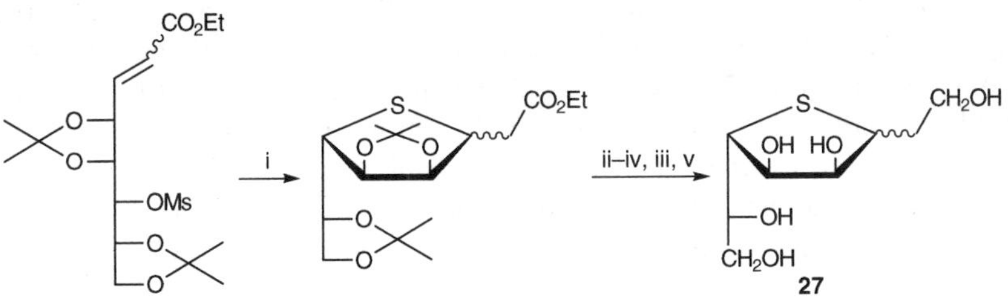

Reagents: i, NaHS, EtOH; ii, TFA, aq. EtOH; iii, Ac_2O, Py; iv, LAH; v, MeO^-, MeOH

Scheme 2

New syntheses of 4′-thionucleosides are covered in Chapter 20.

1.1.3 Other monosaccharide thiosugars. Base-induced transfer of a benzothiazol-2-yl group from sulfone to a neighbouring hydroxyl group with electrophilic trapping of the sulfinate anion represents a new S→O migration. An example is shown in Scheme 3.[15]

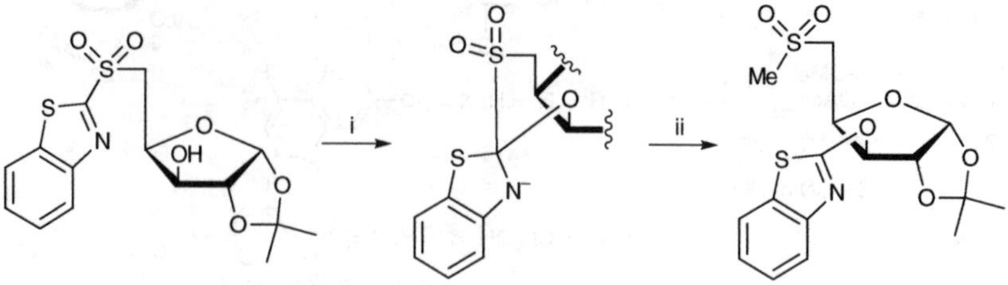

Reagents: i, Bu^{t}OK; ii, MeI

Scheme 3

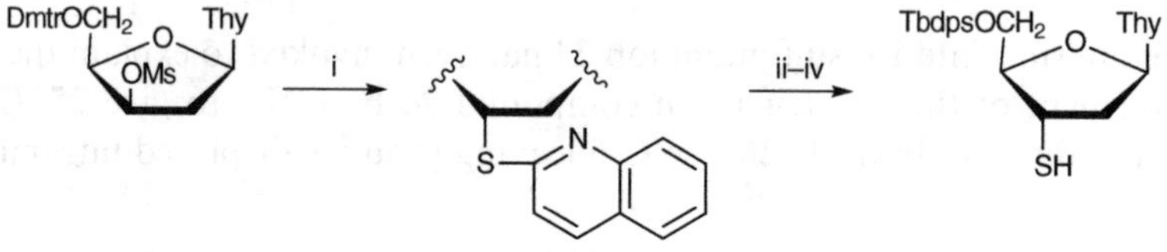

28

Reagents: i, 2-thioquinoline; ii, MsOH; iii, TbdpsCl, Im; iv, $NaBH_3CN$

Scheme 4

A new method for introducing a thio group by use of 2-thioquinoline is illustrated in Scheme 4. Quinolidine sulfides, such as **28**, are stable to aq. NaOH, methanolic NH_3, aq. HCl and mild oxidants; the free thiols are readily released on exposure to sodium cyanoborohydride.[16] Protection of 1-thio sugars as the symmetrical disulfides allowed introduction of an anomeric thiol group prior to functionalization at other positions. Acetylation, silylation and etherification proceeded without complications (*e.g.* **29**→**30**), but acetalation gave unsymmetrical products. Reduction of the disulfides to the thiols was performed with Zn/HOAc (*e.g.* **30**→**31**).[17] When 2-acetamido-3,4,6-tri-*O*-acetyl-2-deoxy-α-D-glucopyranosyl chloride was heated with thiourea in acetone, compound **32** was obtained which on treatment with methanolic diethyl disulfide in the presence of triethylamine, followed by deacetylation, afforded the unprotected, unsymmetrical disulfide **33** in 75% overall yield.[18] Protection of 1-thiosugars as unsymmetrical disulfides has been used in the solid-phase synthesis of thiodisaccharides (see Section 1.2 below).

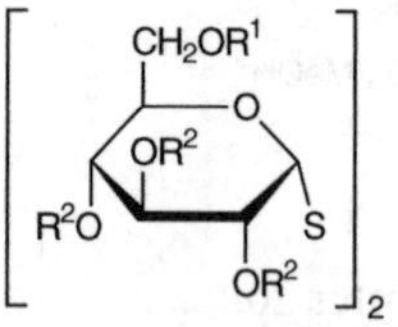

29 $R^1 = R^2 = H$
30 R^1 = Tbdms, R^2 = Ac

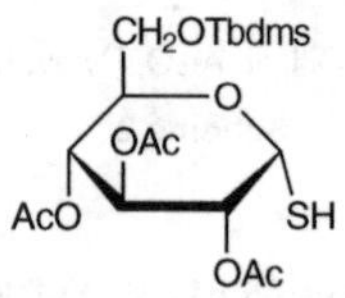

31

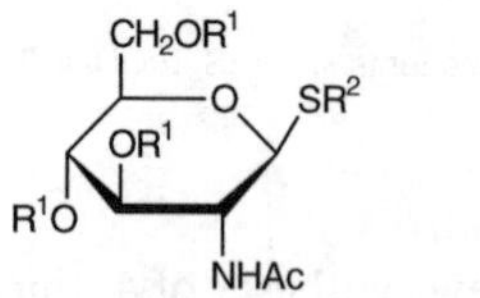

32 R^1 = Ac, R^2 = $C(=NH)NH_3^+\ Cl^-$
33 R^1 = H, R^2 = SEt

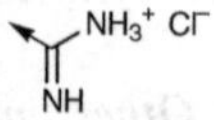

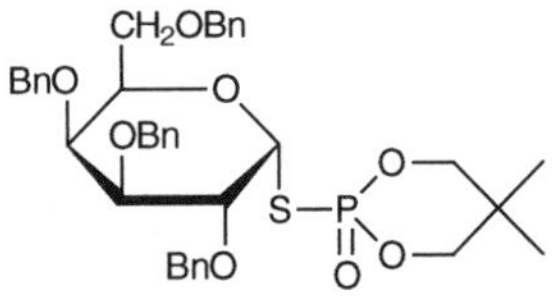

34 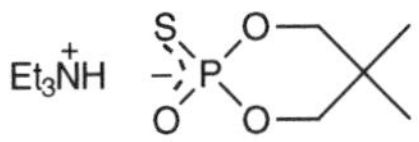 35 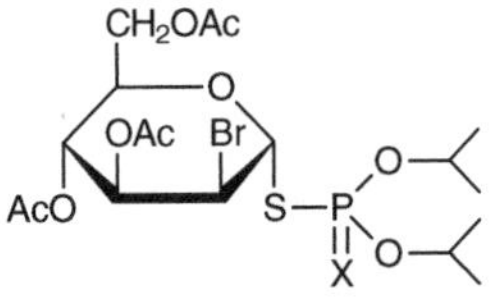 36 X = O 37 X = S

Hexopyranosyl 1-*S*-thiophosphates, such as **34**, have been made by exposure of benzyl-protected free sugars to triethylammonium di-*O*-alkylthiophosphate **35** in the presence of a Lewis acid.[19] The first examples of 1-*S*-thiophosphate derivatives of 2-bromo-2-deoxy-sugars, such as compounds **36**, were obtained analogously by reaction of triethylammonium di-*O*-alkylthiophosphates with 1,2-dibromides or, alternatively, by addition of di-*O*-alkylthiophosphoric acids to 2-bromoglycals,[20] and 1-*S*-dithiophosphate derivatives of 2-bromosugars, *e.g.* **37**, were similarly prepared by use of di-*O*-alkyldithiophosphoric acids or their triethylammonium salts.[21] *S*-Nitrosothiols **38** and **39**, required for kinetic studies on their decomposition, have been prepared by *S*-nitrosation ($NaNO_2$/ aq. $MeOH/H^+$) of the corresponding 1-thiosugars.[22]

S-Glycosides of aminothiols, for example thio-serine derivative **40**, were available from *O*-protected free thiosugars and *N*-protected aminoalcohols under modified Mitsunobu conditions [1,1′-(azodicarbonyl)dipiperidine/tri-methylphosphine].[23] 2-Thiosialic acid attached to Sepharose 4B *via* an aminothiol spacer has been used for the purification of sialic acid-recognizing proteins.[24]

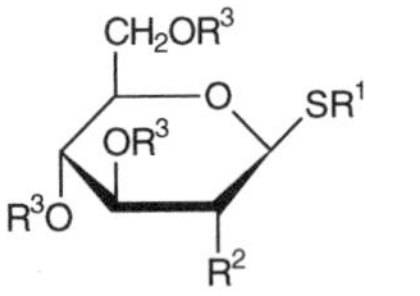

38 R^1 = NO, R^2 = OH, R^3 = H
39 R^1 = NO, R^2 = NHAc, R^3 = Ac
40 R^1 = $CH_2CH(NHBoc)CO_2Me$, R^2 = OAc, R^3 = Ac

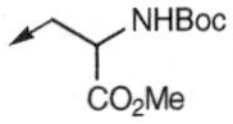

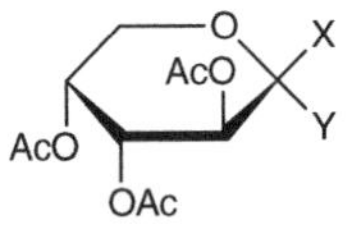

41 X = Br, Y = CO_2Me
42 X = CO_2Me, Y = S-(o-nitrophenyl)

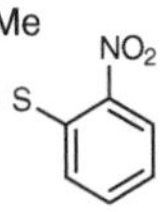

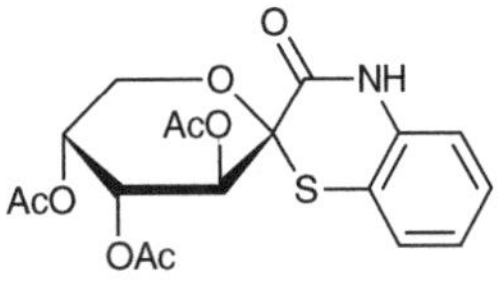

43

Spiro(1,4-benzothiazine-2,2′-pyran) derivative **43** was formed by condensation of the peracetylated glycosyl bromide **41**, derived from methyl β-D-*arabino*-2-hexulopyranosonate, with *o*-nitrothiophenol to furnish **42**, and subsequent acetylation and hydrogenation in alkaline medium.[25]

Reaction of the 1-bromo-1-cyanosugar **44** with thiocyanate ion afforded the epimeric 1-cyano-1-thiocyanates **45**, which surprisingly did not epimerize to the corresponding isothiocyanates even after prolonged heating at 130 °C in the melt. Treatment of the separated isomers with H_2S resulted in the formation of complex mixtures of spirocyclic products, such as **46**.[26] 4′-Thiaspiroglycosides, for example 1-thio-hept-5-ulose derivative **47**, were the products of inverse electron demand hetero-Diels Alder reactions between

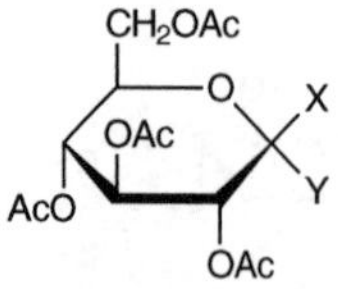

44 X = CN, Y = Br
45 X, Y = CN, SCN
46 X, Y =

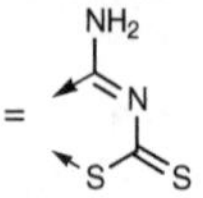

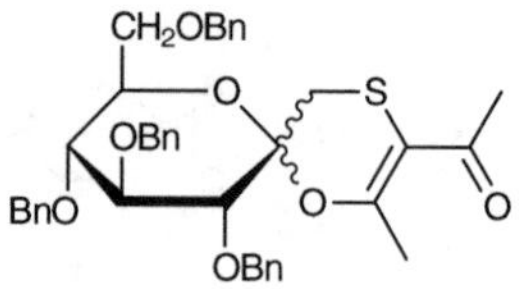

47

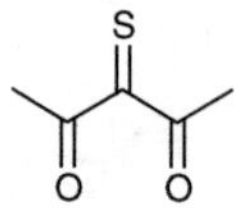

48

protected exoglycals and α,α′-dioxothione **48**.[27] *o*-Thioquinones added similarly to exoglycals, as well as to glycals. The latter reaction was employed in the synthesis of 2-deoxysugars and is referred to in Chapter 12.

Addition of ArSCl to tri-*O*-benzyl-D-glucal gave a mixture of 2-*S*-aryl-D-2-thioglycosyl chlorides which reacted with vinyl ethers in the presence of $SnCl_4$ to furnish *C*-glycosides, by way of sulfonium ions. An example is given in Scheme 5.[28]

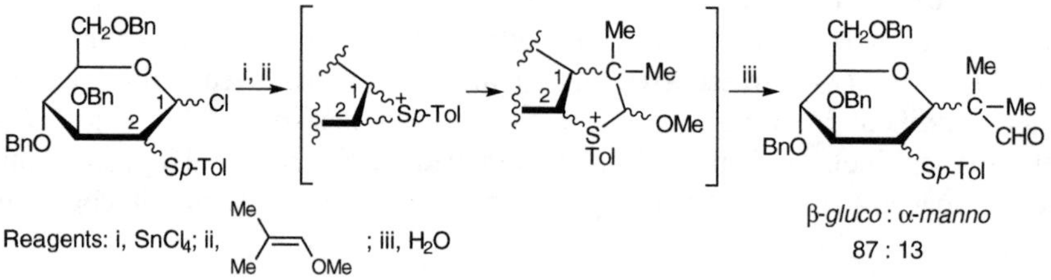

Reagents: i, $SnCl_4$; ii, $Me_2C{=}CHOMe$; iii, H_2O

β-*gluco* : α-*manno*
87 : 13

Scheme 5

The conversion of the *O*-acetate **49** into the corresponding *S*-acetate **50** in 73% yield, with overall retention of configuration, was the only carbohydrate example in a general paper on the Pd(0)-catalysed synthesis of allylic thioacetates.[29] The new *S*-glycosylated diaryl-2-thiolumazine derivative **51** was prepared conventionally by condensation of 6-*O*-tosyl-1,2:3,4-di-*O*-isopropylidene-α-D-galactopyranose with the sodium salt of the heterocyclic thiol.[30]

The synthesis of functionalized cyclohexenyl sulfides and sulfoxides is referred to in Chapter 18 and the asymmetric synthesis of chiral 3′-oxathionucleosides is covered in Chapter 20.

The sulfolipid **52** has been isolated from the brown alga *Dictyota ciliolata*.[31]

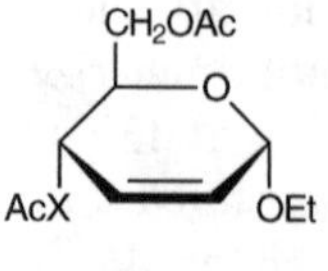

49 X = O
50 X = S

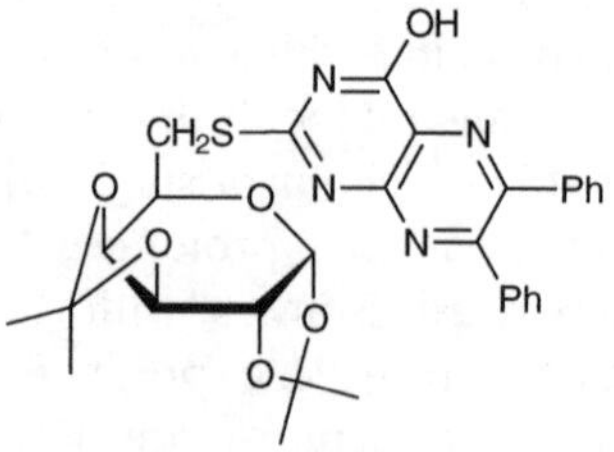

51

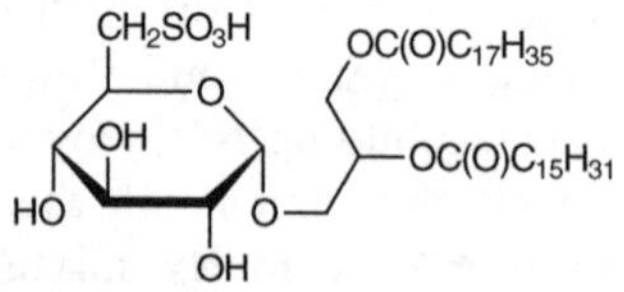

52

1.2 Di- and Oligo-saccharides. – The new procedure for the preparation of dithioacetals described in ref. 1 above has also been applied to cellobiose, lactose, gentiobiose, melibiose, maltobiose and maltotriose.

The solid-supported, *O*-unprotected thiolate **54**, obtained by loading the *S*-protected thiosugar **53** onto a polystyrene resin followed by reduction with

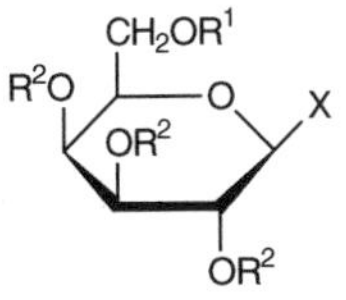

53 $R^1 = R^2 = Ac$, X = SSEt
54 R^1 = Polystyrene resin
$R^2 = H$, X = SNa

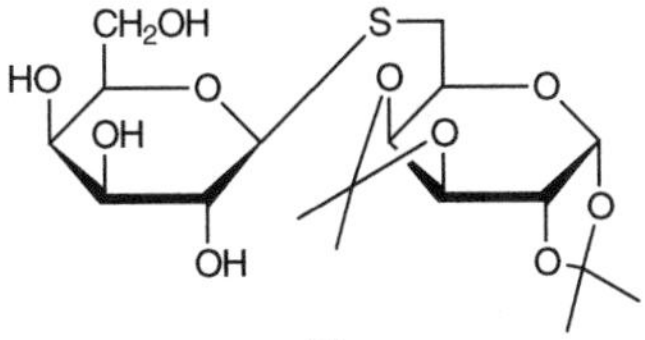

55

dithioerythritol and treatment of the thiol thus liberated with NaOMe in THF, was coupled with 1,2:3,4-di-*O*-isopropylidene-α-D-galactopyranose 6-triflate. The expected, thio-linked disaccharide **55** was obtained in 64% yield after cleavage from the solid support.[18]

Glycosylation of 1,6-anhydrothiosugar **56** with various sugar alcohols promoted by NIS/TfOH gave the expected disaccharides as the disulfides, *e.g.*

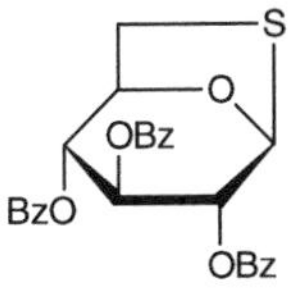

56 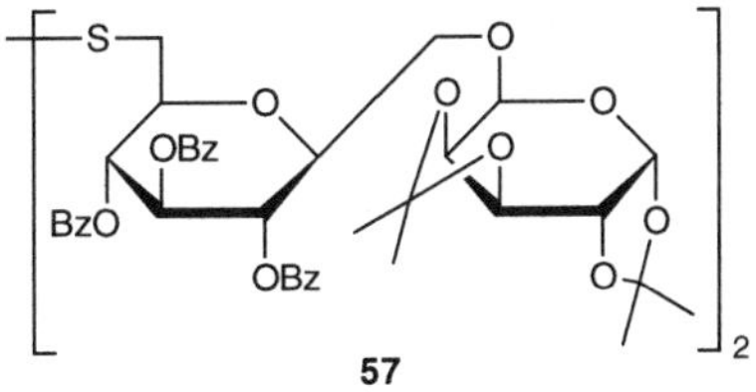 **57**

57, in satisfactory yields. Reduction to the 6′-thiodisaccharides required debenzoylation before treatment with Na/NH_3. Likewise, desulfurization with Raney Nickel to furnish the 6′-deoxydisaccharide could only be effected after debenzoylation.[32] Attempts to synthesize acarbose by application of this

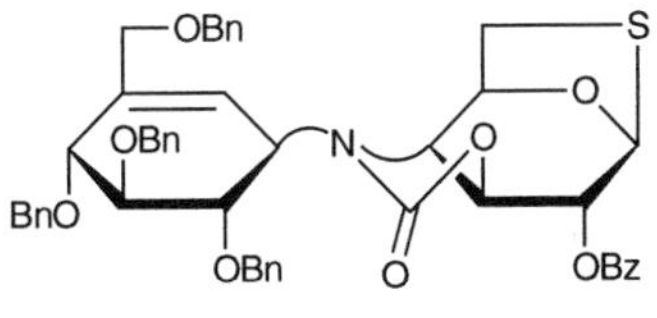

58 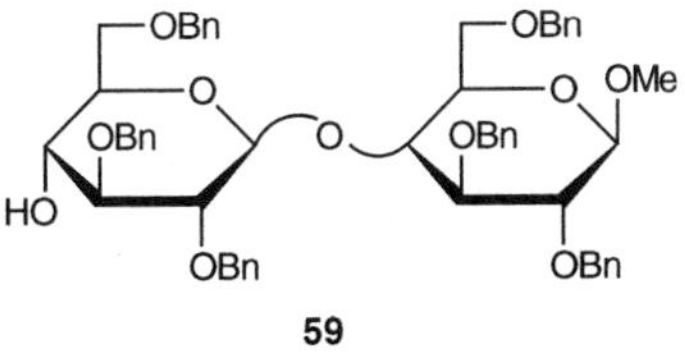 **59**

method to 1,6-anhydrothiosugar **58** and sugar alcohol **59** were unsuccessful.[33] A non-glycosylation strategy for the synthesis of disaccharides with sulfur in the reducing sugar ring has been used for the conversion of lactose into 5-epi-5-thiolactose, as outlined in Scheme 6.[34]

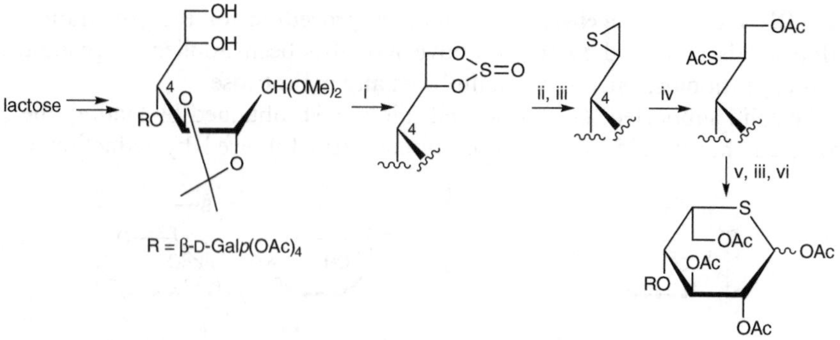

Reagents: i, $SOCl_2$, Et_3N; ii, KSAc; iii, NaOMe; iv, Ac_2O, AcOH, AcONa; v, aq. HOAc; vi, Ac_2O, Py

Scheme 6

Treatment of *S*-phenyl 1,2-dithio-β-D-mannopyranoside with base led to an 1,2-episulfide which underwent *in situ* oligomerization to afford a family of α-(1→2)-linked thio-oligomannosides.[35]

3′,3‴-Di-*S*-β-D-glucopyranosyl-3′,3‴,6,6″,6‴-pentathiogentiotetraose, an all-sulfur-linked, branched hexasaccharide has been prepared, together with a positional isomer and two homologous heptasaccharides, using displacement of good leaving groups at the primary positions by anomeric thiolates for the formation of the interglycosidic bonds.[36] Mono-2- or 3-thiocyclodextrins have been obtained by opening of the epimeric mono-2,3-epoxides with benzylthiol and subsequent de-*S*-benzylation with Na/NH_3.[37] In connection with ^{19}FNMR specroscopic studies on the properties of cyclodextrin host–guest complexes, mono-6-trifluoroethylthio-β-cyclodextrin was synthesized by reaction of the mono-6-tosylate with 2,2,2-trifluoroethanethiol and purification of the product *via* the peracetate.[38] A macrocyclic octa-sialic acid cluster, to be used as host, adsorbate or ligand for sialo-targeting lectins or viruses, has been constructed by attaching eight 2-thio-NeuAc residues *via* amide spacers to calix[4]resorcarene.[39]

The synthesis of a 1-*N*-iminosugar-based UDP-galactose analogue containing a 5′-thioribosyl moiety is referred to in Chapter 18.

2 Seleno- and Telluro-sugars

Se-Glycosyl selenothiophosphates, such as compound **60**, have been synthesized by BF_3 etherate-catalysed reaction of sugar peracetates with di-*O*-alkylthioselenophosphates. The *S*-linked isomers were obtained as minor products.[40,41] Reaction of tri-*O*-acetyl-D-glucal with phenylselenenyl chloride, followed by *in situ* exposure to KSCN, gave the *trans*-addition compounds **61** and **62** in 29 and 33% yield, respectively. Tri-*O*-acetl-D-galactal furnished only the α-D-*talo*-configured addition product in 44% yield. Further transformation of these isothiocyanates to urea derivatives is referred to in Chapter 10.[42]

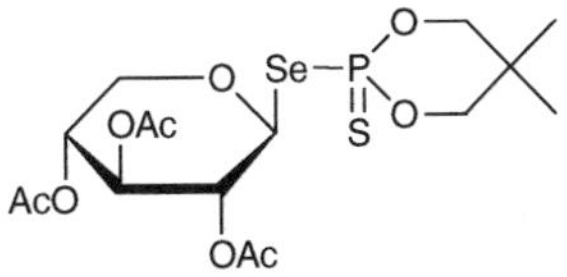

60

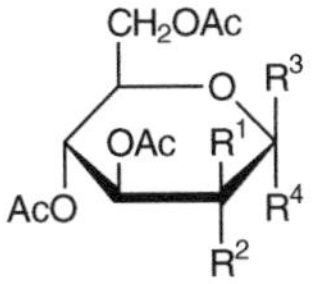

61 R^1 = SePh, R^2 = R^3 = H, R^4 = NCS
62 R^1 = R^4 = H, R^2 = SePh, R^3 = NCS

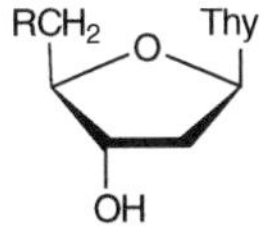

63 R = $OSO_2CH_2CF_3$
64 R = SeCN

Selenium substituents have been introduced by displacement of trifluoroethyl sulfonate groups (which gave better results than trifluoromethyl sulfonates) with Bu_4NSeCN (*e.g.* **63**→**64**); application of this process to the preparation of new nucleoside heteroanalogues is covered in Chapter 20.[43] When 3,4,5-tri-*O*-5-*Se*-benzyl-5-seleno-D-ribose (**65**) was treated with samarium iodide, intramolecular radical substitution at Se took place to give the 5-seleno-D-ribopyranose derivative **66**.[44]

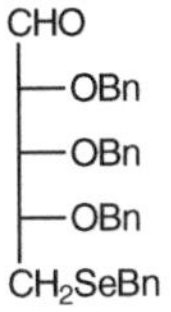

65

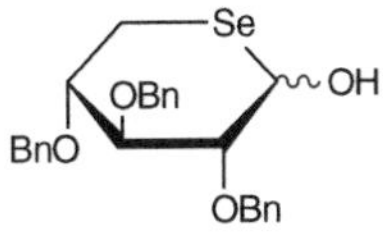

66

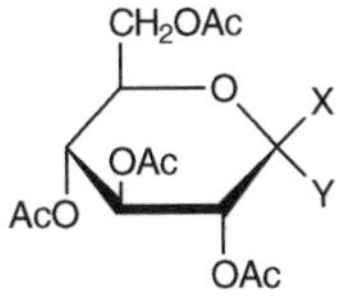

67 X = TeTol, Y = H
68 X = H, Y = TeTol
69 X = Y = H
70 X = H, Y = TolTe(C=N-2,6-dimethylphenyl)
71 X = H, Y = CH=C(R)TeTol
72 X = H, Y = CH=CHR

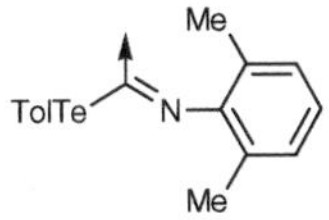

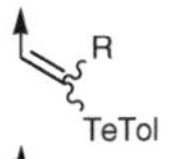

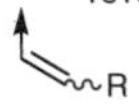

R = alkyl, 4-methoxyphenyl, 3-pyridyl, *etc.*

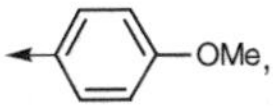

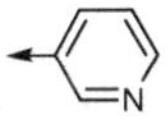

Glycosylation experiments analogous to those carried out with 1,6-anhydrothiosugars (refs. 32 and 33 above) have been carried out with 1,6-anhydroselenosugars with very similar results.

Glycosyl radicals were generated from tellurium glycosides under photolysis or thermolysis conditions. In the absence of any reagents this led to anomerization, *e.g.* **67**→**68**, resulting in a *ca.* 4:1 α/β equilibrium mixture. In the presence of Bu_3SnH, reduction to the 1,5-anhydroalditol took place (**67** or **68**→**69**).[45] Trapping with aromatic isocyanides[46] or alkynes[47] gave *C*-glycosides with Te in the aglycon (**67**→**70** and **71**, respectively), which could be detellurized with Bu_3SnH (*e.g.* **70**→**72**).[47]

References

1 M. Funabashi, S. Arai and M. Shinohara, *J. Carbohydr. Chem.*, 1999, **18**, 333.
2 Y. Hama, H. Nakagawa, T. Sumi, X. Xie and K. Yamaguchi, *Carbohydr. Res.*, 1999, **318**, 154.
3 D. Marek, A. Wadouachi and D. Beaupère, *Synthesis,* 1999, 839.
4 H. Yuasa and H. Hashimoto, *Rev. Heteroat. Chem.*, 1998, **19**, 35 (*Chem. Abstr.*, 1999, **130**, 267 629).
5 E. Bózo, S. Boros and J. Kuszmann, *Carbohydr. Res.*, 1999, **321**, 52.
6 A. Fleetwood and N.A. Hughes, *Carbohydr. Res.*, 1999, **317**, 204.
7 Y. Li, D. Horton, V. Barberousse, S. Samreth and F. Bellamy, *Carbohydr. Res.*, 1999, **316**, 104.
8 E. Bózo, S. Boros and J. Kuszmann, *Pol. J. Chem.*, 1999, **73**, 989 (*Chem. Abstr.*, 1999, **131**, 73 891).
9 Y. Colette, K. Ou, J. Pires, M. Baudry, M. Descotes, J.-P. Praly and V. Barberousse, *Carbohydr. Res.*, 1999, **318**, 162.
10 K.D. Randell, T.P. Frandsen, B. Stoffer, M.A. Johnson, B. Svensson and B.M. Pinto, *Carbohydr. Res.*, 1999, **321**, 143.
11 M.K. Strumpel, J. Buschmann, L. Szilagyi and Z. Gyorgydeak, *Carbohydr. Res.*, 1999, **318**, 91.
12 J.-P. Praly, G. Hetzer and M. Steng, *J. Carbohydr. Chem.*, 1999, **18**, 833.
13 M. Compain-Batissou, L. Mesrari, D. Anker and A. Doutheau, *Carbohydr. Res.*, 1999, **316**, 201.
14 J. Riedner, I. Robina, J.G. Fernández-Bolaños, S. Gómez-Bujedo and J. Fuentes, *Tetrahedron: Asymmetry*, 1999, **10**, 3391.
15 D. Gueyrard, C. Lorin, J. Moravcova and P. Rollin, *J. Carbohydr. Chem.*, 1999, **18**, 317.
16 J. Zhang and M.D. Matteucci, *Tetrahedron Lett.*, 1999, **40**, 1467.
17 M.J. Kiefel, R.J. Thomson, M. Radovanovic and M. von Itzstein, *J. Carbohydr. Chem.*, 1999, **18**, 937.
18 G. Hummel and O. Hindsgaul, *Angew. Chem., Int. Ed. Engl.*, 1999, **38**, 1782.
19 W. Kudelska, *Z. Naturforsch., B: Chem. Sci.*, 1998, **53**, 1277 (*Chem. Abstr.*, 1999, **130**, 110 480).
20 J. Borowiecka, *Heteroat. Chem.*, 1999, **10**, 465 (*Chem. Abstr.*, 1999, **131**, 310 774).
21 J. Borowiecka, *Pol. J. Chem.,* 1999, **73**, 793 (*Chem. Abstr.*, 1999, **130**, 338 304).
22 A.P. Munro and D.L.H. Williams, *Can. J. Chem.*, 1999, **77**, 550.
23 R.A. Falconer, I. Jablonkai and I. Toth, *Tetrahedron Lett.*, 1999, **40**, 8663.
24 S. Abo, S. Ciccotosto, A. Alafaci and M. von Itzstein, *Carbohydr. Res.*, 1999, **322**, 201.
25 J. Andersch, D. Sicker and H. Wilde, *J. Heterocycl. Chem.*, 1999, **36**, 457.
26 E. Ösz, L. Szilágyi, L. Somsák and A. Bényyi, *Tetrahedron*, 1999, **55**, 2419.
27 A. Bartolozzi, G. Capozzi, C. Falciani, S. Mennnnichetti, C. Nativi and A. Paolacci Bacialli, *J. Org. Chem.*, 1999, **64**, 6490.
28 I.P. Smoliakova, M. Han, J. Gong, R. Caple and W.A. Smit, *Tetrahedron*, 1999, **55**, 4559.
29 S. Divekar, M. Sati, M. Soufiaoui and D. Sinoü, *Tetrahedron*, 1999, **55**, 4369.
30 M.A. Martins Alho, C. Ochoa, A. Chana and N.B. D'Accorso, *An. Asoc. Quím. Argent.*, 1998, **86**, 197 (*Chem. Abstr.*, 1999, **130**, 153 881).
31 D.J. Bourne, S.E. Pilchowski and P.T. Murphy, *Aust. J. Chem.*, 1999, **52**, 69.
32 R.V. Stick, D.M.G. Tilbrook and S.J. Williams, *Aust. J. Chem.*, 1999, **52**, 685.

33 R.V. Stick, D.M.G. Tilbrook and S.J. Williams, *Aust. J. Chem.*, 1999, **52**, 885.
34 J. Isac-García, F.-G. Calvo-Flores, F. Hernández-Mateo and F. Santoyo-Gonzales, *Chem. Eur. J.*, 1999, **5**, 1512.
35 S. Knapp and K. Malolanarasimhan, *Org. Lett.*, 1999, **1**, 611 (*Chem. Abstr.*, 1999, **131**, 199 908).
36 V. Ding, M.-O. Contour-Galcera, J. Ebel, C. Ortiz-Mellet and J. Defaye, *Eur. J. Org. Chem.*, 1999, 1143.
37 M. Fukudome, Y. Okabe, D.-Q. Yuan and K. Fujita, *Chem. Commun.*, 1999, 1045.
38 J. Diakur, Z. Zuo and L.I. Wiebe, *J. Carbohydr. Chem.*, 1999, **18**, 209.
39 K. Fujimoto, O. Hayashida, Y. Aoyama, C.-T. Guo, K.I.-P. Jwa Hidari and Y. Suzuki, *Chem. Lett.*, 1999, 1259.
40 W. Kudelska, A. Olczak, M.L. Glowka and S. Jankowski, *Pol. J. Chem.*, 1999, **73**, 487 (*Chem. Abstr.*, 1999, **130**, 252 540).
41 W. Kudelska, *Heteroat. Chem.*, 1999, **10**, 259 (*Chem. Abstr.*, 1999, **130**, 338 302).
42 C. Uriel and F. Santoyo-Gonzáles, *Synthesis*, 1999, 2049.
43 A.M. Belostotskii, J. Lexner and A. Hassner, *Tetrahedron Lett.*, 1999, **40**, 1181.
44 C.H. Schiesser and S.-L. Zheng, *Tetrahedron Lett.*, 1999, **40**, 5095.
45 S. Yamago, H. Miyazoe and J.-i. Yoshida, *Tetrahedron Lett.*, 1999, **40**, 2339.
46 S. Yamago, H. Miyazoe, R. Goto and J.-i. Yoshida, *Tetrahedron Lett.*, 1999, **40**, 2347.
47 S. Yamago, H. Miyazoe and J.-i. Yoshida, *Tetrahedron Lett.*, 1999, **40**, 2343.

12
Deoxy-sugars

Studies of the biosynthesis of the 3,6-dideoxyhexoses found in the lipopolysaccharides of Gram-negative bacteria and of the 2,6- and 4,6-dideoxyhexoses found in cardiac glycosides and macrolide antibiotics have been reviewed.[1] The trisaccharides **1** and **2** as well as the glycal derived from **2** have been isolated from the plant *Marsdenia roylei*.[2,3]

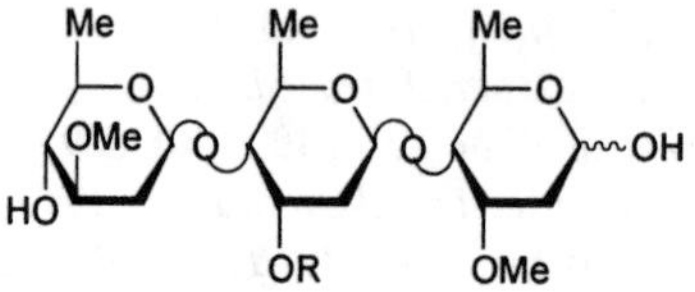

1 R = Me
2 R = H

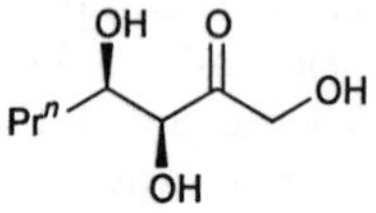

3

A novel one-pot procedure for the synthesis of carbohydrates from glycerol and an aldehyde has utilized a cascade of four enzymatic steps. A kinetically controlled phosphorylation of glycerol followed by glycerol phosphate oxidase-catalysed oxidation to dihydroxyacetone phosphate (DHAP) and an aldol reaction of DHAP with an aldehyde acceptor (*e.g.* butanal) and then, finally, enzymatic dephosphorylation of the aldol adduct gave, for example, the trideoxy-D-*threo*-heptulose **3**.[4] A short chemo-enzymatic synthesis of 4-deoxy-D-fructose 6-phosphate has been described (Scheme 1).[5] The commer-

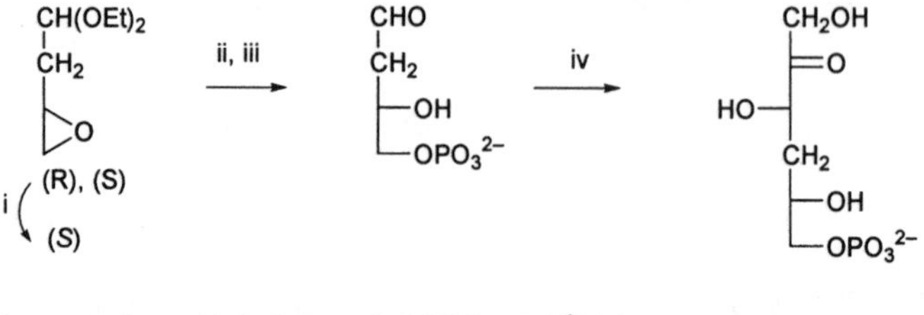

Reagents: i, epoxide hydrolase; ii, K_2HPO_4; iii, H^+/H_2O; iv, transketolase, L-erythrulose

Scheme 1

cially available aldolase antibody 38C2 uses hydroxyacetone as a donor for an aldol reaction with aldehydes affording 1-deoxyalduloses (Scheme 2).[6] An improved method for the preparation of *C*-5 dideuterated 1-deoxy-D-xylulose

Carbohydrate Chemistry, Volume 33

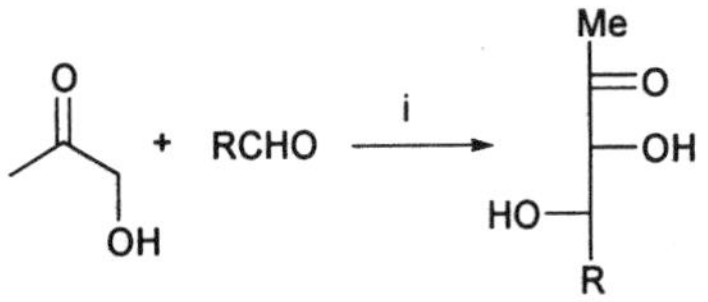

Reagents: i, Antibody 38C2

Scheme 2

from 2,3-*O*-isopropylidene-D-tartrate has been disclosed. A mono-ester of the tartrate was reduced with $LiEt_3BD$, the product treated with MeLi and then subjected to acid hydrolysis to afford the desired material.[7] A similar synthesis of 1-deoxy-D-xylulose has started from 2,3-*O*-isopropylidene-D-*threo*-tetritol.[8] The Lewis acid-catalysed cycloaddition of enone **4** with benzyl (or ethyl) vinyl ether has afforded the mixture **5** and **6** from which both enantiomers of 2,6-dideoxy-6,6,6-trifluoro-*arabino*-hexopyranose were obtained (Scheme 3). The

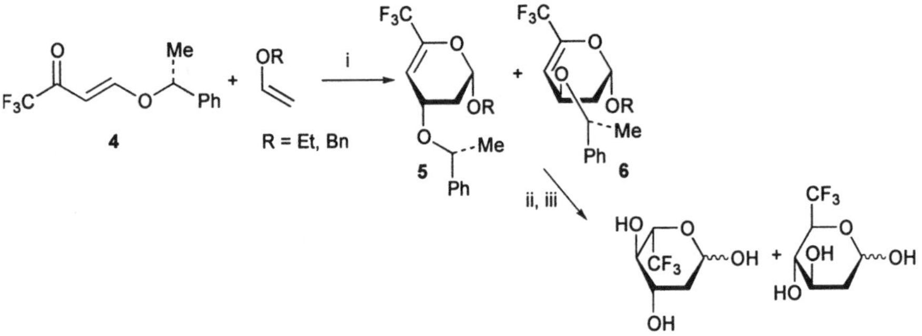

Reagents: i, $TiCl_4$, CH_2Cl_2, –78 °C; ii, BH_3•SMe_2 then $NaOH/H_2O_2$; iii, H_2, Pd/C, then H^+/H_2O if necessary

Scheme 3

enantiomers were separated by chromatography after the hydroboration step.[9] A procedure for the stereoselective synthesis of tetroses and pentoses, which can be adapted for the synthesis of deoxy-sugars, involving consecutive one-carbon chain elongations with high *anti*-selectivity from 2,3-*O*-isopropylidene-D-glyceraldehyde by use of the Li salt of ethyl ethylthiomethyl sulfoxide, is covered in Chapter 2. The synthesis, by standard methods, of phenyl 4-*O*-acetyl-1-thio-2,3,6-trideoxy-6,6,6-trifluoro-3-trifluoroacetamido-α- and β-L-*lyxo*-hexopyranoside and their coupling with daunomycinone has been described.[10] A radical deoxygenation has been effected at C-2 of a glucofuranose derivative by way of a pentafluorophenoxy thiocarbonyl ester,[11] and methyl 2,6-dideoxy-2-fluoro-β-L-talopyranoside has been synthesized from the readily available 2-deoxy-2-fluoro-D-glucose.[12] A deoxy disaccharide analogue of moenomycin is discussed in Chapter 19.

The air-stable crystalline reagent 1,1,2,2-tetraphenyldisilane has proved effective for the radical dehalogenation of bromodeoxy-sugar derivatives.[13] Glycosylation of 2-deoxy-2-iodo- and 2-deoxy-2-bromo-glucopyranosyl trichloroacetimidates has afforded β-glycosides with good selectivity and thus, after dehalogenation, a route to 2-deoxy-β-glycosides.[14] Similarly, glycosylation of 2-deoxy-2-iodo-α-D-manno- and talopyranosyl acetates promoted by TmsOTf gave only α-glycosides and thus an approach to 2-deoxy-α-glycosides.[15] 2-Deoxy-1-*O*-isobutyryl-α- and β-D-glucopyranose have been synthesized from tri-*O*-benzyl-D-glucal and used as donors in transesterification reactions.[16]

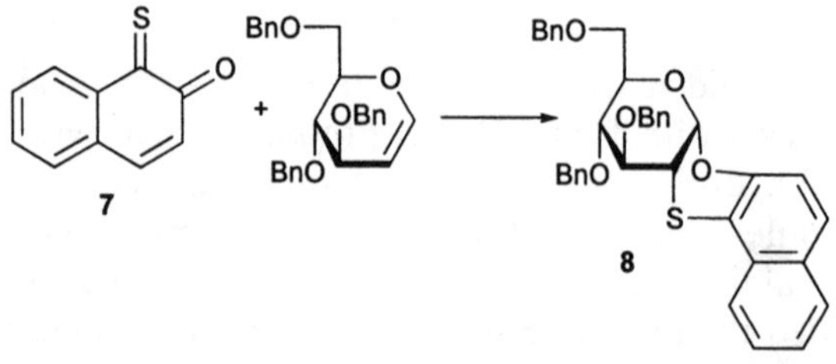

Ortho-thioquinones (*e.g.* **7**) on exposure to glycals gave adducts such as **8** which, after reduction with Raney nickel, offer access to aryl 2-deoxy-α-glycosides.[17] The use of known 2-deoxy-2-thio-derivatives such as **9** as glycosyl donors has afforded, after Raney nickel reduction, 2-deoxy-β-glycosides (Scheme 4). The corresponding β-*manno*-derivative of **9** also gave 2-deoxy-α-

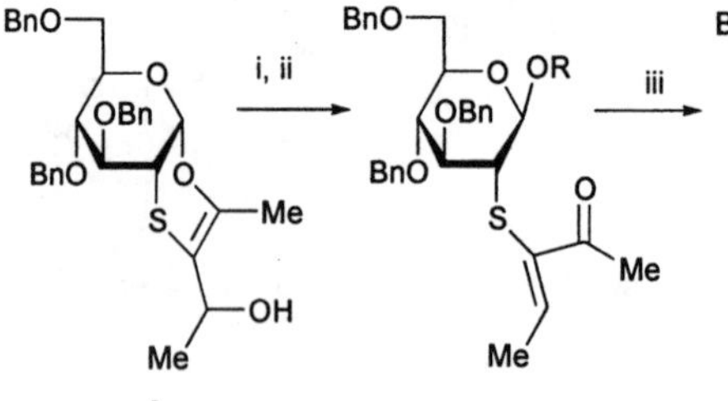

Reagents: i, ROH, CF_3CO_2H; ii, TsOH; iii, Raney Ni

Scheme 4

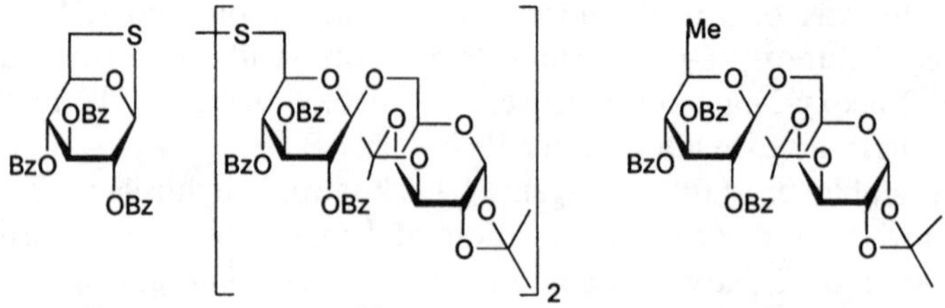

glycosides.[18] Glycosylation of various sugar alcohols by anhydrothiosugar **10** promoted by NIS/TfOH gave the corresponding disulfide-linked disaccharides, *e.g.* **11**. Subsequent reductive removal of the sulfur led to deoxy-disaccharides, *e.g.* **12**.[19] Two new syntheses of 2-deoxy-L-ribose from the inexpensive L-arabinose have been described. One, apparently suitable for large scale synthesis, effected a Barton-McCombie deoxygenation at C-2,[20] whereas the other featured the reduction of a 1,2-*O*-ethylthioorthoester to give a dialkoxyalkyl radical intermediate which produces a 2-deoxy-L-ribose triester.[21]

References

1 T.M. Hallis and H.-W. Liu, *Acc. Chem. Res.*, 1999, **32**, 579.
2 A. Kumar, A. Khare and N.K. Khare, *Phytochemistry*, 1999, **52**, 675.
3 A. Kumar, A. Khare and N.K. Khare, *Phytochemistry*, 1999, **50**, 1353.
4 R. Schoevaart, F. van Rantwijk and R.A. Sheldon, *Chem. Commun.*, 1999, 2465.
5 C. Guerard, V. Alphand, A. Archelas, C. Domuynck, L. Hecquet, R. Furstoss and J. Bolte, *Eur. J. Org. Chem.*, 1999, 3399.
6 D. Shabat, B. List, R.A. Lerner and C.F. Barbas, *Tetrahedron Lett.*, 1999, **40**, 1437.
7 A. Jux and W. Boland, *Tetrahedron Lett.*, 1999, **40**, 6913.
8 B.S.J. Blagg and C.D. Poulter, *J. Org. Chem.*, 1999, **64**, 1508.
9 C.M. Hayman, L.R. Hanton, D.S. Larsen and J.M. Guthrie, *Aust. J. Chem.*, 1999, **52**, 921.
10 K. Nakai, Y. Tagaki, S. Ogawa and T. Tsuchiya, *Carbohydr. Res.*, 1999, **320**, 8.
11 J. Molina Arévalo and C. Simons, *J. Carbohydr. Chem.*, 1999, **18**, 535.
12 S.T. Deal and D. Horton, *Carbohydr. Res.*, 1999, **315**, 187.
13 O. Yamazaki, H. Togo, S. Matsubayashi and M. Yokoyama, *Tetrahedron*, 1999, **55**, 3735.
14 W.R. Roush, B.W. Gung and C.E. Bennett, *Org. Lett.*, 1999, **1**, 891 (*Chem. Abstr.*, 1999, **131**, 272 093).
15 W.R. Roush and S. Narayan, *Org. Lett.*, 1999, **1**, 899 (*Chem. Abstr.*, 1999, **131**, 272 092).
16 G.S. Ghangas, *Phytochemistry*, 1999, **52**, 785.
17 G. Capozzi, C. Falciani, S. Menichetti, C. Narivi and B. Raffaelli, *Chem. Eur. J.*, 1999, **5**, 1748.
18 A. Dios, C. Nativi, G. Capozzi and R.W. Franck, *Eur. J. Org. Chem.*, 1999, 1869.
19 R.V. Stick, D.M.G. Tilbrook and S.J. Williams, *Aust. J. Chem.*, 1999, **52**, 685.
20 W. Zhang, K.S. Ramasamy and D.R. Averett, *Nucleosides Nucleotides*, 1999, **18**, 2357.
21 M.E. Jung and Y. Xu, *Org. Lett.*, 1999, **1**, 1517 (*Chem. Abstr.*, 1999, **131**, 351 547).

13

Unsaturated Derivatives

1 Pyranoid Derivatives

1.1 1,2-Unsaturated Cyclic Compounds and Related Derivatives. – *1.1.1 Syntheses of glycals.* Several new syntheses of glycals from glycosyl halides have been reported. Acetobromo-sugars, both mono- and di-saccharide, are reductively eliminated to glycals using titanocene dichloride (Cp_2TiCl_2) and manganese in THF.[1] This method does not require the separate preparation, using glove-box techniques, of $(Cp_2TiCl)_2$, which is the active species in the reaction. Zinc and aluminium proved less successful in the reduction than manganese. Another report of the use of this reagent to generate glycals, again in good yields, indicated its compatibility with protective groups such as silyl ethers, acetals and esters.[2] Both reports suggest a glycosyl–Ti(IV) intermediate is involved.

Glycosyl halides can also be treated with $Cr(EDTA)^{2-}$ in aqueous DMF at pH 5–7 to give glycals.[3,4] This methodology gives products in good yield and in high purity without the need for purification and has been applied to different sugars protected with a variety of esters. Another method, which uses catalytic amounts of vitamin B_{12} with zinc as a co-reductant in methanol/water gives excellent yields of tri-*O*-acetyl-D-glucal and -D-galactal in 5 minutes from the corresponding tetraacetylglycosyl bromides.[5] However, di-*O*-acetyl-D-rhamnal was obtained in only 45% yield.

1-*C*-Substituted glycals have been synthesized by metathesis.[6] The readily available alkene **1** can be acylated and the carboxyl group methylenated to enol ethers **2** with TMEDA/ $TiCl_4$/Zn. The ring-closing metathesis reaction with Grubbs' catalyst affords the 1-*C*-substituted glycal **3** (Scheme **1**).

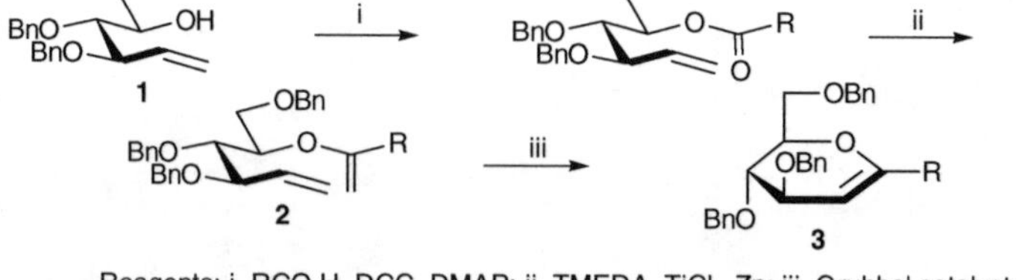

Reagents: i, RCO_2H, DCC, DMAP; ii, TMEDA, $TiCl_4$, Zn; iii, Grubbs' catalyst

Scheme 1

Carbohydrate Chemistry, Volume 33

During the first total synthesis of altohyrtin C (an antitumour macrolide, see Chapter 3) glycals such as **4** and **5** were prepared as precursors.[7]

β-D-Oleandrosyl-(1→4)-β-digitoxosyl-(1→4)-D-cymaral (**6**) has been isolated from dried twigs of *Marsdenia roylei*.[8]

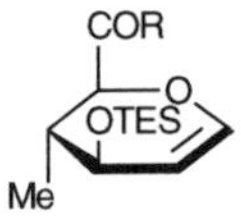

4 R = OMe
5 R = SPh

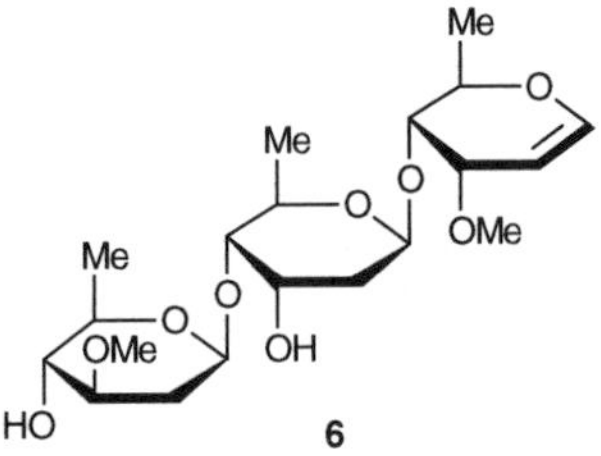

6

1.1.2 Reactions of glycals. Radical addition of *n*-Bu_3SnH to glycals gives 2-deoxyglycosyl stannanes regioselectively but not stereoselectively.[9] The use of $Bu_3Sn(Bu)Cu(CN)Li_2$ for the addition gives the 1-*C*-stannanes with both regio- and stereo-selectivity. For their conversion to *C*-glycosides see Chapter 3.

Tri-*O*-benzyl-D-glucal undergoes addition of aryl sulfenyl chlorides to give 2-arylthio-D-glycosyl chlorides.[10] These compounds can be converted into *C*-glycosides (see Chapters 3 and 11).

Detailed studies of fluorine addition at C-2 with nucleophilic addition at C-1 in the glycal series are given in a large paper concerning the mechanism and applications of electrophilic fluorination by Selectfluor (**7**).[11] Choice of protecting group, solvent and the Selectfluor counter ion affect the fluorination and entry of the nucleophile at C-1 of the glycal. Nucleophiles include carbohydrate secondary alcohols, amino acids, phosphonates and phosphates.

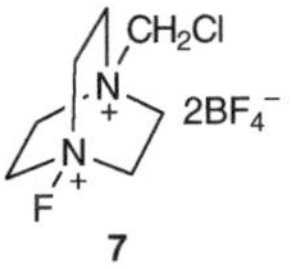

7

The alkoxides derived from protected serine and threonine derivatives undergo conjugate addition to tri-*O*-benzyl-2-nitro-D-galactal with no racemization at the amino acid α-positions.[12]

The unusual nitroenone shown in Scheme 2 undergoes a one-pot conjugate

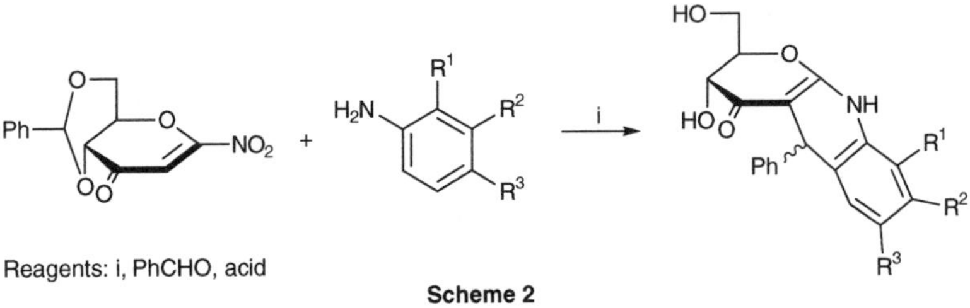

Reagents: i, PhCHO, acid

Scheme 2

addition with various anilines followed by an electrophilic substitution reaction with benzaldehyde and acid to give substituted dihydroquinolines (see also Chapters 14 and 15).[13]

A ring contraction of *O*-substituted glycals to 2,5-anhydroaldose dimethylacetals **8** occurs when a variety of glycals are exposed to thallium(III) nitrate in acetonitrile/methanol.[14] *Gluco*/*galacto* configurations and several protecting groups are tolerated.

The treatment of D-glucal with a catalytic amount of $Sm(OTf)_3$ or $RuCl_2(PPh_3)_3$ in the presence of 1 equivalent of water afforded optically active furandiol **9** in good yield under mild conditions.[15] The reactions also works well with D-galactal to give the same product.

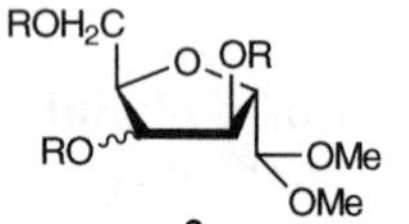

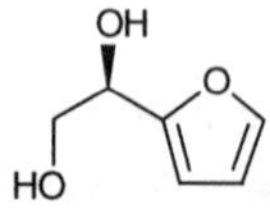

Glucals may be linked to polymers, containing benzylic alcohols, by a two-stage method using dialkyldihalosilanes to prepare intermediate dialkylhalosilyl ethers from the glucal.[16] Any position of the glucal may be linked.

The regioselective *O*-formylation of 3,4,6-tri-*O*-Tbdms-D-glucal under Vilsmeier-Haack conditions has been reported.[17] Formylation occurs at O-6 and other, non-carbohydrate, examples are reported.

Three semi-empirical methods were used to model the electroreductive formation of oxyanions from D-glucal.[18] Reaction of the anions with alkylating or acylating reagents produced ethers or esters with a clear regiochemistry, O-4 > O-3 > O-6. Of the three methods (AMI, MNDO, PM3), the best suited was AMI, and the electrochemically induced reactions produced an anion distribution pattern corresponding to the Gibbs free energies.

Tri-*O*-benzyl-D-galactal can be selectively oxidized to the 3-ulose by PhI(O-H)OTs. The carbonyl group was reduced under Luche conditions to give the hydroxy glycal of predominantly D-*lyxo* configuration (10:1) thus providing an effective 3-*O*-deblocking protocol.[19] Several other examples were given but they are less selective than the D-galactal 3-*O*-deblocking.

D-Glucal and D-galactal can be oxidized to their respective 1,5-anhydrohex-1-en-3-uloses in excellent yields by treatment with a Pd precipitate in CH_3CN or DMF, under an atmosphere of ethylene.[20,21]

A Pd-catalysed formation of dienes (2-vinylglycals) from 2-bromo-D-glucals and various ethylene derivatives has been reported.[22,23] The dienes may be used to generate tricyclic compounds through Diels-Alder reactions (see Chapter 14).

Addition of dibenzyl phosphite to the aldehyde **10** results in a mixture of isomeric α-hydroxy phosphonates, which may be separated.[24]

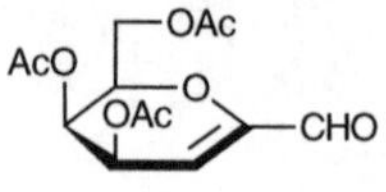

1.2 2,3-Unsaturated Cyclic Compounds. – *1.2.1 Syntheses involving allylic rearrangements of glycals.* Indium(III) chloride has been used to induce the formation of allyl 2,3-unsaturated *C*-glycosides from per-acetylated glycals and allyltrimethylsilane.[25] Tri-*O*-acetyl-D-galactal gave only the α-allyl compound **11** in 78% yield, while tri-*O*-acetyl-D-glucal afforded a 95% yield of the corresponding compounds with a 9:1 α:β ratio. Two other examples are given. In a related reaction alkynyl 2,3-unsaturated *C*-glycosides (**12**) were produced when tri-*O*-acetyl-D-glucal was treated with alkynyltrimethylsilanes and a Lewis acid.[26] When 3-trimethylsilylpropagyl acetate was used the reaction failed, but 3-TbdpsO-1-Tms-propyne was successful, with $SnCl_4$ as Lewis acid, giving **12** (X = $CH_2OTbdps$) in 85% yield.

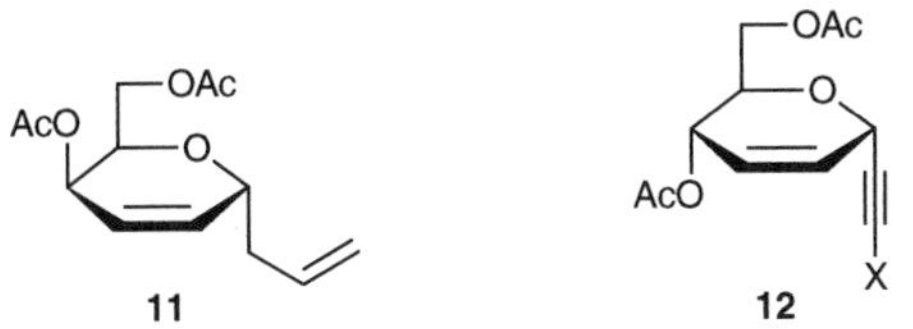

AcO OAc O 11

OAc O AcO X 12

2,3-Unsaturated glycosyl nitriles can be prepared from unprotected glycals by treatment with trimethylsilyl cyanide and $Pd(OAc)_2$ in acetonitrile.[27] With D-glucal a 99% yield of glucosyl nitrile was obtained with a 3:1 α:β ratio. In another paper it was reported that 2,3-unsaturated thioglycosides could be obtained by treatment of tri-*O*-acetyl-D-glucal with thiols in the presence of $LiBF_4$ in acetonitrile. Several aromatic and aliphatic thiols were used and yields were in the range of 56–72%.[28]

A *C*-glycoside fragment of angucyclines, a group of quinoid antibiotics, was prepared by way of the reaction of the glycosyl naphthalene **13** with either diacetyl-L-fucal, to give the unsaturated disaccharide **14**, or with acetylrhodinal to give the addition product **15**.[29] Low yields were observed in all reactions, and removal of the sugar protecting group on **13** before the addition reaction gave, unsurprisingly, a mixture of regioisomers of **14** or of **15**.

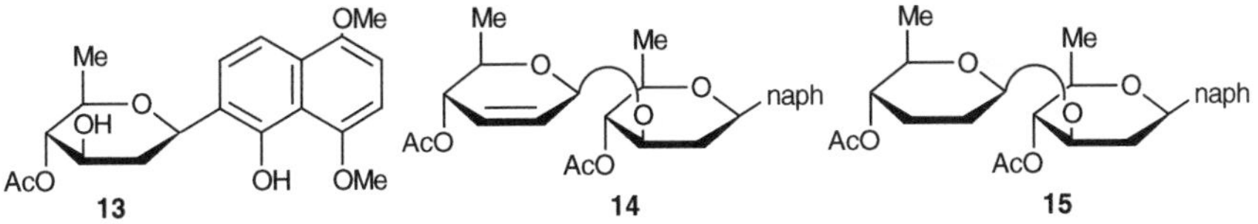

Me O OH AcO OMe OH OMe 13

Me O AcO Me O O AcO naph 14

Me O AcO Me O O AcO naph 15

The per-acetate of 2-hydroxy-D-glucal may be dimerized with I_2 or $BF_3.OEt_2$ in aqueous acetone to give, by way of a di-2,3-unsaturated compound and after hydrogenation and deacetylation, α,α-3,3-dideoxytrehalose.[30] Similarly the corresponding D-galactal compound gave the α-D-*lyxo*/α-D-*lyxo* analogue.

When 2,3-unsaturated compounds of type **16** are treated with Pd catalysts, Heck-type cyclizations take place to give bicyclic glycals or occasionally enol ethers (Scheme 3).[31]

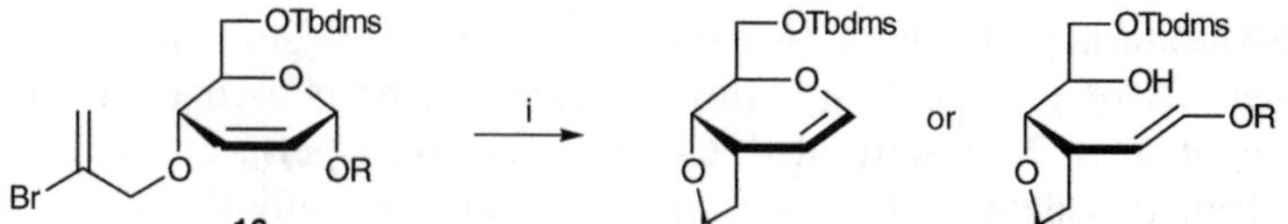

Reagents: i, $Pd(OAc)_2$

Scheme 3

1.2.2 Other syntheses and reactions. The synthesis of methyl 2-deoxy-2,3-dehydro-*N*-acetyl-neuraminate (**17**) and its 4-epimer has been reported.[32] Also the synthesis of 2,7-dideoxy-7-fluoro-2,3-didehydrosialic acid (**18**) has been reported and, in addition to inhibiting sialidase, it has been found to inhibit the binding of influenza virus H1 hemagglutinin to ganglioside GM 3.[33]

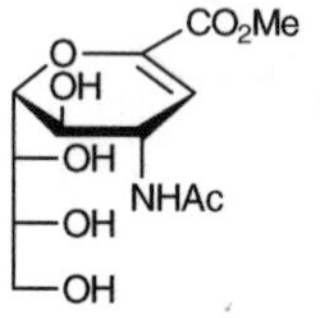

Addition of (methoxycarbonylmethyl)dimethylphosphonate to the enulose **19** gave either 1,4- or 1,2-addition (Scheme 4) depending on the choice of base and solvent.[34] The use of (ethoxycarbonylmethyl)diethylphosphonate led to 1,4-addition only without phosphonate–phosphate rearrangement (see also Chapters 7 and 14).

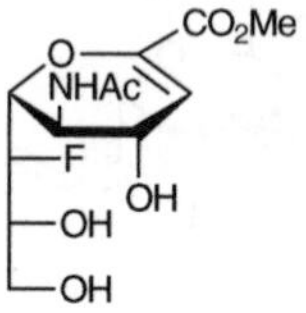

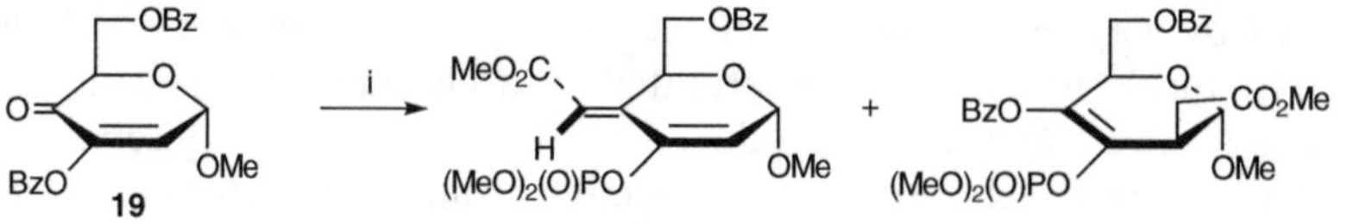

Reagents: i, $MeOP(O)CH_2CO_2Me$

Scheme 4

The hydrosilylation of several phenylthioalkynes (Scheme 5) with a catalytic amount of a dicobalthexacarbonyl alkyne complex has been reported.[35] These reactions proceed quickly and in good yield. The regioselective addition of nucleophiles to the products is also reported (see Chapter 24).

Palladium(0) catalysed π-allyl chemistry has been used to generate tertiary

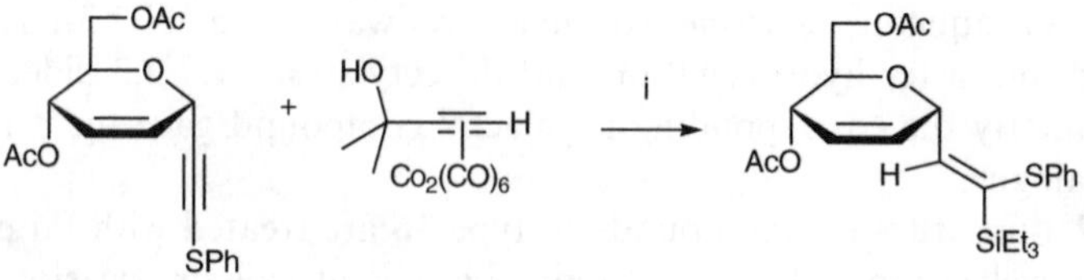

Reagents: i, Et_3SiH

Scheme 5

amines **20** from cyclohexyl 4,6-di-*O*-acetyl-2,3-dideoxy-α-D-*erythro*-hex-2-enoside (see Chapter 9 for additional reactions).[36] In a related reaction palladium(0) catalysed π-allyl chemistry has been used to generate the thioacetate **21** from ethyl 4,6-di-*O*-acetyl-2,3-dideoxy-α-D-*erythro*-hex-2-enoside in 73% yield (see Chapter 11 for additional details).[37]

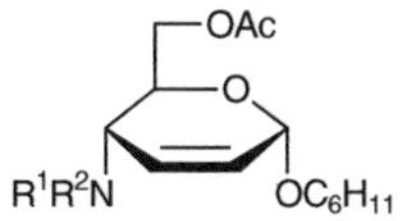

In an unusual approach, the hetero-Diels-Alder reaction of aldehydes with dienes has been used to generate unsaturated, 'doubly-reducing' disaccharides (Scheme 6).[38]

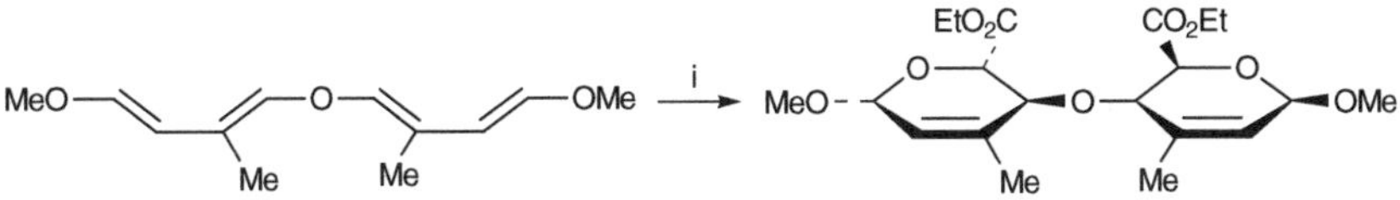

Reagents: i, $OHCCO_2Et$

Scheme 6

Unsaturated tetrahydrofurans and cyclopentanes are available by application of palladium catalysed metallo-ene cyclizations to 2,7-dienes derived from 2,3-unsaturated sugars (Scheme 7).[39]

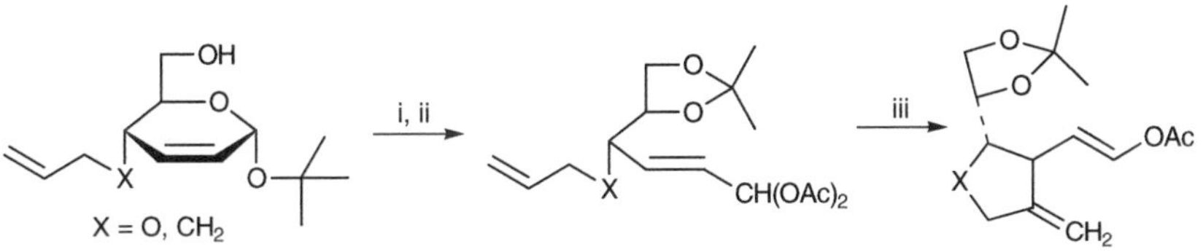

Reagents: i, Me_2CO, H^+; ii, I_2, Ac_2O; iii, $Pd(PPh_3)_4$, HOAc

Scheme 7

1.3 3,4-Unsaturated Cyclic Compounds. – Levoglucosenone is readily converted into the illustrated, known thio sugar (Scheme 8), and hence to an α-oxoketene dithioacetal and further still to the 'push-pull' butadiene (see Chapter 14 for additional reactions).[40]

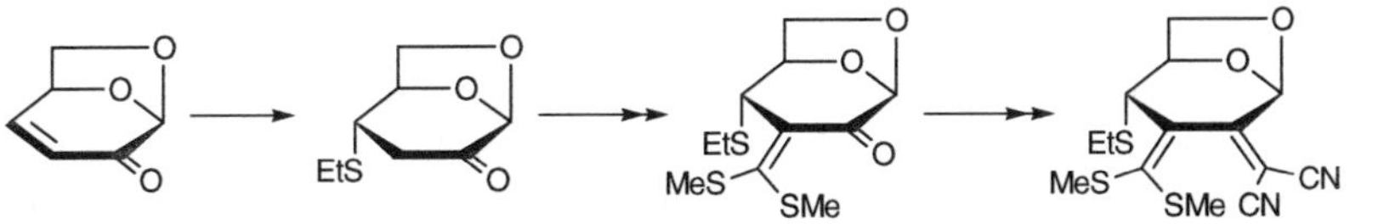

Scheme 8

1.4 4,5-Unsaturated Cyclic Compounds. – The unsaturated uronic acid **22** has been readily prepared from isopropyl 2,3,4-tri-*O*-acetyl-2-amino-2-deoxy-glucuronoside and used as an inhibitor of bacterial and viral sialidases.[41]

1.5 5,6-, 6,7-Unsaturated Cyclic Compounds and Exocyclic Glycals. – Methyl 2,3-di-*O*-benzyl-6-deoxy-4-*O*-methanesulfonyl-α-D-*xylo*-hex-5-enoside is an intermediate in the syntheses of the core of Zaragozic acid (see also Chapters 5 and 24).[42]

Both anomers of methyl 2,3-di-*O*-benzyl-4-*O*-(2,3,4-tri-*O*-benzyl-6-deoxy-β-D-*xylo*-pyrano-5-enosyl)-D-glucoside undergo reductive cyclization upon treatment with $Ti(O^iPr)Cl_3$ or Al^iBu_3 to give the pseudodisaccharides **23**.[43] The reaction is also successful for making the β-phenylthioglycoside and the 1,6-linked analogues of **23**, the latter giving a single, 5*R*, diasteroisomer unlike **23**, which are formed as mixtures.

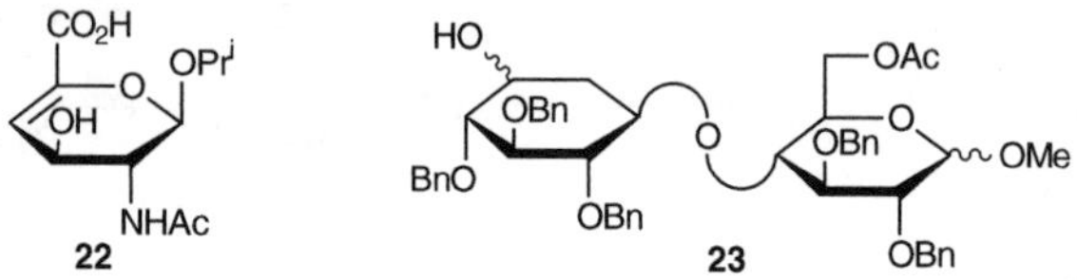

The known exoglycals **24** were prepared in good yields as the starting point for conversion to the *C*-glycosides **25** using the Ireland-Claisen rearrangement as the key step.[44] The tetrayne **26** has been prepared and tested as an inhibitor of *Staphylococcus aureus* but was found to be only moderately effective.[45]

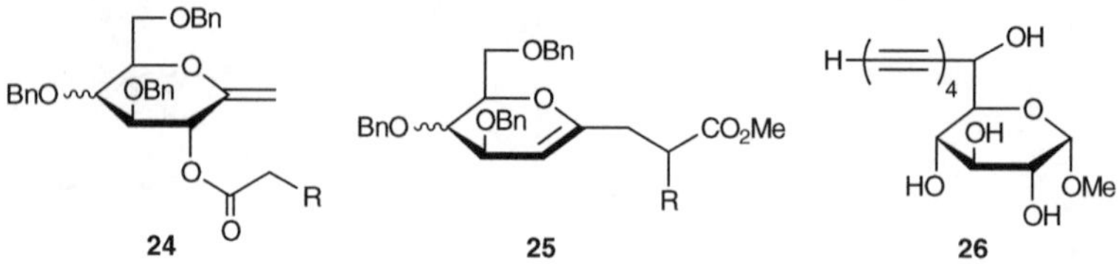

2 Furanoid Derivatives

2.1 1,2-Unsaturated Cyclic Compounds. – A new method for the formation of furanoid glycals relies on the oxidative elimination of selenium from 1,4-anhydro-3,5-di-*O*-benzyl-2-deoxy-2-phenylseleno-D-pentitols.[46] These compounds are in turn derived from the $PhSe^+$ mediated cyclization of some hydroxy alkenes. If the 5-*O*-benzyl group is replaced by *tert*-butyldiphenylsilyl the yields rise from 62–74% to 82–95%.

2.2 2,3-, 3,4- and 5,6-Unsaturated Cyclic Compounds. – Furanoid glycals have been converted into 1,4-anhydro-1-*C*-aryl-2,3-dideoxy-α-D-*glycero*-pent-2-enitols by treatment with an aryl Grignard reagent and a nickel(0) catalyst.[47]

2-Acylamino-2-deoxy-D-glucofuranoses have been converted into the hex-2-

enofuranoses **27** (see also Chapter 9).[48] A series of 2-deoxyribonucleosides were chain extended at C-5 by oxidation to the aldehydes followed by Wittig reactions with $Ph_3P{=}CHC(O)Ph$. Yields were in the range 38–63% (see also Chapter 20).

An unsaturated dinucleotide isostere **28** has been prepared by the palladium-catalysed coupling of a vinyl bromide with a phosphite.[50]

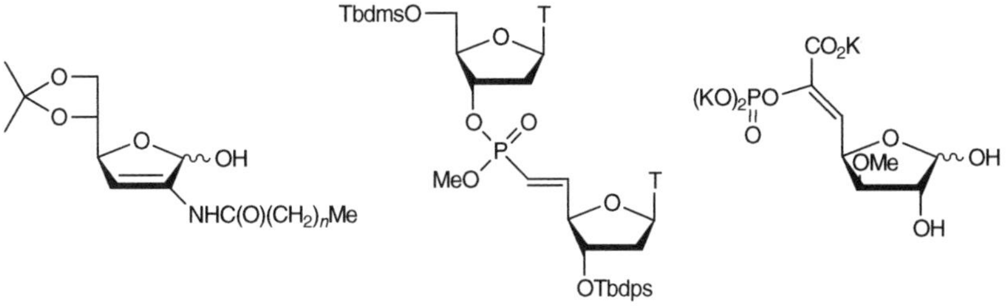

27 $n = 4–6$ **28** **29**

The glycos-4-yl phosphoenolpyruvic acid derivative **29** has been prepared.[51]

A new and efficient method for the transformation of 1,2-diols to alkenes has been reported (Scheme 9).[52] The diols are converted to cyclic thiocarbonates, heated under reflux with iodomethane to form iodomethylthiocarbonates, which are treated with phenyllithium to give the alkene. The last two steps proceed with 96% yield.

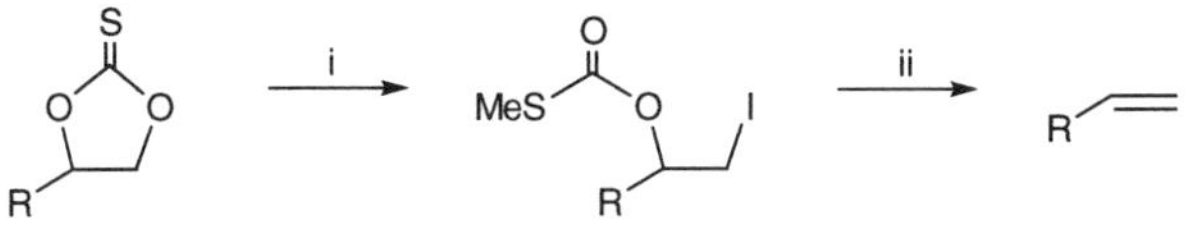

Reagents: i, MeI; ii, PhLi

Scheme 9

Further to their previous work[53,54] on iodonium ion mediated cyclization of C-5 *C*-allylated isopropylidene furanose **30** to give the THF-THP compound **31**, Mootoo *et al.* have reported on the stereochemistry of the key THP cyclization step. Application of the same type of reaction to the synthesis of THP analogues of pseudomonic acids has also been reported.[55]

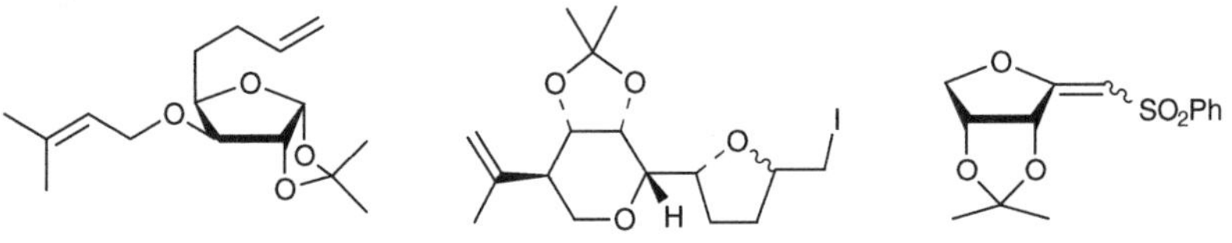

30 **31** **32**

2.3 Exocyclic Glycals. – Treatment of 2,3-*O*-isopropylidene erythronolactone with phenylsulfonylmethyllithium results in an adduct which, when exposed to $BF_3.OEt_2$, loses water to give the exocyclic glycals **32**.[56]

3 Septanoid Derivatives

An unusual ring expansion reaction of ribosyluracil derivative **33** which occurs on treatment with methylenetriphenylphosphorane in THF has been reported (Scheme 10).[57] A deuterium labelling experiment found deuterium at C-4′ of the dihydrooxepine ring and a ring cleavage between C-3′ and C-4′ of **33** was suggested.

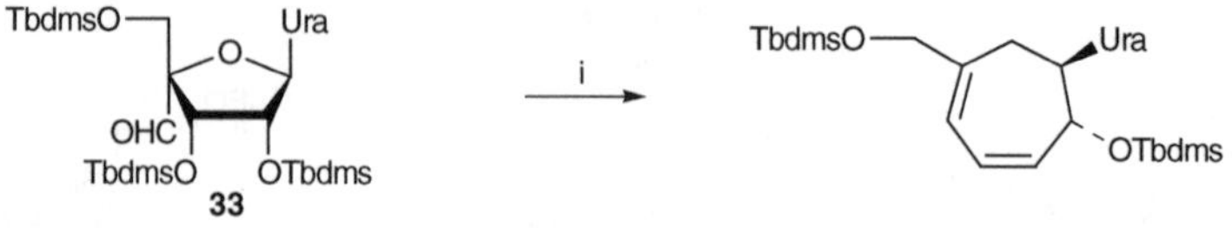

Reagents: i, $Ph_3P=CH_2$

Scheme 10

4 Acyclic Derivatives

The enone **34** was the starting point for the synthesis of the 1,6-diene **35**, the clean ring-closing metathesis of which, using Schrock's catalyst, gave the cyclopentanoid system **36** (see Chapter 18 for conversion of **36** into cyclitols).[58]

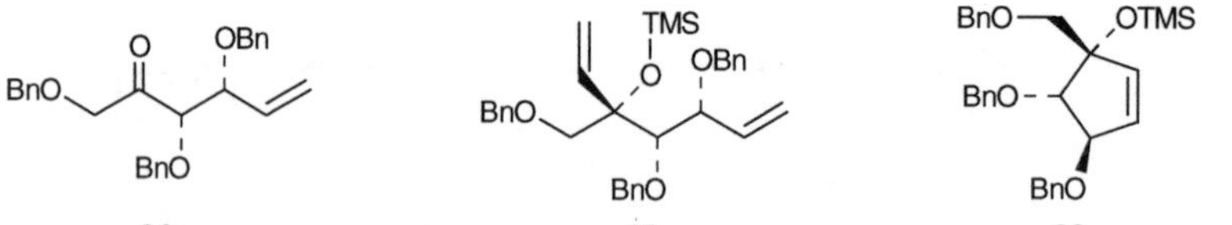

2,3,5-Tri-*O*-benzyl-D-arabinose has been converted, *via* the aldehyde **37** and Horner-Emmons olefination, into the *Z*-unsaturated ester **38**. This was subsequently transformed by deprotection at O-6, followed by triflation, azide displacement and 1,3-dipolar addition/N_2 extrusion, into the unsaturated azasugar **39**.[59]

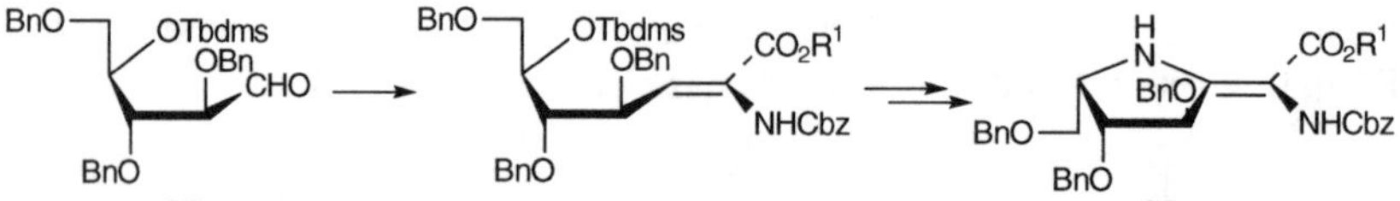

References

1 T. Hansen, S.N. Krintel, K. Daasbjerg and T. Skrydstrup, *Tetrahedron Lett.*, 1999, **40**, 6087.
2 R.P. Spencer, C.L. Cavallaro and J. Schwartz, *J. Org. Chem.*, 1999, **64**, 3987.
3 G. Kovács, K. Tóth, Z. Dinya, L. Somsák and K. Micskei, *Tetrahedron*, 1999, **55**, 5253.

4 *Cf. Tetrahedron Lett.*, 1996, **37**, 1293; see Vol. 30, p. 173, ref. 6; p. 219, ref. 30.
5 C.L. Forbes and R.W. Franck, *J. Org. Chem.*, 1999, **64**, 1424.
6 D. Calimente and M.H.D. Postema, *J. Org. Chem.*, 1999, **64**, 1770.
7 D.A. Evans, B.W. Trotter, P.J. Coleman, B. Cote, L.C. Dias, H.A. Rajapakse and A.N. Tyler, *Tetrahedron*, 1999, **55**, 8671.
8 A. Kumar, A. Khare and N.K. Khare, *Phytochemistry*, 1999, **50**, 1353.
9 D.H. Braithwaite, C.W. Holzapfel and D.B.G. Williams, *S. Afr. J. Chem.*, 1998, **51**, 162 (*Chem. Abstr.*, 1999, **130**, 223 501).
10 I.P. Smoliakova, M. Han, J. Gong, R. Caple and W.A. Smit, *Tetrahedron*, 1999, **55**, 4559.
11 S.P Vincent, M.D. Burkart, C.-Y. Tsai, Z. Zhang and C.-H. Wong, *J. Org. Chem.*, 1999, **64**, 5264.
12 G.A. Winterfeld, Y. Ito, T. Ogawa and R.R. Schmidt, *Eur. J. Org. Chem.*, 1999, 1167.
13 G. Scheffler, M. Justus, A. Vasella and H.P. Wessel, *Tetrahedron Lett.*, 1999, **40**, 5845.
14 D.H. Braithwaite, C.W. Holzapfel and D.B.G. Williams, *J. Chem. Res. (S)*, 1999, 108.
15 M. Hayashi, H. Kawabata and K. Yamada, *Chem. Commun.*, 1999, 965.
16 K.A. Savin, J.C.G. Woo and S.J. Danishefsky, *J. Org. Chem.*, 1999, **64**, 418.
17 S. Koeller and J.-P. Lellouche, *Tetrahedron Lett.*, 1999, **40**, 7043.
18 C.H. Hamann, S. Pleus, R. Koch and K. Baghorn, *J. Carbohydr. Chem.*, 1999, **18**, 1051.
19 A. Kirschning, U. Hary, C. Plumeier, M. Ries and L. Rose, *J. Chem. Soc., Perkin Trans. 1*, 1999, 519.
20 M. Hayashi, K. Yamada and O. Arikita, *Tetrahedron Lett.*, 1999, **40**, 1171.
21 M. Hayashi, K. Yamada and O. Arikita, *Tetrahedron*, 1999, **55**, 8331.
22 M. Hayashi, K. Amano, K. Tsukada and C. Lambeth, *J. Chem. Soc., Perkin Trans. 1*, 1999, 239.
23 M. Hayashi, K. Tsukada, H. Kawabata and C. Lambeth, *Tetrahedron*, 1999, **55**, 12287.
24 V. Kolb, F. Amann, R.R. Schmidt and M. Duszenka, *Glycoconjugate J.*, 1999, **16**, 537.
25 R. Gosh, D. De, B. Shown and S.B. Maiti, *Carbohydr. Res.*, 1999, **321**, 1.
26 M. Isobe, R. Saeeng, R. Nishizawa, M. Konobe and T. Nishikawa, *Chem. Lett.*, 1999, 467.
27 M. Hayashi, H.Kawabata and O. Arikita, *Tetrahedron Lett.*, 1999, **40**, 1729.
28 B.S. Babu and K.K. Balasubramanian, *Tetrahedron Lett.*, 1999, **40**, 5777.
29 K. Krohn and C. Bäuerlein, *J. Carbohydr. Chem.*, 1999, **18**, 807.
30 F.W. Lichtenthaler and B. Wonder, *Carbohydr. Res.*, 1999, **319**, 47.
31 K. Bedjeguelal, L. Joseph, V. Bolitt and D. Sinou, *Tetrahedron Lett.*, 1999, **40**, 87.
32 S. Li, H. Cui and O. Haruo, *Zhongguo Yoawu Huaxue Zazhi*, 1997, **7**, 167 (*Chem. Abstr.*, 1999, **130**, 14 126).
33 T. Sato, F. Ohtake, Y. Ohira and V. Okahata, *Chem. Lett.*, 1999, 145.
34 O.M. Moradei, C.M. du Mortier and A.F. Cirelli, *J. Carbohydr. Chem.*, 1999, **18**, 709.
35 M. Isobe, R. Nishizawa, T. Nishikawa and K. Yoza, *Tetrahedron Lett.*, 1999, **40**, 6927.
36 T.M.B. de Brito, L.P. da Silva, V.L. Siqueira and R.M. Srivastava, *J. Carbohydr. Chem.*, 1999, **18**, 609.

37 S. Divekar, M. Safi, M. Soufiaoui and D. Sinoü, *Tetrahedron*, 1999, **55**, 4369.
38 A. Guillam, L. Toupet and J. Maddaluno, *J. Org. Chem.*, 1999, **64**, 9348.
39 C.W. Holzapfel, L. Marcus and F. Toerien, *Tetrahedron*, 1999, **55**, 3467.
40 M. Gómez, J. Quincoces, B. Kuhla, K. Peseke and H. Reinke, *J. Carbohydr. Chem.*, 1999, **18**, 57.
41 P. Florio, R.J. Thomson, A. Alafaci, S. Abo and M. von Itzstein, *Bioorg. Med. Chem. Lett.*, 1999, **9**, 2065.
42 C. Taillefumier, M. Lakhrissi and Y. Chaleur, *Synlett*, 1999, 697.
43 A.J. Pearce, M. Sollogoub, J.-M. Mallet and P. Sinaÿ, *Eur. J. Chem.*, 1999, 2103.
44 T. Vidal, A. Haudrechy and Y. Langlois, *Tetrahedron Lett.*, 1999, **40**, 5677.
45 H.-J., Park and N.-D. Sung, *HanGuk Nonghwa Hakhoechi*, 1998, **43**, 258 (*Chem. Abs.*, 1999, **130**, 38 580).
46 F. Bravo, M. Kassou and S. Castillón, *Tetrahedron Lett.*, 1999, **40**, 1187.
47 M. Tingoli, B. Panunzi and F. Santacroce, *Tetrahedron Lett.*, 1999, **40**, 9329.
48 J.G. Fernandez-Bolanos, U. Ulgar, I. Robina and J. Fuentes, *Carbohydr. Res.*, 1999, **322**, 284.
49 D. Crich and X.-S. Mo, *Synlett*, 1999, 67.
50 S. Abbas and C.J. Hayes, *Synlett*, 1999, 1124.
51 P. Coutrot, C. Grison, M. Tabyaoui, B. Tabyaoui and S. Dumarçay, *Synlett*, 1999, 792.
52 M. Adiyaman, Y.-J. Jung, S. Kim, G. Saha, W.S. Powell, G.A. FitzGerald and J. Rokach, *Tetrahedron Lett.*, 1999, **40**, 4019.
53 S. Elvey and D.R. Mootoo, *J. Am. Chem. Soc.*, 1992, **114**, 9685.
54 W. Shan, P. Wilson, W. Liang and D.R. Mootoo, *J. Org. Chem.*, 1994, **59**, 7986.
55 N. Khan, H. Xiao, B. Zhang, X. Cheng and D.R. Mootoo, *Tetrahedron*, 1999, **55**, 8303.
56 A. Alzérreca, E. Hernandez, E. Mangual and J.A. Prieto, *J. Heterocycl. Chem.*, 1999, **36**, 555.
57 M. Nomura, K. Endo, S. Shuto and A. Matsuda, *Tetrahedron*, 1999, **55**, 14847.
58 O. Sellier, P.V. de Weghe and J. Eustache, *Tetrahedron Lett.*, 1999, **40**, 5859.
59 Y. Konda, T. Sato, K. Tsushima, M. Dodo, A. Kusunoki, M. Sakayanagi, N. Sato, K. Takeda and Y. Harigaya, *Tetrahedron*, 1999, **55**, 12723.

14
Branched-chain Sugars

R

1 Compounds with a C–C–C Branch-point

O

Branch at C-2 or C-3. – The synthesis of tetradeuterated 2-*C*-methyl-D-erythritol derivative **4** has been reported in connection with the investigation of the non-mevalonate pathway for isoprenoid biosynthesis. Palladium(II)-catalysed deutereostannation of **1** (that gave the stannyl derivative **2**) and cyanocuprate coupling of the derived vinyl iodide **3** with $(CD)_3CuLi.LiCN$ were the key steps in the reaction (see also Chapter 18). Conversion of the stannyl derivative **2** to the iodide **3** was achieved by treatment with iodine in ether.[1]

ROH_2C—≡—CH_2OR
1
R = Tbdps

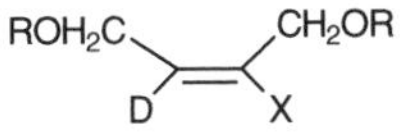

ROH_2C CH_2OR D X
2 X = $SnBu_3$, R = Tbdps
3 X = I

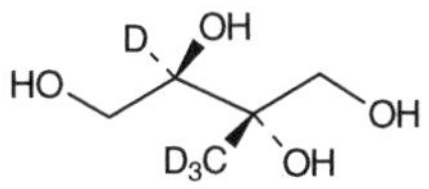

D OH HO OH D_3C OH
4

A new branched-chain aldose, 2-*C*-(hydroxymethyl)-D-allose, and its molybdic acid-catalysed tautomerization to sedoheptulose (in admixture with 2,7-anhydrosedoheptulose) have been reported. The branched chain allose was obtained by a base-catalysed addition of formaldehyde to 2,3:5,6-di-*O*-isopropylidene-β-D-allofuranose followed by acid hydrolysis of the aldol product (see also Chapter 2).[2] A reported one-step synthesis of D-hamamelose [2-*C*-(hydroxymethyl)-D-ribose] involves isomerization of D-fructose in mild acidic solution in the presence of catalytic amount of molybdic acid.[3,4] Preparation of other similarly branched aldoses from the corresponding ketose sugars has also been likewise carried out (see also Chapter 2).[4] Copper triflate-catalysed intramolecular [2+2]photoannulation of branched chain aldosides to produce tetracyclic compounds, *e.g.* **6** from **5**, has been described along with a mechanistic rationalization of the process (see also Chapter 24).[5] See ref. 25 for a 2′-ethynyl nucleoside derivative (**36**).

Carbohydrate Chemistry, Volume 33

Reaction of 1,2-*O*-isopropylidene-5-*O*-methyl-α-D-*erythro*-pent-3-ulofuranose with lithium acetylide in THF led to formation of the expected 3-ethynyl derivative **7** which, on reaction with iodonium ion-producing reagents (for example NIS/TsOH) underwent intramolecular cyclization with the loss of 5-*O*-methyl group to afford furo[3,4-*b*]furan derivative **8** (see also Chapter 24).[6] Stereoselective nucleophilic addition of cyanide to the 5-*O*-benzoyl-1,2-*O*-isopropylidene-α-D-*erythro*-pentofuranose-3-ulose to give the *xylo* cyanohydrin **9** and its conversion to **10** and other analogous isocarbonucleoside derivatives have been reported (see also Chapter 10).[7,8] Structural characterization of some of these derivatives by 2-D NMR and X-ray crystallography have also been described (see also Chapter 22).[7,8] In pursuits of potent new anti HIV-1 agents (see Vol. 32, Chapter 14, ref. 13) 3-*spiro*-branched ribofuranoses **11** and **12** have been prepared in one-pot procedures from cyanohydrin **9** by reaction with chlorosulfonyl isocyanate or sulfamoyl chloride, respectively (see also Chapter 10). NMR studies and theoretical

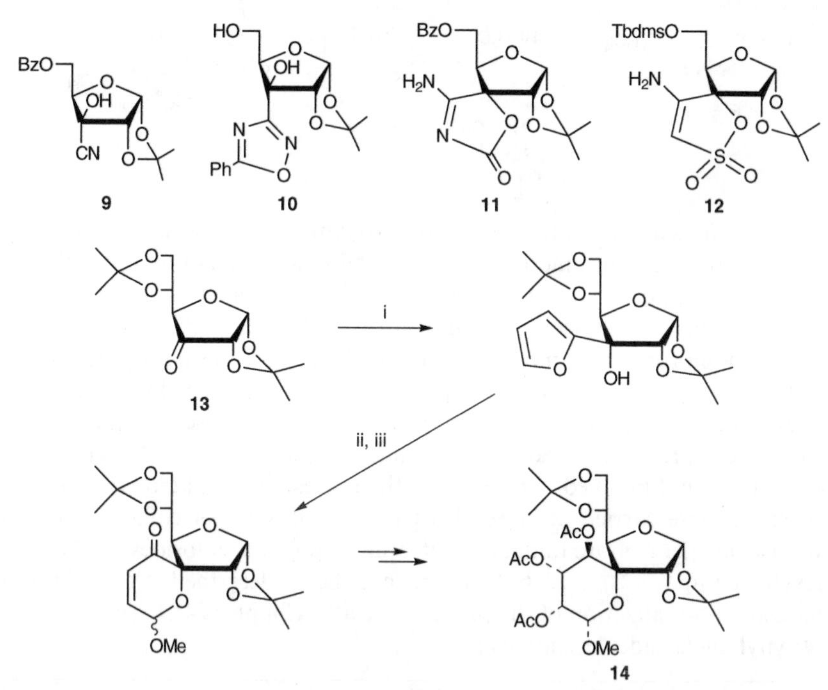

Reagents: i, furan, BuLi, THF, –78 °C; ii, NBS, aq. THF, –5 °C; iii, Ag_2O–MeI, CH_2Cl_2

Scheme 1

calculations on these bicyclic compounds have also been reported (see also Chapter 21).[9]

The diastereoselective synthesis of *spiro* carbon-linked disaccharide **14** from the known 3-ulose derivative **13** has likewise been reported (Scheme 1).[10] Moreover, addition of diethyl (1-bromo-2-propenyl)phosphonate to **13**, and other 3-ulose sugar derivatives, to give *e.g.* allyl phosphonate derivative **15** and related compounds has also been described.[11] (See *Tetrahedron Asymm.*, 1997, **8**, 1411 and also Chapter 2.)

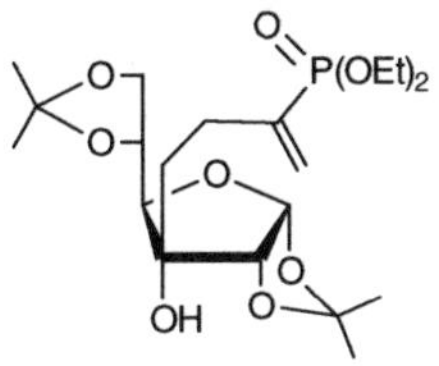

15

A practical approach to the synthesis of methyl L-mycaroside **17**, a subunit of the kedarcidin chromophore, from ethyl (*S*)-lactate *via* the (*E*)-alkene intermediate **16** (see also Chapter 9) has been described.[12] The conversion involved diastereoselective dihydroxylation as the key step, as shown in Scheme 2.

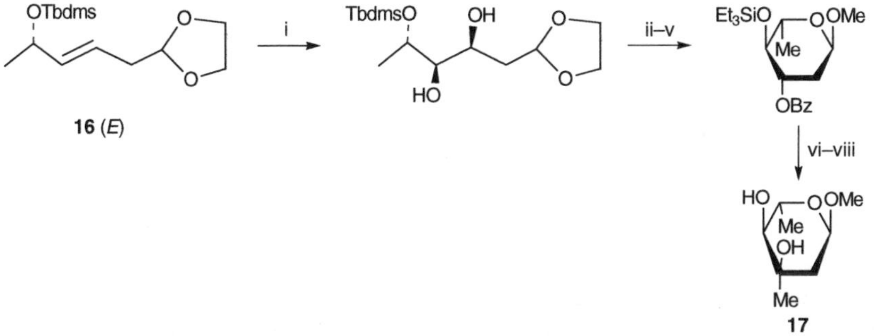

16 (*E*) **17**

Reagents: i, AD-mix-α, $MeSO_2NH_2$, aq. Bu^tOH; ii, H^+, MeOH; iii, Bu^n_2SnO, $PhCH_3$, reflux; iv, BzCl, $CHCl_3$; v, Et_3SiCl, imidazole, DMF; vi, DIBAL-H, Et_2O, –78 °C; vii, Des-Martin periodinate oxdn.; viii, MeMgBr, Et_2O

Scheme 2

Among branched-chain sugar disaccharides reported are the kelampayosides A and B (**18**, **19** respectively), natural products isolated from *Cinnamomium cassia*. Their synthesis was based on a glycosylation protocol involving chemoselective NIS/TfOH-mediated glycosylation (for making the disaccharide unit) and the $BF_3.Et_2O$-mediated trichloroacetimidate method (for connecting the aromatic aglycon moiety) applied to the appropriately protected sugar building blocks (see also Chapter 3).[13] Synthesis of a series of other C-3′ branched-chain sugar disaccharides (*e.g.* **20**) by addition of MeLi or AllMgBr to the corresponding 3′-uloses and their subsequent enzymatic conversion to trisaccharide derivatives have also been described (see also Chapter 4).[14]

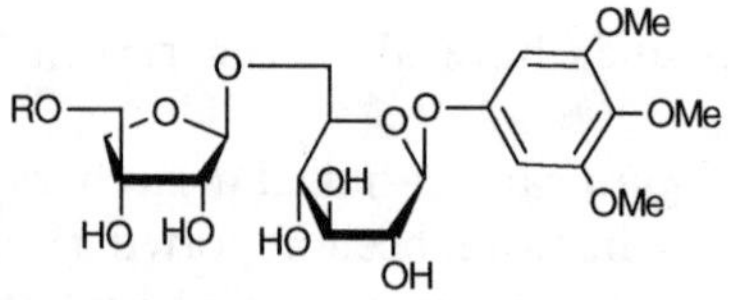

18 R = H
19 R = C(O)CH=CH-[3,4-(OH)$_3$]Ph for B

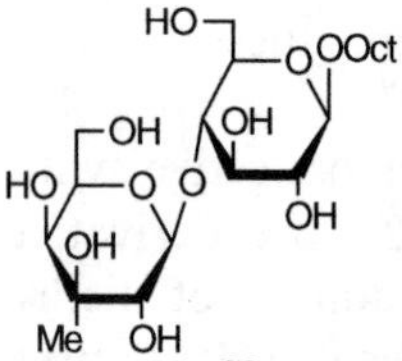

20

1.2 Branch at C-4 or C-5. – Diastereoselective acylation of 3-*O*-benzyl-4-*C*-hydroxymethyl-1,2-*O*-isopropylidene-α-D-*erythro*-pentofuranose using vinyl acetate and either *Pseudomonas cepacia*- or *Candida rugosa*-derived lipase has been studied in organic solvents. The results showed complementarity in diastereoselectivity, the two enzymes giving **21** and **22**, respectively (see also Chapter 7).[15]

Synthesis and allyl ketene acetal Claisen rearrangement of D-mannose-derived acetal **25** (as well as other ketene acetals prepared by various other methods) to generate C-4 branched-chain sugar derivatives have been reported (Scheme 3). X-ray structures of some of these products are noted in Chapter

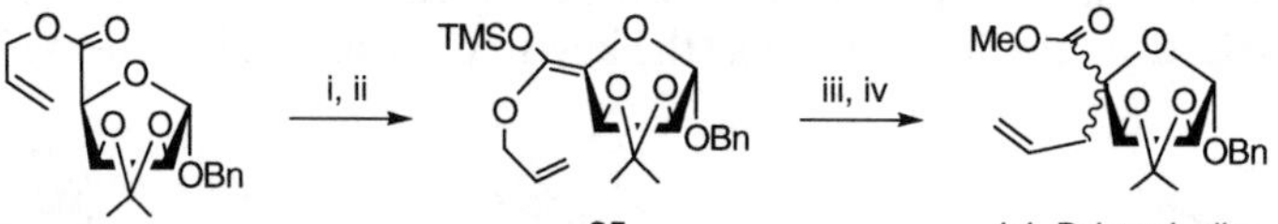

25

4:1, D-*lyxo*-:L-*ribo*-

Reagents: i, LDA, THF/HMPA; ii, TMSCl; iii, TBAF, THF; iv, MeOH/PhH, TMSCHN$_2$

Scheme 3

22.[16] In a new approach to the bicyclic core of zaragozic acids, C-4 branched furanoside **23** has been prepared by a CeCl$_3$-promoted aldol reaction of 2,3-*O*-isopropylidene glyceraldehyde and the corresponding L-*xylo*-furanosiduronate. LiAlH$_4$ reduction followed by acid treatment gave rise to the desired bicyclic core of zaragozic acids (see also Chapter 24).[17] Diastereofacial selectivity observed in the nucleophilic 1,2-addition of MeMgX (X = Cl/I) and MeLi to the ulose derivative **24** has been discussed in terms of the structure of the metal chelate formed and substituent/solvent effects.[18]

Conversion of L-arabinose to the C-5 branched sugar noviose, and its transformation to a series of coumarin inhibitors of gyrase B, *e.g.* **26**, have

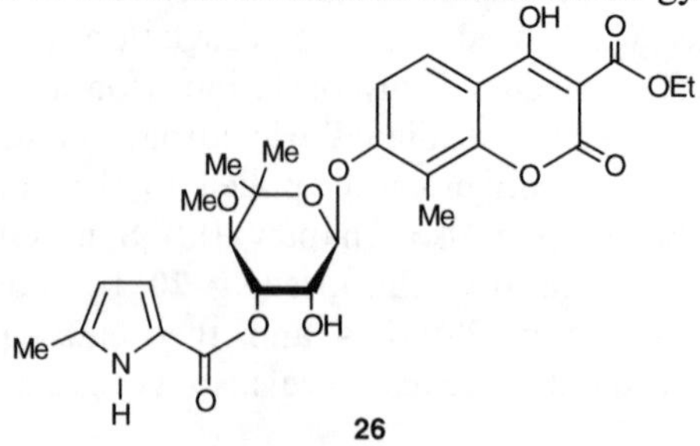

26

been reported. These compounds showed good *in vitro* super coiling inhibitory activity as well as antibacterial activity against vancomycin and teicoplanin-resistant *Enterococci*.[19]

2 Compounds with a C–C(R)(N)–C Branch-point

Full characterization of 3-deoxy-3-*C*-methyl-3-nitro-α-D- and β-L-glucopyranosides, products of Baer reaction of the dialdehyde of IO_4^- oxidation of methyl α-D-hexopyranosides with $MeNO_2$ obtained as a 1:1 mixture, has been achieved by X-ray diffraction analysis (see also Chapter 22). Separation of the isomers was accomplished *via* their 4,6-*O*-isopropylidene derivatives. Further functionalization of the separated compounds by DAST has also been described (see also Chapter 8 and 10).[20] Following the recent synthesis of a model vancomycin aryl glycoside (see *Angew. Chem. Int. Ed. Engl.*, 1998, **37**, 1871), the total synthesis of vancomycin has now been reported in which the C-3 branched glycosyl fluoride **27** has been used as the glycosyl donor (see also Chapters 3 and 19).[21]

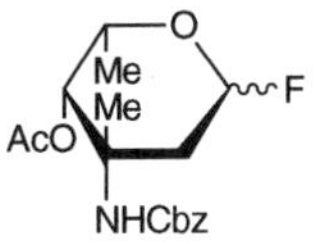

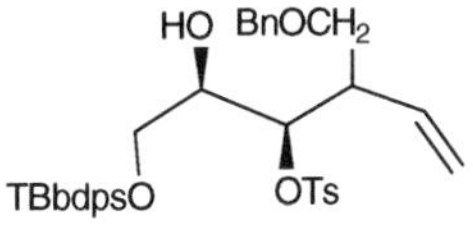

3 Compounds with a C–C(R)(H)–C Branch-point

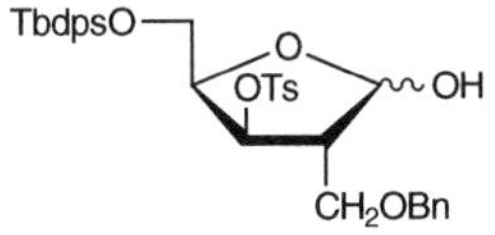

3.1 Branch at C-2. – Synthesis of the C-2 branched pentose derivative **29**, and other branched pentose derivatives, was achieved from the partially protected vinyl alditol **28**, or corresponding vinyl compounds, by ozonolysis.[22]

Fructose 1,6-bisphosphate aldolase isolated from *Staphylococcus carnosus* has been applied in the synthesis of 'bicyclic sugars' such as **31** from the dialdehyde **30**, as shown in Scheme 4.[23] It is important to note that this

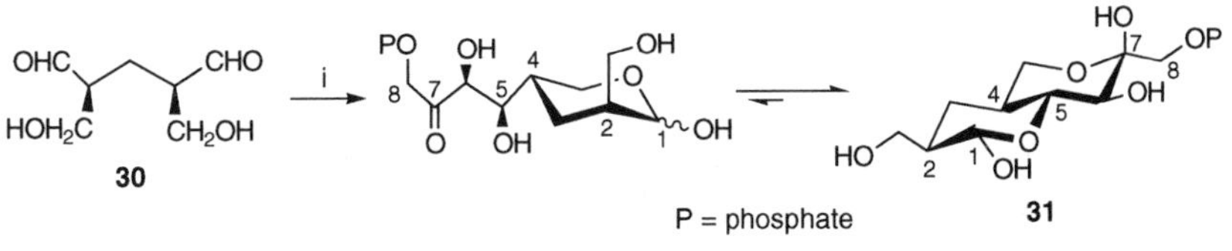

Reagents: i, aldolase from *S. carnosus*, $HOCH_2C(O)CH_2OP$

Scheme 4

transformation could not be carried out with the well-known rabbit enzyme since it is denatured by the dialdehyde.

The impact of solvent and counter ion on the regio- and stereochemical aspects of Horner-Wadsworth-Emmons reactions of 2-en-4-uloses with metal enolates of dimethyl [(methoxycarbonyl)methyl]phosphonate (**32**) or diethyl [(ethoxycarbonyl)methyl]phosphonate has been studied. Thus, branched-chain derivatives **34** (formed by rearrangements of the 1,2-addition product) and **35** (1,4-adduct-rearranged product) were obtained in various proportions from the enulose **33** as shown in Scheme 5 (see also *Tetrahedron*, 1997, **53**, 7397). Evidence of phosphonate–phosphate rearrangement was presented (see also Section 5 and Chapter 7 and 13).[24]

Sovent	Base	**34** (%)	**35** (%)
Polar	NaH	73	—
Non-polar	NaH	28	47
All	LiBr	No reaction	
All	KO^{t}Bu	70	—

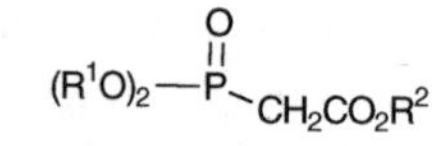

32 R^1 = Me, R^2 = Me

Reagents: i, **32**, base, solvent

Scheme 5

2-Deoxy-2′-*C*-ethynyl-modified nucleotide **37** has been synthesized from the protected **36** (obtained from the 2-ulose derivative by base-catalysed addition of TMSC≡CH in THF). Substitution of two or three of the ethynyl-modified guanosines for deoxyguanosine within $d(CG)_3$ or $d(GC)_3$ led to a Z-DNA-like conformation of the resulting duplex, independent of salt concentration.[25]

Base = Gua or Cyt

36 **37**

Application of SmI_2-promoted *C*-glycosylation has been noted in the stereospecific synthesis of α-1,2-*C*-mannobioside derivative **41**. The mannosyl pyridylsulfone **38** and the 2-*C*-formyl glycoside **39** (see *Tetrahedron Lett.*, 1993, **34**, 6247) were coupled in a remarkably efficient reaction to give **40** which was then converted to **41** (Scheme 6). Conformational analysis carried out by NMR spectroscopy has also been reported (see also Chapter 21).[26]

Studies on the reaction of nitrone **42** with furanone **43**a (see *Chem. Papers*, 1997, **51**, 163) have been extended to its substituted derivatives **43**b–**43**f.

38 39 40 41

Reagents: i, SmI_2, THF

Scheme 6

Acetate **43**b, for example, gave 2-*C*-branched products **44** and **45** together with the 3-*C*-branched compound **46** in the ratio 64:25:11 (see also Section 3.2).[27]

42

43 a–f
R = H, Ac, Bz, Piv, Tbdms, Tbdps respectively

44 **45** **46**

Cycloaddition reactions of dichloroketene with glycals give bicyclic compounds **47**. Dechlorination of these with Bu_3SnH/AIBN and subsequent ring opening with NaOMe/MeOH of the derived products have also been described (see also Chapter 3).[28] Dithionite-mediated addition (a radical 'domino reac-

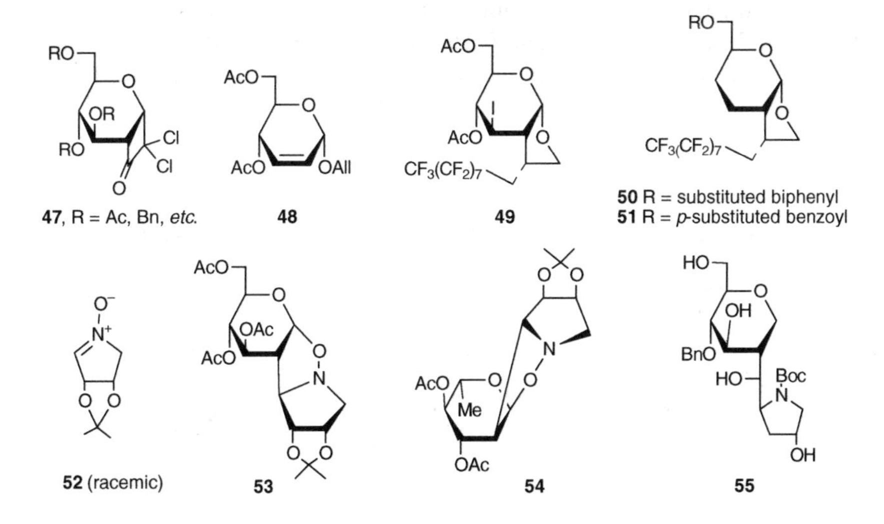
47, R = Ac, Bn, *etc.* **48** **49** **50** R = substituted biphenyl **51** R = *p*-substituted benzoyl

52 (racemic) **53** **54** **55**

tion') of 1-iodoperfluorooctane to the enopyranoside **48** has been carried out to yield pyrano[1,2-*b*]furan derivative **49**. Subsequent transformation into *O*-alkylated/acylated compounds **50/51** respectively has also been described. These compounds acted as non-amphiphilic chiral mesogens (see also Chapter 3 and 24).[29] Other cycloaddition reactions of glycals reported are those of the racemic pyrroline *N*-oxide **52** with tri-*O*-acetyl-D-glucal and di-*O*-acetyl-D-rhamnal. These reactions proceeded with double asymmetric induction to give **53** ('matched' bottom-*exo-anti* product) and **54** ('matched' top-*exo-anti* product) as the respective main products.[30] However, as these reactions proceeded with only partial kinetic resolution other cycloadducts were also produced.

Convergent synthesis of (1,2)- and (1,4)- (see Section 3.3) *C*-linked imino disaccharides (*e.g.* **55**) has been reported following Nozaki-Kishi coupling of a hydroxyproline-derived carbaldehyde with, for example, levoglucosenone enol triflate (for the preparation of **55**).[31] Reaction of 2-*C*-(dicyanomethylene)-substituted 1,6-anhydro hexopyranose derivative **56** with aryl isothiocyanates to furnish, *e.g.*, tricyclic compound **57** has been reported (Scheme 7). Also see Sections 3.2 and 5 as well as Chapter 10).[32] Pauson-Khand reactions of 1,6-

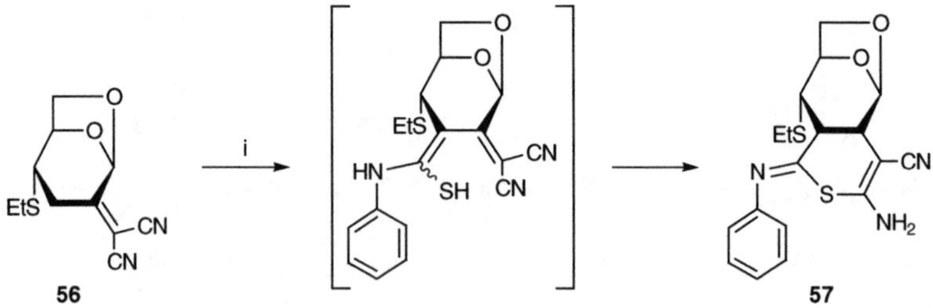

Reagents: i, PhNCS

Scheme 7

enyne **58** and similar enynes, with $Co_2(CO)_8$ have been reported.[33,34] The product **59** obtained from **58** has been extensively studied and forms a well defined intermediate toward the total synthesis of iridoid glycosides. (See also *Tetrahedron Lett.*, 1994, **35**, 5059, *J. Org. Chem.*, 1996, **61**, 7666 and Sections 3.2 and 5). Several such reactions leading to iridoid aglycons have been described.[33]

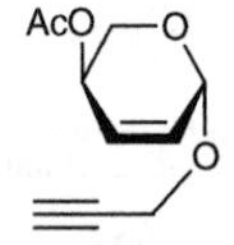

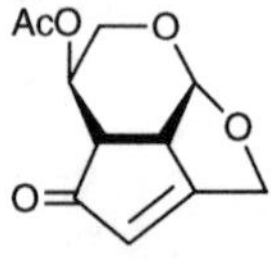

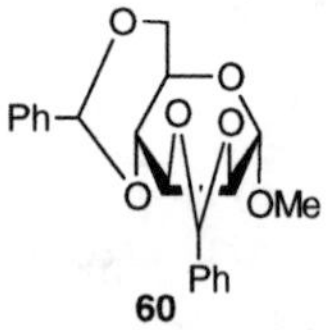

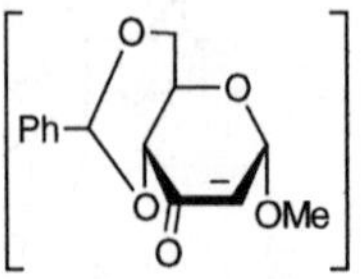

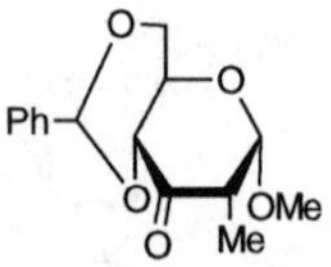

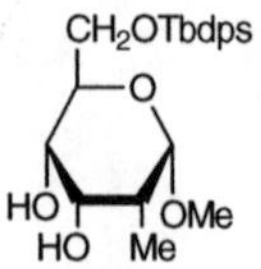

The 2-*C*-branched hexopyranoside **63**, obtained from the acetal derivative **60** *via* **62** formed by deprotonation at C-3 using BuLi followed by *C*-methylation of the resulting enolate **61**, has been described as an intermediate in the total synthesis of the macrolide soraphen $A_{1\alpha}$.[35] Synthesis of C1–C8 fragment **66** of (−)-discodermolide, the antipode of the marine natural product (+)-discodermolide noted for its potent immunosuppressive and anti-cancer activities, has been carried out following aldol coupling of **64** and **65** (see also Section 3.3).[36]

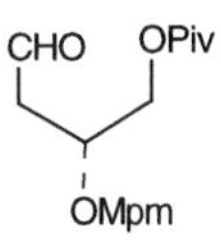

64 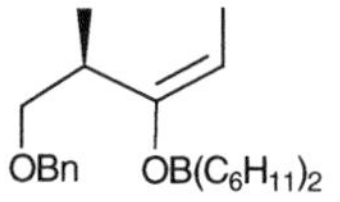 **65** 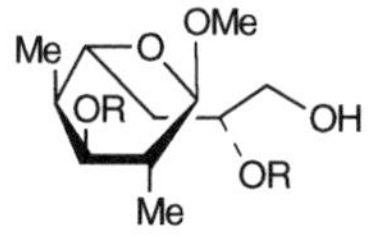 **66** R = Tbdms

3.2 Branch at C-3. – Carbohydrate *N*-oxides have been used as precursors of, among other functionalized carbohydrate derivatives, branched-chain sugars.[37] Thus compound **67**, prepared *in situ* from the corresponding 3-deoxy-3-(4-morpholinyl)-derivative by MCPBA oxidation, gave **69** as a mixture of isomers on treatment with hot pyridine. The reaction may proceed *via* 1,5-diene **68** which undergoes *in situ* Claisen rearrangement producing **69** (mixture of isomers at C-3) as the final products. This mixture, however, on brief treatment with silica gel gave equatorially substituted branched-chain sugar **69** specifically.[37]

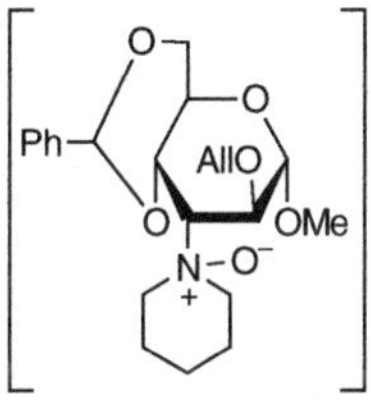

67 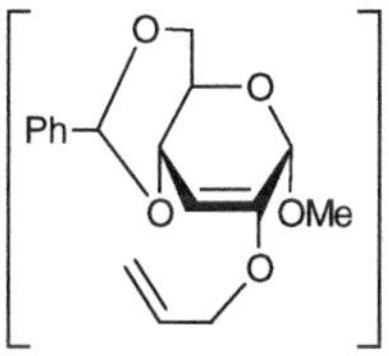 **68** 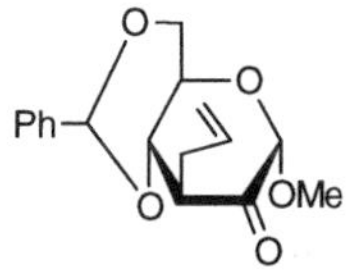 **69**

As described for the synthesis of (1,2)-*C*-linked disaccharides, convergent synthesis of (1,3)-*C*-linked aza disaccharides has been reported.[38] Two carbocyclic guanosine analogues, **70** and **71**, with an electron withdrawing fluoro or hydroxy substituent in the 4′-position, have been reported (see also Chapter 20).[39] They were also evaluated as anti-HIV-1 and anti-HSV-1 agents but showed no activity (see also *AIDS*, 1997, **11**, 157 and *Nucleosides Nucleotides*, 1992, **11**, 1739). Conversion of D-xylose to the C-3 branched-chain sugar **72** has been described, and the latter was converted to nucleosides.[40] Palladium(0)-catalysed Heck-type cyclization of aryl *erythro*- or *threo*- 2,3-unsaturated hexopyranosides has been further exemplified (see also *J. Org. Chem.*, 1997, **62**, 1341 and 6827 and also Chapters 13 and 24).[41,42] Synthesis of bicyclic pyrrolazines (*e.g.* **73**) has likewise been reported from 2,3-anhydropyranosides (see also Chapter 9).[43] Reaction of 3-iodolevoglucosenone with the sodium derivative of ethyl cyanoacetate (or ethyl acetoacetate or acetylacetone) at −60 °C has been reported to give tetrasubstituted cyclopropane derivatives, presumably of structure **74**.[44]

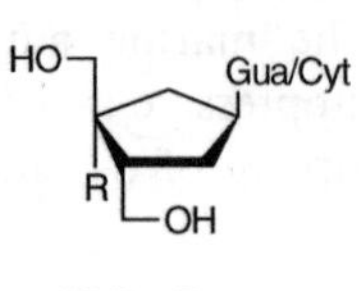

70 R = F
71 R = OH

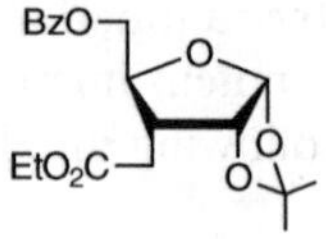

72

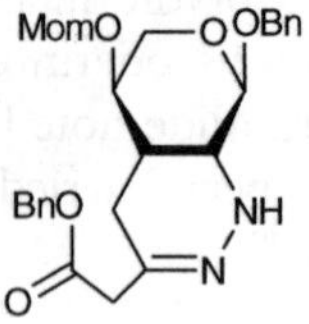

73

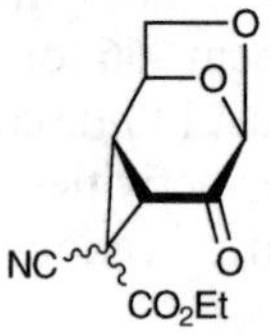

74

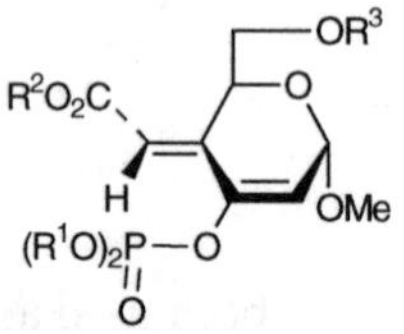

R^1, R^2 = Me, Et; R^3 = H, Tbdps or Bz
75

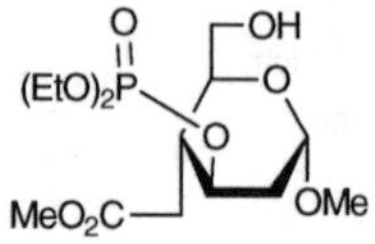

76

3.3 Branch at C-4 or C-5. – Phosphate enol esters **75** (for their synthesis see *Tetrahedron*, 1997, **53**, 7397) have been subjected to various conditions of catalytic hydrogenation to afford partially or fully saturated branched-chain carbohydrate derivatives. One such product, compound **76**, has been described as a precursor to thromboxane analogues.[45] A seven-step (20% overall yield) synthesis of pseudomonic acid C analogues **81** involves an efficient ene reaction between **77** and trioxane **78**, to give **79**, followed by its intramolecular silyl-mediated Sakurai cyclization with the acetal **80** as key

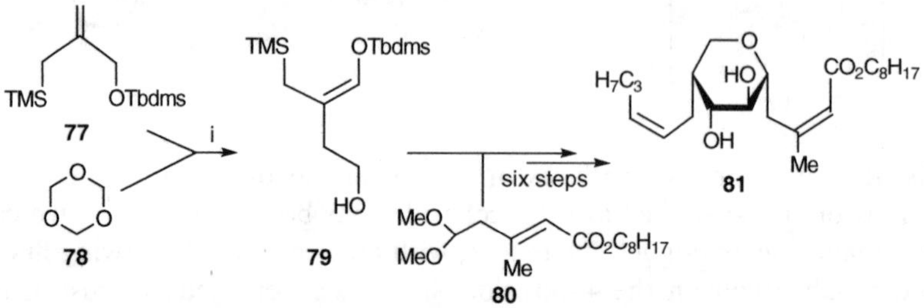

Reagents: i, Yamamoto's Al-reagent

Scheme 8

steps (Scheme 8).[46] Further to the work on the iodonium ion-mediated cyclization of C-5 allylated isopropylidene furanose **82** to give the THF-THP ether **83** (see *J. Am. Chem. Soc.*, 1992, **114**, 9685 and *J. Org. Chem.*, 1994, **59**, 7986), the stereochemistry of the key THP cyclization and its application to the synthesis of THP analogues of pseudomonic acids have been studied (see also Chapter 13).[47] Bicyclic oxazines (*e.g.* **86**) have been prepared in high

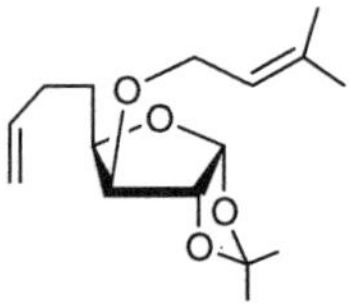

82

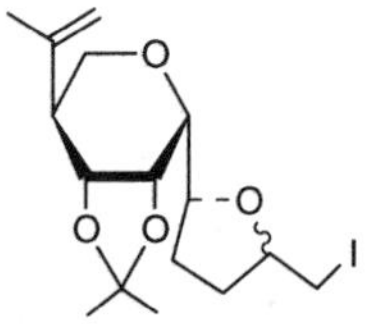

83

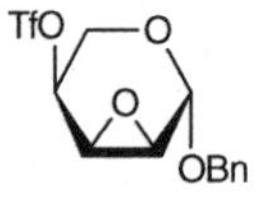

84

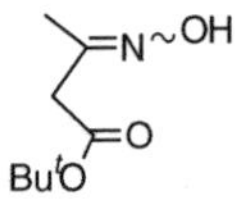

85

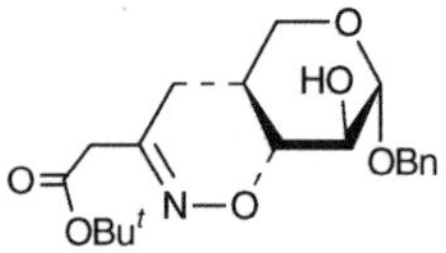

86

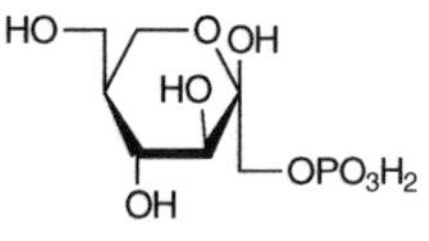

87

yield from sugar epoxides (*e.g.* **84**) and oximes (**85**), pre-treated with BuLi (see also Chapter 9).[43]

Coupling of dihydroxyacetone phosphate with 3-hydroxy-2-hydroxymethylpropanal in the presence of rabbit muscle aldolase is an efficient method to synthesize branched-chain hexulose phosphate **87**.[48] Further derivatization of **87** was also described. (See also Chapters 2, 3, and 7).[48]

4 Compounds with a C–C(R)(R)–C Branch-point

Conversion of methyl 4,6-*O*-benzylidene-2-*O*-methyl-α-D-*ribo*-hexopyranosid-3-ulose to the *xylo*-hexopyranoside derivative **88** has been reported. The starting material was first condensed with nitromethane and the resulting product was subjected to Michael-type addition using Me_2CuLi. Reduction of the nitro group in the Michael adduct followed by deamination gave **88**.[49] A palladium(II)-promoted domino process [using $PdCl_2(MeCN)_2$ and $CuCl_2$] has been suggested for the conversion of pseudo glycals **89** to 3-*C*-geminally substituted chiral dihydropyran derivatives **90**.[50]

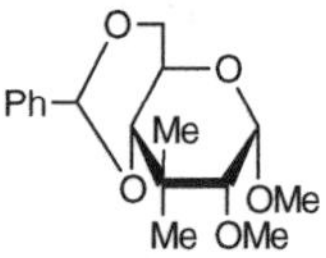

88

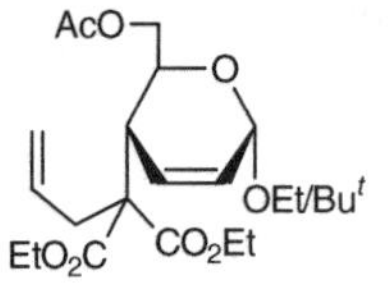

89

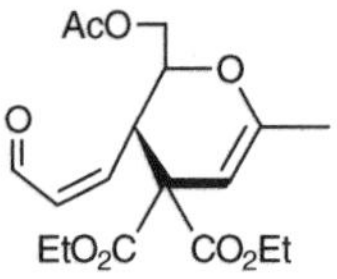

90

5 Compounds with a C–C(=R)–C or C=C(R)–C Branch-point

Push-pull alkenes, such as the oxoketene dithioacetal **92**, have been prepared from **91**, in turn obtained from levoglucosenone (see Vol. 30, p. 196, ref. 47

and Vol. 26, p. 162, ref. 47), by treatment with CS_2–MeI and NaH and were used as precursors of annelated pyranosides (*e.g.* **93**) (see also Chapter 13).[51] 4-Oxo-Kdn2en (**94**), prepared from Kdn2en methyl ester by permanganate oxidation, has been converted to 4-carbethoxyethylene-Kdn2en derivative **95** by Wittig reaction followed by alkaline hydrolysis (see also Chapter 16).[52] Unsaturated branched-chain carbohydrates (*e.g.* **98**) have been efficiently synthesized by a palladium-catalysed Heck-type reaction in aqueous[53] or organic[54] media using the glycal derivatives **96** and vinyl compounds **97** as starting materials. One of the attractions of the former method is that the reaction can be done with unprotected substrates. The conjugated dieno-pyranoside **98** was subsequently subjected to Diels-Alder reactions to produce chiral carbocyclic systems (see also Chapters 3 and 13).[54]

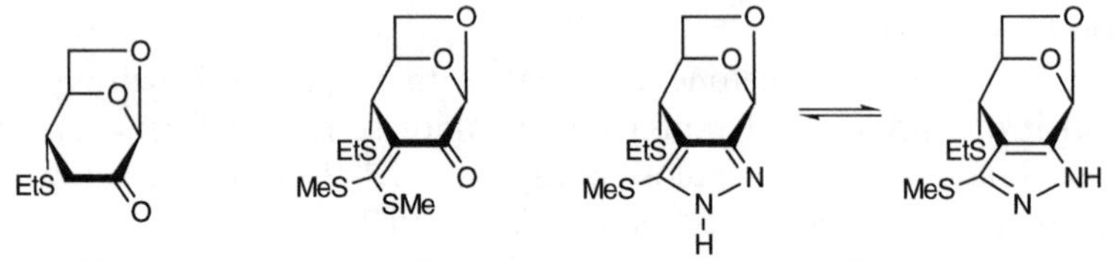

91 **92** **93**

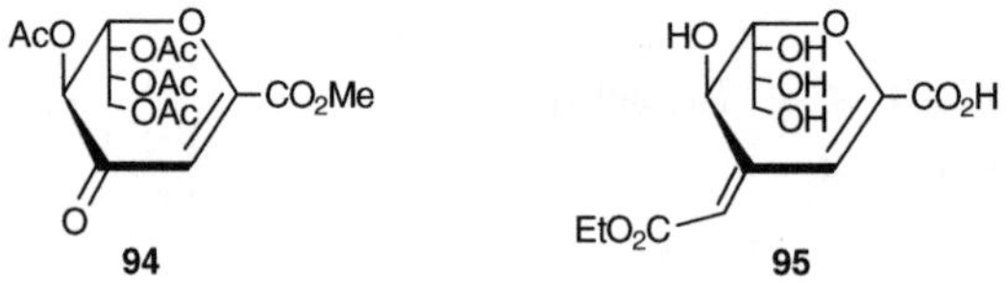

94 **95**

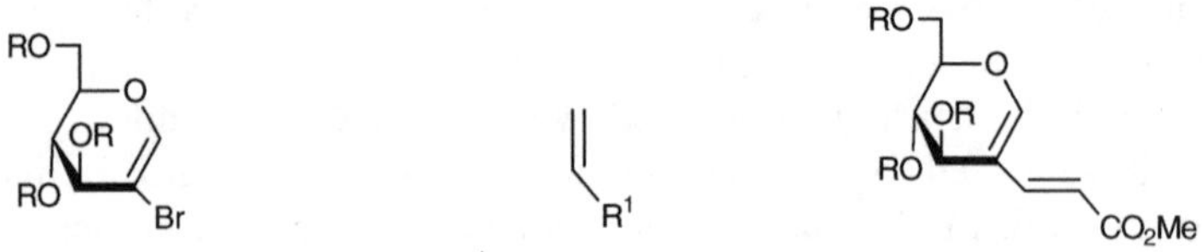

96 R = H or Ac **97** R^1 = H, CN or CO_2Me **98** R = H or Ac

A one-pot reaction has been reported for the conversion of enantiopure nitroenone **99** to dihydropyridine derivatives (*e.g.* **100**) which could be efficiently transformed into quinoline derivatives (*e.g.* **101**) by treatment with CAN (Scheme 9) (see also Chapters 13 and 15).[55]

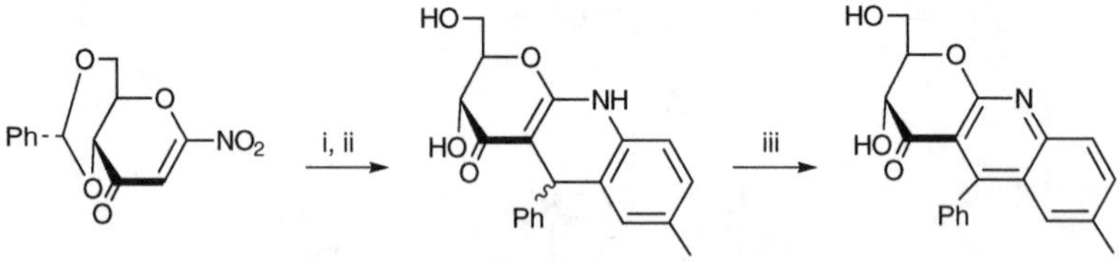

99 **100** **101**

Reagents: i, *p*-toluidine, CH_2Cl_2, base; ii, PhCHO, TFA; iii, CAN. DMF

Scheme 9

References

1 L. Cheron, J.F. Haeffler, C. Pale-Grosdemange and M. Rohmer, *Tetrahedron Lett.*, 1999, **40**, 8369.
2 Z. Hricoviniova-Bilikova and L. Petrus, *Carbohydr. Res.*, 1999, **320**, 31.
3 Z. Hricoviniova, M. Hricovini and L. Petrus, *Chem. Pap.*, 1998, **52**, 692 (*Chem. Abstr.*, 1999, **130**, 139547).
4 Z. Hricoviniova-Bilikova, M. Petrusova, A.S. Serianni and L. Petrus, *Carbohydr. Res.*, 1999, **319**, 38.
5 D.J. Holt, W.D. Barker, P.R. Jenkins, S. Ghosh, D.R. Russell and J. Fawcett, *Synlett.*, 1999, 1003.
6 E. Djuardi and E. McNelis, *Tetrahedron Lett.*, 1999, **40**, 7193.
7 M. Zhang, H. Zhang, L. Ma, J. Min and L. Zhang, *Carbohydr. Res.*, 1999, **318**, 157.
8 M.L. Zhang, Y.X. Cui, L.T. Ma, L.H. Zhang, Y. Lu, B. Zhao and Q.T. Zheng, *Chin. Chem. Lett.*, 1999, **10**, 117 (*Chem. Abstr.*, 1999, **131**, 157 873).
9 M.J. Camarasa, M.L. Jimeno, M.J. Perez-Perez, R. Alvarez and S. Velazquez, *Tetrahedron*, 1999, **55**, 12187.
10 G.V.M. Sharma, V.G. Reddy and P.R. Krishna, *Tetrahedron Lett.*, 1999, **40**, 1783.
11 R. Csuk and C. Schröder, *J. Carbohydr. Chem.*, 1999, **18**, 285.
12 M.J. Lear and M. Hirama, *Tetrahedron Lett.*, 1999, **40**, 4897.
13 H.I. Duynstee, M.C. de Koning, G.A. Van der Marel and J.H. van Boom, *Tetrahedron*, 1999, **55**, 9881.
14 X. Qian, K. Sujino, A. Otter, M.M. Paleic and O. Hindsgaul, *J. Am. Chem. Soc.*, 1999, **121**, 12063.
15 S.K. Sharma, S. Roy, R. Kumar and V.S. Parmar, *Tetrahedron Lett.*, 1999, **40**, 9145.
16 B. Werschkun and J. Thiem, *Synthesis*, 1999, 121.
17 P. Fraisse, I. Hanna, J.Y. Lallemand, T. Prange and L. Ricard, *Tetrahedron*, 1999, **50**, 11819.
18 M. Ghosh, R. Kakarla and M.J. Sofia, *Tetrahedron Lett.*, 1999, **40**, 4511.
19 P. Laurin, D. Ferrow, M. Klich, C. Dupuis-Hamelin, P. Mauvais, P. Lassaigne, A. Bonnefoy and B. Musicki, *Bioorg. Med. Chem. Lett.*, 1999, **9**, 2049.
20 A.T. Carmona, P. Borachero, F. Cabrera-Escribano, M. J. Dianez, M.D. Estrado, A. Lopez-Casto, R. Ojeda, M. Gomez-Guiller and S. Perrez-Garrido, *Tetrahedron Asymmetry*, 1999, **10**, 1751.
21 K.C. Nuolaon, H.J. Mitchell, N.F. Jain, T. Bando, R. Hughes, N. Winssinger, S. Natarajan and A.E. Koumbis, *Chem. Eur. J.*, 1999, **5**, 2648.
22 A. Chattopadhyay, B. Dhotare and S. Hassarajani, *J. Org. Chem.*, 1999, **64**, 684.
23 M.T. Zannetti, C. Walter, M. Knorst and W.D. Fessner, *Chem . Eur. J.*, 1999, **5**, 1882.
24 O.M. Moradei, C.M. du Mortier and A.F. Cirelli, *J. Carbohydr. Chem.*, 1999, **18**, 709.
25 R. Buff and J. Hunziker, *Synlett*, 1999, 905.
26 O. Jarrelon, T. Skrydstrup, J.F. Espinosa, J. Jiménez-Barbero and J.M. Beau, *Chem. Eur. J.*, 1999, **5**, 430.
27 V. Ondrus, M. Orsag, L. Fisera and N. Pronayova, *Tetrahedron,* 1999, **55**, 10425.
28 K.N. Cho, J. Oh, T. Yoon, K.H. Chum and J.E.N. Shin, *J. Korean Chem. Soc.*, 1999, **43**, 375 (*Chem. Abstr.*, 1999, **131**, 286 707).

29. M. Hein and R. Miethchen, *Eur. J. Org. Chem.*, 1999, 2429.
30. F. Cardona, S. Valenza, A. Goti and A. Brandi, *Eur. J. Org. Chem.*, 1999, 1319.
31 Y.H. Zhu and P. Vogel, *Chem. Commun.*, 1999, 1873.
32 M. Gómez, J. Quincoces, K. Peseke, M. Michalik and H. Reinke, *J. Carbohydr. Chem.*, 1999, **18**, 851.
33 J. Marco-Contelles and J. Ruiz, *J. Chem. Res. (S)*, 1999, 260.
34 J. Marco-Contelles and J. Ruiz-Caro, *J. Org. Chem.*, 1999, **64**, 8302.
35 S. Abel, D. Faber, O. Hüller and B. Giese, *Synthesis*, 1999, 188.
36 S.A. Filla, J.J. Song, L. Chen and S. Masamune, *Tetrahedron Lett.*, 1999, **40**, 5449.
37 B. Ravindran and T. Pathak, *J. Org. Chem.*, 1999, **64**, 9715.
38 Y.H. Zhu and P. Vogel, *J. Org. Chem.*, 1999, **64**, 666.
39 J. Watchtmeister, A. Muhlmann, B. Classon and B. Samuelsson, *Tetrahedron*, 1999, **55**, 10761.
40 D.M. Lu, J.M. Min and L.H. Zhang, *Carbohydr. Res.*, 1999, **317**, 193.
41 K. Bedjeguelal, L. Joseph, V. Bolitt and D. Sinou, *Tetrahedron Lett.*, 1999, **40**, 87.
42 K. Bedjeguelal, V. Bolitt and D. Sinou, *Synlett*, 1999, 762.
43 R.A. Al-Gawasmeh, T.H. Al-Tel, R.J. Abdel-Jalil and W. Voelter, *Chem. Lett.*, 1999, 541.
44 F.A. Valeev, E.V. Gorobets and M.S. Miftakhov, *Russ. Chem. Bull.*, 1999, **48**, 152 (*Chem. Abstr.*, 1999, **131**, 59 072).
45 O. Moradei, C.M. du Mortier, A.F. Cirelli and J. Thiem, *J. Carbohydr. Chem.*, 1999, **18**, 15.
46 I.E. Marko and J.M. Plancher, *Tetrahedron Lett.*, 1999, **40**, 5259.
47 N. Khan, H. Xiao, B. Zhang, X. Cheng and D.R. Mootoo, *Tetrahedron*, 1999, **55**, 8303.
48 S. David, *Eur. J. Org. Chem.*, 1999, 1415.
49 N. Kawauchi, *Carbohydr. Lett.*, 1998, **3**, 199 (*Chem. Abstr.*, 1999, **130**, 95 729).
50 C.W. Holzapfel and L. Marais, *J. Chem. Res. (S)*, 1999, 190.
51 E. Gomez, J. Quincices, B. Kuhla, K. Peseke and H. Reinke, *J. Carbohydr. Chem.*, 1999, **18**, 57.
52 X.-L. Sun, T. Kai, N. Satok, H. Takayanagi and K. Furuhata, *J. Carbohydr. Chem.*, 1999, **18**, 1131.
53 M. Hayashi, K. Amano, K. Tsukada and C. Lamberth, *J. Chem. Soc., Perkin Trans. 1*, 1999, 239.
54 M. Hayashi, K. Tsukada, H. Kawabata and C. Lamberth, *Tetrahedron*, 1999, **55**, 12287.
55 G. Scheffler, M. Justus, A. Vasella and H.P. Wessel, *Tetrahedron Lett.*, 1999, **40**, 5845.

15
Aldosuloses and Other Dicarbonyl Compounds

1 Aldosuloses

Three new *C*-glycosidic flavonoids bearing 6-deoxy-L-*ribo*-hexos-3-ulosyl moieties have been isolated from aerial parts of *Cassia occidentalis*[1] and two new β-D-*ribo*-hex-3-ulopyranoside iridoid glycosides were isolated from the bluebeard plant.[2] The metabolism, biological activity and methods for analysis of 3-deoxyglucosone have been reviewed.[3]

A new procedure for the selective oxidation of primary alcohols to the corresponding aldehydes in the presence of secondary alcohols uses H_2O_2/methyltrioxorhenium(VII)/HBr/TEMPO in HOAc. The parameters can be adjusted to afford the corresponding carboxylic acids.[4] The oxidation of carbohydrate stannylene acetals to osulose derivatives has been achieved in improved yields by using 1,3-dibromo-5,5-dimethylhydantoin in place of bromine or NBS. The product ratios were unchanged.[5] The enzymatic oxidation of some disaccharides using *Agrobacterium tumefaciens* to 3-osulose derivatives has been studied,[6] and similar results were obtained using a pyranose-2-oxidase to make 2-osuloses.[7]

Oxidation and reduction of the aldos-5-ulose derivative **1** afforded **2**, which gave the previously unknown L-*ribo*-hexos-5-ulose on acid hydrolysis.[8] The rigid L-fucose mimics **3** and **4** have been prepared in connection with the synthesis of potential fucosyltransferase inhibitors.[9,10]

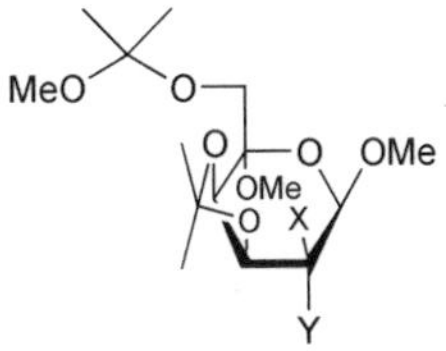

1 X = H, Y = OH
2 X = OH, Y = H

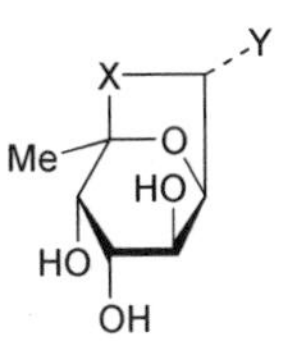

3 X = O, Y = NH_2
4 X = NAc, Y = OMe

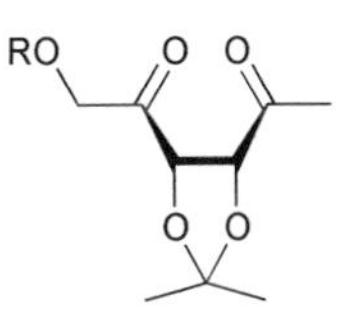

5 R = Tr or Tbdps

2 Other Dicarbonyl Compounds

The 2,5-diketoses **5**, useful for preparing carbocyclic compounds, epimerize and hence racemize on standing at room temperature.[11] The chain extension of pento-1,5-dialdofuranoses with one-carbon Grignard reagents and the isolation of unexpected products is discussed in Chapter 2.

Carbohydrate Chemistry, Volume 33

References

1 T. Hatano, S.Mizuta, H. Ito and T.Yoshida, *Phytochemistry*, 1999, **52**, 1379.
2 S. Hannedouche, I. Jacquemond-Collet, N. Fabre, E. Stanislas and C. Moulis, *Phytochemistry*, 1999, **51**, 767.
3 T. Niwa, *J. Chromatogr. B*, 1999, **731**, 23.
4 W.A. Herrmann, J.P. Zoller and R.W. Fischer, *J. Organomet. Chem.*, 1999, **579**, 404 (*Chem. Abstr.*, 1999, **131**, 185 164).
5 P. Söderman and G. Widmalm, *Carbohydr. Res.*, 1999, **316**, 184.
6 K. Buchholz and E. Stoppok, *Schriftenr. "Nachwachsende Rohst."*, 1998, **10**, 259 (*Chem. Abstr.*, 1999, **130**, 110 490).
7 J. Volc, C. Leitner, P. Sedmera, P. Halada and D. Haltrich, *J. Carbohydr. Chem.*, 1999, **18**, 999.
8 P.L. Barili, M.C. Bergonzi, G. Berti, G. Catelani, F. D'Andrea and F. De Rensis, *J. Carbohydr. Chem.*, 1999, **18**, 1037.
9 K.H. Smelt, Y. Blériot, K. Biggadike, S. Lynn, A.L. Lane, D.J. Watkin and G.W.J. Fleet, *Tetrahedron Lett.*, 1999, **40**, 3255.
10 K.H. Smelt, A.J. Harrison, K. Biggadike, M. Müller, K. Prout, D.J. Watkin and G.W.J. Fleet, *Tetrahedron Lett.*, 1999, **40**, 3259.
11 J.B. Rodriguez, *Tetrahedron*, 1999, **55**, 2157.

16
Sugar Acids and Lactones

1 Aldonic and Lactones

2-Deoxy-D-*erythro*-pentono-γ-lactone (in addition to 1-deoxy-D-ribitol, -xylitol, -glucitol, -threitol and 2-deoxy-D-ribitol) has been isolated from commercial fennel (prepared from the fruit of *Foeniculum vulgare*). This is the first report of its occurrence in nature.[1]

A review on the electrolytic and catalytic oxidation processes for producing sodium gluconate from glucose has appeared.[2] The oxidation of aldoses to aldonolactones by Cr(VI) occurs by two separate paths, one involving reduction of Cr(VI) to Cr(III), and the other its reduction to Cr(V) and then to Cr(III),[3] and the mechanism of the oxidation of D-glucose, D-mannose, D-fructose, D-arabinose and D-ribose to the corresponding aldonic acids with Chloramine B has been proposed.[4]

Per-*O*-benzylated free sugars have been used as starting materials for the preparation of the corresponding aldonolactones with perruthenate/4-methylmorpholine *N*-oxide as oxidizing agent, and the products were olefinated using Wittig reagents to give *exo*-alkenes and then hydrogenated to produce *C*-glycosides, both types of products being extended-chain aldonic acid derivatives. Alternatively, the Wittig products were dihydroxylated using OsO_4 to give extended-chain ulosonic acid derivatives.[5] The oxidation of 2,3-unsaturated glycosides with hydrogen peroxide and molybdenum trioxide to the corresponding 2,3-unsaturated aldono-δ-lactones has been reviewed, and in the same paper the conjugate addition of *N*-substituted hydroxylamines and hydrazines to these products was shown to proceed at C-3 and *anti*- to the C-6 groups, with the adducts undergoing rearrangements to give 3-substituted isoxazolidin-5-ones or 5-substituted pyrazolidin-3-ones, respectively.[6]

Heating 1-*C*-(2-thiazolyl)aldo-furanoses or -pyranoses in boiling toluene resulted in the elimination of thiazole and formation of the corresponding aldonolactones. Model furyl- and thienyl-ketofuranoses and various thiazolyl alcohols are stable to these conditions.[7] An improved synthesis of 3-deoxy-D-*arabino*-hexono-γ-lactone (**1**) and its use in the preparation of the natural product leptosphaerin [(2-acetamido-2,3-dideoxy-D-*erythro*-hex-2-enono-1,4-lactone (**2**)] have been developed (Scheme 1).[8]

Based on the product of reaction of D-glucose with Meldrum's acid [3,6-anhydro-2-deoxy-D-*glycero*-D-*ido*-octono-1,4-lactone (**3**) (*Carbohydr. Res.*

Carbohydrate Chemistry, Volume 33

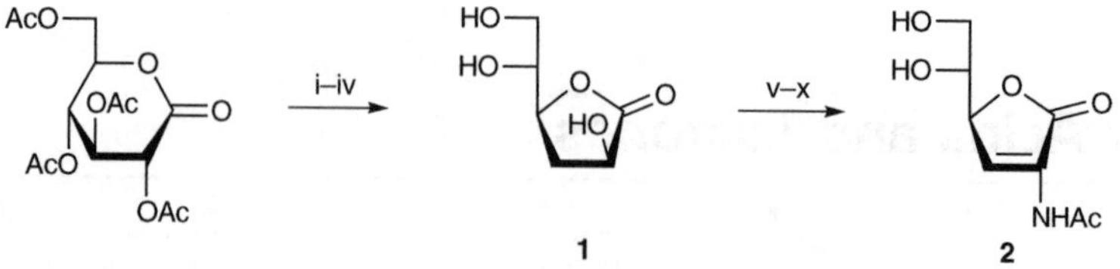
AcO
OAc
AcO
OAc
i–iv
HO
HO
HO
1
v–x
HO
HO
NHAc
2

(isomerized to 1,4-lactone after deacetylation)

Reagents: i, Et_3N/CH_2Cl_2; ii, H_2, Pd/C, EtOAc; iii, NaOMe; iv, H^+, resin; v, acetone, H_2SO_4, $MgSO_4$; vi, py, MsCl; vii, NaN_3, MeCN; viii, NaOMe, MeOH; ix, Py, Ac_2O; x CF_3CO_2H (aq)

Scheme 1

1992, **225**, 159)], the syntheses of (+)-goniofufurone **4** and (+)-7-*epi*-goniofufurone **5** have been achieved,[9] and further access to these potential antitumour compounds was opened from *C*-glycosides of 2,3:5,6-di-*O*-isopropylidene-D-mannofuranose.[10]

Amide and hydrazide phosphates **6,7** were synthesized from D-arabinono-γ-lactone 5-phosphate as competitive inhibitors of yeast phosphoglucose isomerase,[11] and the natural hunger substance **8** (3-deoxy-D-*threo*-pentono-1,4-lactone) was obtained from non-carbohydrate starting materials by use of an asymmetric aldol coupling and subsequent chromatographic isolation of the diastereomer **9.**[12] The Diels-Alder adduct **10,** which was made by use of a chiral salenCo(II) complex as a catalyst, gave access to the ethyl 2,6-anhydro-3-deoxy-D-heptanoates **11** and **12.**[13]

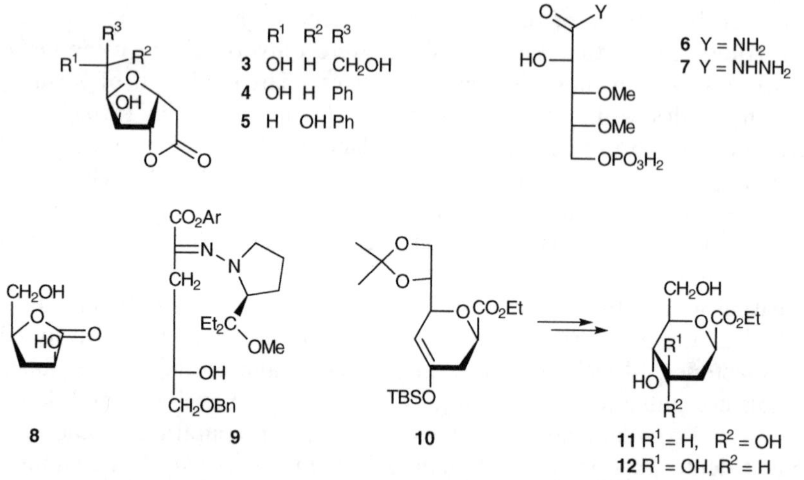
R¹ R² R³
3 OH H CH2OH
4 OH H Ph
5 H OH Ph
6 Y = NH2
7 Y = NHNH2
CH2OH
CO2Ar
CH2
Et2C
OMe
OH
CH2OBn
TBSO
CO2Et
8
9
10
11 R¹ = H, R² = OH
12 R¹ = OH, R² = H

2 Ulosonic Acids

A facile synthesis of NeuNAc was accomplished using an allylic substitution as the key step (Scheme 2), and similar methods were also used to produce KDN and KDO.[14] KDO and its 2-deoxy analogue have been made from diene **13** by

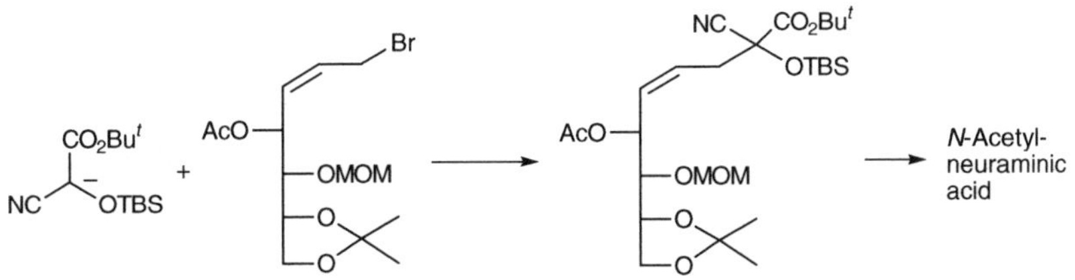

Scheme 2

Sharpless hydroxylation followed in the former case by hydroxylation at C-2 with LDA/MoO_5Py.HMPA (Scheme 3).[15] The sequence illustrated in Scheme 4 has been used to prepare **14** which is an inhibitor of KDO-8-phosphate synthase.[16] By use of fructose-1,6-bisphosphate aldolase, analogues **15** of KDO have been made as indicated in Scheme 5.[17] The synthesis of the heptulosonic acid derivative **16** was achieved using a stereoselective 1,4-addition of benzylamine to an acyclic enone as the key step (Scheme 6), and this strategy was also applied to the synthesis of the 4-*N*-acetylamino analogue of NeuNAc.[18]

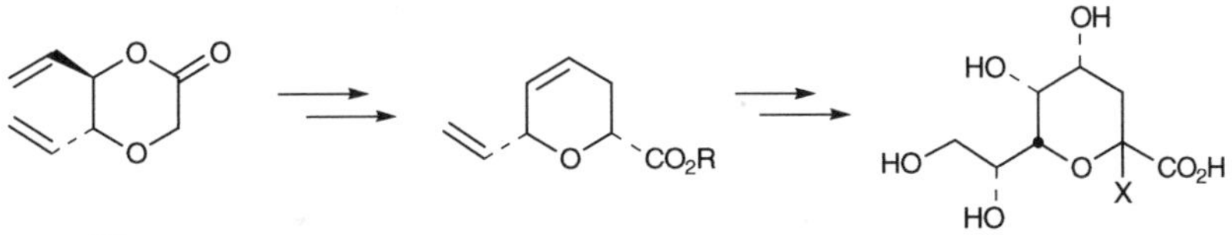

13

2-deoxy-β-KDO X = α-H
KDO X = α-OH

Scheme 3

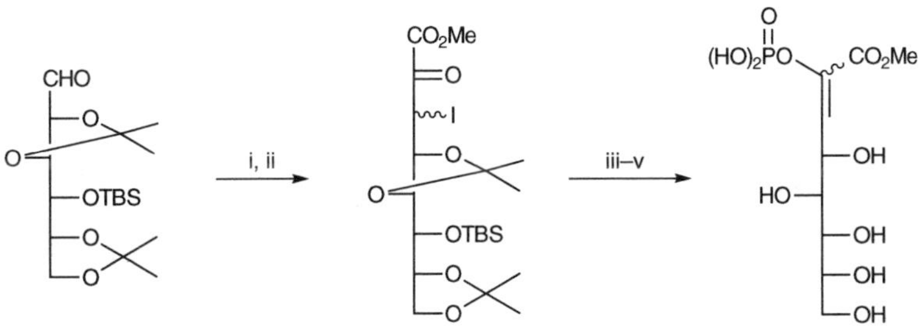

14

Reagents: i, $CHBr_2CO_2Me$, MeOK; ii, MgI_2; iii, $P(OMe)_3$, iv, TmsBr; v, MeOH

Scheme 4

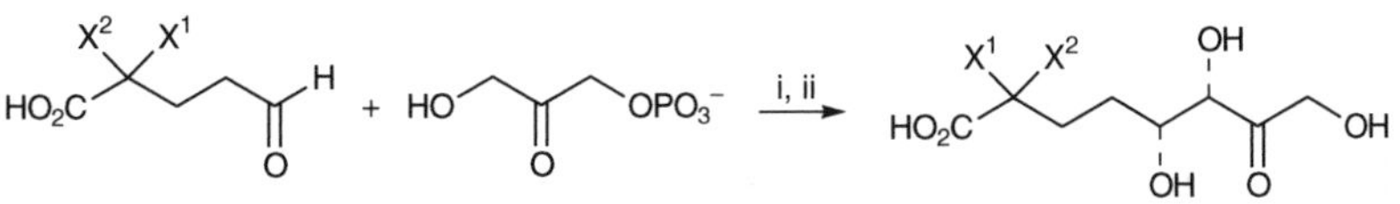

15

X^1, X^2 = O, CH_2, H,OMe

Reagents: i, aldolase; ii, phosphatase

Scheme 5

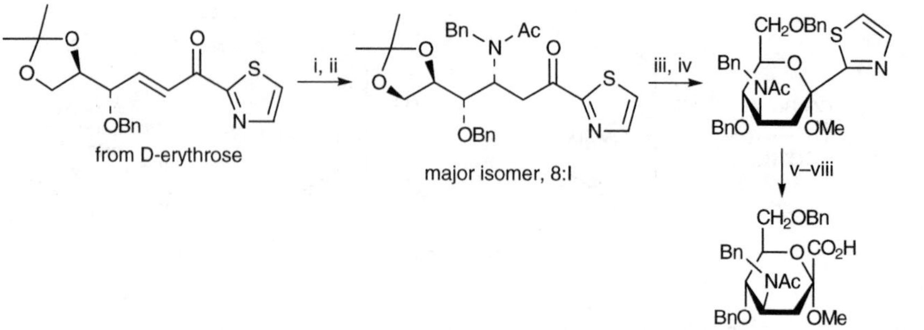

from D-erythrose

major isomer, 8:1

16

Reagents: i, $BnNH_2$; ii, Ac_2O; iii, HCl,AcOH; iv, BnBr, NaH; v, MeOTf; vi, $NaBH_4$; vii, $CuCl_2$, CuO; viii, Ag_2O

Scheme 6

Compounds **17** and **18**, on reaction in the presence of boron trifluoride, gave the expected products **19**, but mainly the fluorinated **20** and **21**, the process involving a Mukaiyama aldol reaction using the aldehyde shown and vinylation.[19] A route from L-serine to galantinic acid (**23**), an amino acid of the peptide antibiotic galantin I, involved the intermediate 3-ulosonic acid derivative **22**.[20]

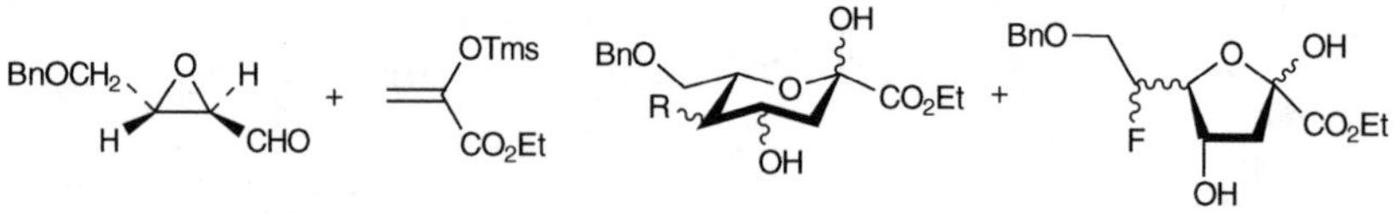

17 **18** **19** R = OH **20** R = F **21**

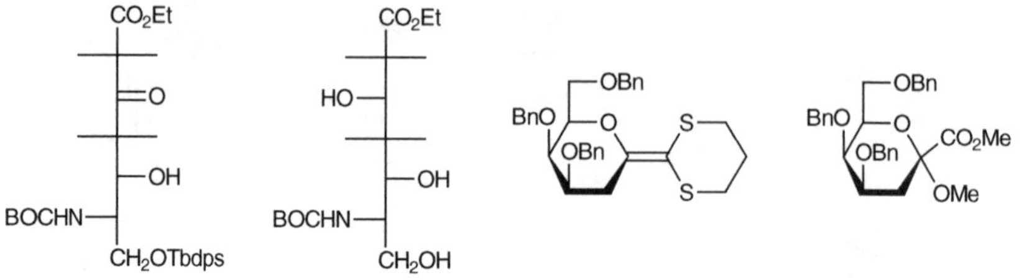

22 **23** **24** **25**

In a further synthesis of ald-2-ulosonic acids 2-oximolactones were involved and were cleaved by use of nitrosylsulfuric acid.[21,22]

A stereospecific construction of α-glycosides **25** of ulosonic acids was developed by a TMSOTf-promoted addition of alcohols, including saccharides, to the ketene dithioacetal **24**. Hydrolysis of the dithiyl residues and oxidation gave the required products.[23,24]

Hydrogenation of methyl D-*arabino*-hex-2-ulopyranosidonate **26** has led to a mixture of *spiro*-pyrido[3,2-*b*][1,4]oxazin-2,2-pyrans **27** (Scheme 7),[25] and several analogues have been made similarly.[26,27] The related glycosylidene-*spiro*-heterocycles **28** were produced following an unprecedented incorporation

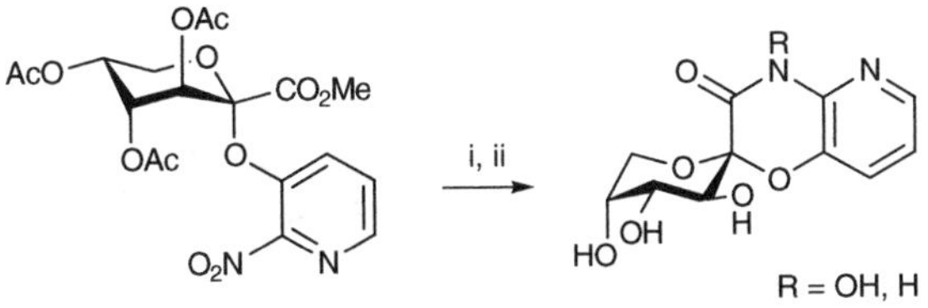

Reagents: i, H_2/Pt–C; ii, NaOMe/MeOH

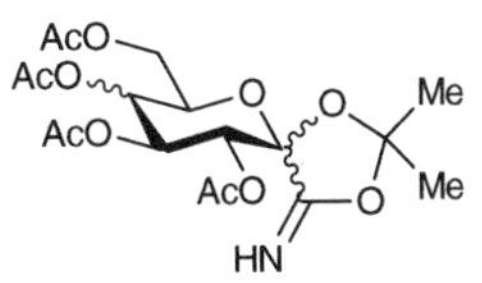

Scheme 7

of a molecule of the solvent acetone during Koenigs-Knorr-like reactions of *C*-(1-bromo-β-D-glycopyranosyl)formamides.[28]

Calix[4]resorcarene-based macrocyclic octa-*O*-sialyl derivative **29** forms complexes with guests such as Rose Bengal in water. The derivative also forms a closely packed monolayer on an SPR chip, and a sialo-targeting lectin is readily adsorbed onto the saccharide residues exposed to water. It also inhibits the haemagglutination of human erythrocytes and the cytopatheic effect of Madine–Darby canine kidney (MDCK) cells mediated by human influenza A virus at concentrations of 5–50 μM.[29] The preparation of the NeuNAc glycoside **30**, a colorimetric substrate for the neuraminidase from influenza A or B, has been reported.[30] In the same series derivative **31** was synthesized from **32** then coupled to interleukin 1α by the acylazide method.[31] Selective protection of thioglycoside **33** has been accomplished to give the 4-, 5-, 6-, 7-,8- and 9-mono-hydroxy derivatives.[32]

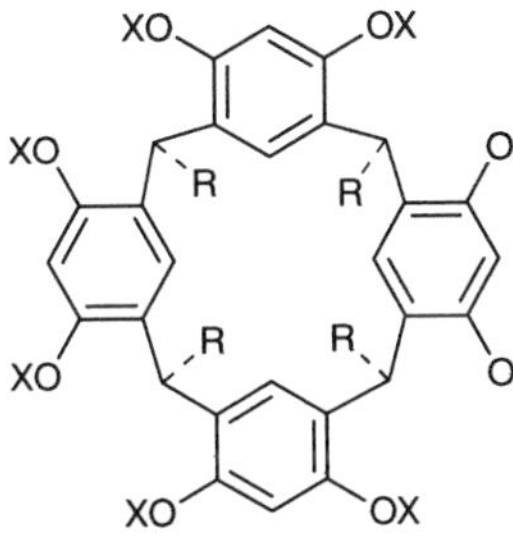

29 R = $(CH_2)_{10}CH_3$
X = $NeuNAcSCH_2OCHNCH_2CH_2$

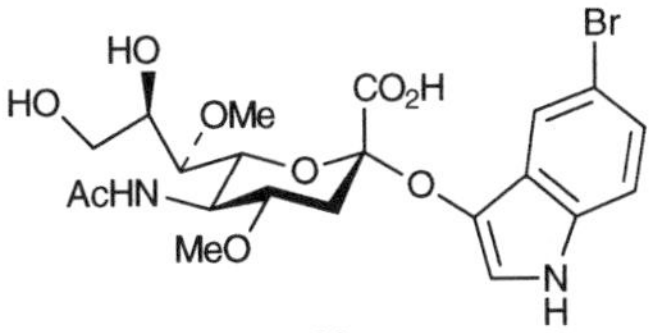

30

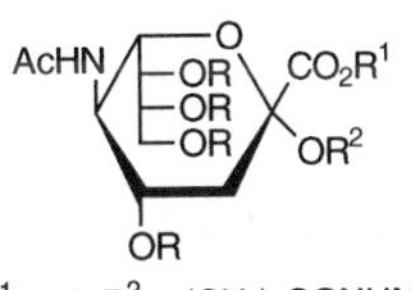

31 R = H, R^1 = K, R^2 = $(CH_2)_8CONHNH_2$
32 R = Ac, R^1 = Me, R^2 = $(CH_2)_8CONHNHCO_2Bn$

The synthesis of the D-*glycero*-D-*ido*-hept-4-ulosonate derivative **34** has been accomplished from diethyl 4-oxopimelate. A complex set of aliphatic changes were involved, with Sharpless oxidation being the key to the development of enantioselectivity.[33]

The preparation of 2,3,4,6-tetra-*O*-benzyl-D-*xylo*-hex-5-ulosonamide was achieved followed by its cyclization to the 5-amino-5-deoxy-D-glucono- and D-ido-1,5-lactams.[34]

In the area of unsaturated derivatives methyl 2-deoxy-2,3-dehydro-*N*-acetylneuraminate (Neu5Ac2enMe) and its 4-epimer have been synthesized.[35] For

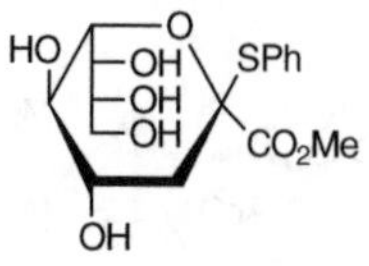

33

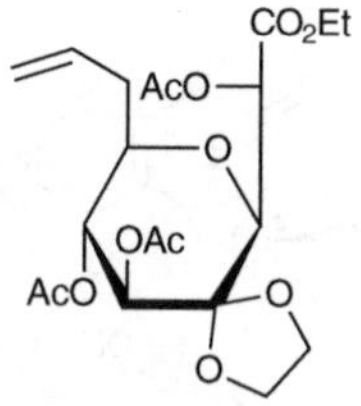

34

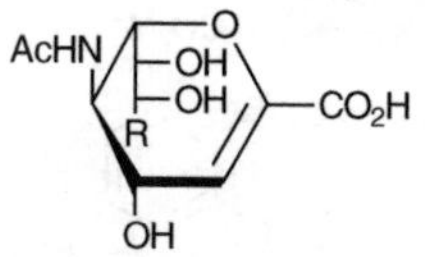

35 R = CH_2NH_2
36 R = $CH_2NHCO(CH_2)_3CO_2H$

the preparation of the 9-amino-neuraminic acid analogue **35** standard methods were used. The acid was then attached to a linker to give **36** which was conjugated with proteins and used to generate a monoclonal antibody to NeuNAc.[36] A one step conversion of glycosides of NeuNAc derivatives into their 2,3-unsaturated (NeuNAc2en) analogues, which provided six examples all with 4-*C*-substitution, has also been described (see Scheme 8).[37]

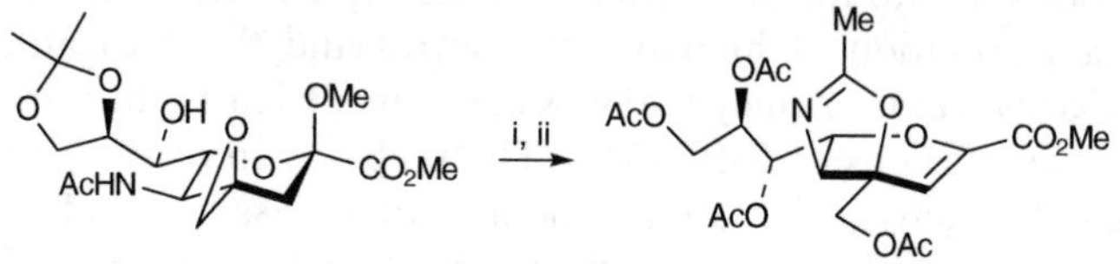

Reagents: i, H_2SO_4, Ac_2O, HOAc; ii, $NaHCO_3/H_2O$

Scheme 8

The 7-deoxy-7-fluoro sialidase inhibitor **37**, which also inhibits effectively the binding of influenza virus H1 haemaglutinin to ganglioside GM3, has been prepared. This is alleged to be the first demonstration of inhibition of sialidase

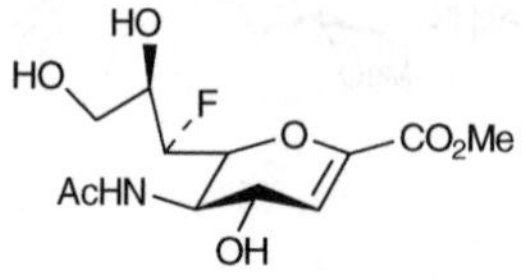

37

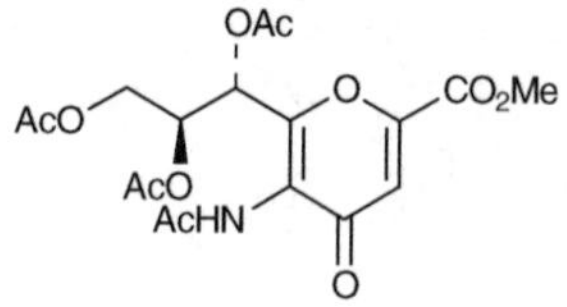

38

and haemaglutinin binding.[38] The sialic acid analogue **38**, containing a γ-pyrone ring, was prepared from a 2,3-unsaturated derivative,[39] and *N*-acetyl-4-deoxyneuraminic acid **39** was synthesized from NeuNAc (Scheme 9).[40] The 4-hydroxyimino (**40**) and the 4-carbethoxymethylene (**41**) derivatives of KDN2en have been reported.[41]

Boon's group has described the chemical synthesis of the α-(2→8)-linked NeuNAc dimer,[42] and the formation of the α-NeuNAc-(2→8)-α-NeuNAc-(2→8)-NeuNAc-1,9;1,9-dilactone **42** was accomplished from the NeuNAc trimer, and the product was selectively hydrolysed to give mono-lactones.[43] Sulfonomethyl *C*-glycosidic analogues of aldo-2-ulosonic acids have been prepared as SiaLe^x mimics.[44]

Many other references to NeuNAc and its derivatives and oligomers are noted in Chapters 3 and 4.

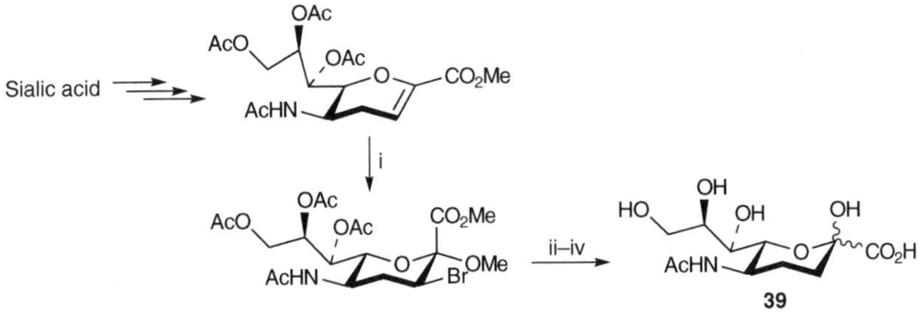

Reagents: i, NBS, MeOH; ii, H_2, Pd/C, Et_3N; iii, NaOMe/MeOH, then H_2O; iv, influenza viral sialidase

Scheme 9

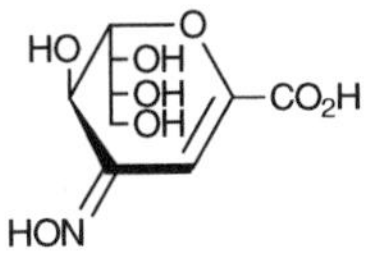

3 Uronic Acids

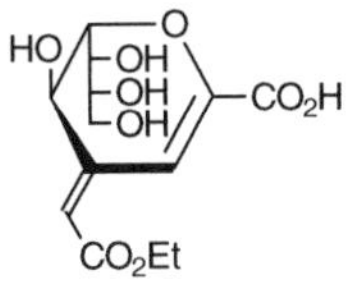

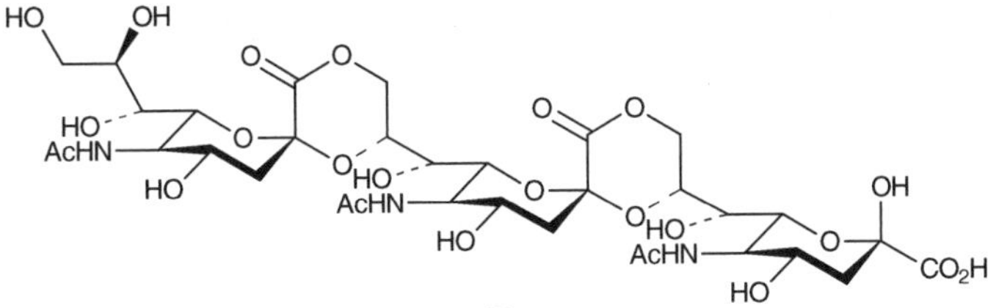

A new selective oxidation procedure, with applicability to carbohydrates – especially to the synthesis of uronic acid derivatives – uses H_2O_2, methyltrioxorhenium, HBr and TEMPO in acetic acid to oxidize primary alcohols to carboxylic acids. The parameters can also be adjusted to give the corresponding aldehydes.[45] A potential additional method of synthesis of uronic acids, *via* nitriles, uses hypervalent cyano-silicates ($^tBu_3SiCN/Bu_4NF$) as effective sources of the cyanide.[46] A synthesis of α-L-ido- and α-L-altro-pyranosiduronic acids from ester **43** has been reported (Scheme 10),[47] and a route involving seven steps, from trehalose to L-iduronic acid has been developed (Scheme 11).[48]

The Gonzalez group reported a facile method for the formation of uronic and aldulosonic acids by addition of lithium dianions of carboxylic acids to the aldehydes **44** and **45**. Acids such as acetic, propanoic, crotonic, sorbic, phenylacetic and 3,3-dimethylacrylic acids were used, and with the first of these, the products were **46** and **47**.[49] An inhibitor of bacterial and viral sialidases, unsaturated uronic acid **48**, was synthesized by a three-step strategy from the glucosamine uronoside triacetate.[50]

An investigation into the properties of various oleanolic acid glycosides

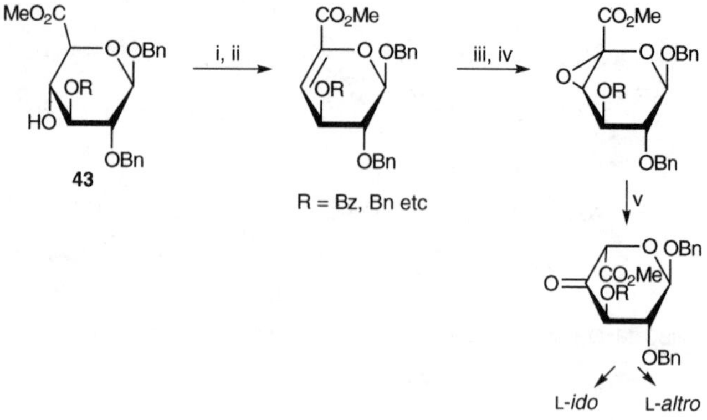

Reagents: i, BzCl, Py; ii, DBU; iii, NBS, H_2O; iv, Ag_2O, DMF; v, Lewis acid

Scheme 10

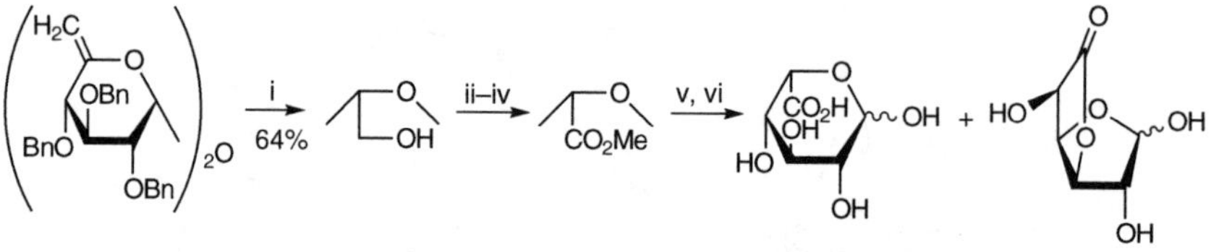

Reagents: i, hydroboration; ii, Swern; iii, Jones; iv, CH_2N_2; v, H_2, $Pd(OH)_2/C$; vi, Amberlite (H^+), H_2O

Scheme 11

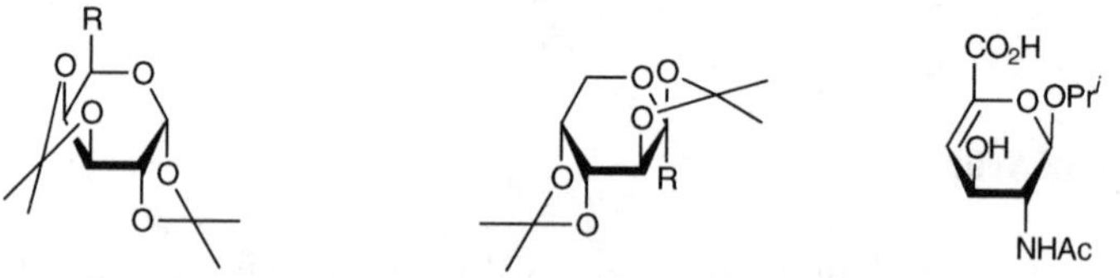

44 R = CHO
46 R = $CH(OH)CH_2CO_2H$

45 R = CHO
47 R = $CH(OH)CH_2CO_2H$

48

obtained from herbs found saponins containing a glucuronic acid residue, which often was further glycosylated with glucose, xylose or arabinose residues. The study sought inhibitors of 'gastric emptying', *i.e.* digestion and absorption of a meal.[51]

In an investigation into antibody-directed enzyme prodrug therapy (ADEPT), the three prodrugs **49–51** of 5-fluorouracil were prepared. These target colorectal cancer and rely on β-glucuronidase activity release.[52] Similar chemistry, also for ADEPT study and reliant on β-glucuronidase, was used to release 9-aminocamptothecin.[53]

The glucuronide **52** of an azetidinone-based cholesterol absorption inhibitor has been prepared using glucuronyl transferases derived from bovine and canine liver microsomes and UDPGlcA.[54] Compound **53**, a synthon for a library of compounds related to the F unit of moenomycin A, was made to permit attachment to phospholipids through the anomeric centre, glycosyla-

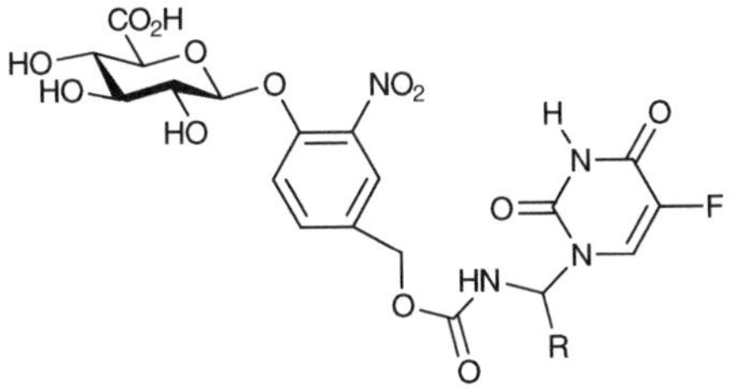

49 R = H
50 R = COOK

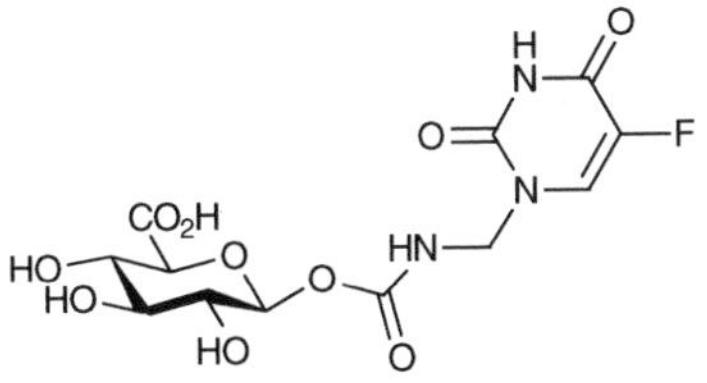

51

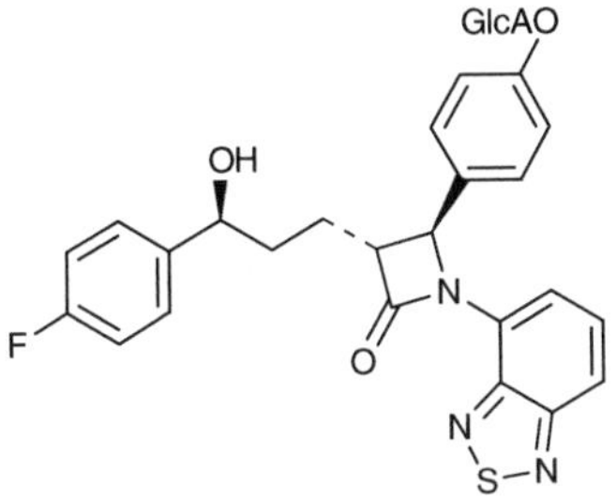

52

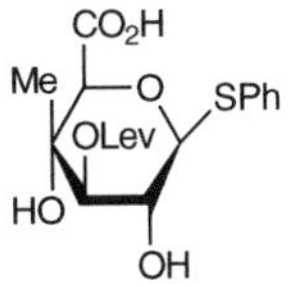

53

tion at O-2, hydrolysis and carbamoylation at O-3 and attachment to solid supports *via* C-6.[55]

4 Aldaric Acids

The syntheses of two novel vinyl monomers for use in the preparation of glycopolymeric inhibitors of β-glucuronidase, *N*-(*p*-vinylbenzyl)-6-D-glucaramide and potassium *N*-(*p*-vinylbenzyl)-6-D-glucaramid-1-ate, have been reported.[56]

5 Ascorbic Acids

A review of L-ascorbic acid biosynthesis in plants and analogues in fungi has been compiled.[57] In the first reported synthesis of L-ascorbic acid from a non-carbohydrate source (Scheme 12) a microbial oxidation was performed to obtain the chiral diol **54**.[58]

A range of compounds derived from dehydro-L-ascorbic acid to have been reported are: hydrazones such as **55**,[59] derivatives of lactam **56**[60] and **57**,[61] the last of these having cytostatic activity against several malignant cell lines as well as antiviral properties.

Dactylose A and B, 1-deoxy-1-(4-hydroxyphenyl)-L-sorbose **59** and -L-tagatose **60** respectively, have been isolated from the roots of *Dactylorhizia hatagirea*, a Nepalese crude drug. It was suggested that they are biosynthesized from L-ascorbic acid and 4-hydroxybenzyl alcohol occurs *via* the

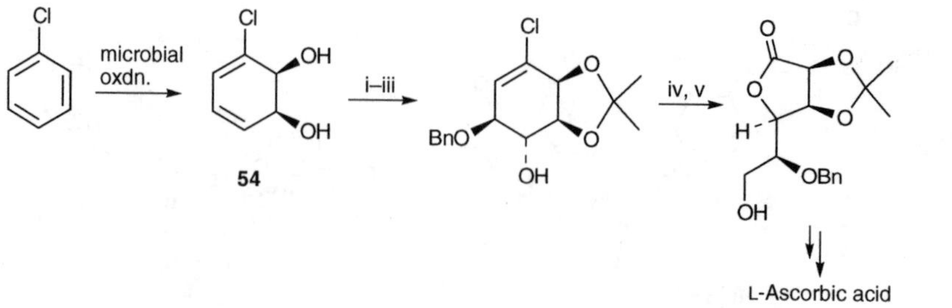

L-Ascorbic acid

Reagents: i, $Me_2C(OMe)_2$, H^+; ii, mCPBA; iii, BnOH, TfOH; iv, O_3, MeOH; v, $NaBH_3CN$

Scheme 12

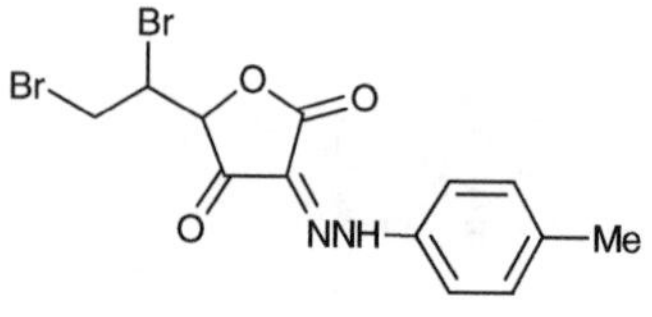

55

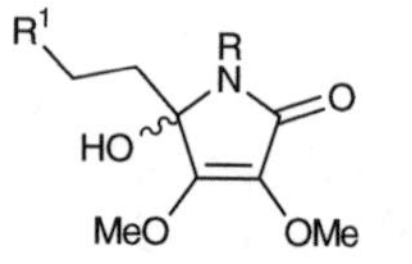

56 R = *n*-butyl, benzyl or cyclohexyl
R^1 = OH or NHR

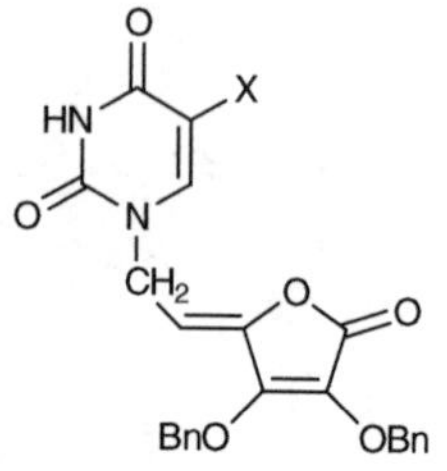

57 X = H, F, Cl, Br, I, F or CF_3

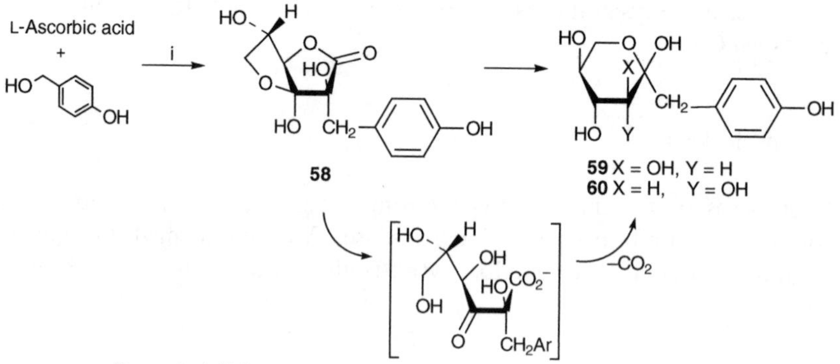

58

59 X = OH, Y = H
60 X = H, Y = OH

Reagents: i, H_2O

Scheme 13

branched-chain lactone **58**, a process that was reproduced in the laboratory (Scheme 13).[62]

References

1. J. Kitajima, T. Ishikawa, Y. Tanaka and Y. Ida, *Chem. Pharm. Bull.*, 1999, **47**, 988.
2. D. Huang, L. Yu, G. Wang and J. He, *Henan Huagong*, 1999, 35 (*Chem. Abstr.*, 1999, **131**, 144 752).
3. S. Signorella, V. Daier, S. Garcia, R. Cargnello, J.C. Gonzalez, M. Rizzotto and L.F. Sala, *Carbohydr. Res.*, 1999, **316**, 14.

4 M.P. Raghavendra, K.S. Rangappa, D.S. Mahadevappa and D.C. Gowda, *Indian J. Chem., Sect B: Org. Chem. Incl. Med. Chem.*, 1998, **37B**, 783 (*Chem. Abstr.*, 1999, **130**, 110 466).
5 J. Xie, A. Molina and S. Czernecki, *J. Carbohydr. Chem.*, 1999, **18**, 481.
6 I. Panfil, D. Mostowicz and M. Chmielewski, *Pol. J. Chem.*, 1999, **73**, 1099 (*Chem. Abstr.*, 1999, **131**, 144 751).
7 A. Dondoni and A. Marra, *Heterocycles*, 1999, **50**, 419 (*Chem. Abstr.*, 1999, **130**, 153 885).
8 C. Pedersen, *Carbohydr. Res.*, 1999, **315**, 192.
9 R. Bruns, A. Wernicke and P. Koll, *Tetrahedron*, 1999, **55**, 9793.
10 H.B. Mereyala, R.R. Gadikota, A Joe, S.K. Arora, S.G. Dastidar and S. Agarwal, *Bioorg. Med. Chem.*, 1999, **7**, 2095.
11 R. Hardré and L. Salmon, *Carbohydr. Res.*, 1999, **318**, 110.
12 D. Enders, H. Sun and F.R. Lensink, *Tetrahedron*, 1999, **55**, 6129.
13 L.-S. Li, Y. Wu and YA. Wu, *J. Carbohydr. Chem.*, 1999, **18**, 1067.
14 T. Takahashi, H. Tsukamoto, M. Kurosaki and H. Yamada, *Tennen Yuki Kagobutsu Toronkai Koen Yoshishu*, 1997, **39th**, 49 (*Chem. Abstr.*, 1999, **130**, 352 485).
15 S.D. Burke and G.M. Sametz, *Org. Lett.*, 1999, **1**, 71.
16 P. Coutrot, S. Dumarcay, C. Finance, A Tabyaoui, B. Tabyaoui and C. Grison, *Bioorg. Med. Chem. Lett.*, 1999, **9**, 949.
17 C. Guérard, C. Demuynck and J. Bolte, *Tetrahedron Lett.*, 1999, **40**, 4181.
18 A. Dondoni, A. Marra and A. Boscarato, *Chem. Eur. J.*, 1999, **5**, 3562.
19 Y. Ruland, C. Zedde, M. Baltas and L. Gorrichon, *Tetrahedron Lett.*, 1999, **40**, 7323.
20 J.S.R. Kumar and A. Datta, *Tetrahedron Lett.*, 1999, **40**, 1381.
21 Y.M. Mikshiev, B.B. Paidak, V.I. Kornilov and Y.A. Zhdanov, *Russ. J. Gen. Chem.*, 1998, **68**, 1168 (*Chem. Abstr.*, 1999, **130**, 182 704.)
22 Y.M. Mikshiev, V.I. Kornilov B.B. Paidak and Y.A. Zhdanov, *Russ. J. Gen. Chem.*, 1999, **69**, 476 (*Chem. Abstr.*, 1999, **131**, 257 776).
23 J. Mlynarski and A. Banaszek, *Pol. J. Chem.*, 1999, **73**, 973 (*Chem. Abstr.*, 1999, **131**, 73901).
24 J. Mlynarski and A. Banaszek, *Tetrahedron*, 1999, **55**, 2785.
25 J. Andersch, D. Sicker and H. Wilde, *Tetrahedron Lett.*, 1999, **40**, 57.
26 J. Andersch, D. Sicker and H. Wilde, *J. Heterocycl. Chem.*, 1999, **36**, 457.
27 J. Andersch, D. Sicker and H. Wilde, *Carbohydr. Res.,* 1999, **316**, 85.
28 L. Somsak, K. Kovacs, V. Gyollai and E. Osz, *Chem. Commun.*, 1999, 591.
29 K. Fujimoto, O. Hayashida, Y. Aoyama, C.-T. Guo, K.I.-P. Jwa Hidari and Y. Suzuki, *Chem. Lett.*, 1999, 1259.
30 A. Liav, J.A. Hansjergen, K.E. Achyuthan and C.D. Shimasaki, *Carbohydr. Res.*, 1999, **317**, 198.
31 T. Chiba, K. Moriya, S. Nabeshima, H. Hayashi, Y. Kobayashi, S. Sasayama and K. Onozaki, *Glycoconjugate J.*, 1999, **16**, 499.
32 S. Akai, T. Nakagawa, Y. Kajihara and K.-I. Sato, *J. Carbohydr. Chem.*, 1999, **18**, 639.
33 S. Lemaire-Audoire and P. Vogel, *Tetrahedron Asymm.*, 1999, **10**, 1283.
34 R. Kovarikova, M. Ledvina and D. Saman, *Collect. Czech. Chem. Commun.*, 1999, **64**, 673 (*Chem. Abstr.*, 1999, **131**, 5452).
35 S. Li, H. Cui and O. Harvo, *Zhongguo Yoawu Huaxue Zazhi*, 1997, **7**, 167 (*Chem. Abstr.*, 1999, **130**, 14 126).

36 H. Kamei, Y. Kajihara and Y. Nishi, *Carbohydr. Res.*, 1999, **315**, 243.
37 G.B. Kok, D. Groves and M. von Itzstein, *J. Chem. Soc., Perkin Trans. 1*, 1999, 2109.
38 T. Sato, F. Ohtake, Y. Ohira and V. Okahata, *Chem. Lett.*, 1999, 145.
39 H.C. Ooi, S.M. Marcuccio, W.R. Jackson and D.F. O'Keefe, *Aust. J. Chem.*, 1999, **52**, 1127.
40 H.C. Ooi, S.M. Marcuccio and W.R. Jackson, *Aust. J. Chem.*, 1999, **52**, 937.
41 X.-L. Sun, T. Kai, N. Sato, H. Takayanagi and K. Furuhata, *J. Carbohydr. Chem.*, 1999, **18**, 1131.
42 A.V. Demchenko and G.-J. Boons, *Chem. Eur. J.*, 1999, **5**, 1278.
43 M.-C. Cheng, C.-H. Lin, K.-H. Khoo and S.H. Wu, *Angew. Chem. Int. Ed. Engl.* 1999, **38**, 686.
44 A. Borbás, G. Szabovik, Z. Antal, P. Herczegh, A. Agócs and A. Lipták, *Tetrahedron Lett.*, 1999, **40**, 3639.
45 W.A. Herrmann, J.P. Zoller and R.W. Fischer, *J. Organomet. Chem.*, 1999, **579**, 404 (*Chem. Abstr.*, 1999, **131**, 185 164).
46 E.D. Soli, A.S. Manoso, M.C. Patterson, P. DeShong, D.A. Favor, R. Hirschmann and A.B. Smith III, *J. Org. Chem.*, 1999, **64**, 3171.
47 H.G. Bazin, M.W. Wolff and R.J. Linhardt, *J. Org. Chem.* 1999, **64**, 144.
48 H. Hinou, H. Kurosawa, K. Matsuoka, D. Terunuma and H. Kuzuhara, *Tetrahedron Lett.*, 1999, **40**, 1501.
49 Z. Gonzalez and A. Gonzalez, *Carbohydr. Res.*, 1999, **317**, 217.
50 P. Florio, R.J. Thomson, A. Alafaci, S. Abo and M. von Itzstein, *Bioorg. Med. Chem. Lett.*, 1999, **9**, 2065.
51 H. Matsuda, Y. Li, T. Murakami, J. Yamahara and M. Yoshikawa, *Bioorg. Med. Chem.*, 1999, **7**, 323.
52 R. Madec-Lougerstay, J.-C. Florent and C. Monneret, *J. Chem. Soc., Perkin Trans. 1*, 1999, 1369.
53 Y.-L. Liu, S.R. Roffler and J.-W. Chern, J. *Med. Chem.*, 1999, **42**, 3623.
54 P. Reiss, D.A. Burnett and A. Zaks, *Bioorg. Med. Chem.*, 1999, **7**, 2199.
55 R. Kakarla, M. Ghosh, J.A. Anderson, R.G. Dulina and M.J. Sofia, *Tetrahedron Lett.*, 1999, **40**, 5.
56 K. Hashimoto, R. Ohsawa, N. Imai and M. Okada, *J. Polym. Sci. Part A: Polym. Chem.*, 1999, **37**, 303.
57 F.A. Loewus, *Phytochemistry*, 1999, **52**, 193.
58 M.G. Banwell, S. Blakey, G. Harfoot and R.W. Longmore, *Aust. J. Chem.*, 1999, **52**, 137.
59 M.A. El-Sekily, M.E. Elba and F.S. Fouad, *J. Chem. Res. (S)*, 1999, 296
60 M.A. Khan and H. Adams, *Carbohydr. Res.*, 1999, **322**, 279.
61 S. Raic-Malic, A. Hergold-Brundic, A. Nagl, M. Grdisa, K. Pavelie, E. De Clercq and M. Mintas, *J. Med.Chem.*, 1999, **42**, 2673.
62 H. Kizu, E. Kaneko and T. Tomimori, *Chem. Pharm. Bull.*, 1999, **47**, 1618.

17
Inorganic Derivatives

1 Carbon-bonded Phosphorus Derivatives

Some 'P-in-the-ring' 2-amino-2-deoxy-D-mannopyranose, D-glucopyranose and L-idopyranose derivatives have been synthesized by standard methods[1,2] and the conformations of some 5-deoxy-5-*C*-(butylphosphinyl)xylopyranoses are discussed in Chapter 22. A phostone analogue **3** of L-fucose triacetate has been prepared from the glycal **1** *via* **2** (Scheme 1).[3] Aldehydes **4** and **5** have

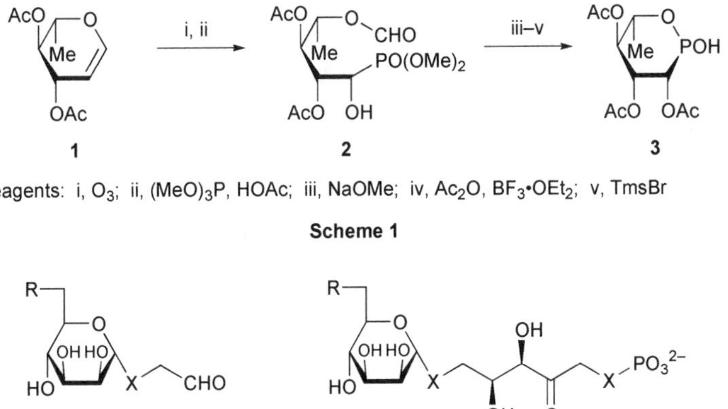

Reagents: i, O_3; ii, $(MeO)_3P$, HOAc; iii, NaOMe; iv, Ac_2O, $BF_3 \cdot OEt_2$; v, TmsBr

Scheme 1

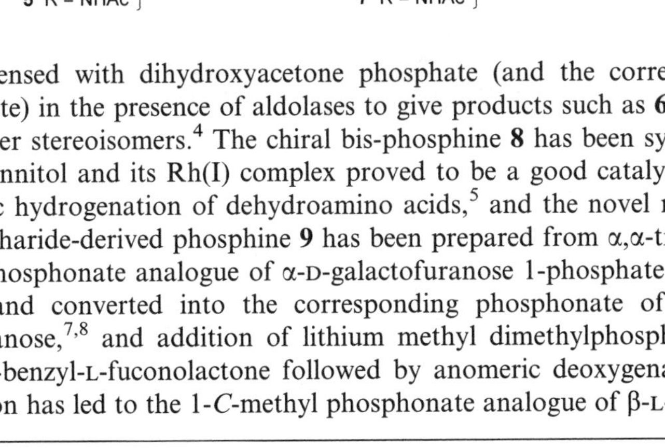

4 R = OH, **5** R = NHAc; X = O, CH_2

6 R = OH, **7** R = NHAc; X = O, CH_2

been condensed with dihydroxyacetone phosphate (and the corresponding phosphonate) in the presence of aldolases to give products such as **6** and **7** as well as other stereoisomers.[4] The chiral bis-phosphine **8** has been synthesized from D-mannitol and its Rh(I) complex proved to be a good catalyst for the asymmetric hydrogenation of dehydroamino acids,[5] and the novel non-reducing disaccharide-derived phosphine **9** has been prepared from α,α-trehalose.[6] The C-1 phosphonate analogue of α-D-galactofuranose 1-phosphate has been prepared and converted into the corresponding phosphonate of UDP-D-galactofuranose,[7,8] and addition of lithium methyl dimethylphosphonate to 2,3,4-tri-*O*-benzyl-L-fuconolactone followed by anomeric deoxygenation and deprotection has led to the 1-*C*-methyl phosphonate analogue of β-L-fucose 1-

phosphate.[9] Two phosphonate analogues **10** and **11** of methyl mannoside 6-phosphate have been prepared by standard methods. The isostere **11** showed high affinity for mannose 6-phosphate receptors whereas **10** did not.[10] A Reformatsky reaction of diethyl (1-bromo-prop-2-enyl)phosphonate with uloses or dialdose derivatives is covered in Chapter 14.

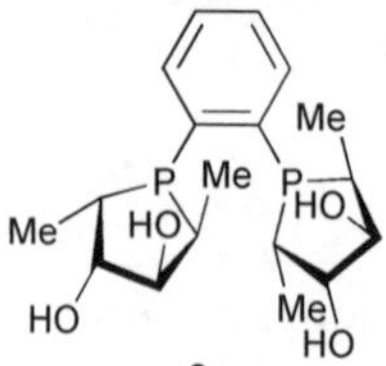

8

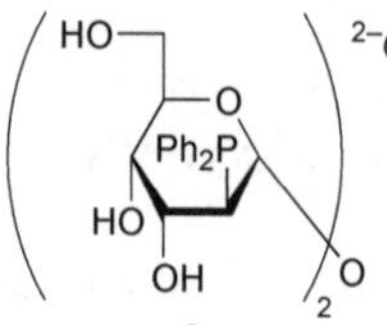

9

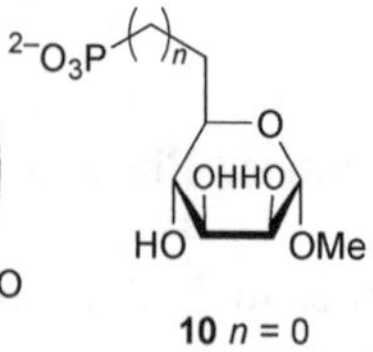

10 *n* = 0
11 *n* = 1

2 Other Carbon-bonded Derivatives

A titanium reagent, $[Cp_2Ti(III)Cl]_2$, has been utilized for the high yielding conversion of *O*-protected glycosyl halides into glycals. Presumably the reaction goes through a glycosyl–Ti(IV) intermediate.[11] Chromium(II) complexes have also been used to effect the same conversions.[12] Some chromium furanosylidene complexes such as **12**–**14** have been prepared using previously reported methods (Vol. 31, p. 218).[13-15] A 1-β-tributylstannyl derivative of *N*-acetylglucosamine has been treated with butyllithium and then an electrophile to give a β-*C*-glycoside *via* an apparently configurationally stable lithiated derivative.[16] Treatment of glycals with Bu_3SnH or with a tributylstannyl cuprate has afforded 1-tributylstannyl derivatives as mixtures of anomers.[17]

Some bis-glycosides of but-2-yne-1,4-diol as well as propargyl glycosides have been converted at the alkyne function into carbohydrate carboranes by standard methods.[18] The glycosyl borinate **15** has been described.[19]

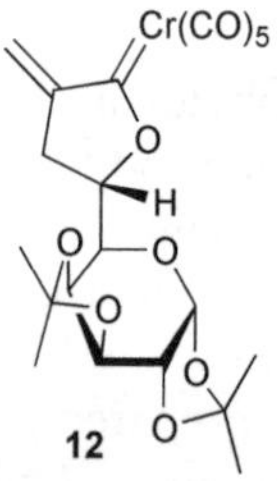

12

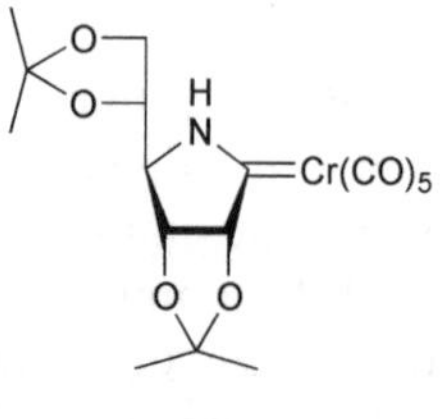

13

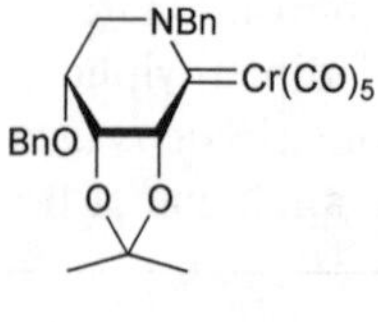

14

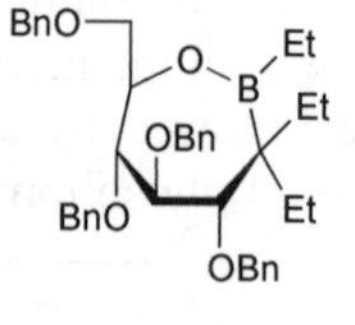

15

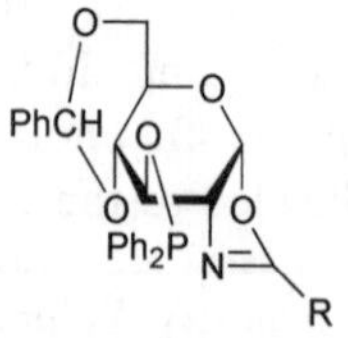

16 R = Pr^i, Bn, Ph, Me, Bu^t

3 Oxygen-bonded Derivatives

The phosphinites **16** derived from glucosamine have been used as chiral ligands for palladium-catalysed asymmetric alkylations.[20] Complex formation of eight monosaccharides with Me_2Sn^{2+} cations in aqueous solution has been studied. The stability and composition of the complexes were determined by potentiometric studies while ^{13}C NMR measurements led to assignment of the participating hydroxy groups.[21] A study by FTIR reflection spectroscopy of aqueous solutions of some monosaccharides containing salts (NaCl, KCl, $MgCl_2$, $CaCl_2$) showed that Ca^{2+} and Mg^{2+} ions effect significant changes to the C–OH bond deformation and C–O stretching regions.[22]

An NMR spectroscopic analysis of the borate ester of methyl β-D-apiofuranoside indicated that two diastereoisomers **17** and **18** are present in approximately equal amounts.[23] It has been claimed that borate ester formation from ascorbic acid involves the 3- and 5-OH groups forming a six-membered ring even though ^{13}B NMR analysis suggested the presence of a five-membered ring.[24] In an extension of previous work (Vol. 30, p. 220), some porphyrin-containing arylboronic acids have been coupled with D-glucose and L-fucose by making the 4,6- and 3,4-*O*-boronates, respectively, and the fluorescence spectra have been studied.[25]

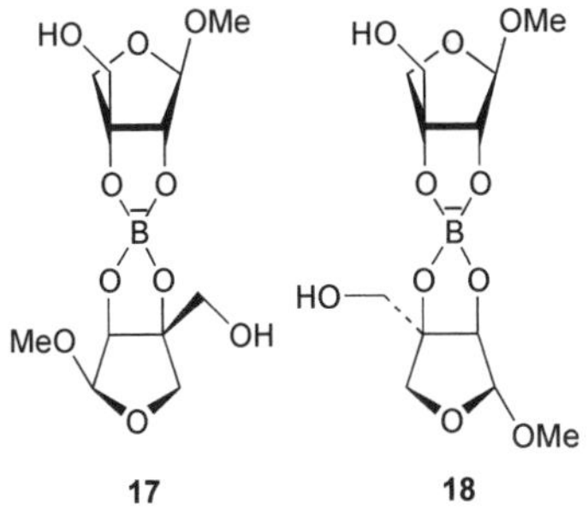

HO O OMe O O B O O MeO OH O **17**

HO O OMe O O B O O HO O OMe **18**

Stannylene acetals of diols have been usefully oxidized to hydroxyketone products using 1,3-dibromo-5,5-dimethylhydantoin in place of NBS or bromine.[26]

References

1 T. Hanaya, Y. Fujii and H. Yamamoto, *J. Chem. Res.*, (*S*), 1998, 790 (*Chem. Abstr.*, 1999, **130**, 125 312).
2 T. Hanaya, Y. Fujii, S. Ikejiri and H. Yamamoto, *Heterocycles*, 1999, **50**, 323 (*Chem. Abstr.*, 1999, **130**, 153 884).
3 S. Hanessian and O. Rogel, *Bioorg. Med. Chem.*, 1999, **9**, 2441.
4 C.-C. Li, F. Moris-Varas, G. Weitz-Schmidt and C.-H. Wong, *Bioorg. Med. Chem.*, 1999, **7**, 425.
5 W. Li, Z. Zhang, D. Xiao and X. Zhang, *Tetrahedron Lett.*, 1999, **40**, 6701.

6 K. Yonehara, T. Hashizume, K. Ohe and S. Uemura, *Tetrahedron: Asymmetry*, 1999, **10**, 4029.
7 J. Kovensky, M. McNeil and P. Sinaÿ, *J. Org. Chem.*, 1999, **64**, 6202.
8 J. Kovensky, A.F. Cirelli and P. Sinaÿ, *Carbohydr. Lett.*, 1999, **3**, 271.
9 A.J. Norris and T. Toyokuni, *J. Carbohydr.. Chem.*, 1999, **18**, 1097.
10 C. Vidil, A. Morère, M. Garcia, V. Barragan, B. Hamdaoui, H. Rochefort and J.-L. Montero, *Eur. J. Org. Chem.*, 1999, 447.
11 R.P. Spencer, C.L. Cavallaro and J. Schwartz, *J. Org. Chem.*, 1999, **64**, 3987.
12 G. Kovács, K. Tóth, Z. Dinya, L. Somsák and K. Micskei, *Tetrahedron*, 1999, **55**, 5253.
13 B. Weyershausen, M. Nieger and K.H. Dötz, *J. Org. Chem.*, 1999, **64**, 4206.
14 W.-C. Haase, M. Nieger and K.H. Dötz, *Chem. Eur. J.*, 1999, **5**, 2014.
15 K.H. Dötz, M. Klumpe and M. Nieger, *Chem. Eur. J.*, 1999, **5**, 691.
16 B. Wesyermann, A. Walker and N. Diedrichs, *Angew. Chem., Int. Ed. Engl.*, 1999, **38**, 3384.
17 D.H. Braithwaite, C.W. Holzapfel and D.B.G. Williams, *S. Afr. J. Chem.*, 1998, **51**, 162 (*Chem. Abstr.*, 1999, **130**, 223 501).
18 G.B. Giovenzana, L. Lay, D. Monti, G. Pamisano and L. Panza, *Tetrahedron*, 1999, **55**, 14123.
19 A. Vasella, W. Wenger and T. Rajamannar, *Chem. Commun.*, 1999, 2215.
20 K. Yonehara, T. Hashizume, K. Mori, K. Ohe and S. Uemura, *Chem. Commun.*, 1999, 415.
21 L. Nagy, N. Buzns, H. Baratne Jankovics, T. Gajda, E. Kuzmann, A. Vertes and K. Burger, *Acta Pharm. Hung.*, 1999, **69**, 9 (*Chem. Abstr.*, 1999, **131**, 185 140).
22 H. Kodad, R. Mokhlisse, E. Davin and G. Mille, *Can. J. Anal. Sci. Spectrosc.*, 1998, **43**, 129 (*Chem. Abstr.*, 1999, **131**, 59 051).
23 T. Ishii and H. Ono, *Carbohydr. Res.*, 1999, **321**, 257.
24 N. Obi, M. Katayama, J. Sano, Y. Kojima, Y. Shigemitsu and K. Takada, *New J. Chem.*, 1998, **22**, 933 (*Chem. Abstr.*, 1999, **130**, 52 657).
25 M. Takeuchi, S. Yoda, Y. Chin and S. Shinkai, *Tetrahedron Lett.*, 1999, **40**, 3745.
26 P. Söderman and G. Widmalm, *Carbohydr. Res.*, 1999, **316**, 184.

18
Alditols and Cyclitols

Recent work in the field covered by this chapter has been reviewed in a book on carbohydrate mimics.[1]

1 Alditols and Derivatives

1.1 Alditols. – 1-Deoxy-D-threitol, -D-ribitol, -D-xylitol, and -D-glucitol, as well as 2-deoxy-D-*erythro*-pentitol and 3-deoxy-D-*threo*-pentitol, have been isolated from commercial fennel, their first reported occurrences in nature.[1a] The rate of oxidation of D-mannitol and D-glucitol by hexacyanoferrate(III) in alkaline media was found to be directly proportional to [alditol] and $[OH^-]$.[2] A number of 5-modified D-xyloses and 5,6-modified D-glucoses have been tested as substrates for yeast aldose reductase and were found to be better substrates than the unmodified sugars. The overall results support the view that aldose reductase binds aldoses in the acyclic form.[3] Two clay-supported nickel catalysts have been evaluated in the hydrogenation of glucose to glucitol.[4] Better yields have been achieved in this reduction by use of an ultrafine, amorphous Co–B alloy, prepared by chemical reduction of $Co(OAc)_2$ with KBH_4. Mo- and W-dopants further increased the activity of the catalyst.[5]

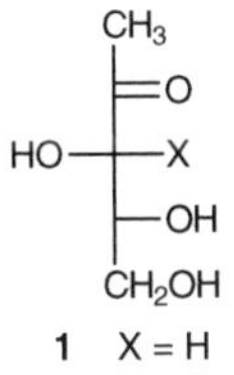

1 X = H
1a X = D

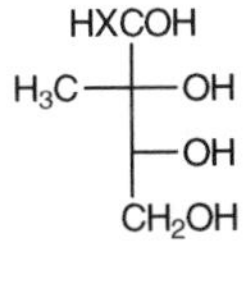

2 X = H
2a X = D

The stereochemical course of the reduction-step in the enzymic formation of 2-*C*-methyl-D-erythritol from 1-deoxy-D-xylulose (**1**→**2**) in leaves of higher plants, which involves skeletal rearrangement, has been studied by use of the easily accessible [3-^{2}H]-labelled precursor **1a**, which rearranged to **2a** with the label in the H_{si}-position of C-1.[6] The relevant steps in the chemical synthesis of [3,5,5,5-$^2H_4$]-2-*C*-methylerythritol (**8**), a substrate designed for the elucidation

Carbohydrate Chemistry, Volume 33

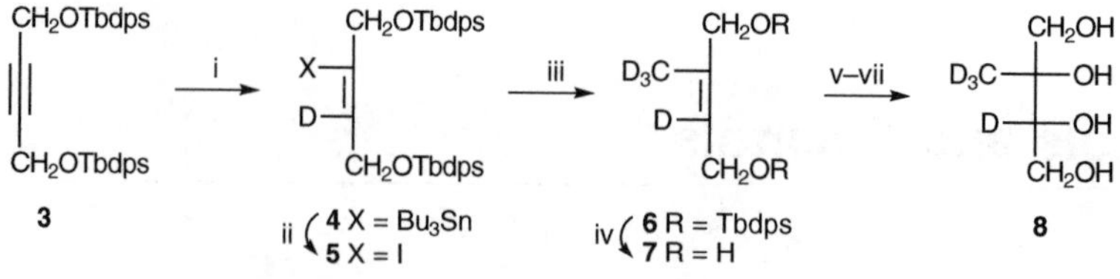

Reagents: i, Bu_3SnD, $PdCl_2(PPh_3)_2$; ii, I_2, Et_2O; iii, $(D_3C)_2CuLi.LiCN$, Et_2O; iv, Bu_4NF; v, Ac_2O, DMAP, Et_3N; vi, AD-mix β, H_2O, Bu^tOH; vii; Resin (OH^-), MeOH

Scheme 1

of the mevalonate-independent isoprenoid biosynthesis, were a Pd(II)-catalysed *syn*-deuterostannation (**3**→**4**), the introduction of the trideuteromethyl group by cyanocuprate coupling (**5**→**6**) and the enantioselective dihydroxylation of *cis*-2-butene intermediate **7**, as shown in Scheme 1.[7]

A novel synthetic route to alditols based on reductive radical fragmentation of *O*-*N*-phthalimidyl glycosides, involving cleavage of the C-1–C-2 bond following formation of an anomeric alkoxy radical, is shown in Scheme 2.[8] The well-known zinc-mediated fragmentation of methyl 5-deoxy-5-iodopentofuranosides which furnishes enals, *e.g.* **9**, has been coupled with *in situ* alkylation to give alditols, such as **10**.[9]

A range of hexitols carrying long-chain *n*-alkyl ether groups, required for a

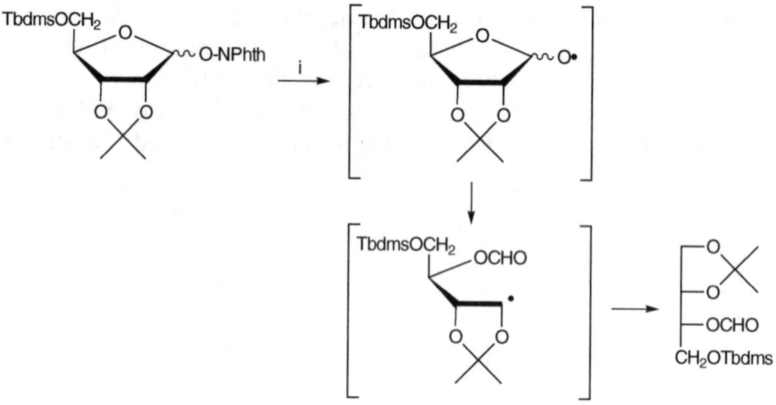

Reagents: i, Bu_3SnH, AIBN

Scheme 2

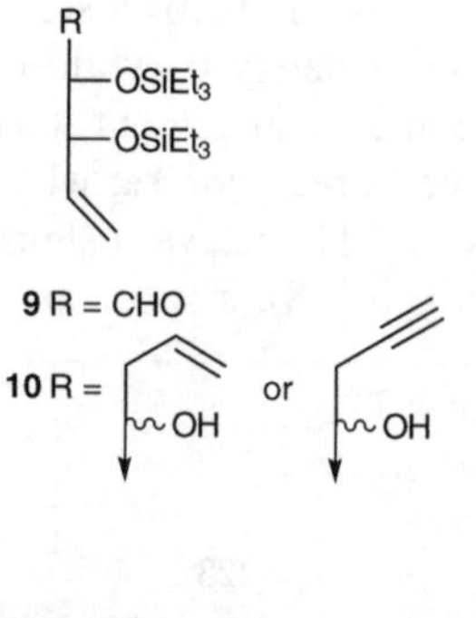

study of their liquid crystal properties, have been prepared conventionally by borohydride reduction of suitably *O*-alkylated hexopyranoses.[10]

1.2 Anhydro-alditols. – An investigation of the mechanism of the reductive cleavage of methyl glycopyranosides with silanes in the presence of a Lewis acid catalyst concluded that acyclic oxonium ions are the sole intermediates; Bu^tMe_2SiH gave the best results and β-glycosides reacted twice as fast as their α-anomers to give 1,5-anhydro-alditols.[11] When 1,2-*O*-isopropylidene-furanoses, *e.g.* 5-deoxy-α-D-*ribo*-hexofuranose derivative **11**, were exposed to $EtSi_3H$ in the presence of $BF_3.OEt_2$ or TmsOTf, deoxygenation at the anomeric centre took place with formation of 1,4-anhydroalditols, in this case **12**.[12,13] Anomeric deoxygenation has also been applied to the previously known pyranulose glycoside **13** to obtain the precursor **14** of a 'direct-linked' *C*-disaccharide.[14]

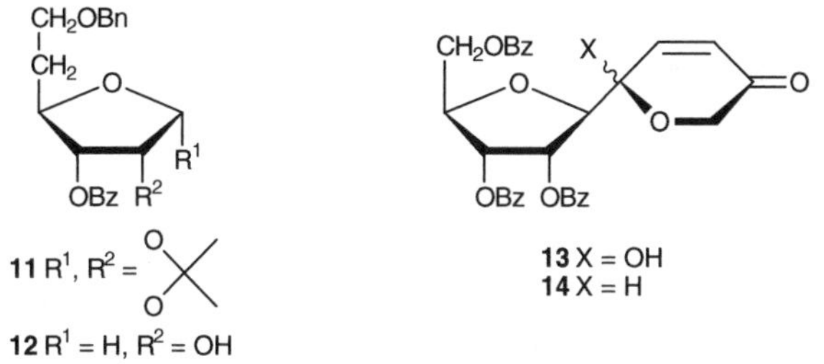

11 R^1, R^2 = (isopropylidenedioxy)

12 R^1 = H, R^2 = OH

13 X = OH

14 X = H

1-*C*-Silylated 2,5-anhydro-hexitols have been synthesized in moderate yields by stereocontrolled, aciD-promoted cyclization of 3,4-di-*O*-benzyl-1-deoxy-1-*C*-silyl-hexitols, obtained from suitably protected *aldehydo*-pentoses (D-ribose, D-arabinose or D-xylose) by reaction with $Me_2PhSiCH_2MgCl$

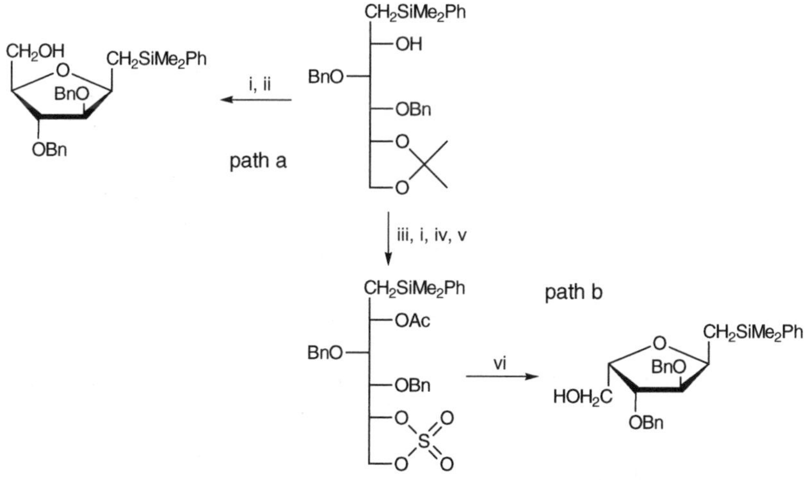

path a

path b

Reagents: i, aq. HOAc; ii, $BF_3.OEt_2$; iii, Ac_2O, Py; iv, $SOCl_2$, Py, EtOH; v, $RuCl_3$, $NaIO_4$;
v, $RuCl_3$, $NaIO_4$; vi, LiOMe, MeOH, then H^+

Scheme 3

(Scheme 3, path a).[15] In a more efficient modification of this synthesis, 5,6-cyclic sulfates were used instead of 5,6-diols, cyclization taking place under basic conditions and inversion occurring at C-5 (Scheme 3, path b).[16]

The isoster **15** of β-L-fucopyranose 1-phosphate (**16**) has been prepared from 2,3,4-tri-*O*-benzyl-β-L-fucono-1,5-lactone by condensation with $LiCH_2$-$P(O)(OMe)_2$, followed by a two-step deoxygenation (SO_2Cl–Py, then H_2–Pd/C) with concomitant deprotection; the methylene isoster of β-L-rhamnopyranose 1-phosphate was analogously prepared.[17] Asymmetric hydration-cyclization of terminal *meso*-diepoxides derived from pentitols or hexitols in the presence of chiral (salen)Co(III)OAc gave access to 1,4- and 2,5-anhydro-alditols, respectively, with e.e. ⩾ 92.5% (for example **17**→**18**).[18]

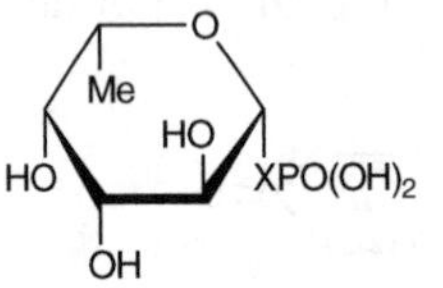

15 X = CH_2
16 X = O

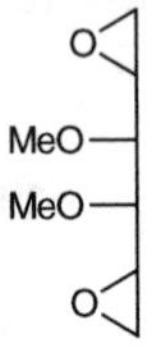

17

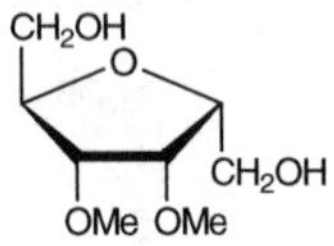

18

The ethyl esters **20** of 2,6-anhydro-3-deoxy-D-*gluco*- and -D-*allo*-heptonic acid have been synthesized from the known Diels Alder adduct **19** (see Y.-J. Hu *et al.*, *J. Org. Chem.*, 1998, **63**, 2456) by asymmetric hydroxylation, followed by stereoselective reduction of the resulting hydroxyketone to a diol, side-chain-cleavage and deprotection.[19]

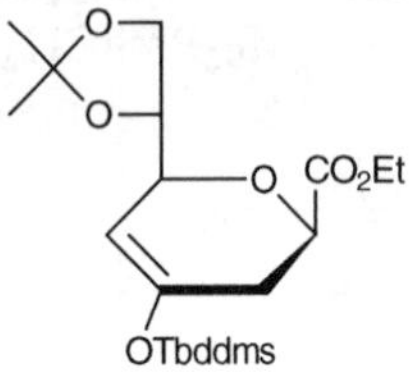

19

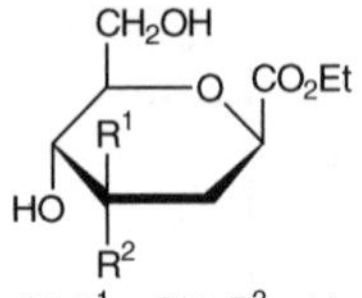

20 R^1 = OH, R^2 = H
or R^1 = H, R^2 = OH

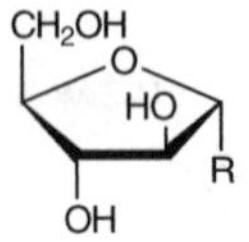

21 R = CHO
22 R = CH_2NH_2

Reductive amination of aldehyde **21** afforded 1-amino-2,5-anhydro-1-deoxy-D-mannitol (**22**), as well as various *N*-aryl derivatives. Aldehyde **21** was obtained by the well-established nitrous acid deamination of D-glucosamine which proceeds with concomitant ring contraction.[20,21] An improved procedure has been presented for this reaction.[20] 3,7-Anhydro-1-anilino-1,2-dideoxy-D-*glycero*-D-*ido*-octitol was similarly made from 2-(*C*-α-D-glucopyranosyl)ethanal.[21]

Selective radical-chain epimerization at C-2 and C-5 of 1,4:3,6-dianhydro-D-glucitol on exposure to tri-*t*-butoxysilanethiol as protic polarity-reversal catalyst and 2,2-di-(*t*-butylperoxy)butane as radical initiator gave an equilibrium mixture of the D-*gluco*-, D-*manno*-, and L-*ido*-epimers (which have one, two and zero *endo*-hydroxyl groups, respectively) in the ratio 25:1:40.[22] Application of ring-closing metathesis-osmylation to δ-unsaturated allyl ethers furnished oxygenated oxepanes, *e.g.* **24** from **23**.[22a]

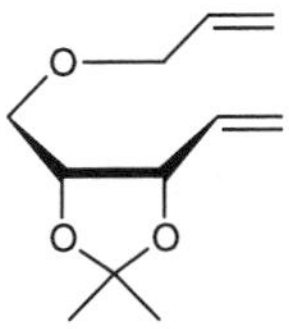

23 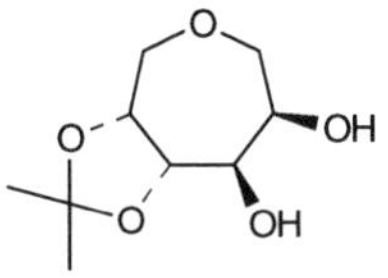 **24** 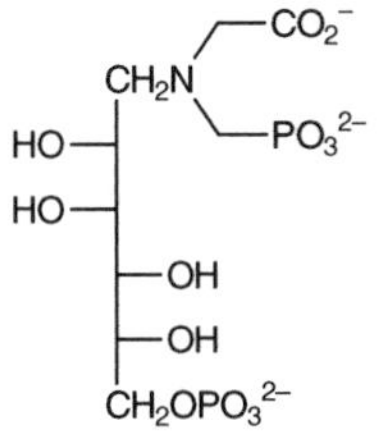 **25**

1.3 Acyclic Amino- and Imino-alditols. – 1-Deoxy-1-*N*-alkylaminoglucitols, obtained by reductive amination of D-glucose with methyl-, ethyl- or butylamine using hydrogenation over Raney nickel at elevated temperature and pressure, have been employed for resolving racemic carboxylic acids.[23] The *N*-functionalized 1-amino-1-deoxy- D-mannitol 6-phosphate **25**, an inhibitor of Kdo 8-phosphate synthase, has been prepared by reductive amination of D-mannose 6-phosphate with the herbicide glyphosate.[24]

The conversion of D-mannitol to a chiral bis-phospholane by way of ditosylate **26** and cyclic sulfate **27** is covered in Chapter 24. The enantiopure, C_2-symmetric bis(cyclic isothiourea) derivative **28** and its regioisomer with sulfur at the primary positions have been elaborated from commercial 3,4-*O*-isopropylidene D-mannitol in eight and ten standard reaction steps, respectively.[25] The potential inhibitors **30** of lumazine synthase, the penultimate enzyme of riboflavine biosynthesis, have been prepared by coupling of D-ribitylamine with chlorides **29**.[26,27] Two bis(D-ribityl-lumazine) analogues have also been made in a similar manner.[28]

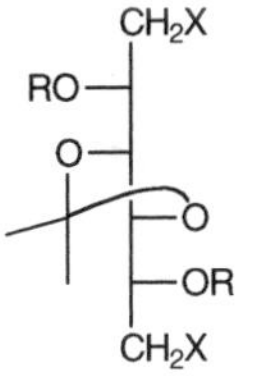

26 X = OTs, R = H
27 X = H, R,R = SO_2

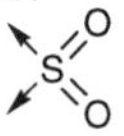

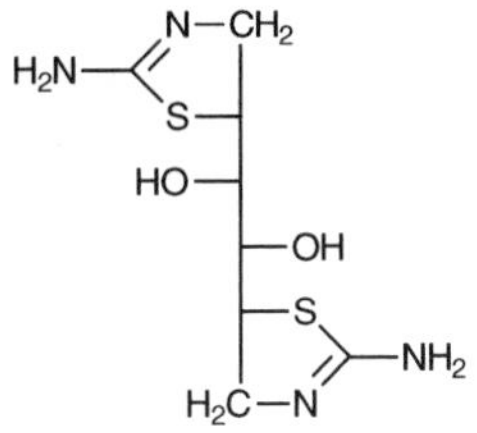

28

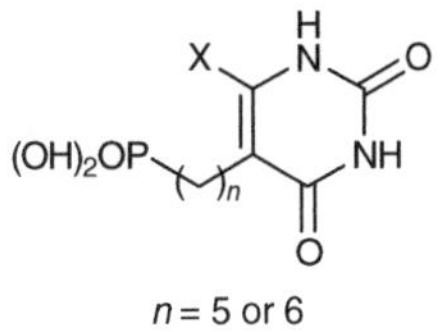

n = 5 or 6

29 X = Cl
30 X = CH_2NH–(CHOH)$_3$–CH_2OH

Acid hydrolysis of 3′-deoxy-3′-thymin-1-yl-thymidine (**31**) and its 3′-uracil-1-yl analogue **32** led to loss of the anomeric base moieties and release of the

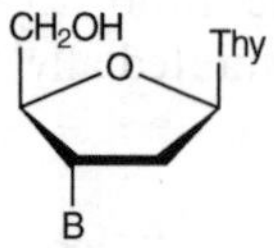

31 B = Thy
32 B = Ura

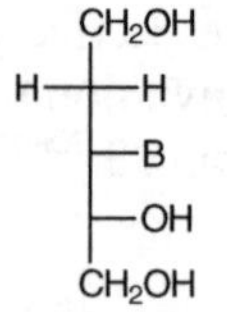

33 B = Thy
34 B = Ura

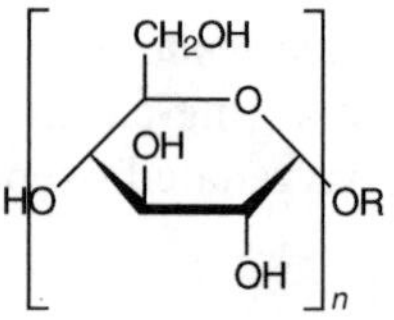

n = 0–3

35 R =

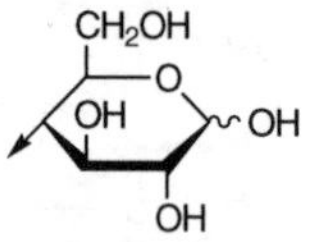

36 R =

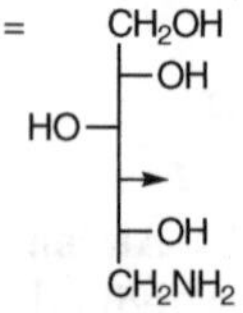

corresponding free sugars, which were reduced to alditols **33** and **34**, respectively.[29]

Enzymic transglycosylation of 6-azido-6-deoxy-D-glucose with α-cyclodextrin and subsequent hydrolysis with β-amylase gave mono- to tetra-saccharide glycosides **35**, which on reduction with $NaBH_4$ furnished malto-oligosaccharide-imino-alditols **36**.[30]

A number of acyclic aminoalditols which were prepared as intermediates in the synthesis of cyclic imino compounds are referred to in Section 1.4 below.

1.4 Cyclic Imino-alditols. – The azasugar isolated from *Aglaonema treublii* has now been shown to be α-homo-D-*allo*-nojirimycin (**37**), rather than the α-D-*gulo*-isomer as previously reported (see N. Asano *et al.*, *J. Nat. Prod.*, 1997, **60**, 98).[31] The new iminoalditols **38** and **39** have been extracted from the stalks

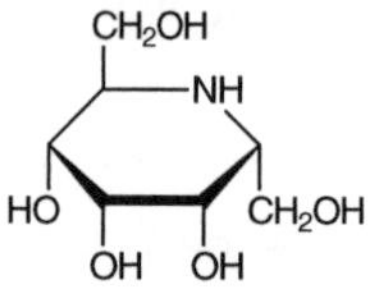

37

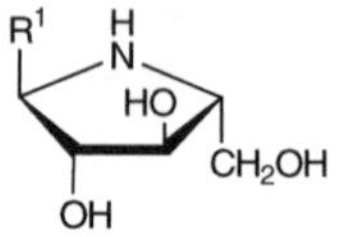

38 R[1] =

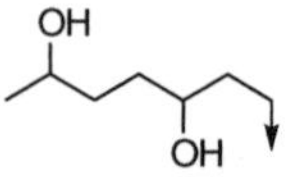

39 R[1] = β-D-xylosyl-O

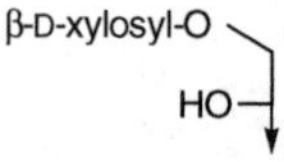

and unripe fruits of the bluebell (*Hyacinthoides non scripta*) together with four similar, known compounds. Compound **39** is an inhibitor of β-D-glucosidase and lactase.[32] Two new pyrrolidine alkaloids, named broussonetines K and L, have been isolated from the branches of *Broussonetia kazinoki,* which have previously yielded ten related compounds (see Vol. 30, Chapter 18, refs. 41 and

42; Vol. 31, Chapter 18, ref. 68).[33] The inhibition of glycosidases by a series of iminoalditols with structures **40** has been evaluated in detail by capillary electrophoresis. Compound **40** (R^1 = Me, R^2 = NHAc, X = H), in particular, was found to be a very potent β-*N*-acetylhexosaminidase inhibitor.[34]

As usual, numerous dideoxy-iminitol derivatives have been synthesized as potential glycosidase inhibitors. In the following sections, syntheses are grouped according to the method used for forming the nitrogen-containing rings.

1.4.1 New approaches to the formation of nitrogen-containing rings. The cyclic nitrone **42** has been prepared from allyl 4,6-*O*-benzylidene-α-D-glucopyranoside in 12 steps by way of a 6-deoxy-α-D-altropyranoside and oxime **41**, as a versatile advanced intermediate for the synthesis of azasugars. The ring-closure **41**→**42** took place on exposure to excess of NH_2OH–Et_3N. Compound **42** was converted, for example, to piperidine **43**, a potent α-L-fucosidase inhibitor.[35]

Under deacetylation conditions with methoxide, thioureas **44** rearranged to

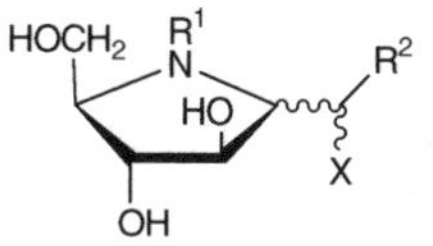

R^1 = H, Me, Et or Bu
R^2 = CH_2OH, CH_2NHAc or NHAc
X = H or OH
40

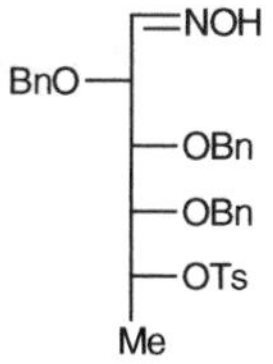

41

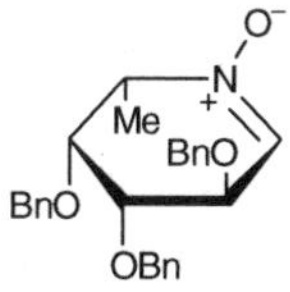

42

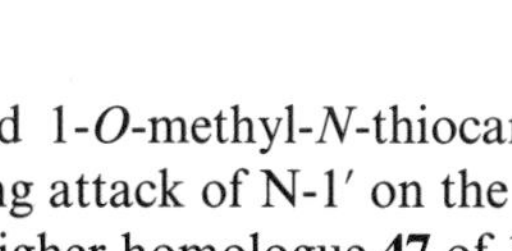

43

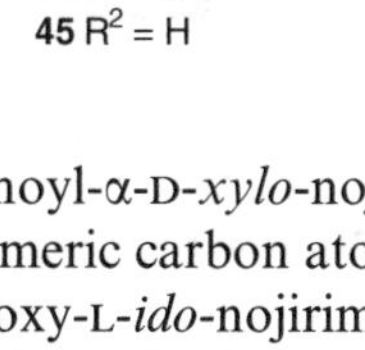

R^1 = Me, Ph
44 R^2 = Ac
45 R^2 = H

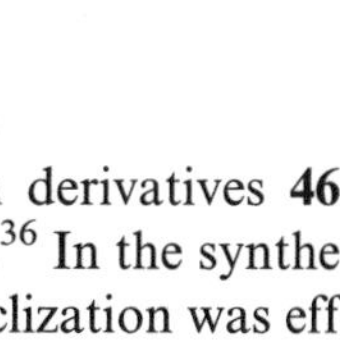

46 R = Me, Ph

afford 1-*O*-methyl-*N*-thiocarbamoyl-α-D-*xylo*-nojirimycin derivatives **46**, following attack of N-1′ on the anomeric carbon atom of **45**.[36] In the synthesis of the higher homologue **47** of 1-deoxy-L-*ido*-nojirimycin, cyclization was effected by intramolecular Michael addition of a benzylamino group, introduced by reductive amination, to an α,β-unsaturated ester side-chain, as shown in Scheme 4.[37]

N,N-Disubstituted amines have been used as precursors in several azasugar syntheses, with ring-closures involving C–C-bond formation. The methyl ester of *N*-benzenesulfonylalanine, for example, was *N*-alkylated with dibromide **48** to give **49**. Cyclization took place upon bromo-lithium exchange, and further

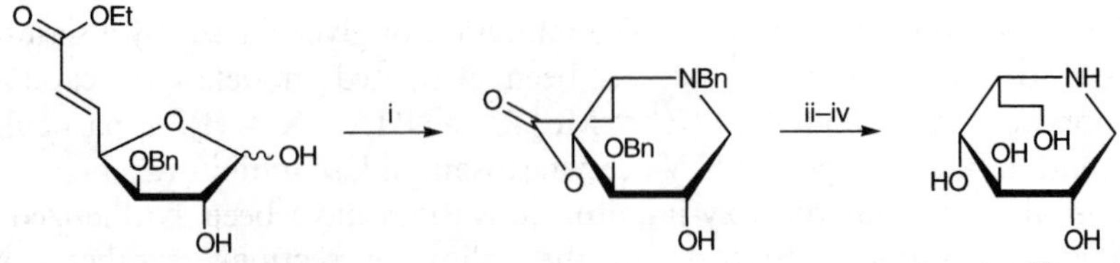

Reagents: i, $BnNH_2$, $NaCNBH_3$; ii, Ac_2O, Py; iii, LAH; iv, Pd/C, HCO_2NH_4

Scheme 4

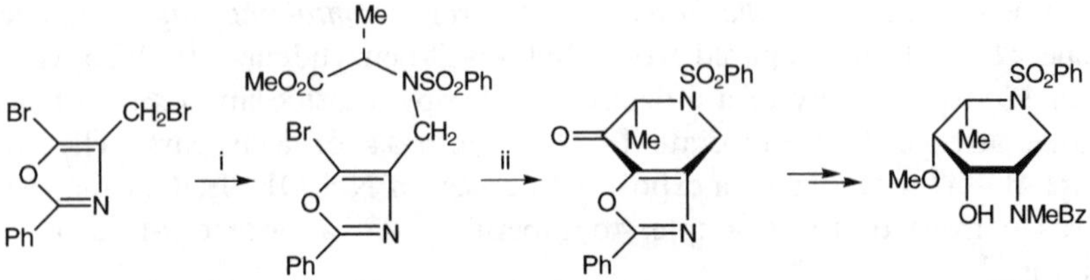

Reagents: i, *N*-Benzenesulfonyl-L-alanine methyl ester, K_2CO_3, ii, BuLi

Scheme 5

elaboration led to the l-*talo*-configured product **50**, as shown in Scheme 5.[38] In another instance, diesters **51**, available by a published procedure, were ring-closed by Dieckmann condensation. Enolsilylation of the resulting ketones **52**, then hydrogenation of the double bond over Raney-nickel, provided 2,4,5-trisubstituted piperidines **53** with high diastereoselectivity.[39] The β-keto acid **52a**, which is commercially available, was transformed in a similar manner to the (±)-*lyxo*-configured azasugar acid **54** and its (±)-*ribo* isomer; the hydroxymethyl group was introduced by direct alkylation of the dianion derived from **52a**.[40]

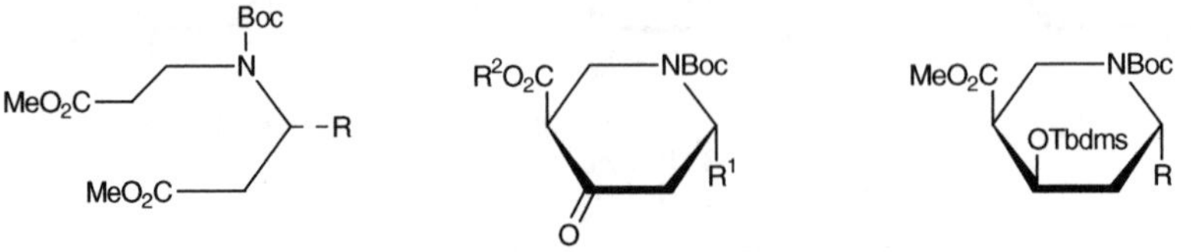

51 R = Bn, Pr or $(CH_2)_{14}Me$

52 R^1 = Bn, Pr or $(CH_2)_{14}Me$
R^2 = Me
52a $R^1 = R^2$ = H

53 R = Bn, Pr or $(CH_2)_{14}Me$

The enantiopure, endocyclic enecarbamate **55** has been converted into the *C*-aryl azasugar **56** and to the pyrrolidine alkaloids codonopsine (**57**) and codonopsinine (**58**); the aromatic moieties were introduced by modified Heck arylations and the double bonds were dihydroxylated *via* the epoxides.[41]

1.4.2 Ring-closure by nucleophilic displacement. A new route to 2,5-anhydro-2,5-imino-hexitols from C-1-unprotected pentofuranoses made use of Don-

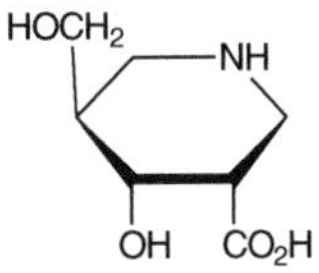

54

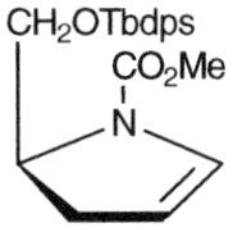

55

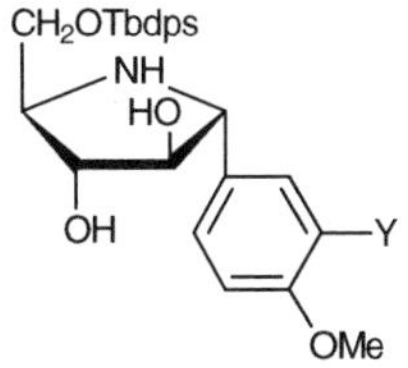

56 X = OH, Y = H
57 X = H, Y = OMe
58 X = Y = H

doni's 'aminohomologation' to construct 1-thiazolyl-2-amino-5-sulfonates, which cyclized spontaneously. The thiazolyl moiety was transformed into a hydroxymethyl group following the usual protocol. As an example, the preparation of 2,5-dideoxy-2,5-imino-D-glucitol (**60**) from D-arabinofuranose derivative **59** is shown in Scheme 6.[42]

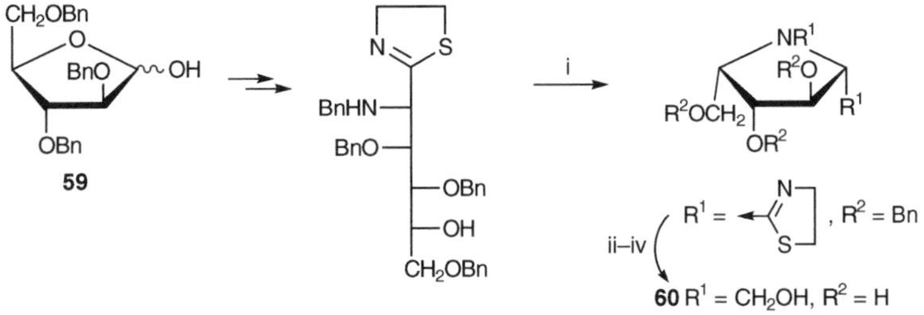

59

60 $R^1 = CH_2OH$, $R^2 = H$

Reagents: i, Tf_2O, Py ii, TfOMe, then $NaBH_4$, then $HgCl_2$; iii, $NaBH_4$; iv, H_2, $Pd(OH)_2/C$

Scheme 6

Methyl 2,3:5,6-di-*O*-isopropylidene-D-galactonate, the minor product formed (20%) when D-galactono-1,4-lactone was treated with dimethoxypropane and TsOH in acetone–methanol, was readily converted into the 4-azido-1-mesylate **61**, which cyclized on acetal hydrolysis and reduction of the azide, furnishing 1,4-dideoxy-1,4-imino-D-glucitol (**62**).[43]

On exposure to ammonia or amines, compound **63** underwent lactone opening followed by displacement of the triflate by the C-2 amino group to give iminosugar amides **64**. These were readily reduced (BH_3) and deprotected, furnishing iminosugar amines **65**. Use of an aminouridine-derived amine gave the iminosugar peptidonucleosides **66** as analogues of UDP Gal*f*.[44]

1-Deoxy-D-*manno*-nojirimycin and 1-deoxy-D-*altro*-nojirimycin were prepared from key-intermediate **68** by dihydroxylation of the double bond under the appropriate conditions, then treatment with aq. NaOH to cleave the oxazolidinone ring and remove the silyl group. Compound **68** was formed when mesylate **67**, obtained in a lengthy reaction sequence from *N*-benzyl-4-methoxycarbonyl oxazolidinone, was exposed to NaH in DMF.[45] Compound **71**, the product of a *syn*-aldol condensation between L-threose derivative **69** and the stannylated bis-lactim ether **70**, was cyclized *via* mesylate **72**, as shown in Scheme 7. The ring-closed product **73** was readily transformed to 1-deoxy-D-*galacto*-nojirimycin. A synthesis of 1-deoxy-D-*talo*-

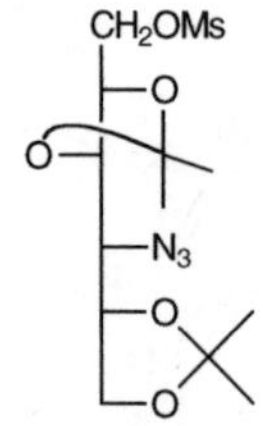

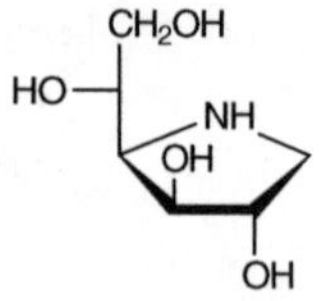

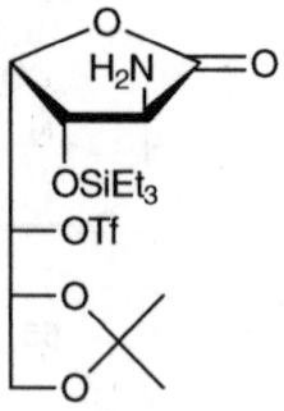

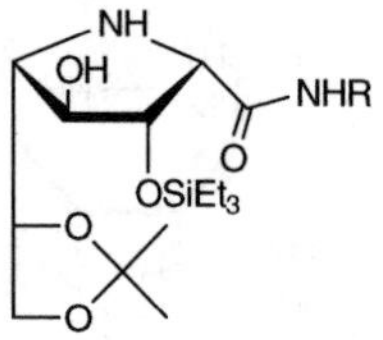

61 **62** **63** **64** R = H, Me, Bn

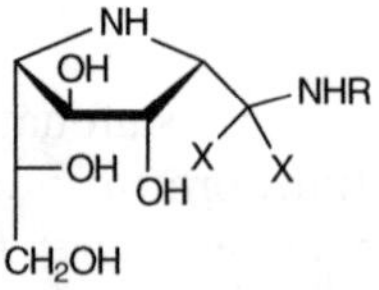

65 X = H, R = H, Me, Bn

66 X,X = O, R =

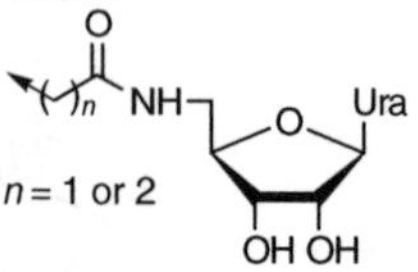

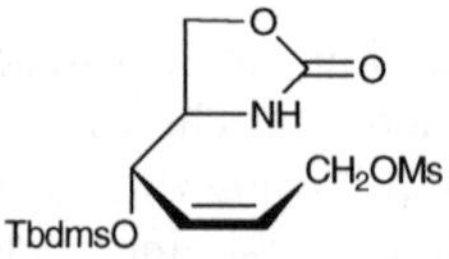

n = 1 or 2

67 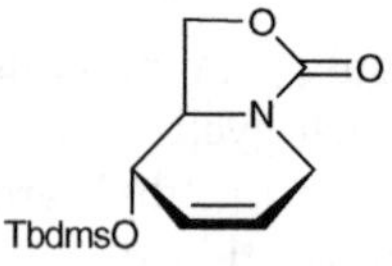 **68**

69 **70** **71** **72** **73**

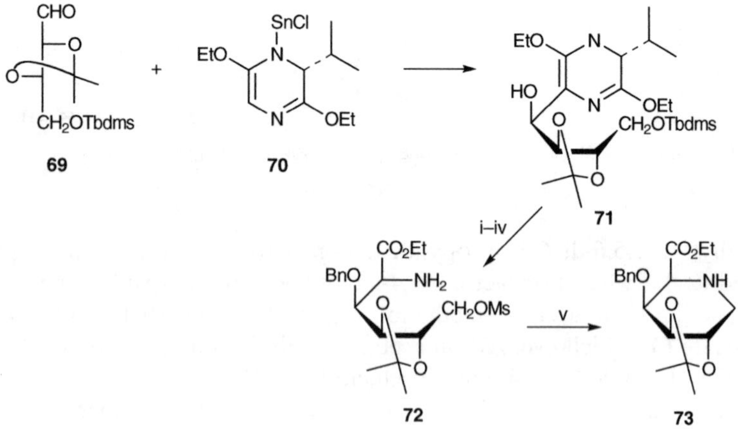

Reagents: i, BnBr, NaH; ii, Bu_4NF; iii, MsCl, Et_3N; iv, HCl, EtOH; v, Et_3N, DMSO

Scheme 7

nojirimycin from an epimer of adduct **71** is covered in Section 1.4.3 (ref. 56) below.[46]

The protected, L-rhamnose-derived 2-azido-2,7-dideoxy-heptono-1,5-lactone acetal **74**, was converted into the 6-*O*-triflate, then ring-closed by hydrogenation to give bicycle **75**. This was used to generate 5-*epi*-β-homo-*rhamno*-nojirimicin (**76**) and a series of related compounds by nucleophilic lactone-opening and subsequent reduction and deprotection. A reaction sequence without inversion of configuration at C-6, involving ring-formation by use of an aza-Wittig reaction and leading to α-homo-*rhamno*-nojirimicin derivatives, is covered in Section 1.4.3 below.[47]

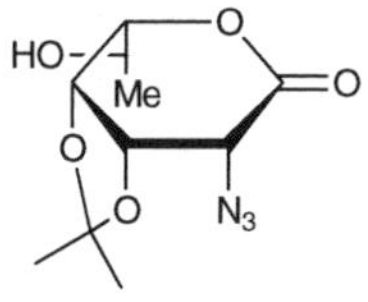

74 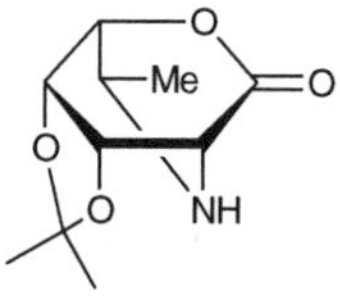 75 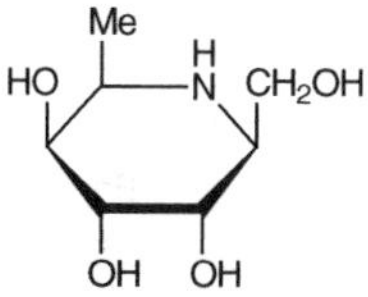 76

Azasugars with seven-membered rings have been made from easily available glycosylenamine derivatives **77** (D-*gluco*-, D-*galacto*- or D-*manno*-configuration). The mesylates were displaced by the enamine-nitrogens to furnish the bicyclic 1,6-aza-anhydrosugars **78**, after oxidative removal of the *N*-substituent. Reductive or hydrolytic cleavage of the oxygen-bridge gave 1,6-dideoxy-1,6-iminohexitols **79** and their 1-hydroxy analogues **80**, respectivey.[48] *N*-Benzylated, C_2-symmetric 1,6-dideoxy-1,6-iminohexitols have been prepared from 1,2:5,6-di-*O*-isopropylidene-D-mannitol *via* dimesylated hexitol intermediates,[49] and the solid-supported, C_2-symmetric 1,6-dideoxy-1,6-iminohexitol **82** was formed by opening of L-iditol 1,5-bis-epoxide **81** with a resin-bound amine (Rink resin).[50]

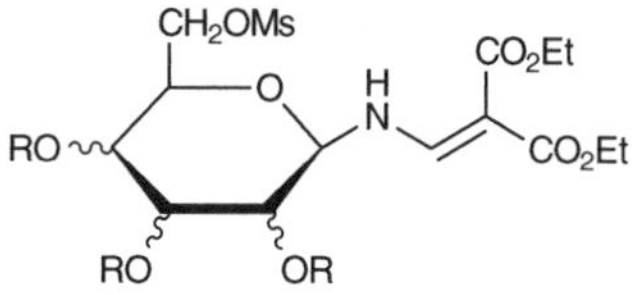

R = Ac or Bz
77

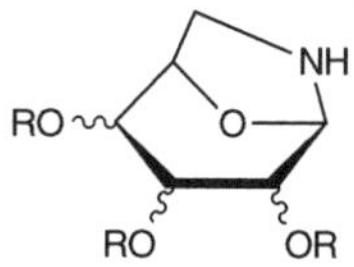

R = Ac or Bz
78

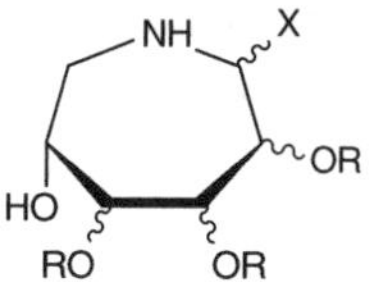

R = Ac or Bz
79 X = H
80 X = OH

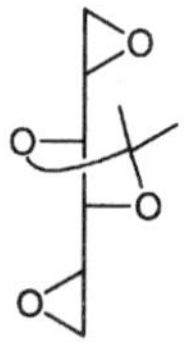

81

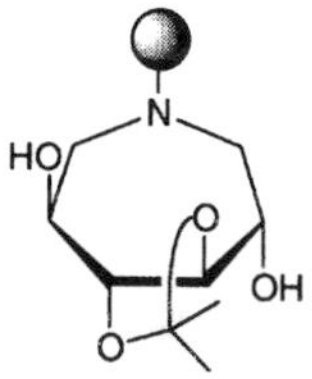

82 = solid support

1.4.3 Ring-closure by reductive amination. A review on the synthesis of 4- and 5-amino-aldoses and 5- and 6-amino-ketoses by chemical and enzymic methods and their intramolecular reductive aminations to form iminoalditols has been published.[51]

In an eight-step synthesis of nojirimycin (**84**), 2,3,4,6-tetra-*O*-benzyl-α-D-glucopyranose was first transformed into 5-aminoaldehyde **83** *via* a 5-oxime. Exposure to hydrogenating conditions caused simultaneous de-*N*-protection, intramolecular reductive amination and de-*O*-protection.[52] The same tetrabenzylglucose was used in a straightforward preparation of 1-deoxynojirimycin (**85**) and its *N*-butyl derivative **86** by consecutive reduction to alditol **87**,

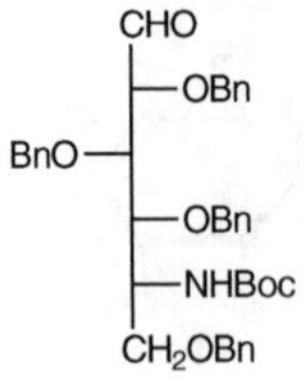

CHO, OBn, BnO, OBn, NHBoc, CH_2OBn

83

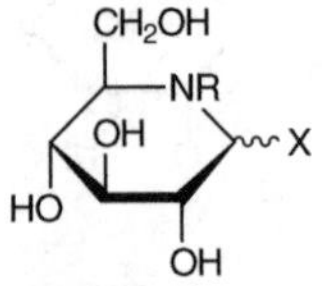

CH_2OH, NR, OH, X, HO, OH

84 R = H, X = OH
85 R = X = H
86 R = Bu, X = H

CH_2OH, OBn, BnO, OBn, OH, CH_2OBn

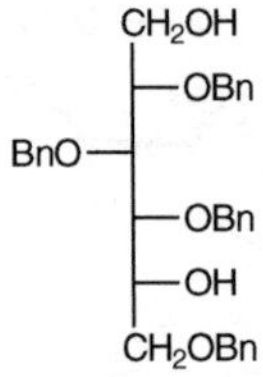

87

oxidation to the corresponding aldos-5-ulose and *in situ* reductive amination, using NH_3 and $BuNH_2$, respectively, then debenzylation.[53]

1-Deoxy-D-*galacto*-nojirimycin (**90**) and 1-deoxy-L-*altro*-nojirimycin (**91**) were formed when azides **88** and **89**, available from D-galactose in six and four steps, respectively, underwent intramolecular reductive amination on exposure to H_2–Pd/C,[54] and 1-deoxy-L-*fuco*-nojirimycin (**92**) was similarly obtained from the 5-imino-5-deoxy-L-fucose derivative **93**, which was constructed in a lengthy reaction sequence from methyl L-alaninate.[55] The L-erythrose-derived epimer **94** of compound **71** (see Section 1.4.2 above) ring-closed with concomitant loss of the auxiliary on selective oxidation of the primary hydroxyl group, followed by hydrogenation over Pd/C, furnishing the 1-deoxy-D-*talo*-nojirimycin precursor **95**.[56]

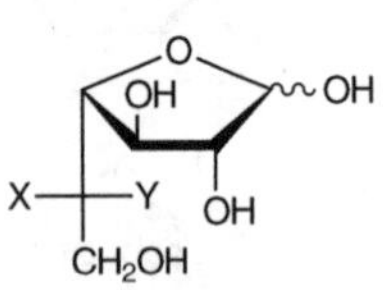

88 X = H, Y = N_3
89 X = N_3, Y = H

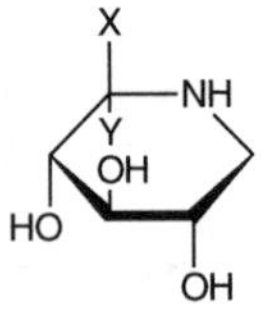

90 X = CH_2OH, Y = H
91 X = H, Y = CH_2OH

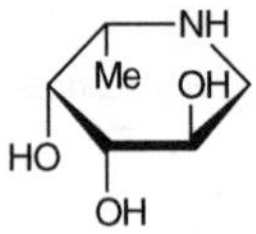

92

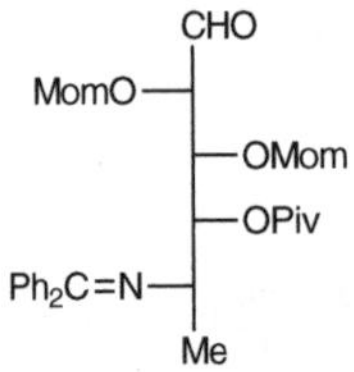

93

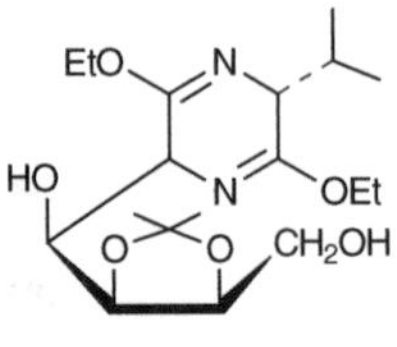

94

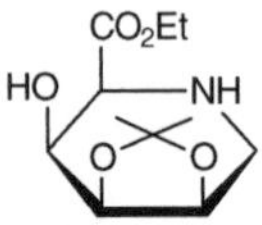

95

In a chemo-enzymic synthesis of the hydroxymethyl derivative **97** of isofagomine, the hexulose 1-phosphate **96** carrying an *N*-formamidomethyl branch was made by an aldolase-catalysed process from small molecules, and cyclization was effected chemically by reductive amination, as shown in Scheme 8. Racemization of the starting aldehyde under the reaction conditions necessitated separation of the end product **97** from its epimer **97a** by preparative HPLC.[57]

A modification of the conventional reductive amination (aza-Wittig reac-

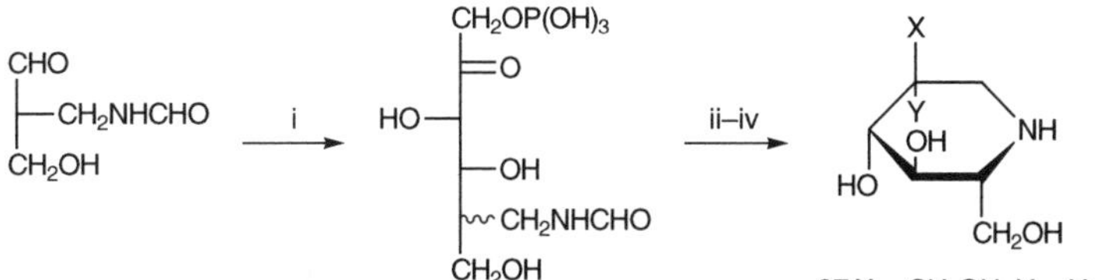

96

97 X = CH_2OH, Y = H

97a X = H, Y = CH_2OH

Reagents: i, Dihydroxyacetone phosphate, rabbit muscle aldolase; ii, phosphatase; iii, aq. HCl; iv, H_2, Pd/C

Scheme 8

tion) was used in the synthesis of β-homo-*rhamno*-nojirimicin (**98**) from the protected L-rhamnose-derived 2-azido-2,7-dideoxy-1,5-lactone **74** (see Section 1.4.2 above), as shown in Scheme 9.[47]

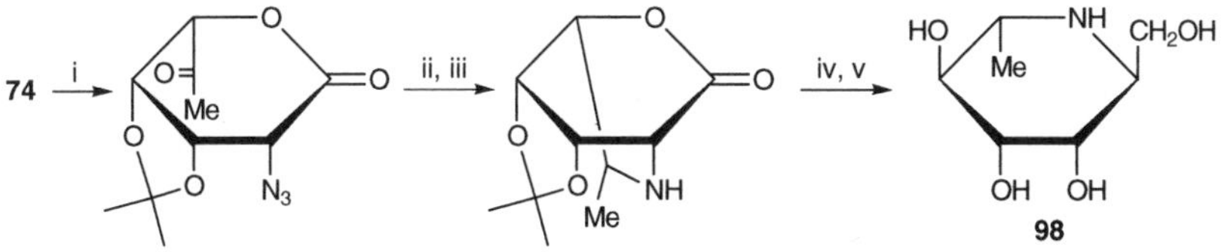

98

Reagents: i, PCC; ii, $P(OMe)_3$; iii, $NaCNBH_3$; iv, $LiBEt_3H$; v, HCl, MeOH

Scheme 9

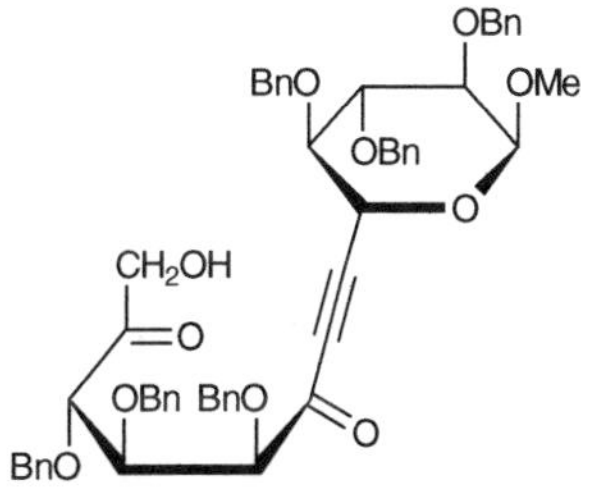

99

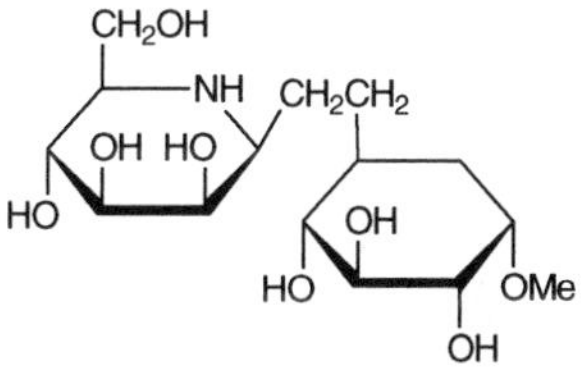

100

The synthesis of *C*-linked aza-β-(1→6)-disaccharides, *e.g.* aza-Man-β-(1→6)-*C*-α-D-GlcOMe (**100**) was based on the double reductive amination of the acetylenic sugar diketone **99** with ammonium formate.[58]

1.4.4 Ring-closure by lactam formation. All four monoamino-monodeoxy-D-tetronolactams **102** were available by reaction of either the D-threonic acid-derived bromo-*cis*-epoxide **101** or its D-erythronic aciD-derived *trans*-isomer with liquid ammonia. The lactams were readily converted into the 4-amino-3-hydroxypyrrolidines **103** by reduction with BH_3.[59]

5-*O*-Benzyl-1,4-dideoxy-1,4-imino-D-lyxitol (**106**) has been prepared from the bicyclic lactam **105**, which was formed by condensation of keto-acid **104**

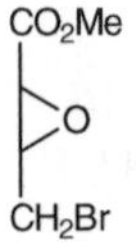

101 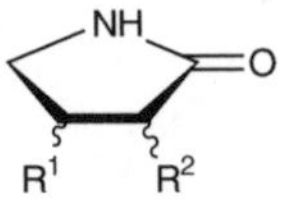 102 R^1, R^2 = OH, NH_2 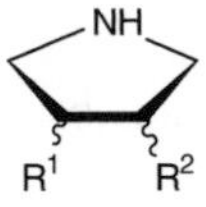 103 R^1, R^2 = OH, NH_2

with (*S*)-phenylglycinol under dehydrating conditions. Dihydroxylation was achieved by first generating α,β-unsaturation, followed by stereoselective osmylation, then simultaneous carbonyl reduction, C–O-4 cleavage with inversion at the angular position and *N*-deprotection by use of BH_3-BBN. This procedure was adapted to the synthesis of 1-deoxy-L-*manno*-nojirimycin and 1-deoxy-L-*rhamno*-nojirimycin from the homologous keto-acid **104a**, by incorporating an additional allylic hydroxylation step.[60]

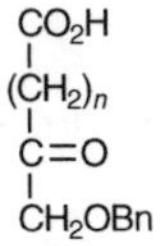

104 $n = 2$
104a $n = 3$ 105 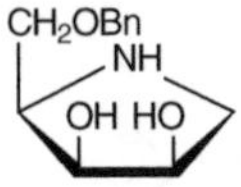 106

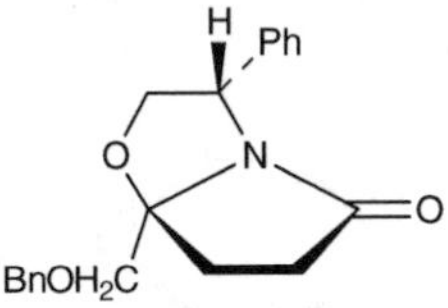

Lactams **108** were formed on oxidation of alcohol **107**, obtained from diethyl D-tartrate by use of a modified Strecker synthesis as the key-operation. The 3,4-*trans*- and 3,4-*cis*-isomers were separately processed to afford nectrisine (**109**), a fungal metabolite and potent glycosidase inhibitor, and its 4-*epi*-isomer, respectively.[61]

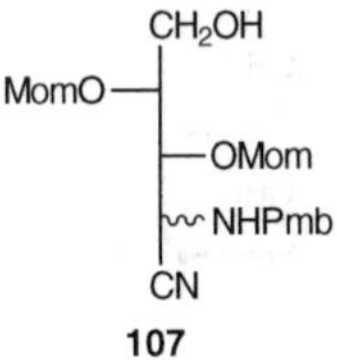

107 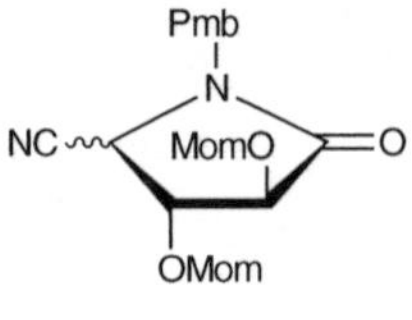 108 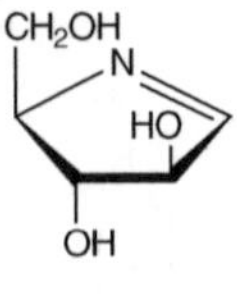 109

2-Acylamino-1,2-dideoxynojirimycin derivatives, *e.g.* compound **112**, have been prepared as haptens for raising catalytic antibodies that can hydrolyse the glycosidic bonds of Lipid A. Opening of 3,4,6-tri-*O*-benzyl-2-benzyloxycarbonylamino-2-deoxy-D-gluconolactone with ammonia, then oxidation and treatment of the resulting keto-amide **110** with $NaCNBH_3$, gave the 2-amino-

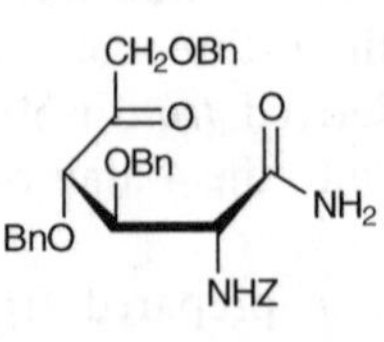

110 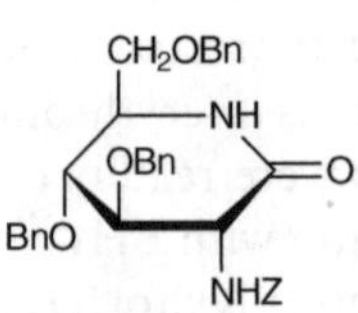 111 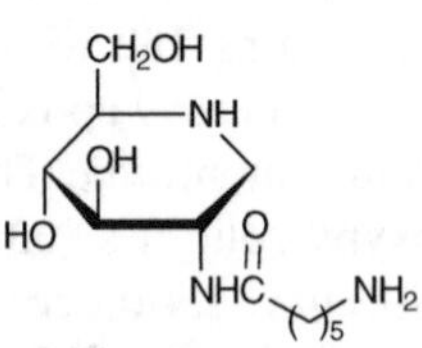 112

2-deoxy-lactam derivative **111**. Deoxygenation at C-1 (BH_3) was followed by removal of the benzyloxycarbonyl group, introduction of the acyl-chains by PyBOP-mediated condensation with the appropriate *N*-protected free ω-aminoacid, and deprotection.[62]

1.4.5 Ring-formation by cycloaddition. Use of [2-^{13}C]penta-2,4-dienoic acid, derived from [2-^{13}C]malonic acid, as starting diene in the known Diels Alder approach to racemic azafagomine (see Vol. 32, Chapter 18, ref. 42) gave the [3-^{13}C]-labelled compound.[63] Enantiopure (+)-calystegine B_2 (**113**) has been prepared following the procedure employed for making the racemic compound (see Vol. 30, p. 236, ref. 95) except for the use of a sugar-derived nitroso dienophile instead of a simple, achiral one in the initial cycloaddition reaction.[64]

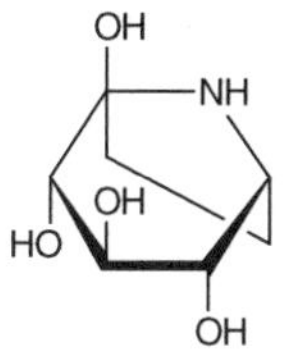

OH NH OH HO OH CH₂OAc O AcO O N AcO O O Me O AcO O N AcO O O

113 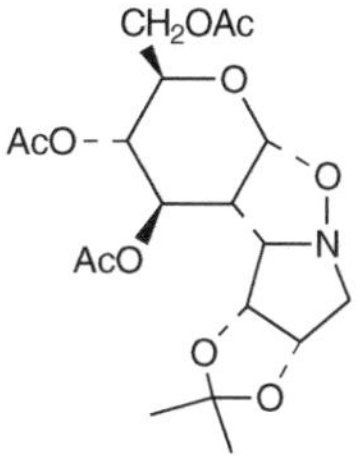 **114** 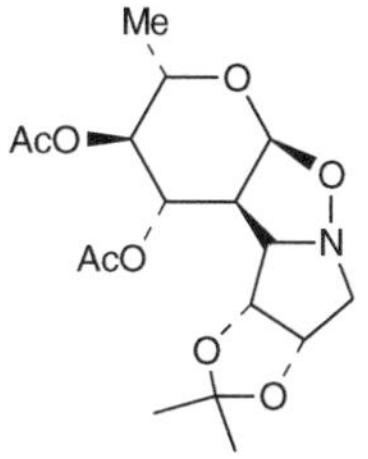 **115**

In a parallel kinetic resolution experiment, racemic isopropylidene-protected *cis*-dihydroxypyrroline *N*-oxide was exposed simultaneously to triacetyl-D-glucal and diacetyl-L-rhamnal. The cycloaddition products **114** and **115** deriving from 'matched' interactions were obtained in 1:1 ratio with exclusion of the disfavoured alternatives. Reductive cleavage of the N–O bond and deprotection furnished a pseudo-iminodisaccharide from **114**.[65] Tandem Wittig reaction/intramolecular [2+3] cycloadditon applied to azidoaldose **116** provided the stereoisomeric triazoline adducts **117** which were converted into the all-*cis*-β-hydroxypiperidine derivative **118** and several

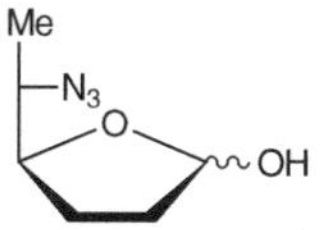

Me N₃ O OH N N Me HO H CO₂Et NH Me HO CH₂CO₂Et

116 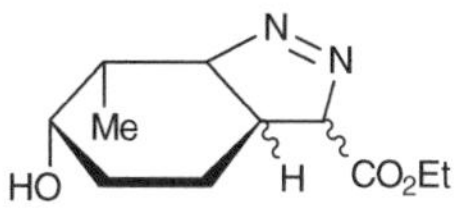 **117** 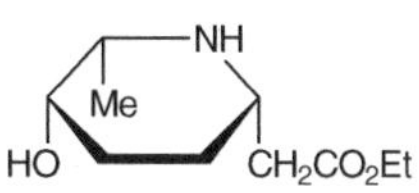 **118**

closely related compounds by thermolysis with loss of nitrogen, then modification of the side-chain by conventional procedures.[66] *cis*-Dihydroxylation of the known carbocyclic azide **119**, which was obtained by a cycloaddition route, followed by diol cleavage, gave hemiketal **120**. This was converted by consecutive reduction of the free aldehyde group, deacetylation and reductive amination into the (±)-monohydroxytetra(hydroxymethyl)-piperidine **121**.[67]

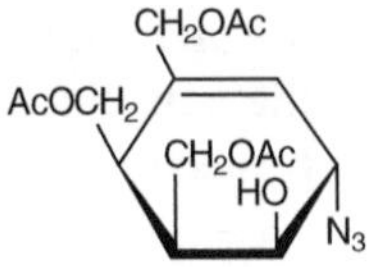

1.4.6 Ring-closure by other methods. The new azirdines **123** have been synthesized by extension of aldehyde **122** with $PPh_3{=}CBrCO_2Et$, then Michael addition of ammonia or benzylamine to the resulting α-bromoacrylates. Their reactivity has been explored, together with that of several known aziridines of similar structure.[68]

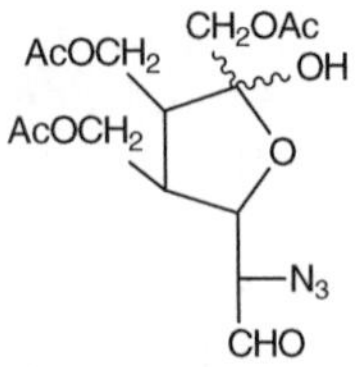

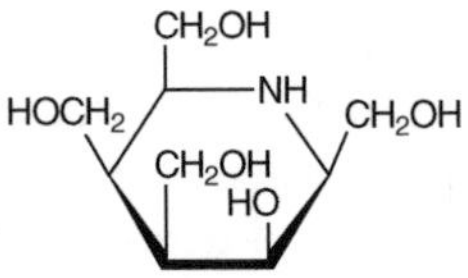

Several references to the preparation of 1-deoxyazasugars can be found in a *Tetrahedron Report* on the synthesis and application of optically active α-furfurylamine derivatives.[69] A new, concise, stereoselective preparation of enantiopure β-hydroxyfurfurylamine derivative **124** from vinyl furan in five steps, including a Sharpless asymmetric dihydroxylation, in 65% overall yield has been developed. Compound **124** is a key-starting material in the synthesis of 1-deoxyazasugars [*e.g.* 1-deoxy-D-*manno*-nojirimycin and 1-deoxy-L-*altro*-nojirimycin (**91**), see Vol. 31, p. 235, ref. 84] by way of its oxidation product **125**, and has now been employed to prepare 1-deoxy-D-*gulo*-nojirimycin (**126**) and 1-deoxy-D-*talo*-nojirimycin (**127**).[70]

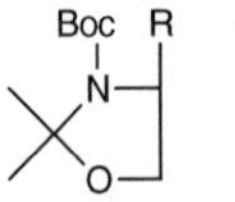

122 R = CHO
123 R = CO_2Et

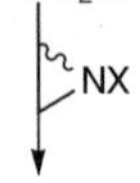

X = H or Boc

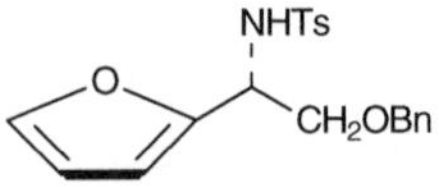

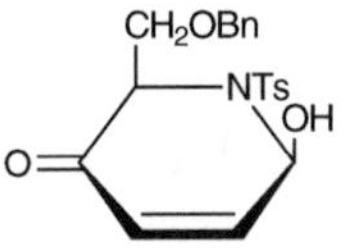

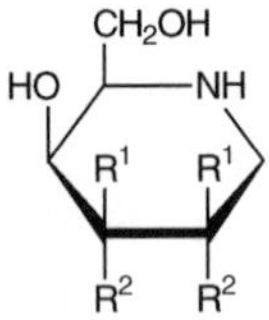

126 R^1 = H, R^2 = OH
127 R^1 = OH, R^2 = H

1.4.7 Modification of known azasugars. A number of 1-aryl-1,4-dideoxy-1,4-imino-D-ribitols **129** and homonucleosides, such as **130**, have been produced as transition state analogue inhibitors of protozoan nucleoside hydrolases. The

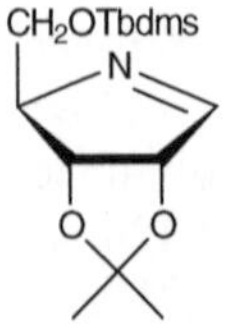

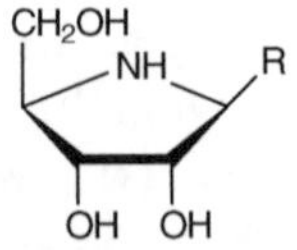

129 R = aryl, heteroaryl
130 R = CH_2-aden-9-yl

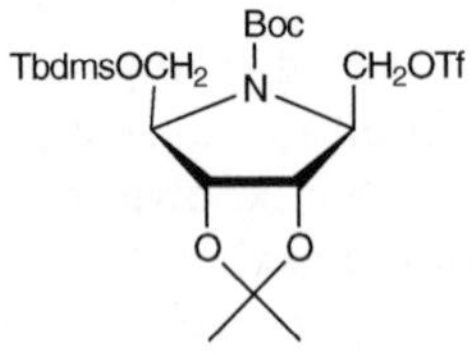

former were accessed by addition of organometallic species to imine **128**, the latter by displacement of the sulfonyloxy group from the 2,5-dideoxy-2,5-imino-D-allitol derivative **131** by adenine, followed by deprotection.[71]

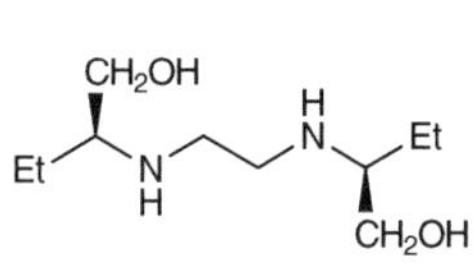

132

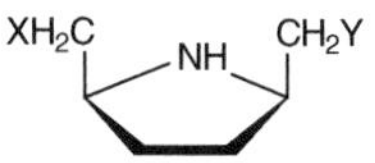

133 X, Y = H, HN–CH(Et)(CH$_2$OH)

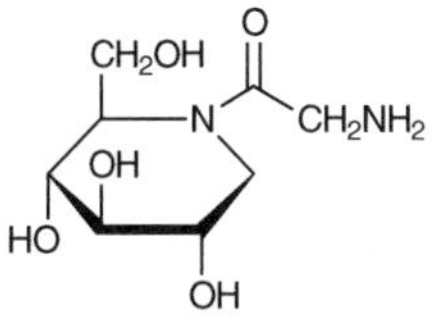

134

Analogues **133** of the anti-tuberculosis agent ethambutol (**132**) have been prepared by standard procedures from 2,5-dideoxy-2,5-imino-D-glucitol.[72] *N*-^{13}C-Methyl-1-deoxynojirimycin was synthesized by alkylation of 1-deoxynojirimycin (**85**) with [^{13}C]-methyl iodide for use in isotope-edited NMR studies of the binding site of an α-glucosidase. These led to the design and synthesis of *N*-glycyl-1-deoxynojirimycin (**134**) as a novel, more potent inhibitor, by TBTU-mediated coupling of 1-deoxynojirimycin tetra-*O*-acetate with *N*-Boc-glycine, followed by deprotection.[73] Synthesis of D-nojirimycin-δ-lactam (**137**) from the known *meso*-imide **135** involved asymmetric reduction with bis(2,6-dimethylphenoxy)borane in the presence of a thiazazincolidene complex, and acetylation of the new hydroxy-group to give **135a**. The side-chain was introduced by substitution of the new acetoxy group with propargyltrimethylsilane to give the allenic intermediate **136**. This was ozonolysed, then reduced with $NaBH_4$.[74]

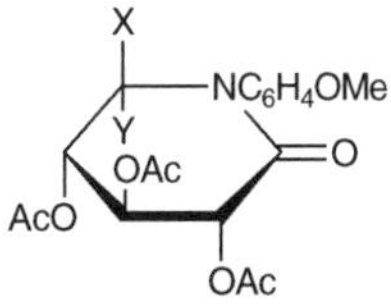

135 X, Y = O
135a X = OAc, Y = H
136 X = H, Y = allenyl

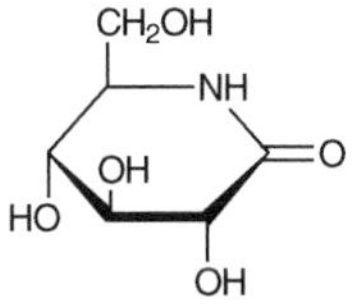

137

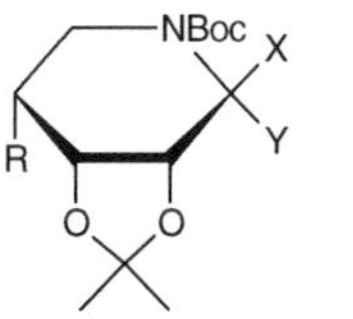

138 X, Y = O, R = OTbdms
139 X = H, Y = OH, R = Me

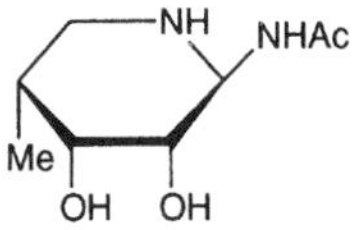

140

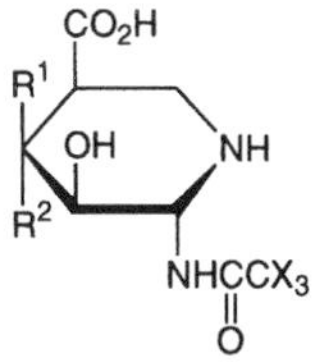

141 R^1 = OH, R^2 = X = H
142 R^1 = H, R^2 = OH, X = H or F

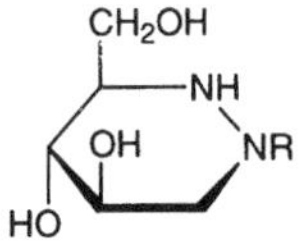

143 R = H
144 R = CH$_2$C(O)-tripeptide-(solid support)

● = solid support

gem-Diamino compound **140**, a new type of glycosidase inhibitor, was prepared from the known lactam **138** by first opening the ring reductively and closing it again to the aminal **139**, after replacement of the 4-OTbdms group by methyl. Introduction of the amino group was achieved under Mitsunobu conditions.[75] Siastatin B (**141**), isolated from *Streptomyces* cultures, has been transformed into analogues **142** by standard procedures.[76] Enzyme-mediated resolution of an intermediate in the published synthesis of racemic 1-azafagomine **143** (see Vol. 32, Chapter 18, ref. 42) gave access to the bioactive (−)-enantiomer in pure form.[77] Racemic 1-azafagomine has been alkylated at N-1 with a solid-supported library of tripeptides chloroacetylated at the terminal amino group (**143**→**144**) in the assembly of the first combinatorial library of azasugar glycosidase inhibitors,[78,79] and the iminosugar-based UDP-galactose analogue **146** has been constructed as a potential inhibitor of α-(1→3)-galactosyltransferase by reductive amination of iminosugar **145** with the appropriate aldehyde.[80]

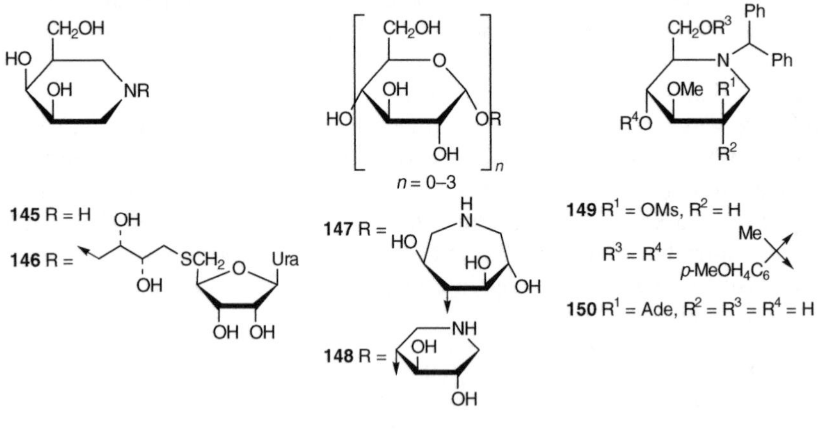

145 R = H

146 R = (OH, OH, SCH_2, furanose, Ura, OH OH)

n = 0–3

147 R = (HO, N-H, HO, OH)

148 R = (NH, OH, OH)

149 R^1 = OMs, R^2 = H, R^3 = R^4 = (Me, p-$MeOH_4C_6$)

150 R^1 = Ade, R^2 = R^3 = R^4 = H

Enzymic transglycosylation of 6-azido-6-deoxy-D-glucose to α-cyclodextrin and subsequent hydrolysis with β-amylase gave mono- to tetra-saccharide glycosides **35** (see Section 1.3 above) which furnished compounds **147** and **148** on exposure to H_2–Pd/C directly or after oxidative degradation by one carbon, respectively.[30] The protected azasugar nucleoside analogue **150** was obtained by displacement of the mesyloxy group of *N*-benzhydryl-1-deoxynojirimycin derivative **149** by adenine in the presence of NaH. It was used for preparing

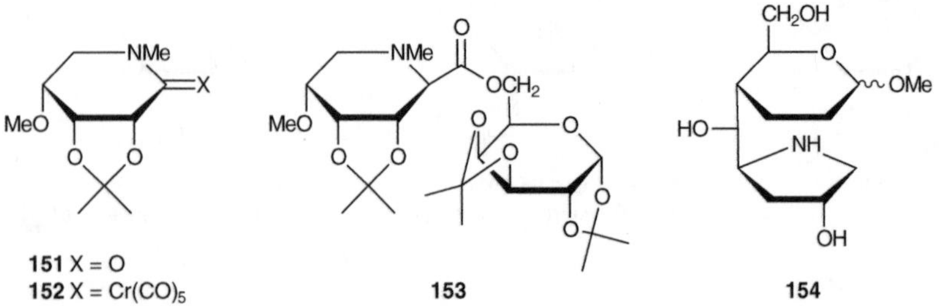

151 X = O
152 X = $Cr(CO)_5$

153

154

modified oligonucleotides.[81] Chromium iminoglycosylidenes **152**, available by treatment of the corresponding lactams **151** with $K_2Cr(CO)_5$–TmsCl, underwent photoinduced reactions with glycosyl acceptors to give, for example, galactos-6-yl 2,6-imino-D-allonate **153**.[82]

The synthesis of methyl glycosides with iminoalditol branches, for example compound **154**, is referred to in Chapter 14.

2 Cyclitols and Derivatives

2.1 Cyclopentane Derivatives. – Previously published work on the synthesis of trehazoline and trehazoline analogues has been reviewed.[83]

Metathesis using Schrock's catalyst converted the branched diolefin **155** into a cyclopentanoid system, which was dihydroxylated and deprotected to furnish cyclitols **156**.[84,85]

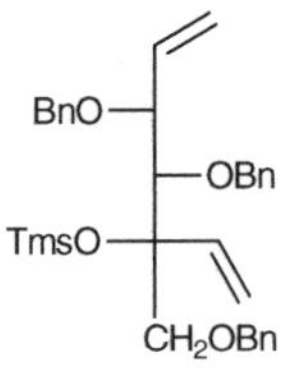

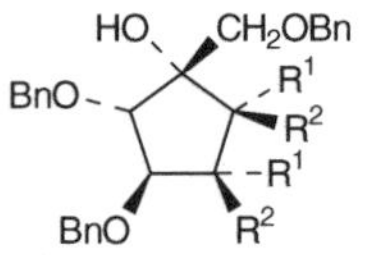

R^1 = H, R^2 = OH or R^1 = OH, R^2 = H

155 **156** **157**

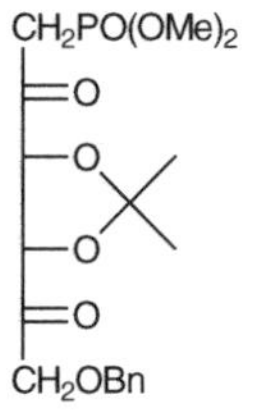

The branched cyclopentenone **158**, obtained by intramolecular Wittig-Horner reaction of the D-ribono-1,4-lactone-derived phosphonate **157**, served as key-intermediate for the synthesis of modified carbocyclic nucleosides.[86] Highly selective radical cyclization has been achieved by use of Bu_3SnH and 2,2′-azobis(2,4-dimethyl-4-methoxyvaleronitrile) as initiator. In

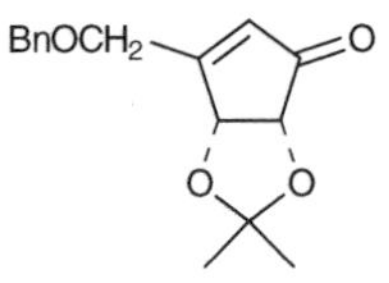

158 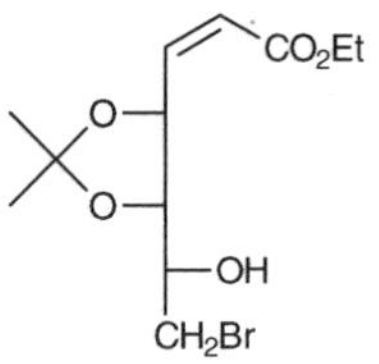 **159** 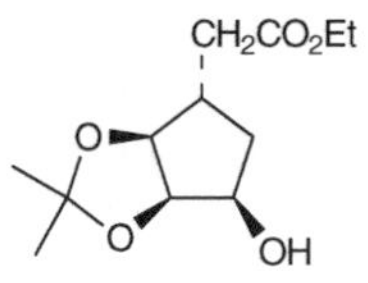 **160**

the transformation **159**→**160**, for example, the *syn/anti* ratio was 2:98, with a combined yield of 85%, compared with a 1:6 ratio when AIBN was used.[87] Radical cyclization of a chitobiose-derived oxime ether (**161**→**162**) has been employed in the preparation of an analogue of the chitinase inhibitor allosamidin.[88] The high-yielding, SmI_2-mediated cyclization of the di-*O*-isopropylidene-keto-oxime **163** was the key-step in the synthesis of an epimer of trehazolamine.[89]

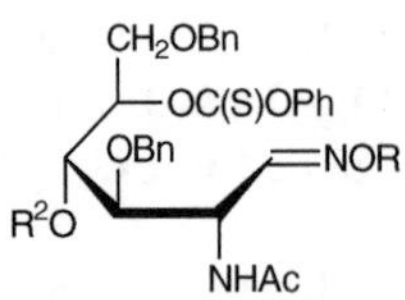

161 R^1 = Me, Bn; R^2 =

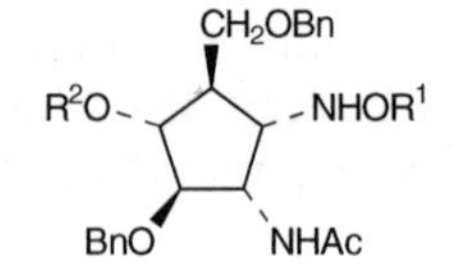

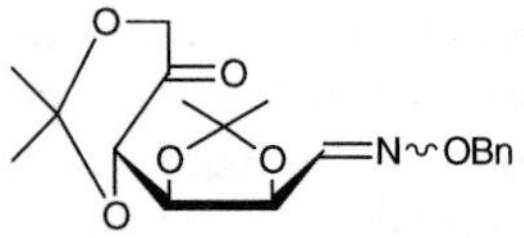

162

163

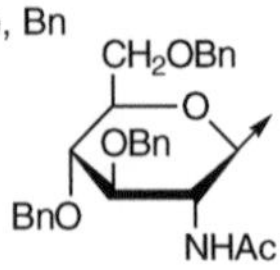

Bicyclic isoxazolidines and isoxazolines produced by intramolecular cycloadditions of carbohydrate-derived, δ-unsaturated nitrile oxides,[90] and nitrones[90] or oximes,[91] respectively, were further transformed into various aminocyclopentitols. A novel route to densely functionalized carbocycles involving formation and Raney-nickel cleavage of intermolecular nitrile oxide addition products of pent-4-enopyranosides is referred to in Chapter 24.

The sugar-derived bicyclic lactone **164** served as starting material for the synthesis of cyclitols **165** and aminocyclitol **166**. The stereochemistry of the functionalization of the double bond was controlled by choice of method (osmylation or epoxidation/epoxide-opening) and protecting groups.[92]

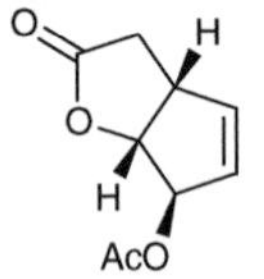

164

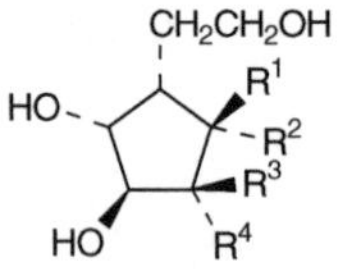

165 $R^1 = R^3$ = OH, $R^2 = R^4$ = H
or $R^1 = R^3$ = H, $R^2 = R^4$ = OH
or $R^1 = R^4$ = H, $R^2 = R^3$ = OH

166 $R^1 = R^3$ = OH, R^2 = H, R^4 = $NH_3^+Cl^-$

Bi- and tri-cyclic systems available by cycloaddition routes from simple molecules were the starting materials in a number of cyclitol syntheses: the precursors **169** and **170** of carbocyclic 2′-deoxynucleosides and homocarba-

167

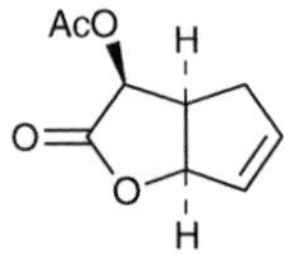

168

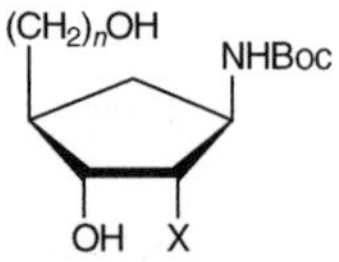

169 X = H, *n* = 1
170 X = OH, *n* = 2

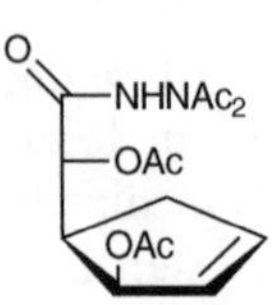

171

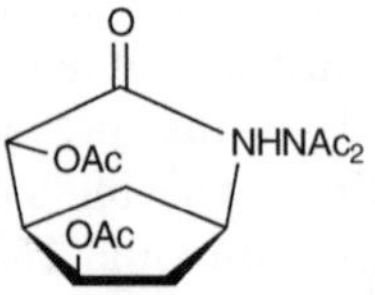

172

nucleosides, for example, were elaborated from the tricyclic epoxy-lactam **167**[93] and bicyclic lactone **168**,[94] respectively, both readily available in resolved form. In the latter case, the crucial reaction step was the unprecedented Pd(0)-catalysed cyclization of a hydrazine derivative (**171**→**172**). Grob-type fragmention of the norbornanone-7-mesylate **173** to give the cyclopentene derivative **174** was the key-step in the synthesis of all-*cis*-cyclopentane pentaol,[95] of two new analogues of trehazolamine[96] and of the natural product salpanitol (**175**).[96] Several inconsistencies in the spectral data of the synthetic **175** with those reported for the natural product have been pointed out. β-Aminoketone **177** was formed by the highly regio- and stereo-selective reduction (Li–NH_3) of the enaminone **176** with concomitant loss of the auxiliary; cycloreversion of **177** afforded aminocyclopentenol **178**, after exposure to $NaBH_4$.[97]

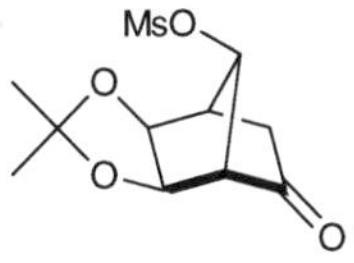

173

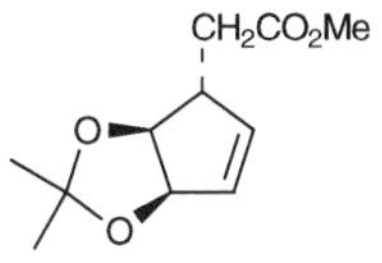

174

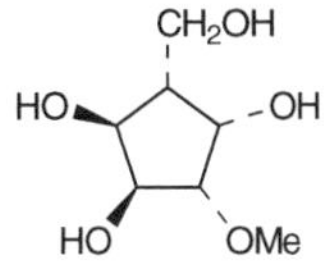

175

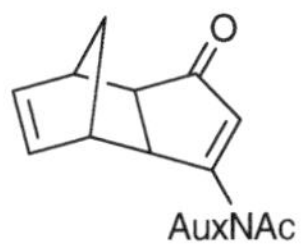

Aux = chiral auxiliary

176

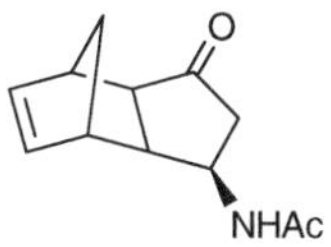

177

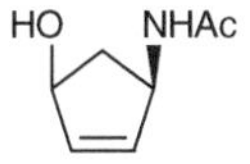

178

The 1-, 2- and 3-deoxy analogues of mannostatin were prepared from all-*cis*-5-acetamido-1,2,3,4-cyclopentanetetrol by conventional deoxygenation of suitably protected derivatives and displacement of an appropriately placed triflate by a sulfur nucleophile.[98] Opening of epoxide **180**, obtained in two steps from

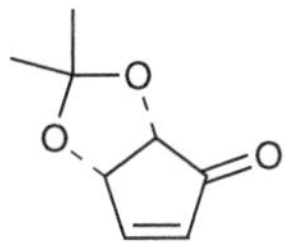

179

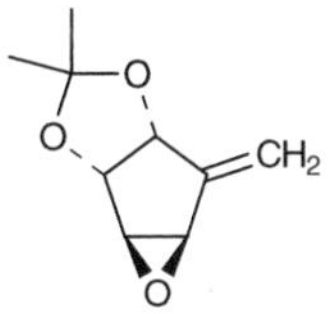

180

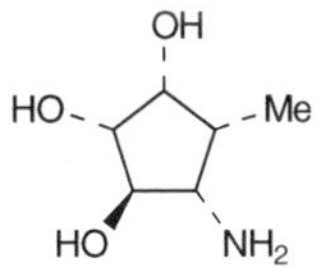

181

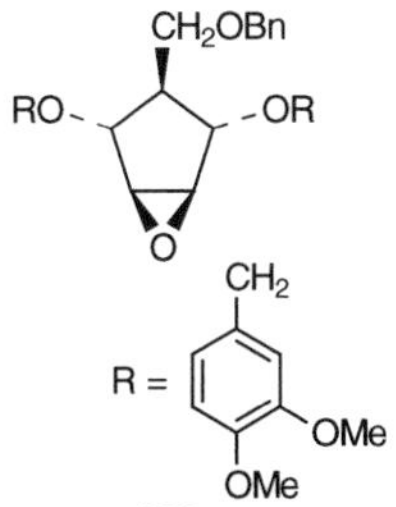

182

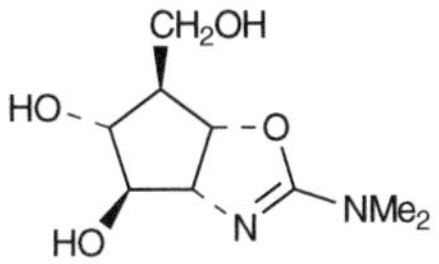

183

the known acetal-protected 4,5-dihydroxy-cyclopent-2-enone **179**, with ammonia proceeded with high selectivity to furnish, after three further standard steps, aminocyclitol **181.**[99] A new synthesis of allosamizoline (**183**) involved asymmetric opening of the known *meso*-epoxide **182** by use of $TmsN_3$ in the presence of a (salen)–Cr(III) complex (94% e.e.).[100]

2.2 Inositols and Related Compounds. – L-Bornesitol (1-*O*-methyl-L-*myo*-inositol) has been found as a major soluble carbohydrate in sweet pea petals.[101] A new oligosaccharin signalling molecule isolated in micromolar quantities from cell-suspension cultures of *Rosa* sp. has been identified as α-D-Man*p*-(1→4)-α-D-GlcA*p*-(1→2)-*myo*-inositol.[102]

Several synthetic pathways to conduritol C and conduritol F have been compared and evaluated for efficiency.[103] Cyclizing metathesis under the influence of Grubbs' catalyst has been applied to the preparation of conduritols from a range of sugar-derived dienes: thus, compound **184**, obtained from 2,3,4,5-tetra-*O*-benzyl-galactitol by Swern oxidation, then Wittig reaction, furnished perbenzyl-conduritol A; perbenzyl-conduritols E and F were obtained analogously from tetra-*O*-benzyl-D-mannitol and -D-glucitol, respectively.[104] A mixture of diene **185** and its 6-epimer, prepared by sequential Wittig, Swern and Grignard reactions from 2,3,4-tri-*O*-benzyl-D-xylose, ring-closed to afford derivatives of conduritols B and F; benzylation of the free hydroxyl groups and dihydroxylation (OsO_4) of the double bonds then gave tetra-*O*-benzyl-*myo*-and -L-*chiro*-inositol, respectively.[105] Diene **188**, available in 6 steps from enal **186** *via* **187,** which underwent a $PdCl_2$-mediated allylic rearrangement, gave access to conduramine derivative **189.**[106]

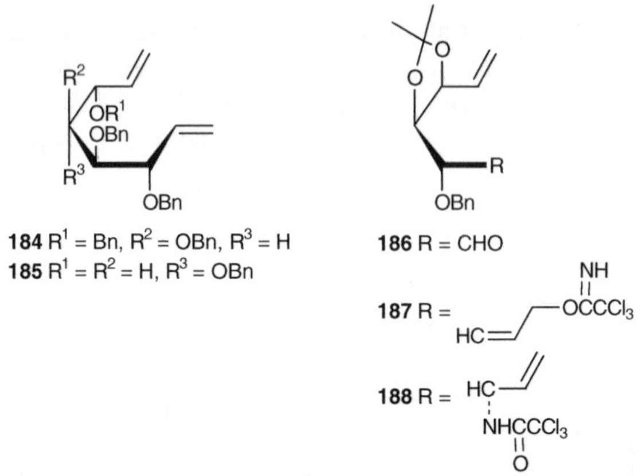

184 R^1 = Bn, R^2 = OBn, R^3 = H
185 R^1 = R^2 = H, R^3 = OBn

186 R = CHO
187 R = HC=CH–CH₂–OC(=NH)CCl₃
188 R = HC(NHC(=O)CCl₃)–CH=CH₂

189

The amino-tetrahydroxycyclohexane ring of the biologically active alkaloid 7-deoxypancratistatin (**190**) was formed by radical cyclization of an oxime elaborated from D-gulonolactone,[107] and the functionalized cycloheptane ring of compound **191** was constructed by intramolecular nitrone-olefin cycloaddition of a 7-alkenyl-*N*-benzyl-nitrone derived from 2,3:5,6-di-*O*-isopropylidene-D-mannofuranose.[108] By use of four enzymes, D-glucose was

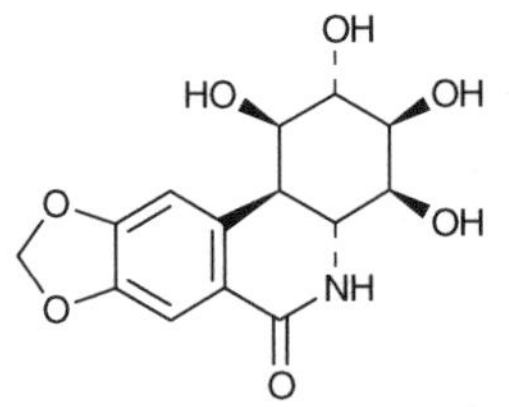

190

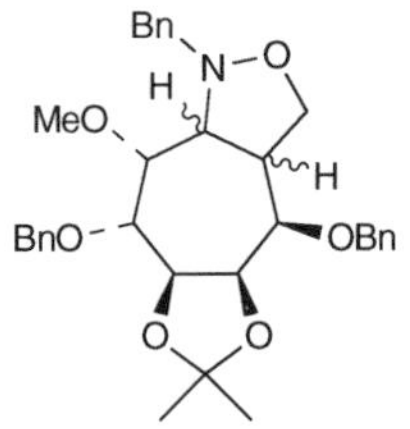

191

transformed into *myo*-2-inosose, which aromatized on exposure to dilute sulfuric acid.[109]

The tricyclic, allylic hydroxy-acetate **192** served as a rigid equivalent of chiral *cis*-1,4-dihydroxycyclohexa-2,5-diene, allowing the stereoselective introduction of hydroxyl groups and hence the systematic synthesis of the six epimeric conduritols A–F.[110] The known product **193** of microbial oxidation, followed by acetonation, of bromobenzene was elaborated to aminofluoroinositol **194** and its enantiomer by way of mono-epoxides,[111] and a similar starting material has been employed in the synthesis of (−)-*gala*-quercitol.[112]

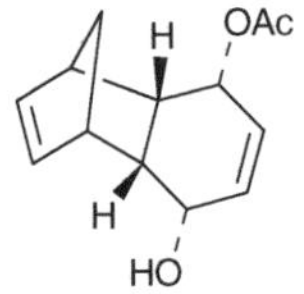

192

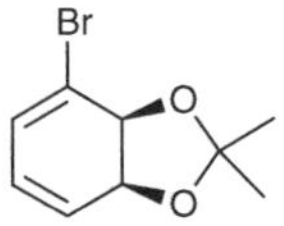

193

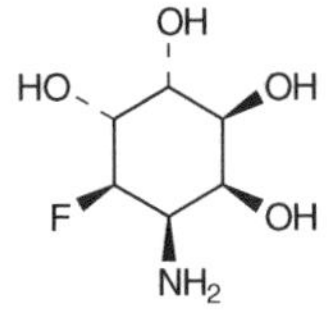

194

myo-Inositol has been converted, in standard multi-step reaction sequences, to three mono-deoxy-inositols,[113] three fluorinated deoxy-inositols,[114] and by way of conduritols to a range of isomeric inositols.[115] The known functionalized cyclohexanones **195** have been transformed into the corresponding vinylic sulfides **196** and sulfoxides **197** as potential α-glycosidase inhibitors.[116] The allylic dithiocarbonate **198** and allylic trichloroacetimidate **199** rearranged

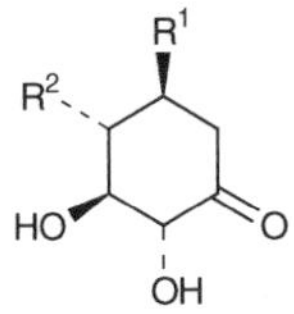

195 $R^1 = R^2 = H$, or
$R^1 = H$, $R^2 = OH$, or
$R^1 = CH_2OH$, $R^2 = H$

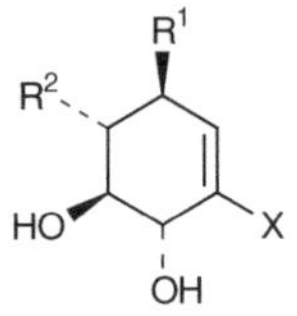

196 X = SPh
197 X = S(O)Ph

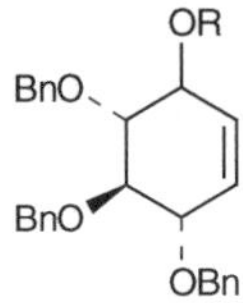

198 R = C(S)SMe
199 R = $C(NH)CCl_3$

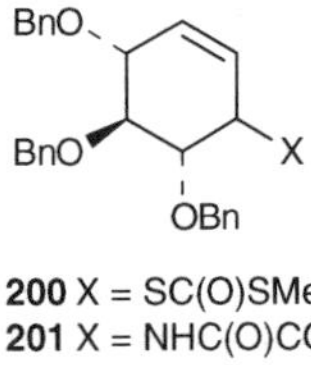

200 X = SC(O)SMe
201 X = $NHC(O)CCl_3$

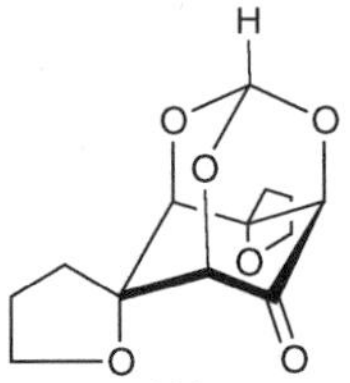

202

to the 'condurithiol' derivative **200** and trichloroacetamide **201**, respectively, on heating in refluxing xylene.[117] Several allylic trichloroacetimides obtained in a similar way were used to study the *syn/anti* selectivity of their dihydroxylations.[118] The regioselective behaviour of the three contiguous free hydroxyl groups of 3,4,5-tri-*O*-benzyl-*myo*-inositol in stannylene-activated alkylations and acylations has been examined,[119] and the epimerization of a cyclohexanepentol derivative has been achieved by the method of Miethchen (see Vol. 32, Chapter 6, ref. 8; Vol. 30, p. 99, refs. 17–19).[120]

The stereochemical course of the hydride reduction and of 1,2-additions to the rigid and sterically hindered inosose derivative **202** have been investigated.[121]

2.3 Carbasugars. – Intramolecular aldol condensation of aldehydo-amide **203**, furnishing precursor **204** of 1-amino-2-deoxy-5a-carba-β-D-gulopyranose, exemplifies a new and potentially versatile synthetic route to carbasugars.[122] Another novel entry to 5a-carbasugars was based on the 6-*exo*-dig radical cyclization of phenylacetylides derived from acetaL-protected free hexoses. Compound **205**, for example, ring-closed to give **206**, which was readily converted into 5a-carba-β-D-mannopyranose.[123]

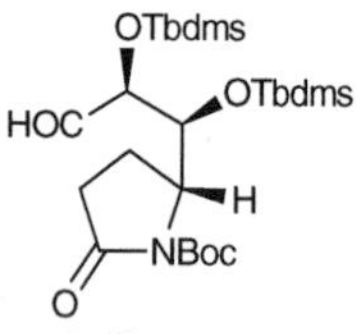

203 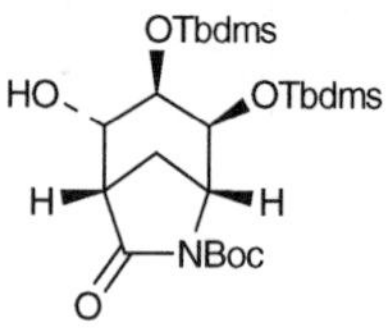 **204** 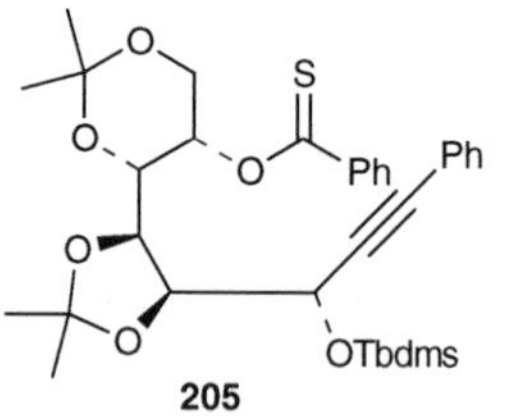 **205**

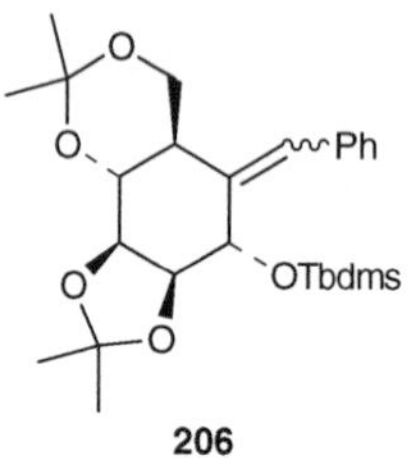

206 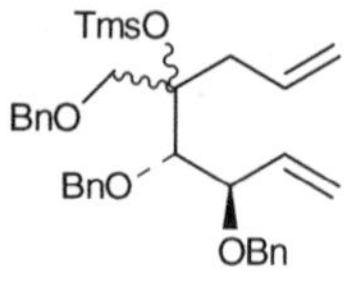 **207** 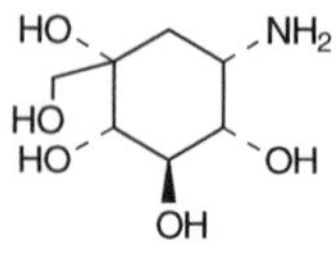 **208**

Ring-closing metathesis/oxyamination of sugar-derived dienes has been made use of in short syntheses of valiolamine (**207**→**208**)[124] and valienamine (**209**→**210**). The latter transformation involved the [3,3]-sigmatropic rearrangement of a cyanate generated *in situ*, as shown in Scheme 10.[125]

Another synthesis of valienamine and its 2-epimer started from quinic acid,[126] and ring-expansion of 5-amino-2 ,3,4-trihydroxycyclopentanone derivatives with diazomethane afforded the α-*allo*- and α-*galacto*-analogues of valiolamine.[127]

[1-^{13}C]-Valienone and a related ^{13}C-labelled cyclitol have been synthesized from [1-^{13}C]-D-glucose by the method of Fukase and Horii (see Vol. 26, p. 201, refs. 111, 112) for biosynthetic studies on acarbose. The same method has been

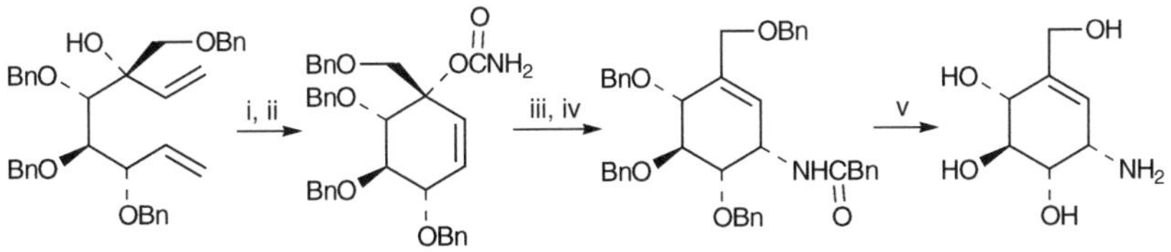

Reagents: i, Grubbs' catalyst; ii, $Cl_3CC(O)NCO$, then K_2CO_3, aq. MeOH; iii, Ph_3P, Et_3N, CBr_4; iv, BnOH; v, Na, NH_3, THF

Scheme 10

used to prepare a series of C-6-deuterated cyclitols related to 2-*epi*-valienamine from D-mannose by reductively dehalogenating the intermediate **211** with Bu_3SnD,[128] and a [7-^{3}H]-label was introduced by reduction of valienamine 6-carbaldehyde with $NaBT_4$.[129]

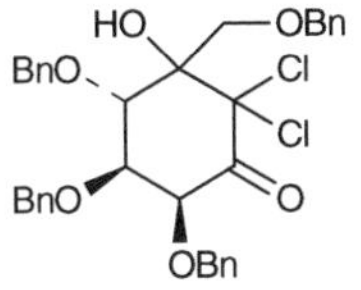

The rigidity of unsaturated di- and tri-cyclic systems, such as the Diels Alder adduct **212** of a substituted 2-pyrone and vinylene carbonate or dicyclopentadienone, has been exploited for the stereoselective introduction of hydroxyl groups, giving access to (±)-2-*epi*-validamine,[130] the rancinamycin II derivative **213** and hence 5a-carba-α-D-talopyranose,[131] cyclitols **214** with three contiguous, oxidized, one-carbon side-chains,[132] novel annulated carbasugars with hydrindane,[133] decaline,[134] and diquinane[134] frameworks, such as compound **215**, and five polyoxygenated cyclohexenylmethanol derivatives such as tonkinenin A (**216**), whose structure was thereupon revised.[135]

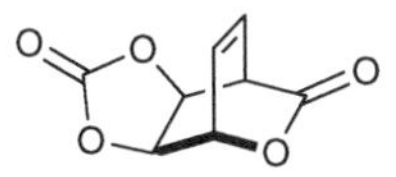

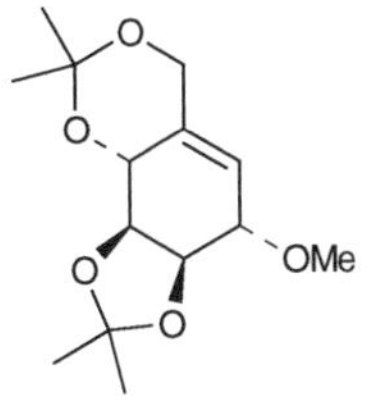

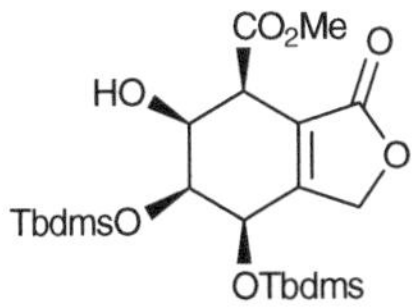

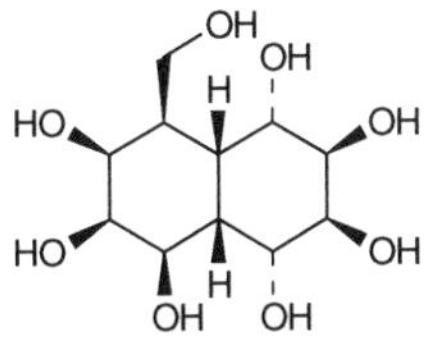

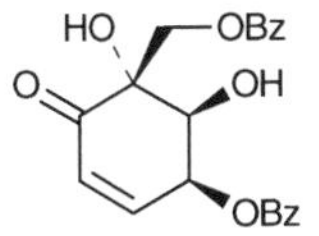

A *de novo* asymmetric synthesis of the carbaglucose epoxide (+)-cyclophellitol was based on hydroxyl-directed epoxidation of homoallylic alcohol **218**, available by [2,3]-sigmatropic rearrangement of the *C*-lithiated conduritol B derivative **217**, which in turn was obtained in a lengthy reaction sequence from benzoquinone.[136]

A Diels Alder approach was also used in the synthesis of dicarboxycyclo-

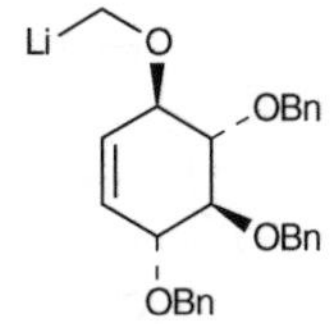

217 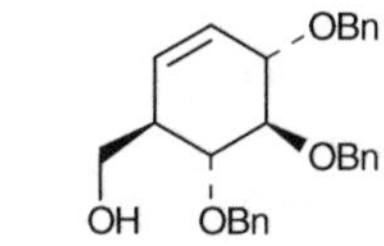 218 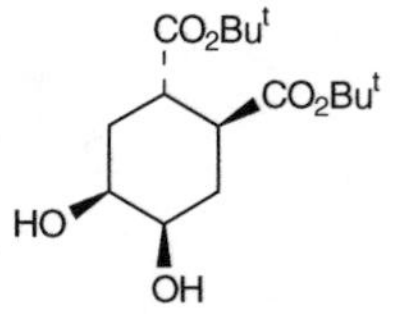 219

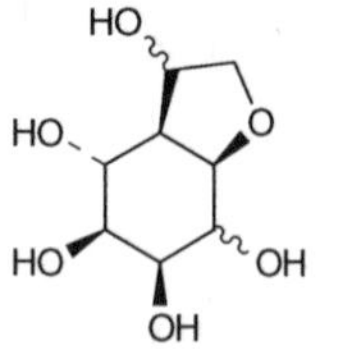

220 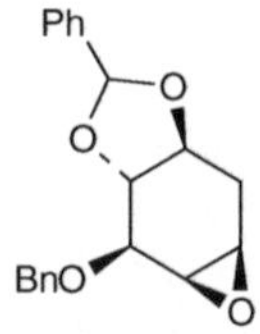 221 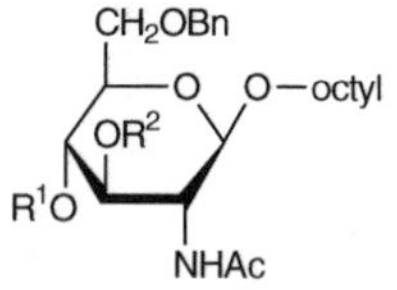 222 $R^1 = H, R^2 = Bn$
223 $R^1 = Bn, R^2 = H$

hexane diol **219** required for incorporation into a pseudotetrasaccharide that mimicks ganglioside GM1 (see Chapter 4).[137] Conformationally restricted, bicyclic analogues **220** of 5a-carba-α- and -β-D-mannopyranose have been synthesized in many steps from a known 1,2-anhydro-5a-carba-β-D-mannopyranose derivative as key components of modified trisaccharides required for enzyme studies.[138] 5′a-Carba-disaccharides linked (1→4) and (1→6)-have been obtained from the corresponding 5′,6′-unsaturated disaccharides by Ti(O-Pr^i)Cl_3-promoted reductive cyclization.[139] The interglycosidic ether linkages in octyl 5′a-carba-β-lactosaminide and the corresponding isolactosaminide were formed by opening of the epoxide ring of 1,2-anhydro-5a-carba-β-D-mannopyranose derivative **221** with the oxyanions generated from octyl glycosides **222** and **223**, respectively.[140] The displacement of a triflate group at C-4″ of a trisaccharide by tetra-*O*-benzylvalienamine in an attempted synthesis of methyl ascarboside proceeded with only 4% yield.[141]

2.4 Quinic and Shikimic Acid Derivatives. – [7-^{12}C]Ethyl shikimate was obtained by cyclization of the [7-^{12}C] labelled phosphonate **224** on treatment with Na–EtOH, then hydrolysis. It was further converted into [7-^{12}C]chorismic acid.[142] Methyl shikimate, its 3-epimer and 3-amino-3-deoxy analogue have been prepared from quinic acid *via* the butane-2,3-bisacetaL-protected oxidation product **225**, which was readily dehydrated.[143] 2,3-Anhydroquinic acid, 3-

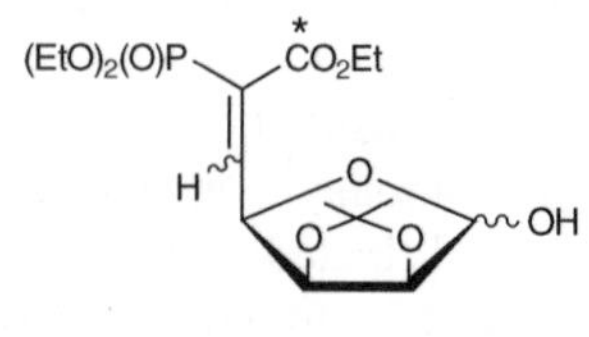

224 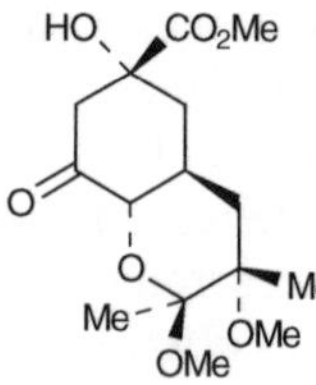 225 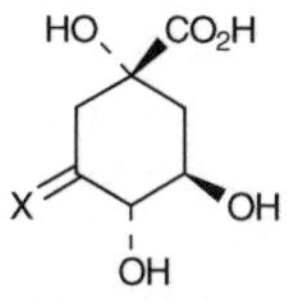 226 X = NOH
227 X = CH_2

deoxyquinic acid, oxime **226** and *exo*-methylene analogue **227**, synthesized following in the main previously reported protocols, are the first inhibitors of type II dehydroquinase.[144]

2.5 Inositol Phosphates. – *2.5.1 Monophosphates.* *Chiro*-inositol-containing phospholipids, in addition to the normal, *myo*-inositol-containing ones, have been found in soyabean seedlings.[145] *myo*-Inositol 1-thiophosphate with a ^{17}O label in the thiophosphate group has been made to investigate the mechanism of hydrolysis by bovine *myo*-inositol monophosphatase.[146]

A series of 6-deoxyphosphatidylinositol analogues **228** have been prepared conventionally, the cyclitol moieties being available from a D-galactose-derived 2-deoxyinosose.[147] A *myo*-inositol 1,2-cyclic phosphate glycosylated at O-6 with α-D-Man-(1→4)-α-D-GlcNH$_2$[148] and the O-6-glycosylated inositol phosphate derivatives **229**[149] and **230**[150] have also been synthesized by standard methods as subunits of GPI membrane anchors. The pseudotrisaccharide **230** has been labelled by attaching a fluorescent or a radioactive tag at the primary position of the glucosamine moiety.[151]

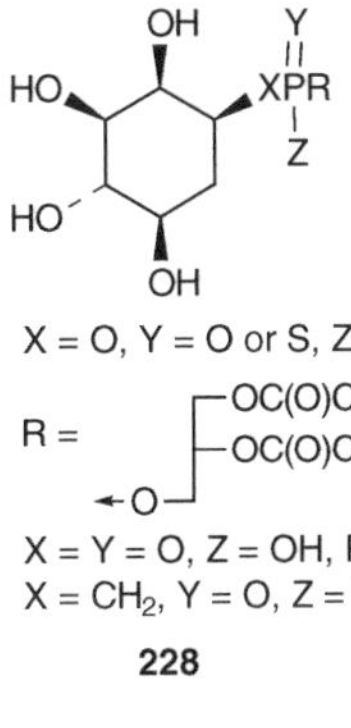

X = O, Y = O or S, Z = OH
R = glyceryl 1,2-di-OC(O)C$_{15}$H$_{31}$, or
X = Y = O, Z = OH, R = Bu, or
X = CH$_2$, Y = O, Z = R = OEt

228

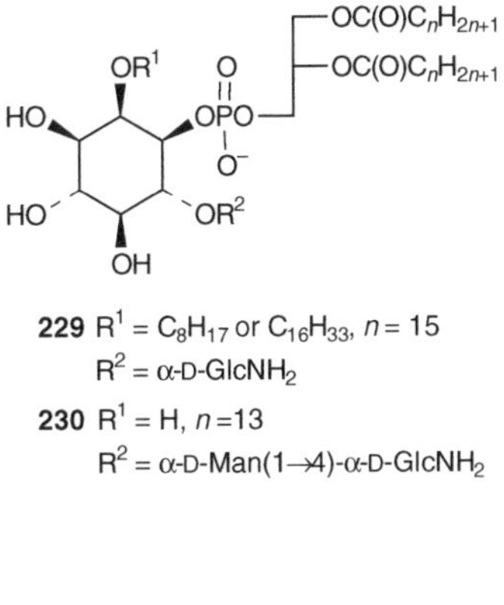

229 R^1 = C$_8$H$_{17}$ or C$_{16}$H$_{33}$, *n* = 15
R^2 = α-D-GlcNH$_2$

230 R^1 = H, *n* = 13
R^2 = α-D-Man(1→4)-α-D-GlcNH$_2$

2.5.2 Phosphates with more than one ester substituent. 1-*O*-Benzoyl-2,3,4,5-tetra-*O*-benzyl-*scyllo*-inositol, obtained from 1,4,5,6-tetra-*O*-benzyl-*myo*-inositol by Mitsunobu inversion, was transformed into all twelve *scyllo*-inositol phosphates by lengthy protecting group manipulations.[152] Standard protecting group manipulations and an oxidation-reduction sequence have been employed to prepare several mono-, bis-, and tris-phosphates of 6-deoxy-D-*myo*- and 3-deoxy-L-*chiro*-inositol,[153] as well as 6-deoxy-D-*myo*-inositol 1,3,4,5-tetrakisphosphate,[154] all from a deoxyinosose available by Ferrier carbocyclization of a D-galactose derivative.

A benzoquinone-derived, resolved conduritol B derivative (see ref. 136 above) was dihydroxylated, then phosphorylated to furnish D-*myo*-inositol 1,4,5-trisphosphate.[155] A different route, also starting from benzoquinone, proceeded by way of resolved cyclohexene bisepoxides, conduritol bisphosphates and inositol tetrakisphosphates to D-*myo*-inositol 3,4,5-trisphosphate and its optical antipode D-*myo*-inositol 1,5,6-trisphosphate. The conversion from the tetrakis- to the tris-phosphates was carried out enzymically.[156]

A number of 6-modified *myo*-inositol 1,4,5-trisphosphates (6-*epi*-, 6-deoxy-, 6-amino-6-deoxy- and 6-deoxy-6-fluoro-) have been obtained from the parent trisphosphate by conventional procedures.[157] Synthesis of the bicyclic inositol 1,4,5-trisphosphate analogues **232**, required for exploring the structural basis of the biological activity of adenophostins involved condensation of the resolved *myo*-inositol butane-2,3-diacetal **231** with 3-chloro-2-chloromethyl-1-propene to form a seven-membered ring.[158]

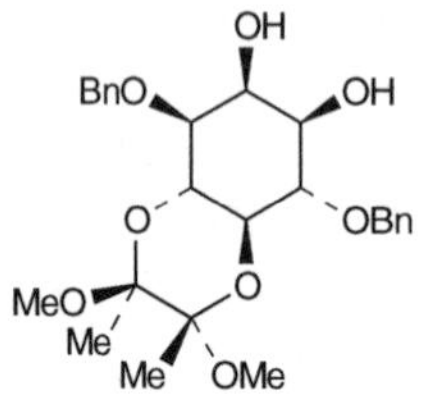

OH BnO OH O OBn MeO O Me Me OMe
231

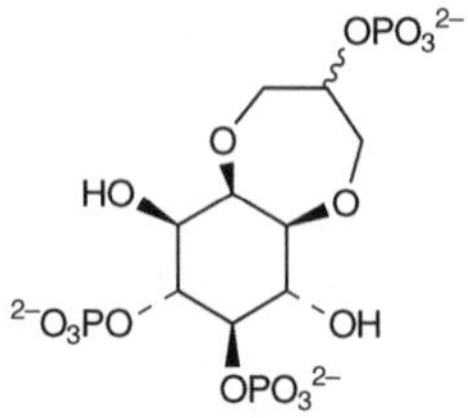

OPO_3^{2-} O HO O $^{2-}O_3PO$ OH OPO_3^{2-}
232

Both enantiomers of *myo*-inositol 1,3,4,5-tetrakisphosphate were prepared from racemic 2,6-di-*O*-benzyl-*myo*-inositol by enzyme mediated resolution, then chemical phosphorylation and deprotection.[159] D- and L-*chiro*-inositol 1,3,4,6-tetrakisphosphate have been synthesized from D-pinitol and L-quebrachitol, respectively *via* selectively dibenzylated (Bu_sSnO–BnBr) intermediates.[160] D,L-*myo*-Inositol 1,3,4,5-tetrakisphosphate with a 2-aminoethyl-1-phospho tether at the 6-position is a potential precursor of affinity probes for purification and structural studies of the receptor proteins.[161]

The conformational behaviour of *myo*-inositol hexakisphosphate is referred to in Chapter 21.

Phosphodiesters with the phosphate linking *myo*-inositol 1,4-bisphosphate and *myo*-inositol 1,4,5-trisphosphate, respectively, through the O-1 phosphate to fluorescein have been made, the former being required for continuous assay of a phospholipase,[162] the latter for investigating binding to the IP_3 receptor.[163]

In a new route to L-α-phosphatidyl-D-*myo*-inositol 4,5-bisphosphate and a short-chain glyceryl lipid analogue thereof, the cyclitol moiety was derived from dehydroshikimic acid, thus avoiding the need for resolution.[164] The 4,5-bisphosphates of some of compounds **228**, whose cyclitol moiety was obtained from D-galactose, have also been made.[147] Phosphatidylinositol-3-phosphate, -3,4-bisphosphate, -3,5-bisphosphate and -3,4,5-trisphosphate have been elaborated from *myo*-inositol orthoformate intermediates using camphor acetals for protection–resolution.[165] Two phosphatidyl-*myo*-inositol 3,4,5-triphosphates containing unsaturated lipid moieties (arachidonoyl, linolenoyl) have been synthesized from *myo*-inositol. Except for the novel use of 9-fluorenylmethyl as phosphate protecting group, the methods used were previously reported.[166]

References

1 *Carbohydrate Mimics*, ed. Y. Chapleur, Wiley, Weinheim, 1998.
1a J. Kitajiama, T. Ishikawa, Y. Tanaka and Y. Ida, *Chem. Pharm. Bull.*, 1999, **47**, 988.
2 H.S. Singh, G.R. Verma, A. Gupta and A. Mittal, *J. Indian Chem. Soc.*, 1999, **76**, 392 (*Chem. Abstr.*, 1999, **131**, 286 722).
3 P. Hadwiger, P. Mayr, A. Tauss, A.E. Stütz and B. Nidetzky, *Bioorg. Med. Chem. Lett.*, 1999, **9**, 1683.
4 E. Hermoza Guerra and M. Del Rosario Sun Kou, *Bol. Soc. Quim. Peru,* 1998, **64**, 161 (*Chem. Abstr.*, 1999, **130**, 125 275).
5 H. Li, H. Li, W. Wang and J.-F. Deng, *Chem. Lett.*, 1999, 629.
6 D. Arigoni, J.-L. Giner, S. Sagner, J. Wungsintaweekul, M.H. Zenk, K. Kis, A. Bacher and W. Eisenreich, *Chem. Commun.,* 1999, 1127.
7 L. Charon, J.-F. Hoeffler, C. Pale-Grosdemange, and M. Rohmer, *Tetrahedron Lett.,* 1999, **40**, 8369.
8 A. Martín, M.S. Rodríguez and E. Suárez, *Tetrahedron Lett.,* 1999, **40**, 7525.
9 L. Hyldtoft, C.S. Poulsen and R. Madsen, *Chem. Commun.*, 1999, 2101.
10 P. Bault, S. Bachir, L. Spychala, P. Godé, G. Goethals, P. Martin, G. Ronco and P. Villa, *Carbohydr. Polym.*, 1998, **37**, 299 (*Chem. Abstr.*, 1999, **130**, 153 902).
11 C.K. Lee and E.J. Kim, *Carbohydr. Res.*, 1999, **320**, 223.
12 G.J. Ewing and M.J. Robins, *Org. Lett.*, 1999, **1**, 635 (*Chem. Abstr.*, 1999, **131**, 228 895).
13 X. Zheng and V. Nair, *Nuclosides Nucleotides,* 1999, **18**, 1961.
14 M. Yuasa, S. Kanazawa, N. Nishimura, T. Higuchi and I. Maeba, *Carbohydr. Res.*, 1999, **315**, 98.
15 F.L. van Delft, A.R.P.M. Valentijn, G.A. van der Marel and J.H.van Boom, *J. Carbohydr. Chem.*, 1999, **18**, 165.
16 F.L. van Delft, A.R.P.M. Valentijn, G.A. van der Marel and J.H.van Boom, *J. Carbohydr. Chem.*, 1999, **18**, 191.
17 A.J. Norris and T. Toyokuni, *J. Carbohydr. Chem.*, 1999, **18**, 1097.
18 M. Kamada, T. Satoh, T. Kakuchi and K. Yokota, *Tetrahedron: Asymmetry*, 1999, **10**, 3667.
19 L.-S. Li, Y. Wu and Y.-L. Wu, *J. Carbohydr. Chem.*, 1999, **18**, 1067.
20 S. Claustre, F. Bringaud, L. Azéma, R. Baron, J. Périé and M. Willson, *Carbohydr. Res.*, 1999, **315**, 339.
21 A.A.H. Abdel-Rahman, E.S.H. El Ashry and R.R. Schmidt, *Carbohydr. Res.*, 1999, **315**, 106.
22 H.-S. Dang and B.P. Roberts, *Tetrahedron Lett.,* 1999, **40**, 4271.
22a J.C.Y. Wong, P. Lacombe and C.F. Sturino, *Tetrahedron Lett.,* 1999, **40**, 8751.
23 S. Mukhopadhyay, G.K. Gandi and S.B. Chandalia, *Indian J. Chem. Technol.*, 1999, **6**, 107 (*Chem. Abstr.*, 1999, **131**, 88 106).
24 S. Du, H. Faiger, V. Belakhov and T. Baasov, *Bioorg. Med. Chem.*, 1999, **7**, 2671.
25 L. Gauzy, Y. Le Merrer, J.-C. Depezay, D. Damour-Barbalat and S. Mignani, *Tetrahedron Lett.,* 1999, **40**, 3705.
26 M. Cushman, J.T. Mihalic, K. Kis and A. Bacher, *Bioorg. Med. Chem. Lett.*, 1999, **9**, 39.
27 M. Cushman, J.T. Mihalic, K. Kis and A. Bacher, *J. Org. Chem.*, 1999, **64**, 3838.
28 M. Cushman, F. Mavandadi, D. Yang, K. Kugelbrey, K. Kis and A. Bacher, *J. Org. Chem.*, 1999, **64**, 4635.

29 A.M. Costa, M. Faja and J. Vilarossa, *Tetrahedron,* 1999, **55**, 6635.
30 R. Uchida, A. Nasu, S. Tokutake, K. Kasai, K. Tobe and N. Yamaji, *Chem. Pharm. Bull.*, 1999, **47**, 187.
31 O.R. Martin, P. Compain, H. Kizu and N. Asano, *Bioorg. Med. Chem. Lett.*, 1999, **9**, 3171.
32 A. Kato, I. Adachi, M. Miyauchi, K. Ikeda, T. Komae, H. Kizu, Y. Kameda, A.A. Watson, R.J. Nash, M.R. Wormald, G.W.J. Fleet and N. Asano, *Carbohydr. Res.*, 1999, **316**, 95.
33 N. Shibano, S. Nakamura, N. Motoya and G. Kusano, *Chem. Pharm. Bull.*, 1999, **47**, 472.
34 M. Takebayashi, S. Hiranuma, Y. Kanie, T. Kajimoto, O. Kanie and C.-H. Wong, *J. Org. Chem.*, 1999, **64**, 5280.
35 A. Peer and A. Vasella, *Helv. Chim. Acta,* 1999, **82**, 1044.
36 M.I. García-Moreno, C. Ortiz Mellet and J.M. García Fernández, *Tetrahedron: Asymmetry*, 1999, **10**, 4271.
37 V.N. Desai, N.N. Saha and D.D. Dhavale, *Chem. Commun.*, 1999, 1719.
38 S. Swaleh and J. Liebscher, *Tetrahedron Lett.,* 1999, **40**, 2099.
39 D. Ma and H. Sun, *Tetrahedron Lett.,* 1999, **40**, 3609.
40 P. Bach, A. Lohse and M. Bols, *Tetrahedron Lett.,* 1999, **40**, 367.
41 D.F. Oliveira, E.A. Severino and C.R.D. Correia, *Tetrahedron Lett.,* 1999, **40**, 2083.
42 A. Dondoni and D. Perrone, *Tetrahedron Lett.*, 1999, **40**, 9375.
43 D.D. Long, R.J.E. Stetz, R.J. Nash, D.J. Marquess, J.D. Lloyd, A.L. Winters, N. Asano and G.W.J. Fleet, *J. Chem. Soc.*, *Perkin Trans. 1*, 1999, 901.
44 R.E. Lee, M.D. Smith, L. Pickering and G.W.J. Fleet, *Tetrahedron Lett.,* 1999, **40**, 8689.
45 K. Asano, T. Hakogi, S. Iwama and S. Katsumura, *Chem. Commun.,* 1999, 41.
46 M. Ruiz, T.M. Ruanova, V. Ojea and J.M. Quintela, *Tetrahedron Lett.,* 1999, **40**, 2021.
47 J.P. Shilvock, J.R. Wheatley, R.J. Nash, A.A. Watson, R.C. Griffiths, T.D. Butters, M. Müller, D.J. Watkin, D.A. Winkler and G.W.J. Fleet, *J. Chem. Soc.*, *Perkin Trans. 1*, 1999, 2735.
48 J. Fuentes, D. Olano and M.A. Pradera, *Tetrahedron Lett.,* 1999, **40**, 4063.
49 H.-S. Moon and D. Koh, *Kong-op Hwahak*, 1998, **9**, 914 (*Chem. Abstr.*, 1999, **130**, 139 544).
50 L. Gauzy, Y. Le Merrer, J.-C. Depezay, F. Clerc and S. Mignani, *Tetrahedron Lett.,* 1999, **40**, 6005.
51 M.H. Fechter, A.E. Stütz and A. Tauss, *Curr. Org. Chem.*, 1999, **3**, 269 (*Chem. Abstr.*, 1999, **131**, 73 845).
52 S. Moutel and M. Shipman, *J. Chem. Soc.*, *Perkin Trans. 1*, 1999, 1403.
53 C.R.R. Matos, R.S.C. Lopes and C.C. Lopes, *Synthesis*, 1999, 571.
54 C. Uriel and F. Santoyo-González, *Synlett*, 1999, 593.
55 R. Polt, D. Sames and J. Chruma, *J. Org. Chem.*, 1999, **64**, 6147.
56 M. Ruiz, V. Ojea and J.M. Quintela, *Synlett*, 1999, 204.
57 M. Schuster, *Bioorg. Med. Chem. Lett.*, 1999, **9**, 615.
58 M.A. Leeuwenburgh, S. Picasso, H.S. Overkleeft, G.A. van der Marel, P. Vogel and J.H. van Boom, *Eur. J. Org. Chem.*, 1999, 1185.
59 G. Limberg, I. Lundt and J. Zavilla, *Synthesis*, 1999, 178.
60 A.I. Meyers, C.J. Andres, J.E. Reseke and C.C. Woodall, *Tetrahedron,* 1999, **55**, 8931.

61 Y.T. Kim, A. Takasuki, N. Kogoshi and T. Kitahara, *Tetrahedron,* 1999, **55**, 8353.
62 R.J.B.H.N. van der Berg, D. Noort, E.S. Milder-Enacache, G.A. van der Marel, J.H. van Boom and H.P. Benschop, *Eur. J. Org. Chem.*, 1999, 2593.
63 S.V. Hansen and M. Bols, *J. Chem. Soc., Perkin Trans. 1*, 1999, 3323.
64 T. Faitg, J. Soulié, J.-Y. Lalemand and L. Ricard, *Tetrahedron: Asymmetry*, 1999, **10**, 2165.
65 F. Cardona, S. Valenza, A. Goti and A. Brandi, *Eur. J. Org. Chem.*, 1999, 1319.
66 C. Herdeis and T. Schiffer, *Tetrahedron,* 1999, **55**, 1043.
67 N. Jotterand and P. Vogel, *J. Org. Chem.*, 1999, **64**, 8973.
68 H. Dollt and V. Zabel, *Aust. J. Chem.*, 1999, **52**, 259.
69 W.-S. Zhou, Z.-H. Lu, Y.-M. Xu and L.-X. Liao, *Tetrahedron*, **55**, 1999, 11959.
70 L.-X. Liao, Z.-M. Wang, H.-X. Zhang and W.-S. Zhou, *Tetrahedron: Asymmetry*, 1999, **10**, 3649.
71 R.H. Furneaux, V.L. Schramm and P.C. Tyler, *Bioorg. Med. Chem.*, 1999, **7**, 2599.
72 R.C. Reynolds, N. Bansal, J. Rose, J. Friederich, W.J. Suling and J.A. Maddry, *Carbohydr. Res.*, 1999, **317**, 164.
73 J.V. Hines, H. Chang, M.S. Gerdeman and D.E. Warn, *Bioorg. Med. Chem. Lett.*, 1999, **9**, 1255.
74 J. Kang, C.W. Lee, G.J. Lim and B.T. Cho, *Tetrahedron: Asymmetry*, 1999, **10**, 657.
75 Y. Nishimura, E. Shitara and T. Takeuchi, *Tetrahedron Lett.,* 1999, **40**, 2351.
76 E. Shitara, Y. Nishimura, F. Kojima and T. Takeuchi, *Bioorg. Med. Chem.*, 1999, **7**, 1241.
77 X. Liang and M. Bols, *J. Org. Chem.*, 1999, **64**, 8485.
78 A. Lohse, K.B. Jensen and M. Bols, *Tetrahedron Lett.*, 1999, **40**, 3033.
79 A. Lohse, K.B. Jensen, K. Lundgren and M. Bols, *Bioorg. Med. Chem. Lett.*, 1999, **9**, 1965.
80 Y.J. Kim, M. Ichikawa and Y. Ichikawa, *J. Am. Chem. Soc.*, 1999, **121**, 5829.
81 K.-E. Jung, K. Kim, M. Yang, K. Lee and H. Lik, *Bioorg. Med. Chem. Lett.*, 1999, **9**, 3407.
82 K. H. Dötz, M. Klumpe and M. Nieger, *Chem. Eur. J.,* 1999, **5**, 691.
83 Y. Kobayashi, *Carbohydr. Res.*, 1999, **315**, 3.
84 O. Sellier, P.V. de Weghe and J. Eustache, *Tetrahedron Lett.*, 1999, **40**, 5859.
85 M. Seepersaud, R. Bucala and Y. Al-Abed, *Z. Naturforsch., B: Chem. Sci.*, 1999, **54**, 565 (*Chem. Abstr.*, 1999, **131**, 19 200).
86 R. Czuk, P. Dörr, M. Kühn, C. Krieger and M.Y. Antipin, *Z. Naturforsch., B: Chem. Sci.*, 1999, **54**, 1068 (*Chem. Abstr.*, 1999, **131**, 310 795).
87 M. Matsugi, K. Gotanda, C. Ohiro, M. Suemura, A. Sano and Y. Kita, *J. Org. Chem.*, 1999, **64**, 6928.
88 S. Takahashi, H. Terayama, H. Koshino and H. Kuzuhara, *Tetrahedron,* 1999, **55**, 14871.
89 S. Bobo, I.S. de Gracia and J.L. Chiara, *Synlett*, 1999, 1551.
90 J.K. Gallos, A.E. Koumbis, V.P. Xiraphaki, C.C. Dellios and E. Coutouli-Argyropoulou, *Tetrahedron,* 1999, **55**, 15167.
91 P.J. Dransfield, S. Moutel, M. Shipman and V. Sik, *J. Chem. Soc., Perkin Trans. 1*, 1999, 3349.
92 S.K. Johansen and I. Lundt, *J. Chem. Soc., Perkin Trans. 1*, 1999, 3615.
93 M. Domínguez and P.M. Cullis, *Tetrahedron Lett.,* 1999, **40**, 5783.

94 C.M. Gonzalez-Alvarez, L. Quintero, F. Santiesteba and J.-L. Fourrey, *Eur. J. Org. Chem.*, 1999, 3085.
95 G. Mehta and N. Mohal, *Tetrahedron Lett.,* 1999, **40**, 5791.
96 G. Mehta and N. Mohal, *Tetrahedron Lett.,* 1999, **40**, 5795.
97 N.G. Ramesh, A.J.H. Klunder and B. Zwanenburg, *J. Org. Chem.*, 1999, **64**, 3635.
98 S. Ogawa and T. Morikawa, *Bioorg. Med. Chem. Lett.*, 1999, **9**, 1499.
99 A. Blaser and J.-L. Reymond, *Helv. Chim. Acta,* 1999, **82**, 760.
100 D.J. Kassab and B. Ganem, *J. Org. Chem.*, 1999, **64**, 1782.
101 K. Ichimura, K. Kohata, Y. Mukosa, Y. Yamaguchi, R. Goto and K. Suto, *Biosci. Biotech. Biochem.*, 1999, **63**, 189.
102 C.K. Smith, C.M. Hewage, S.C. Fry and I.H. Sadler, *Phytochemistry*, 1999, **52**, 387.
103 T. Hudlicky, D.A. Frey, L. Karoniak, C.D. Claeboe and L.E. Brammer, Jr., *Green Chem.*, 1999, **1** 57 (*Chem. Abstr.*, 1999, **131**, 59 057).
104 J.K. Gallos, T.V. Koftis, V.C. Sarli and K.E. Litinas, *J. Chem. Soc., Perkin Trans. 1*, 1999, 3075.
105 A. Kornienko and M. d'Alarcao, *Tetrahedron: Asymmetry*, 1999, **10**, 827.
106 H. Ovaa, J.D.C. Codee, B. Lastdrager, H.S. Overkleft, G.A. van der Marel and J.H. van Boom, *Tetrahedron Lett.,* 1999, **40**, 5063.
107 G.E. Keck, S.F. McHardy and J.A. Murry, *J. Org. Chem.*, 1999, **64**, 4465.
108 J. Marco-Contelles and E. de Opazo, *Tetrahedron Lett.,* 1999, **40**, 4445.
109 C.A. Hansen, A.B. Dean, K.M. Draths and J.W. Frost, *J. Am. Chem. Soc.*, 1999, **121**, 3799.
110 M. Honzumi, K. Hiroya, T. Taniguchi and K. Ogasawara, *Chem. Commun.,* 1999, 1985.
111 K.A. Oppong, T. Hudlicky, F. Yan, C. York and B.V. Nguyen, *Tetrahedron,* 1999, **55**, 2875.
112 N. Maezaki, N. Nagahashi, R. Yoshigami, C. Iwata and T. Tanaka, *Tetrahedron Lett.,* 1999, **40**, 3781.
113 S. Ogawa, S. Uetsuki, Y. Tezuka, T. Morikawa, A. Takahashi and K. Sato, *Bioorg. Med. Chem. Lett.*, 1999, **9**, 1493.
114 A.D. da Silva, A. Antonio, A. Benicio and S.D. Gero, *Tetrahedron Lett.,* 1999, **40**, 6531.
115 S.-K. Chung and Y.-Y. Kwon, *Bioorg. Med. Chem. Lett.*, 1999, **9**, 2135.
116 A. Lubineau and I. Billault, *Carbohydr. Res.*, 1999, **320**, 49.
117 M.J. McDonough, R.V. Stick and D.M.G. Tilbrook, *Aust. J. Chem.*, 1999, **52**, 143.
118 T.J. Donohue, K. Blades, M. Helliwell, P.R. Moore, J.J.G. Winter and G. Stemp, *J. Org. Chem.*, 1999, **64**, 2980.
119 G. Anilkumar, Z.J. Jia, R. Kraehmer and B. Fraser-Reid, *Tetrahedron,* 1999, **55**, 3591.
120 M. Frank R. Miethchen and H. Reinke, *Eur. J. Org. Chem.*, 1999, 1259.
121 L.A. Paquette and J. Tae, *Tetrahedron Lett.*, 1999, **40**, 5971.
122 G. Rassu, L. Auzzas, L. Pinna, F. Zanardi, L. Battistini and G. Casiraghi, *Org. Lett.*, 1999, **1**, 1213 (*Chem. Abstr.*, 1999, **131**, 322 845).
123 A.M. Gomez, G.O. Danelón, E. Moreno, S. Valverde and J.C. López, *Chem. Commun.,* 1999, 175.
124 O. Sellier, P. van de Weghe, D. Le Nouen, C. Strehler and J. Eustache, *Tetrahedron Lett.,* 1999, **40**, 853.

125 P. Kapferer, F. Sarabia and A. Vasella, *Helv. Chim. Acta,* 1999, **82**, 645.
126 T.K.M. Shing., T.-Y. Li and S.H.-L. Kok, *J. Org. Chem.*, 1999, **64**, 1941.
127 S. Ogawa, C. Uchida and T. Ohhira, *Carbohydr. Lett.*, 1999, **3**, 277 (*Chem. Abstr.*, 1999, **130**, 296 923).
128 T. Mahmud, I. Tornus, E. Egelkrout, E. Wolf, C. Vy, H.G. Floss and S. Lee, *J. Am. Chem. Soc.*, 1999, **121**, 6973.
129 S. Lee, I. Tornus, H. Dong and S. Grober, *J. Labelled Compd. Radiopharm.*, 1999, **42**, 361 (*Chem. Abstr.*, 1999, **130**, 312 000).
130 K. Afarinkia and F. Mahmood, *Tetrahedron,* 1999, **55**, 3129
131 O. Arjona, F. Iradier, R. Medel and J. Plumet, *Tetrahedron: Asymmetry*, 1999, **10**, 3431.
132 N. Jotterand and P. Vogel, *Tetrahedron Lett.,* 1999, **40**, 5499.
133 G. Mehta and D.S. Reddy, *Tetrahedron Lett.,* 1999, **40**, 9137.
134 G. Mehta, P.S. Reddy, S.S. Ramesh and V. Tatu, *Tetrahedron Lett.,* 1999, **40**, 9141.
135 K. Hiroya and K. Ogasawara, *Chem. Commun.,* 1999, 2197.
136 B.M. Trost and E.J. Hembre, *Tetrahedron Lett.,* 1999, **40**, 219.
137 A. Bernardi, G. Boschin, A. Checchia, M. Lattanzio, L. Manzoni, D. Potenza and C. Scolastico, *Eur. J. Org. Chem.*, 1999, 1311.
138 S. Ogawa, M. Ohno and T. Ohhira, *Heterocycles*, 1999, **50**, 57 (*Chem. Abstr.*, 1999, **130**, 153 901).
139 A.J. Pearce, M. Sallogoub, J.-M. Mollet and P. Sinaÿ, *Eur. J. Org. Chem.*, 1999, 2103.
140 S. Ogawa, N. Matsunaga, H. Li and M. Palcic, *Eur. J. Org. Chem.*, 1999, 631.
141 R.V. Stick, D.M.G. Tilbrook and S.J. Williams, *Aust. J. Chem.*, 1999, **52**, 895.
142 D.J. Gustin and D. Hilvert, *J. Org. Chem.*, 1999, **64**, 4935.
143 C. Alves, M.T. Barros, C.D. Maycock and M.R. Ventura, *Tetrahedron,* 1999, **55**, 8443.
144 M. Frederickson, E.J. Parker, A.R. Hawkins, J.R. Coggins and C. Abel, *J. Org. Chem.*, 1999, **64**, 2612.
145 Y. Hong and Y. Pak, *Phytochemistry*, 1999, **51**, 861.
146 C.M.J. Fauroux, M. Lee, P.M. Cullis, K.T. Douglas, S. Freeman and M.G. Gove, *J. Am. Chem. Soc.*, 1999, **121**, 8385.
147 M. Vieira de Almeida, J. Cleophax, A. Gateau-Olesker, G. Prestat, D. Dubreuil and S.D. Gero, *Tetrahedron*, 1999, **55**, 12997.
148 C.H. Jaworek, P. Calias, S. Iacobussi and M. d'Alarcao, *Tetrahedron Lett.,* 1999, **40**, 667.
149 A. Crossman Jr., J.S. Brimacombe, M.A.J. Ferguson and T.K. Smith, *Carbohydr. Res.*, 1999, **321**, 42.
150 H. Dietrich, J.F. Espinosa, J.L.Chiara, J. Jimenez-Barbero, Y. Leon, J.-M. Mato, F.H. Cano, C. Foces-Foces and M. Martin-Lomas, *Chem. Eur. J.*, 1999, **5**, 320.
151 T.G. Mayer, R. Weingart, F. Münstermann, T. Kawada, T. Kurzchalia and R.R. Schmidt, *Eur. J. Org. Chem.*, 1999, 2563.
152 S.-K. Chung, Y.-U. Kwon, Y.-T. Chang, K.-H. Sohn, J.-H. Shin, K.-H. Park, B.-J. Hong and I.-H. Chung, *Bioorg. Med. Chem.*, 1999, **7**, 2577.
153 M. Vieira de Almeida, D. Dubreuil, J. Cleophax, C. Verre-Sebrié, M. Pipelier, G. Prestat, G. Vass and S.D. Gero, *Tetrahedron*, 1999, **55**, 7251.
154 D. Dubreuil, J. Cleophax, M. Vieira de Almeida, C. Verre-Sebrié, M. Pipelier, G. Vass and S.D. Gero, *Tetrahedron*, 1999, **55**, 7573.

155 B.M. Trost, D.E. Patterson and E.J. Hembre, *J. Am. Chem. Soc.*, 1999, **121**, 10834.

156 S. Adelt, O. Plettenburg, R. Stricker, G. Reiser, H.-J. Altenbach and G. Vogel, *J. Med. Chem.*, 1999, **42**, 1262.

157 S. Ballereau, P. Guedat, S.N. Poirier, G. Guillemette, B. Spiess and G. Schlewer, *J. Med. Chem.*, 1999, **42**, 4824.

158 A.M. Riley and B.V.L. Potter, *Tetrahedron Lett.,* 1999, **40**, 2213.

159 K. Laumen and O. Ghisalbe, *Biosci. Biotech. Biochem.*, 1999, **63**, 1374.

160 C. Liu, R.S. Davis, S.R. Nahorski, S. Ballereau, B. Spiess and B.V.L. Potter, *J. Med. Chem.*, 1999, **42**, 1991.

161 S.-K. Chung, L.M. Zhao and Y Ryan, *Bull. Korean Chem. Soc.*, 1999, **20**, (*Chem. Abstr.*, 1999, **131**, 272 106).

162 A.V. Rukavishnukov, T.O. Zaikova, G.B. Birrell, J.F.W. Keana and O.H. Griffith, *Bioorg. Med. Chem. Lett.*, 1999, **9**, 1133.

163 T. Inoue, K. Kikuchi, K. Hirose, M. Iino and T. Nagano, *Bioorg. Med. Chem. Lett.*, 1999, **9**, 1697.

164 J.R. Falck, U.M. Krishna and J.H. Capdevila, *Tetrahedron Lett.,* 1999, **40**, 8771.

165 G.F. Painter, S.J.A. Grove, I.H. Gilbert, A.B. Holmes, P.R. Raithby, M.L. Hill, P.T. Hawkins and L.R. Stephens, *J. Chem. Soc., Perkin Trans. 1*, 1999, 923.

166 W. Watanabe and M. Nakatomi, *Tetrahedron*, 1999, **55**, 9743.

19
Antibiotics

1 Aminoglycosides and Aminocyclitols

An account of some of Wong's recent work on mimics of complex carbohydrates and their recognition by receptors includes a brief discussion of studies on the binding of aminoglycosides and analogues to RNA.[1] A report from the same laboratory describes the synthesis of the neamine analogue **1** (paromamine) and the three compounds in which the amino-group at C-2′ is successively moved to each of the other positions in the hexopyranose unit. These were made by reaction of appropriate azido-substituted phenylthioglucosides with an acceptor made from 2-deoxystreptamine by desymmetrization, either chemical or enzymic. None of these compounds was as effective as neamine itself in binding to RNA from *E. coli*. Subsequently, a series of neamine derivatives of type **2** were made, where the substituents R contained other amino-functions linked by short chains. The syntheses of **2** involved the degradation of neomycin B to give a pseudodisaccharide in which the 5-hydroxyl group was selectively made available for reaction, and used the device of protection of the amino groups by conversion into azides (see Vol. 30, p. 149). The antibiotic activities of these compounds and other aminoglycoside antibiotics were compared using a new assay.[2] Other workers have

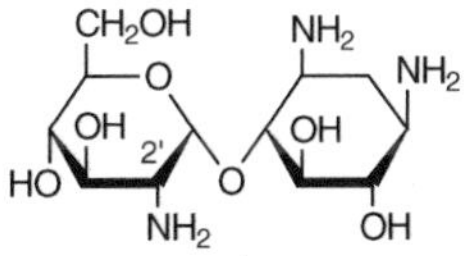

1

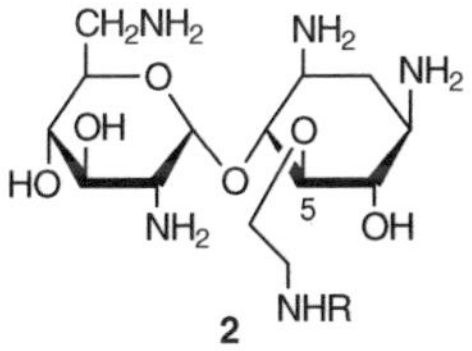

2

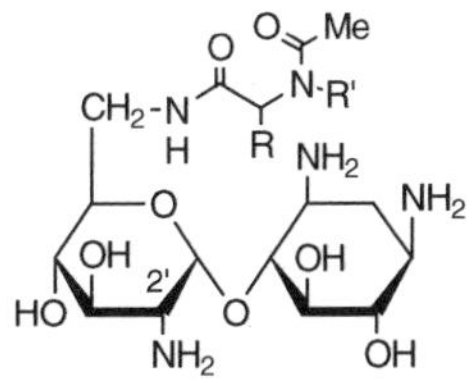

3

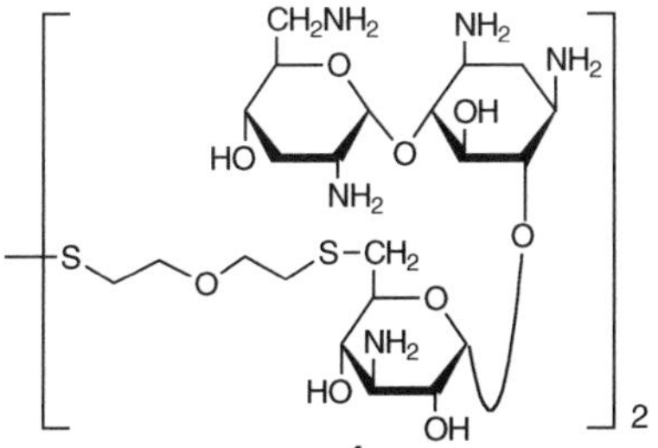

4

Carbohydrate Chemistry, Volume 33

prepared libraries of neamine derivatives modified at C-6′ either by reductive amination to give 6′-*N*-alkyl- or -aryl-derivatives, or by Ugi condensation of a known 6′-isocyanide to give compounds of type **3**.[3]

There has been an expanded account of work in Tor's laboratory on the synthesis and RNA binding of dimeric aminoglycosides, where the two units are connected by a flexible aliphatic chain (see Vol. 31, p. 254). The compounds investigated are illustrated by the tobramycin dimer **4**, and similar dimers of neomycin B and kanamycin A were also made, along with mixed dimers of tobramycin–neomycin and kanamycin–tobramycin. The dimers inhibit the *Tetrahymena* ribozyme between 20- and 1200-fold more effectively than the monomeric compounds, and the inhibition curves of the dimers suggest the presence of at least two high-affinity binding sites within the ribozyme's three-dimensional fold.[4] Neomycin B, kanamycin A, and the neomycin dimer have also been shown to bind to $tRNA^{Phe}$.[5] Paromamycin has been linked, *via* its CH_2NH_2 group and short spacers, to the intercalators thiazole orange and pyrene. The conjugates showed stronger binding to the A-site in a truncated model of prokaryotic 16S rRNA than did paromamycin itself, and the binding was best with short spacers.[6]

The copper complex of kanamycin, at concentrations as low as picomolar levels, has been shown to degrade an RNA aptamer known to have a high affinity for neomycin, but was without effect on random RNA and DNA molecules, thus showing potential as a novel antiviral agent.[7]

One of the major mechanisms for bacterial resistance to aminoglycoside antibiotics involves aminoglycoside 3′-phosphotransferase, which catalyses the phosphorylation of kanamycin A (**5**, R = H) by ATP. In an ingenious approach to circumvent this resistance, the hydrate **5** (R = OH) was prepared from kanamycin, and found in a coupled system of phosphotransferase, pyruvate kinase, kanamycin A, ADP and excess phosphoenol pyruvate (PEP) to lead to consumption of PEP, assumed to be due to phosphorylation of kanamycin at the 3′-β-hydroxyl group to give an unstable intermediate which decomposed to regenerate the ketone/hydrate. The hydrate was effective against an engineered *E. coli* which contained the phosphotransferase.[8] A gene segment from an actinomycete, conferring multiple resistance to aminoglycoside antibiotics with a 6′-amino-group, has been cloned. Use of cell-free extracts of *Streptomyces lividans* carrying the cloned gene showed that it coded for an aminoglycoside 6′-acetyl transferase capable of acetylating all the tested aminoglycosides at C-6′. Acetylated arbekacin still had some antibiotic activity, and acetylation of neomycin did not result in inactivation.[9]

The crystal structure of amikacin has been determined, revealing that the orientation between the three rings of the molecule is maintained by intramolecular hydrogen bonds.[10]

2-Deoxy-*scyllo*-inosose synthase, which effects the carbocyclization of glucose-6-phosphate to form **6**, has been purified and characterized. It was found to require Co^{2+}, and is distinct from dehydroquinate synthase, although the mechanism of action of both enzymes is similar.[11]

A review has appeared covering the synthesis and biological activity of

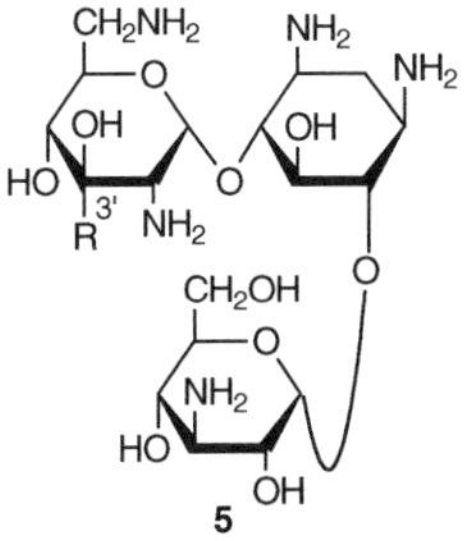

5

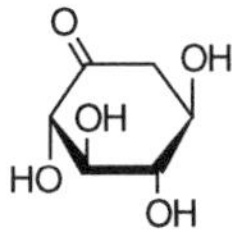

6

natural aminocyclopentitol glycosidase inhibitors, dealing with mannostatin, trehazolin, allosamidin and their analogues.[12]

2 Macrolide Antibiotics

Erythromycin E, the structure of which contains an orthoester involving the cladinose unit (part-structure **7**) has been isolated as a metabolite of *Nocardia brasiliensis*.[13]

An alternative method for mono-*N*-demethylation of the desosamine unit of macrolide antibiotics has been developed, and applied to 4″-deoxyerythromycin B. The procedure consists of treatment of the 2′-*O*-acetylated antibiotic with 1-chloroethyl chloroformate to give the urethane **8**, followed by methanolysis to give the mono-*N*-methylated compound **9**, this latter step also leading to the formation of the 6,9-enol ether in the macrolide ring.[14]

Treatment of amphotericin B with acyl perfluorides results in *N*-acylation of the aminosugar moiety, the resultant *N*-perfluoroacyl derivatives having lower acute toxicity and high antifungal activity.[15]

An account of research on the biosynthesis of deoxysugars includes a review of work on the 2,6- and 4,6-dideoxyhexoses found in macrolide antibiotics.[16]

The structure of colubricidin A, a novel macrolide antibiotic from a *Streptomyces* species has been determined, mainly by NMR methods; it

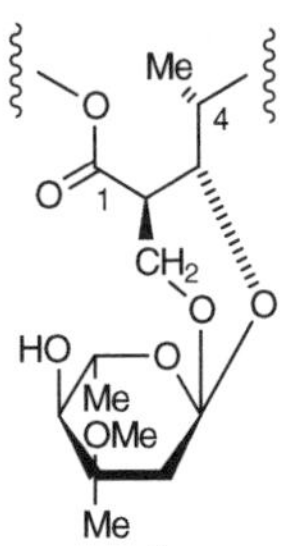

7

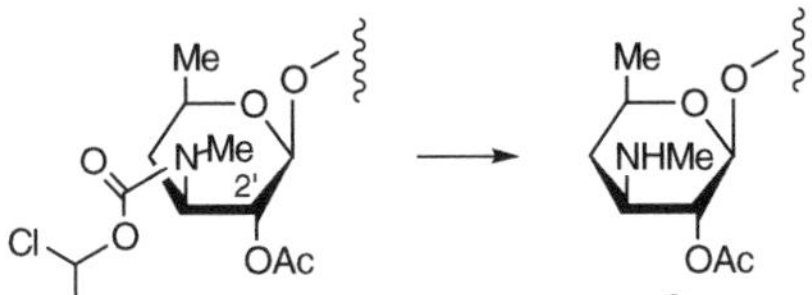

8

9

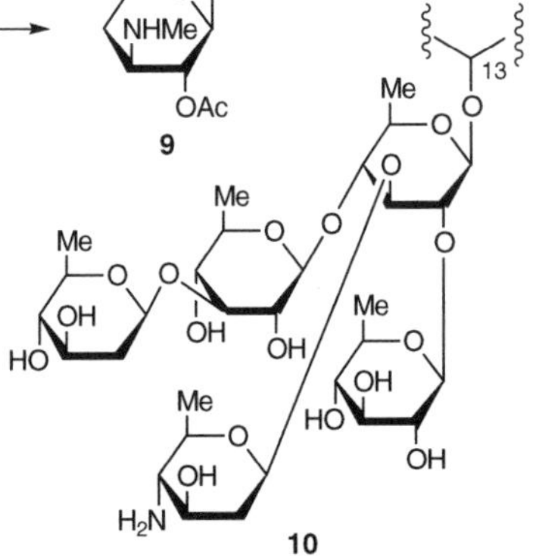

10

contains a 34-membered macrolide and six deoxyhexose units, five of which are linked as indicated in **10**, with an additional acylated 4-amino-2,4,6-trideoxyglucose unit elsewhere in the structure.[17]

3 Anthracyclines and Other Glycosylated Polycyclic Antibiotics

The daunorubicin analogue **11** has been prepared by glycosylation using a glycosyl iodide,[18] and 5′-demethyl-5′-trifluoromethyl-daunorubicin and -doxorubicin (**12**, X = H and OH respectively) have been prepared using a phenylthioglycoside as sugar donor;[19] for the synthesis of the sugar units, see Chapter 8.

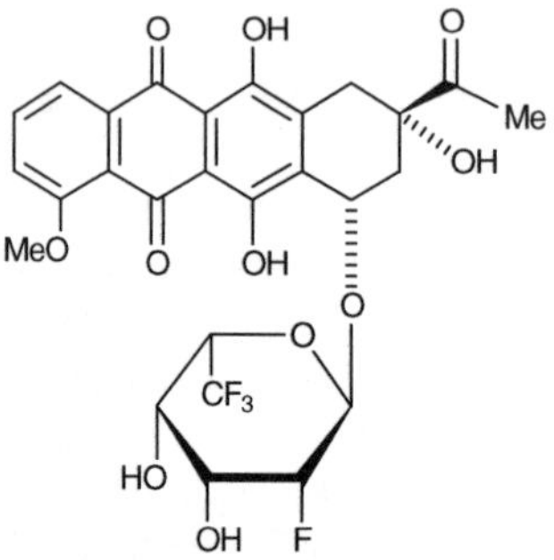

O OH O Me OH MeO O OH O O CF_3 HO OH F

11

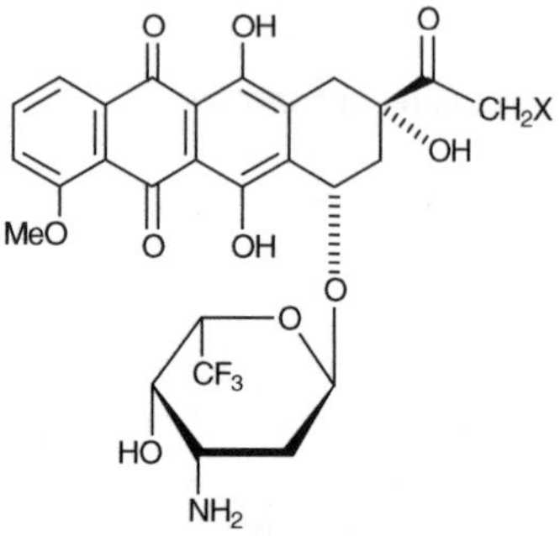

O OH O CH_2X OH MeO O OH O O CF_3 HO NH_2

12

There have been further reports on the development of conjugates of anthracyclines for targetted drug-delivery. The doxorubicin-spacer-glutamate compounds **13** (X = O, NH) have been prepared, these compounds liberating doxorubicin by action of a carboxypeptidase expressed in a mammalian carcinoma cell line,[20] whilst Scheeren and co-workers have described somewhat similar conjugates of daunorubicin and doxorubicin linked through a carbamate to *p*-aminobenzyl alcohol, which was *N*-acylated with a tripeptide; these compounds released the anthracycline on incubation with tumour-associated human plasmin.[21] The same team have also reported the synthesis of anthracycline-spacer-glucuronate conjugates **14** (X = H, OH), together with related compounds with substituents in the spacer unit, and with β-D-glucopyranosyl and β-D-galactopyranosyl units in place of the uronate, the glycopyranosyl carbamates being made by the β-selective method of Vol. 30, p. 114. These compounds were designed to liberate daunorubicin or doxorubicin by β-glucuronidase, known to be present at high levels in necrotic tumour tissue, and, based on favourable results *in vitro*, **14** (X = OH) was carried forward to a phase I trial.[22]

The antineoplastic activity of doxorubicin has been shown to be enhanced by a factor of ten in the presence of β-cyclodextrin.[23] The solution structures of the Fe(III) complexes of idarubicin have been studied.[24]

Two daunosamine units have been linked together by a spacer to give **15**.

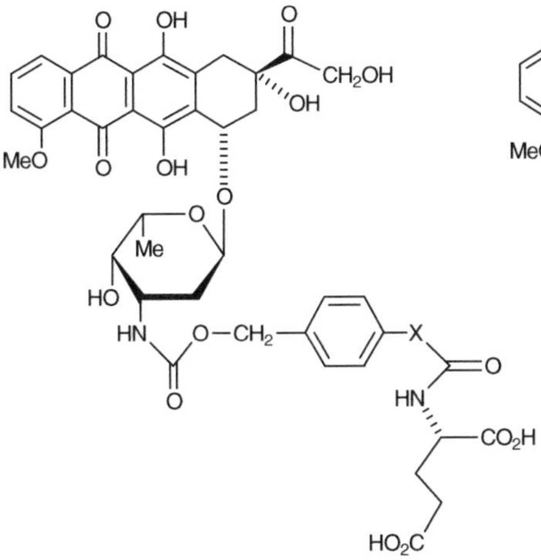

13

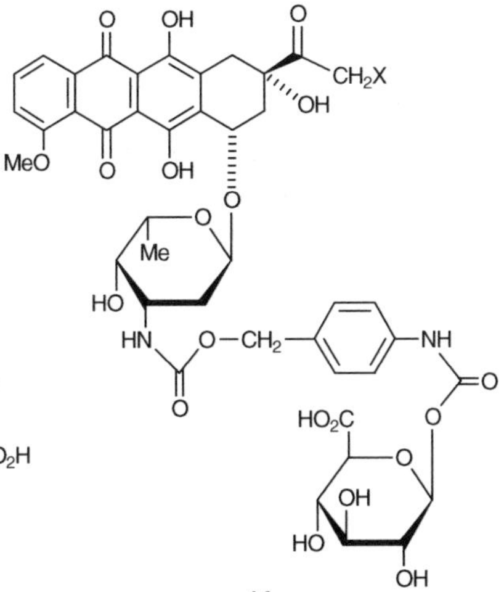

14

This compound, designed as a divalent probe to study aminosugar–polynucleotide interactions, was made either through double glycosylation of monosilylated 1,4-butanediol, or by olefin metathesis using an allyl glycoside followed by hydrogenation.[25]

A significant achievement has been the synthesis of olivomycin A (**16**) in Roush's laboratory, building on earlier work leading to olivin, and studies reported in previous volumes which led to the development of methods for the stereoselective synthesis of 2-deoxyglycosides. In the route to **16**, the C ring was firstly attached, followed by a E–D unit, in both cases the β-link being made by use of an α-trichloroacetimidate, with an α-oriented phenylthio group

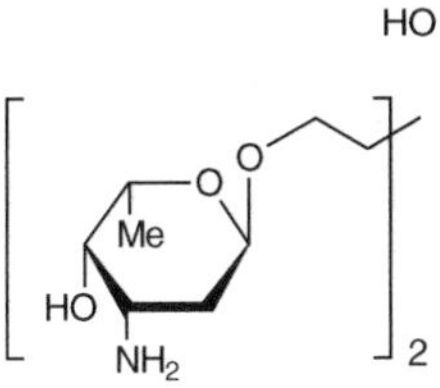

15

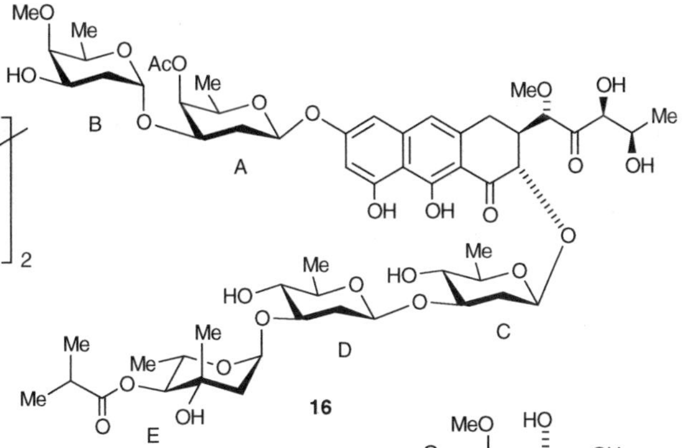

16

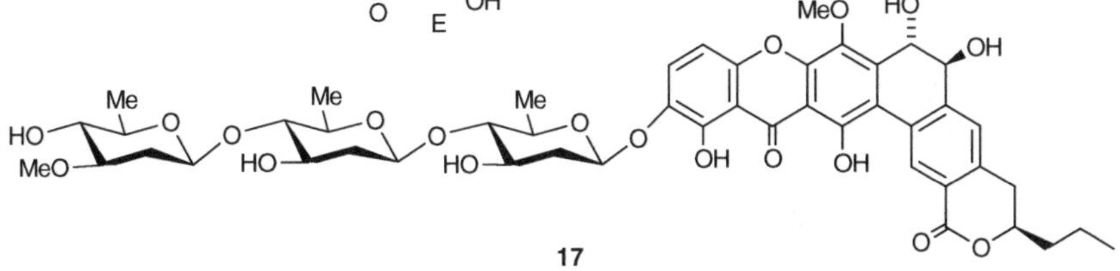

17

at C-2. The B–A unit was added at a later stage, using Mitsunobu coupling (Vol. 29, p. 20).[26] The full stereostructure of the cytotoxic antibiotic FD-594 has now been determined by X-ray crystallography to be as in **17**, and CD and NMR studies showed that the compound displayed interesting solvent-dependent atropisomerism in the dihydroxylated ring of the aglycon.[27]

A paper on the synthesis of the *C*-glycoside fragment of the angucyclines is mentioned in Chapter 3.

4 Nucleoside Antibiotics

The new nucleoside analogue futalosine (**18**) has been isolated from a *Streptomyces* species. 6-*O*-Methylfutalosine methyl ester was prepared, and found to display cell growth inhibitory properties.[28]

A more efficient and higher-yielding procedure for the synthesis of toyocamycin has been developed, in which 4-amino-6-bromo-5-cyanopyrrolo[2,3-*d*]pyrimidine was coupled to the sugar unit, followed by hydrogenolysis to remove the bromine.[29]

The structure **19** has been proposed for the anti-HIV antibiotic EM2487 from a *Streptomyces* species, although stereochemistry of the aglycon and absolute stereostructure of the sugar were not defined.[30]

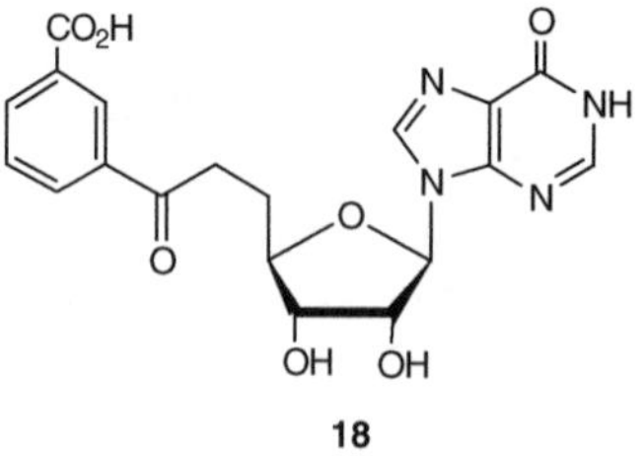

18 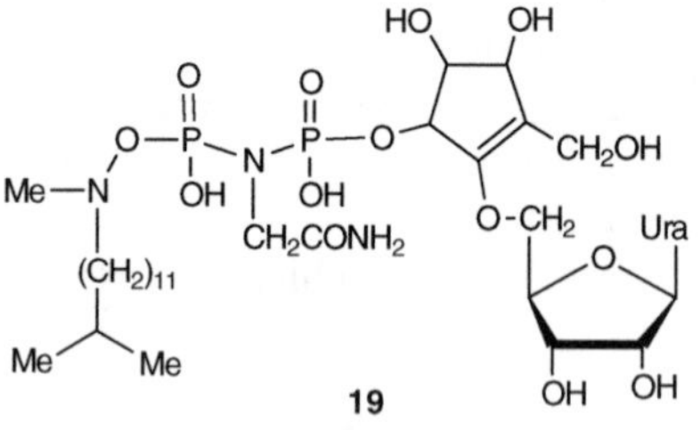 **19**

In a stereocontrolled approach to polyoxin C, the trifluoroacetimidate **21**, prepared as indicated in Scheme 1 from the Horner-Emmons product **20**, underwent [3,3]-sigmatropic rearrangement to establish the C-5′ stereocentre of thymine polyoxin C (**22**).[31] In an alternative approach to the same target, the allylic alcohol **23** was converted highly stereoselectively into epoxyalcohol

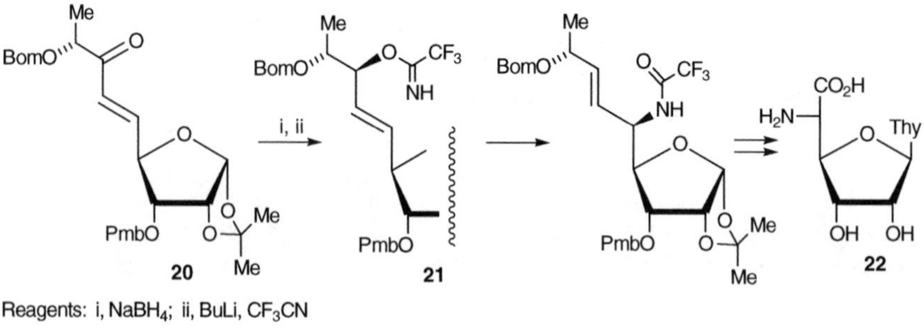

20 **21** **22**

Reagents: i, $NaBH_4$; ii, BuLi, CF_3CN

Scheme 1

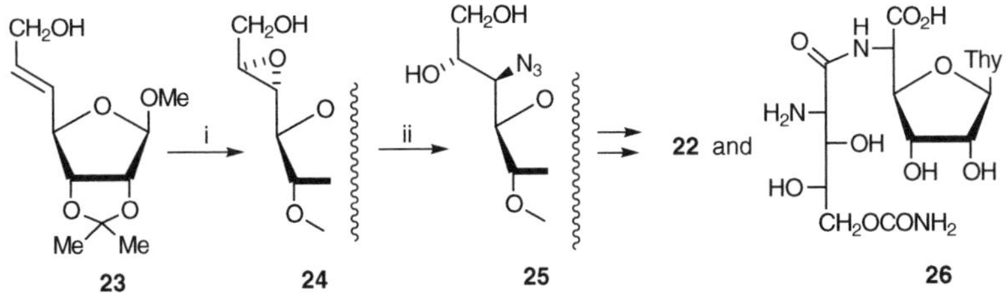

Reagents: i, MCPBA; ii, $Ti(OPr^i)_2(N_3)_2$

Scheme 2

24 using MCPBA, a result ascribed to reagent delivery through coordination to O-4. Regioselective attack of azide gave **25**, convertible to thymine polyoxin C (**22**). Polyoxin J (**26**) was also synthesized, the required aminoacid being prepared by a stereocontrolled route also involving the regioselective opening of an allylic epoxide with the titanium-based reagent of Scheme 2.[32] Other workers have also described syntheses of polyoxin J and polyoxin L (the uracil analogue) by coupling uracil and thymine polyoxin C to the aminoacid.[33] A genetically-engineered mutant of *Streptomyces tendae*, unable to produce the side-chain of nikkomycins I, J, X (**27**, X = N) and Z (the uracil analogue), accumulates nikkomycin C_X and nikkomycin C_Z (uracil polyoxin C). Cell cultures of the mutant fed with benzoic acid produce the biologically-active nikkomycin B_X (**27**, X = CH) and nikkomycin B_Z (the uracil analogue), which have significantly higher stability at pH 5.5 than do nikkomycins X and Z.[34]

New types of liposidomycins (see Vol. 32, pp. 248–249), inhibitors of bacterial peptidoglycan biosynthesis, have been reported, the structure of the fatty acid components being determined by tandem mass spectrometry.[35]

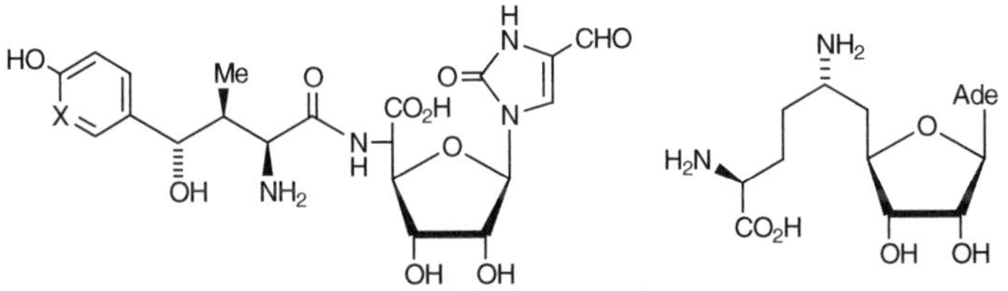

A new formal synthesis of sinefungin (**28**) has been developed, in which the chiral centre at C-6′ was created by azide opening of an epoxide formed by asymmetric epoxidation, and the centre at C-9′ was formed by asymmetric hydrogenation of an enamide using a rhodium catalyst, to give an intermediate in a previous synthesis of sinefungin from the same laboratory (Vol. 30, p. 259).[36]

The total synthesis of herbicidin B (**32**) has been reported. A key step (Scheme 3) involved the reductive coupling of the 2-keto-phenylthio glycoside **29** with aldehyde **30**. The choice of protecting groups for **29** was vital to ensure that the conformation of **31** was such that hydrogenation occurred to give the correct configuration at C-6′, particularly since, although it was possible to get

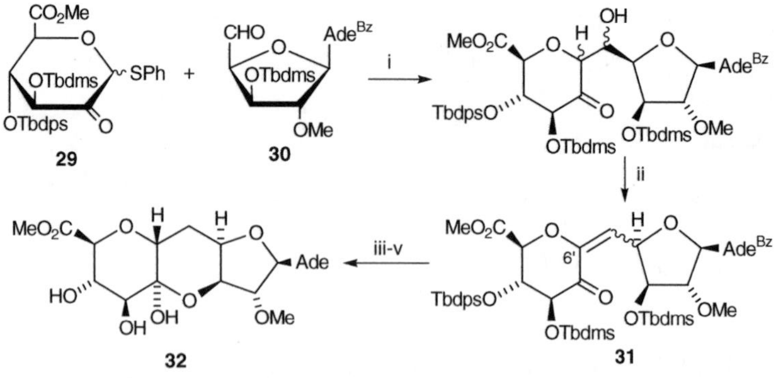

Reagents: i, SmI_2; ii, $MeO_2C\text{-}N\text{-}SO_2NEt_3$; iii, Pd/C, $NH_4^+ HCO_2^-$; iv, SmI_2, MeOH; v, TBAF

Scheme 3

predominantly the correct configuration at C-6′ in the linkage of the two building blocks, direct deoxygenation at C-5′ was not fruitful.[37]

Condensation of the aldehyde **33** with an iminophosphorane was used to prepare the enamide **34**, related to mureidomycin, but the analogue did not inhibit bacterial peptidoglycan biosynthesis.[38]

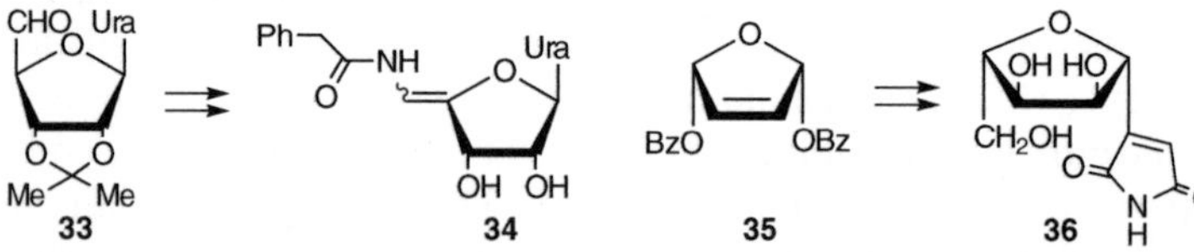

A total synthesis of the L-enantiomer **36** of the *C*-nucleoside antibiotic showdowmycin has been reported, starting from the dihydrofuran **35** and using π-allyl-palladium intermediates to introduce sequentially suitable precursors for both the substituents, and with the use of a chiral bis-phosphine ligand in the first step to induce asymmetry. The method is potentially applicable to other *C*-nucleosides in either enantiomeric series.[39] The desamino analogue **37** of formycin has been made by deamination of the natural antibiotic or from formycin B, and it proved to be a good herbicide. Its corresponding 5′-phosphate was a strong inhibitor of plant AMP deaminase, and other inhibitors of this enzyme have been previously shown be have herbicidal activity. The same workers also prepared carbocyclic nebularine (desamino-aristeromycin, **38**) by a similar deamination, but it proved to be inactive.[40]

Also in the area of carbocyclic nucleoside antibiotics, a new synthesis of (−)-aristeromycin (**40**) relied upon a highly stereoselective dihydroxylation of the cyclopentene **39**.[41] The intermediate **41**, derived from an aldol condensation of the cyclopentanone, itself available from D-ribose by previously-known methods, with ethyl cyanoacetate, has been converted to the carbocyclic *C*-nucleoside 9-deaza-aristeromycin (**42**). The required β-stereochemistry was obtained by reduction of **41** using Bu_3SnH-AIBN.[42] In a ten-step synthesis of

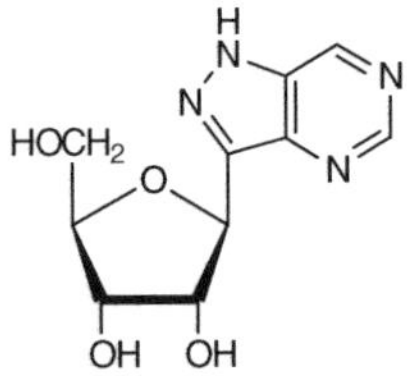

37 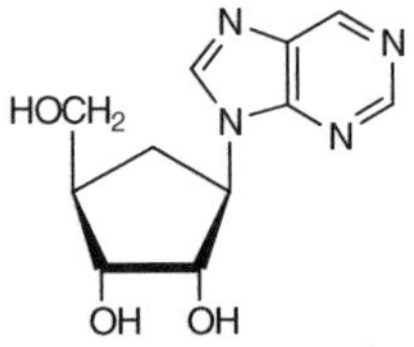 38 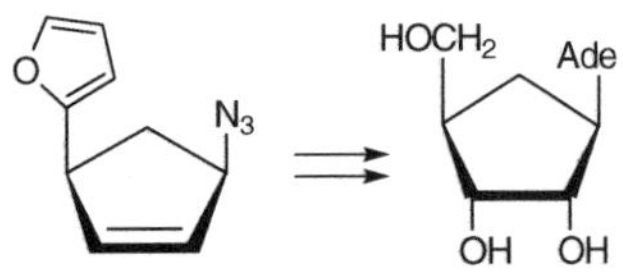 39 40

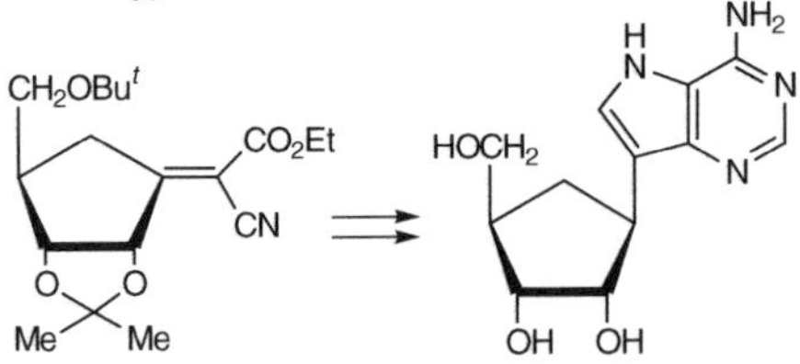

41 42 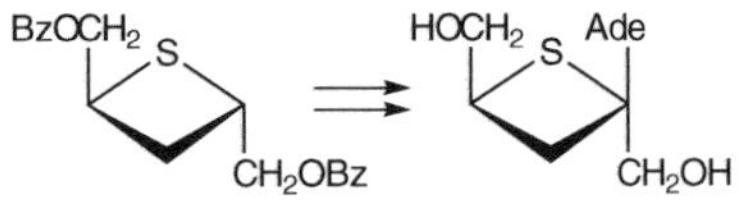 43 44

2′,3′-dideoxyaristeromycin from cyclohexenone, the heterocycle was introduced by a Mitsunobu reaction.[43]

The 3′-thio analogue **44** of oxetanocin has been prepared; a key step in the route was the linkage of the thietane **43** to 6-chloropurine using Pummerer-type chemistry in a manner similar to that previously used to make thietane nucleosides (Vol. 30, pp. 301–302).[44] Carbocyclic oxetanocin and the thymine analogue have been incorporated into modified oligonucleotides (hexadecamers with all units modified except for the one at the 3′-end). The oligomer of carbocyclic oxetanocin formed a stable triple helix with uridine oligoribonucleotide under physiological conditions.[45]

Some references to hexopyranosyl indolocarbazoles (rebeccamycin, *etc.*) and related systems are given in the next section.

5 Other Types of Carbohydrate Antibiotics

Further analogues of lincomycin, including 7-amino-7-deoxy-lincomycin (**45**, R = H), alkylated derivatives (**45**, R = alkyl) and ureas, have been prepared by carrying out a Staudinger reaction and then further chemistry on a previously-

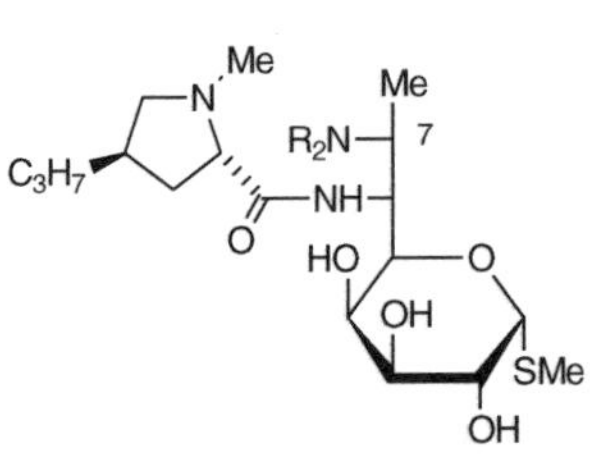

45 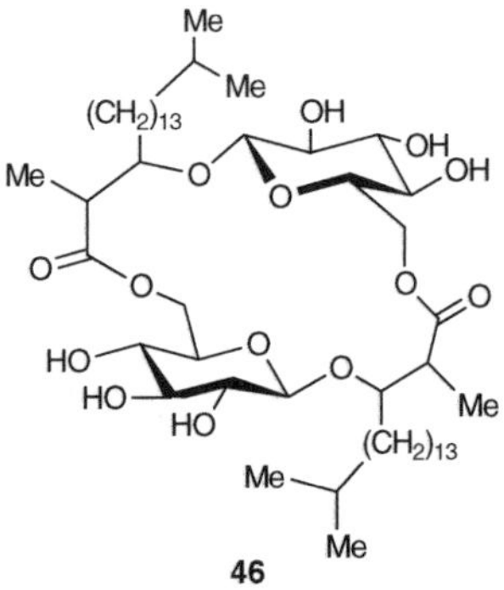 46

reported 7-azidocompound (Vol. 30, pp. 262–263).[46] It has been shown that on prolonged storage at 40 °C, the semisynthetic lincosaminide pirlimycin, which has a chlorine atom at C-7, undergoes a cyclization in which chloride is displaced by O-4.[47]

Glucolipsin A, a glucokinase activator from *Streptomyces purpurogeniscleroticus* and *Nocardia vacinii*, has been assigned the structure **46**, and glucolipsin B is closely related, with a different chain length in one of the sidechains.[48]

Complete spectroscopic characterizations of novobiocin (**47**), isonovobiocin (carbamyl on O-2″), 2″-*O*-carbamyl- and decarbamyl-novobiocin, and novobiocin 2″,3″-carbonate have been carried out, some of these compounds not having been completely characterized before.[49] Workers at Hoechst Marion Roussel have devised a new synthesis of noviose (Chapter 14), and made a series of novobiocin analogues of type **48**, with a reversed amide unit, some of which had good gyrase B inhibitory and antibacterial activity. The pyrrole ester unit was introduced by reaction of a 2′,3′-carbonate with 2-methylpyrrylmagnesium bromide, and the glycosylation of the phenol was carried out with the required α-selectivity by a Mitsunobu reaction.[50] The same team has also prepared active analogues with substituted aminomethyl groups at C-4 and with C-3 unsubstituted, and with substituted 2-aminoethyl groups at O-4, with C-3 either unsubstituted or chlorinated.[51] Bioactive analogues have also been made in which 4-*O*-methyl-L-rhamnose (with the same acyl substituent at O-3′) replaces noviose, and the coumarin has a variety of substituents at C-3, -4, and -5.[52]

Glucosylation of the phenolic group of the antitumour agent duocarmycin B1 was carried out in an effort to improve its pharmacological profile.[53]

A series of novel polyketides, TMC-151 A-F, have been isolated from

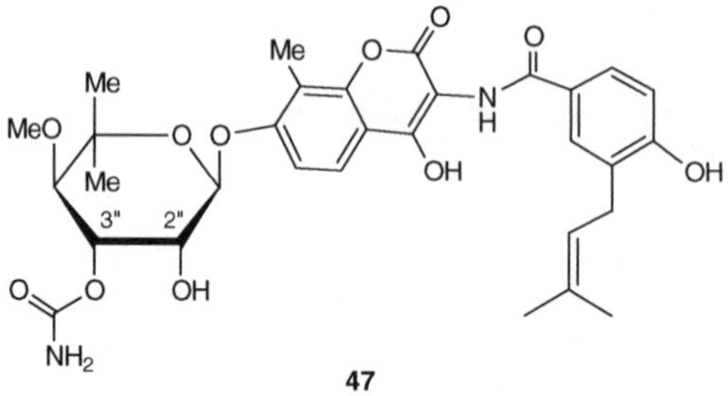

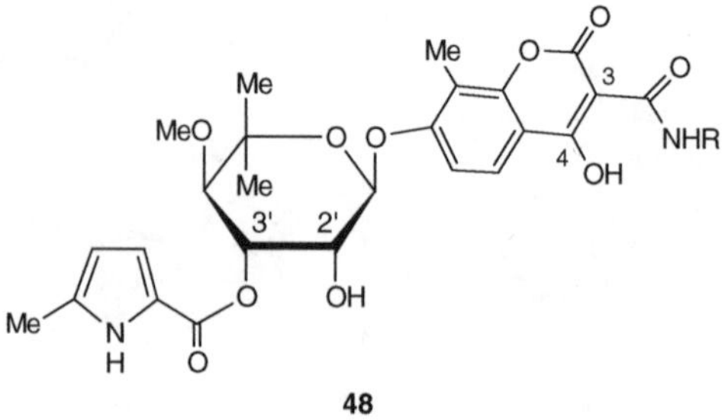

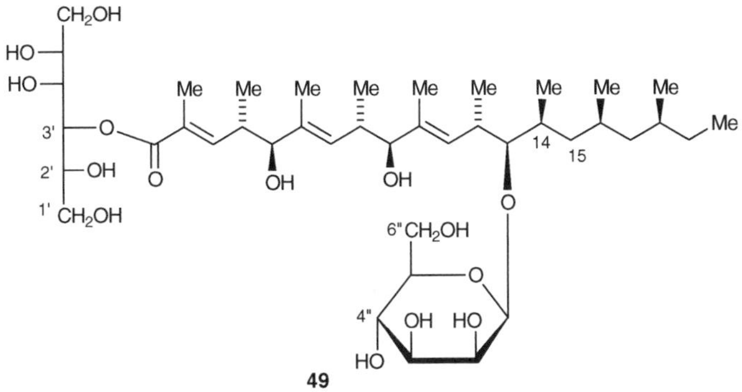

49

Gliocladium catenulatum fermentation broth. TMC-151A has structure **49**, whilst the others differ in the position through which the D-mannitol unit and the polyketide chain are linked (O-1′ or O-2′ instead of O-3′), by the presence of D-arabinitol (1′-*O*-acylated) instead of D-mannitol, and/or by an acetyl group at O-6″ of the mannose unit.[54] A closely-related series of compounds TMC-171 A–C and TMC-154 was also isolated, in which an extra double bond is present between C-14 and C-15; TMC-171A is the dehydrogenated form of TMC 151A, TMC-171B and TMC 171C are acylated at O-2′ and O-1′ respectively and TMC-154 has a 1′-*O*-acylated D-arabinitol unit and is additionally methylated at O-4″.[55] Others have isolated roselipins 1A and 1B from another fungus; although full stereochemical details are not available, the roselipins seem to be very similar, if not identical, to the TMC class of compounds, with D-arabinitol as the alditol involved, and differing in the end (O-1′ or O-5′) that is acylated.[56, 57]

A number of papers have dealt with chemistry of the disaccharide-phospholipid degradation product of moenomycin A, in which seems to reside most of the biological activity of the antibiotic in inhibition of the transglycosylation step of bacterial peptidoglycan biosynthesis. After initial work exploring the applicability of non-2-ynyl glycosides, removable by regioselective Hg(II) catalysed hydration and subsequent β-elimination, for protection of O-1 of the glucuronamide unit,[58] Welzel and colleagues have prepared the analogue **50** and a related species with a shorter alkyl chain, which lack the methyl group at C-4 of the moenuronamide unit in the natural product. The glucuronamide unit was derived from D-glucurono-3,6-lactone. Analogue **50** had full activity against the transglycosylase *in vitro*, but the compound with the shorter chain was inactive. Taken with earlier results, these data allowed the development of a general view of the structural requirements for antibiotic activity amongst compounds of this type.[59] An analogue related to **50**, but lacking the hydroxyl group at C-4, has also been prepared, using photochemical reduction of the *m*-trifluorobenzoate in the presence of *N*-methylcarbazole and isopropanol. This deoxygenated compound was not an inhibitor of the transglycosylase.[60] The analogue of **50** in which the GlcNAc unit is replaced with glucose has also been made, and was a poor transglycosylase inhibitor.[61] A combinatorial approach has been used to produce 1300 disaccharides of the type **51** (R = H

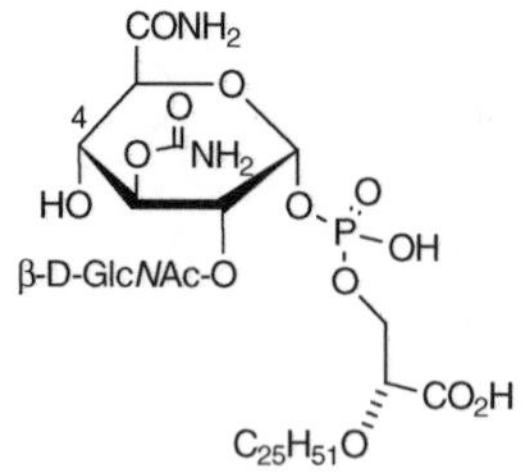

50 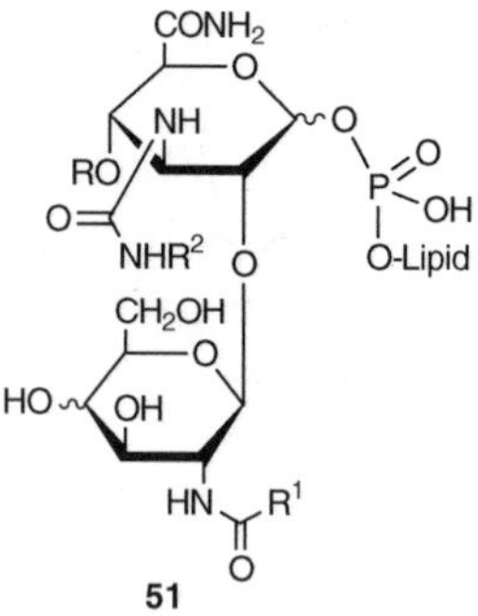 **51**

or Me, D-*gluco*- and D-*galacto*-in the lower ring), in which R^1, R^2 and the lipid were varied, and in another series with glucuronamide as the top ring and D-glucosamine as the bottom unit. Some of the compounds were effective against vancomycin-resistant enterococci (VRE).[62]

Chemical modifications have been carried out on ziracin, an antibiotic of the everninomycin group. These included specific alkylations of sugar hydroxyl groups, deoxygenations, and reduction of the nitro group of the evernitrose moiety.[63]

A review has been published on the work of Williams's Cambridge group on the structures and mode of action of the vancomycin group of glycopeptide antibiotics, and recent developments in semisynthetic compounds to counter VRE.[64] Another extensive review from Nicolaou and his collaborators has dealt with the chemistry, including total synthesis, biology and medicine of the glycopeptide antibiotics.[65] An account has been given, in a series of four papers, of the total synthesis of vancomycin by the Nicolaou group. The last paper covers the attachment of the disaccharide unit to the aglycon by stepwise glycosylation, firstly with a glucopyranosyl trichloroacetimidate in which O-2 was selectively protected with an allyloxycarbonyl group, and then, after release of O-2, by attachment of a vancosamine unit made by total synthesis (*Angew. Chem. Int. Ed. Engl.*, 1998, **37**, 1871).[66] An alternative way of making vancomycin from its aglycon involved glucosylation using donor **52** activated by Tf_2O in the presence of 2,6-di-*t*-butyl-4-methylpyridine and $BF_3.Et_2O$, the latter suppressing the orthoester formation which had previously proved problematical in glycosylations by the sulfoxide method in cases where O-2 bears an unhindered ester. After removal of the ester (Ph_3P, aqueous dioxan), the vancosamine unit was attached using a sulfoxide donor, previously shown to permit α-selective attachment of this sugar (Vol. 32, p. 23).[67]

Recognition that self-association of two glycopeptide molecules enhances binding to their target led to two molecules of vancomycin being linked together *via* amides at the vancosamine units. The self association of the dimers was studied by surface plasmon resonance technology.[68] Vancomycin has been linked, again using the amino group of the vancosamine, to a norbornene unit *via* a spacer. Subsequent metathesis polymerization using Grubbs's catalyst gave a multivalent vancomycin polymer, which showed significant enhancement of activity against VRE.[69]

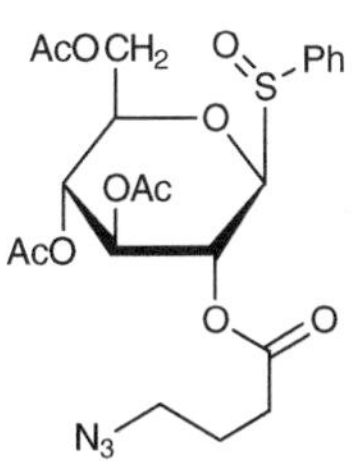

52

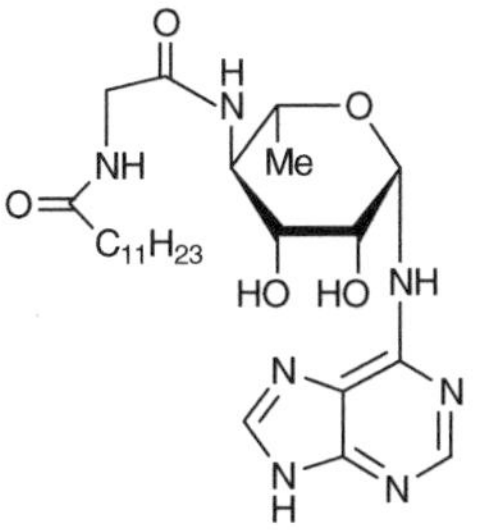

53

A review has appeared covering the synthesis and mechanism of action of the antitumour glycopeptide bleomycin.[70]

A full account has been given of the synthesis of the L-rhamnose analogue **53** of spicamycin (see Vol. 32, p. 251), and the work was extended to the synthesis of the enantiomer from methyl α-D-mannopyranoside. Whilst **53** had high cytotoxicity towards human myeloma cells, the enantiomer was inactive, as was the analogue in which the nucleobase was replaced by a methoxy group.[71] The D-mannosyl-compound **54** and the D-*gluco*-isomer have been made with β-selectivity by Pd-catalysed coupling of tetra-*O*-benzyl-β-D-glycopyranosylamines with *N*-Sem-6-chloropurine, followed by deprotection.[72]

Hydantocidin 5′-phosphate is an inhibitor of adenylosuccinate synthetase, and so is hadacidin (*N*-formyl-*N*-hydroxyglycine), an aspartate analogue. When the two inhibitors were linked together in the structure **55**, made by Mitsunobu coupling of a hydantocidin derivative with a protected derivative of the δ-hydroxy-aminoacid, a highly effective bi-substrate inhibitor ensued,

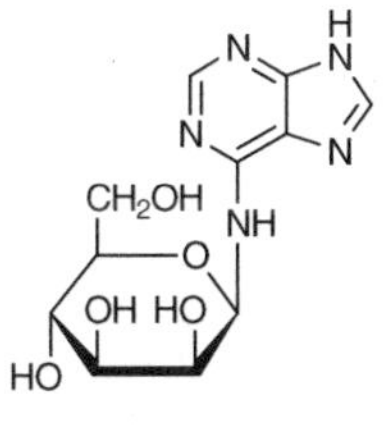

54

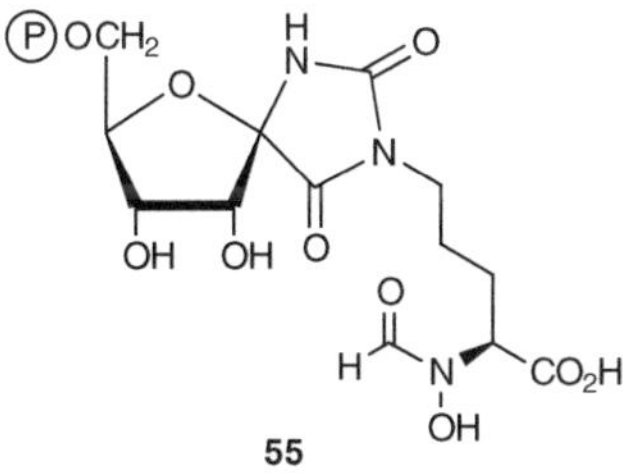

55

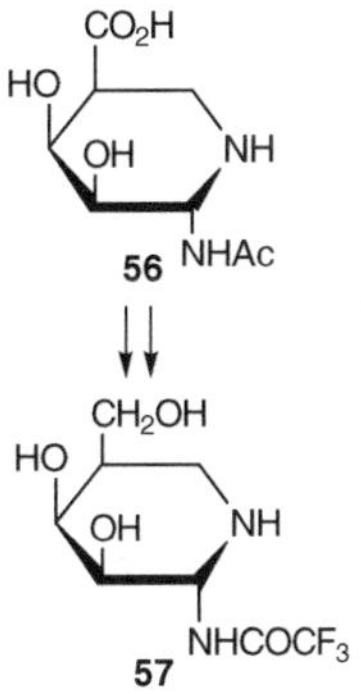

56

57

with the side-chain epimer being much less inhibitory. Some modelling was done in support of the idea that the trimethylene chain adopts a conformation so as to place the adenylate and aspartate analogues in the correct regions of space to overlap with the positions of the two natural substrates.[73]

Siastatin B (**56**) has been converted into the D-galactose analogue **57**, which turned out to be a new type of galactosidase inhibitor.[74]

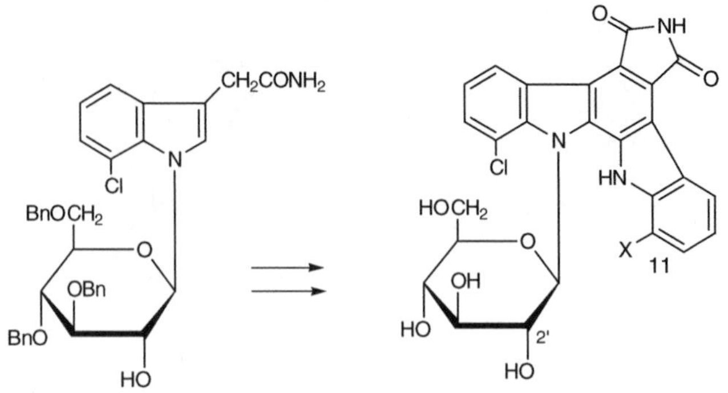

58 **59**

In the area of indolocarbazole antibiotics, the *N*-glycosylindole **58** was made by linking the indole to the 1,2-anhydrosugar, in a manner previously described by others (*J. Org. Chem.*, 1993, **58**, 343), and then used to prepare rebeccamycin (**59**, X = Cl) and 11-dechlororebeccamycin (**59**, X = H).[75] Rebeccamycin analogues with a halogenoacetyl substituent at O-2′ have been prepared, and, although the expected increased interaction with DNA was not observed, the compounds behaved as typical topoisomerase I inhibitors.[76] ED-110 (**60**) has been prepared, and so have the other three isomers in which the hydroxyl groups are repositioned symmetrically in the benzene rings, the

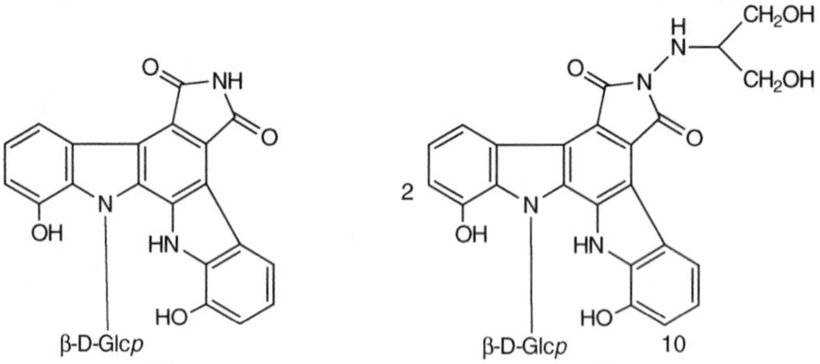

60 **61**

compounds having potent topoisomerase inhibitory activity.[77] A set of substituted *N*-aminocompounds related to NB-506 (the *N*-amino-derivative of **60**) has been reported, and the example shown in **61** was significantly active against human stomach cancer cells.[78] The analogue of **61** in which the phenolic groups are at C-2 and C-10 has also been reported.[79] Holyrine A (**62**) and holyrine B (**63**) have been isolated from a marine actinomycete, and are possible intermediates in the biosynthesis of staurosporine (**64**). Also isolated was K252d (Vol. 20, p. 193), which is possibly an earlier intermediate in the pathway.[80]

The benzopyranopyrrole antibiotic pyralomycin 2c (**65**) has been prepared by Mitsunobu condensation of the heterocycle to tetra-*O*-Mom-β-D-glucopyranose.[81]

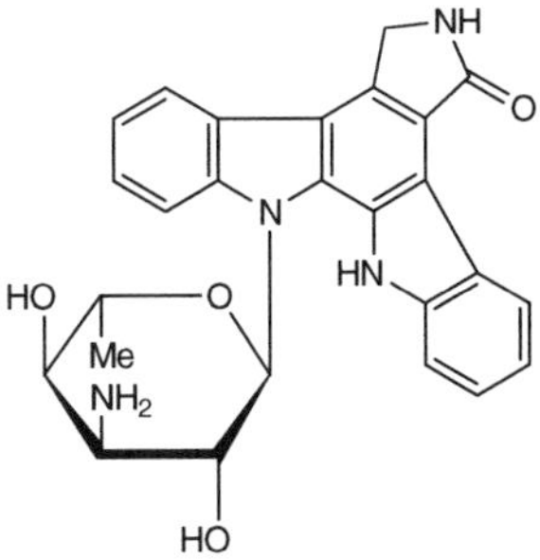

62

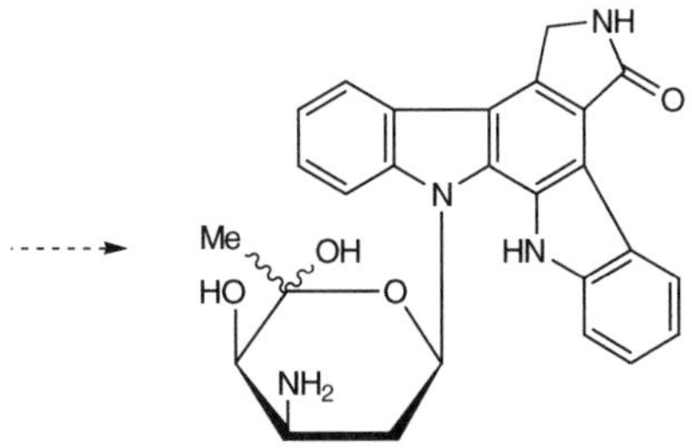

63

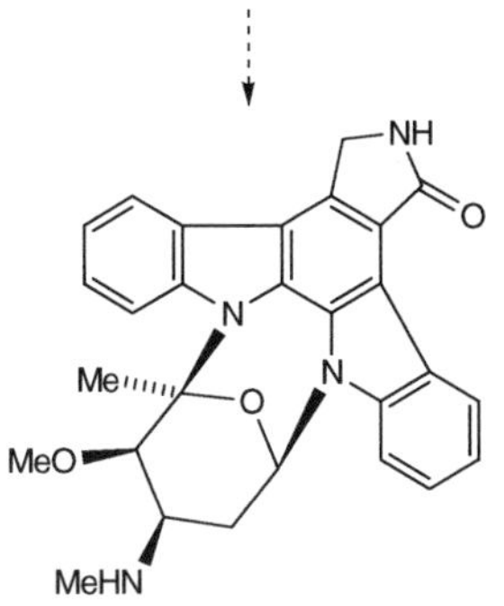

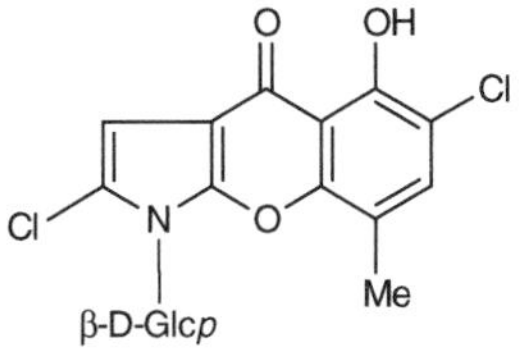

65

References

1 C.-H. Wong, *Acc. Chem. Res.*, 1999, **32**, 376.
2 W.A. Greenberg, E.S. Priestley, P.S. Sears, P.B. Alper, C. Rosenbohm, M. Hendrix, S.-C. Hung and C.-H. Wong, *J. Am. Chem. Soc.*, 1999, **121**, 6527.
3 C.L. Nunns, L.A. Spence, M.J. Slater and D.J. Berrisford, *Tetrahedron Lett.*, 1999, **40**, 9341.
4 K. Michael, H. Wang and Y. Tor, *Bioorg. Med. Chem.*, 1999, **7**, 1361.
5 S.R. Kirk and Y. Tor, *Bioorg. Med. Chem.*, 1999, **7**, 1979.
6 J.B.-H. Tok, J. Cho and R.R. Rando, *Tetrahedron*, 1999, **55**, 5741.
7 A Steedhara, A. Patwardhan and J.A. Cowan, *Chem. Commun.*, 1999, 1147.
8 J. Haddad, S. Vakulenko and S. Mobashery, *J. Am. Chem. Soc.*, 1999, **121**, 11922.
9 C.-B. Zhu, A. Sunada, J. Ishikawa, Y. Ikeda, S. Kondo and K. Hotta, *J. Antibiot.*, 1999, **52**, 889.
10 R. Bau and I. Tsyba, *Tetrahedron*, 1999, **55**, 14839.
11 F. Kudo, Y. Hosomi, H. Tamegai and K. Kakinuma, *J. Antibiot.*, 1999, **52**, 81.
12 A. Berecibar, C. Grandjean and A. Siriwardena, *Chem. Rev.*, 1999, **99**, 779.

13 Y. Mikami, K. Yazawa, A. Nemoto, H. Komaki, Y. Tanaka and U. Gräfe, *J. Antibiot.*, 1999, **52**, 201.
14 J.E. Hengeveld, A.K. Gupta, A.H. Kemp and A.V. Thomas, *Tetrahedron Lett.*, 1999, **40**, 2497.
15 Y.D. Shenin, V.V. Belakhov, L.I. Shatik and R.A. Araviyskii, *Antibiot. Khimister.*, 1998, **43**, 8 (*Chem. Abstr.*, 1999, **131**, 59 033).
16 T.M. Hallis and H.-W. Liu, *Acc. Chem. Res.*, 1999, **32**, 579.
17 F. Kong, D.Q. Liu, J. Nietsche, M. Tischler and G.T. Carter, *Tetrahedron Lett.*, 1999, **40**, 9219.
18 K. Nakai, Y. Takagi and T. Tsuchiya, *Carbohydr. Res.*, 1999, **316**, 47.
19 K. Nakai, Y. Takagi, S. Ogawa and T. Tsuchiya, *Carbohydr. Res.*, 1999, **320**, 8.
20 I. Niculescu-Duvaz, D. Niculescu-Duvaz, F. Friedlos, R. Spooner, J. Martin, R. Marais and C.J. Springer, *J. Med. Chem.*, 1999, **42**, 2485.
21 F.M.H. de Groot, A.C.W. de Bart, J.H. Verheijen and H.W. Scheeren, *J. Med. Chem.*, 1999, **42**, 5277.
22 R.G.G. Leenders, E.W.P. Damen, E.J.A. Bijsterveld, H.W. Sheeren, P.H.J. Houba, I.H. van der Meulen-Muileman, E. Boven and H.J. Haisma, *Bioorg. Med. Chem.*, 1999, **7**, 1597.
23 A. Al-Omar, S. Abdou, L. de Robertis, A. Marsura and C. Finange, *Bioorg. Med. Chem. Lett.*, 1999, **9**, 1115.
24 M.M.L. Fiallo, H. Drechsel, A. Garnier-Suillerot, B.F. Matzanke and H. Koslowski, *J. Med. Chem.*, 1999, **42**, 2844.
25 A. Kirschning and G.-w. Chen, *Tetrahedron Lett.*, 1999, **40**, 4665.
26 W.R. Roush, R.A. Hartz and D.J. Gustin, *J. Am. Chem. Soc.*, 1999, **121**, 1990.
27 T. Eguchi, K. Kondo, K. Kakinuma, H. Uekusa, Y. Ohashi, K. Mizoue and Y.-F. Qiao, *J. Org. Chem.*, 1999, **64**, 5371.
28 N. Hosokawa, H. Naganawa, T. Kasahara, S. Hattori, M. Hamada, T. Takeuchi, S. Yamamoto, K.S. Tsuchiya and M. Hori, *Chem. Pharm. Bull.*, 1999, **47**, 1032.
29 A.R. Porcari and L.B. Townsend, *Nucleosides, Nucleotides,* 1999, **18**, 153.
30 H. Takeuchi, N. Asai, K. Tanabe, T. Kozaki, M. Fujita, T. Sakai, A. Okuda, N. Naruse, S. Yamamoto, T. Sameshima, N. Heida, K. Dobashi and M. Baba, *J. Antibiot.,* 1999, **52**, 971.
31 A. Chen, E.J. Thomas and P.D. Wilson, *J. Chem. Soc., Perkin Trans. 1*, 1999, 3305.
32 A.K. Ghosh and Y. Wang, *J. Org. Chem.*, 1999, **64**, 2789.
33 K. Uchida, K. Kato and H. Akita, *Synthesis*, 1999, 1678.
34 C. Bormann, A. Kálmánczhelyi, R. Süssmuth and G. Jung, *J. Antibiot.,* 1999, **52**, 102.
35 Y. Esumi, Y. Suzuki, K. Kimura, M. Yoshihama, T. Ichikawa and M. Uramoto, *J. Antibiot.,* 1999, **52**, 281.
36 A.K. Ghosh and Y. Wang, *J. Chem. Soc., Perkin Trans. 1*, 1999, 3597.
37 S. Ichikawa, S. Shuto and A. Matsuda, *J. Am. Chem. Soc.*, 1999, **121**, 10270.
38 C.A. Gentle and T.D.H. Bugg, *J. Chem. Soc., Perkin Trans. 1*, 1999, 1279.
39 B.M. Trost and L.S. Kallander, *J. Org. Chem.*, 1999, **64**, 5427.
40 S.D. Lindell, B.A. Moloney, B.D. Hewitt, C.G. Earnshaw, P.J. Dudfield and J.E. Dancer, *Bioorg. Med. Chem. Lett.*, 1999, **9**, 1985.
41 Y. Tokoro and Y. Kobayashi, *Chem. Commun.*, 1999, 807.
42 B.K. Chun and C.K. Chu, *Tetrahedron Lett.*, 1999, **40**, 3309.
43 M.J. Comin and J.B. Rodriguez, *An. Asoc. Quím. Argent.*, 1998, **86**, 131 (*Chem. Abstr.*, 1999, **130**, 153 910).

44 E. Ichikawa, S. Yamamura and K. Kato, *Tetrahedron Lett.*, 1999, **40**, 7385.
45 N. Katagiri, Y. Morishita, I. Oosawa and M. Yamaguchi, *Tetrahedron Lett.*, 1999, **40**, 6835.
46 F. Sztaricskai, G. Batta, Z. Dinya, M. Hornyák, E. Röth, R. Masuma and S. Omura, *J. Antibiot.*, 1999, **52**, 1050.
47 F.W. Crow, J.R. Blinn, C.G. Chidestere, A.M. Cooper, W.K. Duholke, J.W. Hallberg, G.E. Martin, R.F. Smith and T.J. Thamann, *J. Heterocycl. Chem.*, 1999, **36**, 1049.
48 J. Qian-Cutrone, T. Ueki, S. Huang, K.A. Mookhtiar, R. Ezekiel, S.S. Kalinowski, K.S. Brown, J. Golik, S. Lowe, D.M. Pirnik, R. Hugill, J.A. Veitch, S.E. Klohr, J.L. Whitney and S.P. Manly, *J. Antibiot.*, 1999, **52**, 245.
49 F.W. Crow, W.K. Duholke, K.A. Farley, C.E. Hadden, D.A. Hahn, B.D. Kaluzny, C.S. Mallory, G.E. Martin, R.F. Smith and T.J. Thamann, *J. Heterocycl. Chem.*, 1999, **36**, 365.
50 P. Laurin, D. Ferroud, M. Klich, C. Dupuis-Hamelin, P. Mauvais, P. Lassaigne, A. Bonnefoy and B. Musicki, *Bioorg. Med. Chem. Lett.*, 1999, **9**, 2079.
51 P. Laurin, D. Ferroud, L. Schio, M. Klich, C. Dupuis-Hamelin, P. Mauvais, P. Lassaigne, A. Bonnefoy and B. Musicki, *Bioorg. Med. Chem. Lett.*, 1999, **9**, 2875.
52 D. Ferroud, J. Collard, M. Klich, C. Dupuis-Hamelin, P. Mauvais, P. Lassaigne, A. Bonnefoy and B. Musicki, *Bioorg. Med. Chem. Lett.*, 1999, **9**, 2881.
53 A. Asai, S. Nagamura, E. Kobayashi, K. Gomi and H. Saito, *Bioorg. Med. Chem. Lett.*, 1999, **9**, 2995.
54 J. Kohno, M. Nishio, M. Sakurai, K. Kawano, H. Hiramatsu, N. Kameda, N. Kishi, T. Yamashita, T. Okuda and S. Komatsubara, *Tetrahedron*, 1999, **55**, 7771.
55 J. Kohno, Y. Asai, M. Nishio, M. Sakurai, K. Kawano, H. Hiramatsu, N. Kameda, N. Kishi, T. Okuda and S. Komatsubara, *J. Antibiot.*, 1999, **52**, 1114.
56 S. Omura, H. Tomoda, N. Tabata, Y. Ohyama, T. Abe and M. Namikoshi, *J. Antibiot.*, 1999, **52**, 586.
57 N. Tabata, Y. Ohyama, H. Tomoda, T. Abe, M. Namikoshi and S. Omura, *J. Antibiot.*, 1999, **52**, 815.
58 D. Weigelt, R. Krähmer, K. Brüschke, L. Hennig, M. Findeisen, D. Müller and P. Welzel, *Tetrahedron*, 1999, **55**, 687.
59 N. El-Abadla, M. Lampilas, L. Hennig, M. Findeisen, P. Welzel, D. Müller, A. Markus and J. van Heijenoort, *Tetrahedron*, 1999, **55**, 699.
60 S. Riedel, A. Donnerstag, L. Hennig, P. Welzel, J. Richter, K. Hobert, D. Müller and J. van Heijenoort, *Tetrahedron*, 1999, **55**, 1921.
61 F.-T. Ferse, K. Floeder, L. Hennig, M. Findeisen, P. Welzel, D. Müller and J. van Heijenoort, *Tetrahedron*, 1999, **55**, 3749.
62 M.J. Sofia, N. Allanson, N.T. Hatzenbuhler, R. Jain, R. Kakarla, N. Kogan, R. Liang, D. Liu, D.J. Silva, H. Wang, D. Gange, J. Anderson, A. Chen, F. Chi, R. Dulina, B. Huang, M. Kamau, C. Wang, E. Baizman, A. Branstrom, N. Bristol, R. Goldman, K. Han, C. Longley, S. Midha and H.R. Axelrod, *J. Med. Chem.*, 1999, **42**, 3193.
63 A.K. Ganguly, J.L. McCormick, A.K. Saksena, P.R. Das and T.-M. Chan, *Bioorg. Med. Chem. Lett.*, 1999, **9**, 1209.
64 D.H. Williams and B. Bardsley, *Angew. Chem., Int. Ed. Engl.*, 1999, **38**, 1172.
65 K.C. Nicolaou, C.N.C. Boddy, S. Bräse and N. Winssinger, *Angew. Chem., Int. Ed. Engl.*, 1999, **38**, 2097.

66 K.C. Nicolaou, H.J. Mitchell, N.F. Jain, T. Bando, R. Hughes, N. Winssinger, S. Natarajan and A.E. Koumbis, *Chem. Eur. J.*, 1999, **5**, 2648.
67 C. Thompson, M. Ge and D. Kahne, *J. Am. Chem. Soc.*, 1999, **121**, 1237.
68 M. Adamczyk, J.A. Moore, S.D. Rege and Z. Yu, *Bioorg. Med. Chem. Lett.*, 1999, **9**, 2437.
69 H. Arimoto, K. Nishimura, T. Kinumi, I. Hayakawa and D. Uemura, *Chem. Commun.*, 1999, 1361.
70 D.L. Boger and H. Kai, *Angew. Chem., Int. Ed. Engl.*, 1999, **38**, 449.
71 A. Martin, T.D. Butters and G.W.J. Fleet, *Tetrahedron: Asymmetry*, 1999, **10**, 2343.
72 N. Chida, T. Suzuki, S. Tanaka and I. Yamada, *Tetrahedron Lett.*, 1999, **40**, 2573.
73 S. Hanessian, P.-P. Lu, J.-Y. Sanceau, P. Chemla, K. Gohda, R. Fonne-Pfister, L. Prade and S.W. Cowan-Jacob, *Angew. Chem., Int. Ed. Engl.*, 1999, **38**, 3160.
74 E. Shitara, Y. Nishimura, F. Kojima and T. Takeuchi, *J. Antibiot.*, 1999, **52**, 348.
75 M.M. Faul, L.L. Winneroski and C.A. Krumrich, *J. Org. Chem.*, 1999, **64**, 2465.
76 P. Moreau, F. Anizon, M. Sancelme, M. Prudhomme, C. Bailly, D. Sevère, J.-F. Riou, D. Fabbro, T. Meyer and A.-M. Aubertin, *J. Med. Chem.*, 1999, **42**, 584.
77 D.E. Zembower, H. Zhang, J.P. Lineswala, M.J. Kuffel, S.A. Aytes and M.M. Ames, *Bioorg. Med. Chem. Lett.*, 1999, **9**, 145.
78 M. Ohkubo, K. Kojiri, H. Kondo, S. Tanaka, H. Kawamoto, T. Nishimura, I. Nishimura, T. Yoshinari, H. Arakawa, H. Suda, H. Morishima and S. Nishimura, *Bioorg. Med. Chem. Lett.*, 1999, **9**, 1219.
79 M. Ohkubo, T. Nishimura, T. Honma, I. Nishimura, S. Ito, T. Yoshinari, H. Arawaka, H. Suda, H. Morishima and S. Nishimura, *Bioorg. Med. Chem. Lett.*, 1999, **9**, 3307.
80 D.E. Williams, V.S. Bernan, F.V. Ritacco, W.M. Maiese, M. Greenstein and R.J. Andersen, *Tetrahedron Lett.*, 1999, **40**, 7171.
81 K. Tatsuta, M. Takahaski and N. Tanaka, *Tetrahedron Lett.*, 1999, **40**, 1929.

20
Nucleosides

1 General

The neuroactive glyconucleoside disulfate HF-7 (**1**), produced by the funnel-web spider, has been shown by synthesis to have the illustrated structure. A regioselective synthesis used as an intermediate the orthoester **2**; stereoselective fucosylation was followed by base-sugar linkage and sulfation. The regioisomers L-fucosylated at O-2′ and O-5′, and the isomer with D-fucose attached at O-3′, were also prepared for comparison with the natural material which was available in extremely small amounts.[1]

Both 5-(carbamoylmethyl)uridine (**3**) and its 2-thiono-analogue have been isolated from human urine,[2] and the condensed tricyclic nucleosides of phenylalanine tRNAs, wyosine, wybutosine (**4**, R = H) and β-hydroxywybutosine (**4**, R = OH) have been isolated in sufficient amounts for unambiguous structural confirmation and comparison with synthetic samples.[3]

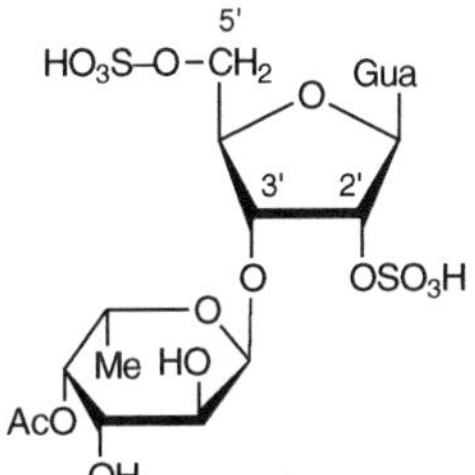

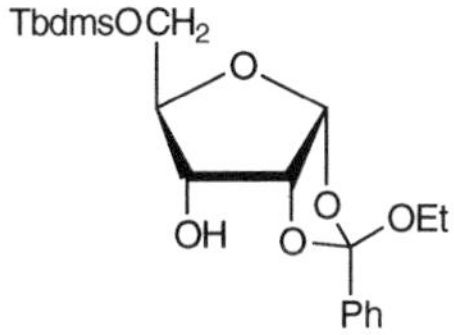

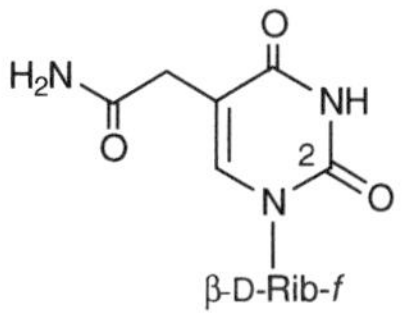

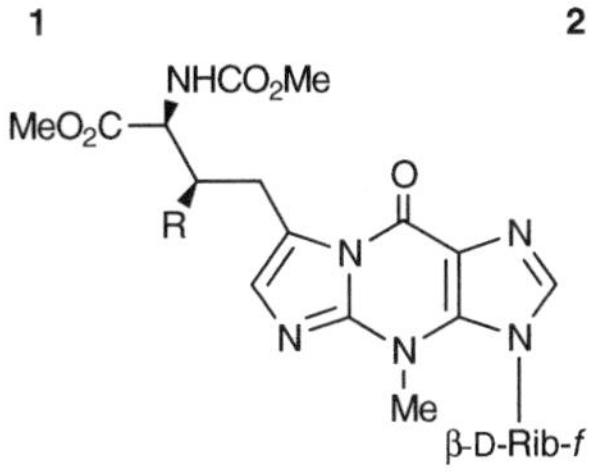

A review on recent results relevant to the possible origin of life includes sections on nucleobases, nucleosides and nucleotides.[4] Cook has given a perspective and review on the prospects for oligonucleotides as drugs,[5] whilst

Carbohydrate Chemistry, Volume 33

Secrist *et al.* have discussed their work on gene therapy of cancer, involving the delivery of *E. coli* nucleoside phosphorylase, followed by a relatively non-toxic nucleoside prodrug which is cleaved by the enzyme to a toxic compound.[6] Robins has reviewed mechanism-based inhibition of ribonucleotide reductases,[7] and the anti-HIV activity of AZT and a variety of other nucleoside analogues have been correlated with their electrostatic potential distributions, electron-density shapes and conformations to give a structure–activity profile that it is hoped will contribute to the development of QSAR in the area.[8] An account has been given of attempts in Townsend's laboratory to improve the stability and anti-HCMV activity of 2,5,6-trichloro-1-(β-D-ribofuranosyl)benzimidazole (TCRB).[9] Carell and co-workers have reviewed work on the synthesis of DNA lesions and oligonucleotides containing such structures.[10] Reviews have also appeared discussing nucleosides and nucleic acids containing carboranes[11] and the synthesis of L-nucleosides and the effects on structure and function of nucleic acids and oligonucleotides of incorporating L-ribose units.[12] The proceedings of the XIII International Round Table on nucleosides, nucleotides and their biological applications (some 370 short articles) have been published.[13]

2 Synthesis

A review has appeared which covers the topic of nucleoside synthesis by the use of glycosyl transferases.[14] A paper has discussed the lack of stereochemical control in the synthesis of β-nucleosides. The effects of different conditions were investigated, and the use of low temperatures and $SnCl_4$ as catalyst were recommended for high yields and stereoselectivities.[15] The synthesis of the IMP dehydrogenase inhibitor 5-ethynyl-1-β-D-ribofuranosylimidazole-4-carboxamide has been discussed, and some of the SAR data for this enzyme has been reviewed.[16]

Glycosylation of silylated pyrimidines with aryl 1-thioribofuranosides and 2-deoxy-1-thioribofuranosides has been carried out under electrolytic conditions in the presence of catalytic bromine or NBS; in the ribofuranosyl series, good β-selectivity was found in the synthesis of 5-fluorouridine if *O*-acetyl protection was used, whilst *O*-benzyl protection gave mostly α-nucleosides.[17] L-Ribonucleosides have been prepared by Vorbrüggen-type coupling with 2,3,5-tri-*O*-benzoyl-β-L-ribofuranosyl acetate, which was made from L-xylose, with inversion of stereochemistry at C-3.[18] The specifically-deuteriated nucleosides **5** (B = Ura, Ade, Cyt, Gua), required for incorporation into RNA to facilitate NMR studies, have been prepared from the appropriately-labelled sugar (Chapter 2).[19] [1,3,NH_2-$^{15}N_3$]-Adenosine, [2-^{13}C-1,3,NH_2-$^{15}N_3$]-adenosine and the same isotopomers of guanosine, 2′-deoxyadenosine and 2′-deoxyguanosine, have been prepared from appropriately-labelled purines using enzymatic transglycosylation.[20]

Reaction of 6,7-dichloroimidazo[4,5-*b*]quinoline-2-one with tri-*O*-benzoyl-ribofuranosyl acetate in the presence of $SnCl_4$ gave the N^3-riboside, whilst

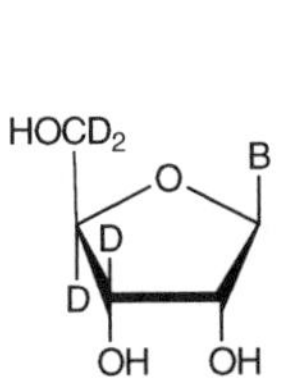

5

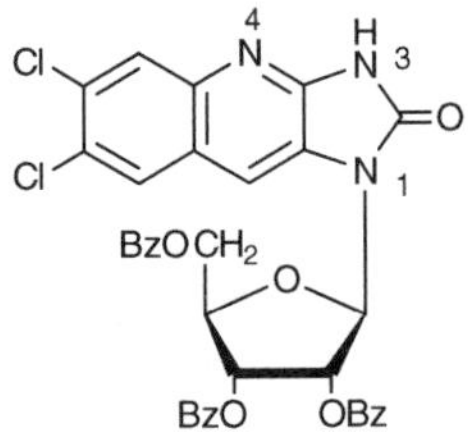

6

under silyl-Hilbert-Johnson conditions the N^1-regioisomer **6** was obtained. Under these latter conditions, reaction was thought to proceed *via* initial reaction at N^4, followed by the formation of a 1,4-diribosylated species and then loss of the N^4-substituent, since the protected N^4-riboside and the 1,4-disubstituted material could be isolated in small amounts under appropriate conditions.[21] 5-Nitro-1-β-D-ribofuranosylimidazole (**7**) could be obtained with some regioselectivity through the reaction of the silver salt of 4(5)-nitroimidazole with tri-*O*-acetyl-D-ribofuranosyl bromide. Phosphorylation at O-5′ and

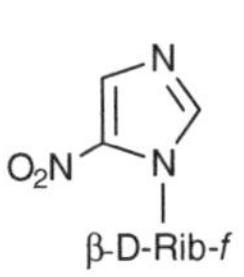

7 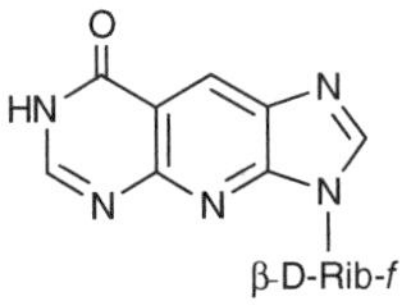 8 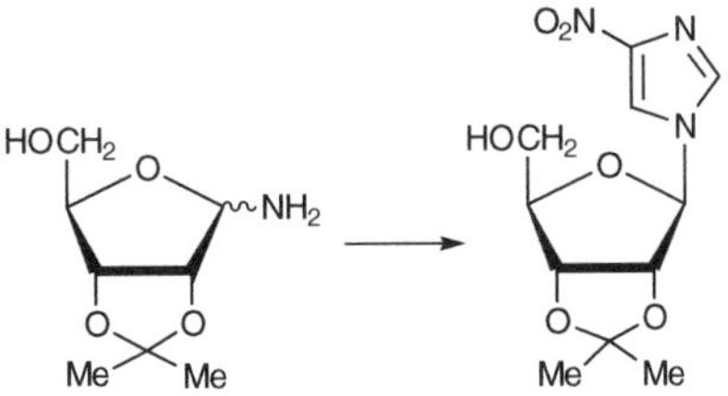 9 10

reduction of the nitrogroup gave the intermediate AIR in purine biosynthesis, and the 5-aminoimidazole could be elaborated to bicyclic and tricyclic systems such as **8**.[22] Reaction of the glycosylamine **9** with 1,4-dinitroimidazole gave the 5-nitroimidazole nucleoside **10**.[23] 3-Deaza-3-halopurine ribonucleosides such as **11** have been prepared from imidazole nucleosides; if the halogen is Cl, Br or I, the compounds exist in an *anti*-orientation about the glycosidic link, but 3-fluorocompounds can undergo free rotation.[24] The first stable benzoborauracil nucleoside **12** has been obtained, mostly as the α-anomer shown, by interaction of a substituted glycosylamine with methyl isocyanate.[25]

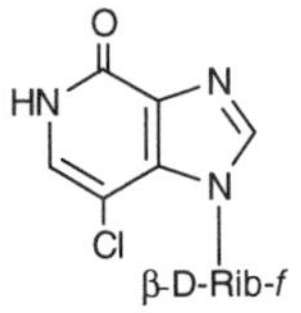

11 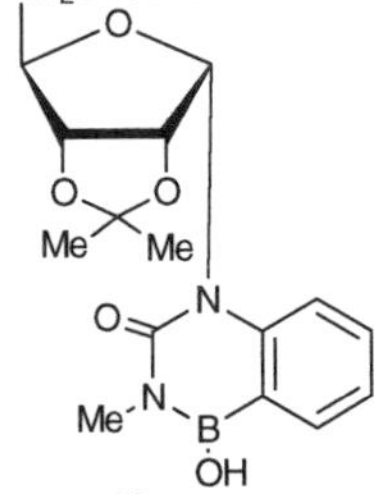 12 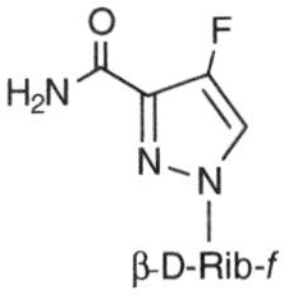 13

Standard coupling procedures have been used to make β-D-ribofuranosyl derivatives of 5-*o*-carboranyluracil (L-*ribo*-, D-*arabino*- and 2′-deoxy-compounds were also described),[26] 6-(trifluoromethyl)- and 6-(heptafluoropropyl)-purines,[27] 4-fluoropyrazole-3-carboxamide (the ribavirin analogue **13**),[28] 3,4-diaryl-4,5-dihydro-1,2,4-triazole-5-thiones (ribosylation at N-1),[29] 3-cyano-4,6-diarylpyridine-2-thiones,[30] and halogenated 4-quinolone-3-carboxylic acids.[31]

The synthesis of β-D-xylofuranosyl nucleosides has been reviewed,[32] and a paper from the same laboratory has described the synthesis of β-D-xylofuranosyluracil derivatives by condensations of silylated bases in the presence of $SnCl_4$.[33]

2-Amino-6-aziridino-9-β-D-arabinofuranosylpurine has been prepared by sugar-base condensation,[34] and arabinofuranosyl pyrimidine nucleosides of type **14** have been made from 5-bromouridine, inversion of configuration at C-2′ being carried out *via* a 2,2′-anhydronucleoside. The 4-methoxycompound **15**

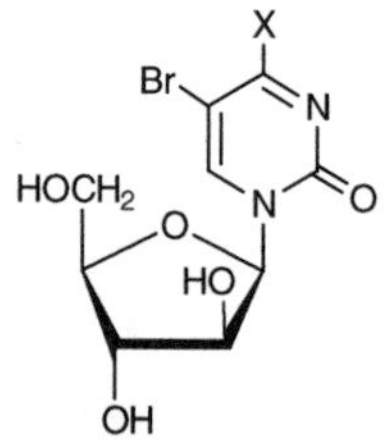

14 X = NR^1R^2
15 X = OMe

was also reported.[35] 9-(β-L-Arabinofuranosyl)adenine, which showed no antiviral activity, has been prepared from L-xylose by a method in which a β-L-xylofuranosyl intermediate was prepared and then subjected to stereochemical inversion at both C-2′ and C-3′ by a method previously used in the D-series (Vol. 18, p. 193).[36] A new route to ara-guanosine is mentioned in Section 5 below. The β-L-xylofuranosylbenzimidazole **16** gave the anhydronucleoside **17**, the structure of which was confirmed by X-ray crystallography, when treated as indicated in Scheme 1. Reaction with an amine than gave the β-L-lyxofuranosyl system **18**.[37]

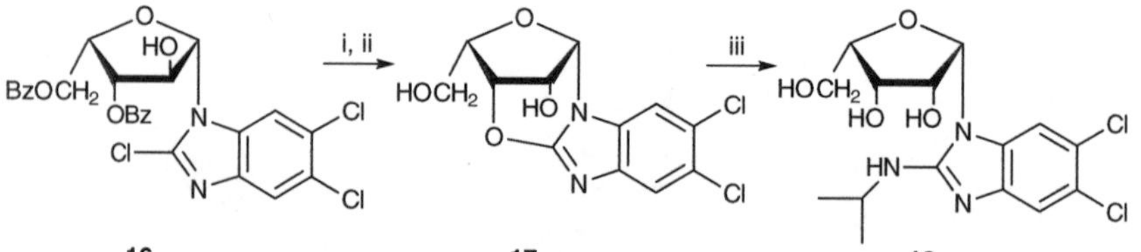

16 **17** **18**

Reagents: i, Tf_2O, py; ii, Na_2CO_3, EtOH, H_2O; iii, Pr^iNH_2

Scheme 1

A ribopyranosyl nucleoside involving tryptamine as base has been prepared by the indoline method and incorporated (2′→4′ links) into oligonucleotides, as part of an investigation into pyranosyl-RNA supramolecules containing non-hydrogen-bonding base pairs.[38] Xylopyranosyl nucleosides of type **19** (R = various alkyl/aryl) have been prepared conventionally,[39] and 2,3,4,6-tetra-*O*-benzyl-D-glucopyranose has been coupled to 6-chloropurine to give the β-nucleoside under modified Mitsunobu conditions, in which a basic group was incorporated into the triphenylphosphine and hence the resultant oxide, and di-*t*-butyl azodicarboxylate was used so as to give an acid-labile by-product.[40] Some glucosyl triazoles of type **20** ($n = 3$ or 4) have been prepared by

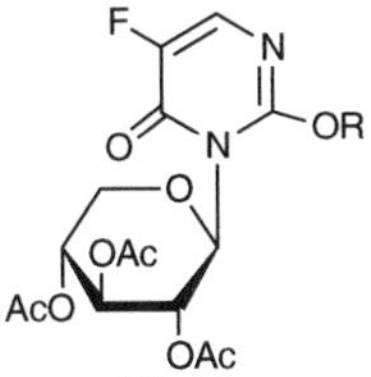

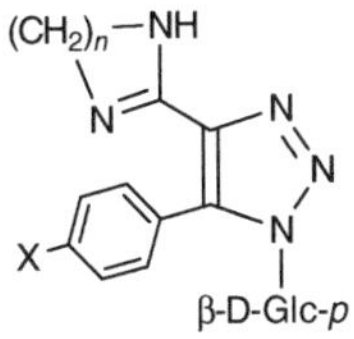

cycloadditions involving a glucosyl azide,[41] and β-D-galactopyranosyl nucleosides of substituted 1,2,4-triazol-3-thiones have been made conventionally,[42] as have both β-D-gluco- and -galactopyranosyl derivatives of 3-cyano-pyridin-2-ones and -pyridin-2-thiones,[43] 2-methylthiopyrimidin-6-ones,[44] and indano[1,2-*b*]pyridinethiones.[45] β-D-Glucopyranosyl nucleosides of 6-arylvinyl-1,2,4-triazines have also been described, with glucosylation occurring at both N-2 and N-4.[46]

3 Anhydro- and Cyclo-nucleosides

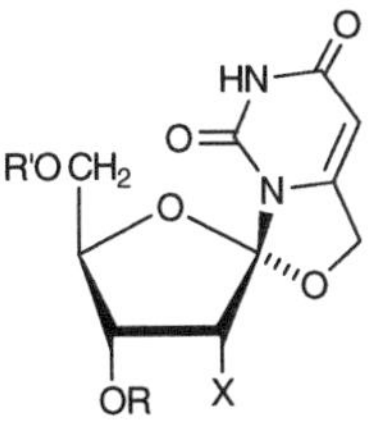

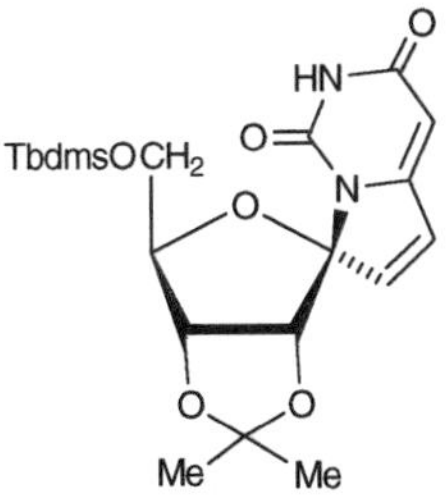

Kittaka and colleagues have given a full account of their work on the preparation of spirocyclic compounds of type **21** [X = H or O(protected)] through oxidative cyclization of 6-hydroxymethyluridine derivatives (see Vol. 31, pp. 269–270).[47] The same team have also given a full and extended account of the formation of spirocycles such as **22** through reductive radical cyclization of 6-(2,2-dibromoethenyl)uridines (see Vol. 30, pp. 270–271). Whilst the β-isomer **22** was the major product formed in the *ribo*-series, treatment of *arabino*-compound **23** with Bu_3SnH-AIBN gave mostly the α-compound **24**.[48] Very similar studies have also been underway in Chatgilialoglu's laboratory (see Vol. 30, p. 271 and Vol. 31, pp. 269–270), and a full account has been given of this work, covering cyclizations to give compounds of type **21** and also to form alkenyl-bridged systems related to **22**.[49] Treatment of the nitrile **25** with organolithium reagents at low temperature, followed by quenching after 20 minutes, gives the spiro-compounds **26** (R = Me, Bu^t); immediate quench of the reaction leads to the 1′-acyl-nucleosides, as previously reported (Vol. 32, pp. 298–299).[50]

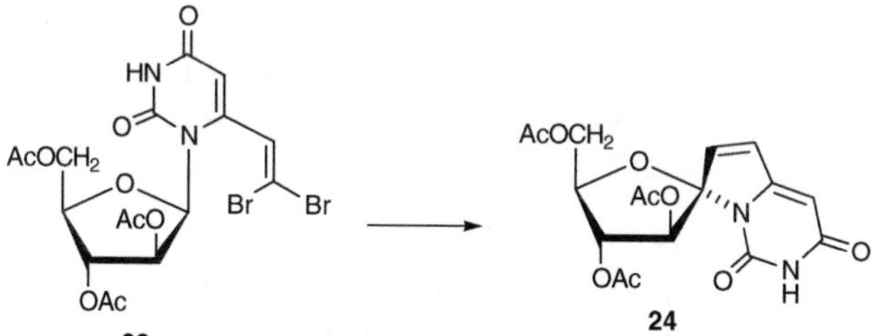

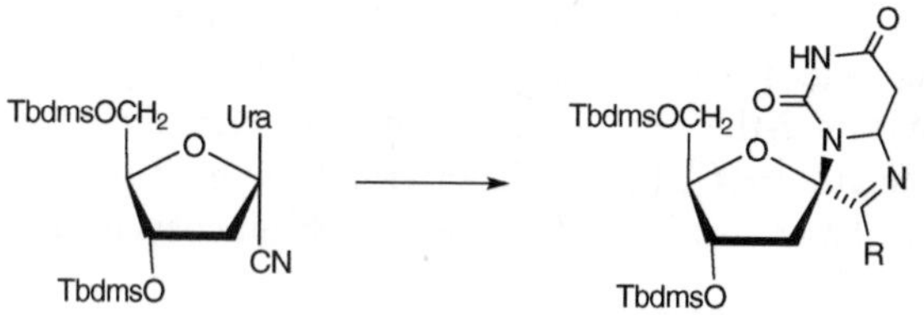

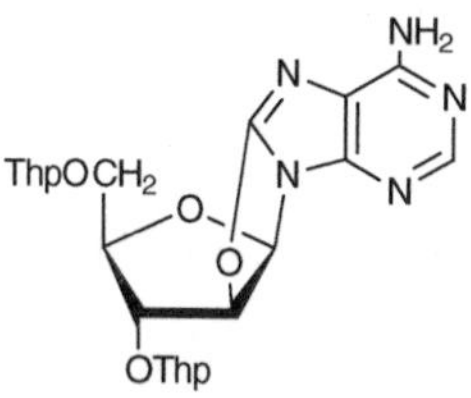

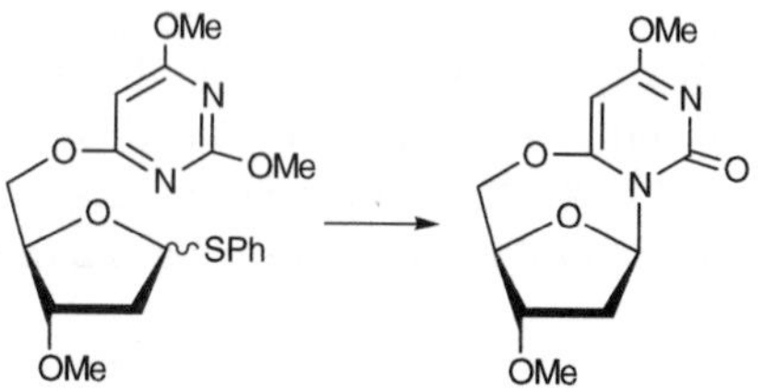

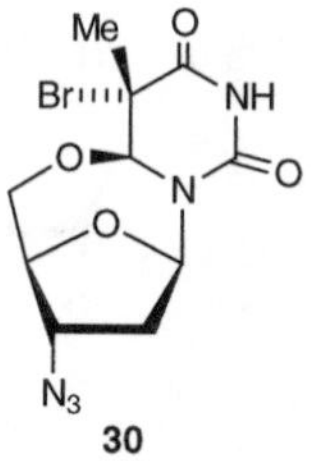

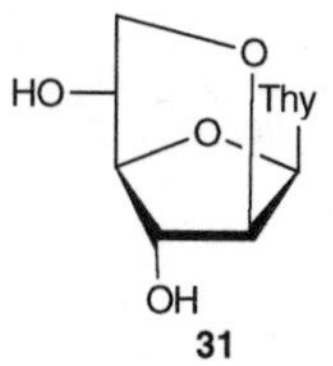

The 8,2′-anhydrocompound **27** was obtained when the 2′-*O*-acetylated derivative of 8-bromo-ara-A was treated with ammonia in methanol.[51] Intramolecular glycosidation of **28**, induced by dimethyl(thiomethyl)sulfonium tetrafluoroborate, led to the anhydronucleoside **29**.[52] Treatment of AZT with NBS in DMF led to the cyclized product **30** as the major isomer, together with the other *trans*-adduct.[53]

The 2′,6′-anhydro-altrofuranosylnucleoside **31** has been prepared by intramolecular displacement of a 2′-*O*-tosyl group; the furanose ring adopts an *S*-type conformation, from NMR measurements.[54]

4 Deoxynucleosides

Conventional base-sugar coupling procedures have been used in a new synthesis of 3-deaza-2′-deoxycytidine, suitably derivatized for incorporation into oligonucleotides,[55] 4-*O*-methyl-5-(2-hydroxyethyl)-2′-deoxyuridine,[56] triazole- and pentafluorophenyloxy-substituted 2′-deoxy-pyrimidine ribosides,[57] 5-(carboranylalkylmercapto)-2′-deoxyuridines,[58] 7-deaza-2′-deoxyinosine, where nucleobase anion glycosylation gave a higher yield than previously obtained under different conditions (Vol. 19, p. 200),[59] 7-halogenated-7-deaza-2′-deoxyinosines[60] and 7-halogenated 8-aza-7-deaza-2′-deoxyguanosines (**32**, X = Br, I), which had a high-*anti*-conformation as demonstrated by NMR and X-ray crystallography.[61] Two regioisomeric 2′-deoxyribonucleosides were obtained from 5-(3,3-diethyl-1-triazeno)pyrazole-4-carbonitrile, and these were subsequently hydrolysed to the corresponding carboxamides.[62]

A synthesis of 2′-deoxy-[8-^{13}C-9,NH_2-$^{15}N_2$]-adenine has been described, in which enzymic transglycosylation (thymidine, thymidine phosphorylase and purine nucleoside phosphorylase) was used to link the labelled base to 2-deoxyribose.[63] Other workers have used a similar approach to prepare 2′-deoxy-[1,3,NH_2-$^{15}N_3$]-adenosine, 2′-deoxy-[2-^{13}C-1,3,NH_2-$^{15}N_3$]-adenosine and the same isotopomers of 2′-deoxyguanosine.[20] An automated radiosynthesis of 2-[^{11}C]-thymidine has been described.[64]

A review has been published concerning synthetic methodology for the preparation of 2′,3′-dideoxynucleosides and their analogues.[65] A chemoenzymatic synthesis of 2′,3′-dideoxyinosine (ddI) from 2′-deoxyadenosine involves deamination with adenosine deaminase, regioselective acetylation at O-5′ using *Candida antarctica* lipase, and chemical deoxygenation.[66] The analogues **33** (X = CH or N) of ddI, which should not be substrates for purine nucleoside phosphorylase, have been prepared using conventional deoxygenations,[67] 2′,3′-Dideoxythymidine has been prepared from 2′-deoxythymidine in four steps,[68] and 3′-deoxythymidine has been prepared stereoselectively from D-xylose.[69] A full account has been given of the synthesis of β-L-2′,3′-didehydro-2′,3′-dideoxypurine nucleosides **34** (B = Ade, Gua, Hx, 2-F-Ade), and the reduced analogues, with D-glutamic acid acting as the source of the sugar unit (see Vol. 30, p. 275).[70] The didehydro-dideoxy analogue **35** of 5-fluorocytosine, which has anti-HIV activity, has been prepared by a procedure in which base–sugar coupling was directed with β-selectivity using a 2-α-phenylselenyl group in the sugar, which subsequently gave rise to the alkene after oxidation. Bromo- and iodo-analogues of **35** were also reported, as was the uracil compound and the enantiomer of **35**, which was bioactive.[71] The d4 analogue of 8-bromoadenosine has been prepared and converted into other compounds by nucleophilic displacement of the bromine.[72]

5′-Deoxythymidine has been prepared by reductive removal of a 5′-bromo-substituent using 1,1,2,2-tetraphenyldisilane and AIBN.[73]

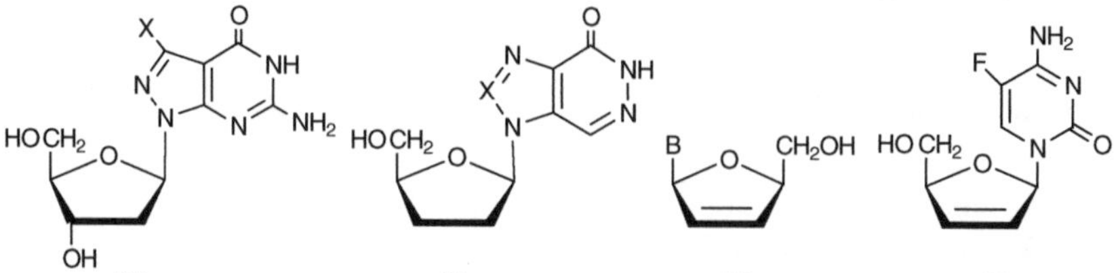

32 **33** **34** **35**

5 Halogenonucleosides

There has been a report on the synthesis of 2′-bromo-2′-deoxyadenosine and 2′-deoxy-2′-iodoadenosine (**36**, B = Ade, X = Br, I) from ara-A by inversion of configuration at C-2′. The same paper also describes a direct small-scale synthesis of *ara*-guanosine from guanosine, and the use of the product to make 2′-bromo-2′-deoxy- and 2′-deoxy-2′-iodoguanosine (**36**, B = Gua, X = Br, I).[74] 8-Bromo-2′-deoxy-2′-fluoroadenosine (*e.g.* **36**, B = 8-Br-Ade, X = F) has been made from 8-bromoadenosine by a double-inversion procedure, and the 8-oxo-compound was also described.[51] In a significantly improved route to 2′-deoxy-2′-fluoro-arabinofuranosyl purine nucleosides the selectively-protected nucleoside **37** was prepared using benzoylation of a 2′,3′-dibutylstannylidene derivative, and subjected to fluorination with inversion using DAST, the presence of the chloro-substituent reducing the tendency for competitive intramolecular participation by N-3. The product was converted to the fluorinated ara-A **38** (X = OH), and also to the 3′-deoxy-compound FddA **38** (X = H).[75] The 8-aza-7-deazapurine nucleoside **39** (X = OMe) has been prepared, together with its N^2-regioisomer and the corresponding α-compounds, by base-sugar linkage; the amino-compound **39** (X = NH_2) was made from the methoxy-derivative, and could itself be converted into the inosine analogue using adenosine deaminase.[76]

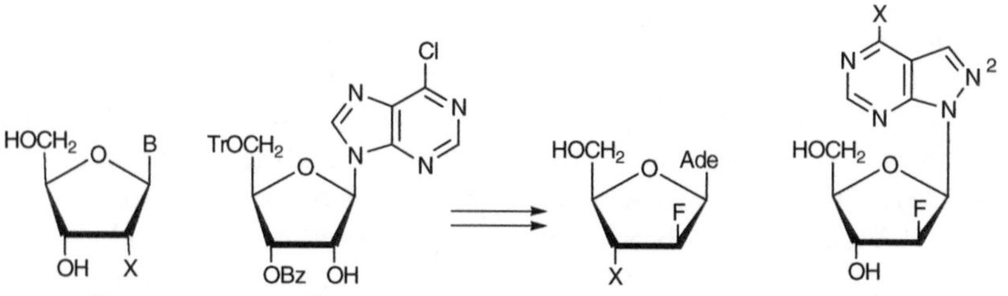

36 **37** **38** **39**

An expanded account has been given of the synthesis of 2′-fluoro-2′,3′-unsaturated L-nucleosides **40** (see Vol. 32, p. 264, and Vol. 24, pp. 230–231 for the D-series) and the α-anomers, in a synthesis that uses isopropylidene-L-glyceraldehyde as the source of chirality for the modified sugar unit, now including cases with both pyrimidine and purine bases. The β-anomers with B = Cyt, 5-F-Cyt and Ade had moderate-to-potent anti-HIV activity.[77] A new synthesis of the anti-HBV agent L-FMAU (**41**) has been reported, in which the sugar unit is derived from L-arabinose (Chapter 8), prior to silyl coupling with the base (for an earlier route from L-xylose, see Vol. 30, pp. 118 and 276).[78] A more comprehensive study has been reported (see Vol. 29, p. 275 for earlier work) on the synthesis of 2′,3′-dideoxy-2′-fluoro-L-*threo*-pentofuranosyl nucleosides of type **42**, and the α-anomers, with a range of purine and pyrimidine bases, the sugar being prepared from L-xylose (Chapter 8). Only the cytidine analogue had moderate anti-HIV and anti-HBV activity.[79] The fluorinated thionucleoside **43** has been prepared using bis(2-methoxyethyl)aminosulfur trifluoride, a more thermally-stable variant on DAST.[80]

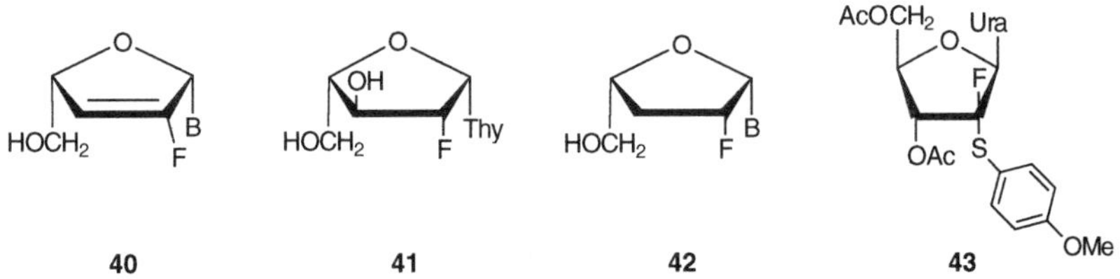

40 **41** **42** **43**

3′-Deoxy-3′-fluoro-derivatives of 5-amino-1-(β-D-ribofuranosyl)imidazole-4-carboxamide and 4-amino-1-(β-D-ribofuranosyl)imidazole-5-carboxamide have been prepared, but did not show antiviral or antitumour activity.[81] A new route to 2′,3′-dideoxy-3′-fluoronucleosides involves the formation of **44** (X = OMe) from D-xylose, its conversion into the phenylselenyl glycoside **44** (X = SePh), and linkage with 6-chloropurine under Mitsunobu conditions to give the 2′-phenylselenonucleoside **45** as a 1:1.8 mixture of α- and β-anomers. The phenylselenyl group could then be reductively removed.[82] The 3′-deoxy-3′-fluorothymidine derivative **46** could be formed with good stereoselectivity by coupling silylated thymine with the furanosyl diethyl phosphite in the presence

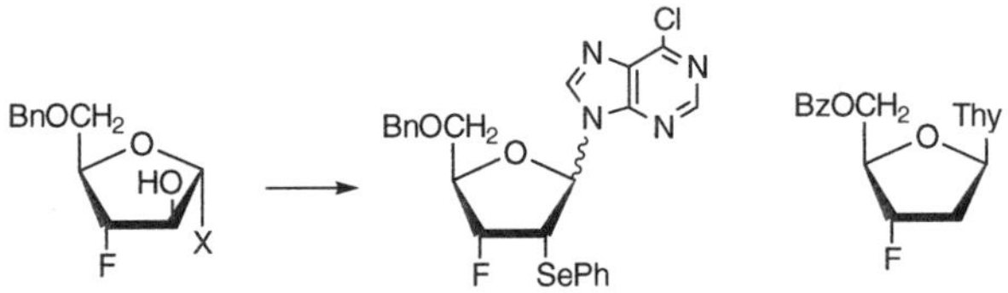

44 **45** **46**

of TmsOTf, provided propionitrile was used as solvent, but the case with the fluorine ‘up’ was not stereoselective.[83]

In a new approach to 3′-deoxy-3′-halogenated xylofuranosyl pyrimidine nucleosides (Scheme 2), the cyclic sulfate **47** was prepared, *N*-nitration being carried out to prevent anhydronucleoside formation. This intermediate could then give rise to the chloro- and bromo-derivatives **48** (X = Cl, Br). Use of NaI in the ring opening gave the iodo-analogue with concomitant denitration, but reaction with tetrabutylammonium azide was unsuccessful due to competing attack at C-4. The 2′,3′-ribofuranosyl epoxide reported by the same group last year (see Vol. 32, p. 258) did not react cleanly with nucleophiles due, it was thought, to the same problem.[84]

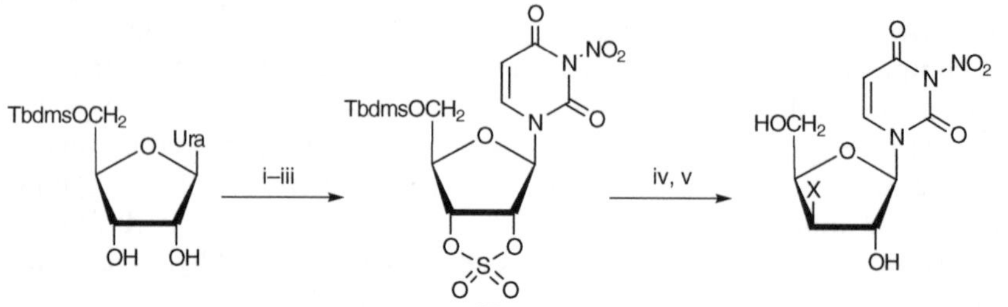

Reagents: i, $SOCl_2$, py; ii, CF_3COONO_2; iii, oxone, $RuCl_3$; iv, Bu_4NX (X = Cl, Br); v, H_2SO_4, H_2O, THF

Scheme 2

The *gem*-difluorinated pyranosyl systems **49** and **50**, which can be regarded as ring-expanded analogues of gemcytabine, have been prepared, along with α-anomers, through coupling of fluorinated sugar units to silylated thymine or uracil. The difluoromethylene groups were introduced by reaction of appropriately protected uloses with DAST. The compounds did not have anti-HIV activity.[85]

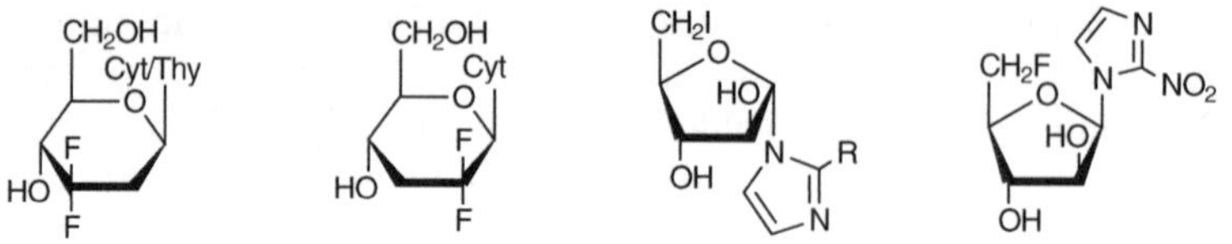

Iodoaminoimidazole arabinoside (IAIA, **51**, R = NH_2), a potential reductive metabolite of iodoazomycin arabinoside (IAZA, **51**, R = NO_2), which is used when labelled with ^{123}I to mark hypoxic tissue, has been prepared by base–sugar coupling followed by iodination, in a study which confirmed the α-configuration of IAZA.[86] The related fluorinated 2-nitroimidazole nucleoside **52** has been prepared with deuterium or tritium labels at C-5′.[87]

A number of papers discussed in later sections mention compounds in which other structural features are present in addition to halogen atoms.

6 Nucleosides with Nitrogen-substituted Sugars

The 2′,3′-dideoxy-2′-dimethylamino-nucleosides **53** and **54** have been prepared as racemates by coupling thymine to 'sugar' units prepared using cycloaddition chemistry.[88] Conjugate addition of *N*-methylhydroxylamine to unsaturated lactone **55** was used in a route to the nucleoside analogues **56** (B = Ura, Thy, Cyt), which were produced along with their α-anomers.[89]

A further synthesis of AZT from D-xylose has been described,[90] and 2′,3′-dideoxy-3′-[5-(4-fluorophenyl)-2-tetrazolyl]thymine has been reported.[91]

A study of the thermolysis of sugar-modified uridine- *N*-oxides has led to novel nucleoside analogues. The *N*-oxide of 3′-deoxy-3′-morpholino-arabinofuranosyluracil generated double bonds in the sugar ring without much specificity, whereas *N*-oxides of 2′,3′-epiminouridines gave only 2′,3′-didehydro-2′,3′-dideoxyuridine. Treatment of **57** with MCPBA gave the *O*-morpholino-anhydronucleoside **58**, whilst the cyclonucleoside **59** gave the oxazolidine **60**.[92]

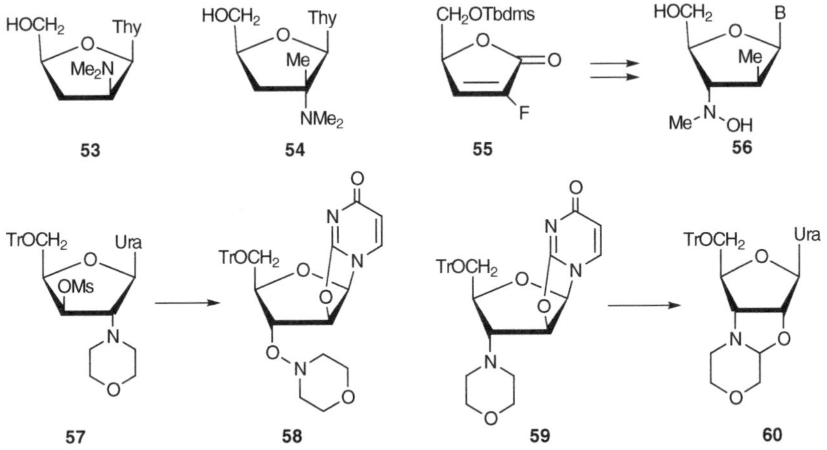

Intramolecular glycosidation of **61** using dimethyl(thiomethyl)sulfonium tetrafluoroborate, followed by reaction with tetrabutylammonium azide, gave the 5′-azido-2′,5′-dideoxynucleoside **62** ($X = N_3$).[93] A route to 5′-deoxy-5′-*N*-hydroxylaminonucleosides has been developed in Miller's laboratory. For example, treatment of adenosine with BocNHOCbz under Mitsunobu conditions gave **63** in good yield, and this could be deprotected to give the *N*-alkylhydroxylamine. The method can also be applied to pyrimidine systems, although uridine gave considerable amounts of the 2,5′-anhydrocompound.[94]

The analogue **64** of *S*-adenosylmethionine has been made *via* alkylation of a 5′-deoxy-5′-*N*-methylamino-derivative of adenosine, and the *N,N*-dimethyl compound was also prepared; this latter has a positive charge at the position corresponding to the sulfonium centre of SAM, a feature which **64** possesses only at sufficiently low pH values.[95] The sulfamoyl derivative **65** of 5′-amino-5′-deoxyadenosine has been prepared as an analogue of tyrosyl adenylate and a potential inhibitor of prokaryotic tyrosyl t-RNA synthetases,[96] whilst a report from Fleet's laboratory describes the synthesis of the peptidomimetic **66** of UDP-Gal*f*, a potential inhibitor of UDP-galactopyranose mutase.[97]

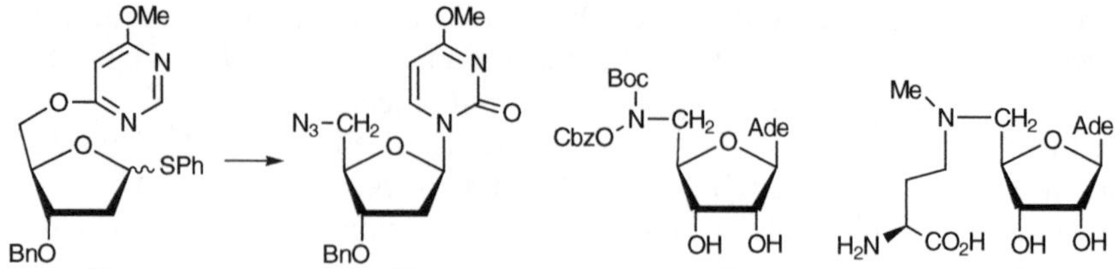

The building block **67** has been prepared for use in the synthesis of peptide nucleic acid (PNA)-deoxynucleic guanidine (DNG) chimeras, which show good binding with DNA.[98]

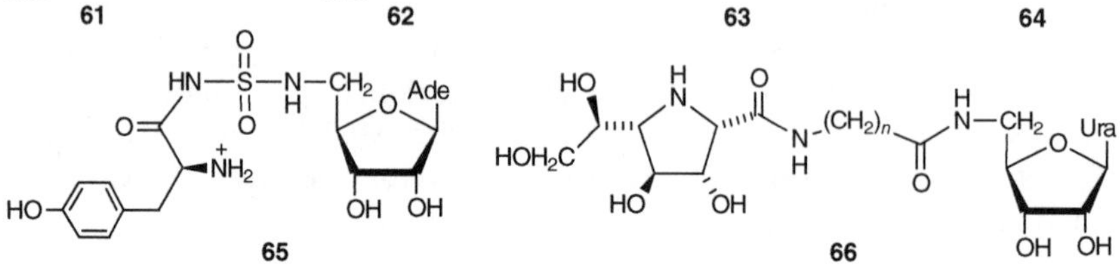

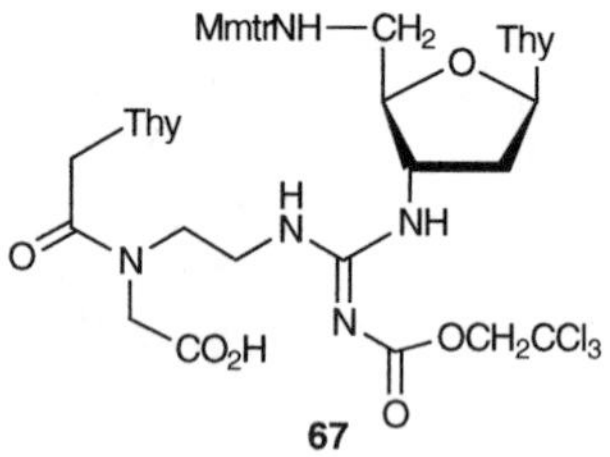

The kinetics of hydrolysis of diaminoanalogues of 2′- or 3′-deoxyadenosine and of 9-(2- and 3-deoxy-β-D-*threo*-pentofuranosyl)adenine have been studied in buffers of various pH values. The rate of hydrolysis at acid pH was found to be related to the position and configuration of the amino group on the sugar moiety, and the diamino compounds in this study were found to be more stable than corresponding monoaminated nucleosides.[99]

The photolytic removal of an *N*-tosyl group in aqueous acetonitrile has been employed in the synthesis of some new 5′-amino-5′-deoxy-analogues of AZT.[100] Some references to nucleoside analogues with pyrrolidine rings are given in Section 15 below.

7 Thio- and Seleno-nucleosides

A practical synthesis of 3′-thioguanosine (**68**) has been developed from the nucleoside itself, which also permitted **68** to be converted into an appropriately-protected 3′-thiophosphoramidite for incorporation into modified oligonucleotides.[101] The 2-thioquinoline unit has been developed as a masked thiol, and applied in the synthesis of the 3′-thiothymidine derivative **70**. The thioether **69** was prepared by displacement of a mesyloxy group by 2-thioquinoline in the presence of DBU; after exchange of protecting groups, the thiol could be liberated by reduction with $NaBH_3CN$ in glacial acetic acid, followed by the addition of water to hydrolyse the presumed thioaminal intermediate.[102] The use of this chemistry in making 3′-thioformacetal internucleotidic links is mentioned below (Section 13).

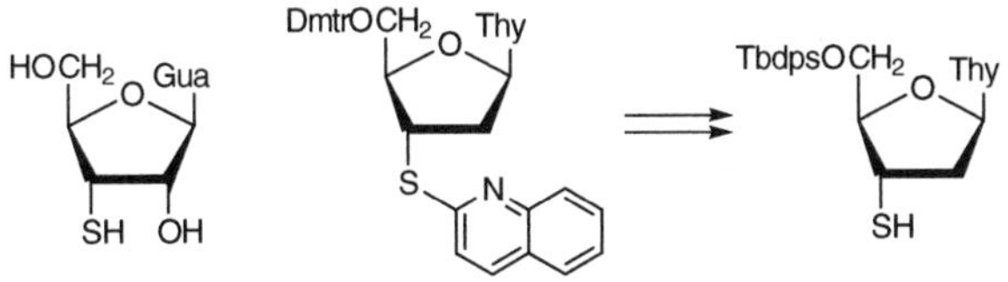
HOCH2
Gua
SH OH
DmtrOCH2
Thy
S
N
TbdpsOCH2
Thy
SH

68 **69** **70**

A number of further reports have been concerned with the synthesis of 4′-thionucleosides ('sulfur-in-ring' systems). A paper from Matsuda's laboratory has described an investigation of the suitability of the Pummerer reaction for the stereoselective synthesis of 4′-thionucleosides . When the *meso*-sulfoxide **71** was treated with thymine, TmsOTf and triethylamine, the *trans*-nucleoside analogue **72** was obtained in 70% yield, and with 30:1 stereoselectivity, the 2,4-

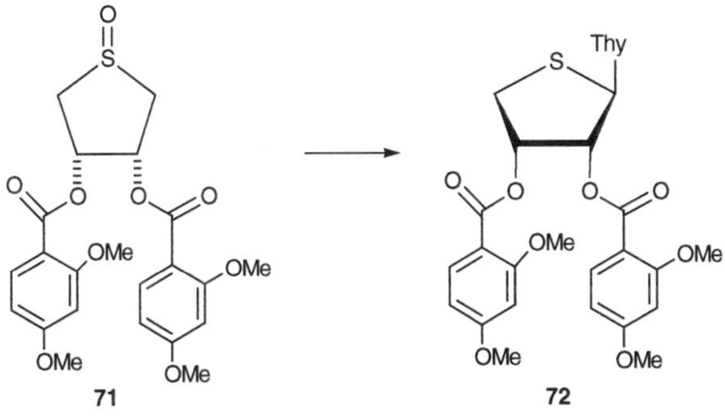
OMe
OMe
OMe
OMe
Thy
S
OMe
OMe
OMe
OMe

71 **72**

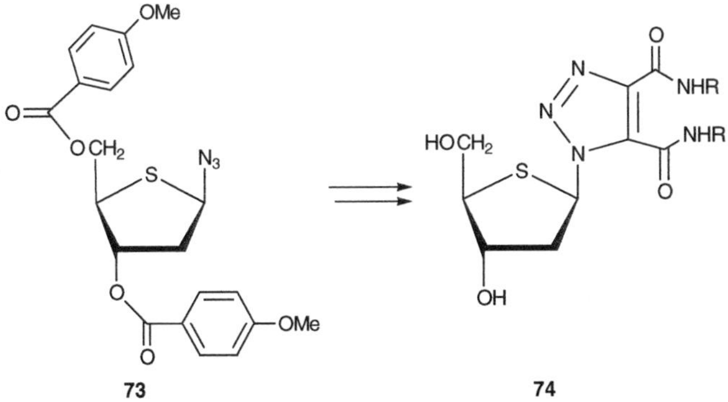
OMe
OCH2
S
N3
OMe
NHR
NHR
HOCH2
S
N
OH

73 **74**

dimethoxybenzoyl unit proving better in terms of both yield and stereocontrol than a number of other substituted benzoyl groups that were investigated.[103] A number of 2′-deoxy-4,5-disubstituted-4′-thio-imidazole nucleosides have been prepared by base–'sugar' coupling,[104] and triazoles of type **74** have been made using the cycloaddition of DMAD with the β-glycosyl azide **73**, where use of *p*-methoxybenzoyl protection for O-3′ and O-5′ ensured good β-selectivity in the formation of **73** from the 1-*O*-acetyl-thiosugar.[105] The kinetics of the acid-catalysed hydrolysis of purine and cytosine 2′-deoxy-4′-thionucleosides has been studied; all proved to be more stable than the oxygen analogues, presumably due to the lowered stability of the intermediate carbocation, and other evidence from pH–rate profiles was presented to support the contention that the mechanisms were similar in the oxygen and sulfur cases.[106]

The fluorinated 4′-thiocytosine **76** (4′-thioFAC) is an orally-active anti-tumour agent. A new synthesis of this compound has been reported, in which the fluorinated epoxide **75**, prepared from 1,2:5,6-di-*O*-isopropylidene-α-D-

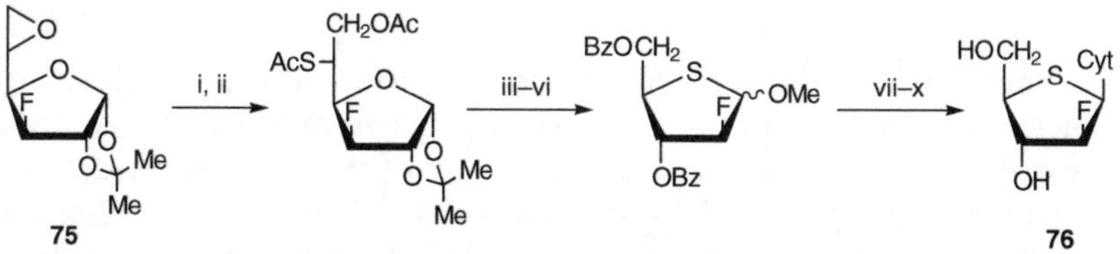

Reagents: i, thiourea, MeOH; ii, KOAc, Ac_2O, AcOH; iii, H_2O, TFA; iv, $NaIO_4$; v, MeOH, HCl; vi, BzCl, Py; vii, Ac_2O, AcOH, H_2SO_4; viii, HBr, AcOH; ix. silylated *N*-Ac-Cyt; x, NH_3, MeOH

Scheme 3

allofuranose, is manipulated as outlined in Scheme 3. The use of a glycosyl bromide in the base–sugar coupling, based on early work on the synthesis of the oxygen analogue, was important in ensuring reasonable β-selectivity.[107, 108] The route has been extended to the preparation of the guanosine analogue of **76**, and also the 2′-azido-analogue (4′-thiocytarazid).[108]

The analogue **77** of BVDU has been prepared *via* the uracil derivative, onto which the C-5 substituent was attached (for earlier routes to similar compounds, see Vol. 29, p. 279 and Vol. 28, p. 275).[109] The 3′-*C*-methyl-4′-thio-apionucleoside **78** has been prepared, together with its α-anomer, by a sequence in which the quaternary chiral centre at C-3′ was established with high enantiomeric purity using a Johnson-Claisen rearrangement carried out on an allylic alcohol, the chirality of which arose from isopropylidene-D-glyceraldehyde.[110] A synthesis of the thietane analogue of oxetanocin is mentioned in Chapter 19.

Intramolecular glycosidation of **61** induced by dimethyl(thiomethyl)sulfonium tetrafluoroborate, followed by reaction with aryl thiols and Hünig's base led to 5′-arylthionucleosides **62** (X = ArS).[93] The 3′-deoxy-5′-thionucleoside **79**, a potential inhibitor of cytoplasmic thymidine kinase, has been prepared either from the diol (Vol. 26, p. 249) or by base–sugar condensation.[111] Various 5′-alkylthio-N^6-(aryl-substituted)benzyl-adenosines have been prepared, which

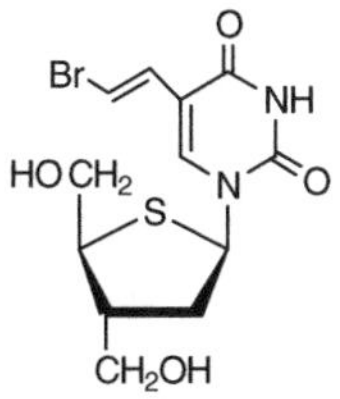

77 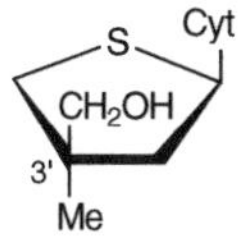 78 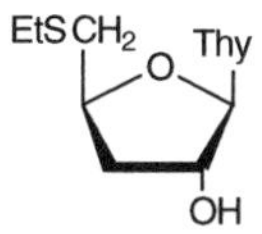 79 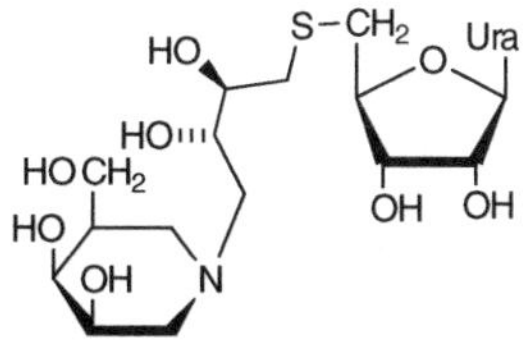 80

combine a substructure to target the adenosine A_3 receptor with the alkylthio-substituent known to give partial agonism for the A_1 receptor (Vol. 32, pp. 270–271).[112] An approach to the rational design of inhibitors of α-1,3-galactosyl transferase led to the synthesis of the UDP-Gal analogue **80**, by a procedure in which a 5′-tosylate was displaced from a uridine derivative by the thiolate of a monothiothreitol derivative. The analogue was indeed a potent inhibitor of α-1,3-galactosyl transferase, which transfers galactose from UDP-Gal by a double-inversion mechanism, but it did not inhibit β-1,4-galactosyl transferase, which operates by a direct displacement.[113]

The nucleoside selenocyanates **81** and **82** have been obtained in good yields by displacements of trifluoroethyl sulfonates (tresylates) using tetrabutylammonium selenocyanate. The 3′-epimer of **82** could only be obtained from the 'down'-tresylate in 20% yield, the major product being the 2,3′-anhydronucleoside, which was unreactive towards tetrabutylammonium selenocyanate.[114]

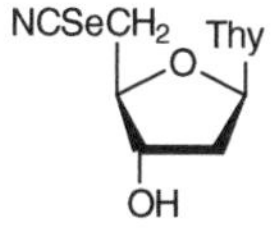

81 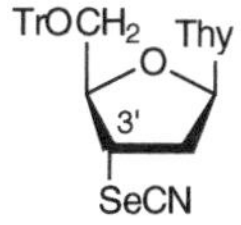 82

8 Nucleosides with Branched-chain Sugars

Wengel has reviewed work, primarily that carried out in his laboratory, on the synthesis of 3′-*C*- and 4′-*C*-branched nucleosides, the incorporation of these into oligonucleotides, and the development of 'locked nucleic acid' (LNA).[115] A report on the use of radical chemistry in the synthesis of medium-sized rings includes references to applications in the branched-chain nucleoside area of radical cyclizations using silicon tethers.[116]

The 2′-deoxy-2′-methylene-analogues of 6-azauridine and 6-azacytidine have been prepared from 6-azauridine by oxidation and Wittig methylenation of the 3′,5′-*O*-Tipds derivative, whilst the 2′-deoxy-2′-methylene analogues of 5-azacytidine and 3-deazaguanosine, together with their α-anomers, were made by chemical transglycosylation of a 2′-deoxy-2′-methyleneuridine derivative.[117] 2′-Deoxy-2′-β-*C*-ethynylguanosine (**83**) has been prepared by a method like that used for similar analogues with other bases (see Vol. 29, p. 276), and

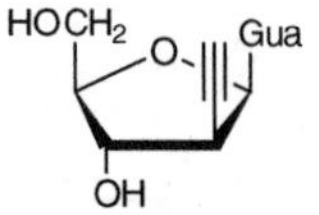

HOCH2 O Gua OH 83 · BzOCH2 O Ura OBz X 84 ⇛ HOCH2 O Ura OH C=C=CH2 H 85 · HOCH2 O Ura BzO OH CF2H 86

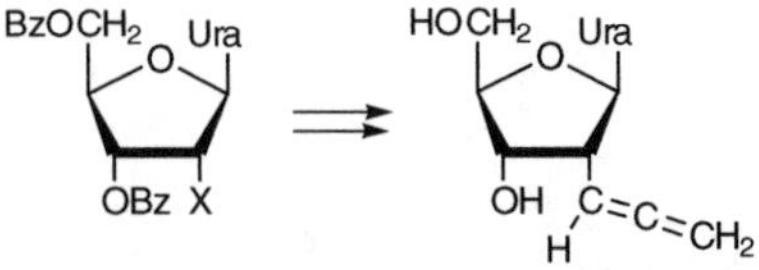

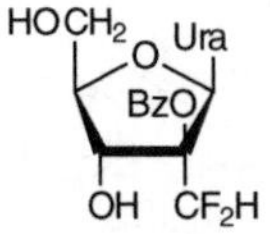

incorporated into oligonucleotides, where substitution of two or three units of **83** for deoxyguanosine in $d(CG)_3$ or $d(GC)_3$ leads to a Z-DNA-like conformation.[118]

Reaction of the halides **84** (X = Br or I) with propargyltriphenylstannane in the presence of AIBN gave, after deprotection, 2′-α-*C*-allenyl-2′-deoxyuridine (**85**) in good yield, and use of allyltributylstannane gave rise to the α-*C*-allyl derivative.[119] 2′-β-*C*-Methylcytidine has been made by base-sugar coupling in a manner similar to that described previously (see, *e.g.*, Vol. 31, p. 277); it was incorporated into 3′,5′-oligonucleotides which were found to be subject to base-catalysed degradation in the same way as is RNA itself.[120] The *C*-difluoromethyl nucleoside **86** was prepared by addition of the anion of $PhSO_2CF_2H$ to a 3′,5′-protected-2′-one, and was incorporated into a hammerhead ribozyme.[121]

3′-*C*-Methylene- and 2′-*C*-methyl-3′-*C*-methylene-3′-deoxythymidine have been prepared by sequences in which the base is linked at a late stage to a unit prepared using asymmetric epoxidation.[122] The enone **87** has been prepared in six steps from 5-*O*-benzyl-1,2-*O*-isopropylidene-α-D-xylofuranose.[123] The protected nucleoside **88** has been made by base–sugar linkage in a manner similar to that used earlier by others (Vol. 27, p. 256), and was incorporated into oligonucleotides (with the 2′-*O*-Mem group still present). The oligomers had lower affinity for complementary RNA and DNA than did the natural oligomers (for similar structures involving 3′-hydroxymethylthymidine see Vol. 30, p. 290).[124] Workers at Novartis have prepared 3′-α-carboxymethyl-3′-deoxy-2′-*O*-methylribonucleosides **89** (B = Thy, Ade, Gua), for incorporation into amide-modified oligonucleotides, by sequences involving the nucleosides as starting materials,[125] whilst others, with a similar aim in mind, have described the synthesis of **90** by a route which involves base–sugar linkage at a late stage, after introduction of the side-chain.[126] Yet

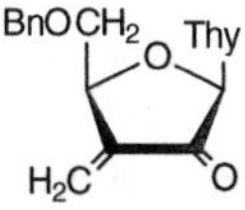

BnOCH2 O Thy H2C O 87 · DmtrOCH2 O Thy HOCH2 OMem 88 · HOCH2 O B CH2 OMe CO2Me 89

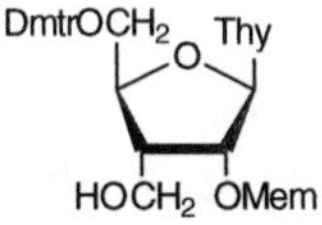

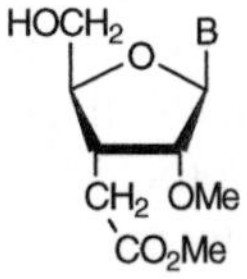

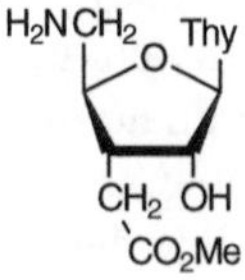

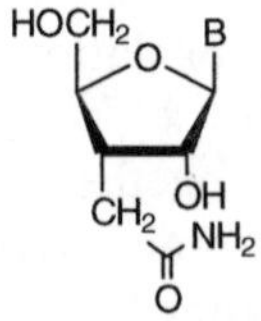

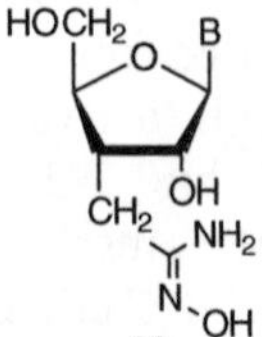

H2NCH2 O Thy CH2 OH CO2Me 90 · HOCH2 O B CH2 OH NH2 O 91 · HOCH2 O B CH2 OH NH2 N OH 92

others have described the synthesis of amides and *N*-hydroxyamidines of types **91** and **92**.[127]

The 3′-*C*-acylated thymidines **94** (R = Bu^t, Ph, Me) were made from the silylated cyanohydrin **93**, the major product from addition of TmsCN to the 3′-ketone, through reaction with organolithium reagents. The C-3′-epimers were also prepared from the known 5′-*O*-Dmtr-3′-*C*-hydroxymethylthymidine (Vol. 28, p. 278). The *C*-pivaloyl compounds on photolysis gave radicals at C-3′ which abstracted hydrogen from Bu_3SnH to give 5′-*O*-Dmtr-thymidine and its C-3′-epimer, the efficiency of radical production being independent of the stereochemistry at C-3′.[128]

The 3′-*C*-allyl-2′-*O*-methylnucleoside **95** has been made from a previously-reported intermediate (Vol. 31, pp. 277–278); units of both it and the known bicyclic compound **96** were incorporated into oligonucleotide dodecamers together with thymidine, but the modified oligonucleotides showed reduced T_m values.[129]

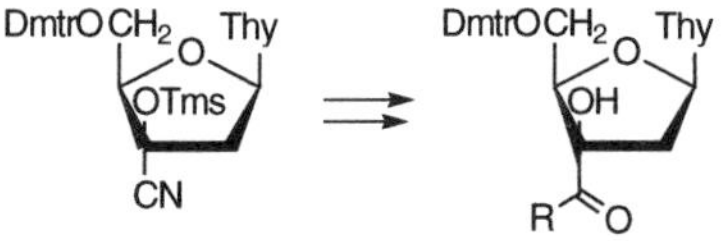

93 **94** 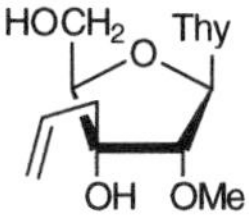 **95** 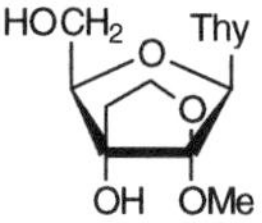 **96**

The spirocyclic nucleoside analogue TSAO-T has been linked from N-3 through a spacer to the base unit of AZT or d4U, to give a hybrid which combines a non-nucleoside reverse transcriptase inhibitor with a nucleoside RT inhibitor. The dimer with the d4 structure and a trimethylene spacer was ~10 times more inhibitory to HIV-1 than was an AZT heterodimer.[130]

There continues to be considerable interest in nucleoside analogues branched at C-4′, and bicyclic compounds derived from them. The 4′-*C*-methyl analogues **97** (R = H, OH) of BVDU have been prepared by base–sugar linkage, using a 'down' acetoxy group at C-2′, subsequently removed or inverted, to control stereochemistry. The 2′-deoxycompound **97** (R = H) had strong anti-varicella zoster activity *in vitro*.[131] A synthesis of 4′-*C*-methyl-2′,3′-dideoxythymidine involves an asymmetric synthesis of the sugar unit using a chiral auxiliary, followed by attachment of the base.[132] A previously-reported intermediate with a 4′-(2-hydroxyethyl) substituent (Vol. 32, pp. 273–274) was used to prepare the branched thymidine derivatives **98** (R = Et, vinyl), and the bicyclic compound **99**, whilst **98** (R = alkynyl) was made from a 4′-*C*-formyl compound. Some of these derivatives showed potent anti-HSV-I and anti-HIV

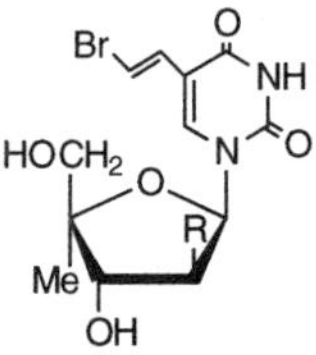

97 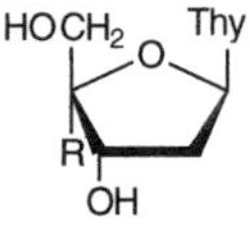 **98** 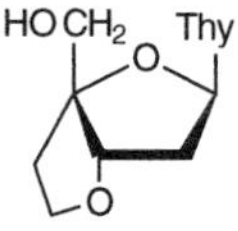 **99** 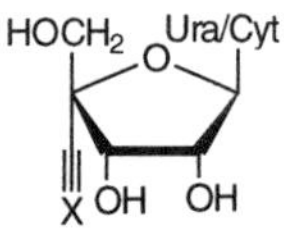 **100**

activity without cytotoxicity.[133] The same group has extended this work to compounds of type **98** (R = vinyl, alkynyl, CH=CHCl, *etc.*) with cytosine as base, and has also described compounds such as **100** (X = CH or N) with the 2′-hydroxyl group present.[134] In a variant on a previous route to such compounds (Vol. 32, p. 273), treatment of **101** under radical atom-transfer conditions [$(Bu_3Sn)_2$, *hν*], followed by desilylation, led to 4′-*C*-vinylthymidine (**98**, R = vinyl) in good yield.[135] Other workers have also prepared the alkynes **100** (X = CH), and the thymine analogue, by an alternative procedure in which a branched-chain sugar (Chapter 14) was prepared and then coupled with the bases at a late stage.[136] This work was extended to the synthesis of 2′-deoxycompounds **102** (X = H), and the *arabino*-compounds **102** (X = OH), where the inversion of configuration was carried out *via* anhydronucleoside intermediates.[137] 2′-Deoxy-4′-*C*-methoxycarbonylnucleosides **103** have been prepared by a route where the base was attached at a late stage to a 'sugar' unit derived from L-tartaric acid, using a stereoselective alkylation of the enolate of dimethyl 2,3-*O*-cyclopentylidene-L-tartrate with $BnOCH_2Cl$ to establish the quaternary chiral centre.[138] The 4′-*C*-aminomethyl compounds **104** (X = OMe or F) have been prepared by adaptations of previous syntheses in the area, and were incorporated into oligodeoxynucleosides. The resultant oligomers displayed enhanced binding towards complementary DNA as well as RNA, as compared with 4′-branched compounds without the electronega-

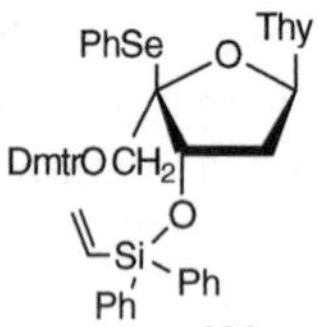

PhSe Thy DmtrO CH2 O Si Ph Ph
101

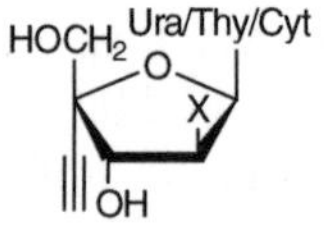

HOCH2 Ura/Thy/Cyt X OH
102

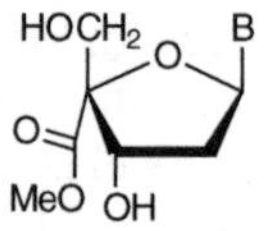

HOCH2 B O MeO OH
103

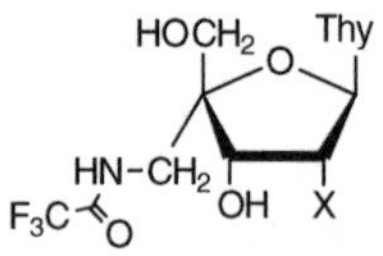

HOCH2 Thy HN-CH2 F3C O OH X
104

tive substituent at C-2′, which may well reflect the effect of the substituents in directing the sugar ring towards a C-3′-*endo* conformation.[139]

In the area of bicyclic nucleosides derived from 4′-branched compounds, Imanishi and Obika have reviewed syntheses and conformational analysis of such systems.[140] The monomer **105** for constructing *xylo*-LNA has been made by a sequence in which the cyclic ether link is put in place by displacement of a tosylate after attachment of the thymine, directed by a 2′-*O*-acetyl group,[54] and both **105** and the α-L-LNA monomer **106** have been incorporated into oligonucleotides. The incorporation of **106** led to strongly enhanced binding to complementary RNA, as for LNA itself, and superposition of models indicates close spatial similarity between the thymine units and O-3′ and O-5′ of α-L-LNA monomer **106** and the LNA monomer, which exist in locked 3′-*exo* and 3′-*endo* conformations respectively.[141] An abasic LNA monomer **107** has been prepared and incorporated into oligonucleotides. It induced the same lowering of T_m as did the introduction of a normal abasic unit. Thus effects mediated by the base are vital for the properties of LNA.[142] A report from Imanishi's laboratory has also described the synthesis of the locked AZT analogue **108**, and also the corresponding amine which can potentially be incorporated into phosphoramidate-linked oligomers.[143] The same group have also described a route to 3′-*O*,4′-*C*-methyleneribonucleosides **109** which is applicable to all bases (see Vol. 31, p. 279 for an earlier less general approach), and shown that the compounds have an *S*-conformation (C-2′-*endo*).[144] The uridine and 5-methyluridine analogues have been incorporated into oligonucleotides with

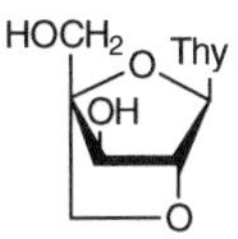

HOCH2 O Thy OH O
105

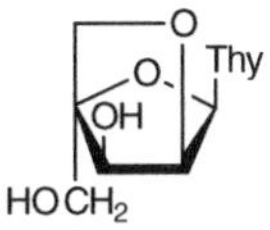

O O Thy OH HOCH2
106

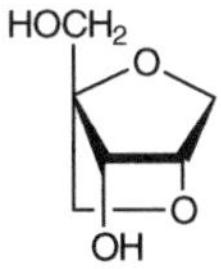

HOCH2 O O OH
107

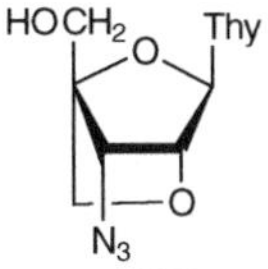

HOCH2 O Thy O N3
108

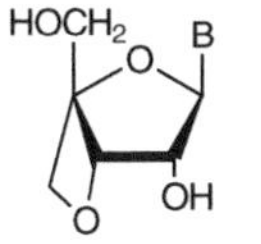

HOCH2 O B O OH
109

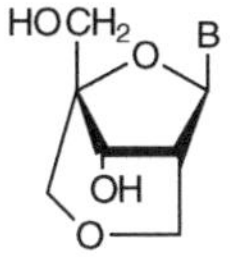

HOCH2 O B OH O
110

2′,5′-links, and the resultant structures were found to hybridize satisfactorily with complementary RNA, but not DNA.[145]

A novel type of conformationally-locked nucleoside **110** has been reported by workers at ICN Pharmaceuticals. These analogues can be prepared from methyl 3′,5′-*O*-Tipds-α-D-arabinofuranoside by a sequence involving introduction of a C-2 hydroxymethyl group through hydroboration of an exocyclic alkene and a C-4 hydroxymethyl group through a Cannizzaro reaction, formation of the bicyclic system, and attachment of the base at a late stage by a method that, at least in the case of pyrimidines, could be carried out with β-selectivity.[146] When **110** (B = Thy) was incorporated into oligodeoxynucleotides, significantly increased hybridization to complementary RNA was observed.[147]

9 Nucleosides of Unsaturated Sugars, Aldosuloses and Uronic Acids

A review has appeared covering the preparation and reactions of 4′,5′-unsaturated nucleosides.[148] A study has been made of the polymerization of the 4′,5′-unsaturated compound **111**, and the hydrolysis of the polymer to produce hypoxanthine.[149] The potentially-useful intermediate **113** has been prepared in high yield by treatment of the iodocompound **112** with base (Scheme 4). Reaction of **113** with a variety of nucleophiles gave the 3′-substituted com-

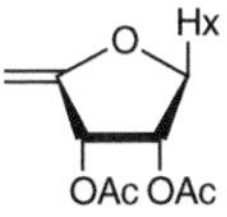

O Hx OAc OAc
111

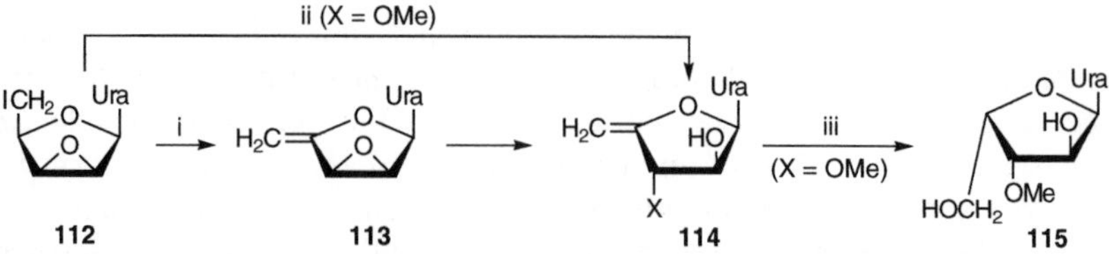

Reagents: i, LiHMDS, DMF; ii, NaOMe, MeOH, reflux; iii, BH_3.THF, then H_2O_2, NaOH

Scheme 4

pounds **114** [X = OMe, Me, BnNH, $CH(CO_2Me)_2$, OBz, SBz, N_3, SPh] in good yields, and **114** (X = OMe) could be prepared directly from **112** in high yield. Hydroboration–oxidation of **114** (X = OMe) gave mostly the α-L-xylofuranosyl nucleoside **115**.[150] Some references to 2′,3′-didehydro-2′,3′-dideoxynucleosides are mentioned in Section 3, along with the saturated analogues.

When the 4′-*C*-formyl compound **116** was treated with CH_2=PPh_3, in an approach to the 4′-*C*-vinyl compound (see above), the unusual ring-expanded compound **117** (X = H) was obtained instead. Use of CD_2=PPh_3 gave the deuteriated product **117** (X = D), and a reasonable mechanism was proposed.[151]

The bicyclic hemiacetal **118** and some related compounds have been prepared from di-*O*-isopropylidene-α-D-glucofuranose. Purification of such hemiacetals proved problematical, however, although systems in which 3′-OH was protected could be fully characterized.[54]

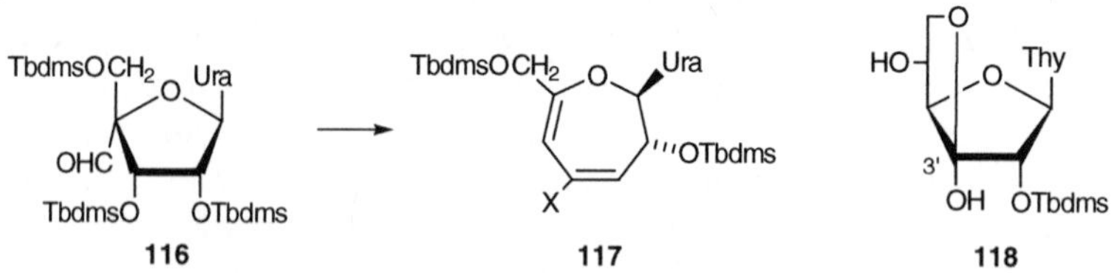

2′,3′-*O*-Isopropylidene ribonucleosides can be oxidized cleanly to their 5′-carboxylic acids using TEMPO and [bis(acetoxy)iodo]benzene.[152]

Bisubstrate-analogue inhibitors of protein kinase have been prepared by linking oligomers of arginine to adenosine-5′-carboxylic acid *via* a spacer, usually an ω-aminoacid, and some proved to be effective inhibitors of protein kinases A and C.[153] A range of 5′-*N*-substituted carboxamidoadenosines, and some analogous thioamides, have been prepared as agonists for adenosine receptors,[154] and some further 5′-amides of 2-(1-hexynyl)adenosine-5′-carboxylic acid have been described, but with lower affinity for all adenosine receptor subtypes as compared to the *N*-ethyl-compound HENECA.[155]

10 *C*-Nucleosides

A new synthesis of pseudouridine has been reported, in which 2,4-dimethoxy-5-lithiopyrimidine was added to a ribonolactone derivative, followed by reduction of the resultant hemiacetal using triethylsilane in the presence of $BF_3.Et_2O$ and demethylation using NaI.[156] A similar approach has been used for the synthesis of some 2′-deoxy-aryl-*C*-nucleoside analogues.[157]

The 2′-deoxy-2′-tritiated compound **119** has been prepared by reduction of a thiocarbonylimidazolide with tritiated tributylstannane,[158] and 2′-deoxy-9-deazaguanosine, a pyrrolo[3,2-*d*]pyrimidine *C*-nucleoside, has been made by Friedel-Crafts deoxyribosylation of the heterocycle.[159]

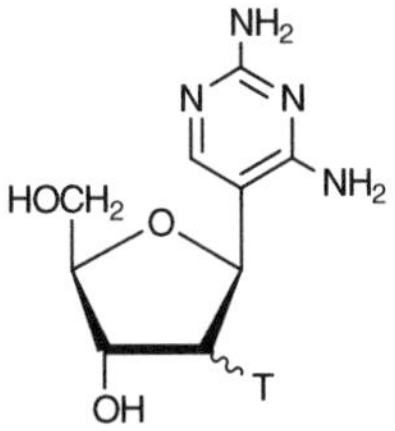

An improved synthesis of tiazofurin has been reported, which avoids some of the by-products obtained in earlier syntheses and which is suitable for large-scale production.[160] A number of thiazole-*C*-nucleosides of type **120** have been synthesized as aminoacyl adenylate mimics, and hence potential inhibitors of aminoacyl-tRNA synthetases. Several of these sulfamates displayed potent inhibitory activity, with good selectivity for bacterial enzymes over the human equivalents.[161] 2′-Deoxy-*C*-nucleosides with fluorescent terthiophene and benzoterthiophene as bases have been prepared by reaction of a 1-chlorosugar with Grignard reagents.[162] A new approach to pyrazole *C*-nucleosides involves the elaboration of species of type **121** from tri-*O*-benzoyl-β-D-ribofuranosyl cyanide, and reaction with benzylhydrazine to give **122**. Treatment of **121** with amidines gave a route to β-D-ribofuranosyl-pyrimidines.[163] The 5-hydroxy-pyran-4-one *C*-nucleoside **123** has been prepared from earlier-reported furanoid intermediates,[164] and 3-(β-D-xylopyranosyl)pyridazine and (β-D-xylopyranosyl)-4,5-dihydro-1*H*-6-pyridazone have been made *via* the intermediacy of 2-(2′,3′,4′-tri-*O*-benzyl-β-D-xylopyranosyl)furan.[165]

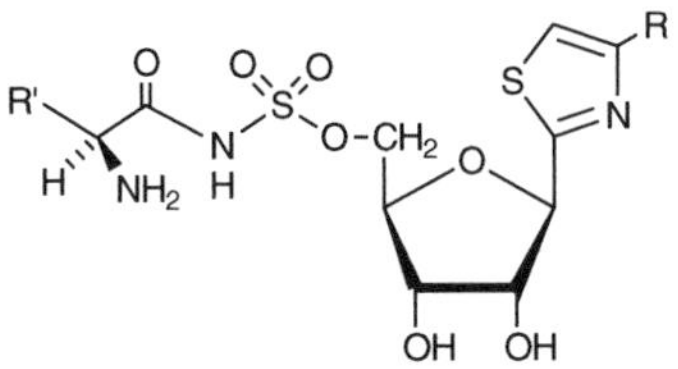

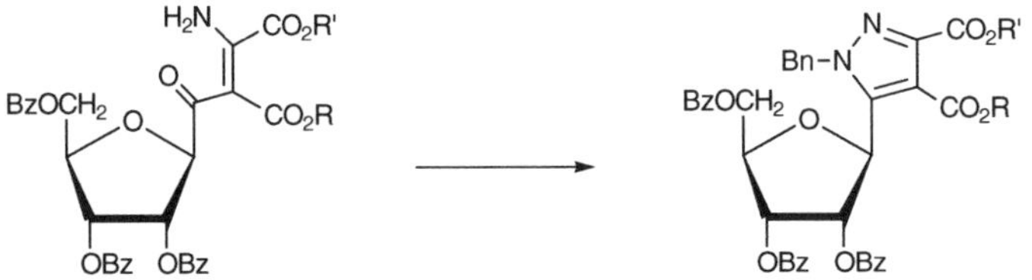

The *C*-glycoside **124**, made by Wittig reaction-cyclization from the protected ribose, has been elaborated into the *C*-nucleoside **125** through reaction with an amino-1,2,4-triazine,[166] and the related system **127** has been prepared from 2,5-anhydro-3,4,6-tri-*O*-benzoyl-D-allonic acid (**126**) by condensation with a hydrazino-1,2,4-triazine.[167] A route has been reported to 1,2,4-triazole-*C*-

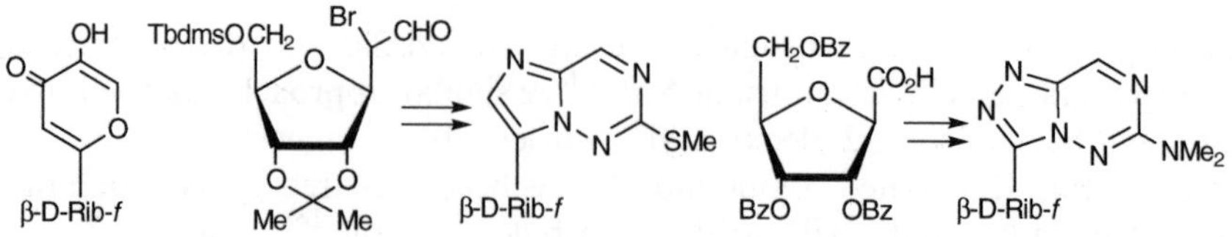

123 124 125 126 127

nucleosides of type **128** from tri-*O*-benzoyl-β-D-ribofuranosyl cyanide, using cycloaddition reactions of 1-aza-2-azoniaallene cations, the method being complementary to related work described earlier (Vol. 32, p. 277).[168]

The showdomycin analogue 3-(1′,2′-*O*-isopropylidene-β-L-threofuranos-4′-yl)maleimide has been prepared from D-glucose.[169] All four stereoisomers of **129** have been prepared, and the one illustrated ('imifuramine') is a novel histamine H_3 agonist.[170]

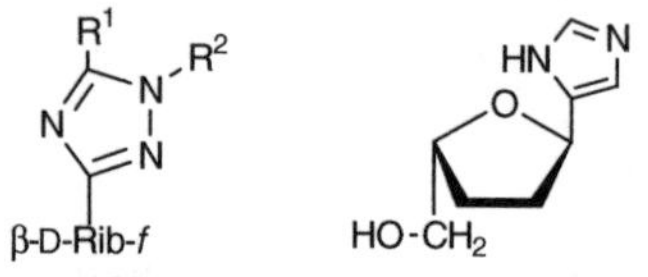

128 129

A study has been made of the charge distribution in some nucleosides and analogues. In particular, thymidine and the 2,4-difluorotoluene-5-yl-*C*-nucleoside analogue were studied. It was concluded that the 2,4-difluorotoluene-5-yl system is the closest to thymidine in shape and molecular electrostatic environment, which helps to explain why it is a good replacement for thymidine in polymerase reactions.[171]

11 Carbocyclic Nucleoside Analogues

The epoxide **130**, prepared highly diastereoselectively from 2-azabicyclo[2.2.1]-hept-5-en-3-one, can be converted efficiently into the amine **131** ($X = NH_2$), which was then transformed conventionally into carbacyclic thymidine **131** (X = Thy). Modifications of this chemistry led regioselectively to the precursor **132** for 3′-deoxy-carbocyclic nucleosides, and to the aminotriol **133**, useful for the synthesis of carbocyclic analogues of arabinofuranosyl nucleosides. Since both enantiomers of 2-azabicyclo[2.2.1]hept-5-en-3-one are commercially available, this chemistry offers considerable versatility.[172] Various 2′-deoxy-carbocyclic purine nucleosides such as **134** and the corresponding 2′,3′-*ribo*-epoxide, have been made by Pd-catalysed coupling of the base to a cyclopentenyl acetate, followed by modification of the resultant 2′,3′-didehydrodideoxy systems. Regio- and stereo-selective additions of NBS and AgOAc across the 2′,3′-ene was a key reaction in this work.[173] The conformationally-restricted analogue **135** of 2′-deoxy-carba-adenosine has been prepared as a racemate by a method similar to that used in the pyrimidine series (Vol. 32, p. 278), and the guanosine analogue was also reported.[174]

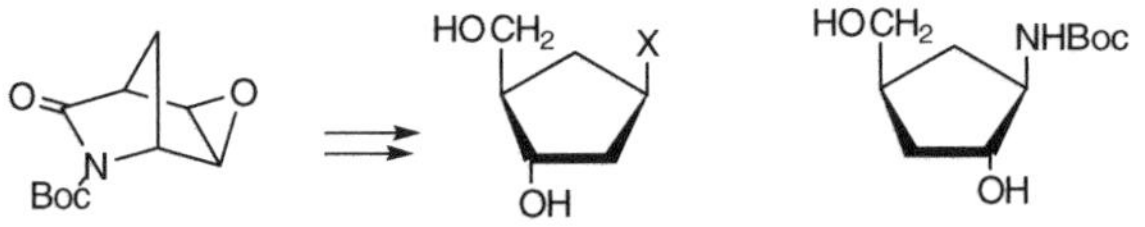

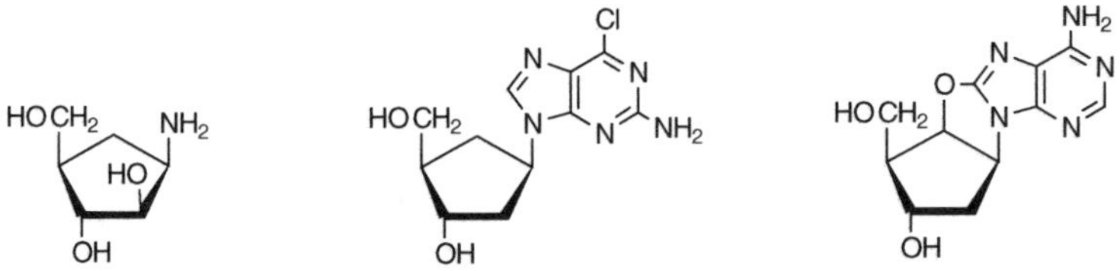

The carbocyclic analogues of d4C and its 5-fluoro-derivative **136** have been made by Pd(0)-catalysed coupling; they were converted to their 5′-triphosphates, which proved to be inhibitors of HIV-1 reverse transcriptase.[71]

An extended account has been given of the use of cyclopentenone **137** (Vol. 24, p. 302) to make carbocyclic purine nucleosides of L-configuration (**138**, B = Ade, Gua, Hx, 6-mercaptopurine) (see Vol. 31, pp. 260–261),[175] and this work has been extended to the synthesis of d4 systems **139**, and the corresponding reduced compounds. The d4A analogue (**139**, B = Ade) had potent anti-HBV activity and moderate anti-HIV activity.[176]

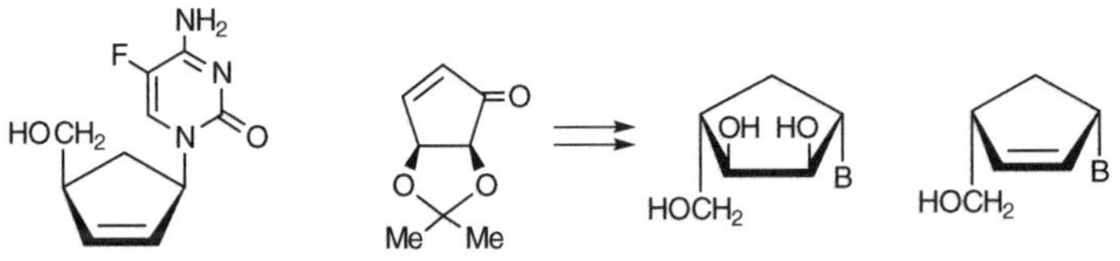

The carbocyclic *C*-nucleoside **140**, an analogue of pyrazofurin, has been prepared using the cycloaddition of an enantiopure cyclopentyldiazomethane to methyl propiolate, and the pyrazolopyridine **141** has been made using a tetrazole-to-pyrazole rearrangement.[177]

The first example of a 2′,3′-methano-carbocyclic nucleoside is the adenosine analogue **142**, prepared as a racemate through the diastereoselective addition of methylene carbene to the *N*-Boc derivative of 2-azabicyclo[2.2.1]hept-5-en-3-one, again illustrating the applicability of this building block in the chemistry of carbocyclic nucleosides.[178] The two carbocyclic guanosine analogues **143** (X = OH, F) have been reported, the synthetic route being based on that used earlier (Vol. 26, p. 248) for the preparation of the compound lacking the group X.[179] A full account has also been given of the synthesis of the conformationally-locked cyclopropa-fused compounds **144** (B = Ade, Gua, Ura, Cyt) through reaction of a 1′,2′-cyclic sulfite either with NaN_3 (in the pyrimidine

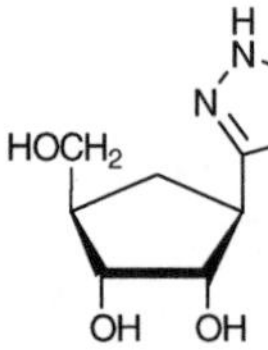

140

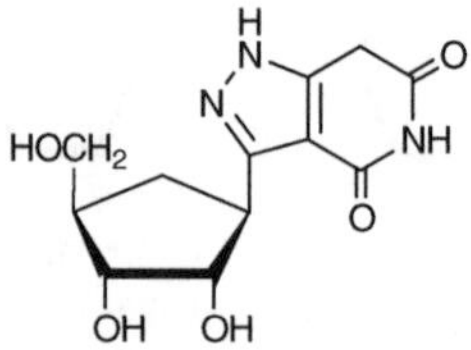

141

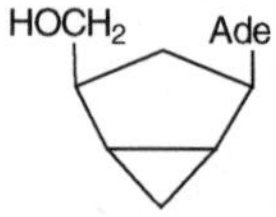

142

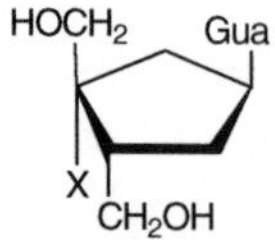

143

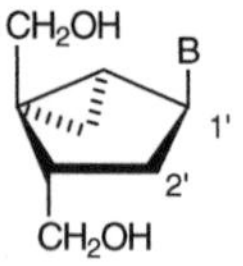

144

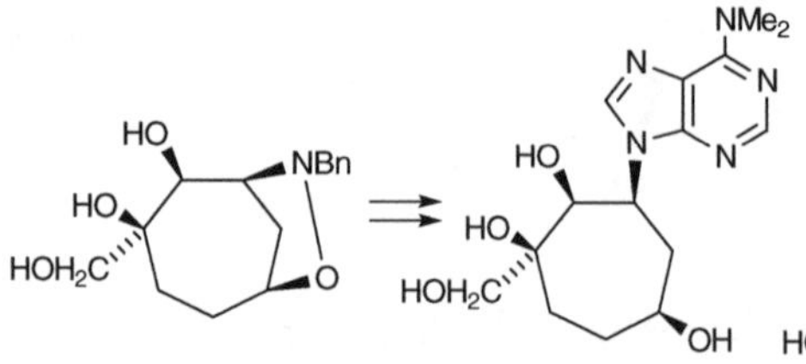

145 146

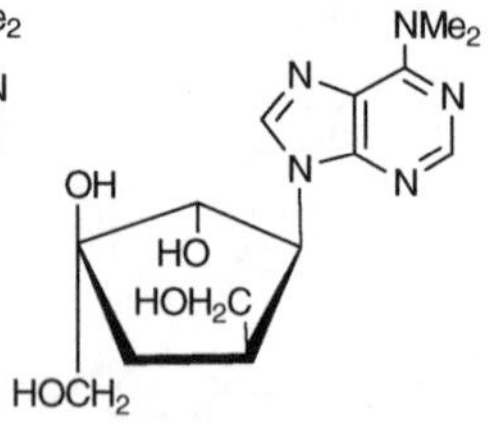

147

series) or with the anion of an intact heterocycle (for the purine cases) (see Vol. 30, p. 288).[180] The intramolecular nitrone–alkene cycloadduct **145** (Chapter 24) has been converted into the nucleoside analogue **146**, and an earlier-reported cycloadduct has been similarly manipulated to give **147.**[181]

Both the 5′-nor-analogue of carbocyclic thymidine, and the 3′-deoxy-regio-isomer **148**, have been made as pure enantiomers through hydroboration of the 2′-ene,[182] and the analogues of carbocyclic sangivamycin and toyocamycin, in the L-series (**149**, R = $CONH_2$ and CN), have been reported, but did not display the anti-trypanosomal activity of **149** (R = H) (Vol. 31, p. 261).[183] Linkage of base units to the bicyclic compound **150** under Pd(0) catalysis gave rise to the aristeromycin analogues **151** (X = CH, N), and the same approach was also used to make the pyrazolopyrimidine **152**, where the hydroxylation of the 2′-ene occurred to give the all-*cis*-isomer.[184] Carbocyclic azanoraristeromycin (**151**, X = CH) has been linked to a siderophore *via* an L-alanyl spacer (see also Vol. 29, p. 285), in an attempt to promote microbially-selective drug delivery, and one of the conjugates proved to be active against viruses sensitive to inhibitors of SAH hydrolase.[185] Some other references to aristeromycin and

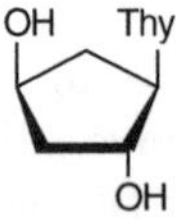

148

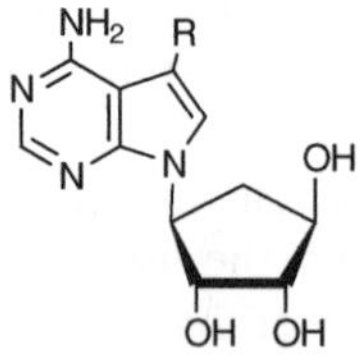

149

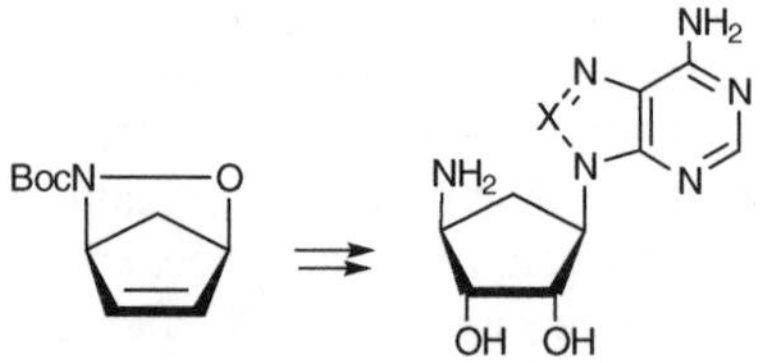

150 151

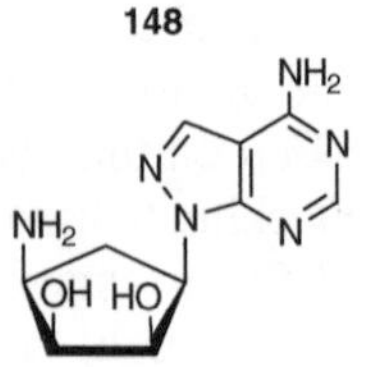

152

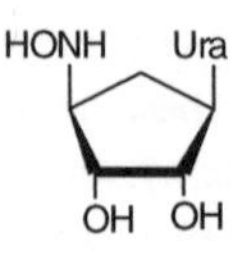

153

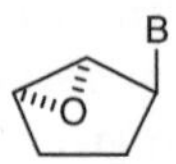

154

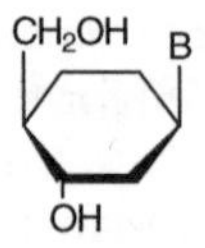

155

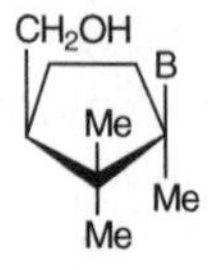

156

its close analogues are given in Chapter 19. Hydroxylamino-compounds such as **153** have also been reported, again by linking the bases to a π-allylpalladium intermediate.[186] The racemic nor-dideoxycarbanucleosides **154** (B = Ade, Gua) with the cyclopentane ring locked in an *N*-type conformation, have been prepared, and had moderate antiviral activity.[187]

Work reported previously (Vol. 32, pp. 279–280) on the synthesis of cyclohexane nucleoside analogues has now been extended to the synthesis of systems of type **155**, and conformational studies of these compounds were carried out.[188] Nucleoside analogues **156** (B = Ura, Ade, Hx, 8-aza-Ade) have been made by building up the heterocycle from a chiral amine.[189]

Some cyclopropyl nucleoside analogues are mentioned in Section 15 below.

12 Nucleoside Phosphates and Phosphonates

12.1 Nucleoside Mono- and Di-phosphates, Related Phosphonates and Other Analogues. – 2′-Deoxynucleoside 3′-*H*-phosphonates, dimethoxytritylated at O-5′, can be converted efficiently into aryl deoxynucleosidyl-*H*-phosphonates using diphenyl phosphorochloridate and pyridine as a coupling system.[190] Two simple and efficient methods have been reported for the conversion of 5′-*O*-Dmtr-deoxynucleosides into their 3′-*H*-phosphonothioate monoesters; in one method, commercially-available methyl phosphoramidites are treated with H_2S, followed by *O*-demethylation,[191] whilst alternatively the 5′-*O*-Dmtr-deoxynucleoside is treated with diphenyl *H*-phosphonate and pyridine, followed by reaction with H_2S, Et_3N and TmsCl.[192] Two groups have developed ways of removing 3′-*H*-phosphonate groups[193] or 3′-*H*-phosphonate diesters and phosphoramidites[194] using polyhydroxy alcohols, these methods being of value in recovering unreacted materials in oligonucleotide synthesis. A new route has been described for the synthesis of 3′-phosphotriester intermediates for solution-phase preparation of oligonucleotide phosphorothioates.[195] Lipophilic thymidine glycopyranosidyl phosphates such as **157** have been reported as models for drug delivery systems across cellular membranes; similar compounds involving α-methyl mannoside and also free glucose, and also systems linked at O-5′, were prepared as well.[196]

The methylenephosphonate analogues **158** of 2′-deoxynucleoside-3′-phosphates have been made using intramolecular glycosidation.[197]

Using N^6-methyl-2′-deoxyadenosine-3′,5′-bisphosphate as a lead compound, various other bisphosphates with additional substituents at C-2 and C-8 were prepared, along with bisphosphates such as **159** (X = S, CH_2), as $P2Y_1$ receptor antagonists.[198]

Analogues of TDP have been prepared as inhibitors of TDPG-4,6-dehydratase; these included cases with a substituent at C-5 (replacing the methyl group) which gave the possibility for immobilization and use in affinity columns, and cases in which the diphosphate was replaced by a methylene diphosphonate or $O(PO_2)OCH_2CO_2H$ unit, to good effect in maintaining inhibition similar to that shown by TDP itself.[199] The 2′- and 3′-deoxy analogues of tyrosinyl adenylate, an inhibitor of tyrosyl tRNA synthetases, have been reported, along with compounds in which the adenine unit was replaced with other non-heterocyclic moieties.[96]

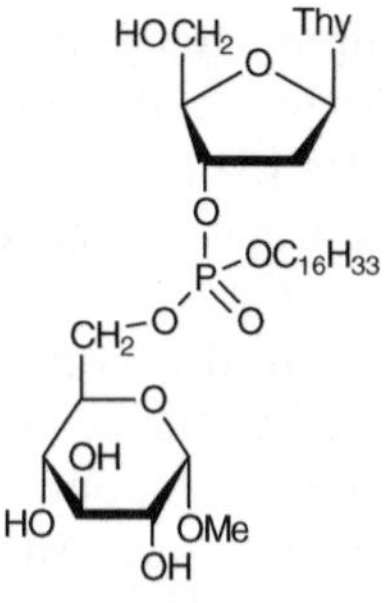

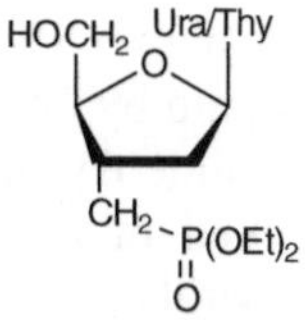

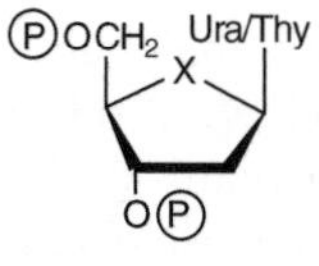

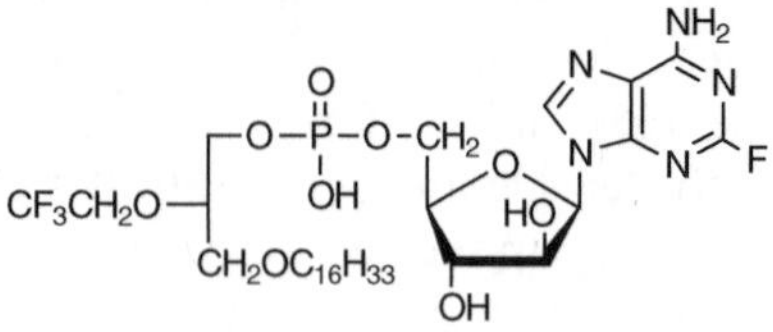

A range of phospholipid analogues of the antiproliferative drug fludarabine, such as **160**, have been prepared, and compounds with a diphosphate link between sugar and lipid unit were also reported.[200] A chemoenzymatic approach has been used to make an analogue of coenzyme A in which the amide bond of the pantothenic acid unit is replaced by a thioester.[201]

Diadenylated systems such as **161** [X = CH(OH) or PO(OH)] have been prepared as new non-isopolar analogues of diadenosine tri- and tetra-phosphates.[202] Sensitive nucleopeptides, such as **162**, the linkage region of the adenovirus 2 nucleoprotein, have been made by a chemoenzymatic route.[203] The formation of aminoacyladenylates such as **163** (X = Y = O) has been monitored by ^{31}P-NMR, and the best coupling reagents were defined for the formation of aminoacyladenylates, thio-compounds **163** (X = S, Y = O) and the phosphonate analogue **163** (X = O, Y = CH_2).[204]

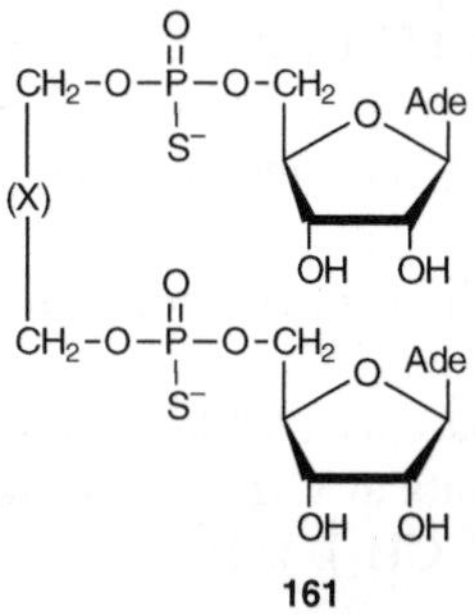

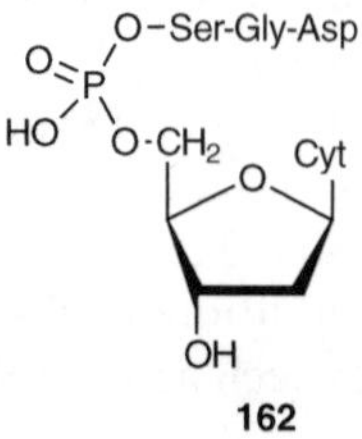

The 5′-phosphates of chiral carbocyclic ribonucleotides involving all the principal nucleobases have been prepared.[205]

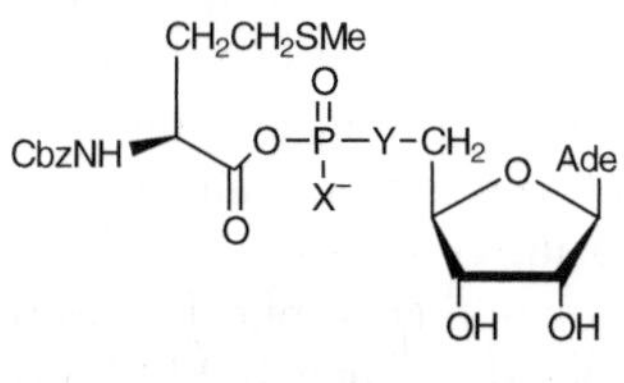

In the area of prodrugs for nucleoside 5′-phosphates, there has been an expanded account of the formation and properties of *cyclo*Sal phosphotriesters of ddA (see Vol. 31, p. 285) and d4A.[206] A similar derivative **164** has

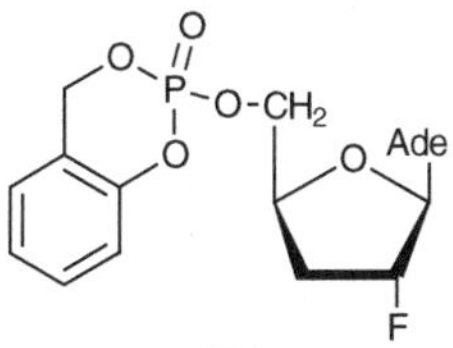

164 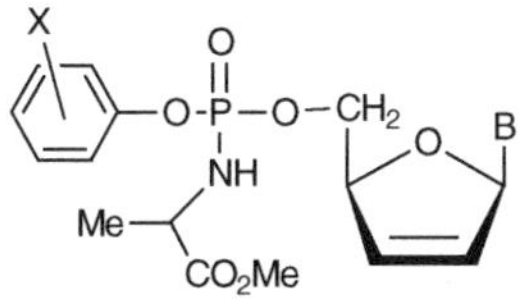 165 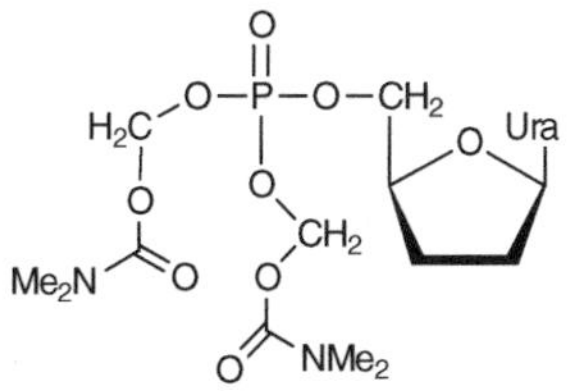 166

been prepared for 2′,3′-dideoxy-2′-fluoroadenosine, and also for the *arabino*-isomer. The *cyclo*Sal derivative **164** showed a level of anti-HIV activity higher than that for the C-2′-epimer, even though the *ribo*-nucleoside itself is inactive. This supports the idea that the monophosphate is produced, bypassing a metabolic blockade, due possibly to the ribonucleoside existing primarily in a northern conformation.[207] There have been further and fuller reports on phosphoramidate prodrugs of d4T (**165**, B = Thy), most of which were more potent than the parent (X = H),[208, 209] and Hansch-type QSAR analysis has been applied to these systems.[209] Similar derivatives of d4A (**165**, B = Ade), and the dihydrocompounds, have also been reported.[210] The triester **166** of ddUMP has been made as a potential membrane-permeable prodrug.[211] The 2-(glucosylthio)ethyl group has been proposed as a potential biolabile phosphate protecting group removable by glucosidases, and as an alternative to the SATE group. This idea was applied to AZT, through the synthesis of **167** (R = H, Ac), but, somewhat surprisingly, it was found that these compounds released free AZT and not its monophosphate.[212]

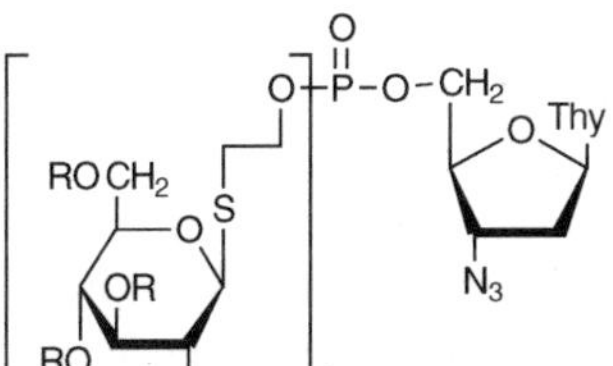

167 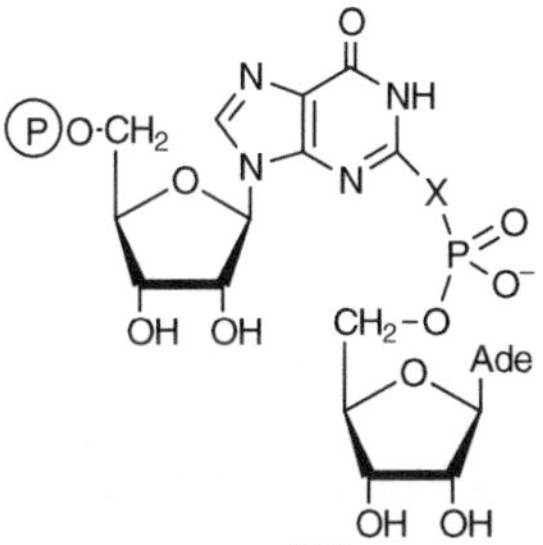 168

5′-Phosphoramidates and 5′-diphosphates of 2′-*O*-allyl-β-D-arabinofuranosyl-uracil, -cytosine and -adenine have been reported, made as potential inhibitors of ribonucleotide reductase.[213]

An interesting phosphonate that has been reported is **168** (X = CH_2), an analogue of the presumed intermediate **168** (X = O) in the conversion of XMP into GMP by GMP synthetase. A key step in the synthesis of **168** (X = CH_2) involved linking a suitably-protected 2-phosphonomethylinosine, made from AICA riboside, to N^6-dibenzoyl-2′,3′-*O*-isopropylideneadenosine under Mitsunobu conditions. The phosphonate inhibited GMP synthetase with K_i 0.56 mM.[214]

12.2 Cyclic Monophosphates and Their Analogues. – The coumarinylmethyl phosphates **169** (B = Ade, Gua) have been prepared, and a detailed study was carried out of the photolysis of these compounds to give cyclic AMP/GMP.[215] Details have also been given of the synthesis of the adenosine derivative **169** (B = Ade), and some related compounds, by a slightly different procedure (see Vol. 27, p. 263).[216] The cyclic phosphates **170** (B = Cyt, 5-F-Cyt, Ade) have been made, either from the known diols, which in the pyrimidine cases are potent antivirals, or, for the fluorocytosine compound, by attaching the base after the cyclic phosphate had been constructed. None of these compounds showed antiviral activity.[217]

2′-Thiocytosine has been incorporated into a tetranucleotide with a 2′,5′-thiophosphate link; on base treatment, this underwent hydrolysis to give the dinucleotide **171** terminated with a 2′-*S*,3′-*O*-cyclic phosphorothiolate.[218]

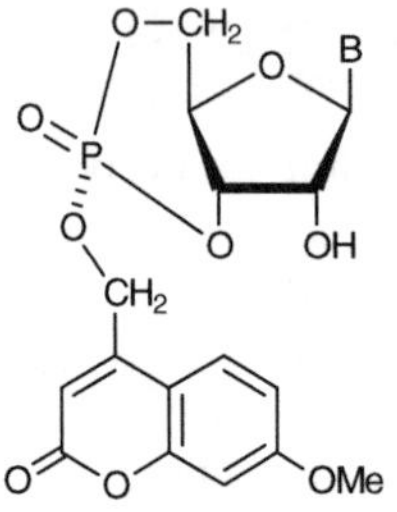

169 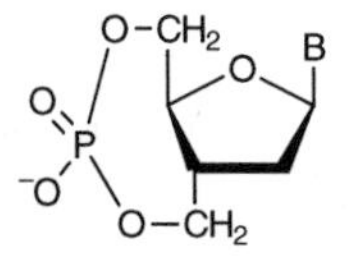 170 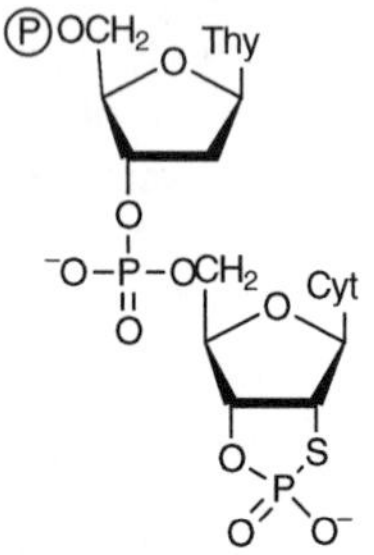 171

12.3 Nucleoside Triphosphates and Their Analogues. – 2′-Deoxynucleoside diphosphates and triphosphates have been prepared from the monophosphates using nucleoside mono- and di-phosphate kinases,[219] and nucleoside diphosphate kinase was used by others to make 2′-deoxy-8-hydroxy-GTP and 2′-deoxy-isoguanosine-5′-triphosphate from the diphosphates.[220] 2′-Deoxy-isoguanosine-5′-triphosphate and 2′-deoxy-5-methylisocytidine-5′-triphosphate have also been made by chemical methods,[221] and the 5′-triphosphate of a ring-expanded ('fat') nucleoside (Vol. 28, p. 264) has also been reported, and was found to inhibit T7 RNA polymerase.[222] 2′-Deoxy-2′-iodo- and 2′-bromo-2′-deoxy-ATP and -GTP have been made from the nucleosides, and the β,γ-imido-triphosphates were also reported, all these systems being potential phasing tools for X-ray crystallography of ATP/GTP binding proteins.[74] New fluorescent derivatives of GTP have been prepared, with the fluorophore attached by a spacer to N-2, and again β,γ-imido-triphosphates were also made; the interaction of these compounds with some GTP-binding proteins was studied.[223] The 5′-triphosphates of 2-butyloxy- and 2-butylthio-adenosine, and of 8-butyloxy-, 8-butylthio- and 8-butylamino-adenosine, have been made and investigated as regards their activation of the P2Y$_1$ receptor,[224] and the α-phosphorothioate analogues of some 2-(alkylthio)adenosines have been similarly investigated, the diastereomers at phosphorus being separated[225]

ATP analogues such as **172**, made by coupling an AMP derivative with

$Cl_2C(PO_3H_2)_2$ using carbonyldiimidazole, have been investigated as antagonists of the platelet P2T receptor.[226] In an extension of earlier work (Vol. 30, pp. 295–629 and Vol. 28, p. 291), a series of phosphonates **173** (X = CF_2, CCl_2, CBr_2), and some related compounds, related to the L-enantiomer of carbovir triphosphate, have been prepared,[227] and the related species **174** (X = CF_2, CHF, CH_2) have also been described.[228] 5-(3-Aminopropen-1-yl)-5′-*O*-(β,γ-diphosphoryl-α-phosphonomethyl)-2′-deoxyuridine and 5-[3-(*N*-biotinyl-6-aminohexanoylamino)propen-1-yl]-5′-*O*-(β,γ-diphosphoryl-α-phosphonomethyl)-2′-deoxyuridine have been synthesized and evaluated as substrates for DNA polymerases α and δ.[229] The enzymically-stable analogue **175** of AZT-triphosphate has been prepared; the functionalization at N-3 permitted linkage to a carrier protein for subsequent immunization of rabbits to raise specific antibodies against AZT-triphosphate.[230] The 5′-triphosphate of 3′-amino-3′-deoxythymidine, and the α,β:β,γ-bis-methylene analogue, which was coupled to the protein KLH, have also been made by the same team in order to develop a specific immunometric assay for AZT-triphosphate.[231] Various other related species, such as the α-anomer of **175** and the branched phosphonate-phosphinate **176** have also been made.[232]

Following from work on the synthesis of cyclic phosphoramidates (*Angew.*

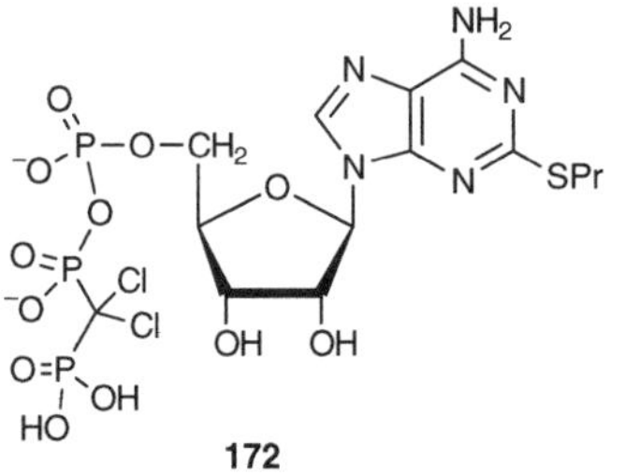

172

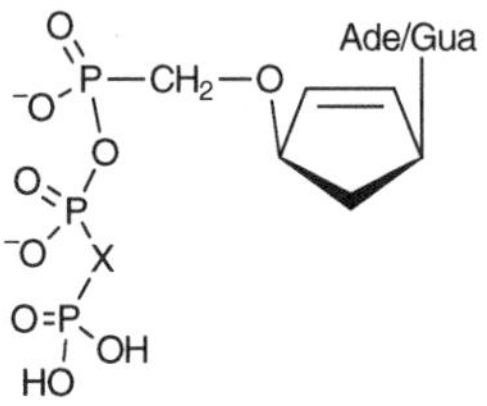

173

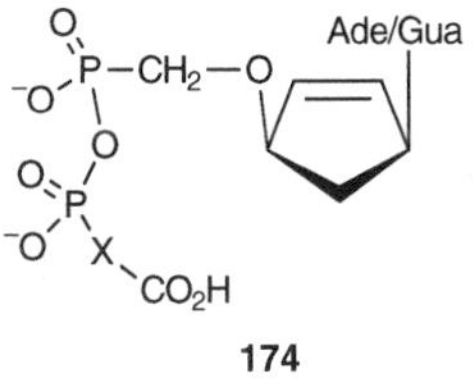

174

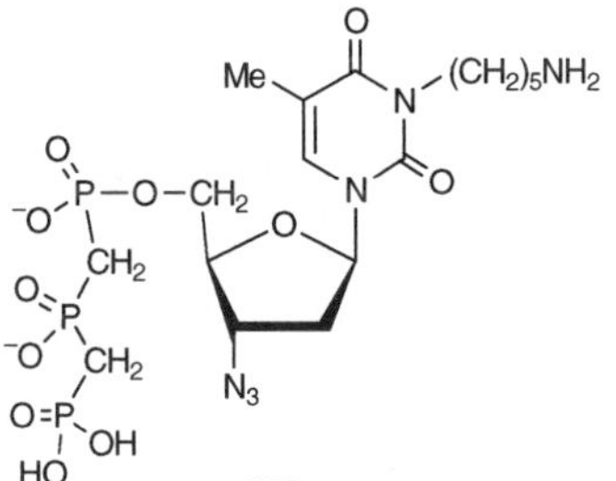

175

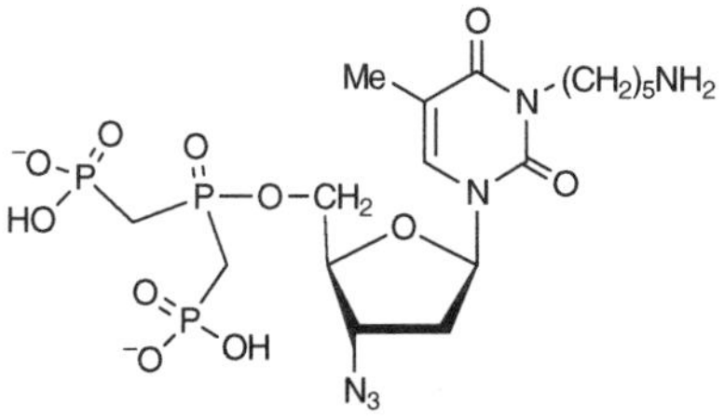

176

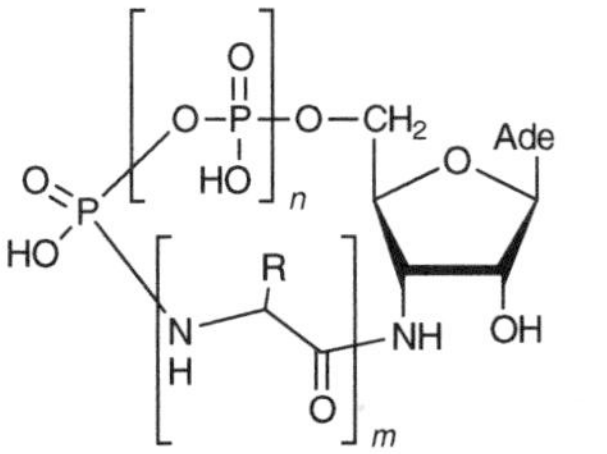

177

Chem. Int. Ed. Engl., 1994, **33**, 1395), various cyclic peptide-nucleotide hybrids of type **177** ($m = 1$–5, $n = 0$–2) have now been made by a sequence of attachment of an oligopeptide to the 3′-amino-3′-deoxynucleoside, 5′-phosphorylation and macrocyclization using a water-soluble carbodiimide.[233]

12.4 Nucleoside Mono- and Di-phosphosugars and Their Analogues. – A survey has been given of the biosynthetic pathways involving nucleotide sugars that are important for the *in vitro* synthesis of mammalian conjugates, including a summary and evaluation of the large-scale enzymic syntheses of these compounds.[234] High yielding enzymic syntheses of GDP-D-[^{3}H]-arabinose and GDP-L-[^{3}H]-fucose have been described.[235]

A new method of synthesis of nucleoside 5′-glycosyl phosphates involves the coupling of nucleosides with acylated glycosyl *H*-phosphonates, followed by oxidation and deacylation.[236] During the synthesis of CMP-Neu5Ac and its derivatives using phosphite triesters as intermediates (Vol. 31, p. 206), dimethyl-dioxiran has been developed as a useful mild oxidant for the intermediate phosphites.[237] The 3-fluorosialic acid derivative **178** can be prepared in high yield from the 2-ene using selectfluor, and could then be used in a synthesis of CMP-3-fluoro-Neu5Ac (**179**), where again a phosphite triester was used as an intermediate (with Bu^{t}OOH as oxidant).[238]

Some analogues of UDPG have been made in order to investigate the mechanism of UDPG dehydrogenase, which oxidizes C-6 of the glucose moiety to a carboxylic acid *via* the aldehyde. The ketone **180** proved to be a competitive inhibitor; the 6*S*-alcohol (for the synthesis of the sugar see Vol. 31, p. 6) was a slow substrate but the epimer was not, agreeing with previous evidence that the *pro-R* hydrogen is lost in the first oxidation.[239]

A short review has highlighted work from Schmidt's laboratory on phosphonate analogues of CMP-Neu5Ac as potent sialyltransferase inhibitors,[240]

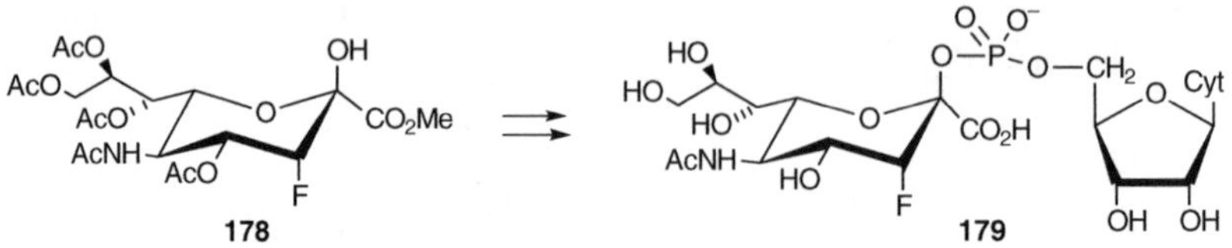

178 **179**

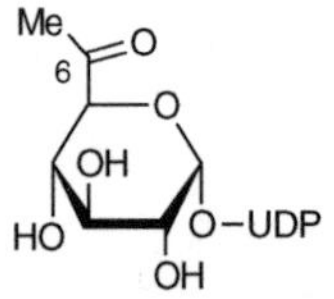

180 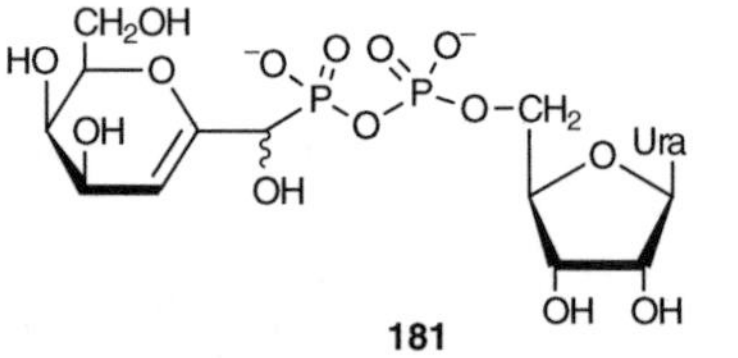 **181**

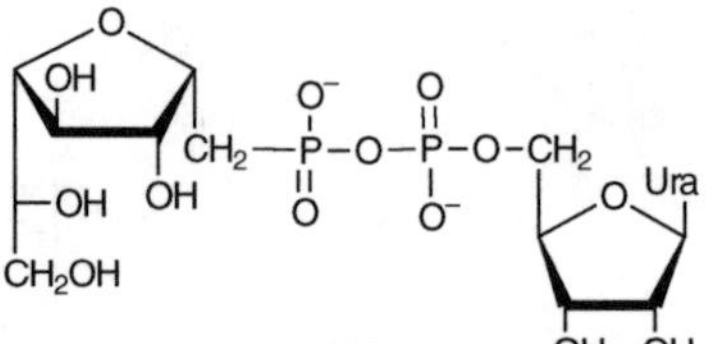

182

whilst Schmidt and colleagues have used similar principles to design the phosphonate **181** (two separate isomers) as an α-galactosyltransferase inhibitor.[241] The phosphono-analogues of GDP-*N*-acetyl-α-D-mannosamine and TDP-α-L-rhamnose have been prepared by treatment of the phosphono-analogues of the sugar-1-phosphates with the appropriate NMP-morpholidate,[242] as was the phosphonate analogue **182** of UDP-α-D-galactofuranose, of potential interest as an antimycobacterial agent.[243]

12.5 Small Oligonucleotides and Their Analogues. – *12.5.1 3′→5′-Linked systems; methodology and modified internucleotidic links.* A review has appeared discussing fluorescent oligonucleotides, including cases where the fluorophore is linked to the sugar unit.[244]

A one-pot, cost-effective synthesis of deoxyribonucleoside phosphoramidites has been described, in which the 5′-*O*-Dmtr-protected nucleoside is treated with bis(diisopropylamino)chlorophosphine in the presence of Et_3N to give the phosphorobisamidite, which, without isolation, is treated with the alcohol (*e.g.* 2-cyanoethanol).[245] Improved routes have been developed for making deoxyribo- and ribo-nucleoside phosphoramidites with allylic protecting groups, including at phosphorus.[246] In an extension of earlier work (Vol. 31, pp. 287–288), Beaucage and colleagues have developed the use of the 4-[*N*-methyl-*N*-trifluoroacetyl-amino]butyl group (as in **183**) as an alternative to the 2-cyanoethyl unit, so as to prevent possible alkylation of nucleobases by acrylonitrile. The internucleosidic phosphodiester is liberated on treatment with ammonia, with *N*-methylpyrrolidine as a byproduct.[247] The 2-(4-nitrophenyl)ethyl group has also been used as a cyanoethyl replacement, during the synthesis by phosphoramidite methodology of oligomers of 2′-*O*-methyl-ribonucleosides.[248] Some phosphoramidites of xylofuranosyl nucleosides have been made, using a redox process to invert the stereochemistry at C-3′ of the normal nucleoside, and incorporated into ribozymes at specific sites.[249]

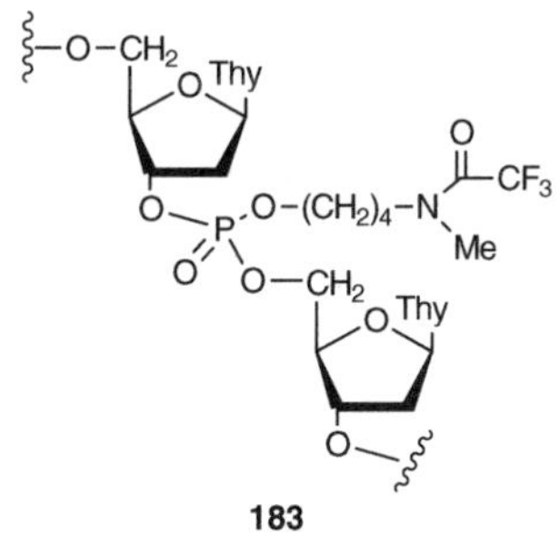

183

There have been reports describing procedures for the synthesis of cyclic oligoribonucleotides, including those with four or fewer nucleotide units.[250–252]

A method reported earlier (Vol. 31, p. 288) from Sekine's laboratory has now been applied to the synthesis of *H*-phosphonate oligonucleotides, made without the need for protection of the bases.[253] Reese and Song have also given a full account of their solution-phase approach to oligonucleotides and their phosphorothioate analogues by an *H*-phosphonate approach (see Vol. 31, p. 288),[254] and the use of bis(trimethylsilyl)peroxide and *N,O*-bis(trimethylsilylacetamide) in the presence of TmsOTf has been advocated as an effective method for the oxidation of dinucleoside *H*-phosphonates to the phosphodiesters.[255] Workers at Isis Pharmaceuticals have introduced the use of phenylacetyl disulfide as an effective and economical sulfurizing reagent in the solid-phase synthesis of oligodeoxyribonucleotide phosphorothioates,[256] and diethyldithiocarbonate disulfide as a similarly useful reagent for converting bis(deoxynucleoside) phosphite triesters into phosphorothioate triesters.[257] It has been found that attempted synthesis of dinucleotide phosphorothioates by the displacement of a sulfonate from C-3′ of a xylofuranosyl nucleoside by a nucleoside 5′-phosphorothioate monoester led to both *O*- and *S*-alkylation and much elimination.[258] Wang and Just have given a full account of their elegant approach to the diastereoselective synthesis of dinucleosidyl phosphorothioates using an indole derivative as a chiral auxiliary (see Vol. 27, pp. 288–289),[259] and the same group has also shown (Scheme 5) that if the intermediate **184** from this work is treated with DBU, it undergoes a β-elimination to give **185**, which on treatment with 3′-*O*-Tbdps-thymidine gives the phosphonate **186**, presumably by displacement of the indole unit, followed by recombination.[260]

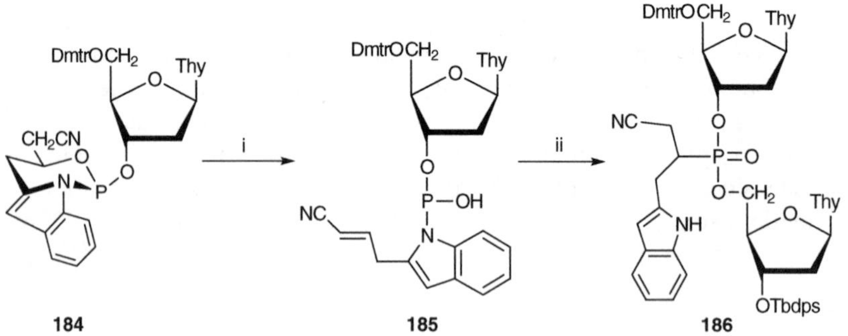

184 185 186

Reagents: i, DBU; ii, 3'-*O*-Tbdps-thymidine

Scheme 5

Both diastereomers of protected bis(deoxyadenosyl)methylphosphonate have been prepared, and incorporated into pentamers at specific sites in connection with studies on phosphodiester hydrolysis by *Serratia* endonuclease.[261]

An interesting new type of phosphonate analogue is **187**, produced by the reaction of the *H*-phosphonate diester with pyridine and trityl chloride in the presence of DBU. The reaction was shown to be stereospecific, probably proceeding with retention of configuration.[262]

A new method for introducing amidate linkages into dinucleotides, used for the synthesis of **188**, involves treating the 5′-*O*-protected dinucleotide, linked at O-3′ to a solid support, with tosyl chloride–pyridine, and then reaction of the mixed anhydride with butylamine.[263] Oligonucleotides containing the functio-

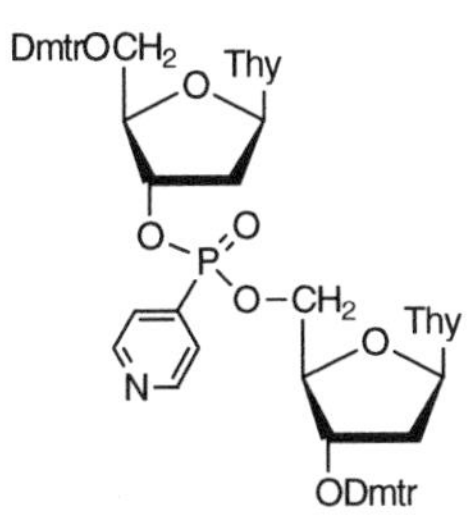

187

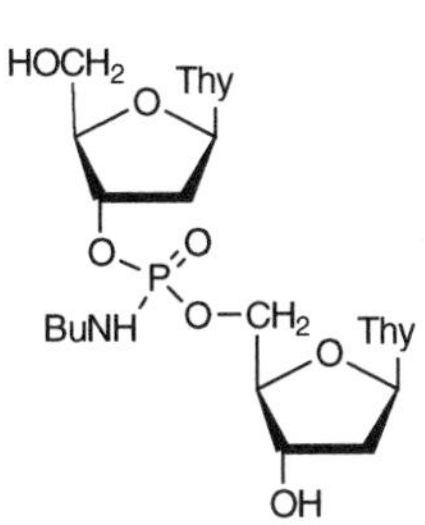

188

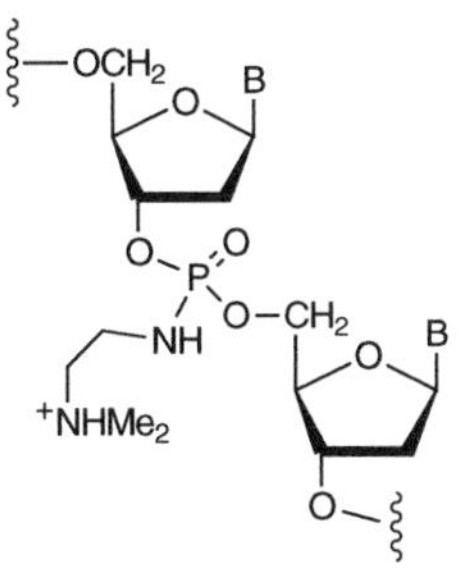

189

nalized phosphoramidate unit **189** have been made, using reaction of *H*-phosphonates with iodine and *N,N*-dimethyl-1,2-diaminoethane to introduce the phosphoramidate at each required position.[264] A method has been reported for preparing N3′→P5′ phosphoramidates of type **190** from 3′-amino-2′,3′-dideoxynucleosides, *via* their *N*-(2-oxo-1,3,2-oxathiaphospholane) deriva-

190

191

tives,[265] and a report from Gryaznov's laboratory describes the synthesis of oligonucleotide N3′→P5′ phosphoramidothioates **191**, which form highly stable duplexes with complementary RNA, as do the corresponding phosphoramidates.[266] A route to P3′→N5′ dinucleoside phosphoramidates and phosphoramidothioates is outlined in Scheme 6, the aryl *H*-phosphonate **192** being formed *in situ* from the *H*-phosphonate monoester and the phenol using diphenyl chlorophosphate as coupling reagent.[267]

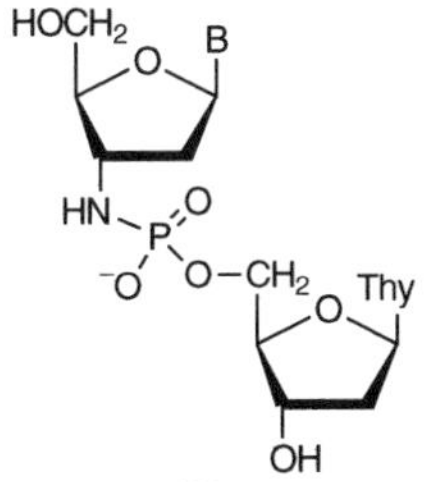

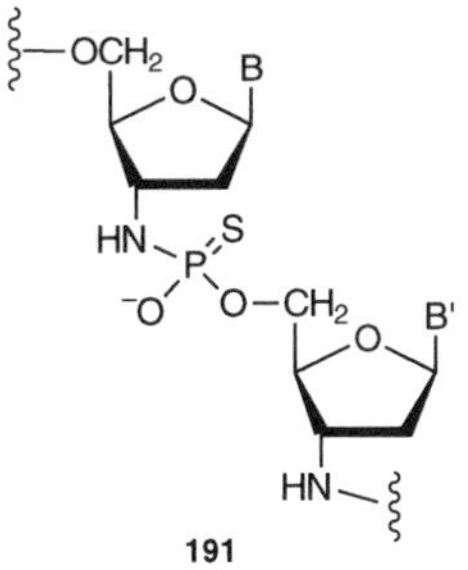

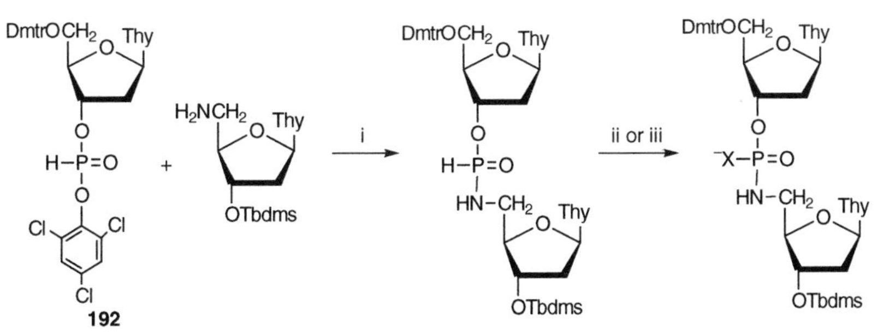

192

Reagents: i, Et_3N; ii, Et_3N, I_2, H_2O, Py (X = O); iii, S, Et_3N (X = S)

Scheme 6

There have been further reports on dinucleoside boranophosphates. The R_P-isomer **193** of dithymidine boranophosphate, and its epimer, have been made from the sugar-protected dinucleoside *H*-phosphonates, by silylation to the dinucleosidyl trimethylsilyl phosphite, boronation with DIPEA.BH_3, and deprotection, the method proceeding with retention of configuration.[268] This *H*-phosphonate approach has been extended to the ribonucleoside series with the preparation of diuridine 3′,5′-boranophosphate (for an earlier synthesis using a phosphoramidite approach, see *Tetrahedron Lett.,* 1995, **36**, 745).[269] Dithymidine 3′,5′-boranophosphorothioate (**194**), the first example of this class of nucleotide mimetic, has been made by a sequence involving boronation of a protected dinucleosidyl *p*-nitrophenyl phosphite triester, followed by displacement of *p*-nitrophenolate by Li_2S. Dithymidine boranophosphorothioate proved to be highly stable to acidic, basic or phosphodiesterase-catalysed degradation, as well as having higher lipophilicity than the boranophosphate.[270]

Tri- and tetra-thymidines containing a phosphorofluoridate link at each internucleosidic linkage have been made from the corresponding phosphoroselenoates by treatment with triethylamine tris(hydrofluoride), the products having very limited hydrolytic stability.[271] It has been demonstrated that the previously reported (Vol. 28, p. 288) sensitivity of dithymidine 3′,5′-phosphorofluoridate and -phosphorothiofluoridate to spleen and snake venom phosphodiesterases is due to initial chemical hydrolysis, and that both enzymes will only cleave a P–F bond in a phosphorofluoridate that is negatively charged.[272]

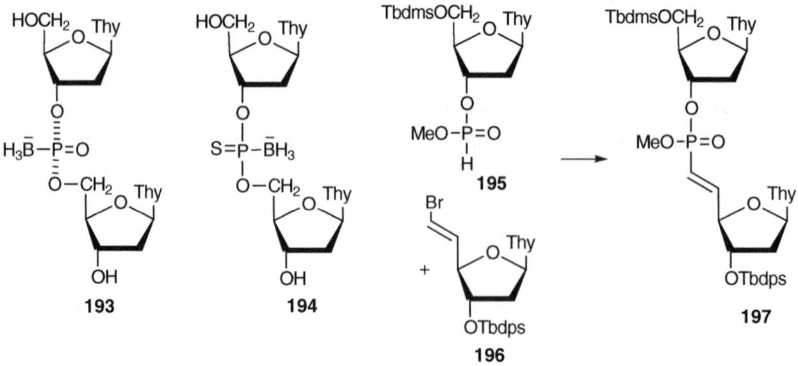

HOCH2 Thy O H3B–P=O O–CH2 Thy OH **193**

HOCH2 Thy O S=P–BH3 O–CH2 Thy OH **194**

TbdmsOCH2 Thy O MeO–P=O H **195** + Br Thy OTbdps **196** → TbdmsOCH2 Thy O MeO–P=O Thy OTbdps **197**

A Pd-catalysed coupling of the *H*-phosphonate **195** with the alkenyl bromide **196** was used to make the unsaturated phosphonate **197**.[273]

A number of papers discussed in earlier sections refer to the incorporation of nucleoside analogues into oligonucleotides, and some papers in Section 14 below discuss protecting groups of use in oligonucleotide synthesis.

12.5.2 5′→5′-Linked systems. Various symmetrical phosphoramidates **198** of AZT have been prepared by a method that involved forming the bis-nucleosidyl *H*-phosphonate and oxidative amination (RNH_2, CCl_4, Et_3N).[274] In

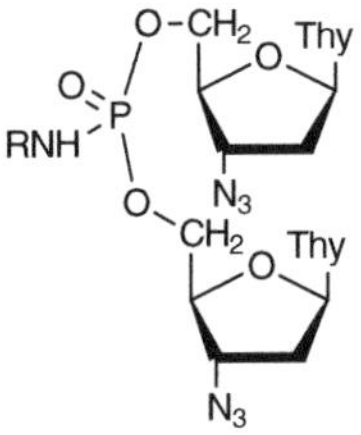

198

syntheses of amphiphilic dinucleoside phosphate derivatives, N^4-palmitoyl-2′,3′-dideoxycytidine 5′-*H*-phosphonate was condensed with AZT and ddI to give the 5′→5′ dinucleoside phosphates, which had anti-HIV activity.[275]

In a new synthesis of symmetrical dinucleoside 5′,5′-pyrophosphates, moderate to good yields were obtained by treatment of the 5′-monophosphates with common activating agents such as TsCl and *p*-nitrophenyl chloroformate.[276] A practical synthesis of nicotinamide mononucleotide has been described, along with a high-yield coupling with AMP-morpholidate to give NAD^+.[277] The radiolabelled photoactive NAD^+ analogue [α-^{32}P]-nicotinamide-8-azidoadenine dinucleotide has been made by a chemoenzymatic approach,[278] and the transglycosylation activity of NAD glycohydrolase has been used to replace the nicotinamide unit of NAD^+ with various diols.[279] The phosphonate analogue of thiazole-4-carboxamide adenine dinucleotide (TAD) in which the central P–O–P link has been replaced by P–CH_2–P has been prepared, and caused depleted guanine nucleotide pools in tiazofurin-resistant tumour cells.[280]

A study has been reported of the hydrolysis of 2′-deoxyguanosine-5′-phosphate 2-methylimidazolide, which gives rise to di(dideoxyguanosine)-5′,5′-diphosphate ($G_2{}^P$), along with other products. The rate of $G_2{}^P$ formation increased in the presence of poly-C, implying a template-directed dimerization.[281]

A synthesis of nucleoside 5′-diphosphate imidazolides involves the reaction of ADP with imidazole in the presence of 2,2′-dipyridyl disulfide, Ph_3P and Et_3N. The phosphorimidazolide bond is more stable than in the corresponding monophosphate derivative, and reaction of adenosine 5′-diphosphate imidazolide with AMP in the presence of magnesium ions gave diadenosine 5′,5′-triphosphate, a reaction that could be used to cap oligoribonucleotides carrying 5′-phosphates.[282]

A review from Sih's laboratory on the bioorganic chemistry of cyclic ADP-ribose (cADPR) includes methodology for the synthesis of analogues of cADPR,[283] whilst Matsuda's group have reported a more efficient synthesis of cyclic IDP-carbocyclic ribose (see Vol. 32, p. 289).[284]

12.5.3 2′→5′-Linked systems. Derivatives of the 2′,5′-adenylate trimer, selectively deoxygenated at C-3′ in each unit, and at all three positions, and with a palmitoylamino group replacing the 5′-OH, have been prepared by standard means. Some of these compounds had powerful inhibitory activity against HIV syncytia formation.[285] Similar derivatives, but with *N*-palmitoyl-3′-

amino-3′-deoxyadenosine at the 2′-end, have also been reported, and the compound with the central adenosine unit deoxygenated at C-3′ had powerful HIV-RT inhibitory activity.[286]

13 Oligonucleotide Analogues with Phosphorus-free Linkages

The 2-quinolinyl thioether **69** has been converted using previous methods (Vol. 29, p. 293) into the 5′-*O*-diphenylphosphinyloxymethyl derivative, which was coupled (DBU) with **70** to give, after reductive-hydrolytic removal of the quinoline unit, the 3′-thioformacetal dinucleotide analogue **199**.[102] The Gilead group has also used similar methods, but with dimethoxytrityl protection of the thiol, for the solution-phase synthesis of a pentathymidine unit with four contiguous 3′-thioformacetal links, which was then incorporated into a longer oligonucleotide by conventional solid-phase phosphoramidite methods to give a product with slightly improved complementarity properties.[287] Other workers have described the synthesis of 5′-thioformacetal analogues such as **200** by the reaction of a 3′-*O*-methylthiomethyl nucleoside with sulfuryl chloride/base, followed by a 5′-thionucleoside.[288]

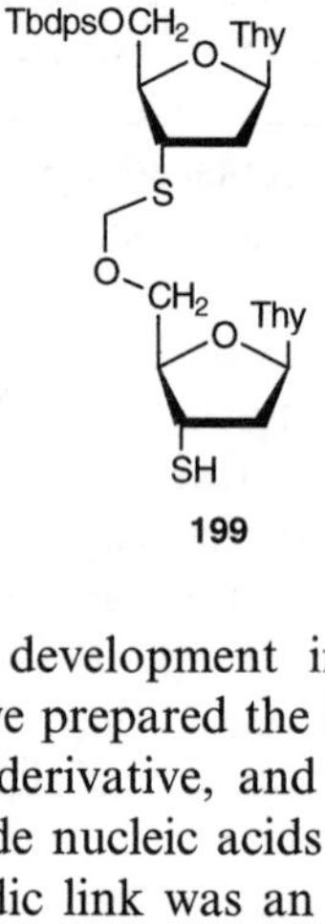

TbdpsOCH2 Thy S O CH2 Thy SH
199

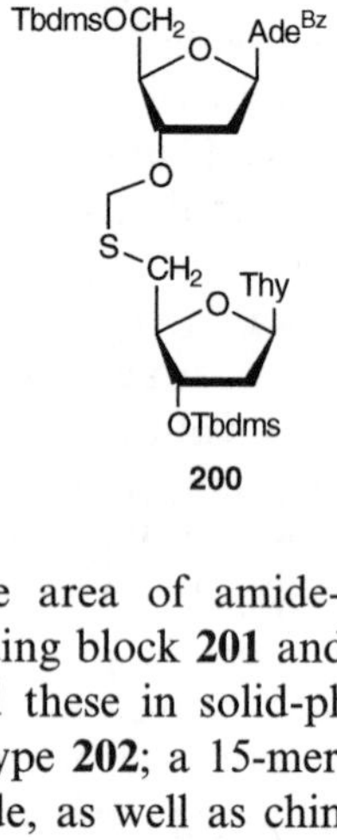

TbdmsOCH2 Ade^{Bz} O S CH2 Thy OTbdms
200

In a significant development in the area of amide-linked systems, the Novartis group have prepared the building block **201** and an analogous base-protected cytidine derivative, and used these in solid-phase synthesis of 2′-deoxyribo-polyamide nucleic acids of type **202**; a 15-mer was made in which every internucleotidic link was an amide, as well as chimeric structures with regions of polyamides and regions of phosphodiester links. These analogues had similar affinities towards both complementary RNA and DNA as did the normal oligodeoxynucleotides.[289] The amide-linked system **203**, the compound with Thy and Gua reversed, and the bis-Thy analogue have been prepared by modifications of previous work in this area; these were made as nucleomimetics of a sequence recognized by the HIV nucleocapsid protein NCp7, and **203** was indeed recognized by the protein.[290] Dimers of type **204** (R = Tbdms) have been prepared by DCC-induced amide formation.[291] The 2′→5′-amide-linked building blocks **205** (R = H, OTbdms) have been made and used to incorporate these units into deoxynucleotide dodecamers, but a significant destabilizing effect on duplex formation was found.[292]

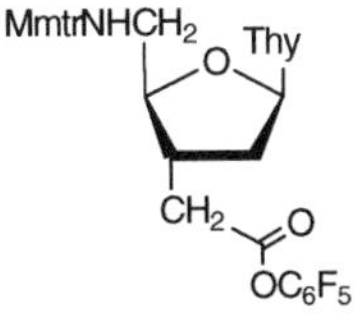

201

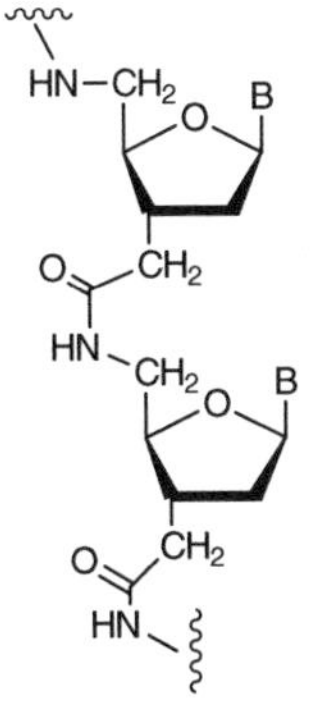

202

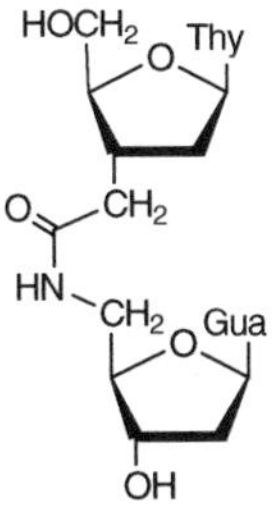

203

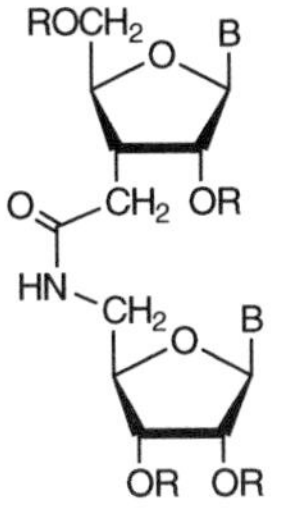

204

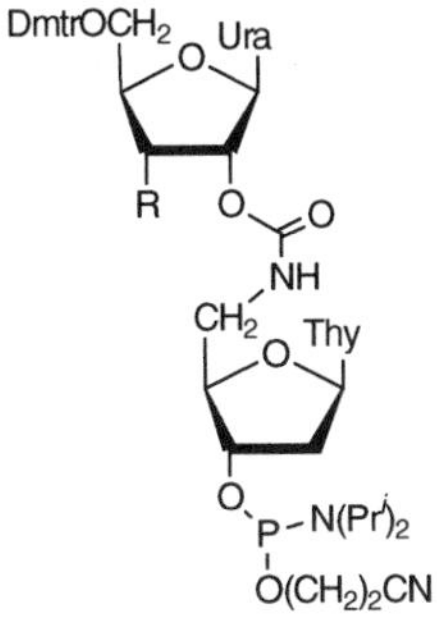

205

A new type of analogue is the *N*-acylsulfamide **206**, made by coupling a 'top' sulfamide with a 'bottom' uronic acid, but oligonucleotides containing this replacement showed considerably reduced affinity for complementary sequences.[293] The hydroxylamines **207** (X = H, OH), and a similar 2′→5′-linked compound, were made by reaction of 'lower' hydroxylamines with 'upper' aldehydes, and reduction of the resultant nitrones, and the *N*-hydroxyurea **208**, and the compound with the other nitrogen hydroxylated, were also described. These systems easily gave rise to aminoxyl radicals.[294] A dinucleoside carbonate, suitable for incorporation into oligonucleotides, has been prepared by reaction of 5′-*O*-Mmtr-thymidine with 1,1′-carbonyldiimidazole, followed by subsequent treatment with 3′-*O*-allyl-thymidine and deallylation.[295] The carbazoyl-linked nucleoside dimers **209** have been made from the carbazoyl nucleosides previously prepared (Vol. 32, pp. 292–293) by a chemoenzymatic route.[296]

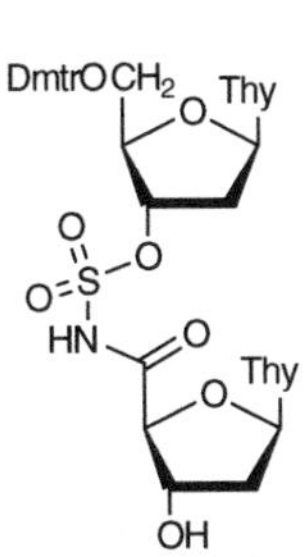

206

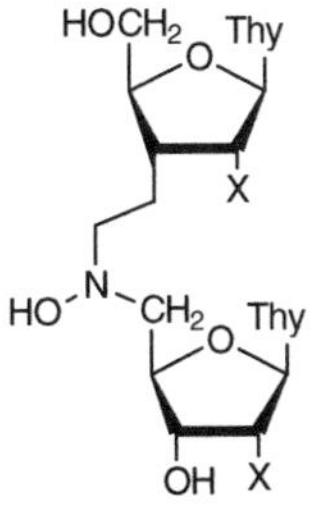

207

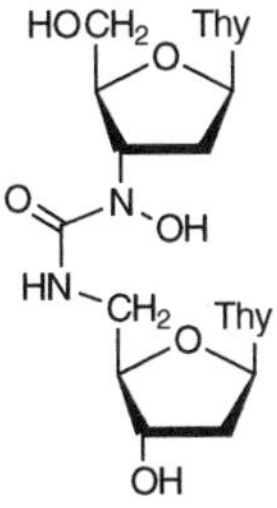

208

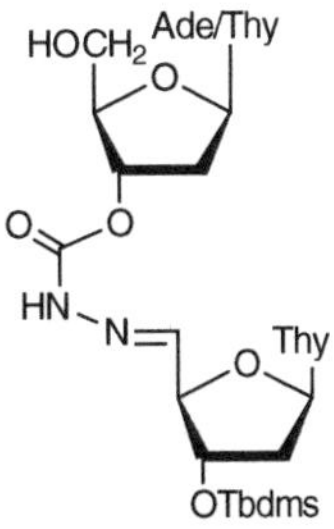

209

14 Ethers, Esters and Acetals of Nucleosides

14.1 Ethers. – The synthesis of methyl ethers of 2′-deoxy-cytidine and -adenosine can be carried out by direct methylation (Me_2SO_4, KOH), as an alternative to methods involving protecting groups. Mixtures of both mono-*O*-methylated compounds and the 3′,5′-di-*O*-methyl compounds were obtained, but these could be separated by chromatography.[297] 2′-*O*-(2-Methoxyethyl)-uridine has been obtained in high yield from uridine when the 2,2′-anhydro-nucleoside was treated with aluminium 2-methoxyethoxide,[298] and 2′-*O*-cyanomethyl ribonucleosides have been prepared and used as precursors for oligoribonucleotides containing 2′-*O*-carbamoylmethyl groups.[299] An azobenzene unit has been ether-linked at O-2′ of a ribonucleoside by reaction with a benzylic bromide. This was incorporated into a modified oligonucleotide, and on photoirradiation the T_m of this with complementary DNA was significantly changed by *cis–trans* isomerization of the azobenzene.[300]

Following from earlier work (Vol. 31, p. 294), 2′-*O*-Tbdms-nucleoside-3′-*H*-phosphonates **210** (B = protected Ade or Gua), prepared regioselectively, can be treated with glycerol to effect removal of the *H*-phosphonate, thus opening a route to selectively silylated purine nucleosides for use in oligonucleotide synthesis by the phosphoramidite method (see also Section 12.1 above).[301]

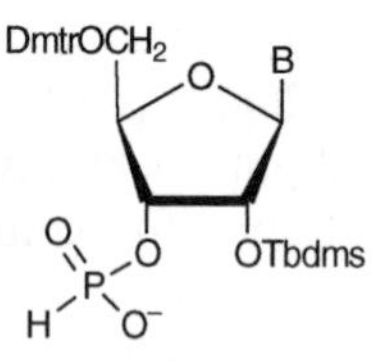

DmtrOCH2 B O OTbdms O P H O⁻ O

210

HOCH2 Ade/Thy O O OMe O—(CH2)5NH-Dansyl O2N

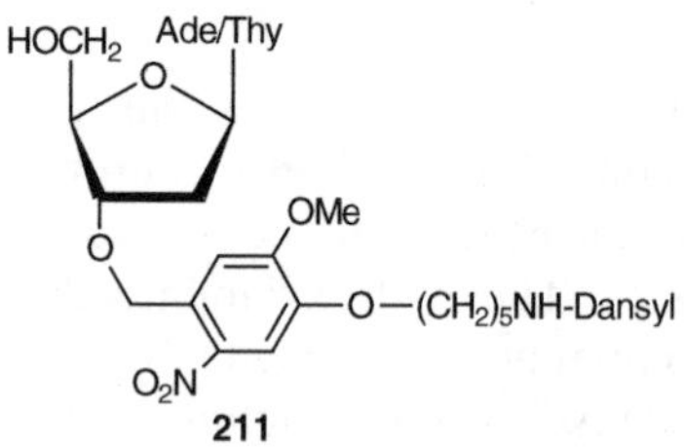

211

Nucleosides with ethers at O-3′ that are both photolabile and fluorescent, as in **211**, have been made and converted into their triphosphates for use in combinatorial DNA sequencing.[302, 303]

Treatment of deoxynucleosides with DmtrCl, imidazolium mesylate and Hünig's base gave high yields of the 5′-*O*-Dmtr ethers without any complications due to reaction of amino groups in the bases.[304] The use of a polymer formed from 4-vinylpyridine has also been advocated for use in the synthesis of 5′-*O*-dimethoxytritylated nucleosides.[305] The reagent *O*-(benzotriazol-1-yl)-*N,N,N′,N′*-tetramethyluronium tetrafluoroborate (TBTU) can cleave silyl and dimethoxytrityl groups, and permits the selective cleavage of a Dmtr group from a protected deoxynucleoside in the presence of a Tbdms group, whilst the same reagent selectively removed a Tbdms group from O-5′ in the presence of a 3′-*O*-Tbdps unit.[306] The Dmtr group can also be removed from the 5′-position of a nucleoside under aprotic neutral conditions using stannous chloride.[307] The photochemical cleavage of pixyl groups from nucleosides (Vol. 32, ref. 266) has now been extended to the cleavage of 5′-*O*-(*S*-pixyl) groups.[308]

14.2 Esters. – The esterification of nucleosides to solid supports using uronium or phosphonium coupling reagents and DMAP has been described.[309] A number of 5-bromo-2′-deoxyuridine and thymidine analogues which have aminoacid and peptide residues in the 3′- and 5′-*O*-positions have been prepared and tested as antiviral agents.[310]

A study has been made of the benzoylation of 6-chloro-9-(β-D-ribofuranosyl)purine, and conditions were found, using benzoic anhydride and an amine in aqueous MeCN, to obtain the 3′-*O*-monobenzoate in reasonable yield and with ~2:1 selectivity over the 2′-*O*-benzoate. Alternative conditions gave the 2′,3′-di-*O*-benzoyl compound in 81% yield.[311]

Use of the appropriate lipase, with vinyl acetate as an acetyl donor, permits the conversion of inosine into either the 3′- or 5′-*O*-acetyl ester.[66] Enzymic acylation, using an immobilized preparation of *Candida antarctica* lipase, was also used to make the 5′-*O*-acetyl derivative of the anti-leukaemic agent 2-amino-9-β-D-arabinofuranosyl-6-methoxypurine,[312] and *Candida antarctica* lipase was used in the synthesis of 5′-*O*-acryloyl nucleosides, which were subsequently polymerized.[313] Selective 5′-*O*-acylation of guanosine, 2′-deoxyadenosine and thymidine can be carried out using *N*-hydroxysuccinimidyl esters of benzoic and *p*-nitrobenzoic acids,[314] whilst 5′-*O*-benzoyl-2′-deoxyuridine has been made using a variant on the Mitsunobu reaction which gives more easily separated by-products.[40]

The antitumour agent gemcytabine (2′-deoxy-2′,2′-difluorocytidine) has been condensed separately at each of O-3′, O-5′ and the amino-group through a succinoyl spacer with an isoquinoline unit which targets peripheral benzodiazepine receptors on mitochondria, and which is known to be overexpressed in human brain tissues. In this work, methods were developed to place a Boc group selectively on either O-3′ or O-5′, or both.[315] AZT has been linked through an ester at O-5′ with HIV protease inhibitors. The conjugates have excellent antiviral activities compared with the individual components, possibly because of good cell-penetration followed by hydrolysis to give anti-HIV agents of two different types.[316] Conjugates of siderophores and 5-fluorouridine, ester-linked at O-5′, have been prepared to target delivery of the drug to pathogenic microorganisms (see also Section 11).[317] In an approach to inhibitors of methionyl-tRNA synthase, the ester **212** (X = O) has been prepared, as have the nitrogen-containing systems **212** (X = NH, NOH, NHO). They inhibited the aminoacylation activity of the enzymes from several bacteria, but showed growth inhibitory activity only on *E. coli*.[318] The bisphosphonate alendronate has been coupled to thymidine through a urethane at O-5′.[319] 5′-*O*-[*N*-(Cyclohexylcarbonyl)glycyl]- and 5′-*O*-[*N*-(cyclohexylpro-

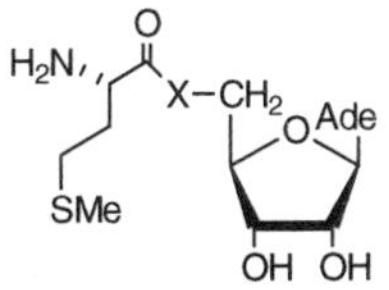

212 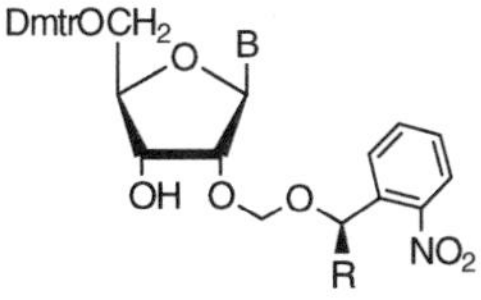 **213** 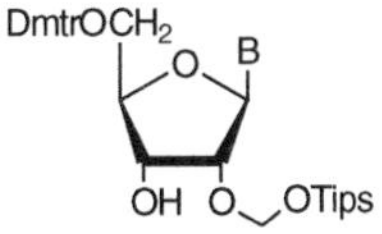 **214**

pionyl)glycyl]-2′,3′-*O*-isopropylideneuridine have been made as potential inhibitors of UDP-glucuronosyltransferase.[320]

The 2-(levulinyloxymethyl)-5-nitrobenzoyl and 2-(levulinyloxymethyl)-benzoyl protecting groups have been developed for the 5′-positions of nucleoside 3′-phosphoramidites.[321]

14.3 Acetals. – The 2′-*O*-[(2-nitrobenzyl)oxy]methyl (Nbom) group [as in **213** (R = H)] has been developed in Pitsch's laboratory for use in the synthesis of oligoribonucleotides by the phosphoramidite method; it can be put in place by regioselective alkylation of a 2′,3′-*O*-dibutylstannylene derivative, and removed by photolysis, along with 2-nitrobenzyloxycarbonyl groups used to protect bases.[322] The same team has also developed the (*R*)-Npeom group [as in **213** (R = Me)], again photolabile, and the triisopropylsilyloxymethyl (Tom) group (see **214**), removable by fluoride ion.[323]

O-(Benzotriazol-1-yl)-*N,N,N′,N′*-tetramethyluronium tetrafluoroborate (TBTU) has been used for the selective removal of a Thp ether from a nucleoside in the presence of a Tbdms group.[306]

Treatment of 3′,5′-*O*-Tipds-2′-ketouridine with primary alcohols gives stable hemiketals with a single diastereomer formed, the configuration being confirmed by X-ray crystallography.[324]

O-Linked 2,3-unsaturated glycosyl derivatives of nucleosides have been made by Ferrier reaction of a glycal ester in the presence of the nucleoside,[325] and pyrimidine 5′-*O*-β-D-ribofuranosyl nucleosides have been obtained by ribosylation of 2′,3′-di-*O*-acetylribonucleosides or 3′-*O*-Tbdps-2′-deoxyribonucleosides.[326] The modified nucleoside 5-(β-D-glucopyranosyloxymethyl)-2′-deoxyuridine has been prepared through the reaction of a protected form of 5-(hydroxymethyl)-2′-deoxyuridine with tetra-*O*-benzoyl-α-D-glucopyranosyl trichloroacetimidate, and use of a non-participatory protecting group at O-2 of the sugar gave rise to the α-anomer.[327] Some other references to glycosylated nucleosides are given in Chapter 3.

15 Other Types of Nucleoside Analogues

Chain-extended deoxynucleosides **215** (B = Ade, Gua, Thy, Cyt) have been prepared from the parent nucleoside by a one-pot procedure involving selective oxidation of the primary alcohol with *o*-iodoxybenzoic acid (IBX) and subsequent Wittig reaction.[328] In another approach to chain-extended compounds, the *C*-allylated thymidine **216** was prepared diastereoselectively from the 5′-aldehyde by reaction with allyltrimethylsilane in the presence of $BF_3.Et_2O$. By hydroboration, **216** was converted into the hydroxypropyl compound, and by further manipulation into the *C*-hydroxyhexyl compound **217**. Attempts to introduce a longer chain more directly by the use of more complex allylsilanes were not always straightforward, but the bromocompound **218** could be obtained in good yield.[329] When the dibromoalkene **219** was treated with LDA in THF, followed by paraformaldehyde, an unexpected

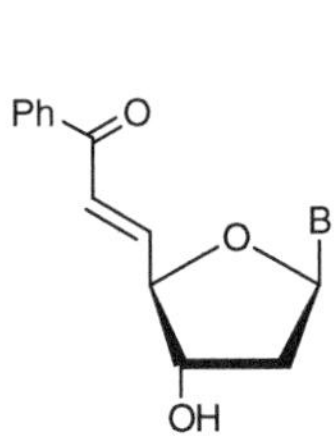

215 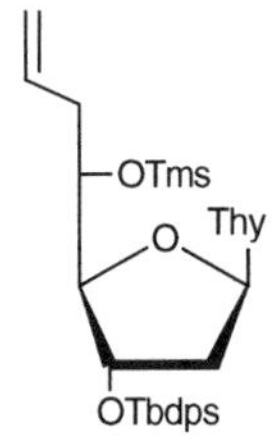 216 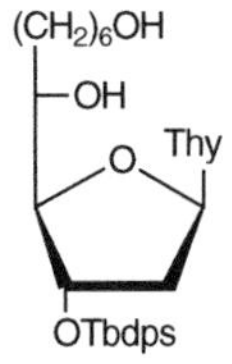 217 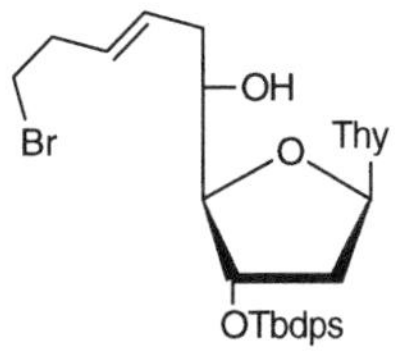 218

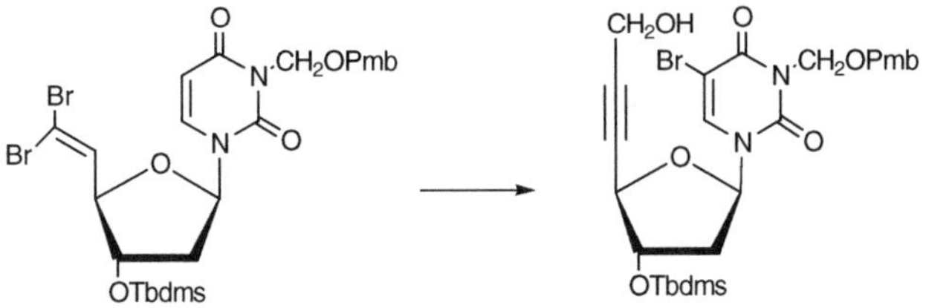

219 220

bromination of the uracil unit took place in addition to the expected reaction on the sugar unit, to give **220**. This product was exploited for the synthesis of a porphyrin-functionalized nucleoside, a meso-position of the porphyrin ring being attached to C-5 of the uracil *via* a phenylacetylene spacer.[330]

The bicyclic nucleoside analogues **221** (B = Ura, Cyt, Ade) have been prepared from 3,6-anhydro-1,2-*O*-isopropylidene-α-D-glucofuranose, itself made in a new way from 5,6-cyclic sulfates of 3-substituted mono-isopropylidene-α-D-glucofuranoses (Chapter 5).[331] Another type of bicyclic nucleoside analogue which has been reported is the amino-compound **222**, made from AZT by chain extension, followed by reduction of the azide and cyclization. The 5′-epimer was also prepared.[332]

A route to the *S,S*-isomer **223** of iso-ddA, and the corresponding thymidine derivative, has been developed, starting from *O*-Tbdps-(*S*)-glycidol.[333] Nair has prepared phosphonates **224** (B = Ade, Ura, Thy) related to iso-dideoxynucleosides, but which can also be regarded as cyclic analogues of the anti-HIV acyclonucleoside phosphonate PMEA, by a route that began with (*S*)-1,2,4-butanetriol.[334] Other reports from the same laboratory have described

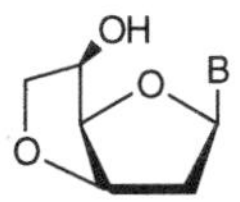

221 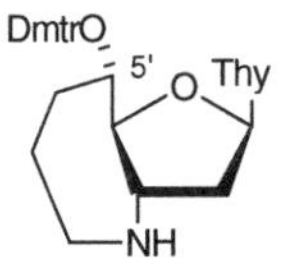 222 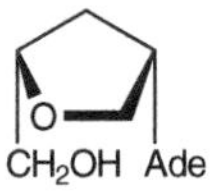 223

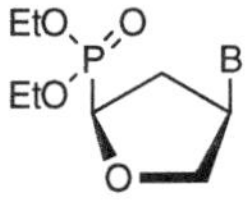

224 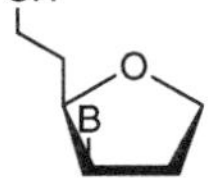 225 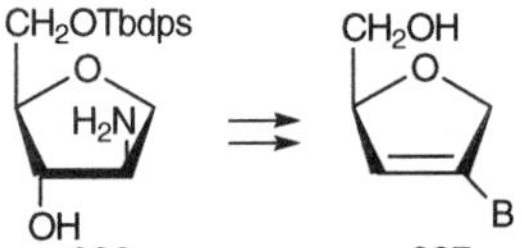 226 227

the syntheses of **225** (B = Ade, Ura, Thy), by a route from di-*O*-isopropylidene-α-D-allofuranose, with the bases being attached to an anhydroalditol (Chapter 18) at a late stage,[335] and of the unsaturated compounds **227** (B = Ade, Ura, Thy, Cyt), the synthesis of which involved the key aminoalcohol **226**, obtained from a 2,3-cyclic sulfite of 1,4-anhydro-D-ribitol, the amino group of **226** being used to elaborate the base units.[336]

In the area of 1,5-anhydrohexitol nucleosides, the conformations of pyrimidine systems **228** (X = I, Cl, Me, Et, CF_3), which have potent activity against HSV-1, have been studied. The authors concluded that these compounds adopted a different conformation when bound to thymidine kinase than that which was preferred in solution or in the solid state.[337] There has been a more substantial account of the synthesis of compounds **230** (B = Ade, Gua, Ura, Cyt) with a D-*altro*-configuration (see Vol. 32, p. 297), which were made by opening of the epoxide ring of **229** using salts of the bases.[338] These compounds have been incorporated into 'D-altritol nucleic acids (ANA)', which hybridize strongly and sequence-selectively with RNA in an anti-parallel manner, and are superior to HNA (hexitol nucleic acids), lacking the hydroxyl group at C-3′.[339] Somewhat similar chemistry was used to prepare 1,5-anhydro-2,4-dideoxy-D-mannitol nucleosides **231** (B = Ade, Ura) from 1,5-anhydro-4,6-*O*-benzylidene-D-glucitol. The pyranoid rings of these nucleoside analogues were shown by NMR to adopt $^4C_1$ conformations, with axial base units.[340] A similar piperidine (iminoalditol) nucleoside analogue **232** has been made from deoxynojirimycin; it was incorporated into oligonucleotides through its phosphoramidite, and hybridization studies were reported.[341]

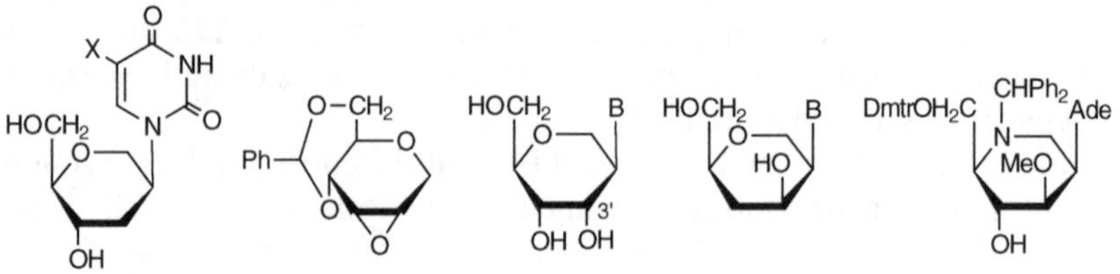

228 229 230 231 232

In this area of 'N-in-ring' nucleoside analogues, a review has been published on the synthesis and biological activity of *N*- and *C*-azanucleosides.[342] The L-*ribo*-compound **233** was made from methyl 2,3-*O*-isopropylidene-α-D-lyxopyranoside by introduction of nitrogen (azide) at C-4 with inversion of configuration, and was used to make 4′-aza-L-nucleosides **234** (B = various pyrimidines). In the thymine series, 2′-deoxy, 2′,3′-dideoxy- and d4-species were subsequently prepared by fairly standard deoxygenation procedures.[343] The pyrrolidine *C*-nucleoside **236** has been made as a ~1:1 mixture of epimers, by a sequence in which the ring was formed by treatment with Ph_3P of the azidoketone **235**, prepared in a number of steps from 2-deoxyribose and the lithiated heterocycle, followed by reduction of the resultant cyclic imine with cyanoborohydride.[344] Earlier work on derivatives of 1,4-dideoxy-1,4-imino-D-ribitol as inhibitors of nucleoside hydrolases (Vol. 31, p. 229) has been extended. In addition to further examples of 1-aryl-iminoribitols (Chapter 18), the synthetic route, involving additions of organometallic reagents to an imine, has been adapted to the preparation of nucleoside analogues **237** (B = Ade, Gua, Hx), which proved to be effective inhibitors of the purine-specific trypanosomal 'IAG nucleoside hydrolase'.[345] Some pyrrolidinone thymine nucleoside analogues such as **238** have been prepared from L-malic acid.[346] The morpholino-thymidine **239** has been made from 5-methyluridine by periodate cleavage

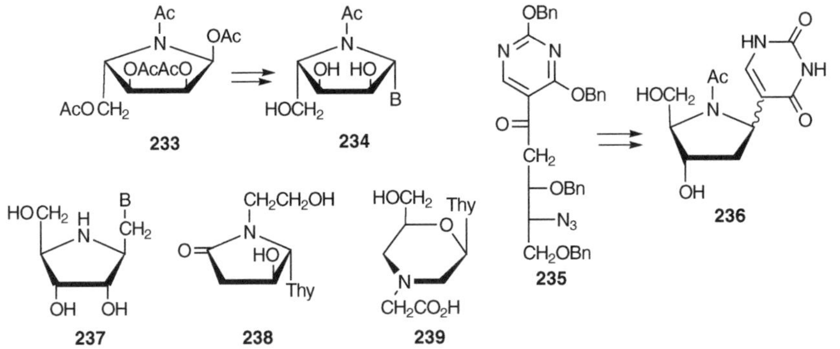

followed by reductive amination. It was converted into its triphosphate, which was recognised by Taq polymerase, and stopped DNA elongation in a similar manner to dideoxynucleotides.[347]

In the area of dioxolanyl nucleosides and related systems, there has been an account of the synthesis of L-dioxolanyl uracil nucleosides **240** (X = H, all halogens, CF_3), with the 'sugar' unit being derived from L-gulonolactone, which were tested against Epstein-Barr virus, with **240** (X = I) having the highest activity.[348] The dioxolanyl triazole *C*-nucleoside **241**, and its 2'-epimer, have been prepared using isopropylidene-D-glyceraldehyde as the source of the chiral centre at C4'; the enantiomeric compounds were also made from isopropylidene-L-glyceraldehyde, but none of these compounds showed significant antiviral activity.[349] An asymmetric synthesis of oxathiolanyl nucleoside analogues involves the formation of **242**, with the illustrated *E*-isomer as major (82:18) product formed with 60% e.e. by Sharpless oxidation of the racemic sulfide using diethyl L-tartrate. Pummerer-type coupling then gave the nucleoside analogues **243** (B = Thy, Cyt) as 1:1 epimeric mixtures.[350] The synthesis of the L-(+)-isomer **244** of homolamivudine (homo-3TC) has been reported, along with the 5-fluoro-compound (L-homo-FTC), and the uracil and 5-fluorouracil derivatives.[351] The year has also seen the first report on the synthesis of 1,3-dithiolane nucleosides **245** (B = Cyt, 5-F-Cyt, Gua, Hx), which were prepared as racemates, along with the *trans*-isomers. The *cis*-cytosine and 5-fluorocytosine analogues displayed anti-HIV activity, but the levels were significantly less than for the oxathiolanes (3TC and FTC).[352]

An asymmetric synthesis of the isoxazolidine **246** has been reported, which was obtained by cycloaddition between vinyl acetate and a homochiral nitrone bearing a sugar-based auxiliary on nitrogen, followed by linkage to thymine (see also Chapter 24).[353]

A novel type of nucleoside analogue **247** has been reported, based on a benzo[*c*]furan; both *cis*- and *trans*-isomers were made, as racemates.[354]

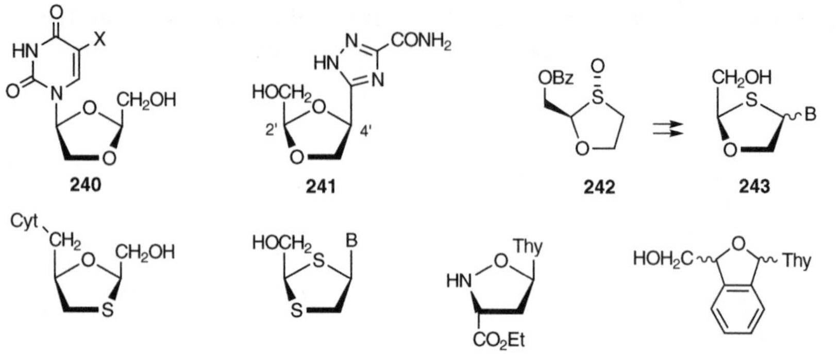

240 241 242 243

244 245 246 247

The 'reversed' cyclonucleosides **249** (X = N, CH) have been prepared by acid-induced cyclization of **248**. These precursors were derived from 5-azido-5-deoxy-1,2-*O*-isopropylidene-α-D-ribofuranose either by a cycloaddition reaction to make **248** (X = N), or by reduction to the amine followed by standard formation of the imidazole using aminocyanoacetamide and triethyl orthoformate, to give **248** (X = CH).[355] Similar studies were also reported for compounds of D-*xylo*-configuration.[356]

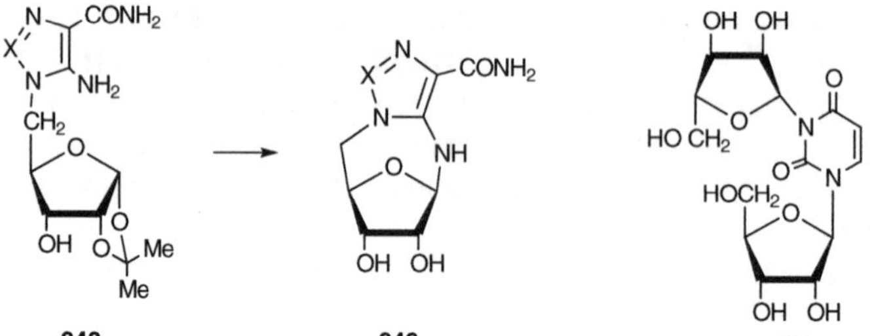

248 249 250

The ribosylated derivative **250** of uridine has been obtained by Mitsunobu condensation between 2′,3′,5′-tri-*O*-acetyluridine and 2,3,5-tri-*O*-acetyl-D-ribofuranose, followed by deacetylation. A similar reaction on inosine gave both N^1- and O^6-ribosylated products, and related glucopyranosylated nucleosides were also described.[357] Treatment of peracetylated 6-thioguanosine with TsOH in chlorobenzene at reflux leads to the formation of the S^6-(tri-*O*-acetylribofuranosyl) derivative, and N^9 to N^7 transglycosylation (Vol. 30, p. 303) was not observed in this case.[358]

The racemic cyclopentane derivative **251** has been prepared,[359] and there have been a number of further reports on cyclopropyl nucleoside analogues. The all-*cis*-trisubstituted cyclopropane **252** has been prepared,[360] as have the thymine analogue **253** and the adenine derivative **254**,[361] and a practical synthesis has been devised for the anti-HSV and anti-VZV agent A-5021 (**255**) (see Vol. 32, p. 297).[362] Racemic difluorinated compounds **256** (R = H) have been made with a variety of bases,[363] and the compounds **256** (R = CH_2OH, B = Thy, 5-F-Ura) have been made from 1,3-di-*O*-benzylglycerol.[364] The phosphonates **257** (B = Ade, 6-chloropurine) have been prepared as constrained analogues of known antiviral phosphonate derivatives of purines.[365]

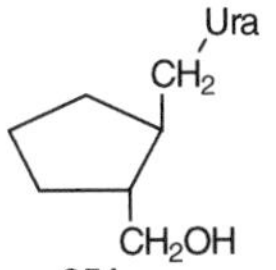

251 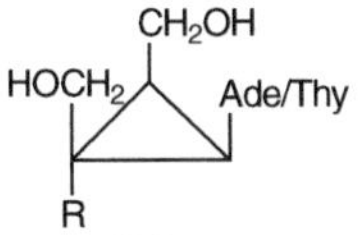 **252** 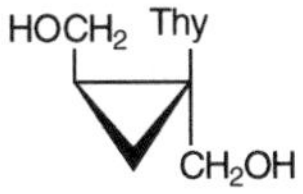 **253** 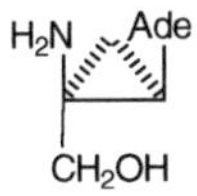 **254**

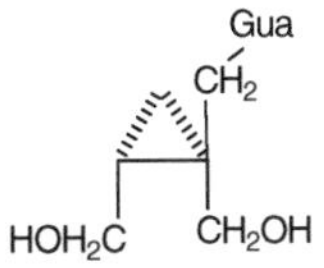

255 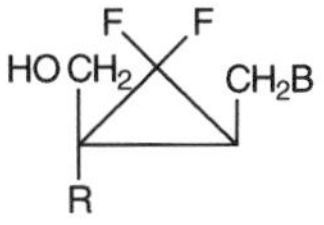 **256** 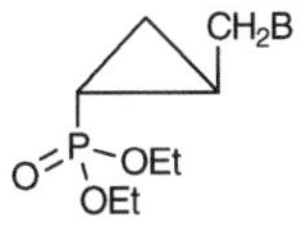 **257**

16 Reactions

A review has been published on the use of potentiometric titration data for the detection of stability constants of Cd(II) and Hg(II) complexes with nucleosides and nucleotides.[366]

The synthesis, structure and chemical properties of nucleoside dialdehyde derivatives have been reviewed.[367] The seconucleoside **258** derived from the antineoplastic agent triciribine has been made, but had no useful bioactivity.[368] The aldehyde **259**, prepared using DIBAL reduction of 2′,3′-*O*-isopropylideneadenosine (Vol. 27, p. 270) has proved to be an irreversible inhibitor of human *S*-adenosyl-L-homocysteine hydrolase, and behaved as a good affinity labelling probe of the enzyme.[369]

The phosphates **260** (R = Me, Tr), the 2′-regioisomers and the 2′-deoxy-compound (R = Tr) have been used to study the mechanism of the uncatalysed methanolysis as a model for ribozyme action.[370] A detailed study has appeared of the hydrolysis and intramolecular transesterification of symmetrical and

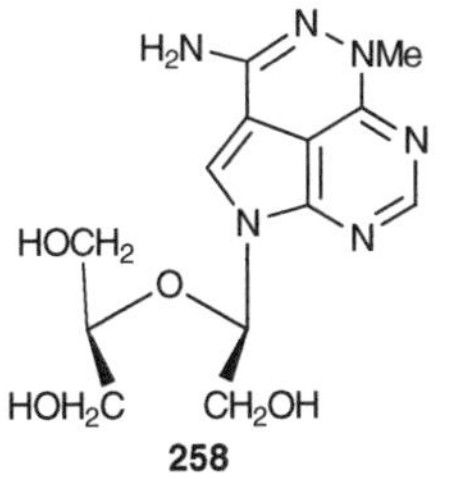

258 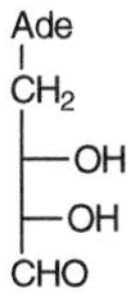 **259** 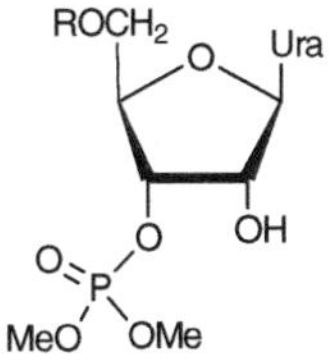 **260**

unsymmetrical dialkyl esters of 5′-*O*-pivaloyl-uridine -3′-phosphates,[371] and of the buffer-catalysed interconversion of ribonucleoside 2′- and 3′-methylphosphonates and 2′- and 3′-alkyl phosphates.[372] The course of the hydrolysis of the dinucleoside thiophosphate **261** has been followed by HPLC over a wide pH range. Cleavage to give thioinosine monophosphates and uridine competed with isomerization to the 2′,5′-isomer, and a mechanistic discussion was given.[373] A kinetic study of the hydrolysis of adenosine 2′,3′-cyclic monophosphate by aqueous Cu(II) terpyridine has concluded that $(Cu.terpyridine.OH)^+$ acts as a nucleophilic catalyst, in contrast to its behaviour in the transesterification of RNA.[374]

In the area of prebiotic chemistry, the conversion of the phosphorylimidazolide **262** (R = H) into 3′,5′-linked oligoguanylates is catalysed by polycytidylate, indicating a template effect. Without the template, other types of link are also formed.[375] Experiments have shown that the oligomerization of racemic **262** (R = Me) on a template of polycytosyl HNA gives the D-polymer.[376] Oligomerization of **262** (R = H, Me), and the 2′-deoxycompounds on a decamer template of polycytosyl HNA or $(Cyt)_{10}$-RNA was effective, but $(dCyt)_{10}$-DNA, which adopts a B-type structure, was ineffective.[377] A study has been made of the reaction of the adenosine analogue of **262** (R = Me) with phosphate to give ADP.[378]

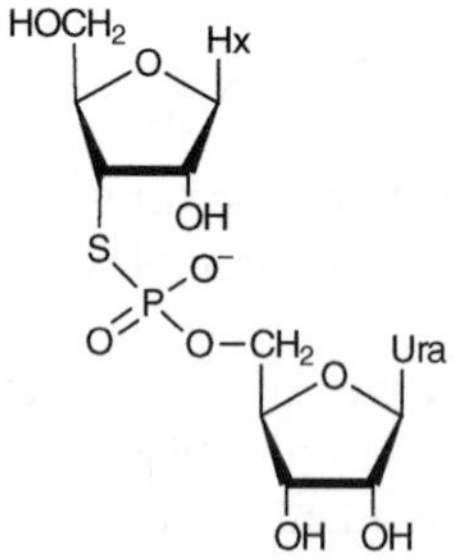

261 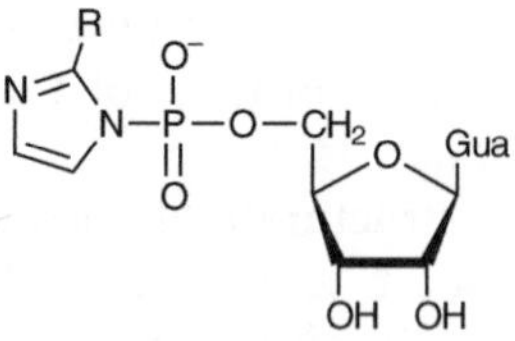 **262**

References

1 J. McCormick, Y. Li, K. McCormick, H.J. Duynstee, A.K. van Engen, G.A. van der Marel, B. Ganem, J.H. van Boom and J. Meinwald, *J. Am. Chem. Soc.*, 1999, **121**, 5661.

2 G.B. Chheda, H.B. Patrzyc, H.A. Tworek and S.P. Dutta, *Nucleosides, Nucleotides*, 1999, **18**, 2155.

3 T. Itaya and T. Kanai, *Tetrahedron Lett.*, 1999, **40**, 8003.

4 M.-C. Maurel and J.-L. Décout, *Tetrahedron*, 1999, **55**, 3141.

5 P.D. Cook, *Nucleosides, Nucleotides*, 1999, **18**, 1141.

6 J.A. Secrist III, W.B. Parker, P.W. Allan, L.L. Bennett Jr., W.R. Waud, J.W. Truss, A.T. Fowler, J.A. Montgomery, S.E. Ealick, A. H. Wells, G.Y. Gillespie, V.K. Gadi and E.J. Sorscher, *Nucleosides, Nucleotides*, 1999, **18**, 745.

7 M.J. Robins, *Nucleosides, Nucleotides*, 1999, **18**, 779.

8 T. Mickle and V. Nair, *Bioorg. Med. Chem. Lett.*, 1999, **9**, 1963.
9 L.B. Townsend, K.S. Gudmundsson, S.M. Daluge, J.J. Chen, Z. Zhu, G.W. Koszalka, L. Boyd, S.D. Chamberlain, G.A. Freeman, K.K. Biron and J.C. Drach, *Nucleosides, Nucleotides*, 1999, **18**, 509.
10 J. Butenandt, L.T. Burgdorf and T. Carell, *Synthesis*, 1999, 1085.
11 Z.J. Lesnikowski, J. Shi and R.F. Schinazi, *J. Organomet. Chem.*, 1999, **581**, 156.
12 H. Urata, *Yakugaku Zasshi*, 1999, **119**, 689 (*Chem. Abstr.*, 1999, **131**, 322 833).
13 *Nucleosides, Nucleotides*, 1999, **18**, Nos. 4/5 and 6/7.
14 A.K. Prasad, S. Trikha and V.S. Parmar, *Bioorg. Chem.*, 1999, **27**, 135 (*Chem. Abstr.*, 1999, **130**, 325 291).
15 R. Schure, A.A. Mar, B. Pease, W. Jones, B. Felt and M.S. Iyer, *Org. Process Res. Dev.*, 1999, **3**, 135 (*Chem. Abstr.*, 1999, **130**, 182 710).
16 N. Minakawa and A. Matsuda, *Curr. Med. Chem.*, 1999, **6**, 615 (*Chem. Abstr.*, 1999, **131**, 185 135).
17 J. Nokami, M. Osafune, Y. Ito, F. Miyake, S. Sumida and S. Torii, *Chem. Lett.*, 1999, 1053.
18 E. Moyroud and P. Strazewski, *Tetrahedron*, 1999, **55**, 1277.
19 A. Trifonova, A. Földesi, Z. Dinya and J. Chattopadhyaya, *Tetrahedron*, 1999, **55**, 4747.
20 J.-L. Abad, B.L. Gaffney and R.A. Jones, *J. Org. Chem.*, 1999, **64**, 6575.
21 Z. Zhu, B. Lippa and L.B. Townsend, *J. Org. Chem.*, 1999, **64**, 4159.
22 M.J. Humphries and C.A. Ramsden, *Synthesis*, 1999, 985.
23 K. Walczak, *Pol. J. Chem.*, 1999, **73**, 799 (*Chem. Abstr.*, 1999, **130**, 338 336).
24 N. Minakawa, N. Kojima and A. Matsuda, *J. Org. Chem.*, 1999, **64**, 7158.
25 J.-C. Zhao, A.H. Soloway, J.C. Beeson, W. Ji, B.A. Barnum, F.-G. Rong, W. Tjarks, G.T. Jordan, J. Liu and S.G. Shore, *J. Org. Chem.*, 1999, **64**, 9566.
26 N.S. Mourier, A. Eleuteri, S.J. Hurwitz, P.M. Tharnish and R.F. Schinazi, *Bioorg. Med. Chem.*, 1999, **7**, 2759.
27 M. Hocek and A. Holy, *Collect. Czech. Chem. Commun.*, 1999, **64**, 229.
28 R. Storer, C.J. Ashton, A.D. Baxter, M.M. Hann, C.L.P. Marr, A.M. Mason, C.-L. Mo, P.L. Myers, S.A. Noble, C.R. Penn, N.G. Weir, J.M. Woods and P.L. Coe, *Nucleosides, Nucleotides*, 1999, **18**, 203.
29 M.A.N. Mosselhi, *Nucleosides, Nucleotides*, 1999, **18**, 2043.
30 A.M. Attia, H.A. Mansour, A.A. Almehdi and M.M. Abbasi, *Nucleosides, Nucleotides*, 1999, **18**, 2301.
31 A.D. Da Matta, C.V.B. Dos Santos, H. De S. Pereira, I.C. De P.P. Frugulhetti, M.R.P. De Oliveira, M.C.B.V. De Souza, N. Moussatch and V.F. Ferreira, *Heteroat. Chem.*, 1999, **10**, 197 (*Chem. Abstr.*, 1999, **130**, 338 335).
32 A.G. Mustafin, M.A. Petrova, I.B. Abdrakhmanov and G.A. Tolstikov, *Bashk. Khim. Zh.*, 1998, **5**, 3 (*Chem. Abstr.*, 1999, **131**, 88 085).
33 A.G. Mustafin, M.A. Petrova, R.F. Khalimov, G.A. Tolstikov and I.B. Abdrakhmanov, *Bashk. Khim. Zh.*, 1999, **6**, 4 (*Chem. Abstr.*, 1999, **131**, 257 777).
34 J.B. Chang, R.Y. Guo, J. He and R.F. Chen, *Chin. Chem. Lett.*, 1999, **10**, 357 (*Chem. Abstr.*, 1999, **131**, 228 917).
35 R. Saladino, M. Mezzetti, E. Mincione, A.T. Palamara, P. Savini and S. Marini, *Nucleosides, Nucleotides,* 1999, **18**, 2499.
36 V. Boudou, J.-L. Imbach and G. Gosselin, *Nucleosides, Nucleotides,* 1999, **18**, 2463.
37 J.-L. Girardet and L.B. Townsend, *J. Org. Chem.*, 1999, **64**, 4169.
38 C. Hamon, T. Brandstetter and N. Windhab, *Synlett*, 1999, 940.

39 C. Sun, Y. Wang, H. Li, Y. Qi, Z. Chen and J. Zhang, *Zhongguo Yaowu Huaxue Zazhi*, 1997, **7**, 84 (*Chem. Abstr.*, 1999, **130**, 52 661).
40 M. Kiankarimi, R. Lowe, J.R. McCarthy and J.P. Whitten, *Tetrahedron Lett.*, 1999, **40**, 4497.
41 X.-M. Chen, Z.-J. Li, Z.-X. Ren and Z.-T. Huang, *Carbohydr. Res.*, 1999, **315**, 262.
42 R. Iqbal, N.H. Rama, K.H. Zamani and M.T. Hussain, *Indian J. Heterocycl. Chem.*, 1999, **8**, 269 (*Chem. Abstr.*, 1999, **131**, 257 778).
43 B.A. Hussain, A.M. Attia and G.E.H. Elgemeie, *Nucleosides, Nucleotides,* 1999, **18**, 2335.
44 A.M. Attia, M.A. Sallam, A.A. Almehdi and M.A. Abbasi, *Nucleosides, Nucleotides,* 1999, **18**, 2307.
45 G.E.H. Elgemeie, O.A. Mansour and N.H. Metwally, *Nucleosides, Nucleotides,* 1999, **18**, 113.
46 A.K. Mansour, Y.A. Ibrahim and N.S.A.M. Khalil, *Nucleosides, Nucleotides,* 1999, **18**, 2265.
47 A. Kittaka, H. Kato, H. Tanaka, Y. Nonaka, M. Amano, K.T. Nakamura and T. Miyasaka, *Tetrahedron*, 1999, **55**, 5319.
48 A. Kittaka, T. Asakura, T. Kuze, H. Tanaka, N. Yamada, K.T. Nakamura and T. Miyasaka, *J. Org. Chem.*, 1999, **64**, 7081.
49 C. Chatgilialoglu, T. Gimisis and G.P. Spada, *Chem. Eur. J.*, 1999, **5**, 2866.
50 C. Chatgilialoglu, C. Ferrari and T. Gimisis, *Tetrahedron Lett.*, 1999, **40**, 2837.
51 T. Maruyama, S. Kozai, T. Manabe, Y. Yazima, Y. Satoh and H. Takaku, *Nucleosides, Nucleotides,* 1999, **18**, 2433.
52 C. Castro, C. Chen and M.E. Jung, *Nucleosides, Nucleotides,* 1999, **18**, 2415.
53 M. Motura, H. Salomón, G. Moroni, M. Wainberg and M.C. Briñón, *Nucleosides, Nucleotides,* 1999, **18**, 337.
54 V.K. Rajwanshi, R. Kumar, M. Kofod-Hansen and J. Wengel, *J. Chem. Soc., Perkin Trans. 1*, 1999, 1407.
55 T. Searls and L.W. McLaughlin, *Tetrahedron*, 1999, **55**, 11985.
56 C.-S. Yu and F. Oberdorfer, *Synthesis*, 1999, 2057.
57 M.P. Wallis, N. Mahmood and W. Fraser, *Farmaco*, 1999, **54**, 83 (*Chem. Abstr.*, 1999, **130**, 338 334).
58 A.J. Lunato, J. Wang, J.E. Woollard, A.K.M. Anisuzzaman, W. Ji, F.-G. Rong, S. Ikeda, A.H. Soloway, S. Eriksson, D.H. Ives, T.E. Blue and W. Tjarks, *J. Med.Chem.*, 1999, **42**, 3378.
59 F. Seela and C. Mittelbach, *Nucleosides, Nucleotides*, 1999, **18**, 425.
60 N. Ramzaeva, C. Mittelbach and F. Seela, *Helv. Chim. Acta*, 1999, **82**, 12.
61 F. Seela, G. Becher, H. Rosemeyer, H. Reuter, G. Kastner and I.A. Mikhailopulo, *Helv. Chim. Acta*, 1999, **82**, 105.
62 J.S. Larsen, M.A. Zahran, E.B. Pedersen and C. Nielsen, *Monatsh. Chem.*, 1999, **130**, 1167.
63 C.C. Orti, R. Michalczyk and L.A. Silks III, *J. Org. Chem.*, 1999, **64**, 4685.
64 C.J. Steel, F. Brady, S.J. Luthra, G. Brown, I. Khan, K.G. Poole, A. Sergis, T. Jones and P.M. Price, *Appl. Radiat. Isot.*, 1999, **51**, 377 (*Chem. Abstr.*, 1999, **131**, 257 784).
65 A. Converso, C. Siciliano and G. Sindona, *Targets Heterocycl. Syst.*, 1998, **2**, 17 (*Chem. Abstr.*, 1999, **131**, 5419).
66 P. Ciuffreda, S. Casati and E. Santaniello, *Bioorg. Med. Chem. Lett.*, 1999, **9**, 1577.

67 J.C. Bussolari and R.P. Panzica, *Bioorg. Med. Chem.*, 1999, **7**, 2373.
68 J. Liu, B. Zhu and H. Chen, *Zhongguo Yaowu Huaxue Zazhi*, 1997, 7, 208 (*Chem. Abstr.*, 1999, **130**, 14 144).
69 S. Lajsic, G. Cetkovic, M. Popsavin, V. Popsavin, B.L. Milic and D. Miijkovic, *Carbohydr. Lett.*, 1999, **3**, 285.
70 P. Wang, P.J. Bolon, M.G. Newton and C.K. Chu, *Nucleosides, Nucleotides*, 1999, **18**, 2819.
71 J. Shi, J.J. McAtee, S.S. Wirtz, P. Tharnish, A. Juodawlkis, D.C. Liotta and R.F. Schinazi, *J. Med. Chem.*, 1999, **42**, 859.
72 M.T. Omar and A.H. El-Masry, *Bull. Natl. Res. Cent. (Egypt)*, 1999, **24**, 1 (*Chem. Abstr.*, 1999, **131**, 157 924).
73 O. Yamazaki, H. Togo, S. Matsubayashi and M. Yokoyama, *Tetrahedron*, 1999, **55**, 3735.
74 M. Gruen, C. Becker, A. Beste, C. Siethoff, A.J. Scheidig and R.S. Goody, *Nucleosides, Nucleotides*, 1999, **18**, 137.
75 T. Maruyama, S. Takamatsu, S. Kozai, Y. Satoh and K. Izawa, *Chem. Pharm. Bull.*, 1999, **47**, 966.
76 A.T. Shortnacy-Fowler, K.N. Tiwari, J.A. Montgomery, R.W. Buckheit Jr. and J.A. Secrist III, *Helv. Chim. Acta*, 1999, **82**, 2240.
77 K. Lee, Y. Choi, E. Gullen, S. Schlueter-Wirtz, R.F. Schinazi, Y.-C. Cheng and C.K. Chu, *J. Med. Chem.*, 1999, **42**, 1320.
78 J. Du, Y. Choi, K. Lee, B.K. Chun, J.H. Hong and C.K. Chu, *Nucleosides, Nucleotides*, 1999, **18**, 187.
79 S.C.H. Cavalcanti, Y. Xiang, M.G. Newton, R.F. Schinazi, Y.-C. Cheng and C.K. Chu, *Nucleosides, Nucleotides*, 1999, **18**, 2233.
80 G.S. Lal, G.P. Pez, R.J. Pesaresi, F.M. Prozonic and H. Cheng, *J. Org. Chem.*, 1999, **64**, 7048.
81 H. Gugliemi, M. Dachtler and K. Albert, *Z. Naturforsch., B: Chem. Sci.*, 1999, **54**, 1055.
82 N. Poopeiko, R. Fernández, M.I. Barrena, S. Castillón, J. Forniés-Cámer and C.J. Cardin, *J. Org. Chem.*, 1999, **64**, 1375.
83 J. Inagaki, H. Sakamoto, M. Nakajima, S. Nakamura and S. Hashimoto, *Synlett*, 1999, 1274.
84 C. Serra, J. Farràs and J. Vilarrasa, *Tetrahedron Lett.*, 1999, **40**, 9111.
85 R. Fernández and S. Castillón, *Tetrahedron*, 1999, **55**, 8497.
86 H.C. Lee, P. Kumar, L.I. Wiebe, R. McDonald, J.R. Mercer, K. Ohkura and K. Seki, *Nucleosides, Nucleotides*, 1999, **18**, 1995.
87 P. Kumar, D. Stypinski, H. Xia, A.J.B. McEwan, H.-J. MacHulla and L.I. Wiebe, *J. Labelled Compd. Radiopharm.*, 1999, **42**, 3 (*Chem. Abstr.*, 1999, **130**, 196 909).
88 U. Chiacchio, A. Rescifina, D. Iannazzo and G. Romeo, *J. Org. Chem.*, 1999, **64**, 28.
89 S. Pan, J. Wang and K. Zhao, *J. Org. Chem.*, 1999, **64**, 4.
90 A.G. Mustafin, R.R. Gataullin, I.B. Abdrakhmanov, L.V. Spirikhin and G.A. Tolstikov, *Russ. Chem. Bull.*, 1998, **47**, 2007 (*Chem. Abstr.*, 1999, **130**, 182 718).
91 V.V. Filichev, A.A. Malin, M.V. Yas'ko, M.B. Shcherbinin and V.A. Ostrovskii, *Russ. J. Org. Chem.*, 1998, **34**, 449 (*Chem. Abstr.*, 1999, **130**, 38 622).
92 S. Bera and T. Pathak, *Tetrahedron*, 1999, **55**, 13051.
93 H. Sugimura and Y. Katoh, *Chem. Lett.*, 1999, 361.
94 H. Li and M.J. Miller, *J. Org. Chem.*, 1999, **64**, 9289.

95 M.J. Thompson, A. Mekhalfia, D.P. Hornby and G.M. Blackburn, *J. Org. Chem.*, 1999, **64**, 7467.
96 P. Brown, C.M. Richardson, L.M. Mensah, P.J. O'Hanlon, N.F. Osborne, A.J. Pope and G. Walker, *Bioorg. Med. Chem.*, 1999, **7**, 2473.
97 R.E. Lee, M.D. Smith, L. Pickering and G.W.J. Fleet, *Tetrahedron Lett.*, 1999, **40**, 8689.
98 D.A. Barawkar and T.C. Bruice, *J. Am. Chem. Soc.*, 1999, **121**, 10418.
99 G. Thoithi, A. Van Schepdael, C. Vinckier, P. Herdewijn, E. Roets and J. Hoogmartens, *Nucleosides, Nucleotides*, 1999, **18**, 1863.
100 W. Urjasz and L. Celewicz, *J. Phys. Org. Chem.*, 1998, **11**, 618 (*Chem. Abstr.*, 1999, **130**, 81 771).
101 J. Matulic-Adamic and L. Beigelman, *Helv. Chim. Acta*, 1999, **82**, 2141.
102 J. Zhang and M.D. Matteucci, *Tetrahedron Lett.*, 1999, **40**, 1467.
103 T. Naka, M. Nishizono, N. Minakawa and A. Matsuda, *Tetrahedron Lett.*, 1999, **40**, 6297.
104 Y. Wang, G. Inguaggiato, M. Jasamai, M. Shah, D. Hughes, M. Slater and C. Simons, *Bioorg. Med. Chem.*, 1999, **7**, 481.
105 G. Inguaggiato, M. Jasamai, J.E. Smith, M. Slater and C. Simons, *Nucleosides, Nucleotides*, 1999, **18**, 457.
106 M.I. Elzagheid, M. Oivanen, R.T. Walker and J.A. Secrist III, *Nucleosides, Nucleotides,* 1999, **18**, 181.
107 Y. Yoshimura, M. Endo and S. Sakata, *Tetrahedron Lett.*, 1999, **40**, 1937.
108 Y. Yoshimura, M. Endo, S. Miura and S. Sakata, *J. Org. Chem.*, 1999, **64**, 7912.
109 E. Ichikawa, S. Yamamura and K. Kato, *Bioorg. Med. Chem. Lett.*, 1999, **9**, 1113.
110 J.H. Hong, M.-Y. Gao and C.K. Chu, *Tetrahedron Lett.*, 1999, **40**, 231.
111 N.A. Al-Masoudi, *Tetrahedron Lett.*, 1999, **40**, 4795.
112 E.W. van Tilburg, J. von Frijtag Drabbe Künzel, M. de Groote, R.C. Vollinga, A. Lorenzen and A.P. IJzerman, *J. Med. Chem.*, 1999, **42**, 1393.
113 Y.J. Kim, M. Ichikawa and Y. Ichikawa, *J. Am. Chem. Soc.*, 1999, **121**, 5829.
114 A.M. Belostotskii, J. Lexner and A. Hassner, *Tetrahedron Lett.*, 1999, **40**, 1181.
115 J. Wengel, *Acc. Chem. Res.*, 1999, **32**, 301.
116 L. Yet, *Tetrahedron*, 1999, **55**, 9349.
117 M.-C. Liu, M.-Z. Luo, D.E. Mozdziesz, T.-S. Lin, G.E. Dutschman, Y.-C. Cheng and A.C. Sartorelli, *Nucleosides, Nucleotides*, 1999, **18**, 55.
118 R. Buff and J. Hunziker, *Synlett*, 1999, 905.
119 M. Ethève-Quelquejeu and J.-M. Valéry, *Tetrahedron Lett.*, 1999, **40**, 4807.
120 X.-Q. Tang, X. Liao and J.A. Piccirilli, *J. Org. Chem.*, 1999, **64**, 747.
121 M. Dunkel and V. Reither, *Bioorg. Med. Chem. Lett.*, 1999, **9**, 787.
122 E. Riehokainen, I.E. Mikerin, N.N. Slobodyan, M.B. Tulebaev and S.E. Severin, *Carbohydr. Res.*, 1999, **320**, 161.
123 C. Len, D. Postel, G. MacKenzie, P. Villa and G. Ronco, *Pharm. Pharmacol. Commun.*, 1999, **5**, 165.
124 L.S. Jeong, J.H. Lee, K.-E. Jung, H.R. Moon, K. Kim and H. Lim, *Bioorg. Med. Chem.*, 1999, **7**, 1467.
125 P. von Matt, T. Lochmann, R. Kesselring and K.-H. Altmann, *Tetrahedron Lett.*, 1999, **40**, 1873.
126 D.-M. Liu, J.-M. Min and L.-H. Zhang, *Carbohydr. Res.*, 1999, **317**, 193.
127 J.M.J. Tronchet, E. Grand, I. Flores Montes, M. Criton, M. Seman and P. Dilda, *Carbohydr. Lett.*, 1999, **3**, 203.

128 S. Körner, A. Bryant-Friedrich and B. Giese, *J. Org. Chem.*, 1999, **64**, 1559.
129 H.M. Pfundheller, A.A. Koshkin, C.E. Olsen and J. Wengel, *Nucleosides, Nucleotides*, 1999, **18**, 2017.
130 S. Velázquez, V. Tuñón, M.L. Jimeno, C. Chamorro, E. De Clercq, J. Balzarini and M.J. Camarasa, *J. Med. Chem.*, 1999, **42**, 5188.
131 K. Kitano, H. Machida, S. Miura and H. Ohrui, *Bioorg. Med. Chem. Lett.*, 1999, **9**, 827.
132 F.A. Luzzio, A.V. Mayorov, M.E. Menes and W.L. Champion Jr., *Nucleosides, Nucleotides,* 1999, **18**, 1977.
133 I. Sugimoto, S. Shuto, S. Mori, S. Shigeta and A. Matsuda, *Bioorg. Med. Chem. Lett.*, 1999, **9**, 385.
134 M. Nomura, S. Shuto, M. Tanaka, T. Sasaki, S. Mori, S. Shigeta and A. Matsuda, *J. Med. Chem.*, 1999, **42**, 2901.
135 I. Sugimoto, S, Shuto and A. Matsuda, *J. Org. Chem.*, 1999, **64**, 7153.
136 R. Yamaguchi, T. Imanishi, S. Kohgo, H. Horie and H. Ohrui, *Biosci. Biotechnol. Biochem.*, 1999, **63**, 736.
137 S. Kohgo, H. Horie and H. Ohrui, *Biosci. Biotechnol. Biochem.*, 1999, **63**, 1146.
138 D. Crich and X. Hao, *J. Org. Chem.*, 1999, **64**, 4016.
139 H.M. Pfundheller and J. Wengel, *Bioorg. Med. Chem. Lett.*, 1999, **9**, 2667.
140 T. Imanishi and S. Obika, *Yuki Gosei Kagaku Kyokaishi*, 1999, **57**, 969 (*Chem. Abstr.*, 1999, **131**, 351 542).
141 V.K. Rajwanshi, A.E. Hakansson, B.M. Dahl and J. Wengel, *Chem. Commun.*, 1999, 1395.
142 L. Kværnø and J. Wengel, *Chem. Commun.*, 1999, 657.
143 S. Obika, J. Ando, T. Sugimoto, K. Miyashita and T. Imanishi, *Tetrahedron Lett.*, 1999, **40**, 6465.
144 S. Obika, K.-i. Morio, Y. Hari and T. Imanishi, *Chem. Commun.*, 1999, 2423.
145 S. Obika, K.-i. Morio, Y. Hari and T. Imanishi, *Bioorg. Med. Chem. Lett.*, 1999, **9**, 515.
146 G. Wang, J.-L. Girardet and E. Gunic, *Tetrahedron*, 1999, **55**, 7707.
147 G. Wang, E. Gunic, J.-L. Girardet and V. Stoisavljevic, *Bioorg. Med. Chem. Lett.*, 1999, **9**, 1147.
148 C. Lamberth, *Org. Prep. Proced. Int.*, 1999, **31**, 379 (*Chem. Abstr.*, 1999, **131**, 351 541).
149 M.J. Han, T.J. Cho, G.W. Lee, K.S. Yoo, Y.D. Park and J.Y. Chang, *J. Polym. Sci., Part A: Polym. Chem.*, 1999, **37**, 3361 (*Chem. Abstr.*, 1999, **131**, 272 121).
150 K. Hirota, H. Takasu, Y. Tsuji and H. Sajiki, *Chem. Commun.*, 1999, 1827.
151 M. Nomura, K. Endo, S. Shuto and A. Matsuda, *Tetrahedron*, 1999, **55**, 14847.
152 J.B. Epp and T.S. Widlanski, *J. Org. Chem.*, 1999, **64**, 293.
153 M. Loog, A. Uri, G. Raidaru, J. Järv and P. Ek, *Bioorg. Med. Chem. Lett.*, 1999, **9**, 1447.
154 M. de Zwart, A. Kourounakis, H. Kooijman, A.L. Spek, R. Link, J.K. von Frijtag Drabbe Künzel and A.P. IJzerman, *J. Med. Chem.*, 1999, **42**, 1384.
155 R. Volpini, E. Camaioni, S. Constanzi, S. Vittori, K.-N. Klotz and G. Cristalli, *Nucleosides, Nucleotides,* 1999, **18**, 2511.
156 P.J. Grohar and C.S. Chow, *Tetrahedron Lett.*, 1999, **40**, 2049.
157 U. Wichai and S.A. Woski, *Org. Lett.*, 1999, **1**, 1173.
158 S. Lutz and S.A. Benner, *Bioorg. Med. Chem. Lett.*, 1999, **9**, 723.
159 E.S. Gibson, K. Lesiak, K.A. Watanabe, L.J. Gudas and K.W. Pankiewicz, *Nucleosides, Nucleotides,* 1999, **18**, 363.

160 K.S. Ramasamy and D. Averett, *Nucleosides, Nucleotides,* 1999, **18**, 2425.

161 X.Y. Yu, J.M. Hill, G. Yu, W. Wang, A.F. Kluge, P. Wendler and P. Gallant, *Bioorg. Med. Chem. Lett.*, 1999, **9**, 375.

162 C. Strässler, N.E. Davis and E.T. Kool, *Helv. Chim. Acta*, 1999, **82**, 2160.

163 C.F. Morelli, M. Manferdini and A.C. Veronese, *Tetrahedron,* 1999, **55**, 10803.

164 S. Kanazawa, S. Mizuno, R. Yamauchi, N. Nishimura and I. Maeba, *Carbohydr. Res.*, 1999, **318**, 180.

165 D.-Q. Sun, J. Zu, L.-T. Ma and L.-H. Zhang, *Gaodeng Xuexiao Huaxue Xuebao,* 1999, **20**, 895 (*Chem. Abstr.*, 1999, **131**, 199 905).

166 P.J. Dudfield, V.-D. Le, S.D. Lindell and C.W. Rees, *J. Chem. Soc., Perkin Trans. 1*, 1999, 2929.

167 P.J. Dudfield, V.-D. Le, S.D. Lindell and C.W. Rees, *J. Chem. Soc., Perkin Trans. 1*, 1999, 2937.

168 Y.A. Al-Soud, W.A. Al-Masoudi, R.A. El-Halawa and N. Al-Masoudi, *Nucleosides, Nucleotides,* 1999, **18**, 1985.

169 M. Chabbi, G. Goethals, G. Ronco and P. Villa, *J. Nat.*, 1997, **9**, 17 (*Chem. Abstr.*, 1999, **130**, 110 512).

170 S. Harusawa, T. Imazu, S. Takashima, L. Araki, H. Ohishi, T. Kurihara, Y. Yamamoto and A. Yamatodani, *Tetrahedron Lett.*, 1999, **40**, 2561.

171 D.B. Silverman, M.C. Pitman and D.E. Platt, *J. Biomol. Struct. Dyn.*, 1999, **16**, 1169 (*Chem. Abstr.*, 1999, **131**, 88123).

172 B.M. Domínguez and P.M. Cullis, *Tetrahedron Lett.*, 1999, **40**, 5783.

173 A. Dhanda, L.J.S. Knutsen, M.-B. Nielsen, S.M. Roberts and D.R. Varley, *J. Chem. Soc., Perkin Trans. 1*, 1999, 3469.

174 H. Urata, H. Miyagoshi, T. Yumoto and M. Akagi, *J. Chem. Soc., Perkin Trans. 1*, 1999, 1833.

175 P. Wang, L.A. Agrofoglio, M.G. Newton and C.K. Chu, *J. Org. Chem.*, 1999, **64**, 4173.

176 P. Wang, B. Gullen, M.G. Newton, Y.-C. Cheng, R.F. Schinazi and C.K. Chu, *J. Med. Chem.*, 1999, **42**, 3390.

177 B. Mohar and J. Kobe, *Nucleosides, Nucleotides,* 1999, **18**, 443.

178 N. Katagiri, Y. Yamatoya and M. Ishikura, *Tetrahedron Lett.*, 1999, **40**, 9069.

179 J. Watchtmeister, A. Muhlmann, B. Classon and B. Samuelsson, *Tetrahedron*, 1999, **55**, 10761.

180 H.R. Moon, H.O. Kim, M.W. Chun, L.S. Jeong and V.E. Marquez, *J. Org. Chem.*, 1999, **64**, 4733.

181 A. Roy, K. Chakrabarty, P.K. Dutta, N.C. Bar, N. Basu, B. Achari and S.B. Mandal, *J. Org. Chem.*, 1999, **64**, 2304.

182 V.R. Hegde, K.L. Seley, X. Chen and S.W. Schneller, *Nucleosides, Nucleotides,* 1999, **18**, 1905.

183 K.L. Seley and S.W. Schneller, *J. Heterocycl. Chem.*, 1999, **36**, 287.

184 M. Cowart, M.J. Bennett and J.F. Kerwin, *J. Org. Chem.*, 1999, **64**, 2240.

185 A. Ghosh, M.J. Miller, E. De Clercq and J. Balzarini, *Nucleosides, Nucleotides,* 1999, **18**, 217.

186 A. Ogawa, S. Shuto, M. Tanaka, T. Sasaki, S. Mori, S. Shigeta and A. Matsuda, *Chem. Pharm. Bull.*, 1999, **47**, 1000.

187 M.J. Comin, C.A. Pujol, E.B. Damonte and J.B. Rodriguez, *Nucleosides, Nucleotides,* 1999, **18**, 2219.

188 J. Wang and P. Herdewijn, *J. Org. Chem.*, 1999, **64**, 7821.

189 M.I. Nieto, J.M. Blanco, O. Caamaño, F. Fernández, X. García-Mera, C. López, J. Balzarini and E. De Clercq, *Nucleosides, Nucleotides,* 1999, **18**, 2253.
190 J. Cieslak, M. Szymczak, M. Wenska, J. Stawinski and A. Kraszewski, *J. Chem. Soc., Perkin Trans. 1*, 1999, 3327.
191 P.H. Seeberger, M.H. Caruthers, D. Benkaitis-Davis and G. Beaton, *Tetrahedron*, 1999, **55**, 5759.
192 I. Kers, A. Kers, J. Stawinski and A. Kraszewski, *Tetrahedron Lett.*, 1999, **40**, 3945.
193 C.B. Reese and C. Visintin, *Tetrahedron Lett.*, 1999, **40**, 6477.
194 W. Wang, Q. Song and R.A. Jones, *Tetrahedron Lett.*, 1999, **40**, 8971.
195 T. Miyashita, K. Mori and K. Shinozuka, *Chem. Lett.*, 1999, 39.
196 E. Belsito, A. Liguori, A. Napoli, C. Siciliano and G. Sindona, *Nucleosides, Nucleotides,* 1999, **18**, 2565.
197 T. Yokomatsu, T. Shimizu, T. Sada and S. Shibuya, *Heterocycles*, 1999, **50**, 21.
198 E. Nandanan, E. Camaioni, S.-Y. Jang, Y.-C. Kim, G. Cristalli, P. Herdewijn, J.A. Secrist III, K.N. Tiwari, A. Mohanram, T.K. Harden, J.L. Boyer and K.A. Jacobson, *J. Med. Chem.*, 1999, **42**, 1625.
199 A. Naundorf and W. Klaffke, *Carbohydr. Res.*, 1999, **318**, 38.
200 H. Brachwitz, J. Bergmann, Y. Thomas, T. Wollny and P. Langen, *Bioorg. Med. Chem.*, 1999, **7**, 1195.
201 R.T. Bibart, K.W. Vogel and D.G. Drueckhammer, *J. Org. Chem.*, 1999, **64**, 2903.
202 J. Baraniak, E. Wasilewska, D. Korczynski and W.J. Stec, *Tetrahedron Lett.*, 1999, **40**, 8603.
203 S. Flohr, V. Jungmann and H. Waldmann, *Chem. Eur. J.*, 1999, **5**, 669.
204 T. Moriguchi, T. Yanagi, T. Wada and M. Sekine, *J. Chem. Soc., Perkin Trans. 1*, 1999, 1859.
205 V.D. Antle and C.A. Caperelli, *Nucleosides, Nucleotides*, 1999, **18**, 1911.
206 C. Meier, T. Knispel, E. De Clercq and J. Balzarini, *J. Med. Chem.*, 1999, **42**, 1604.
207 C. Meier, T. Knispel, V.E. Marquez, M.A. Siddiqui, E. De Clercq and J. Balzarini, *J. Med. Chem.*, 1999, **42**, 1615.
208 A.Q. Siddiqui, C. Ballatore, C. McGuigan, E. De Clercq and J. Balzarini, *J. Med. Chem.*, 1999, **42**, 393.
209 A.Q. Siddiqui, C. McGuigan, C. Ballatore, F. Zuccotto, I.H. Gilbert, E. De Clercq and J. Balzarini, *J. Med. Chem.*, 1999, **42**, 4122.
210 A.Q. Siddiqui, C. McGuigan, C. Ballatore, O. Wedgwood, E. De Clercq and J. Balzarini, *Bioorg. Med. Chem. Lett.,* 1999, **9**, 2555.
211 S.R. Khan and D. Farquhar, *Tetrahedron Lett.*, 1999, **40**, 607.
212 N. Schwenger, C. Périgaud, A.-M. Aubertin, C. Thumann, G. Gosselin and J.-L. Imbach, *Helv. Chim. Acta*, 1999, **82**, 2044.
213 S. Manfredini, P.G. Baraldi, E. Durini, S. Vertuani, J. Balzarini, E. De Clercq, A. Karlsson, V. Buzzoni and L. Thelander, *J. Med. Chem.*, 1999, **42**, 3243.
214 E.J. Salaski and H. Maag, *Synlett*, 1999, 897.
215 B. Schade, V. Hagen, R. Schmidt, R. Herbrich, E. Krause, T. Eckhardt and J. Bendig, *J. Org. Chem.*, 1999, **64**, 9109.
216 T. Furuta and M. Iwamura, *Methods Enzymol.*, 1998, **291**, 50 (*Chem. Abstr.*, 1999, **130**, 182 707).
217 M.G.B. Drew, S. Gorsuch, J.H.M. Gould and J. Mann, *J. Chem. Soc., Perkin Trans. 1*, 1999, 969.

218 M.L. Hamm and J.A. Piccirilli, *J. Org. Chem.*, 1999, **64**, 5700.
219 S. Pal and V. Nair, *Biotechnol. Lett.*, 1998, **20**, 1149 (*Chem. Abstr.*, 1999, **130**, 196 910).
220 H. Kamiya and H. Kawai, *Nucleosides, Nucleotides,* 1999, **18**, 307.
221 S.C. Jurcyk, J.T. Kodra, J.-H. Park, S.A. Benner and T.R. Battersby, *Helv. Chim. Acta*, 1999, **82**, 1005.
222 M. Bretner, D. Beckett, R.K. Sood, D.M. Baldisseri and R.S. Hosmane, *Bioorg. Med. Chem.*, 1999, **7**, 2931.
223 N. Ostermann, M.R. Ahmadian, A. Wittinghofer and R.S. Goody, *Nucleosides, Nucleotides,* 1999, **18**, 245.
224 E. Halbfinger, D.T. Major, M. Ritzmann, J. Ubl, G. Reiser, J.L. Boyer, K.T. Harden and B. Fischer, *J. Med. Chem.*, 1999, **42**, 5325.
225 B. Fischer, A. Chulkin, J.L. Boyer, K.T. Harden, F.-P. Gendron, A.R. Beaudoin, J. Chapal, D. Hillaire-Buys and P. Petit, *J. Med. Chem.*, 1999, **42**, 3636.
226 A. H. Ingall, J. Dixon, A. Bailey, M.E. Coombs, D. Cox, J.I. McInally, S.F. Hunt, N.D. Kindon, B.J. Teobald, P.A. Willis, R.G. Humphries, P. Leff, J.A. Clegg, J.A. Smith and W. Tomlinson, *J. Med. Chem.*, 1999, **42**, 213.
227 A.V. Shipitsin, L.S. Victorova, E.A. Shirokova, N.B. Dyatkina, L.E. Goryunova, R.Sh. Beabealashvilli, C.J. Hamilton, S.M. Roberts and A. Krayevsky, *J. Chem. Soc., Perkin Trans. 1*, 1999, 1039.
228 C.J. Hamilton and S.M. Roberts, *J. Chem. Soc., Perkin Trans. 1*, 1999, 1051.
229 A.M. Murabuldaev, N.F. Zakirova and L.A. Alexandrova, *Bioorg. Khim.*, 1998 **224**, 474 (*Chem. Abstr.*, 1999, **131**, 5461).
230 T. Brossette, A. Valleix, L. Goujon, C. Créminon, J. Grassi, C. Mioskowski and L. Lebeau, *Tetrahedron Lett.*, 1999, **40**, 3391.
231 L. Lebeau, T. Brossette, L. Goujon, C. Créminon, J. Grassi and C. Mioskowski, *Tetrahedron Lett.*, 1999, **40**, 4323.
232 T. Brossette, A. Lefaou, L. Goujon, A. Valleix, C. Créminon, J. Grassi, C. Mioskowski and L. Lebeau, *J. Org. Chem.*, 1999, **64**, 5083.
233 M. Morr, A. Wassmann and V. Wray, *Tetrahedron*, 1999, **55**, 2985.
234 T. Bülter and L. Elling, *Glycoconjugate J.*, 1999, **16**, 147.
235 B.J. Mengeling and S.J. Turco, *Anal. Biochem.*, 1999, **267**, 227.
236 I.A. Ivanova, N.S. Utkina and V.N. Shibaev, *Bioorg. Khim.*, 1998, **24**, 710 (*Chem. Abstr.*, 1999, **131**, 45 037).
237 M.D. Chappell and R.L. Halcomb, *Tetrahedron Lett.*, 1999, **40**, 1.
238 M.D. Burkart, S.P. Vincent and C.-H. Wong, *Chem. Commun.*, 1999, 1525.
239 R.E. Campbell and M.E. Tanner, *J. Org. Chem.*, 1999, **64**, 9487.
240 P.N. Schröder and A. Giannis, *Angew. Chem., Int. Ed. Engl.*, 1999, **38**, 1379.
241 V. Kolb, F. Amann, R.R. Schmidt and M. Duszenko, *Glycoconjugate J.*, 1999, **16**, 537.
242 L. Cipolla, E. Klaps, B. La Ferla, F. Nicotra and W. Schmid, *Carbohydr. Lett.*, 1999, **3**, 173.
243 J. Kovensky, M. McNeil and P. Sinaÿ, *J. Org. Chem.*, 1999, **64**, 6202.
244 C. Wojczewski, K. Stolze and J.W. Engels, *Synlett*, 1999, 1667.
245 A. Eleuteri, D.C. Capaldi, D.L. Cole and V.T. Ravikumar, *Nucleosides, Nucleotides*, 1999, **18**, 1879.
246 S.B. Heidenhain and Y. Hayakawa, *Nucleosides, Nucleotides*, 1999, **18**, 1771.
247 A. Wilk, A. Grajkowski, L.R. Phillips and S.L. Beaucage, *J. Org. Chem.*, 1999, **64**, 7515.
248 H. Cramer and W. Pfleiderer, *Helv. Chim. Acta*, 1999, **52**, 614.

249 J. Matulic-Adamic, A.T. Daniher, C. Gonzalez and L. Beigelman, *Bioorg. Med. Chem. Lett.*, 1999, **9**, 157.
250 R. Micura, *Chem. Eur. J.*, 1999, **5**, 2077.
251 C.B. Reese and Q. Song, *Nucleic Acids Res.*, 1999, **27**, 963.
252 C.B. Reese and Q. Song, *Nucleic Acids Res.*, 1999, **27**, 2672.
253 T. Wada, F. Honda, Y. Sato and M. Sekine, *Tetrahedron Lett.*, 1999, **40**, 915.
254 C.B. Reese and Q. Song, *J. Chem. Soc., Perkin Trans. 1*, 1999, 1477.
255 T. Kato and Y. Hayakawa, *Synlett*, 1999, 1796.
256 Z.S. Cheruvallath, P.D. Wheeler, D.L. Cole and V.T. Ravikumar, *Nucleosides, Nucleotides*, 1999, **18**, 485.
257 A. Eleuteri, Z.S. Cheruvallath, D.C. Capaldi, D.L. Cole and V.T. Ravikumar, *Nucleosides, Nucleotides*, 1999, **18**, 1803.
258 X.-B. Yang, K. Misiura and W.J. Stec, *Heteroat. Chem.*, 1999, **10**, 91 (*Chem. Abstr.*, 1999, **130**, 282 304).
259 J.-C. Wang and G. Just, *J. Org. Chem.*, 1999, **64**, 8090.
260 J.-C. Wang, G. Just, A.P. Guzaev and M. Manoharan, *J. Org. Chem.*, 1999, **64**, 2595.
261 T.K. Srivastava, P. Friedhoff, A. Pingoud and S.B. Katti, *Nucleosides, Nucleotides*, 1999, **18**, 1945.
262 A. Kers and J. Stawinski, *Tetrahedron Lett.*, 1999, **40**, 4263.
263 P.W. Davis and S.A. Osgood, *Bioorg. Med. Chem. Lett.*, 1999, **9**, 2691.
264 J. Robles, V. Ibanez, A. Grandas and E. Pedroso, *Tetrahedron Lett.*, 1999, **40**, 7131.
265 J. Baraniak, D. Kerczyński and W.J. Stec, *J. Org. Chem.*, 1999, **64**, 4533.
266 K. Pongracz and S. Gryaznov, *Tetrahedron Lett.*, 1999, **40**, 7661.
267 I. Kers, J. Stawinski and A. Kraszewski, *Tetrahedron,* 1999, **55**, 11579.
268 Z.A. Sergueeva, D.S. Sergueev and B.R. Shaw, *Tetrahedron Lett.*, 1999, **40**, 2041.
269 K. He, D.S. Sergueev, Z.A. Sergueeva and B.R. Shaw, *Tetrahedron Lett.*, 1999, **40**, 4601.
270 J. Lin and B.R. Shaw, *Chem. Commun.*, 1999, 1517.
271 J. Baraniak, D. Korczynski, R. Kaczmarek and W.J. Stec, *Nucleosides, Nucleotides,* 1999, **18**, 2147.
272 K. Miriura, M. Bollmark, J. Stawinski and W.J. Stec, *Chem. Commun.*, 1999, 2115.
273 S. Abbas and C.J. Hayes, *Synlett*, 1999, 1124.
274 I. Kers, J. Stawinski, J.-L. Girardet, J.-L. Imbach, C. Périgaud, G. Gosselin and A.-M. Aubertin, *Nucleosides, Nucleotides*, 1999, **18**, 2317.
275 H. Schott, P.S. Ludwig, A. Immelmann and R.A. Schwendener, *Eur. J. Med. Chem.*, 1999, **34**, 343 (*Chem. Abstr.*, 1999, **131**, 286 747).
276 K.B. Kim and E.J. Behrman, *Nucleosides, Nucleotides*, 1999, **18**, 51.
277 J. Lee, H. Churchill, W.-B. Choi, J.E. Lynch, F.E. Roberts, R.P. Volante and P.J. Reider, *Chem. Commun.*, 1999, 729.
278 R. Lodaya and J.T. Slama, *J. Labelled Compd. Radiopharm.*, 1999, **42**, 867 (*Chem. Abstr.*, 1999, **131**, 310 790).
279 S. Tono-oka and I. Azuma, *Nucleosides, Nucleotides*, 1999, **18**, 39.
280 H. Ikeda, E. Abushanab and V.E. Marquez, *Bioorg. Med. Chem. Lett.,* 1999, **9**, 3069.
281 A. Kanavarioti and S. Gangopadhyay, *J. Org. Chem.*, 1999, **64**, 7957.
282 H. Sawai, H. Wakai and A. Nakamura-Ozaki, *J. Org. Chem.*, 1999, **64**, 5837.
283 F.-J. Zhang, Q.-M. Gu and C.J. Sih, *Bioorg. Med. Chem.*, 1999, **7**, 653.

284 M. Fukuoka, S. Shuto, N. Minakawa, Y. Ueno and A. Matsuda, *Tetrahedron Lett.*, 1999, **40**, 5361.
285 E.I. Kvasyuk, I.A. Mikhailopulo, J.W. Homan, K.T. Iacono, N.F. Muto, R.J. Suhadolnik and W. Pfleiderer, *Helv. Chim. Acta*, 1999, **82**, 19.
286 H. Schirmeister-Tichy, K.T. Iacono, N.F. Muto, J.W. Homan, R.J. Suhadolnik and W. Pfleiderer, *Helv. Chim. Acta*, 1999, **82**, 597.
287 J. Zhang, J.T. Shaw and M.D. Matteucci, *Bioorg. Med. Chem. Lett.*, 1999, **9**, 319.
288 Y. Ducharme and K.A. Harrison, *Can. J. Chem.*, 1999, **77**, 1410.
289 P. von Matt, A. de Mesmaeker, U. Pieles, W. Zürcher and K.-H. Altmann, *Tetrahedron Lett.*, 1999, **40**, 2899.
290 S. Druillennec, H. Meudal, B.P. Roques and M.-C. Fournié-Zaluski, *Bioorg. Med. Chem. Lett.*, 1999, **9**, 627.
291 M.A. Peterson, B.L. Nilsson, S. Sarker, B. Doboszewski, W. Zhang and M.J. Robins, *J. Org. Chem.*, 1999, **64**, 8183.
292 M.-Y. Chan, R.A. Fairhurst, S.P. Collingwood, J. Fisher, J.R.P. Arnold, R. Cosstick and I.A. O'Neil, *J. Chem. Soc., Perkin Trans. 1*, 1999, 315.
293 J. Zhang and M.D. Matteucci, *Bioorg. Med. Chem. Lett.*, 1999, **9**, 2213.
294 J.M.J. Tronchet, E. Grand, M. Zsely, R. Giovannini and M. Geoffroy, *Carbohydr. Lett.*, 1998, **3**, 161 (*Chem. Abstr.*, 1999, **130**, 95 768).
295 Y. Zhou, L.R. Zhang and L.H. Zhang, *Chin. Chem. Lett.*, 1998, **9**, 791 (*Chem. Abstr.*, 1999, **131**, 299 641).
296 J. Magdalena, S. Fernández, M. Ferrero and V. Gotor, *Tetrahedron Lett.*, 1999, **40**, 1787.
297 B. Kierdaszuk, K. Krawiec, Z. Kazimierczuk, U. Jacobsson, N.G. Johansson, B. Munch-Petersen, S. Eriksson and D. Shugar, *Nucleosides, Nucleotides*, 1999, **18**, 1883.
298 U. Legoburu, C.B. Reese and O. Song, *Tetrahedron*, 1999, **55**, 5635.
299 M. Grøtli, B. Beijer and B. Sproat, *Tetrahedron*, 1999, **55**, 4299.
300 H. Asanuma, T. Yoshida, T. Ito and M. Komiyama, *Tetrahedron Lett.*, 1999, **40**, 7995.
301 Q. Song, W. Wang, A. Fischer, X. Zhang, B.L. Gaffney and R.A. Jones, *Tetrahedron Lett.*, 1999, **40**, 4153.
302 M.B. Welch and K. Burgess, *Nucleosides, Nucleotides*, 1999, **18**, 197.
303 M.B. Welch, C.I. Martinez, A.J. Zhang, S. Jin, R. Gibbs and K. Burgess, *Chem. Eur. J.*, 1999, **5**, 951.
304 M. Kataoka and Y. Hayakawa, *J. Org. Chem.*, 1999, **64**, 6087.
305 N.B. Karalkar, V.G. Akerkar and M.M. Salunkhe, *Indian J. Chem., Sect B: Org. Chem. Incl. Med. Chem.*, 1999, **38B**, 370 (*Chem. Abstr.*, 1999, **131**, 185 180).
306 K.S. Ramasamy and D. Averett, *Synlett*, 1999, 709.
307 A. Khalafi-Nezhad and R. Fareghi Alamdari, *Iran. J. Chem. Chem. Eng.*, 1998, **17**, 58 (*Chem. Abstr.*, 1999, **131**, 45034).
308 M.P. Coleman and M.K. Boyd, *Tetrahedron Lett.*, 1999, **40**, 7911.
309 R.T. Pon, S. Yu and Y.S. Sanghvi, *Bioconjugate Chem.*, 1999, **10**, 1051 (*Chem. Abstr.*, 1999, **131**, 310 787).
310 I.G. Stankova, M.F. Simeonov, V. Maximova, A.S. Galabov and E.V. Golovinsky, *Z. Naturforsch., C: Biosci.*, 1999, **54**, 75 (*Chem. Abstr.*, 1999, **130**, 223 538).
311 S. Kozai, S. Takamatsu, K. Izawa and T. Maruyama, *Tetrahedron Lett.*, 1999, **40**, 4355.
312 M. Mahmoudian, J. Eaddy and M. Dawson, *Biotechnol. Appl. Biochem.*, 1999, **29**, 229 (*Chem. Abstr.*, 1999, **131**, 185 188).

313 A. Khan, D.M. Haddleton, M.J. Hannon, D. Kukulj and A. Marsh, *Macromolecules*, 1999, **32**, 6560.
314 M.T. Mchedlidze, *Vestn. Mosk. Univ., Ser. 2: Khim.*, 1999, **40**, 260 (*Chem. Abstr.*, 1999, **131**, 337 301).
315 Z.-w. Guo and J.M. Gallo, *J. Org. Chem.*, 1999, **64**, 8319.
316 T. Kimura, H. Matsumoto, T. Matsuda, T. Hamawaki, K. Akaji and Y. Kiso, *Bioorg. Med. Chem. Lett.*, 1999, **9**, 803.
317 Y. Lu and M.J. Miller, *Bioorg. Med. Chem.*, 1999, **7**, 3025.
318 J. Lee, S.U. Kang, M.Y. Kang, M.W. Chun, Y.J. Jo, J.H. Kwak and S. Kim, *Bioorg. Med. Chem. Lett.*, 1999, **9**, 1365.
319 M. Lecouvey, C. Dufau, D. El Manouni and Y. Leroux, *Nucleosides, Nucleotides,* 1999, **18**, 2109.
320 D.K. Alargov, R.G. Gugova, P.S. Denkova, G. Muller and E.V. Golovinsky, *Monatsh. Chem.*, 1999, **130**, 937.
321 K. Kamaike, H. Takahashi, K. Morohoshi, N. Kataoka, T. Kakinuma and Y. Ishido, *Acta Biochem. Pol.*, 1998, **45**, 949 (*Chem. Abstr.*, 1999, **130**, 338 341).
322 A. Stutz and S. Pitsch, *Synlett*, 1999, 930.
323 S. Pitsch, P.A. Weiss, X. Wu, D. Ackermann and T. Honegger, *Helv. Chim. Acta*, 1999, **82**, 1753.
324 G.C. Chi, Z.B. Zhang, X.H. Xu and R.Y. Chen, *Chin. Chem. Lett.*, 1998, **9**, 989 (*Chem. Abstr.*, 1999, **131**, 299 640).
325 Z.J. Liu, J.M. Min and L.-H. Zhang, *Gaodeng Xuexiao Huaxue Xuebao,* 1999, **20**, 1400 (*Chem. Abstr.*, 1999, **131**, 337 296).
326 A.A. Rodionov, E.V. Efimtseva, M.V. Fomicheva, N.S. Padyukova and S.N. Mikhailov, *Bioorg. Khim.*, 1999, **25**, 203 (*Chem. Abstr.*, 1999, **131**, 170 545).
327 M. de Kort, E. Ebrahami, E.R. Wijsman, G.A. van der Marel and J.H. van Boom, *Eur. J. Org. Chem.*, 1999, 2337.
328 D. Crich and X.-S. Mo, *Synlett*, 1999, 67.
329 V. Banuls and J.-M. Escudier, *Tetrahedron*, 1999, **55**, 5831.
330 N. Solladié and M. Gross, *Tetrahedron Lett.*, 1999, **40**, 3359.
331 M.P. Molas, M.I. Matheu, S. Castillón, J. Isac-Garcia, F. Hernández-Mateo, F.G. Calvo-Flores and F. Santoyo-González, *Tetrahedron*, 1999, **55**, 14649.
332 G. Wang, *Tetrahedron Lett.*, 1999, **40**, 6343.
333 Y. Díaz, F. Bravo and S. Castillón, *J. Org. Chem.*, 1999, **64**, 6508.
334 V. Nair, *Tetrahedron,* 1999, **55**, 11803.
335 X. Zheng and V. Nair, *Nucleosides, Nucleotides*, 1999, **18**, 1961.
336 S. Bera, T. Mickle and V. Nair, *Nucleosides, Nucleotides*, 1999, **18**, 2379.
337 J. Wouters and P. Herdewijn, *Bioorg. Med. Chem. Lett.*, 1999, **9**, 1563.
338 B. Allart, R. Busson, J. Rozenski, A. Van Aerschot and P. Herdewijn, *Tetrahedron*, 1999, **55**, 6527.
339 B. Allart, K. Khan, H. Rosemeyer, G. Schepers, C. Hendrix, K. Rothenbacher, F. Seela, A. Van Aerschot and P. Herdewijn, *Chem. Eur. J.*, 1999, **5**, 2424.
340 N. Hossain, I. Luyten, K. Rothenbacher, R. Busson and P. Herdewijn, *Nucleosides, Nucleotides*, 1999, **18**, 161.
341 K.-E. Jung, K. Kim, M. Yang, K. Lee and H. Lim, *Bioorg. Med. Chem. Lett.*, 1999, **9**, 3407.
342 M. Yokoyama and A. Momotake, *Synthesis,* 1999, 1541.
343 C.V. Varaprasad, D. Averett, K.S. Ramasamy and J. Wu, *Tetrahedron,* 1999, **55**, 13345.

344 D.C. Kim, K.H. Yoo, D.J. Kim, B.Y. Chung and S.W. Park, *Tetrahedron Lett.*, 1999, **40**, 4825.
345 R.H. Furneaux, V.L. Schramm and P.C. Tyler, *Bioorg. Med. Chem.*, 1999, **7**, 2599.
346 L. Jin, Hong Wu, Huailing Wu, P. Huang, K. Jung and H. Lim, *Chem. Lett.*, 1999, 687.
347 F. Marciacq, S. Sauvaigo, J.-P. Issartel, J.-F. Mouret and D. Molko, *Tetrahedron Lett.*, 1999, **40**, 4673.
348 J.-S. Lin, T. Kira, E. Gullen, Y. Choi, F. Qu, C.K. Chu and Y.-C. Cheng, *J. Med. Chem.*, 1999, **42**, 2212.
349 F. Qu, J.H. Hong, J. Du, M.G. Newton and C.K. Chu, *Tetrahedron*, 1999, **55**, 9073.
350 R. Caputo, A. Guaragna, G. Palumbo and S. Pedatella, *Eur. J. Org. Chem.*, 1999, 1455.
351 N. Khan, S.R. Bastola, K.G. Witter and P. Scheiner, *Tetrahedron Lett.*, 1999, **40**, 8989.
352 N. Nguyen-Ba, W.L. Brown, L. Chan, N. Lee, L. Brasili, D. Lafleur and B. Zacharie, *Chem. Commun.*, 1999, 1245.
353 U. Chiaccio, A. Corsaro, G. Gumina, A. Rescifina, D. Iannazzo, A. Piperno, G. Romeo and R. Romeo, *J. Org. Chem.*, 1999, **64**, 9321.
354 D.F. Ewing, N.-E. Fahmi, C. Len, G. Mackenzie, G. Ronco, P. Villa and G. Shaw, *Nucleosides, Nucleotides,* 1999, **18**, 2613.
355 D.F. Ewing, G. Goethals, G. Mackenzie, P. Martin, G. Ronco, L. Vanbaelinghem and P. Villa, *Carbohydr. Res.*, 1999, **321**, 190.
356 D.F. Ewing, G. Goethals, G. Mackenzie, P. Martin, G. Ronco, L. Vanbaelinghem and P. Villa, *J. Carbohydr. Chem.*, 1999, **18**, 441.
357 L. De Napoli, G. Di Fabio, A. Messere, D. Montesarchio, G. Piccialli and M. Varra, *J. Chem. Soc., Perkin Trans. 1*, 1999, 3489.
358 A. Manikowski and J. Boryski, *Nucleosides, Nucleotides,* 1999, **18**, 2367.
359 L. Santana, M. Teijeira and E. Uriarte, *J. Heterocycl. Chem.*, 1999, **36**, 293.
360 N. Gauvry and F. Huet, *Tetrahedron,* 1999, **55**, 1321.
361 J. Rife and R.M. Ortuno, *Org. Lett.*, 1999, **1**, 1221.
362 T. Onoshi, T. Matsuzawa, S. Nishi and T. Tsuji, *Tetrahedron Lett.*, 1999, **40**, 8845.
363 Y.-L. Qiu and J. Zemlica, *Nucleosides, Nucleotides,* 1999, **18**, 2285.
364 R. Csuk and G. Thiede, *Tetrahedron*, 1999, **55**, 739.
365 J.H. Hah, J.M. Gil and D.Y. Oh, *Tetrahedron Lett.*, 1999, **40**, 8235.
366 L. Lomozik and R. Bregier-Jarzebowska, *Pol. J. Chem.*, 1999, **73**, 927 (*Chem. Abstr.*, 1999, **131**, 59 024).
367 O.M. Gritsenko and E.S. Gromova, *Russ. Chem. Rev.*, 1999, **68**, 241 (*Chem. Abstr.*, 1999, **131**, 144 754).
368 A.R. Porcari, K.Z. Borysko, R.G. Ptak, J.M. Breitenbach, L.L. Wotring, J.C. Drach and L.B. Townsend, *Nucleosides, Nucleotides*, 1999, **18**, 2475.
369 Y. Kitade, M. Nakanishi and C. Yatome, *Bioorg. Med. Chem. Lett.*, 1999, **9**, 2737.
370 C.D. Rousser, G.D. Ivanova, E.K. Bratovanova, N.G. Vassiler and D.D. Petkor, *J. Am. Chem. Soc.*, 1999, **121**, 11267.
371 M. Kosonen, R. Seppanen, O. Wichmann and H. Lönnberg, *J. Chem. Soc., Perkin Trans. 2,* 1999, 2433.
372 E. Maki, M. Oivanen, P. Poijarvi and H. Lönnberg, *J. Chem. Soc., Perkin Trans. 2*, 1999, 2493.

373 M.I. Elzagheid, M. Oivanen, K.D. Klika, B.C.N.M. Jones, R. Cosstick and H. Lönnberg, *Nucleosides, Nucleotides*, 1999, **18**, 2093.
374 L.A. Jenkins, J.K. Bashkin, J.D. Pennock, J. Florian and A. Worshel, *Inorg. Chem.*, 1999, **38**, 3215.
375 A. Kanavarioti, E.E. Baird, T.B. Hurley, J.A. Carruthers and S. Gangopadhyay, *J. Org. Chem.*, 1999, **64**, 8323.
376 I.A. Kozlov, P.K. Polites, S. Pitsch, P. Herdewijn and L.E. Orgel, *J. Am. Chem. Soc.*, 1999, **121**, 1108.
377 I.A. Kozlov, P.K. Polites, A. Van Aerschot, R. Busson, P. Herdewijn and L.E. Orgel, *J. Am. Chem. Soc.*, 1999, **121**, 2653.
378 J. Pereira and A. Cadete, *J. Chem. Soc., Perkin Trans. 2*, 1999, 1143.

21
NMR Spectroscopy and Conformational Features

1 General Aspects

Extensive testing of force fields for carbohydrates has been reported and ways and means of their improvement have been indicated.[1] A combination of molecular mechanics and molecular dynamics simulations has been used to improve the GROMOS force field for ring and hydroxymethyl group conformations and the anomeric effect.[2] A statistical analysis of *N*- and *O*-glycan linkage conformations from crystallographic data has been used to generate a database of 639 glycosidic linkage structures.[3]

A potentially very useful NMR technique has been reported that applies the DPFGSE-TOCSY pulse sequence to selectively remove benzylic proton resonances from ^{1}H NMR spectra of benzyl ether-protected glycosides.[4] Highly concentrated solutions of the cryogenic disaccharides sucrose and trehalose have been shown to exhibit surprisingly well-resolved ^{1}H NMR spectra in deuterium oxide–water mixtures at subzero temperatures. Under such conditions, nearly all homo- and hetero-nuclear coupling constants of OH groups in the carbohydrate can be extracted, which gives valuable information about the conformational preferences of carbohydrate hydroxyl groups in aqueous solution.[5]

^{1}H NMR has been used to characterise the core oligosaccharide of the S-layer glycoprotein from *Aneuinibacillus thermoaerophilus*, identifying for the first time an example of β-linked GalNAc-Thr.[6] The transferase activity of glycosidases has been examined *in situ* by ^{1}H NMR spectroscopy, which permits analysis of reaction progress through analysis of anomeric proton resonances.[7] The glucoamylase inhibitory properties of the anomeric mixture of glycosylamines **1** arises from selective binding of only the α-anomer, as judged by transfer nOe studies.[8]

2 Acyclic Systems

The comprehensive NMR characterisation of novel purine nucleoside analogues **2** and **3** has been reported.[9] Diastereomeric phytosphingosines **4** have been synthesized and characterized spectroscopically to enable the development of protocols for assignment of the relative and absolute configuration of naturally-occurring phytosphingosines.[10]

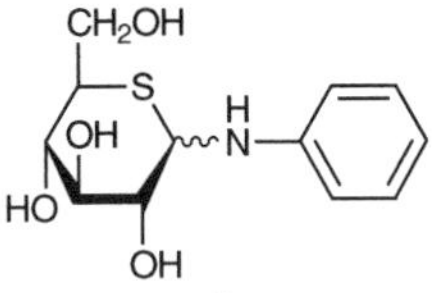

3 Furanose Systems

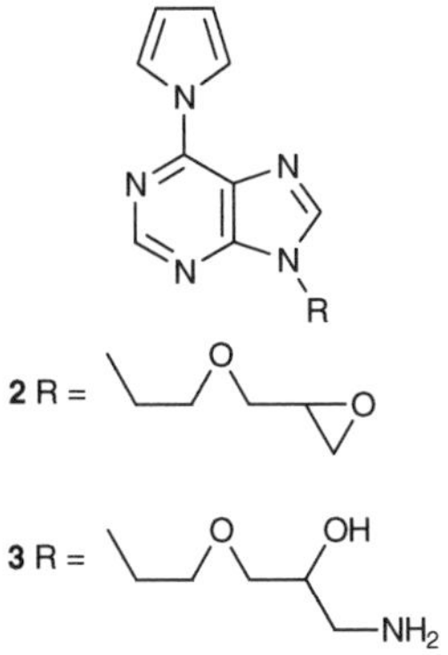

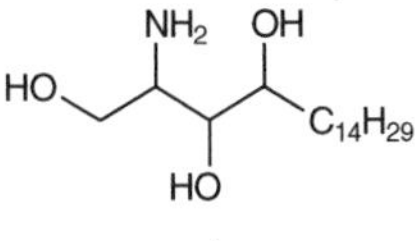

The impact of the 3′-*gauche* effect on the conformation of 3′-*O*-anthraniloyl-adenosine and its 5′-phosphate have been examined.[11] The molecular motions of specifically deuterated thymidines and two C-3-deuterated allofuranoses have been investigated by temperature-dependent relaxation studies. However, with conformationally free nucleosides internal motions are strongly coupled with overall molecular reorientation, which prevents the estimation of the pseudorotation energy barrier.[12] Conformational analysis of nucleoside 5′-deoxy-5′-S-thiosulfates has been reported.[13] Calculated ^{1}H and ^{13}C NMR chemical shifts of various anhydrodeoxythymidines were shown to be in good agreement with experimental data.[14] ^{1}H NMR spectroscopy has been used to study the binding of a variety of monosaccharides and nucleosides with novel pyridine-pyridone-pyridine receptors.[15]

Simulations of the conformations of 2-(4-aminocarbonyl-2-thiazoyl)-1,4-anhydro-L-xylitols, and fluorine derivatives thereof, agree with NMR and X-ray studies that show such compounds prefer to adopt S-type furanose conformations.[16] The structures of a series of C-4 modified pyrimidine nucleosides, such as **5**, have been analysed by 2D NMR and X-ray crystallography.[17] Molecular mechanics calculations have been used to look at the conformation of novel *ribo*-configured reversed cyclonucleoside analogues **6** and to compare them to the corresponding *xylo*-isomers.[18] Detailed NMR studies show that *spiro*-sugar derivatives **7** and **8** exist mainly in the tautomeric forms shown.[19]

Experimental evidence for intramolecular, non-bonded attractive C-F···H-C interactions in branched sugars **9** has been provided by $^{6}J_{C,F}$ and $^{7}J_{H,F}$ coupling constants.[20] Proton NMR studies show that 3′-*O*,4-*C*-methylene ribonucleosides **10** adopt an S-conformation in solution.[21] The complexation of nucleotides **11** with *cis*-platin has been studied by multinuclear NMR

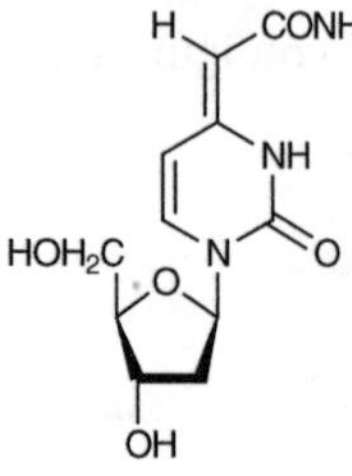

5

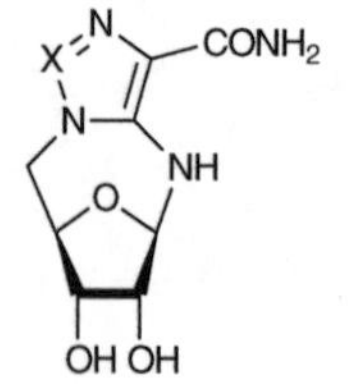

6 X = N or CH

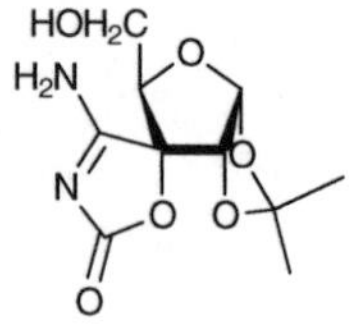

7

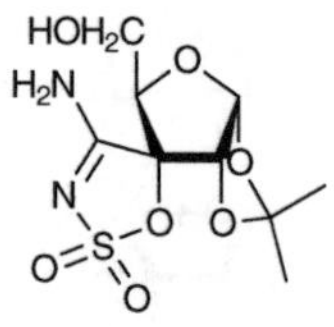

8

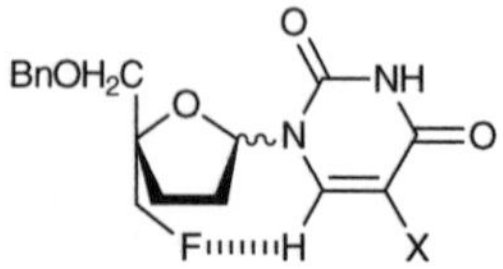

9 X = H, Br or NO_2

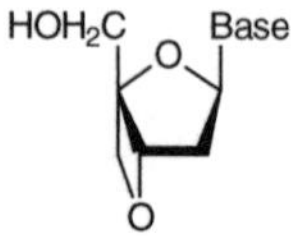

10

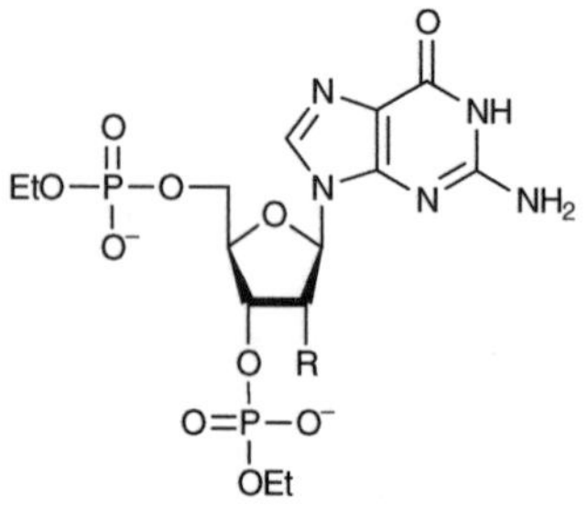

11 R = H or OH

spectroscopy. Platination increases N-type conformers and favours N-1 deprotonation by *ca.* 0.8 pK_a units.[22]

4 Pyranose and Related Systems

1,5-Anhydro-2,4-dideoxy-D-mannitol nucleosides **12** adopt a $^4C_1$ conformation in solution, as judged by proton NMR spectroscopy.[23] The monoglucosyl diacylglycerol analogue **13**, which has an extended conformation in solution as judged by DQF-COSY and HMQC experiments, exhibits the ability to 'self-assemble' into macromolecular structures.[24] Surprisingly, the introduction of a TBS group onto the 3-OH and a TPS onto the 4-OH, as in the D-mannose derivative **14** and the L-rhamnose derivative **15**, results in these bulky sub-

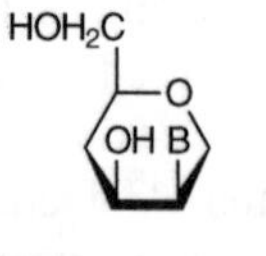

12 B = Adenosine or Uracil

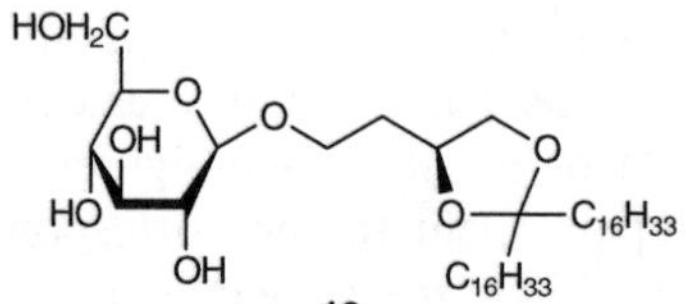

13

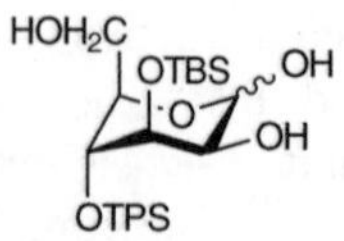

14

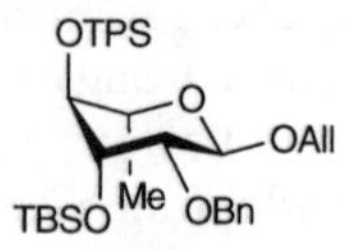

15

stituents sitting in an axial orientation since these molecules adopt $^1C_4$ conformations as shown.[25]

The influence on ^{1}H and ^{13}C chemical shifts of substitution, or replacement, of the amino group on the glucosamine derivative **16** has been investigated.[26] The variable temperature ^{1}H NMR signals of the C-6 protons in **17**, and its 2,3,4,6-tetra-*O*-acetate derivative, have been analysed. These new studies suggest that previous assertions concerning the rotamer population about the C5–C6 bond are invalid.[27] The *p*-methoxybenzyl-assisted β-mannosylation proceeds *via* an intermediate, such as **18**, with formation of a new stereogenic centre; stereochemical assignments have been made by NMR and computation using the DADAS90 program.[28] *N*-D-Glucosyl imidazoles **19**, along with the corresponding 2-deoxy-D-*gluco*- and D-*xylo*-isomers, have been used to provide evidence for the reverse anomeric effect through ^{1}H NMR pH titration experiments.[29]

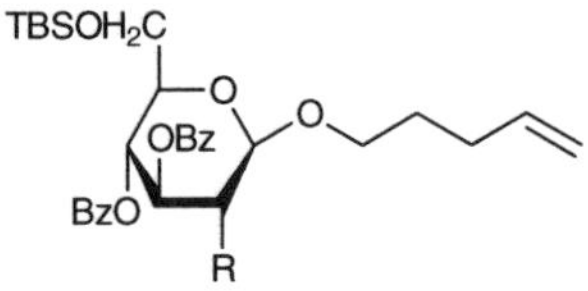

16 R = NH_2, $NHCF_3CO_2H$, NHAc, NPhth, $NhthCl_4$ or N_3

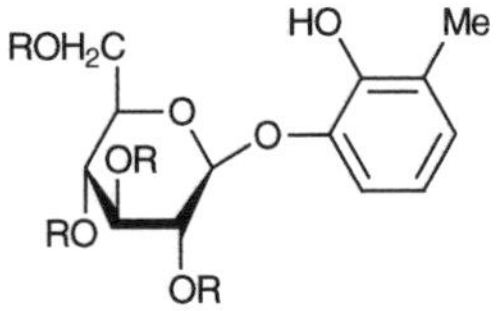

17 R = H, Ac

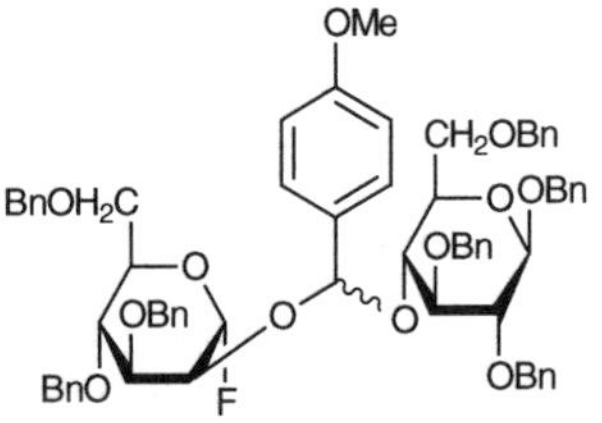

18

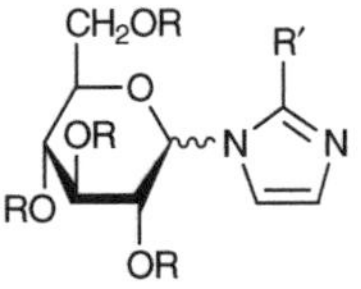

19 R = H or Ac, R′ = H or Me

myo-Inositol hexaphosphate (phytic acid) undergoes a metal ion-dependent conformational inversion from 1ax/5eq to 5ax/1eq, as judged by ^{1}H NMR spectroscopy.[30] Five derivatives of methyl 3,4,6-tri-*O*-acetyl-2-deoxy-2-(3′-dialkylureido)-β-D-glucoside were studied by ^{1}H and ^{13}C NMR. Sterically demanding substituents changed the chemical shifts of H-1,2,3 as well as C-2,3.[31] The crystal structure and ^{13}C NMR analysis of 3,4,6-tri-*O*-acetyl-2-deoxy-2-(3-phenylureido)-β-D-glucopyranoside have been reported.[32] Spin-lattice relaxation times have been used to analyse modes of relaxation of α-D-galacturonic acid monohydrate and of methyl α-D-galacturonic acid methyl ester monohydrate in the solid state.[33] Variable temperature ^{13}C cross-polarization/magic angle spinning (CP/MAS) NMR studies of 3,4-di-*O*-acetyl-1,2,5,6,-tetra-*O*-benzyl-*myo*-inositol show three distinct phases that are readily characterized by changes in the NMR spectrum.[34] Computational studies on a variety of monosaccharides, using novel density functional methods[35] and semi-empirical PM3 calculations,[36] have been reported.

5 Disaccharides

Conformational studies, by NMR and computation, have been reported for inositolphosphoglycans **20–22**; the former two are putative cell signalling molecules,[37] while the latter is a key component of cell surface glycosylphosphatidylinositol (GPI) anchors.[38]

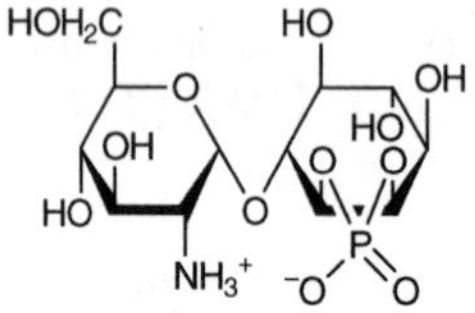

20

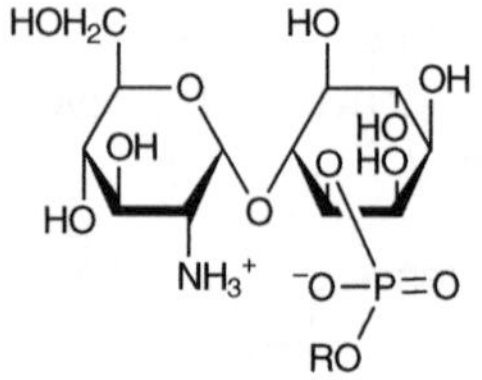

21 R = H

22 R = $CH_2CH(OCOC_{15}H_{31})CH_2OCOC_{15}H_{31}$

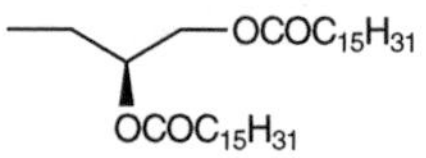

α-L-Rhap-(1,2)-β-D-Glcp-1-OMe
23

β-D-Galp-(1,3)-β-D-Glcp-1-OMe
24

α-L-Rhap-(1,2)-[β-D-Galp-(1,3)]-β-D-Glcp-1-OMe
25

Disaccharide conformational analyses have been conducted by multiple field ^{13}C NMR relaxation studies[39] and with the aid of long range heteronuclear coupling constants.[40] The solvation of sucrose and other carbohydrates in DMSO + H_2O has been probed by NOE measurements.[41] Saccharides **23–25** have been examined by an isotopomer-selected NOE method that gives unequivocal identification of mutually H-bonded hydroxyl groups.[42]

The adiabacity of disaccharide conformational maps has been investigated by using different drivers systematically in combination with an iterative manual searching procedure.[43] Conformational features of the glycosidic bond between two mannosyl units in fragments of galactomannans have been calculated using the MM3 force field.[44] Solid state NMR and *ab initio* calculations have been used to study the conformation of α,α′-trehalose with a view to identifying sources of disorder in glassy trehalose.[45]

The conformations of tethered disaccharides **26** and **27**, generated by intramolecular glycosylation reactions, have been characterized by homo- and hetero-nuclear NMR methods.[46,47] Comparative ^1H NMR studies of hydroxyl protons in galabiosides **28** have been used to investigate weak hydrogen bonding between O-6H and O-2′H in aqueous solution.[48] Conformational analysis of three methyl maltoside analogues, containing S in the non-reducing ring and either O, S or Se in the interglycosidic linkage, have

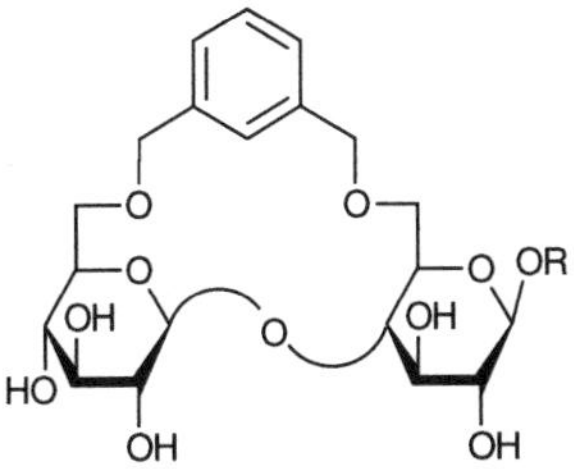

26

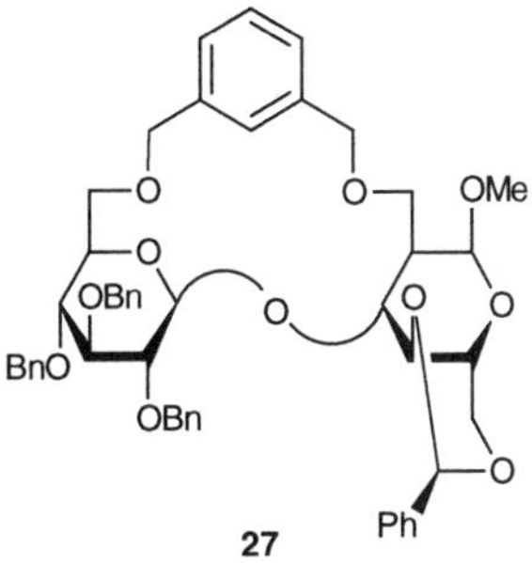

27

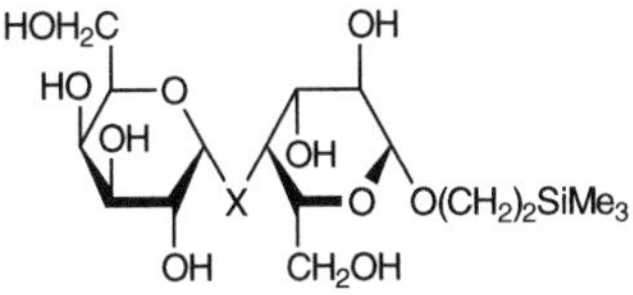

28 X = O or S

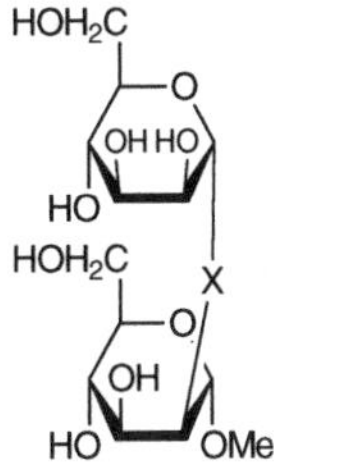

29 X = CH_2 or CH(OH)

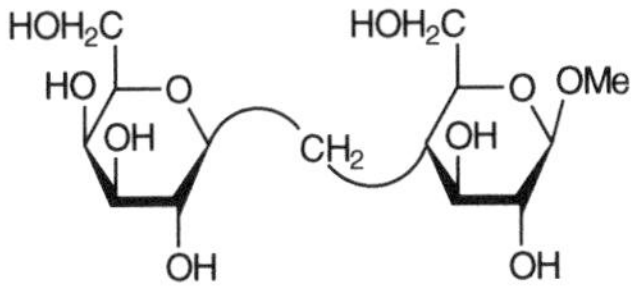

30

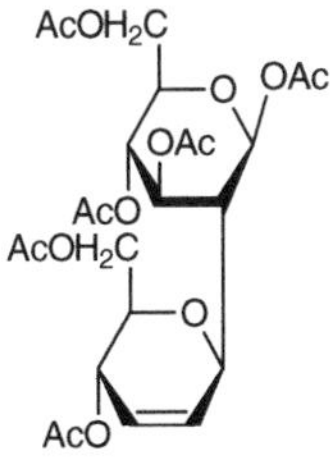

31

been investigated by NOE studies in combination with molecular mechanics calculations. The conformations determined are much the same as for maltose.[49]

The solution conformations of *C*-linked disaccharide analogues **29** have been determined; the results obtained show clear differences to the preferred conformation of the corresponding *O*-linked saccharide.[50] The conformation of *C*-lactoside **30** when bound to bovine heart galectin-1 has been shown by ^{1}H NMR to be the same as found for lactose.[51] ^{1}H NMR studies indicate that the solution conformation of the *C*-disaccharide analogue **31** is the same as the solid state conformation determined by X-ray crystallography.[52]

The conformational behaviour of aza-*C*-glycosides, such as **32** and **33** has been studied by a combination of NMR spectroscopy (*J* and NOE data) and time-averaged restrained molecular dynamics calculations. The conformation of these compounds is dominated by 1,3-*syn*-diaxial interactions, in contrast to *O*-glycosides where the *exo*-anomeric effect governs the conformation.[53]

6 Oligosaccharides

NMR and simulation studies have been reported for four fucosyl lactoses.[54] A series of cellooligosaccharides (dimer through hexamer) have been analysed by ^{13}C CP/MAS NMR spectroscopy and X-ray crystallography.[55] Solution and crystal structures of the trisaccharide **34** show the same conformation.[56] The structure of antineoplastic agent agelagalastatin **35** has been characterized by NMR spectroscopy.[57] The trisaccharide epitope **36** involved in hyperacute rejection during xenotransplantation has been the subject of conformational studies.[58] Conformational studies have also been reported on the trisaccharide **37**, the major repeating unit of a polysaccharide from *Sinorhizobium fredii.*[59] Trisaccharide fragments of the Ogawa and Inaba *Vibrio cholera* serotypes have been studied by NMR and molecular dynamics.[60] The *O*-Acetylated saccharide **38** was show to be a poor mimic of the corresponding acetamide (the tumour-associated globo-H trisaccharide, despite the two molecules adopting similar conformations as judged by ^{1}H NMR and molecular modelling).[61]

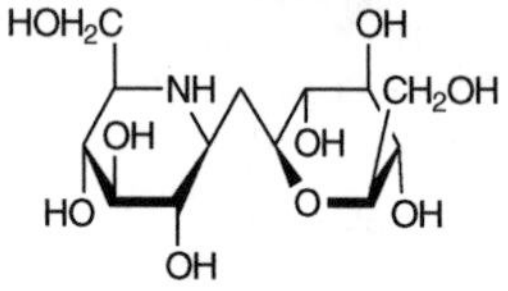

HOH₂C NH OH CH₂OH OH OH HO O OH OH

32

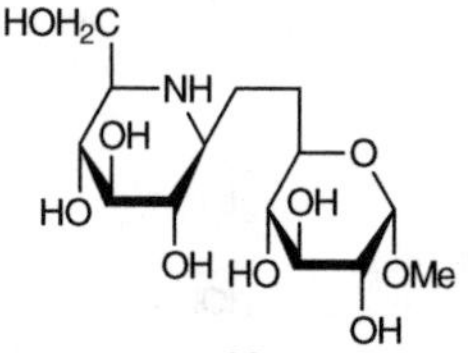

HOH₂C NH OH O HO OH OH HO OMe OH

33

α-L-Fuc*p*-(1,2)-[α-D-Gal*p*-(1,3)]-β-D-Gal*p*-1-OMe

34

α-D-Gal*f*-(1,2)-β-D-Gal*f*-(1,3)]-α-D-Gal*p*-1-OR

R = HO (CH₂)*ₙ*Me HN O OH (CH₂)*ₘ* OH

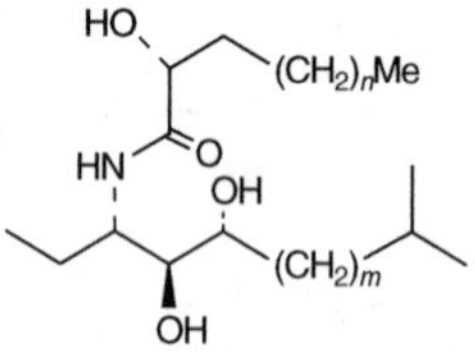

35 *n* = 21 or 20, *m* = 10 or 11

α-D-Gal*p*-(1,3)-β-D-Gal*p*-(1,4)-β-D-Glc*p*-1-NHAc

36

α-D-Gal*p*-(1,2)-β-D-Rib*f*-(1,9)-α-5-*O*-Me-Kdn*p*

37

α-L-Fuc*p*-(1,2)-β-D-Gal*p*-(1,3)-β-D-2-*O*-Ac-Gal*p*-1-OPr

38

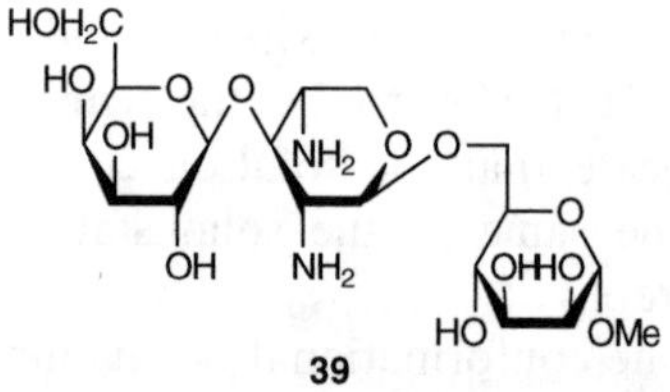

HOH₂C HO O O OH NH₂ O O OH NH₂ O OH HO HO OMe

39

The trisaccharide **39** was shown to change conformation in the presence of Zn^{2+} and Hg^{2+}; this arises from metal ion chelation by the central (hinge) 2,4-diamino-2,4-dideoxy sugar unit which results in this residue flipping to the $^1C_4$ conformation.[62] A number of synthetic lipo-chitooligosaccharides related to

Nod factors have been characterized by NMR spectroscopy and molecular dynamics simulations.[63] Molecular dynamics simulations have been performed on a tetrasaccharide subunit of chondroitin 4-sulfate (CS4). Substructures where water molecules are involved in hydrogen bonds to different sugar rings were found, which may be important for the stabilization of the secondary structure of the CS4 molecule.[64]

Several Lewis glycolipids have been characterized by homo- and heteronuclear methods, with experimental NOE data serving as restraints for molecular dynamics simulations.[65] NMR has also been used to analyse carbohydrate–carbohydrate recognition between Le^x glycoconjugates.[66] The conformation of a Le^x-containing pentasaccharide has been determined in DMSO and methanol; binding constants for Ca^{2+} have been evaluated as 9.5 and 29.6 M^{-1} in these solvents, respectively.[67]

NMR studies on model oligosaccharides acetylated with [^{13}C-carbonyl]-acetic anhydride show additional splittings of the sugar ring protons, which can be used to locate the position of the acetate group. This provides an alternative approach to methylation analysis but it is non-destructive since it does not require the cleavage of glycosidic linkages.[68] A review with 12 references on nuclear spin relaxation in small oligosaccharides has been published.[69] Relaxation and NOE studies have been used to characterize flexibility in the human milk pentasaccharide lacto-*N*-fucopentaose **40**.[70] A large NMR data set and unrestrained molecular dynamics simulations have been used to examine the oligosaccharide $Man_9GlcNAc_2$; considerable internal flexibility did not translate into gross structural changes.[71]

NMR and molecular modelling studies have been used to study two glycopeptides (a tetrasaccharide–dipeptide and an octasaccharide–heptapeptide) from the carbohydrate–protein linkage region of connective tissue proteoglycans.[72] A combined NMR and computer modelling approach has been applied to study and compare the structures of sialyl Lewis x attached to an octapeptide fragment of the mucin MAdCAM-1. The conformation of the carbohydrate moiety was found to be the same as that of free sialyl Lewis x.[73] A recent study concludes that both local protein surface structure and overall tertiary structure influence protein-specific glycosylation. These conclusions arise from NMR and molecular modelling studies on the Ly-6, scavenger receptor and immunoglobulin superfamilies.[74] A review on the use of very high field NMR for the structure determination of large oligosaccharides has been published.[75]

α-L-Fuc*p*-(1,2)-β-D-Gal*p*-(1,3)-β-D-Glc*p*NAc-(1,3)-β-D-Gal*p*-(1,4)-D-Glc*p*

40

α-NeuAc-(2,3)-β-D-Gal*p*-(1,4)-D-Glc*p*

41

Three publications have appeared from Homans and co-workers which review isotopic enrichment and conformational analysis of oligosaccharides,[76] report the measurement of residual dipolar couplings for the ^{13}C-enriched trisaccharide **41** in a liquid crystalline medium,[77] and describe solution structure and dynamics of ^{13}C-enriched sialyl Lewis x and its bound state conformation in association with E-selectin.[78] Isotopic enrichment permits ^{13}C–^{13}C coupling constants to be used as conformational probes.[78]

Structural studies on numerous oligosaccharides have been reported, including: the O-antigens from enteropathogenic *E. coli* 0158,[79] *E. coli* 0116:K$^+$:H10[80] and *E. coli* 065;[81] sulfated fucan from sea urchin egg jelly coat;[82] low molecular weight fucoidin fragments derived by digestion with extracts from a marine mollusc;[83] the exopolysaccharide from *Vibrio diabolicus*, which was isolated from a deep sea vent.[84] Polymer hydration in onion cell wall material has been evaluated by solid-state NMR, which shows at least two different motional regimes for pectin and cellulose domains.[85]

One dimensional HSQC experiments have been used to obtain heteronuclear long-range coupling constants for β-cyclodextrin and 3,6-anhydro-β-cyclodextrin.[86] NMR methods based on a combined INEPT/HMBC sequence have been reported for determination of the position of monosubstitution of β-cyclodextrin.[87] Conformational features of cycooligosaccharides composed of β-(1→3)- and β-(1→6)-linked galactofuranose units have been investigated by Monte Carlo simulations.[88]

7 Other Compounds

(*S*)-α-Methoxyphenylacetic acid was used as an NMR shift reagent to predict, in combination with molecular modelling, the absolute configuration of the sulfinyl centres of a representative pair of epimeric glucopyranosyl sulfoxides.[89] Solution and solid-state structures for the anhydrosugar **42** have been determined by NMR spectroscopy and X-ray crystallography, respectively.[90] NMR and modelling studies on the ^{15}N isotopomer of the anhydrosugar **43** have been used to generate ^{1}H–^{15}N coupling constant data that can be used to

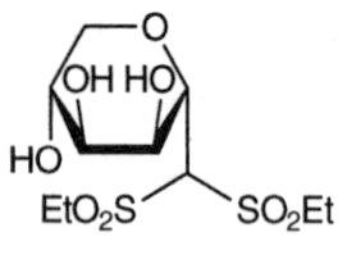

42

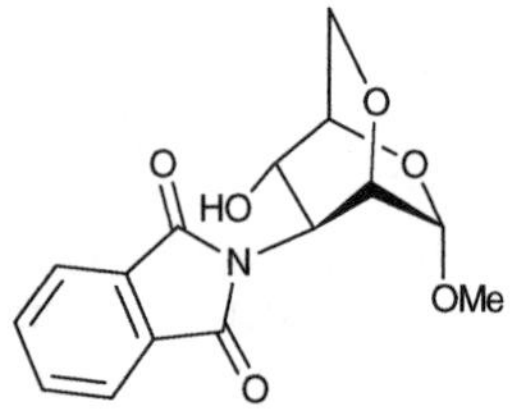

43

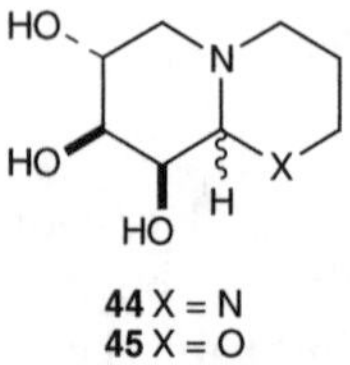

44 X = N
45 X = O

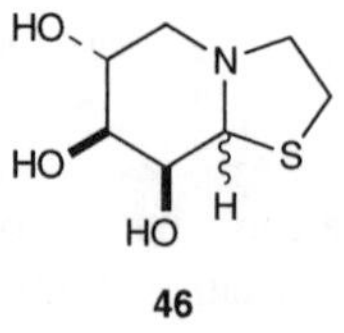

46

define a Karplus equation for analysis of the stereochemistry of aminoglycosides.[91]

NMR has been used to study the conformation of a series of related 6-amino-6-deoxy-hexonolactams.[92] ^{1}H NMR spectra of a series of acetonides obtained during the synthesis of deoxynojirimycin were severely complicated by second order effects.[93] The conformational and configurational (anomeric) preferences of the bicyclic azasugars **44–46** have been investigated by NMR spectroscopy.[94]

Various NMR techniques have been used in the structural and conformational analysis of natural products, including: the oligosaccharide sequence in ginsenosides;[95] the rebeccamycin indolocarbazole glycosides;[96] the protoilludane-type sesquiterpene glycoside, pteridanoside;[97] synthetic diosgenyl saponin analogues;[98] a branched pentasaccharide resin glycoside, soldanelline A.[99]

8 NMR of Nuclei Other than ^{1}H and ^{13}C

^{15}N NMR spectroscopy was used to assess the impact of *N*-substitution on nitrogen chemical shift in a series of *N*-derivatives of 2-amino-2-deoxy-β-D-glucopyranose with dipeptides.[100]

The influence of fluorine substitution in *iso*-C-nucleosides was determined by a combination of ^{1}H, ^{13}C and ^{19}F NMR spectroscopy.[101] A variety of fluorine-substituted cyclodextrins have been synthesized and characterized by ^{19}F NMR spectroscopy,[102] as have the reactions of electrophilic fluorinating agents with glycals.[103]

^{31}P NMR titration experiments have been used to investigate the pH-dependent ionization of phosphate groups in inositol triphosphate analogues.[104]

References

1 K. Rasmussen, *J. Carbohydr. Chem.*, 1999, **18**, 789.
2 S.A.H. Spieser, J.A. van Kuik, L.M.J. Kroon-Batenburg and J. Kroon, *Carbohydr. Res.*, 1999, **322**, 264.
3 A.J. Petrescu, S.M. Petrescu, R.A. Dwek and M.R. Wormald, *Glycobiology*, 1999, **9**, 343.
4 T.J. Rutherford, K.P.R. Kartha, S.K. Readman, P. Cura and R.A. Field, *Tetrahedron Lett.*, 1999, **40**, 2025.
5 G. Batto and K.E. Kövér, *Carbohydr. Res.*, 1999, **320**, 267.
6 T. Wugeditsch, N.E. Zachera, M. Puchberger, P. Kosma, A.A. Gooley and P. Messner, *Glycobiology*, 1999, **9**, 787.
7 P. Spangenberger, V. Chiffoleau-Girard, C. Andre, M. Dion and C. Rabiller, *Tetrahedron Asymm.*, 1999, **10**, 2905.
8a K.D. Randell, T.P. Frandsen, B. Stoffer, M.A. Johnson, B. Svensson and B.M. Pinto, *Carbohydr. Res.*, 1999, **321**, 143.

9 S. Raic-Malic, D. Vikic-Topic and M. Mintas, *Spectrosc. Lett.*, 1999, **32**, 649 (*Chem. Abstr.*, 1999, **131**, 199929).

10 O. Shirota, K. Nakanishi and N. Berova, *Tetrahedron*, 1999, **55**, 13643.

11 P. Acharya, B. Nawrot, M. Sprinzl, C. Thibaudeau and J. Chattopadhyaya, *J. Chem. Soc., Perkin Trans. 1*, 1999, 1531.

12 J. Plavec, P. Roselt, A. Foldesi and J. Chattopadhyaya, *Magn. Reson. Chem.*, 1998, **36**, 732 (*Chem. Abstr.*, 1999, **130**, 38618).

13 A. Orzeszko, A. Niedzwiecka-Kornas, R. Stolarski and Z. Kazimierczuk, *Z. Naturforsch, B: Chem. Sci.*, 1998, **53**, 1191 in English (*Chem. Abstr.*, 1999, **130**, 386262).

14 J. Czernek and V. Sklenar, *J. Phys. Chem. A.*, 1999, **103**, 4089 (*Chem. Abstr.*, 1999, **131**, 88 114).

15 M. Inouye, K. Takahashi and H. Nakazumi, *J. Am. Chem. Soc.*, 1999, **121**, 341.

16 H.-Y. Zhang, L.R. Zhang, L.T. Ma, L.H. Zhang, Y.X. Cui and X.H. Liu, *Gaodeng Xuexiao Huaxue Xuebao*, 1998, **19**, 1767 (in Chinese).

17 S. Blanalt-Feidt, S.O. Doronina and J.-P. Behr, *Tetrahedron Lett*, 1999, **40**, 6229.

18 D.F. Ewing, G. Goethals, G. Mackenzie, P. Martin, G. Ronco, L. Vanbaelinghem and P. Villa, *Carbohydr. Res.*, 1999, **321**, 190.

19 M.-J. Camarasa, M.-L. Jimeno, M.-J. Perez-Perez, R. Alvarez and S. Velazquez, *Tetrahedron*, 1999, **55**, 12187.

20 A. Mele, B Vergani, F. Viani, S. Valdo Meille, A. Farina and P. Bravo, *Eur. J. Org. Chem.*, 1999, 187.

21 S. Obika, K.I. Morio, Y. Hari, T. Imanishi, *Chem. Commun.*, 1999, 2423

22 M. Polak, J. Plavec, A. Trifonova, A. Földesi and J. Chattopadhyaya, *J. Chem. Soc., Perkin Trans. 1*, 1999, 2835.

23 N. Hussain, I. Layten, K. Rothenbacher, R. Busson and P. Herdewijn, *Nucleosides, Nucleotides*, 1999, **18**, 161.

24 J. Song and R.I. Hollingsworth, *J. Am. Chem. Soc.*, 1999, **121**, 1851.

25 H. Yamada, M. Nakatani, T. Ikeda and Y. Marumoto, *Tetrahedron Lett.*, 1999, **40**, 5573.

26 L. Olsson, Z. J. Jia and B. Fraser-Reid, *Pol. J. Chem.*, 1999, **73**, 1091 (*Chem. Abstr.*, 1999, **131**, 144773).

27 G.D. Rockwell, T.B. Grindley and J.-P. Lepoittevin, *J. Carbohydr. Chem.*, 1999, **18**, 51.

28 M. Lergenmüller, T. Nukada, K. Kuramochi, A. Dan, T. Ogawa and Y. Ito, *Eur. J. Org. Chem.*, 1999, 1367.

29 C.L. Perrin, M.A. Fabian, J. Brunckova and B.K. Ohta, *J. Am. Chem. Soc.*, 1999, **121**, 6911.

30 A.T. Bauman, G.M. Chateauneuf, B.R. Boyd, R.E. Brown and P.P.N. Murthy, *Tetrahedron Lett.*, 1999, **40**, 4489.

31 I. Wawer, M. Weychert, J. Klimkiewicz, B. Piekarska-Bartoszewicz and A. Temeriusz, *Magn. Reson. Chem.*, 1999, **37**, 189.

32 Anulewicz, I. Wawer, B. Piekarska-Bartoszewicz and A. Temeriusz, *J. Carbohydr. Chem.*, 1999, **18**, 617.

33 H.R. Tang and P.S. Belton, *Solid State Nucl. Magn. Reson.*, 1998, **12**, 21 (*Chem. Abstr.*, 1999, **130**, 81752).

34 P. Bast, S. Berger and H. Gunther, *Magn. Reson. Chem.*, 1992, **30**, 587 (*Chem. Abstr.*, 1999, **130**, 267693).

35 S.K. Gregurick and S.A. Kafafi, *J. Carbohydr. Chem.*, 1999, **18**, 867.

36 T. Yui, K. Miyawaki, Y. Kawano and K. Ogawa, *J. Carbohydr. Chem.*, 1999, **18,** 905.
37 H. Dietrich, J.F. Espinosa, J.L. Chiara, J. Jiménez-Barbero, Y. Leon, I. Varela-Nieto, J.-M. Mato, F.H. Cano, C. Foces-Foces and M. Martin-Lomas, *Chem. Eur. J.*, 1999, 320.
38 D.K. Sharma, T.K. Smith, C.T. Weller, A. Crossman, J.S. Brimacombe and M.A.J. Ferguson, *Glycobiology*, 1999, **9**, 415.
39 P. Soderman and G. Widmalm, *Magn. Reson. Chem.*, 1999, **37**, 586 (*Chem. Abstr.* 1999, **131**, 257765).
40 T. Rundlöf, A. Kjellberg, C. Damberg, T. Nishida and G. Widmalm, *Magn. Reson. Chem.*, 1998, **36**, 839 (*Chem. Abstr.*, 1999, **130**, 81730).
41 S. Berger, M.D. Diaz and C.L. Hawat, *Pol. J. Chem.*, 1999, **73**, 193 (*Chem. Abstr.*, 1999, **130**, 196868).
42 T. Kozar, N.E. Nifant'ev, H. Grosskurth, U. Dabrowsi and J. Dabrowski, *Biopolymers*, 1998, **46**, 417 (*Chem. Abstr.*, 1999, **130**, 52644k).
43 C.A. Stortz, *Carbohydr. Res.*, 1999, **322**, 77.
44 C.L.O. Petkowicz, F. Reicher and K. Mazeau, *Carbohydr. Polym.*, 1998, **37**, 25 (*Chem. Abstr.*, 1999, **130**, 81734).
45 P. Zhang, A.N. Klymachyov, S. Brown, J.. Ellington and P.J. Grandinetti, *Solid State Nucl. Magn, Reson.*, 1998, **12**, 221 (*Chem. Abstr.*, 1999, **130**, 25244).
46 A. Geyer, U. Huchel and R.R. Schmidt, *Magn. Reson. Chem.*, 1999, **37**, 145 (*Chem. Abstr.*, 1999, **130**, 209879).
47 A. Geyer, M. Müller and R.R. Schmidt, *J. Am. Chem. Soc.*, 1999, **121**, 6312.
48 C. Sandstrom, G. Magnusson, U. Nilsson, L. Kenne, *Carbohydr. Res.*, 1999, **322,** 46.
49 T. Weimar, U.C. Kreis, J.S. Andrews and B.M. Pinto, *Carbohydr. Res.*, 1999, **315**, 222.
50 J.-F. Espinosa, M. Bruix, O. Jarreton, T. Skrydstrup, J.-M. Beau and J. Jiménez-Barbero, *Chem. Eur. J.*, 1999, **5,** 442.
51 J.L. Asensio, J.F. Espinosa, H. Dietrich, F.J. Cañada, R.R. Schmidt, M. Martin-Lomas, S André, H-J. Gabius and J. Jiménez-Barbero, *J. Am. Chem. Soc.*, 1999, **121**, 8995.
52 A.H. Franz, V.V. Zhdankin, V.V. Samoshin, M.J. Inch, V.G. Young and P.H. Gross, *Mendeleev Commun.*, 1999, 45 (*Chem. Abstr.*, 1999, **130**, 338290).
53 J.L. Asenio, F.J. Cañada, A. García-Herrero, M.T. Murillo, A. Fernández-Mayoralas, B.A. Johns, J. Kozak, Z. Zhu, C.R. Johnson and J. Jiménez-Barbero, *J. Am. Chem. Soc.*, 1999, **121**, 11318.
54 Y. Ishizuka, T. Nemoto, M. Fujiwara, K.-i. Fujita and H. Nakanishi, *J. Carbohydr. Chem.*, 1999, **18,** 523.
55 H. Kono, Y. Numata, N. Hagai, T. Erata and M. Takai, *Carbohydr. Res.*, 1999, **322**, 256.
56 A. Otter, R.U. Lemieux, R.G. Ball, A.P. Venot, O. Hindsgaul and D.R. Bundle, *Eur. J. Biochem.*, 1999, **259**, 295 (*Chem. Abstr.*, 1999, **130**, 182675).
57 G.R. Pettit, J.P. Xu, D.E. Gingrich, M.D. Williams, D.L. Doubek, J.C. Chapuis, J.M. Schmidt, *Chem. Commun*, 1999, 915.
58 J. Li, M.B. Ksebati, W. Zhang, Z. Guo, J. Wang, L. Yu, J. Fang and P.G. Wang, *Carbohydr. Res.*, 1999, **315**, 76.
59 M.A. Rodriguez-Carvajal, L. Gonzales, M. Bernabe, J.F. Espinosa, J.L. Espartero, P. Tejero-Mateo, A. Gil-Serran and J. Jiménez-Barbero, *J. Carbohydr. Chem.*, 1999, **18**, 891.

60 L. Gonzales, J.L. Asensio, A. Ariosa-Alvarez, V. Verez-Bencoma and J. Jiménez-Barbero, *Carbohydr. Res.*, 1999, **321**, 88.
61 S. Canevari, D. Colombo, F. Compostella, L. Panza, F. Ronchetti, G. Russo and L. Toma, *Tetrahedron*, 1999, **55**, 1469.
62 H. Yuasa and H. Hashimoto, *J. Am. Chem. Soc.*, 1999, **121**, 5089.
63 Z. Gonzalez, M. Bernabe, J.F. Espinosa, P. Tejero-Mateo, A. Gil-Serrano, N. Montegazza, A. Imberty, H. Driguez and J. Jiménez-Barbero, *Carbohydr. Res.*, 1999, **318**, 10.
64 J. Kaufmann, K. Möhle, H.J. Hoffman and K. Arnold, *Carbohydr. Res.*, 1999, **318**, 1.
65 A. Geyer, G. Hummel, S. Reinhardt and R.R. Schmidt, *Pol. J. Chem.*, 1999, **73**, 181 (*Chem. Abstr.*, 1999, **130**, 196890).
66 A. Geyer, C. Cege and R.R. Schmidt, *Angew. Chem. Int. Ed. Engl.*, 1999, **38**, 1466.
67 B. Henry, H. Desvaux, M. Pristchepa, P. Berthault, Y.-M. Zhang, J.-M. Mallet, J. Esnault and P. Sinaÿ, *Carbohydr. Res.*, 1999, **315**, 48.
68 B. Bendiak, *Carbohydr. Res.*, 1999, **315**, 206.
69 J. Kowalewski, L. Maler and G. Widmalm, *J. Mol. Liq.*, 1998, **78**, 255 (*Chem. Abstr.*, 1999, **130**, 110462).
70 T. Rundlöf, R.M. Venable, R.W. Pastor, J. Kowalewski and G. Widmalm, *J. Am. Chem. Soc.*, 1999, **121**, 11847.
71 R.J. Woods, A. Pathiaseril, M.R. Wormald, C.J. Edge and R.A. Dwek, *Eur. J. Biochem.*, 1998, **258**, 372.
72 P.K. Agrawal, J.-C. Jacquinet, N.R. Krisha, *Glycobiology*, 1999, **9**, 669.
73 W. Wu, L. Pasternack, D.-H. Huang, K.M. Koeller, C.-C Lin, O. Seitz and C.-H. Wong, *J. Am. Chem. Soc.*, 1999, **121**, 2409.
74 P.M. Rudd, M.R. Wormald, P.J. Harvey, M. Devasahayam, M.S.B. McAlister, M.H. Brown, S.J. Davis, A.N. Barclay and R.A Dwek, *Glycobiology*, 1999, **9**, 443.
75 J.D. Duus, P.M. St. Hilaire, M. Meldal and K. Bock, *Pure Appl. Chem.*, 1999, **71**, 755 (*Chem Abstr.*, 1999, **131**, 272082).
76 S.W. Homans, *Biochem. Soc. Trans.*, 1998, **26**, 551 (*Chem. Abstr.* 1999, **130**, 168580).
77 G.R. Kiddle and S.W. Homans, *FEBS Lett.*, 1998, **436**, 128 (*Chem. Abstr.*, 1999, **130**, 14141).
78 R. Harris, G.R. Kiddle, R.A. Field, M.J. Milton, B. Ernst, J.L. Magnani and S.W. Homans, *J. Am. Chem. Soc.*, 1999, **121**, 2546.
79 A.K. Datta, S. Basu and N. Roy, *Carbohydr. Res.*, 1999, **322**, 219.
80 M.R. Leslie, H. Parolis and LA.S. Parolis, *Carbohydr. Res.*, 1999, **321**, 246.
81 M.B. Perry and L.L. MacLean, *Carbohydr. Res.*, 1999, **322**, 57.
82 A.-C.E.S. Vilela-Silva, A.-P Awes, A.R. Valenta, V.D. Vacquier and P.A.S. Mourao, *Glycobiology*, 1999, **9**, 937.
83 R. Daniel, O. Berteau, J. Jozefonvice and N. Goasdone, *Carbohydr. Res.*, 1999, **322**, 291.
84 H. Rougeaux, N. Kervarec, R. Richen and J. Guezennec, *Carbohydr. Res.*, 1999, **322**, 40.
85 S. Hediger, L. Emsley and M. Fischer, *Carbohydr. Res.*, 1999, **322**, 102.
86 P. Forgo and V.T. D'Souza, *Magn. Reson. Chem.*, 1999, **37**, 48 (*Chem. Abstr.*, 1999, **130**, 182679).
87 P. Forgo and V.T. D'Souza, *J. Org. Chem.*, 1999, **64**, 306.

88 H. Gohlke, S. Immel and F.W. Lichtenthaler, *Carbohydr. Res.*, 1999, **321**, 96.

89 P.H. Buist, B. Behrouzian, K.D. MacIsaac, S. Cassel, P. Rollin, A. Imberty, C. Gautier, S. Pérez and P. Genix, *Tetrahedron Asymmetry*, 1999, **10,** 2881.

90 P. Norris and T.R. Wagner, *Carbohydr. Res.*, 1999, **322**, 147.

91 B. Coxon, *Carbohydr. Res.*, 1999, **322**, 120.

92 J. Havlcek, M. Hamernkova and K. Kefurt, *J. Mol. Struct.*, 1999, **482-483**, 311 (*Chem Abstr.*, 1999, **131**, 88104).

93 K. Kato and R.M. Braga, *Magn. Reson. Chem.*, 1999, **37**, 447 (*Chem. Abstr.*, 1999, **131**, 59064).

94 D.A. Berges, N. Zhang and L. Hong, *Tetrahedron,* 1999, **55**, 14251.

95 W. Li, Y. Sa, Y. Bam, D. Zhoug, J. Wang and X. Li, *Bopuxue Zazhi,* 1999, **16**, 269 (*Chem. Abstr.*, 1999, **131**, 185155).

96 E.J. Gilbert, J.D. Chisholm and D.L. Van Vranken, *J. Org. Chem.*, 1999, **64**, 5670.

97 U.F. Castillo, Y. Saragami, M. Alonso-Amelot and M. Ojika, *Tetrahedron,* 1999, **55**, 12295.

98 X.W. Han, H. Yu, X.M. Liu, X. Bao, B. Yu, C. Li and Y.Z. Hui, *Magn. Reson. Chem.*, 1999, **37**, 140 (*Chem. Abstr.*, 1999, **130**, 209863).

99 E.M.M. Gasper, *Tetrahedron Lett*, 1999, **40**, 6861.

100 M. Weychert, J. Klimkiewicz, I. Wawer, B. Piekarska-Bartoszewicz and A. Temeriusz, *Magn. Reson. Chem.*, 1998, **36**, 727 (*Chem. Abstr.*, 1999, **130**, 25251).

101 Y. Cui, H. Zhang, X. Liu, L. Ma and L. Zhang, *J. Chin. Pharm. Sci.*, 1999, **8**, 39 (*Chem Abstr.*, 1999, **130**, 338330).

102 J. Diakur, Z. Zuo and L.I. Wiebe, *J. Carbohydr. Chem.*, 1999, **18,** 209.

103 J. Ortner, M. Albert, H. Weber and K. Dax, *J. Carbohydr. Chem.*, 1999, **18**, 297.

104 M. Felemez, G. Schlewer, D.J. Jenkins, V. Correa, C.W. Taylor, B.V.L. Potter and B. Spiess, *Carbohydr. Res.*, 1999, **322**, 95.

22
Other Physical Methods

1 IR Spectroscopy

The IR and Raman spectra for D,L-threitol and erythritol have been recorded and assigned[1] and the FTIR spectra of uridine, thymidine, ribose, 2-deoxyribose and glucose have been determined.[2] Intramolecular H-bonds indicated that the *syn*-conformations of uridine and thymidine were stabilized by H-bonds involving O-5′H and O-2. The experimental FTIR and FT-Raman spectra of 5-iodo-2′-deoxyuridine, an antiviral nucleoside, have been assigned.[3] From the IR spectra, acquired at regular time intervals, of a solution of D-mannose undergoing mutarotation, the rate constants of the conversion from α- to β-mannose were found to differ slightly from those reported earlier.[4]

Photon correlation spectroscopy and SEM have been used to demonstrate the attachment of per-6-thio-β-cyclodextrin to gold nanoparticles (~12 nm). The FTIR spectrum of the modified nanoparticles strongly resembles that of the free cyclodextrin.[5]

The room temperature Raman and IR spectra of adenosine at pressures between 1 atm and 10 GPa indicated a phase transition near 2.5 GPa.[6] The Raman spectrum of [5′-^{13}C]thymidine has been recorded and compared with that of unlabelled thymidine. Certain absoptions were assigned from the differences observed.[7]

2 Mass Spectrometry

A review on analytical methods for the structural determination of large oligosaccharides focuses on the use of very high field NMR spectroscopy methods as well as MALDI-TOF mass spectrometry.[8]

The CI mass spectra of 10 different glucosides isolated from rape seed have been analysed.[9] A study of the EI mass spectra of 2-amino-2-deoxy sugars and their *N*-alkyl-, *N*-acyl-*N*-alkyl-, and *N*,*N*-dialkyl- derivatives has been reported.[10] α-(2→3)- and α-(2→6)-sialyl linkage types of sialyl-lactose and sialyl-*N*-acetyllactosamine have been analysed by post-source-decay fragmentation techniques using MALDI/TOF mass spectrometry.[11] Under these conditions α-(2→3)-linkages cleave much more readily than do α-(2→6)-linkages.

GlcNAc, GalNAc and ManNAc were differentiated by ESI mass spectro-

Carbohydrate Chemistry, Volume 33

metry when complexed to form $[Co^{III}(DAP)_2HexNAc]Cl_3$.[12] This method has been extended to determine the linkage positions of oligosaccharides.[13]

Eight reducing glucose-containing disaccharides have been analysed by GC- and EI-mass spectrometry as their per-acetylated and per-methylated alditols.[14] From the data the linkage positions and anomeric configurations could be determined.

Nanoflow ESI quadrupole-inlet time of flight mass spectrometry has been use to characterize oligosaccharides derivatized with either 2-(diethylamino)-ethyl 4-aminobenzoate or 2-aminopyridine.[15] Five dimethylated β-cyclodextrins have been analysed by electrospray mass spectrometry and supercritical fluid chromatography.[16]

The MALDI-mass spectral analysis of per-benzoylated sialo-oligosaccharides has been reported. Terminal α-NeuNAc-(2→3)- and α-D-Gal-(2→6)- units gave easily recognizable MS patterns.[17] A computer program has been developed that helps the interpretation of MALDI/TOF postsource decay spectra of *N*-linked oligosaccharides.[18] The program will produce simulated spectra for comparison with those derived from unknown oligosaccharides. Oligosaccharides have been analysed by MALDI/TOF mass spectrometry of their per-succinoyl derivatives.[19] The technique shows good sensitivity down to 100 fmol.

A series of methyl galactotriosides labelled with fluorine at C-3, together with one or two residues specifically deoxygenated, were used as models to examine 'internal residue loss' by MALDI/TOF and ESI mass spectrometry techniques.[20] A means has also been described for determining whether ions observed in the tandem mass spectra of protonated native, or per-*O*-methylated, oligosaccharides originate from 'internal residue loss' or from direct glycosidic linkage fragmentation.[21]

MALDI mass spectra have been recorded on a magnetic sector mass spectrometer set up to acquire spectra in a single scan with high resolution output.[22] The technique has been applied to the biantennary *N*-linked nonasaccharide $[(Gal)_2(Man)_3(GlcNAc)_4]$. A combination of TLC/MALDI-TOF mass spectrometry has been used to identify glycopeptides and the products of complex carbohydrate reactions.[23] The compounds were first separated by TLC, then scraped off the plate and subjected to the mass spectrometric analysis.

A study of the use of 3-methyl-1-phenyl-5-pyrazolone as a derivatization agent in the analysis of oligosaccharides has been reported.[24] This procedure provided an increase in sensitivity for UV and MS detection.

MALDI-TOF/CID has been used to characterize *N*-glycans derived from bile salt-stimulated lipase found in human cells.[25] MALDI-TOF MS and FAB-MS have been used to test the structural heterogeneity of lipooligosaccharides expressed by non-typable *Haemophilus influenzae*.[26] The *N*-glycosylation of carcinoembryonic antigen cell adhesion molecule from rat liver has been characterized by MALDI-TOF MS, fluorescent labelling and 2D HPLC.[27] MALDI/TOF MS has also been used to fingerprint large oligosaccharides linked to ceramide, which have receptor activity for *Helicobacter pylori*.[28] A structural comparison of LPS from two strains of *Helicobacter pylori* has been carried out by FAB-MS and NMR spectroscopy.[29] A structural analysis has been reported of the glycans from the acid phosphatase secreted by *Leishmania donovani* using ES-MS and enzymatic digestion.[30]

Isolation and structural characterization by FAB-MS and GC-EI-MS of glycosphingolipids of *in vitro*-propagated human umbilical vein endothelial cells has been reported,[31] as well as the structural characterization by ESI-tandem MS of gangliosides isolated from mullet milt.[32] The structural heterogeneity in the core oligosaccharide of the S-layer glycoprotein from *Aneurinibacillus thermoaerophilus* has been characterized by LC-ESI-MS and ^{1}H NMR spectroscopy.[33] This is the first example of a β-linkage between *N*-acetyl-D-galactosamine and threonine.

The isolation and structural elucidation (by NMR, MS and chemical studies) of a new resin glycoside containing a branched pentasaccharide, Soldanelline A, from a Portuguese Convolvulaceae have been reported (see also Chapters 4 and 21).[34]

A new novel method for the analysis of short oligonucleotides depends on *in situ* guanine-specific methylation followed by gas-phase fragmentation by electrospray ionization ion-trap mass spectrometry, which permits the detection of the positions of all guanine residues. Rapid depurination at the site of methylation is followed by cleavage of the backbone.[35] Mass spectral studies of 6-substituted-2-methylthio-9-tetrahydrofuranpurine nucleoside analogues have been reported.[36] This technique has been proposed as a suitable method for the structural analysis of a mixture of glucosinolates.

The nucleoside pro-drugs 4-azido-ara-C and 2′-fluoro-2′,3′-dideoxy-4-azido-ara-C have been characterized by mass spectrometry as have the products of their reaction with base.[37]

3 X-Ray and Neutron Diffraction Crystallography

Specific X-ray crystal structures have been reported as follows (solvent molecules of crystallization are frequently not recorded):

3.1 Free Sugars and Simple Derivatives Thereof. – 4-*O*-Benzoyl-2,3-*O*-isopropylidene-α-L-rhamnose.[38]

3.2 Glycosides, Disaccharides and Derivatives Thereof. – A statistical analysis of *N*- and *O*-glycoside linkage conformations from crystallographic data has been reported.[39] A database of 639 glycosidic linkage structures was generated.

Monosaccharide derivatives: methyl 2-azido-4,6-*O*-benzylidene-2-deoxy-β-D-galactopyranoside, methyl 4,6-*O*-benzylidene-2-deoxy-β-D-galactopyranoside,[40] methyl β-D-ribopyranoside, methyl α-D-arabinopyranoside, methyl β-D-xylopyranoside, methyl α-D-lyxopyranoside,[41] *t*-butyl (methyl 2,3,4-tri-*O*-benzyl-7-deoxy-D-*glycero*-α-D-*gluco*-octopyranosid)-uronate (see Chapter 24 for synthesis).[42] *N*-Cbz-*O*-(2,3,4,6-Tetra-*O*-acetyl-β-D-galactopyranosyl)-Lthreonyl-α-aminoisobutyryl-α-aminoisobutyric acid *tert*-butyl ester.[43] The spiro acetal **1**, isolated from twigs and thorns of *Castela polyandra*.[44] The structures of micelles of n-alkyl β-D-glucopyranosides in aqueous solution have been determined with a combination of X-ray and neutron scattering techniques.[45]

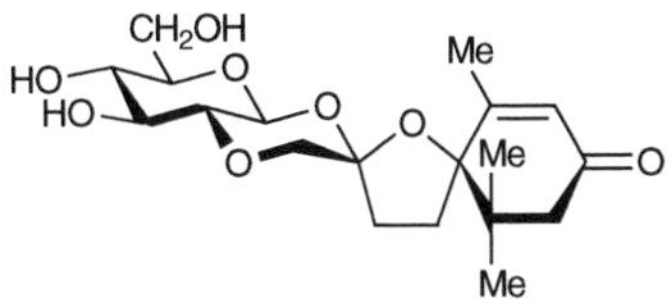

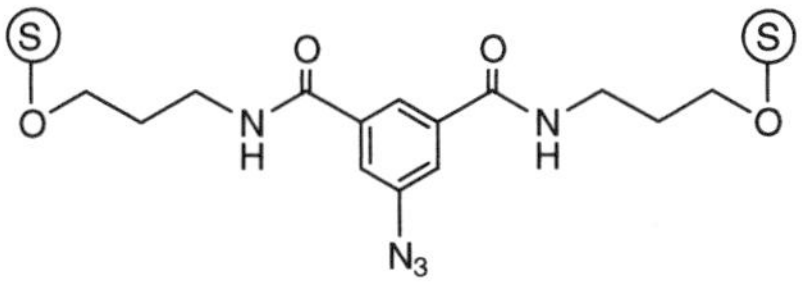

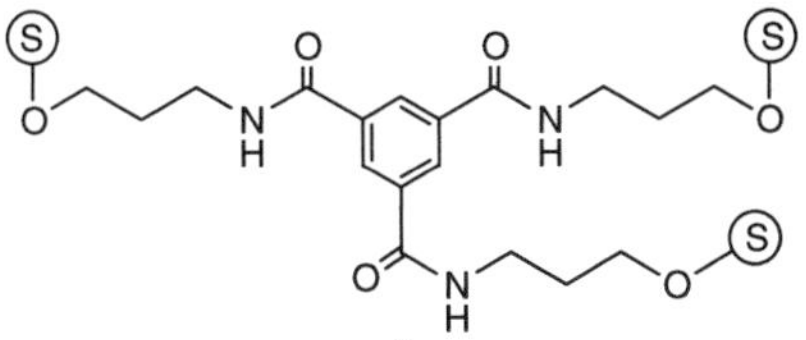

Disaccharide derivatives: methyl β-lactoside,[46] 4-*O*-α-D-glucopyranosyl-D-glucitol[47] (redetermined due to inconsistencies in earlier reports). The interplay of the rate of water removal in the dehydration of α,α-trehalose, and polymorph formation which occurs on the heating of α,α-trehalose, has been characterized by X-ray powder diffraction.[48] The complexes formed between the divalent **2** and trivalent **3** with the plant lectin Concanavalin A have been structurally assigned.[49]

C-Glycoside derivatives: the X-ray structures of the *C*-glycosides **4**,[50] and **5**,[51] the tricyclic *C*-glycoside **6**,[52] and *o*- and *m*-fluorophenyl-*C*-β-D-ribofuranoside[53] have been reported.

3.3 Higher Oligosaccharides. – The following oligosaccharides have been characterized by X-ray crystallography: the new cytotoxic antibiotic FD-594 (**7**) from *streptomyces*,[54] α-D-Glc*p*-(1→4)-α-D-Glc*p*-(1→4)-D-glucitol,[47] α-L-

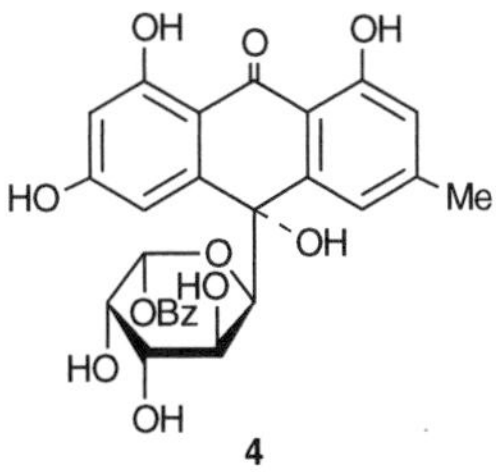

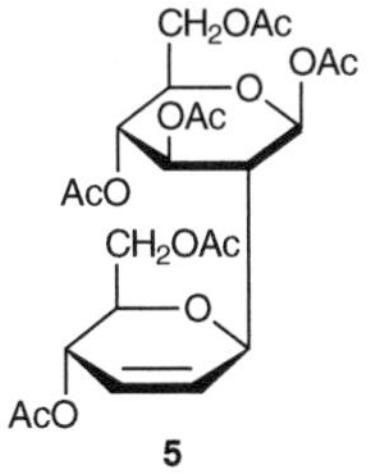

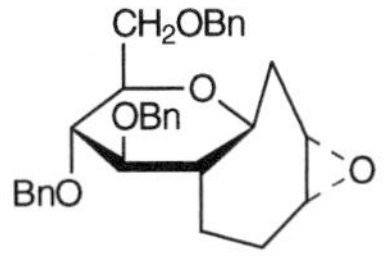

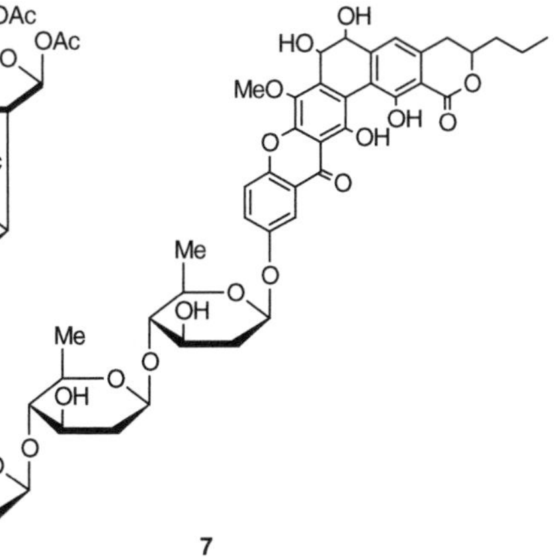

Fuc*p*-(1→2)-[α-D-Gal*p*-(1→3)]-β-D-Gal*p*-OMe.[55] The solution conformation of the last mentioned compound (NMR analysis) is similar to the crystal conformation.

The cellooligosaccharides (from the di- to the hexa-) peracetates have been studied by X-ray analyses as models for cellulose triacetate.[56]

The X-ray structures of heptakis(2,6-di-*O*-methyl)-β-cyclodextrin dihydrate[57] and cyclomaltodecaose[58] have been reported. Cyclodeca- and cyclotetradeca-amylose have been characterized by X-ray crystallography and the results indicate that when cyclodextrins with 6–8 residues are expanded to ten residues and beyond, steric strain builds up as the curvature is reduced and some glucose residues flip conformation.[59]

3.4 Anhydro-sugars. – In the investigation of some rigid bicyclic α-L-fucose derivatives by Fleet and co-workers, the X-ray structures of the compound **8**[60] and the azasugar **9**[61] were reported (see Chapters 5, 9 and 15 for chemistry). The X-ray structure of 3-amino-1,6-anhydro-3-deoxy-β-D-gulose hydrochloride has been determined during work on the synthesis of an analogue of the phytotoxin, tagetitoxin.[62]

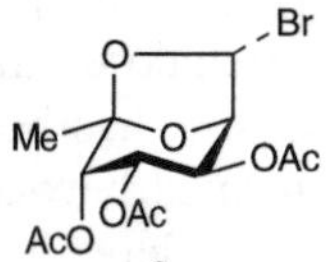

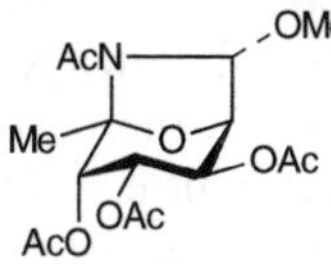

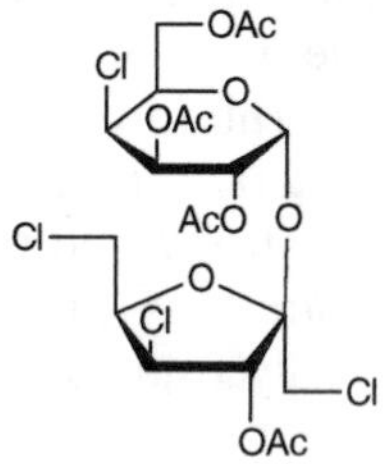

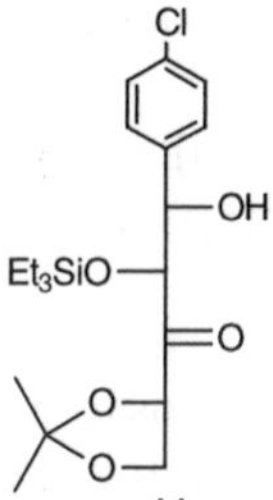

3.5 Halogen-, Phosphorus-, Sulfur-, Selenium- and Nitrogen-containing Compounds. – X-Ray crystallography has been used to determine the structure of the product of sulfuryl chloride treatment of 6-*O*-acetyl-sucrose, after dechlorosulfation and acetylation, as the *tagato*-derivative **10** and not the, C-4′ epimeric, *sorbo*-derivative.[63] The X-ray structure of the ulose **11** has been reported (see Chapter 2 for synthesis).[64] The benzimidazole **12** was produced in an unexpected intramolecular displacement reaction and its structure determined by X-ray crystallography.[65] The X-ray structures of 1-*O*-{2-[(2,6-dichlorophenyl)amino]phenylacetyl}-2,3,4,6-tetra-*O*-acetyl-β-D-glucopyranose and the analogous xylose derivative have been determined.[66]

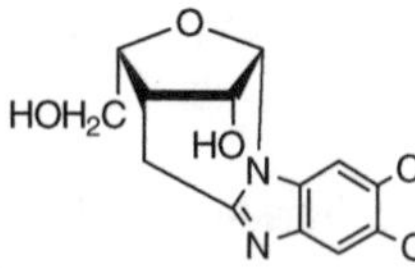

The crystal structure of 1,2,3,4-tetra-*O*-acetyl-5-*C*-(*t*-butylphosphinyl)-5-deoxy-D-xylopyranose[67] has been published, as has that of 1,2:5,6-di-*O*-isopropylidene-3-*O*-(diphenylphosphinyl)-D-*chiro*-inositol.[68] The structures of the thioselenophosphates **13** and **14** have been reported.[69] A study of the crystallization of the bis-thiophosphoryl disulfide **15** to examine inclusion complexes formed with different solvents has been conducted using X-ray crystallography.[70]

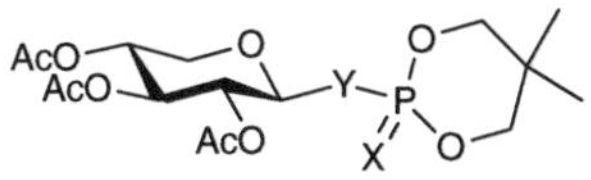

13 Y = S, X = Se
14 Y = Se, X = S

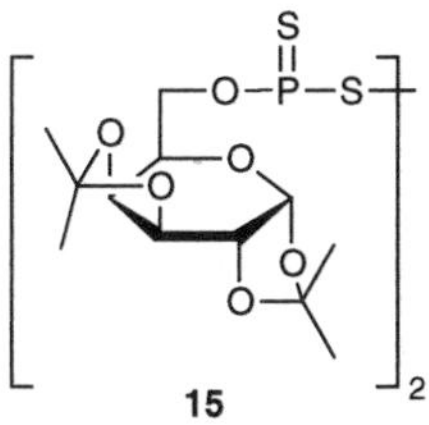

15

Tetra-*O*-acetyl-5-thio-β-D-glucopyranosyl azide has been prepared (see Chapter 11) and characterized by X-ray crystallography,[71] as have tetra-*O*-acetyl-5-thio-β-D-galactopyranosyl azide and isothiocyanate.[72] The X-ray structure of Pirlimycin (**16**), a semisynthetic lincosaminide antibiotic, has been reported[73] and the products formed by reaction of toluenesulfonhydrazide with D-glucose, D-galactose and L-arabinose have been shown to be the pyranosyl glycosylamines in the solid state rather than the open-chain hydrazones.[74] *En route* to a series of spiroaminothiazolines (see Chapter 10) the isothiocyanate **17** has been prepared and the X-ray structure determined.[75] The structures of a series of sulfanamides **18**–**21** have been determined.[76]

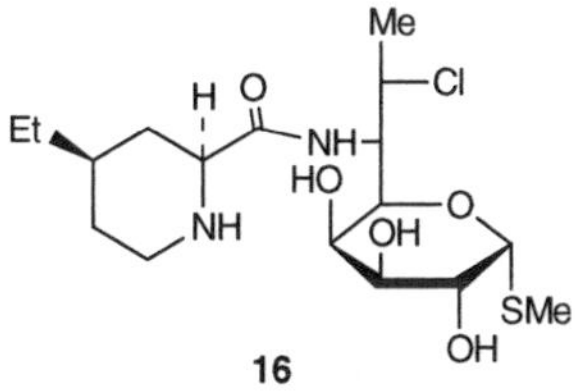

16 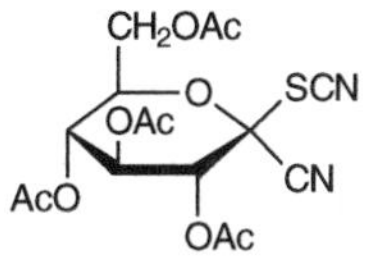 **17** 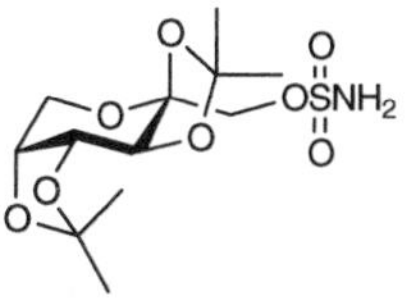 **18**

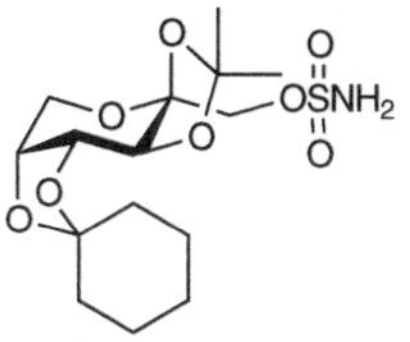

19 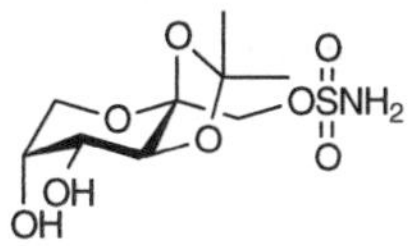 **20** 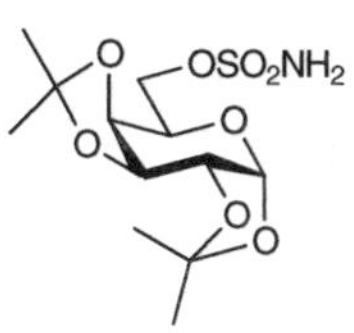 **21**

The diglucosyltriamine **22** has been prepared from D-glucose and diethylenetriamine, and the X-ray structure determined, and similar compounds have been produced from D-galactose, D-allose and L-fucose.[77] A product from a Baer reaction with a dialdehyde (see Chapter 10) has been shown to be a 1:1 mixture of nitrosugars **23** and **24**.[78] In the course of work on the synthesis of azasugars (see Chapter 9) the X-ray structure of methyl 3-*O*-acetyl-4,6-*O*-benzylidene-2-deoxy-2-trichloroacetylamino-α-D-glucopyranose was recorded.[79]

The X-ray structure of the fused imidazole **25** bound to the β-retaining glycosidase Cel5A from *B. agaradhaevens* is consistent with the concept of lateral protonation for these glycosidases.[80] The crystal structure and ^{13}C NMR analysis (see Chapter 10) of methyl 3,4,6-tri-*O*-acetyl-2-deoxy-2-(3-phenylureido)-β-D-glucopyranoside have been recorded.[81] 2-Glycosylchromene derivatives, exemplified by the structure **26,** have been prepared (see Chapter 2).[82]

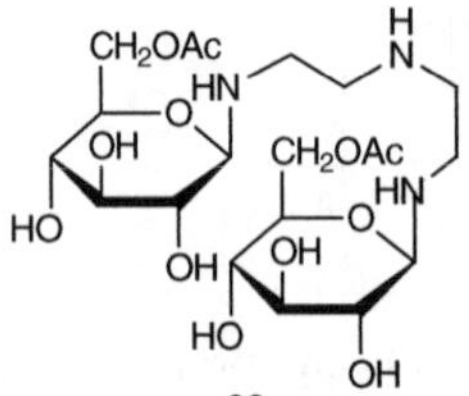

22 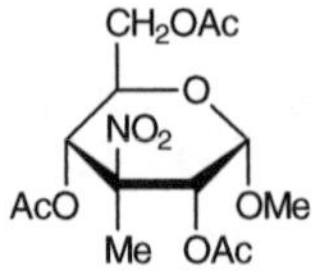 **23** 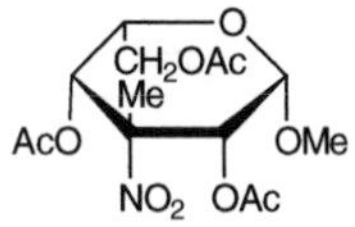 **24**

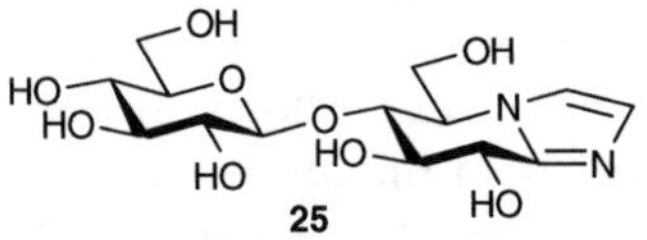

25 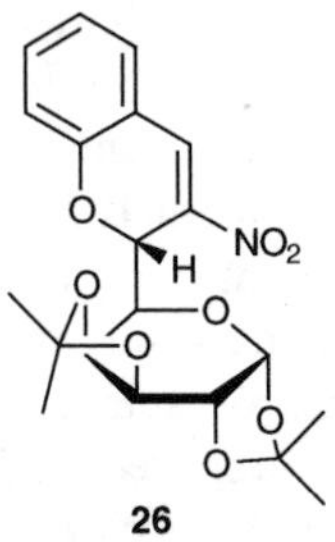 **26** 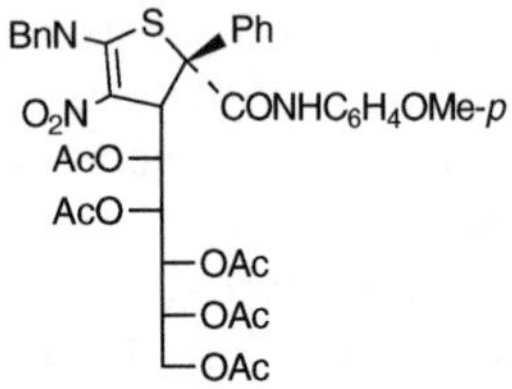 **27**

The structures of 1-deoxy-2,3:5,6-di-*O*-isopropylidene-1-*C*-(piperidin-1-ylthiocarbonylmethyl)-β-D-mannofuranose[83] and the dihydrothiophene **27** have been reported.[84] The structure of 3,6-anhydro-1,1-bis-(ethylsulfonyl)-1-deoxy-D-talitol has been elucidated by NMR spectroscopy and X-ray crystallography[85] The 5-phenyl-1,2,4-oxadiazole derivative of 1,2-*O*-isopropylidene-3-*C*-cyano-5-*O*-benzoyl-α-D-xylofuranose has been prepared (see Chapter 14) and analysed by X-ray crystallography.[86]

Atropisomerism has been investigated in imidazolidin-2-ones and -2-thiones (see Chapter 10), and the structure of imidazolidin-2-one **28** reported,[87] as has that of the xylofuranosylamine **29**.[88] Undecose derivatives have been accessed by cycloaddition of alkenes and nitrile oxides which gives isoxazolines, and the X-ray structure of the specific product **30** has been reported.[89]

3-Deoxy-D-*erythro*-hexos-2-ulose bis(thiosemicarbazone), as the copper

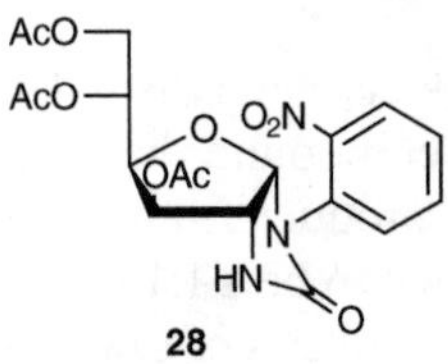

28 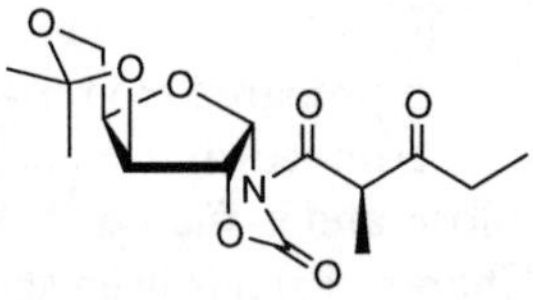 **29** 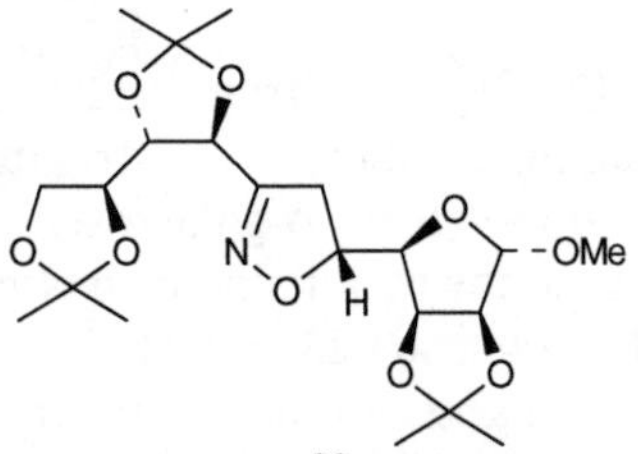 **30**

complex, has been characetized by X-ray powder diffraction analysis.[90] The structure of the aminoaziridine **31** has been reported (see Chapters 16 and 19).[91] The structure of tetrahydropteridine **32**[92] have been reported. During applications of the Staudinger reaction (see Chapter 10) the zwitterion **33** was prepared and the X-ray structure reported.[93]

The crystal structure of amikacin **34**, an aminoglycoside antibiotic used against Gram-negative bacteria, has been reported. The relative orientations of the A, B and C rings of the molecule are maintained by intramolecular hydrogen bonds.[94] The structures of the two bicyclic azasugars **35** and **36** have been reported (see Chapter 10 for chemistry).[95]

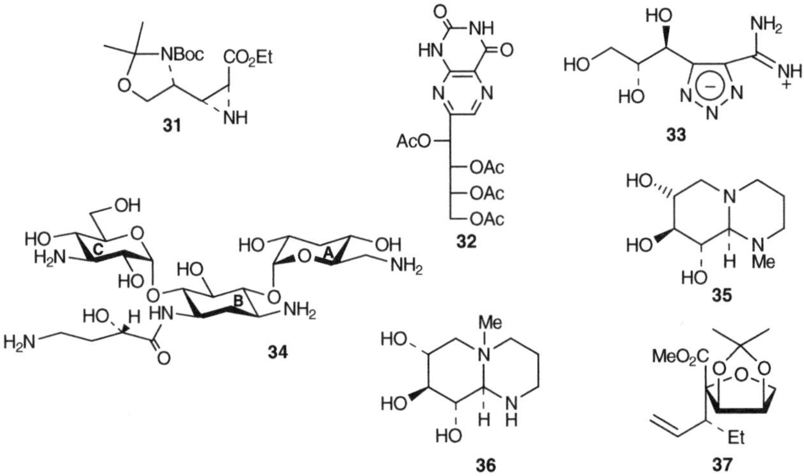

3.6 Branched-chain Sugars. – The X-ray structure of the furanosid derivative **37** has been reported,[96] and nitrosugars **23** and **24** are other examples of compounds containing branch-points to have been characterized crystallographically.

3.7 Sugar Acids and Their Derivatives. – The structure of 3,4,7-tri-*O*-acetyl-2,6-anhydro-D-*glycero*-L-*manno*-heptonic acid has been reported.[62]

3.8 Unsaturated Compounds. – During the course of the synthesis of 5-C-arylpentopyranosides (see Chapters 2 and 24), the enone **38** was prepared and the X-ray structure determined.[97] Likewise the structure of 3,4-*O*-benzylidene-D-galactal has been determined.[40]

3.9 Inorganic Derivatives. – Structures of the following chromiumcarbonyl complexes have been determined crystallographically: the four methyl 4,6-*O*-[(η^6-phenyl) alkylidene-α-D-glucopyranosides **39**–**42**,[98] the Diels-Alder adduct **43** (see Chapter 14 for preparation),[99] the chromium pentacarbonyl iminoglycosylidenes **44** and **45** and their tungsten analogues (see Chapters 17 and 18 for synthesis and application to photoinduced *C*-glycosidation),[100] and the complex **46** with its cyclized derivative **47** (see Chapter 17 for synthesis).[101]

The X-ray structure of the bis-complex of the diamine **48** with Ni(II) has been reported.[102]

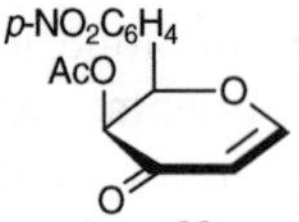

38

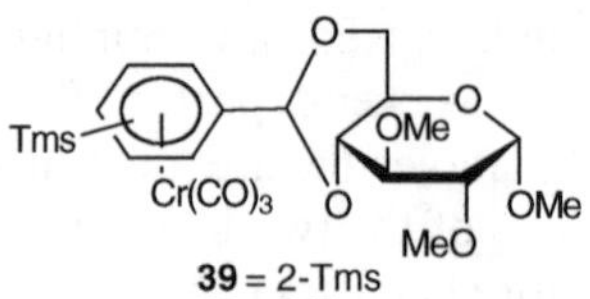

39 = 2-Tms
40 = 3-Tms

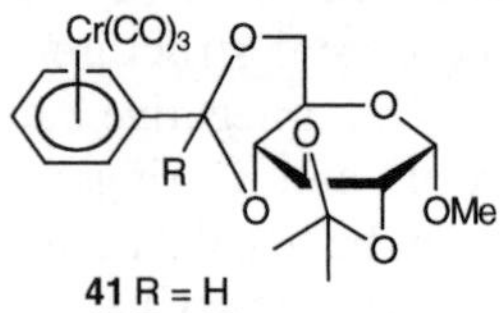

41 R = H
42 R = Me

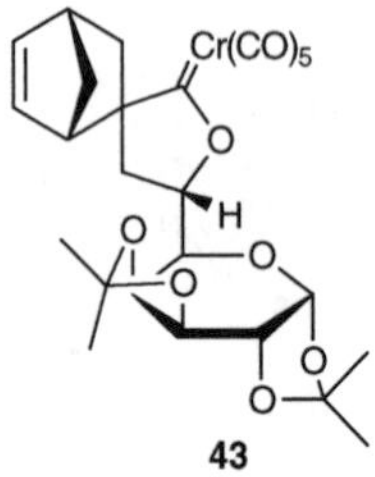

43 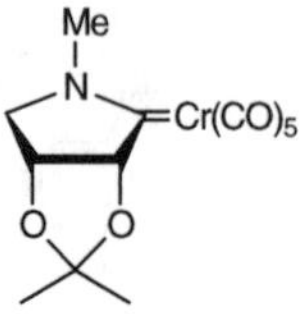 **44** 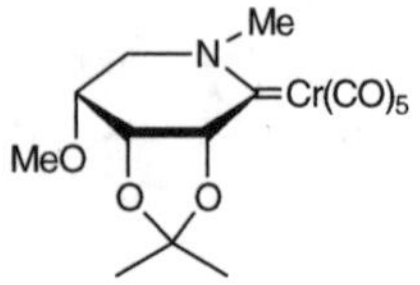 **45**

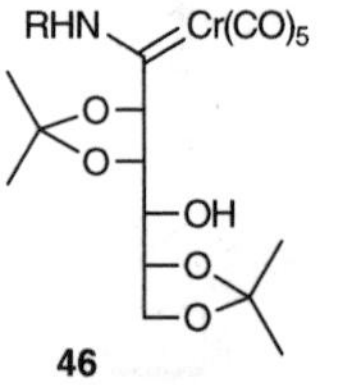

46 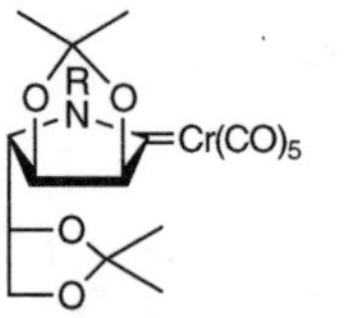 **47** 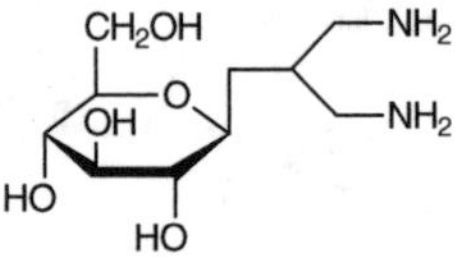 **48**

3.10 Alditols, Cyclitols and Derivatives Thereof. – The structure of 1,2:3,4:5,6-tri-*O*-isopropylidene-D-mannitol[103] has been determined crystallographically as have those of 3,4-anhydro-1,2:5,6-di-*O*-isopropylidene-D-*allo*-inositol[104] and 5-ammonio-1,3-diamino-1,3,5-trideoxy-*cis*-inositol iodide.[105]

3.11 Nucleosides and Their Analogues and Derivatives Thereof. – The X-ray structures of the following have been reported: an octadecahydrated complex between two inosine 5′-monophosphate molecules,[106] 2′-*O*-tosyladenosine,[107] 5′-*O*-tosyladenosine,[108] 2′-deoxy-5-methoxymethyl-*N,N*-dimethylcytidine,[109] N^6-anisoyladenosine,[110] N^3-benzoyl-2′,3′-di-*O*-benzoyluridine,[111] 2′,3′,5′-tri-*O*-benzoylinosine,[112] 2′-deoxy-2′-fluoro-β-D-arabinofuranosyl-5-fluorouracil,[113] and (*S*)-1-benzyl-9-(5′-*O*-tertbutyldimethylsilyl-2′,3′-*O*-benzylidene-β-D-ribofuranosyl)guanine.[114]

Treatment of 3′,5′-*O*-(tetraisopropyldisiloxane-1,3-diyl)-2′-ketouridine with primary alcohols gave single diastereoisomers of stable hemi-ketals. The configuration was confirmed by X-ray diffraction.[115] The structures of cytidinium H-phosphonate monohydrate, bis-2′-deoxycytidinium H-phosphonate and 2′-deoxycytidinium H-phosphate have been studied by X-ray diffraction and FTIR methods. These phosphonates do not form base pairs; however, pleated sheets with alternating cations and anions are formed with additional phosphonate H-bonds between the sheets.[116] The X-ray structure of 5′-*O*-

(guanosine-2′-*O*-phosphonoethyl)cytidine, an isopolar, non-isoteric phosphonate analogue of GpC, has been reported.[117]

The X-ray structures of the following have been reported; N^1-(2,3,5-tri-*O*-acetyl-β-D-ribofuranosyl)-4-oxonicotinamide,[118] 6-amino-5-[*N*-(β-D-gulopyranosyl)amino]-3-methylpyrimidine tetra- and hexa-acetate,[119] 7-deaza-2′-deoxyadenosine,[120] the anomers of 8-aza-7-deaza-2′-deoxyadenosine,[121] 7-deaza-9-(2-deoxy-α-D-ribofuranosyl)-7-iodoadenine,[122] 1-deazaadenosine and 1-deaza-2′-deoxyadenosine,[123] 8-aza-7-bromo-7-deaza-2′-deoxyadenosine and 8-aza-7-deaza-2′-deoxy-7-iodoadenosine.[124] The last two compounds showed high *anti* conformations (see also Chapter 22).

2,6-Dioxabicyclo[3,2,1]octane nucleosides have been prepared (see Chapter 20) and two members, **49** and **50**,[125] have been characterized by X-ray crystallography.[126] X-Ray analysis, as well as NMR spectroscopy (see Chapters 20 and 21), has been used to examine the intermolecular non-bonded attractions of the 4′-*C*-fluoromethyl nucleoside **51** and its anomer.[127] In the course of synthetic work on some nucleoside analogues of type **52**, the ribofuranosyl ring was judged to be in the *S*-conformation by NMR and X-ray analysis (see Chapters 20 and 21).[128] The structure of the modified pyrimidine nucleoside **53** has been reported.[129]

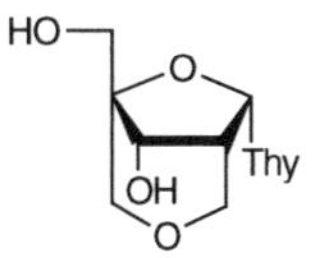

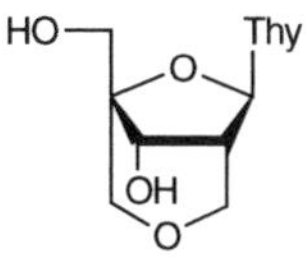

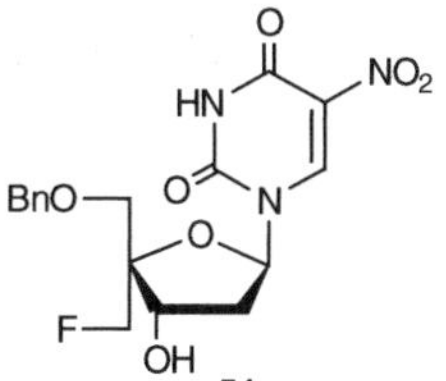

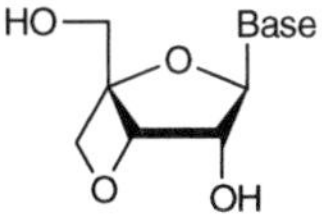

The structure of the L-carbocyclic adenine nucleoside **54** has been determined crystallographically[130] and, similarly, following the synthesis of some L-carbocyclic dideoxy nucleosides, **55**[131] and **56**[132] have been characterized. The X-ray diffraction analyses of 3-nitro-1-(α-D-arabinofuranosyl)-imidazole and 3-nitro-1-(5-deoxy-5-iodo-α-D-arabinofuranosyl)-imidazole were used to confirm their α-configurations (see Chapter 20 for synthesis and chemistry).[133] The structure of the unusual nucleoside **57** was confirmed by X-ray analysis (see also Chapter 10 and 16 for the synthesis of analogues).[134] In the course of some nucleobase chemistry compound **58** was made and analysed by X-ray crystallography.[135] The structural determination of the previously prepared carbocycle **59** has been reported,[136] as has that of the spiro-nucleoside **60**.[137]

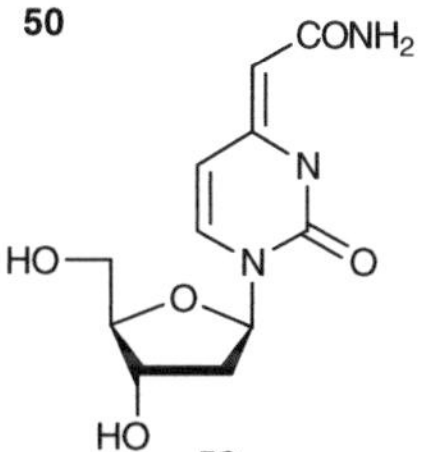

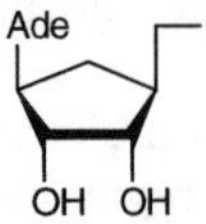
54

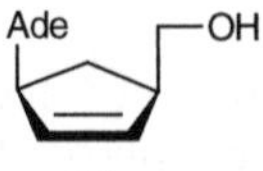
55

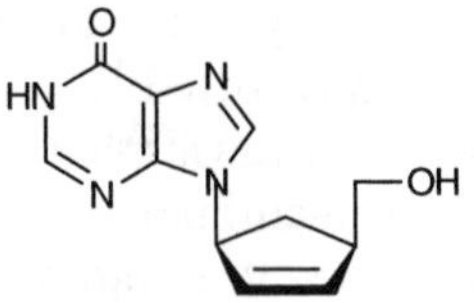
56

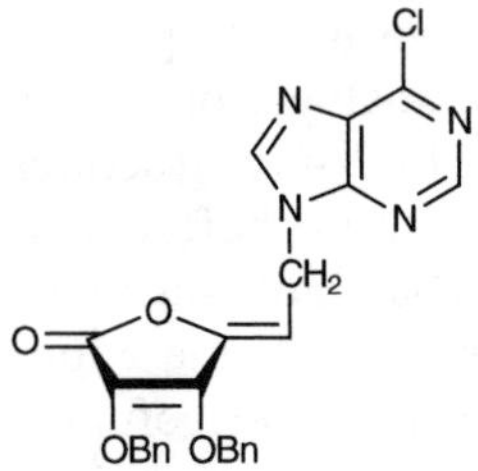
57

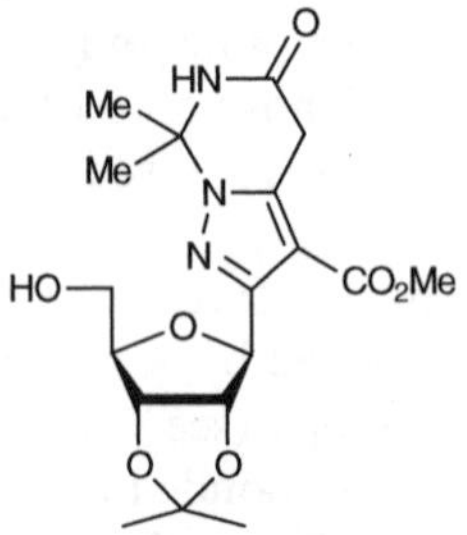
58

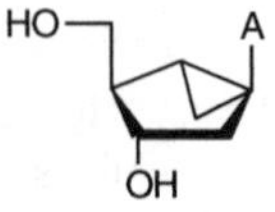
59

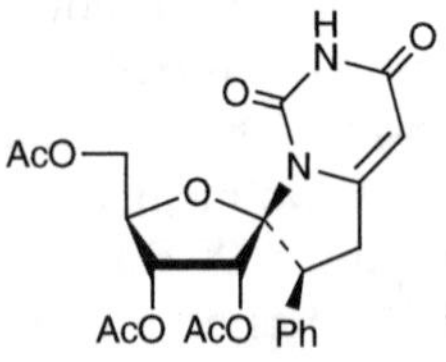
60

4 UV Spectroscopy, Polarimetry, Circular Dichroism, Calorimetry and Other Physical Studies

The UV spectra of some cyclic branched-chain sugar bypyridyls **61**, as well as that of an acyclic precursor, showed that they bind a range of metal ions (Zn, Cu, Ag, Ni).[138] A spectrophotomeric study of the oxidation of *N*-acetylneuraminic acid by periodate ion has been reported.[139] UV resonance Raman spectra of guanosine and seven isotopically substututed analogues (2-^{13}C, 2-^{15}N, 6-^{18}O, 7-^{15}N, 8-^{13}C, 9-^{15}N and 1′-^{13}C) have reported and the data used to assign the Raman bands.[140]

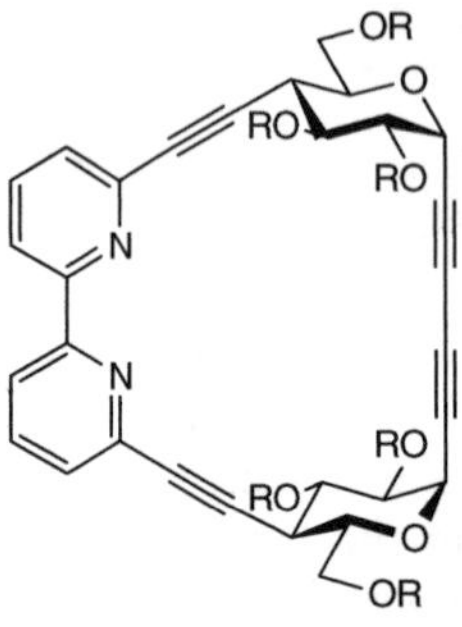
61 R = H, Ac

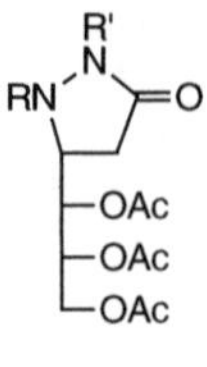
62

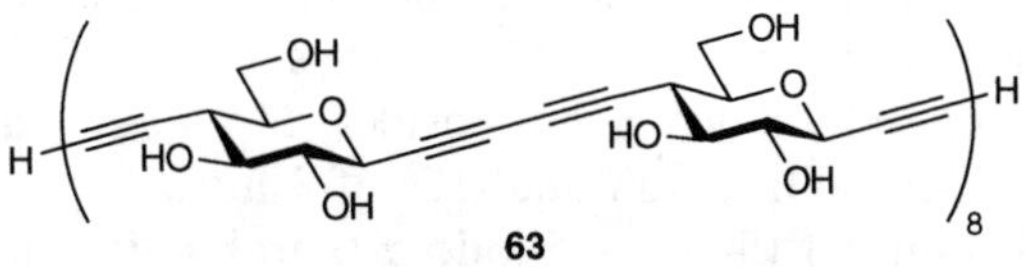
63

A review of CD spectra of carbohydrate complexes with transition metals has appeared.[141] The configurational assignment of carbohydrate-derived 5-substituted pyrazolidin-3-ones **62** has been determined by CD.[142] The sign of the n→π* Cotton effect correlates with the configuration at C-5 of the pyrazolidin-3-one. Vibrational CD has been used to probe the glycosidic linkage in oligosaccharides of D-glucose. The region 1200–900 cm^{-1} can be used to identify α- and β-glycosides, and α-(1→1), α-(1→4) and α-(1→6) linkages, but not β-(1→4) and β(1→6) bonds.[143] The CD spectra of some nucleoside 5′-S-thiosulfates (Bunte salts) have been described and related to their conformations.[144] The absolute stereochemistry of the sesquiterpene glucoside pteridanoside (see Chapter 3) has been determined by CD analysis.[145] Vibrational absorption and CD spectra of several monosaccharides in the 1500–1180 cm^{-1} region have been recorded, and analysed for similarities and differences between anomeric, homomorphic and epimeric pairs of sugars.[146] In analogous fashion, the vibrational absorption and CD spectra of five hexose pentaacetates were measured and interpreted to reflect the orientation of carbonyl groups around the carbohydrate rings.[147] CD has been used to follow the 'dendromercleft' binding of octyl α-L-, α-D- and β-D-glucoside to a dendrimer with an outer layer consisting of nine triethylene glycol units.[148]

High resolution electron microscopy has shown the poly-yne **63** has nanocrystalline domains of parallel chains.[149] At low water contents (~10%) trehalose and rhamnose have been shown to interact directly and hydrophobically with phospholipids (L-α-dipalmitoyl phosphatidylcholine) *via* the hydrophobocity-rich side of the pyranose ring.[150]

5 ESR Spectroscopy

Conformations of some carbohydrates have been studied by ESR spectroscopy of free radicals located on the carbon atoms of the pyranose ring.[151] An ESR study has been published of the free-radicals induced by plasma irradiation of crystalline α- and β-D-glucose.[152] A disaccharide with a hydroxylamine linkage, undergoes spontaneous aerial oxidation to aminyloxy radicals that can be studied by ESR.[153]

References

1 M. Rozenberg, A. Loewenschuss, H.-D. Lutz and Y. Marcus, *Carbohydr. Res.*, 1999, **315**, 89.
2 S.A. Krasnokutski, A. Yu. Ivanov, V. Izvekov, G.G. Sheina and Yu. P. Blagoi, *J. Mol. Struct.*, 1999, **482–483**, 249 (*Chem. Abstr.*, 1999, **131**, 102 486).
3 L. Bailey, R. Navarro and A. Hermanz, *J. Mol. Struct.*, 1999, **480–481**, 465 (*Chem. Abstr.*, 1999, **131**, 88 116).
4 B.V. Grande, H. Kallevik, F.O. Libnau and O.M. Kvalheim, *Chemom. Intell. Lab. Syst.*, 1999, **45**, 7 (*Chem. Abstr.*, 1999, **130**, 237 744).
5 J. Liu, S. Mendoza, E. Román, M.J. Lynn, R. Xu and A.E. Kaifer, *J. Am. Chem. Soc.*, 1999, **121**, 4304.
6 K.C. Martin, D.A. Pinnick, S.A. Lee, A. Anderson, W. Smith, R.H. Griffey and V. Mohan, *J. Biomol. Struct. Dyn.*, 1999, **16**, 1159 (*Chem. Abstr.*, 1999, **131**, 88 122).
7 M. Tsuboi, Y. Takeuchi, E. Kawashima, Y. Ishido and M. Aida, *Spectrochim. Acta*, 1999, **55A**, 1887 (*Chem. Abstr.*, 1999, **131**, 185 183).

8 J.O. Duus, P.M. St. Hilaire, M. Meldal and K. Bock, *Pure Appl. Chem.*, 1999, **71**, 755 (*Chem. Abstr.*, 1999, **131**, 272 082).
9 Y.Q. Deng, *Chin. Chem. Lett.* 1998, **9**, 553 (*Chem. Abstr.*, 1999, **131**, 144 758).
10 J.M. Vega-Perez, J.I. Candela, F. Alcudia and F. Iglesias-Guerra, *Eur. Mass Spectrom.*, 1999, **5**, 191 (*Chem. Abstr.*, 1999, **131**, 337 282).
11 T. Yamagaki and H. Nakanishi, *Glycoconjugate J.*, 1999, **16**, 385.
12 H. Desaire and J.A. Leary, *Anal. Chem.*, 1999, **71**, 1997 (*Chem. Abstr.*, 1999, **130**, 296 919).
13 S. Konig and J.A. Leary, *J. Am. Soc. Mass Spectrom.*, 1998, **9**, 1125 (*Chem. Abstr.*, 1999, **130**, 66 692).
14 I. Faengmark, A. Jansson and B. Nilsson, *Anal. Chem.*, 1999, **71**, 1105 (*Chem. Abstr.*, 1999, **130**, 223 511).
15 W. Mo, H. Sakamoto, A. Nishikawa, N. Kagi, J.I. Langridge, Y. Shimonishi and T. Takao, *Anal. Chem.*, 1999, **71**, 4100 (*Chem. Abstr.*, 1999, **131**, 272 116).
16 A. Salvador, B. Herbreteau and M. Dreux, *J. Chromatogr., A*, 1999, **855**, 645 (*Chem. Abstr.*, 1999, **131**, 337?265).
17 P. Chen, U. Werner-Zwanziger, D. Wiesler, M. Pagel and M. V. Novotny, *Anal. Chem.*, 1999, **71**, 4969 (*Chem. Abstr.*, 1999, **131**, 337 287).
18 Y. Mizumo, T. Sasagawa, N. Dohmae and K. Takio, *Anal. Chem.*, 1999, **71**, 4764 (*Chem. Abstr.*, 1999, **131**, 351 565).
19 Y.H. Ahn, J.S. Yoo and S.H. Kim, *Anal. Sci.*, 1999, **15**, 53 (*Chem. Abstr.*, 1999, **130**, 182 677).
20 V. Kovacik, V. Patoprsty, V. Havlicek and P. Kovac, *Eur. Mass Spectrom.*, 1998, **4**, 417 (*Chem. Abstr.*, 1999, **130**, 296 915).
21 L.P. Brull, V. Kovacik, J.E. Thomas-Oates, W. Heerma and J. Haverkamp, *Rapid Commun. Mass Spectrom.*, 1998, **12**, 1520 (*Chem. Abstr.*, 1999, **130**, 66 703).
22 D.J. Harvey and A.P. Hunter, *Rapid Commun. Mass Spectrom.*, 1998, **12**, 1721 (*Chem. Abstr.*, 1999, **130**, 95 749).
23 P.M. St. Hilaire, L. Cipolla, U. Tedebark and M. Meldal, *Rapid Commun. Mass Spectrom.*, 1998, **12**, 1475 (*Chem. Abstr.*, 1999, **130**, 52 654).
24 X. Shen and H. Perreault, *J. Mass Spectrom.*, 1999, **34**, 502 (*Chem. Abstr.*, 1999, **131**, 32 115).
25 Y. Mechref, P. Chen and M.V. Novotny, *Glycobiology*, 1999, **9**, 227.
26 M.M. Rahman, X.-X. Gu, C.-M. Tsai, V.S.K. Kolli and R.W. Carlson, *Glycobiology*, 1999, **9**, 1371.
27 C. Kannicht, L. Lucka, R. Nuck, W. Renter and M. Gohlka, *Glycobiology*, 1999, **9**, 897.
28 H. Karlsson, L. Johansson, H. Millar-Podraza and K.-A. Karlsson, *Glycobiology*, 1999, **9**, 765.
29 G.O. Aspinall, A.S. Mainkar and A.P. Moran, *Glycobiology*, 1999, **9**, 1235.
30 P.N. Lippert, D.W. Dwyer, F.Li and R.W. Olafson, *Glycobiology*, 1999, **9**, 627.
31 J. Müthing, S. Duvar, D. Heitmann, F.G. Hanisch, U. Neumann, G. Lochnit, R. Geyer and J. Peter-Katalinic, *Glycobiology*, 1999, **9**, 459.
32 J. Zhu, Y.-T. Li, S.-C. Li and R.B. Cole, *Glycobiology*, 1999, **9**, 985.
33 T. Wugeditsch, N.E. Zachara, M. Puchberger, P. Kosma, A.A. Gooley and P. Messren, *Glycobiology*, 1999, **9**, 787.
34 E.M.M. Gasper, *Tetrahedron Lett.*, 1999, **40**, 6861.
35 L. A. Marzilli, J.P. Barry, T. Sells, S.-J. Law, P. Vouros and A. Harsch, *J. Mass Spectrom.*, 1999, **34**, 276 (*Chem. Abstr.*, 1999, **131**, 5 459).

36 A. Tewari, A. Mishra and D.S. Bhakuni, *J. Indian Chem. Soc.*, 1999, **76**, 222 (*Chem. Abstr.*, 1999, **131**, 88 124).

37 P.P. Wang, L.P. Kotra, C.K. Chu and M.G. Bartlett, *J. Mass Spectrom.*, 1999, **34**, 724 (*Chem. Abstr.*, 1999, **131**, 185 184)

38 L. Eriksson, P. Söderman and G. Widmalm, *Acta Crystallogr.*, 1999, **C55**, 1736.

39 A.J. Petreson, S.M. Petreson, R.A. Dwek and M.R. Wormald, *Glycobiology*, 1999, **9**, 343.

40 T.L. Gururaja, P Venugopalan and M.J. Levine, *J. Chem. Crystallogr.*, 1999, **28**, 747 (*Chem. Abstr.*, 1999, **130**, 252 578)

41 A.G. Evdokimov, A.J. Kalb, T.F. Koetzle, W.T. Klooster and J.M.L. Martin, *J. Phys. Chem. A*, 1999, **103**, 744 (*Chem. Abstr.*, 1999, **130**, 182 669).

42 Bartnicka and A. Zamojski, *Tetrahedron*, 1999, **55**, 2061.

43 M. Crisma, M. Gobbo, C. Toniolo and R. Racchi, *Carbohydr. Res.*, 1999, **315**, 334.

44 P.A. Grieco, J. Haddad, M.M. Pineiro-Nunez and J.C. Huffman, *Phytochemistry*, 1999, **51**, 575.

45 R. Zhang, P.A. Marone, P. Thiyagarajan and D.M. Tiede, *Langmuir*, 1999, **15**, 7510 (*Chem. Abstr.*, 1999, **131**, 299 600).

46 R. Stenutz, M. Shang and A.S. Serianni, *Acta Crystallogr.*, 1999, **C55**, 1719.

47 A. Schouten, J.A. Kanters, J. Kroon, P. Looten, P. Duflot and M. Mathlouthi, *Carbohydr. Res.*, 1999, **322**, 274.

48 F. Sussich, F. Princivalle and A. Cesaro, *Carbohydr. Res.*, 1999, **322**, 113.

49 S.M. Dimick, S.C. Powell, S.A. McMahon, D.N. Moothoo, J.H. Naismith and Eric J. Toone, *J. Am. Chem. Soc.*, 1999, **121**, 10286.

50 M.R. Hernandez-Medel, C.O. Ramirez-Corzas, M.N. Rivera-Dominguez, J. Ramirez-Mendez, R. Santillan and S. Rojas-Lima, *Phytochemistry*, 1999, **51**, 1379.

51 A.H. Franz, V.V. Zhdankin, V.V. Samoshin, M.J. Inch, V.G. Young and P.H. Gross, *Mendeleev Commun.*,1999, **45** (*Chem. Abstr.*, 1999, **130**, 338 290).

52 L. Eriksson, S. Guy, P. Perlmutter and R. Lewis, *J. Org. Chem.*, 1999, **64**, 8396.

53 Bats, J. Parsch and J.W. Engels, *Acta Crystallogr.*, 1999, **C55**, i (IUC 9900069 and IUC 9900070).

54 T. Eguchi, K. Kondo, K. Kakinuma, H. Uekusa, Y. Ohashi, K. Mizoue and Y.-F. Qiao, *J. Org. Chem.*, 1999, **64**, 5 371.

55 A. Otter, R.U. Lemieux, R.G. Ball, A.P. Venot, O. Hindsgaul and D.R. Bundle, *Eur. J. Biochem.*, 1999, **259**, 295 (*Chem. Abstr.*, 1999, **130**, 182 675).

56 H. Kono, Y. Numata, N. Nagai, T. Erata and M. Takai, *Carbohydr. Res.*, 1999, **322**, 256.

57 T. Aree, W. Saenger, P. Liebnitz and H. Hoier, *Carbohydr. Res.*, 1999, **315**, 199.

58 T. Endo, H. Nagase, H. Ueda, S. Kobayashi and M. Shiro, *Anal. Sci.*, 1999, **15**, 613 (*Chem. Abstr.*, 1999, **131**, 73 877).

59 J. Jacob, K. Gessler, D. Hoffmann, H. Sanbe, K. Koizumi, S.M. Smith, T. Takaha and W. Saenger, *Carbohydr. Res.*, 1999, **322**, 228.

60 K.H. Smelt, Y. Blèriot, K. Biggadike, S. Lynn, A.L. Lane, D.J. Watkin and G.W.J. Fleet, *Tetrahedron Lett.*, 1999, **40**, 3259.

61 K.H. Smelt, A.J. Harrison, K. Biggadike, M. Müller, K. Prout, D.J. Watkin and G.W.J. Fleet, *Tetrahedron Lett.*, 1999, **40**, 3255.

62 B.R. Dent, R.H. Furneaux, G.J. Gainsford and G.P. Lynch, *Tetrahedron*, 1999, **55**, 6977.

63 C.K. Lee, H.C. Kang and A. Linden, *J. Carbohydr. Chem.*, 1999, **18**, 241.

64 J.A. Marco, M. Carda, E. Falomir, C. Palomo, M. Oiarbide, J.A. Ortiz and A. Linden, *Tetrahedron Lett.*, 1999, **40**, 1065.
65 J.-L. Girardet and L.B. Townsend, *J. Org. Chem.*, 1999, **64**, 4169.
66 E. Kolodziejczyk, K. Suwinska, A.E. Koziol, G. Enright and J. Borowiecka, *Pol. J. Chem.*, 1999, **73**, 367 (*Chem. Abstr.*, 1999, **130**, 209 869).
67 T. Oshikawa, M. Yamashita, K. Seo and Y. Hamauzu, *Heterocycl. Commun.*, 1998, **4**, 393 (*Chem. Abstr.*, 1999, **130**, 139 528).
68 A. Falshaw, G.J. Gainsford and C. Lensink, *Acta Crystallogr.*, 1999, **C55**, 1353.
69 W. Kudelska, A. Olczak, M.L. Glowka and S. Jankowski, *Pol. J. Chem.*, 1999, **73**, 487 (*Chem. Abstr.*, 1999, **130**, 252 540).
70 M.J. Potzebowski, K. Ganicz, W. Ciesielski, A. Skonronska, M.W. Wieczorek, J. Blaszczyk and W. Majzner, *J. Chem. Soc., Perkin Trans. 2*, 1999, 2163.
71 M.K. Strumpel, J. Buschmann, L. Szibgyi and Z. Gyorgydeak, *Carbohydr. Res.*, 1999, **318**, 91.
72 M. Sekti, R. Kassab, H. Parrot Lopez, F. Villain and C. de Rango, *J. Carbohydr. Chem.*, 1999, **18**, 1019.
73 F.W. Crow, J.R. Blinn, C.G. Chidestere, A.M. Cooper, W.K. Duholke, J.W. Hallberg, G.E. Martin, R.F. Smith and T.J. Thamann, *J. Heterocycl. Chem.*, 1999, **36**, 1049.
74 W.H. Ojala, C.R. Ojala and W.B. Gleason, *J. Chem. Crystallogr.*, 1999, **29**, 19 (*Chem. Abstr.*, 1999, **131**, 59 067).
75 E. Ösz, L. Szilágyi, L. Somsak and A. Bényi, *Tetrahedron*, 1999, **55**, 2419.
76 M. Kubickia, P.W. Codding, S.A. Litsterc, M.B. Szkaradzinskac and H.A.R. Bassyounic, *J. Mol. Struct.*, 1999, **474**, 255 (*Chem. Abstr.*, 1999, **130**, 237 750).
77 S.P. Gaucher, S.F. Pedersen and J.A. Leary, *J. Org. Chem.*, 1999, **64**, 4012.
78 A.T. Carmona, P. Barrachero, F. Cabrera-Escribano, M. Jesús Diánez, M. Dolores Estrade, A. López Castro, R. Ojeda, M. Gómez-Guillén and S. Pérez-Garrido, *Tetrahedron Asym.*, 1999, **10**, 1751.
79 T.J. Donohue, K. Blades and M. Helliwell, *Chem. Commun.*, 1996, 1733.
80 A. Varrot, M. Schülein, M. Pipelier, A. Vasella and G.J. Davies, *J. Am. Chem. Soc.*, 1999, **121**, 2621.
81 R. Anulewicz, I. Wawer, B. Piekarska-Bartoszewicz and A. Temeriusz, *J. Carbohydr. Chem.*, 1999, **18**, 617.
82 J.M.J. Tronchet, S. Zerelli and G. Bernardinelli, *J. Carbohydr. Chem.*, 1999, **18**, 343.
83 P. Marchand, S. Masson, D. Rachinel, J.-F. Saint-Clair and M.-T. Averbuch-Pouchot, *Acta Crystallogr.*, 1999, **C55**, 1533.
84 M.J. Diánez, M.D. Estrada, A. López-Castro and S. Pérez-Garrido, *Acta Crystallogr.*, 1999, **C55**, 1020.
85 P. Norris and T.R. Wagner, *Carbohydr. Res.*, 1999, **322**, 147.
86 M.L. Zhang, Y.X. Cui, L.T. Ma, L.H. Zhang, Y. Lu, B. Zhoa and Q.T. Zheng, *Chin. Chem. Lett.*, 1999, **10**, 117 (*Chem. Abstr.*, 1999, **131**, 157 873).
87 M. Avalos, R. Babiano, P. Cintas, F.J. Higes, J.L. Jiménez, J.C. Palacios, G. Silvero and C. Valencia, *Tetrahedron*, 1999, **55**, 4401.
88 J. Kopf, A. Lützen and P. Köll, *Acta Crystallogr.*, 1999, **C55**, 1541.
89 R.O. Gould, K.E. McGhie and R.M. Paton, *Carbohydr. Res.*, 1999, **322**, 1.
90 S. Signorella, C. Palopoli, A. Frutos, G. Escandar, T. Tamase and L.F. Sala, *Can. J. Chem.*, 1999, **77**, 1492.
91 H. Dollt and V. Zabel, *Aust. J. Chem.*, 1999, **52**, 259.

92 J.N. Low, G. Ferguson, J. Cobo, M. Nogueras and A. Sánchez, *Acta Crystallogr.*, 1999, **C55**, iii (IUC 9900163).
93 J. Kovacs, I. Pinter, M. Kajtar-Peredy, G. Argay, A. Kalman, G. Descotes and J.-P. Praly, *Carbohydr. Res.*, 1999, **316**, 112.
94 R.Bau and I. Tsyba, *Tetrahedron*, 1999, **55**, 14839.
95 D.A. Berges, J. Fan, S. Devinek, N. Liu and N.K. Dalley, *Tetrahedron*, 1999, **55**, 6759.
96 B. Werschkun and J. Theim, *Synthesis*, 1999, 121.
97 M. Helliwell, I. M. Phillips, R.G. Pritchard and R.J. Stoodley, *Tetrahedron Lett.*, 1999, **40**, 8651.
98 C.E.F. Rickard, Y. Singh, P.D. Woodgate and Z. Zhao, *Acta Crystallogr.*, 1999, **C55**, 1475.
99 B. Weyershausen, M. Nieger and K.H. Dötz, *J. Org. Chem.*, 1999, **64**, 4206.
100 K.H. Dötz, M. M. Klumpe and M. Nieger, *Chem. Eur. J.*, 1999, **5**, 691.
101 W.-C. Haase, M. Nieger and K.H. Dötz, *Chem. Eur. J.*, 1999, **5**, 2014.
102 S. Vano, Y. Shinohara, K. Mogami, M. Yokogana, T. Tanase, T. Sakakibara, F. Nishida, K. Mochida, I. Kinoshita, M. Doe. K. Ichihara, Y. Naruta, P. Mehrkhodavandi, P. Buglyó, B. Song, C. Orvig and Y. Mikata, *Chem. Lett.*, 1999, 255.
103 P.Liu, T.-B. Wen and J.-K. Cheng, *Acta Crystallogr.*, 1999, **C55**, 1149.
104 A. Falshaw, G.J. Gainsford and C. Lensink, *Acta Crystallogr.*, 1999, **C55**, 958.
105 G.J. Reiss, K. Hegetschweiler and J. Sander, *Acta Crystallogr.*, 1999, **C55**, 123.
106 A.S. Bera, B.P. Mukhopadhyay, K.A. Pal, U. Haldar, S. Bhattacharya and A. Banerjee, *J. Chem. Crystallogr.* 1998, **28**, 509 (*Chem. Abstr.*, 1999, **130**, 168 597).
107 D. Prahadeeswaran and T.P. Seshadri, *Acta Crystallogr.*, 1999, **C55**, 606.
108 D. Prahadeeswaran and T.P. Seshadri, *Acta Crystallogr.*, 1999, **C55**, 389.
109 G.F. Audette, W.M. Zoghaib, S.V. Gupta, J.W. Quail and L.T.J. Delbaere, *Acta Crystallogr.*, 1999, **C55**, 427.
110 S. Kolappan and T.P. Seshadri, *Acta Crystallogr.*, 1999, **C55**, 603.
111 S. Kolappan and T.P. Seshadri, *Acta Crystallogr.*, 1999, **C55**, 604.
112 S. Kolappan and T.P. Seshadri, *Acta Crystallogr.*, 1999, **C55**, 986.
113 A. Hempel, N. Camerman, J. Grierson, D. Mastropaolo and A. Camerman, *Acta Crystallogr.*, 1999, **C55**, 632.
114 S.P. Vincent, C Mioskowski and L. Lebeau, *Nucleosides Nucleotides*, 1999, **18**, 2127.
115 G.C. Chi, Z.B. Zhang, X.H. Xu and R.Y. Chen, *Chin. Chem. Lett.*, 1998, **9**, 989 (*Chem. Abstr.*, 1999, **131**, 299 640).
116 M.D. Bratek-Wiewiórowska, M. Wiewiórowski, M. Alejska, A. Olszewska and K. Wozniak, *Nucleosides Nucleotides*, 1999, **18**, 1825.
117 J. Zachová, I. Císarová, M. Budesinsky, R. Liboska, Z, Tocík and I. Rosenberg, *Nucleosides Nucleotides*, 1999, **18**, 2581.
118 G. Bringmann, M. Ochse, K. Wolf, J. Kraus, K. Peters, E.-M. Peters, M. Herderich, L.A. Assi and F.S.K. Tayman, *Phytochemistry*, 1999, **51**, 271.
119 J.N. Low, J. Cobo, S. Molina, A. Sánchez, M. Norueras and G. Ferguson, *Acta Crystallogr.*, 1999, **C55**, iii (IUC 9900164/1-2).
120 F. Seela, C. Wei, H. Reuter and G. Kastner, *Acta Crystallogr.*, 1999, **C55**, 1335.
121 F. Seela, M. Zulauf, H. Reuter and G. Kastner, *Acta Crystallogr.*, 1999, **C55**, 1947.
122 F. Seela, M. Zulauf, H. Reuter and G. Kastner, *Acta Crystallogr.*, 1999, **C55**, 1560.
123 F. Seela, H. Debelak, H. Reuter, G. Kastner and I.A. Mikhailopulo, *Tetrahedron*, 1999, **55**, 1295.

124 F. Seela, E. Becher, H. Rosemeyer, H. Reuter, G. Kastner and I.A. Mikhailopoulo, *Helv. Chem. Acta*, 1999, **82**, 105.
125 G. Wang, J.-L. Girardet and E. Gunic, *Tetrahedron*, 1999, **55**, 7707.
126 G. Wang, E. Gunic, J.-L. Girardet and V. Stoisavljevic, *Bioorg. Med. Chem. Lett.*, 1999, **9**, 1147.
127 A. Mele, B. Vergan, F. Viani, S. Valdo Meille, A. Farina and P. Bravo, *Eur. J. Org. Chem.*, 1999, 187.
128 S. Obika, K.-I. Mono, Y. Hari and T. Imanishi, *Chem. Commun.*, 1999, 2423.
129 S. Blanalt-Feidt, S.O. Doronina and J.-P. Behr, *Tetrahedron Lett.*, 1999, **40**, 6229.
130 P. Wang, L.A. Agrofoglio, M.G. Newton and C.K. Chu, *J. Org. Chem.*, 1999, **64**, 4173.
131 P. Wang, B. Sullen, M. P. Newton, Y.-C. Cheng, R.F. Schinazi and C.K. Chu, *J. Med. Chem.*, 1999, **42**, 3390.
132 P. Wang, P.J. Bolon, M.G. Newton and C.K. Chu, *Nucleosides Nucleotides*, 1999, **18**, 2819.
133 H.C. Lee, P. Kumar, L.I. Wiebe, R. McDonald, J.R. Mercer, K. Ohkura and K. Seki, *Nucleosides Nucleotides*, 1999, **18**, 1995.
134 S. Raic-Malic, A. Hergold-Brudic, A. Nagl, M. Grdisa, K. Pavelic, E. De Clercq and M. Mintas, *J. Med. Chem.*, 1999, **42**, 2673.
135 M. Prhavc, J. Plavec, J. Kobe, I. Leban and G. Giesker, *Nucleosides Nucleotides*, 1999, **18**, 2601.
136 V.E. Marquez, F. Russ, R. Alonso, M.A. Siddiqui, S. Hernandez, C. George, M.C. Nicklaus, F. Dau and H. Ford, *Helv. Chim. Acta*, 1999, **82**, 2119.
137 A. Kittaka, T. Asakura, T. Kuze, H. Tanaka, N. Yamada, K.T. Nakamura and T. Miyasaka, *J. Org. Chem.*, 1999, **64**, 7081.
138 R. Burli and A. Vasella, *Helv.Chim. Acta*, 1999, **82**, 485.
139 M.-H.E. Spyridaki and P.A. Siskos, *Chim. Chron.*, 1997, **26**, 441 (*Chem. Abstr.*, 1999, **130**, 267 681).
140 A. Tayama, N. Hanada, J. Ono, E. Yoshimitsu and H. Takeuchi, *J. Raman Spectrosc.*, 1999, **30**, 623 (*Chem. Abstr.*, 1999, **131**, 351 600).
141 J. Frelek, M. Geiger and W. Voelter, *Curr. Org. Chem.*, 1999, **3**, 117 (*Chem. Abstr.*, 1999, **130**, 311 977).
142 J. Frelek, I, Panfil, Z. Urbancyk-Lipkowska and M. Chmielewski, *J. Org. Chem.*, 1999, **64**, 6126.
143 P.K. Bose and P.L. Polararapu, *J. Am. Chem. Soc.*, 1999, **121**, 6094.
144 A. Orzeszko, A. Niedzwiecka-Kornas, R. Stolarski and Z. Kazimierczuk, *Z. Naturforsch., B: Chem. Sci.*, 1998, **53**, 1191 (*Chem. Abstr.*, 1999, **130**, 38 626).
145 U.F. Castillo, Y. Sakagami, M. Alonso-Amelot and M. Ojika, *Tetrahedron*, 1999, **55**, 12295.
146 P.K. Bose and P.L. Polararapu, *Carbohydr. Res.*, 1999, **319**, 172.
147 P.K. Bose and P.L. Polararapu, *Carbohydr. Res.*, 1999, **322**, 135.
148 D.K. Smith, A. Zingg and F. Diederich, *Helv. Chim. Acta*, 1999, **82**, 1225.
149 I.V. Bohner, O.-S. Becker and A. Vasella, *Helv. Chim. Acta*, 1999, **82**, 198.
150 H. Nagase, H. Ueda and M. Nakagaki, *Chem. Pharm. Bull.*, 1999, **47**, 607.
151 G.V. Abagyan, A.G. Abagyan and A.S. Apresyan, *Izv.-Natis. Akad. Nauk. Arm., Fiz.*, 1998, **33**, 41 (*Chem. Abstr.*, 1999, **131**, 116 400).
152 Y. Yamauchi, M. Sugito and M. Kuzuya, *Chem. Pharm. Bull.*, 1999, **47**, 273.
153 J.M.J. Tronchet, M. Koufaki, F. Barbalatrey and M. Geoffroy, *Carbohydr. Lett.*, 1999, **3**, 255 (*Chem. Abstr.*, 1999, **130**, 296 922).

23

Separatory and Analytical Methods

1 Chromatographic Methods

The advantages and disadvantages of gas chromatography and HPLC as methods for the simultaneous determination of sugars and acids in various biological sources such as fruits and vegetables have been the subject of a review article.[1]

A comprehensive review on 3-deoxyglucosone (**1**) includes discussion of its analysis by GC–MS and HPLC,[2] and methods for the determination of trehalose, including paper chromatography, TLC and HPLC have been reviewed.[3]

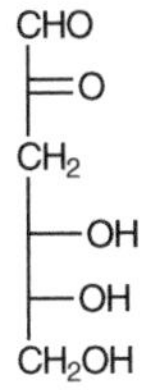

CHO
=O
CH_2
–OH
–OH
CH_2OH

1

Graphitized carbon solid-phase extraction has been used for the preparative-scale purification of sugars. Graded elution with water containing organic solvents permits the separation of groups of oligosaccharides. Acidic sugars are retained by graphitized carbon, whilst neutral and amino-sugars are eluted by water containing an organic modifier, the acidic materials then being eluted by the same eluant containing TFA.[4]

1.1 Gas–Liquid Chromatography. – A review article of general relevance to GLC discusses artefacts that can be obtained during the conversion of compounds into their Tms derivatives, and methods for the prevention of the formation of such artefacts.[5] A review has appeared covering work over the period 1995–1998 on the application of cyclodextrin derivatives as chiral selectors for the direct GLC separation of enantiomers of volatile optically-active components in the essential oil, flavour and aroma areas.[6]

Galactose, a marker of heat treatment, has been analysed in milk as its pentafluorobenzyloxime peracetate by GLC with flame ionization detection, in

Carbohydrate Chemistry, Volume 33

a procedure that did not involve any prederivatization clean-up for elimination of lactose from the samples.[7] A new GC–MS screening method has been developed for the determination of urinary galactose and 4-hydroxyphenyllactic acid in new-born infants.[8] A capillary GC–MS analysis has been developed for urinary sugar and sugar alcohols during pregnancy; the study suggested that changes in the levels of glucose, glucitol, fructose, *myo*-inositol and 1,5-anhydro-D-glucitol might reflect a mild alteration in carbohydrate metabolism not detected by other diabetic indicators.[9]

A method has been developed for the quantitative determination of the monosaccharide composition of glycoproteins and glycolipids, which employs GLC of the per-heptafluorobutyrate derivatives of the methyl glycosides on capillary columns.[10] The major monosaccharide components of the polysaccharides in the root mucilage of maize have been identified as fucose, arabinose, galactose and glucose by analysis of the sugars as their peracetyl derivatives. The gas chromatograms were considerably simpler when direct acetolysis of the polysaccharide was employed, rather than two-step hydrolysis and acetylation, since predominantly one anomer was formed for each monomer present.[11] Bacterial cellular polysaccharides have been analysed by hydrolysis and GC–MS analysis of the alditol acetates of the monomers, with substantial improvements made to sample preparation. Specific species or genera of bacteria can be identified by the presence of particular sugar markers.[12] The contents of D-fructose, D-glucose and sucrose caramels have been analysed by GLC–MS of the Tms or Tms-oxime derivatives of the constituents, namely monosaccharides (D-fructose, D-glucose, anhydrosugars), disaccharide (glucobioses), and pseudodisaccharides (di-D-fructose dianhydrides). Although significant differences emerged in the disaccharide-pseudodisaccharide distribution depending on the caramel source, di-D-fructose dianhydrides were found in all three types of caramel, and these might be useful as specific tracers of the authenticity of the caramels.[13]

Further developments have been reported in the simultaneous quantitation of mono-, di- and tri-saccharides, sugar alcohols and acids in fruits, the sugars being measured as their Tms-oxime ether/esters, prepared in the presence of the fruit matrix.[14]

The elevated levels of 3-deoxyglucosone (**1**) found in the erythrocytes of hemodialysis patients have been measured by GC–CIMS using a selected ion-monitoring method,[15] and a similar technique was employed for the analysis of 1,5-anhydro-D-fructose (**2**), microthecin [2-hydroxy-2-hydroxymethyl-2*H*-pyran-3(6*H*)-one] and 4-deoxy-D-*glycero*-hexo-2,3-diulose (**3**), enzymic degradation products of starch formed in a species of red alga.[16] A GC–MS isotope

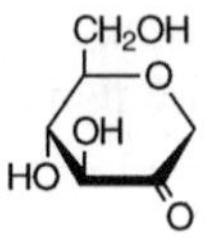

CH2OH O OH HO O HOCH2 O CH2OH OH O

2 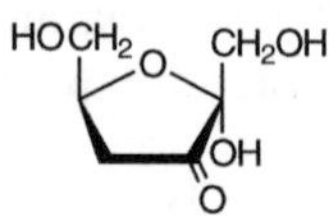 **3**

dilution method has been developed for the rapid analysis of ascorbate in 10 μL samples of plasma.[17]

1.2 High-pressure Liquid Chromatography. – A highly sensitive amperometric detector for the LC analysis of monosaccharides, including aminosugars, has been developed based on the optimization of the nickel content of a Ni–Ti alloy electrode.[18] A new detection methodology, bimodal integrated amperometric detection, has made it possible to detect aminoacids, aminosugars and other carbohydrates selectively following their separation by anion-exchange.[19]

Various ^{14}C- and ^{3}H-labelled isotopomers of D-fructose have been prepared on a microscale by isomerization of the corresponding glucoses in alkali, HPLC using a column of a calcium-loaded sulfonated polymer then being used to purify the fructose.[20]

Isocratic separations of closely-related mono- and di-saccharides have been carried out by high-performance anion exchange chromatography (HPAEC) with pulsed amperometric detection (PAD) using dilute alkaline eluents spiked with barium acetate.[21] HPLC separations of mono, di- and oligo-saccharides have also been carried out on stationary phases prepared from polystyrene-based resin and tertiary amines, with electrochemical detection involving a Ni–Ti working electrode in alkaline eluents.[22] A rapid and sensitive analysis of the disaccharide composition in heparin and heparan sulfate has been carried out, after digestion of the polymers with heparin lyase I, II and III in combination, by reversed-phase ion-pair chromatography on a 2 μm porous silica column.[23] An HPLC-based method for analysis of oligosaccharides in biological fluids uses derivatization with 1-phenyl-3-methyl-5-pyrazolone, and permits evaluation of sialic acid-containing compounds.[24] Reversed and normal-phase HPLC has been used to separate a number of human milk oligosaccharides derivatized by reductive amination with 2-aminoacridone.[25] *Schizosaccharomyces pombe* produces novel $Gal_{0\text{-}2}Man_{1\text{-}3}$ O-linked oligosaccharides, as judged, following release, by a combination of HPAEC–PAD, Bio-gel P4 chromatography and NMR spectroscopy.[26]

A new method has been described that can be performed within a single vessel for the analysis of aldose, hexosamine and sialic acid residues of oligosaccharides and glycoconjugates. Treatment with sialidase or mild acid is followed by the use of NeuNAc aldolase to convert free sialic acid residues into *N*-acylmannosamines. The reaction mixture is then successively subjected to acid hydrolysis, *N*-acetylation and reaction with ethyl *p*-aminobenzoate to give a mixture of monosaccharide derivatives then analysed by reversed-phase HPLC. A slight modification of the method permitted the separate determination of *N*-acetyl- and *N*-glycolyl-neuraminic acids.[27] A review has discussed the use of high pH anion-exchange chromatography to detect differences between the glycosylation patterns of *N*-linked glycoproteins.[28]

The metabolism of alkyl polyglucosides and their determination in waste water has been investigated by HPLC coupled with electrospray MS.[29]

In the area of sialic acid analysis, a new direct reversed-phase HPLC method has been developed for the determination of a series of five sialic acids,[30] and the method has been used to determine NeuNAc, released by acid hydrolysis, and its 2-deoxy-2,3-dehydro-derivative, in serum, urine and saliva.[31] A study has been reported on the quantification of NeuNAc and *N*-glycolylneuraminic acid in serum glycoconjugates before and after surgical treatment of early endometrial cancer, and the relationship of their levels with the progress of surgical therapy. The results suggested that the NeuNAc level in serum could be used as a tumour marker in evaluating the suitability of surgical treatment in this type of cancer.[32] HPAEC with PAD and fluorimetric detection has been evaluated for the analysis of oligo- and poly-sialic acids,[33] and LC-tandem mass spectrometry provides a method for the determination of the sialidase inhibitor zanamivir in human serum.[34]

The method of HPAEC–PAD using alkaline eluents spiked with Ba^{2+}, Ca^{2+} or Sr^{2+} salts has also been applied to the separation of sugar acids including D-gluconic and muramic acids, NeuNAc, and D-galacturonic, D-mannuronic and D-glucuronic acids.[35] An improved method for the simultaneous determination of ascorbic, dehydroascorbic, isoascorbic and dehydroisoascorbic acids in foodstuffs and in biological samples has been described, which employs HPLC separation followed by direct determination of ascorbic acid and indirect fluorimetric determination of dehydroascorbic acid after a post-column derivatization with *o*-phenylenediamine.[36]

LC–Electrospray MS has been used for the determination of ethyl glucuronide, a minor metabolite of ethanol, in human serum.[37] A number of papers have discussed the quantification of morphine and its 3- and 6-β-D-glucuronides using HPLC coupled to electrospray mass spectrometry.[38–40] An immunoaffinity-based extraction procedure has also been devised for morphine and its glucuronides in human blood; specific antisera to morphine, morphine 3-glucuronide and morphine 6-glucuronide were coupled to carbonyldiimidazole-activated tris-acrylgel and used for extraction of morphine and its glucuronides from coronary blood. The resultant extracts were analysed by HPLC with native fluorescence detection.[41] HPLC has been employed for the estimation of the anaesthetic propofol (2,6-di-isopropylphenol), its metabolite 2,6-di-isopropylhydroquinone, and their glucuronides.[42] The metabolism of the opioid-like analgesic tramadol [more active enantiomer, 1*R*,2*R*-2-dimethylaminomethyl-1-(3-methoxyphenyl)-cyclohexan-1-ol] involves *O*-demethylation and conversion into the β-D-glucuronide. Reversed-phase HPLC has been used to separate the diastereomers of the β-D-glucuronides of racemic *O*-demethyltramadol, *N,O*-didemethyltramadol and *N,N,O*-tridemethyltramadol.[43]

A series of 26 sugar and glycerol phosphates have been separated using HPAEC on an alkyl quaternary ammonium column. The capacity, asymmetry and response factors of the compounds were found to vary widely, and secondary acidic dissociation constants of a number of these phosphates were determined in an attempt to explain the separation mechanism.[44]

In the nucleoside and nucleotide areas, reversed-phase HPLC has been used to study levels of rare ribonucleosides produced by post-translational modification of RNA, since abnormal levels of these nucleosides have been suggested as possible tumour markers.[45] 6-Mercaptopurine and its metabolites including the riboside and 6-methylmercaptopurine riboside have been assayed in human plasma by HPLC with diode-array detection,[46] and solid-phase extraction followed by reversed-phase ion-pair HPLC has been used for the simultaneous determination of adenosine, *S*-adenosylhomocysteine and *S*-adenosylmethionine in renal tissue, and for determination of adenosine and *S*-adenosylhomocysteine in urine.[47] The simultaneous determination of deoxyribonucleoside triphosphates has now been accomplished in the presence of much larger amounts of ribonucleoside triphosphates, which may give a method for detecting small changes in dNTP/NTP pools induced by anticancer or antiviral agents, or by diseases.[48] HPAEC has been used for purification of tritiated UDP-galacturonic acid made enzymically from tritiated UTP and glucose-1-phosphate, the last step involving UDP-glucuronic acid 4-epimerase.[49]

Doxorubicin has been determined by an HPLC method that requires few manipulations and is amenable to a large number of samples,[50] and both doxorubicin and doxorubicinol have been determined in the plasma of cancer patients by fluorescence detection,[51] as have idarubicin and idarubicinol in rat plasma.[52] Two reports concerned with the analysis of clindamycin by HPLC using coupled columns[53] and by HPLC–electrospray tandem mass spectrometry have appeared.[54] Streptomycin residues in food can be determined by solid-phase extraction and LC with post-column derivatization and fluorimetric detection,[55] and workers at Schering-Plough have developed an HPLC assay for the everninomycin-type antibiotic SCH 27899.[56]

2 Electrophoresis

2.1 Capillary Electrophoresis. – The association constant between a protein and a carbohydrate can be determined based on the relationship between the delay of migration time of a protein as sample and the concentration of a carbohydrate ligand as additive in capillary zone electrophoresis. However, the sugar ligand must have an electric charge. A method has been described for conversion of neutral carbohydrates into derivatives with a strong negative charge, using reaction with 2-mercaptoethanesulfonate in TFA to give derivatives of type **4**. The process does not cause cleavage of glycosidic links or loss of sialic acid units, and was applied to the determination of the association constants of various carbohydrates and lectins.[57] The same group has also described the determination of association constants between lectins and

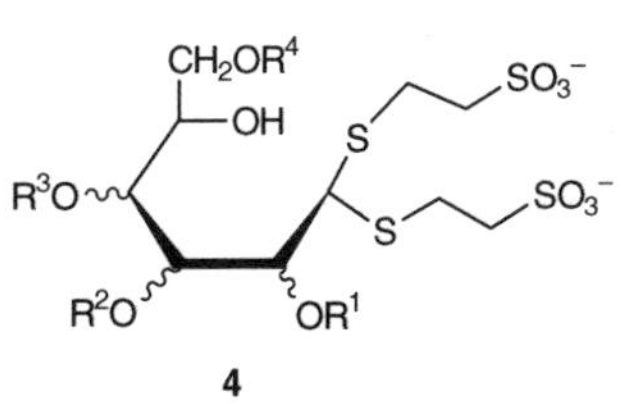

CH_2OR^4 OH SO_3^- S R^3O S SO_3^- R^2O OR^1

4

oligosaccharides, as their 8-amino-1,3,6-naphthalenesulfonate or 1-phenyl-3-methyl-5-pyrazolone derivatives, in the reverse system (oligosaccharide derivative as sample and lectin as additive) by electrophoresis in a capillary coated with linear polyacrylamide.[58] Cyclofructan (cyclic D-fructohexaose, cycloinulohexaose) has been investigated as a new selective complex-forming agent for metal cations in capillary electrophoresis.[59]

Capillary zone electrophoresis has been used for the simultaneous analysis of inorganic anions, organic acids, aminoacids and carbohydrates, of which 32 were included in the study,[60] and for the analysis of the acid hydrolysis products of the dissolved carbohydrates in the effluent from a chlorine-free bleaching plant.[61]

The determination of total glucosinolates in cabbage and rapeseed has been carried out by conversion of the glucosinolates into gluconic acid by the action of myrosinase and glucose oxidase, followed by capillary electrophoresis with laser-induced fluorescence (LIF) detection, after precolumn derivatization with 7-aminonaphthalene-1,3-disulfonic acid.[62] The efficiencies of 3-aminobenzamide and 3-aminobenzoic acid in derivatization of carbohydrates by reductive amination, prior to analysis by capillary electrophoresis with LIF detection, has been investigated with maltose as a model. The 3-substituted aminobenzenes showed good reactivity, although the fluorescence intensities and molar absorbtivities were not as high as for the 2- and 4-substituted anilines.[63] Capillary electrophoresis–LIF detection has also been applied to the analysis, after derivatization with 2-aminoacridone, of sulfated disaccharides produced from enzymic digestion of sulfated chondroitin or dermatan sulfate.[64] An oligosaccharide mixture from partial hydrolysis of dextran has been analysed, after 8-aminonaphthalene-1,3,6-trisulfonate derivatization, using capillary electrophoresis and on-line electrospray ionization quadrupole ion-trap mass spectrometry. As little as 1.6 pmol of dextran octaose could be detected.[65] The manno-oligosaccharide caps of mycobacterial lipoarabinomannan have been examined by capillary electrophoresis–electrospray MS,[66] and the analysis and characterization of cyclodextrins and their inclusion complexes by affinity capillary electrophoresis has been the subject of a review.[67]

Direct measurement of ascorbic acid production in cell suspension cultures of *Arabidopsis thaliana* has been carried out by capillary electrophoresis,[68] and the separation of the hypolipidaemic agent ciprofibrate and its glucuronide has been accomplished. Chiral discrimination between the ciprofibrate enantiomers and the diastereomeric glucuronides was achieved by the use of γ-cyclodextrin as buffer additive.[69]

Two reports have described electrophoretic methods for the assay of the antiviral nucleoside 5-(2-bromovinyl)-2′-deoxyuridine (BVDU) in plasma and urine,[70,71] and the interaction of cisplatin with nucleoside monophosphates, and with di- and tri-deoxynucleotides, has been investigated by capillary electrophoresis, all four common nucleotides and their major platinum adducts being separable in a single run.[72] LIF detection has been applied to capillary electrophoretic analysis of daunorubicin, and the method was found to be more sensitive than HPLC–LIF, the analysis being applied to Kaposy sarcoma tumours.[73]

2.2 Other Electrophoretic Methods. – Micellar electrokinetic capillary chromatography has been developed for the separation of the 3-*O*-glucuronides of entacapone and its *Z*-isomer, the two main urinary metabolites of entacapone, a potent inhibitor of catechol *O*-methyl transferase,[74] and used for the analysis of glucuronide concentrations in urine samples.[75]

Capillary electrochromatography has been investigated as a technique for the analysis of nucleosides, and a system was devised which permitted a baseline separation of adenosine, guanosine, inosine, cytidine, thymidine and uridine in less than 13 minutes.[76] Condensation nucleation light scattering detection has been coupled with a pressurized capillary electrochromatography system to give good reproducibility for a wide range of compounds including carbohydrates, with limits of detection down to the 50 ng ml^{-1} level without the need for derivatization.[77]

3 Other Analytical Methods

Nanofiltration was used for desalination and concentration of CMP-NeuNAc and GDP-mannose, and with appropriate membranes both nucleotide phosphosugars were purified on a gram scale to >95% purity.[78]

6,8-Difluoro-4-methylumbelliferyl β-D-galactopyranoside (**5**) has been used as a fluorogenic substrate for continuous assay of β-galactosidases: the hydrolysis product has a pK_a value lower than does 4-methylumbelliferone and so can be used at pH 7.[79] A fluorimetric assay for cyclic GMP has been developed. Guanosine nucleotides except for cGMP were enzymically phosphorylated to GTP, from which cGMP could be separated on a Sep-Pak aminopropyl cartridge. The purified cGMP was then converted enzymically into GTP which was estimated in a linked enzymic system which produced the fluorescent compound 2-hydroxynicotine aldehyde.[80] The novel symmetrical squarane derivative **6**, with two boronic acid groups, detects carbohydrates in aqueous solutions with a fluorescence intensity increase. The emission maximum is at 645 nm with a γ-band shoulder at 695 nm, making this the first example of a near-IR-emitting carbohydrate sensor.[81]

The porphyrin **7** with four binaphthyl substituents in the meso-positions has been prepared as a single atropisomer; this compound had a high binding

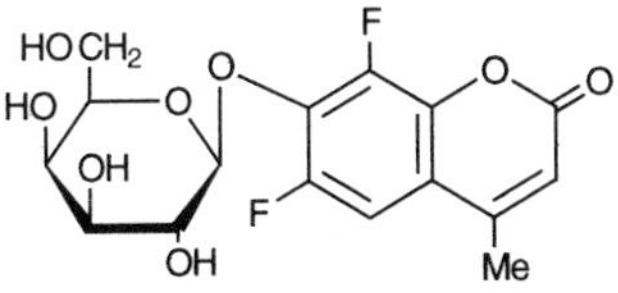

5 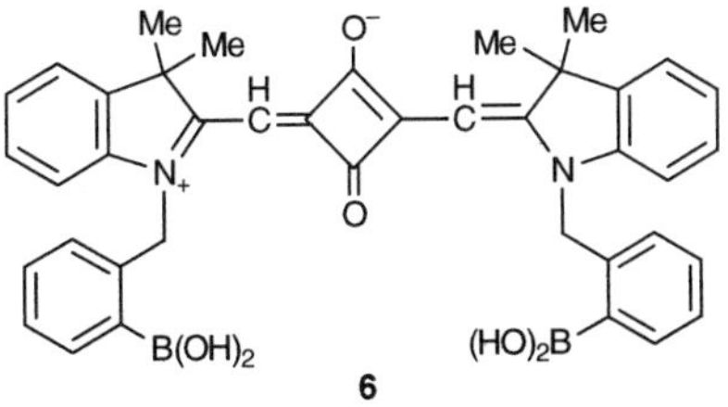 **6**

affinity for di-, and, in particular, tri-saccharides as compared with mono-saccharides.[82]

7

References

1 I. Molnár-Perl, *J. Chromatogr. A,* 1999, **845**, 181.
2 T. Niwa, *J. Chromatogr. B,* 1999, **731**, 23.
3 C. Liu, Z. Yun, J. Lu and W. Mai, *Shipin Yu Fajiao Gongye,* 1998, **24**, 40 (*Chem. Abstr.*, 1999, **130**, 52635).
4 J.W. Redmond and N.H. Packer, *Carbohydr. Res.*, 1999, **319**, 74.
5 J.L. Little, *J. Chromatogr. A,* 1999, **844**, 1.
6 C. Bicchi, A. D'Amato and P. Rubiolo, *J. Chromatogr. A,* 1999, **843**, 99.
7 L.M. Chiesa, L. Radice, R. Belloli, P. Renon and P.A. Biondi, *J. Chromatogr. A,* 1999, **847**, 47.
8 T. Shinka, Y. Inoue, H. Peng, Z.W. Xia, M. Ose and T. Kuhara, *J. Chromatogr. B,* 1999, **732**, 469.
9 M. Tetsuo, C.H. Zhang, H. Matsumoto and I. Matsumoto, *J. Chromatogr. B,* 1999, **731**, 111.
10 J.-P. Zanetta, P. Timmerman and Y. Leroy, *Glycobiology*, 1999, **9**, 255.
11 H.M.I. Osborn, F. Lochey, L. Mosley and D. Read, *J. Chromatogr. A,* 1999, **831**, 267.
12 A. Fox, *J. Chromatogr. A,* 1999, **843**, 287.
13 V. Ratsimba, J.M.G. Fernandez, J. Defaye, H. Nigay and A. Voilley, *J. Chromatogr. A,* 1999, **844**, 283.
14 Z.F. Katona, P. Sass and I. Molnár-Perl, *J. Chromatogr. A,* 1999, **847**, 91.
15 S. Tsukushi, K. Shimokata and T. Niwa, *J. Chromatogr. B,* 1999, **731**, 37.
16 A. Broberg, L. Kenne and M. Pedersén, *Anal. Biochem.*, 1999, **268**, 35.
17 J.C. Deutsch, J.A. Butler, A.M. Marsh, C.A. Ross and J.M. Norris, *J. Chromatogr. B,* 1999, **726**, 79.
18 M. Morita, O. Niwa, S. Tou and N. Watanabe, *J. Chromatogr. A,* 1999, **837**, 17.
19 P. Jandik, A.P. Clarke, N. Avdalovic, D.C. Andersen and J. Cacia, *J. Chromatogr. B,* 1999, **732**, 193.

20 O. Scruel, A. Sener and W.J. Malaisse, *J. Chromatogr. A,* 1999, **847**, 53.
21 T.R.I. Cataldi, C. Campa, M. Angelotti and S.A. Bufo, *J. Chromatogr. A,* 1999, **855**, 539.
22 T. Masuda, Y. Nishimura, M. Tonegawa, K. Kitahara, S. Arai, J. Yamashita and N. Takai, *J. Chromatogr. A,* 1999, **845**, 401.
23 H. Toyoda, H. Yamamoto, N. Ogino, T. Toida and T. Imanari, *J. Chromatogr. A,* 1999, **830**, 197.
24 D. Fu and D. Zopf, *Anal. Biochem.*, 1999, **269**, 113.
25 J. Charlwood, D. Tolson, M. Dwek and P. Camilleri, *Anal. Biochem.*, 1999, **273**, 261.
26 T.R. Gemmill and R.B. Trimble, *Glycobiology*, 1999, **9**, 507.
27 S. Yasuno, K. Kokubo and M. Kamei, *Biosci. Biotechnol. Biochem.*, 1999, **63**, 1353.
28 K.D. Smith, E.F. Hounsell, J.M. McGuire, M.A. Elliot and H.G. Elliot, *Adv. Macromol. Carbohydr. Res.*, 1997, **1**, 65 (*Chem. Abstr.*, 1999, **130**, 38573).
29 P. Eichhorn and T.P. Knepper, *J. Chromatogr. A,* 1999, **854**, 221.
30 M.H.E. Spiridaki and P.A. Siskos, *J. Chromatogr. A,* 1999, **831**, 179.
31 P.A. Siskos and M.H.E. Spiridaki, *J. Chromatogr. B,* 1999, **724**, 205.
32 S. Diamantopoulou, K.D. Stagiannis, K. Vasilopoulos, P. Barlas, T. Tsegenidis and N.K. Karamanos, *J. Chromatogr. B,* 1999, **732**, 375.
33 S.-L. Lin, Y. Inoue and S. Inoue, *Glycobiology*, 1999, **9**, 807.
34 G.D. Allen, S.T. Brookes, A. Barrow, J.A. Dunn and C.M. Grosse, *J. Chromatogr. B,* 1999, **732**, 383.
35 T.R.I. Cataldi, C. Campa and I.G. Casella, *J. Chromatogr. A,* 1999, **848**, 71.
36 M.A. Kall and C. Andersen, *J. Chromatogr. B,* 1999, **730**, 101.
37 M. Nishikawa, H. Tsuchihashi, A. Miki, M. Katagi, C. Schmitt, H. Zimmer, T. Keller and R. Aderjan, *J. Chromatogr. B,* 1999, **726**, 105.
38 A. Dienes-Nagy, L. Rivier, C. Giroud, M. Augsburger and P. Mangin, *J. Chromatogr. A,* 1999, **854**, 109.
39 G. Schanzle, S.X. Li, G. Mikus and U. Hofmann, *J. Chromatogr. B,* 1999, **721**, 55.
40 M. Blanchet, G. Bru, M. Guerret, M. Bromet-Petit and N. Bromet, *J. Chromatogr. A,* 1999, **854**, 93.
41 J. Beike, H. Kohler, B. Brinkmann and G. Blaschke, *J. Chromatogr. B,* 1999, **726**, 111.
42 T.B. Vree, A.J. Lagerwerf, C.P. Bleeker and P.M.R.M. de Grood, *J. Chromatogr. B,* 1999, **721**, 217.
43 P. Overbeck and G. Blaschke, *J. Chromatogr. B,* 1999, **732**, 185.
44 I.C. Schneider, P.J. Rhamy, R.J. Fink-Winter and P.J. Reilly, *Carbohydr. Res.*, 1999, **322**, 128.
45 G. Xu, C. DiStefano, H.M. Liebich, Y. Zhang and P. Lu, *J. Chromatogr. B,* 1999, **732**, 307.
46 Y. Su, Y.Y. Hen, Y.Q. Chu, M.E.C. VandePoll and M.V. Relling, *J. Chromatogr. B,* 1999, **732**, 459.
47 G. Luippold, U. Delabar, D. Kloor and B. Muhlbauer, *J. Chromatogr. B,* 1999, **724**, 231.
48 L.A. Decosterd, E. Cottin, X. Chen, F. Lejeune, R.O. Mirimanoff, J. Biollaz and P.A. Coucke, *Anal. Biochem.*, 1999, **270**, 59.
49 A. Orellana and D. Mohnen, *Anal. Biochem.*, 1999, **272**, 224.
50 L. Alvarez-Cedron, M.L. Sayalero and J.M. Lanao, *J. Chromatogr. B,* 1999, **721**, 271.

51 P. de Bruin, J. Verveij, W.J. Loos, H.J. Kolker, A.S.T. Planting, K. Nooter, G. Stoter and A. Sparreboom, *Anal. Biochem.*, 1999, **266**, 216.
52 O. Kuhlmann, S. Hofmann and M. Weiss, *J. Chromatogr. B,* 1999, **728**, 279.
53 H. Fieger-Buschges, G. Schussler, V. Larsimont and H. Blume, *J. Chromatogr. B,* 1999, **724**, 281.
54 L.L. Yu, C.K. Chao, W.J. Liao, T.Y. Twu, C.M. Liu, T.H. Yang and E.T. Lin, *J. Chromatogr. B,* 1999, **724**, 287.
55 P. Eder, A. Cominoli and C. Corvi, *J. Chromatogr. A,* 1999, **830**, 345.
56 C.C. Lin, C. Korduba and D. Parker, *J. Chromatogr. B,* 1999, **730**, 55.
57 A. Taga, M. Mochizuki, H. Itoh, S. Suzuki and S. Honda, *J. Chromatogr. A,* 1999, **839**, 157.
58 A. Taga, K. Uegaki, Y. Yabusako, A. Kitano and S. Honda, *J. Chromatogr. A,* 1999, **837**, 221.
59 J.C. Reijenga, T.P.E.M. Verheggen and M. Chiari, *J. Chromatogr. A,* 1999, **838**, 111.
60 T. Soga and G.A. Ross, *J. Chromatogr. A,* 1999, **837**, 231.
61 M. Ristolainen, *J. Chromatogr. A,* 1999, **832**, 203.
62 A. Karcher, H.A. Melouk and Z. El Rassi, *Anal. Biochem.*, 1999, **267**, 92.
63 K. Kakehi, T. Funakubo, S. Suzuki, Y. Oda and Y. Kitada, *J. Chromatogr. A,* 1999, **863**, 205.
64 F. Lamari, A. Theocharis, A. Hjerpe and N.K. Karamanos, *J. Chromatogr. B,* 1999, **730**, 129.
65 F.Y. Che, J.F. Song, R. Zeng, K.Y. Wang and Q.C. Xia, *J. Chromatogr. A,* 1999, **858**, 229.
66 B. Monsarrat, T. Brando, P. Londouret, J. Nigou and G. Puzo, *Glycobiology*, 1999, **9**, 335.
67 K.L. Larsen and W. Zimmrmann, *J. Chromatogr. A,* 1999, **836**, 3.
68 M.W. Davey, G. Persieu, G. Bauw and M. Van Montagu, *J. Chromatogr. A,* 1999, **853**, 381.
69 H. Huttemann and G. Blaschke, *J. Chromatogr. B,* 1999, **729**, 33.
70 J. Olgemoller, G. Hempel, J. Boos and G. Blaschke, *J. Chromatogr. B,* 1999, **726**, 261.
71 H.J.E.M. Reeuwijk, U.R. Tjaden and J. van der Greef, *J. Chromatogr. B,* 1999, **726**, 269.
72 A. Zenker, M. Galanski, T.L. Bereuter, B.K. Keppler and W. Lindner, *J. Chromatogr. A,* 1999, **852**, 337.
73 N. Simeon, E. Chatelut, P. Canal, M. Nertz and F. Couderc, *J. Chromatogr. A,* 1999, **853**, 449.
74 P. Lehtonen, L. Malkki-Laine and T. Wikberg, *J. Chromatogr. B,* 1999, **721**, 127.
75 P. Lehtonen, S. Lehtinen, L. Malkki-Laine and T. Wikberg, *J. Chromatogr. A,* 1999, **836**, 173.
76 T. Helboe and S.H. Hansen, *J. Chromatogr. A,* 1999, **836**, 315.
77 W. Guo, J.A. Koropchak and C. Yan, *J. Chromatogr. A*, 1999, **849**, 587.
78 G. Dudziak, S. Frey, L. Hasbach and U. Kragl, *J. Carbohydr. Chem.*, 1999, **18**, 41.
79 K.R. Gee, W.-C. Sun, M.K. Bhalgat, R.H. Upson, D.H. Klaubert, K.A. Latham and R.P. Haugland, *Anal. Biochem.*, 1999, **273**, 41.
80 K. Seya, K.-I. Furukawa and S. Motomura, *Anal. Biochem.*, 1999, **272**, 243.
81 B. Kukrer and E.U. Akkaya, *Tetrahedron Lett.*, 1999, **40**, 9125.
82 O. Rusin and V. Král, *Chem. Commun.*, 1999, 2367.

24
Synthesis of Enantiomerically Pure Non-carbohydrate Compounds

1 Carbocyclic Compounds

A review on the conversion of furanosides and pyranosides into carbocycles has appeared [48 references] covering zirconium- and samarium-mediated contractions of unsaturated carbohydrate derivatives, Ferrier rearrangements and aluminium- and titanium-catalysed processes.[1] A report of a symposium on synthesis of natural products containing a cyclohexane ring from D-glucose has been published.[2]

1.1 Cyclobutane Derivatives. – Copper triflate-catalysed intramolecular cycloaddition reactions of **1**, **2**, **3** and **4** generates the fused cyclobutyl products **5**, **6**, **7** and **8** respectively.[3]

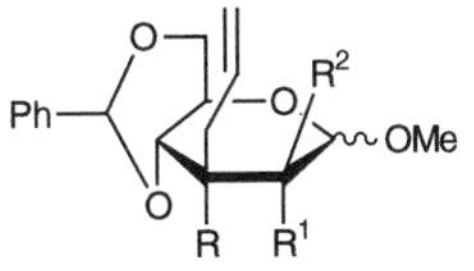

1 $R^1 = CH{=}CH_2$, $R^2 = OH$
2 $R^1 = OH$, $R^2 = CH{=}CH_2$

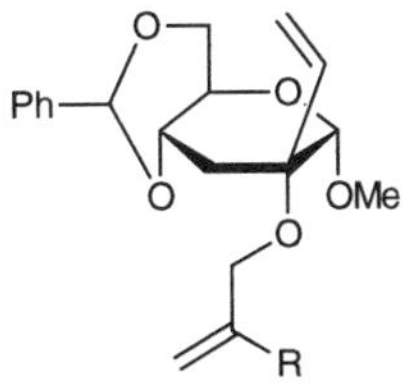

3 R = H, Me

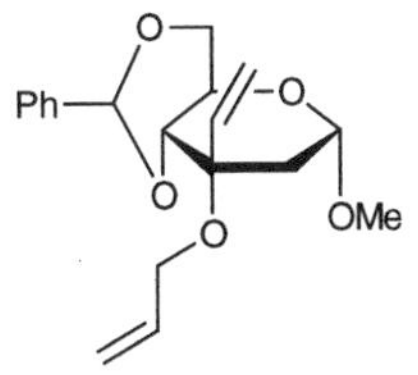

4

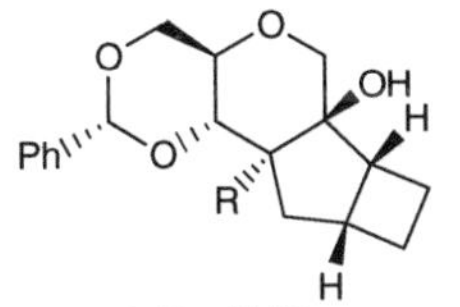

5 R = H, Me

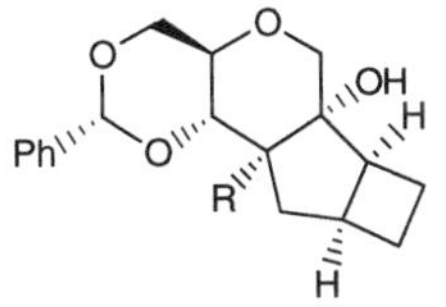

6 R = H, Me

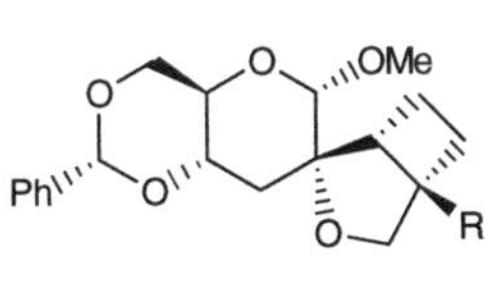

7 R = H, Me

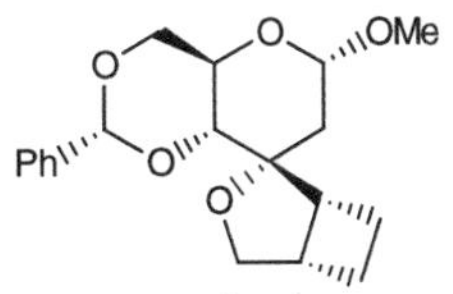

8

Carbohydrate Chemistry, Volume 33

1.2 Cyclopentane Derivatives. – The amino cyclopentane **13** has been prepared from D-ribose in a number of steps (Scheme 1) *via* the known ribose-derived **9**. The key stereocentres were established by a diastereoselective epoxidation of **9** *anti* to the isopropylidene acetal and a diastereoselective reduction of the exocyclic methylene of **12** (ketone **10** being elaborated through Wittig olefination, epoxide ring opening by ammonia and amine protection to provide **12**). The product **13** inhibits L-fucosidase selectively relative to α- or β-glucosidases, mannosidases or galactosidases.[4]

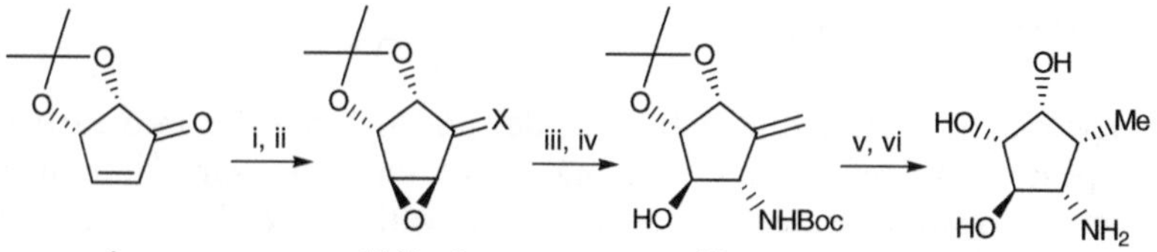

9 → (i, ii) → **10** X = O, **11** X = CH_2 → (iii, iv) → **12** → (v, vi) → **13**

Reagents: i, H_2O_2, OH^-; ii, $Ph_3P{=}CH_2$; iii, NH_3, EtOH; iv, Boc_2O, $NaHCO_3$; v, H_2; vi, HCl

Scheme 1

The cyclopentane derivatives **14** and **15** have been prepared from D-glucose as intermediates towards carbocyclic nucleoside analogues.[5]

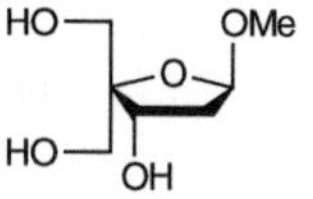

14 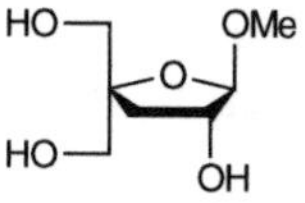 **15**

The exocyclic methylene group of sugar derivative **16** undergoes cycloaddition with nitrile oxides to give the spirocyclic systems **17** as single diastereomers [2,6-$Cl_2C_6H_3$NCO, or *p*-$MeC_6H_4CH{=}NOH$, NCS, Py, or PhNCO in nitro-

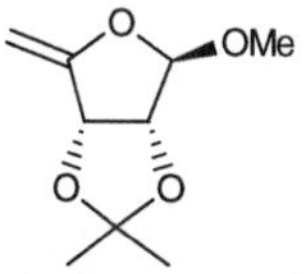

16

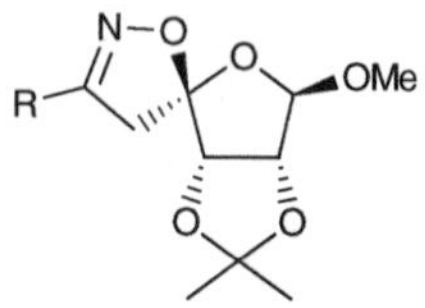

17 R = 2,6,-$Cl_2C_6H_3$
R = *p*-tolyl
R = Me

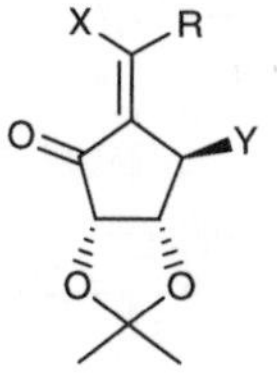

18

R	X	Y	%
2,6,-$Cl_2C_6H_3$	NH_2	OMe(OH)	64 (17)
p-tolyl	NH_2	H	46
Me	NH_2 (OH)	H	17 (23)

ethane]. Under Raney Ni hydrogenation conditions these rearrange to afford the cyclopentanones **18**.[6]

1.3 Cyclohexane and Higher Cycloalkane Derivatives. – The same methodology as described above for cyclopentanones applied to the pyranose-derived substrate **19** gave the cyclohexanones **21**. (In these cases, the intermediate spirocycles **20** were obtained as about 9:1 diastereomeric mixtures).[6]

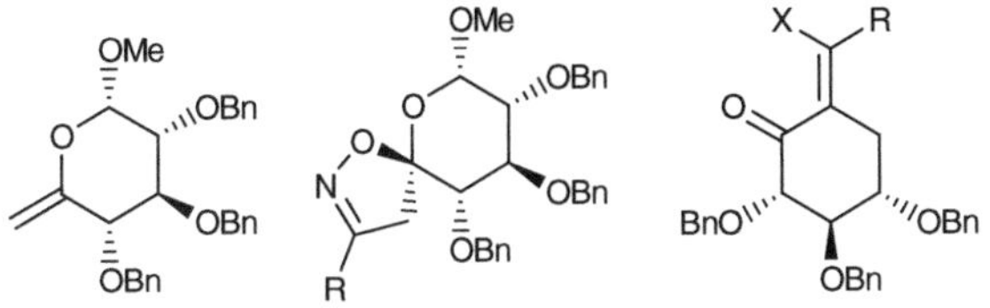

R	X	%
2,6,-$Cl_2C_6H_3$	NH_2	80
p-tolyl	NH_2	46
Me	NH_2	15
Me	OH	35

20 R = 2,6,-$Cl_2C_6H_3$; R = *p*-tolyl; R = Me

A synthesis of D-*myo*-inositol 1,4,5-trisphosphate has been reported starting from D-glucose, which was elaborated to the acyclic precursor **22**. Homologation of the aldehyde of **22** and deprotection and removal of the *O*-trityl group and oxidation at the other terminus afforded the silylmethylated alkyne **22** cyclization of which promoted by CSA then provided cyclohexane **24** bearing an exocyclic allene functionality. Further elaborations convert this to the target trisphosphate *via* the tribenzyl inositol **25** (Scheme 2).[7]

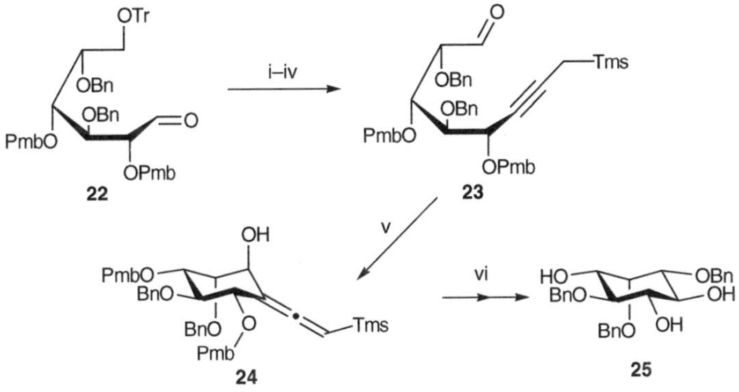

Reagents: i, CBr_4, Ph_3P; ii, Et_3N; iii, BuLi; iv, BuLi, $TmsCH_2OTf$; v, CSA; vi, Swern

Scheme 2

The fungal-derived sesquiterpene FR65814 (**29**) has been synthesized starting from D-glucose by Ferrier rearrangement of **26** to **27** (mercury(II) trifluoroacetate in acetone/water), which was then elaborated to **28** and thence to **29**.[8]

The same group elaborated the known D-glucose-derived 3-deoxy sugar **30** to panuclide A (**31**) in a lengthy synthesis (Scheme 3).

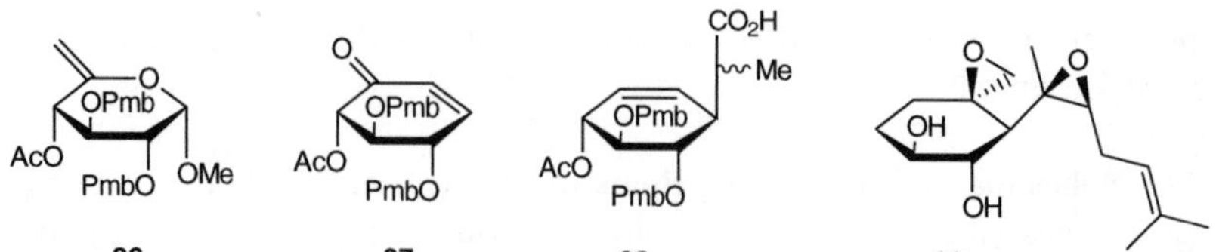

26 27 28 29

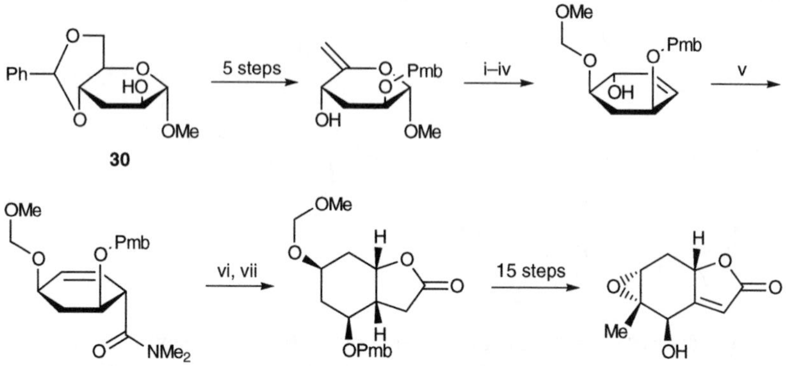

30 → 5 steps → i–iv → v → vi, vii → 15 steps → 31

Reagents: i, $Hg(OCOCF_3)_2$ H_2O, acetone; ii, MsCl, Et_3N; iii, DBU; iv, $NaBH_4$, $CeCl_3$; v, $MeC(OMe)_2NMe_2$, D; vi, I_2, THF-water; vii, Bu_3SnH, AIBN

Scheme 3

The use of intramolecular Diels-Alder reactions for the synthesis of decalin systems from carbohydrate starting materials has been presented in a review. The L-arabinose-derived triene **32** and tetraene **34** undergo cycloaddition reactions (at 0–21 °C) to give **33** and **35** respectively. Similarly, triene **37**, derived from **36**, undergoes cycloaddition to generate **38** (Scheme 4).[9]

D-Glucose was the starting material for a synthesis of cycloheptane-based systems for preparation of nucleoside analogue **41**, key steps being conversion

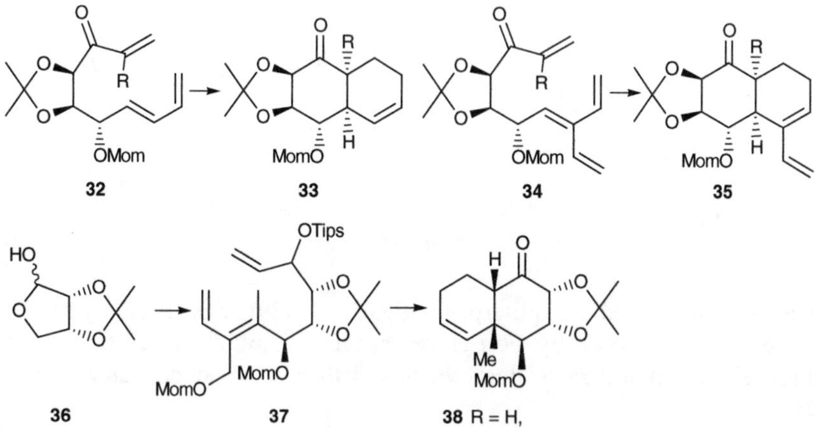

32 33 34 35

36 37 38 R = H, $CH_2CH_2C(CMe{=}CH_2){=}CMe_2$

Scheme 4

of **39** to **40** (BnNHOH; periodate then sodium borohydride), followed by reductive heterocycle opening and base construction. The same chemistry was applied to a synthesis of the analogous five-membered system (see Chapter 22).[10]

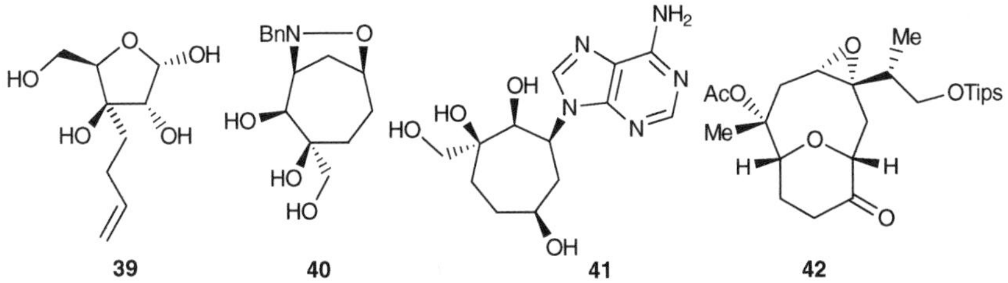

39 **40** **41** **42**

The oxygen-bridged cyclodecane **42** has been synthesized from D-mannitol as an intermediate towards the synthesis of neoliacinic acid.[11]

2 Lactones

2.1 γ-Lactones. – The hydroxylactone harizalactone (**44**), a marine antitumour metabolite, has been synthesized from D-glucose by periodate cleavage of 1,2-isopropylidene glucofuranose and phenyl Grignard addition to the resulting aldehyde to give **43**, radical double deoxygenation of the C3 and C5 hydroxyls, deprotection and a silver carbonate mediated oxidation to the lactone. This synthesis showed that the natural product was in fact the opposite enantiomer, and had previously been misassigned (Scheme 5).[12]

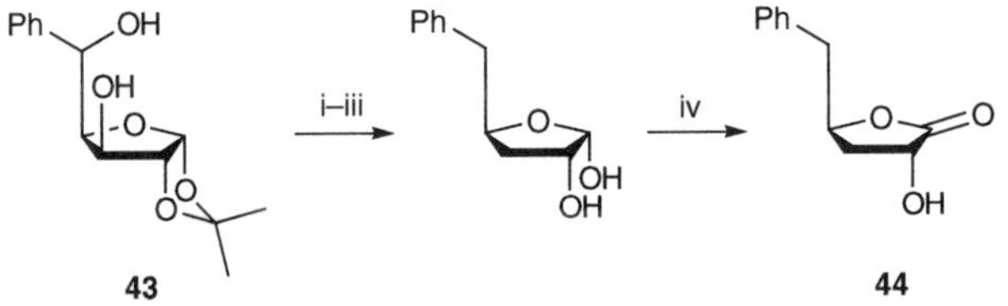

43 **44**

Reagents: i, NaH, CS_2, MeI; ii, Bu_3SnH; iii, aq. AcOH; iv, Ag_2CO_3 on Celite

Scheme 5

D-Ribono-γ-lactone has been converted into unsaturated lactone **45** by samarium iodide-mediated C2 and C3-deacylation of the 2,3-*O*-diacetyl-5-*t*-butyldimethylsilyl ether.[13] Peroxide cleavage of 3,4-*O*-isopropylidene-L-arabinose gives (*R*)-3,4-dihydroxybutanoic acid (**47**) *via* **46** which on acidification yields the lactone product **48** (There is an initial β-elimination of water.)[14]

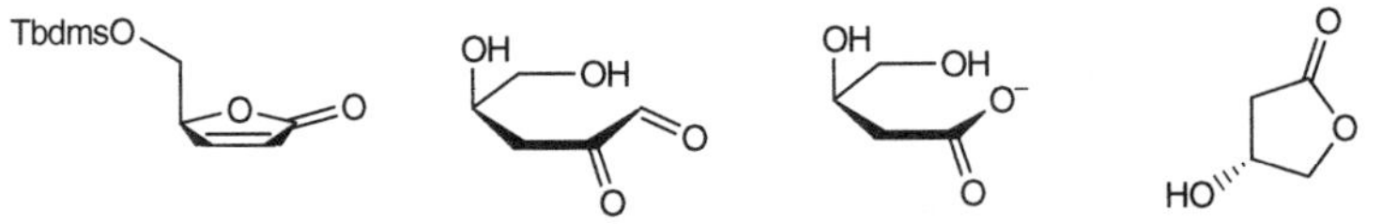

45 **46** **47** **48**

The bicyclic lactone **51** (produced by parasitic wasps in the Braconidae family) have been synthesized from **49**, derived from glucose. The key chemistry involves a *Z*-selective olefination and *in situ* lactonization giving **50** (Scheme 6).[15]

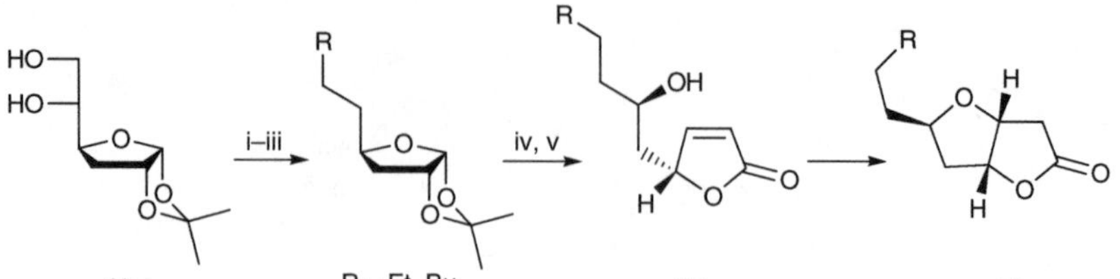

49 R = Et, Bu **50** **51**

Reagents: i, $NaIO_4$; ii, Wittig; iii, Raney Ni; iv, HOAc, H_2O; v, $EtO_2CCH_2P(O)(OCH_2CF_3)_2$

Scheme 6

Both enantiomers of laraconic acids have been prepared from D-mannitol; the synthesis of (−)-methylenolactocin **52**, a constituent of this family, is outlined in Scheme 7. The key ring-forming reaction is an iodolactonization, the iodo function being subsequently removed by radical methods.[16]

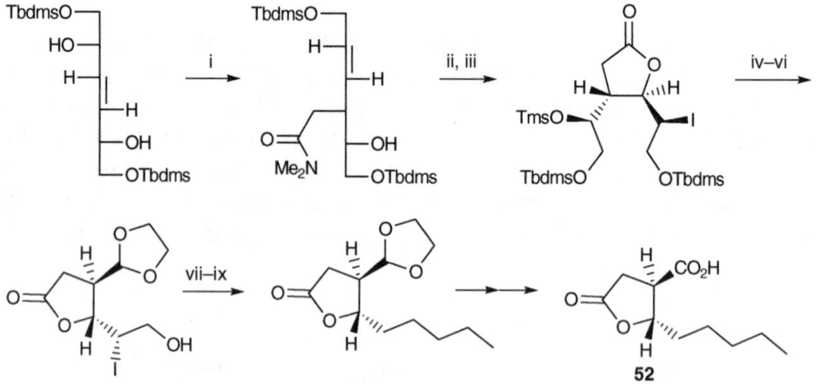

52

Reagents: i, $MeC(OMe)_2NMe_2$; ii, TmsCl, Et_3N; iii, I_2, $NaHCO_3$; iv, CF_3CO_2H; v, $NaIO_4$; vi, TsOH, (2-ethyl-2-methyl-1,3-dioxolane); vii, Bu_3SnH, AIBN; viii, TsCl, Py; ix, Pr_2CuLi

Scheme 7

A fragment (**55**) of the annonaceous acetogenin has been prepared from galactal derivative **53** *via* the anhydride **54** (Scheme 8). A mixture of stereoisomers at the indicated centre was obtained.[17]

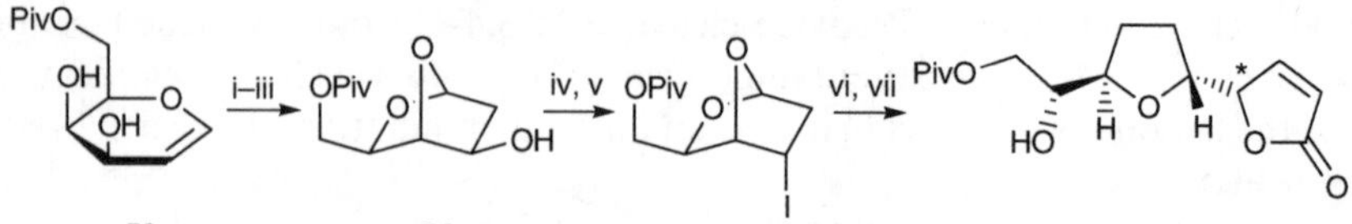

53 **54** **55**

Reagents: i, $(Bu_3Sn)_2O$; ii, NIS; iii, Bu_3SnH, AIBN; iv, Tf_2O, Py; v, Bu_4NI; vi, TmsOTf, 2-trimethylsilyloxyfuran; vii, H_2, Pd-C

Scheme 8

2.2 δ-Lactones. – Full details [Vol. 32, ref. 32] of the synthesis from D-glucose of (2*R*,3*S*,5*R*)-(−)-2,3-dihydroxytetradecanolide (**56**), the enantiomer of a biologically active lactone from the tree pathogen seirdium unicrone, have been reported. (The natural enantiomer was prepared from (*R*)-malic acid).[18] Wittig olefination of 2,4-*O*-isopropylidene-D-xylose provided **57**, which in treatment with DBU and then acetic acid gave the lactone **58**. Similar chemistry was applied to the synthesis of lactones **59** and **60**.[19]

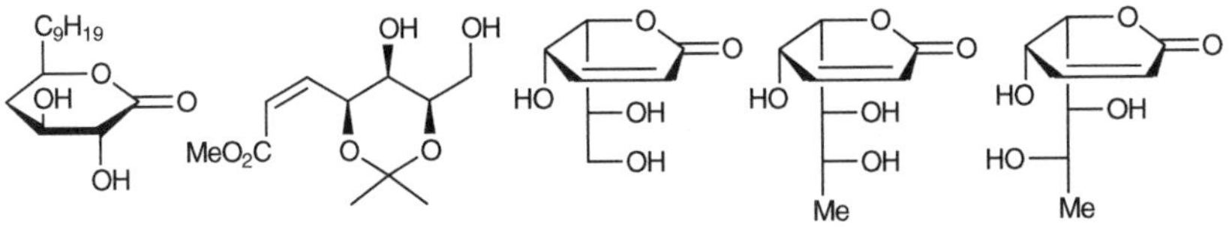

56 **57** **58** **59** **60**

3 Macrolides and Their Constituent Segments

A total synthesis of macrolide sorapohen$A_{1\alpha}$ (**63**) has been reported, which used D-mannose derivative **61** as the starting point. Key chemistry involved methyl substitutions at C2 and C4, and homologation (through acetylene addition to aldehyde) at C6, to provide intermediate **62**, which was then converted to the target in a further 26 steps (Scheme 9).[20]

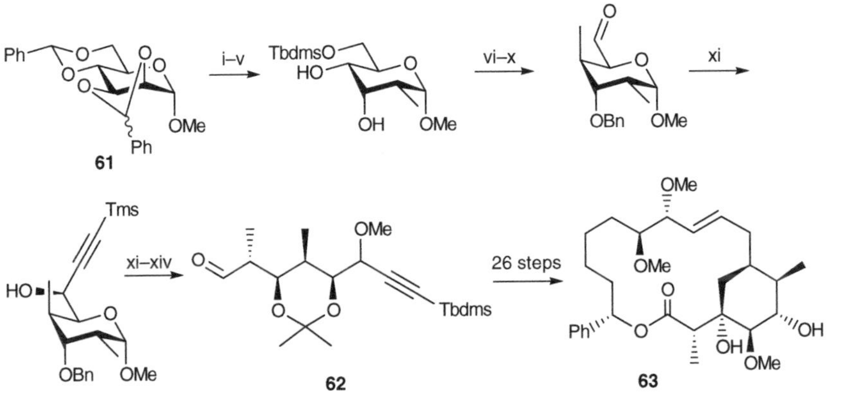

Reagents: i, BuLi; ii, MeI; iii, DIBAL; iv, Pd-C, H_2; v, TbdmsCl, DMAP, Et_3N; vi, $Me_2N^+=C(Cl)_2\ Cl^-$, Et_3N; vii, MeLi, TMEDA; viii, MeMgCl, CuBr; ix, TBAF; x, Swern; xi, Tms-acetylene, MeLi; xii, Ag_2O, MeI; xiii, propanedithiol, BF_3OEt_2; xiv, $(MeO)_2CMe_2$, acetone, CSA

Scheme 9

4 Other Oxygen Heterocycles

4.1 Four-membered *O*-Heterocycles. – The oxetane **65** has been prepared starting from D-glucose-derived **64**, the oxetane ring being formed by displacement of the terminal triflated hydroxyl group (Scheme 10).[21]

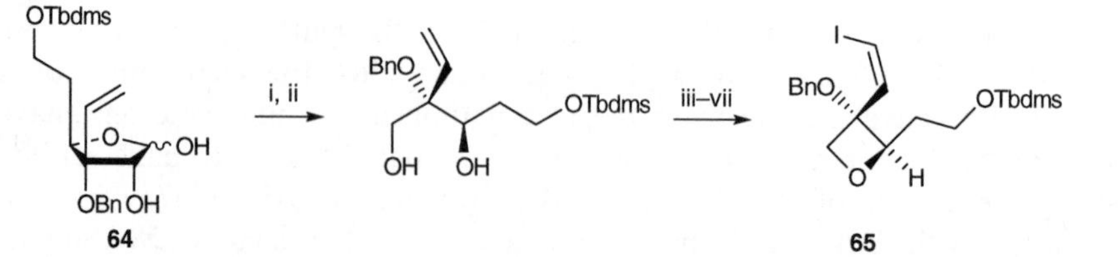

Reagents: i, Dess-Martin; ii, $NaBH_4$; iii, $KN(Tms)_2$; iv, $PhNTf_2$; v, OsO_4, NMO; vi, IO_4^-; vii, $Ph_3P{=}CHI$

Scheme 10

4.2 Five-membered *O*-Heterocycles. – Two novel muscarine analogues **70** and **71** have been prepared from D-glucose. Thus, epoxides **66** and **67** (see Chapter 5 for synthesis) were regioselectively ring-opened with lithium aluminium hydride, and the major isomers **68** and **69** were elaborated (tosylation and then substitution with trimethylamine) to targets **69** and **71**.[22] D-Glucose was also the starting material for the synthesis of tetrahydrofuran derivative **75**, diacetone-D-glucose-derived **72** being iodinated (iodine, triphenylphosphine, imidazole) and the isopropylidene removed (aqueous sulfuric acid) to give **73**, Wittig reaction of which gave a mixture of **75** and the uncyclized olefin **74**.[23]

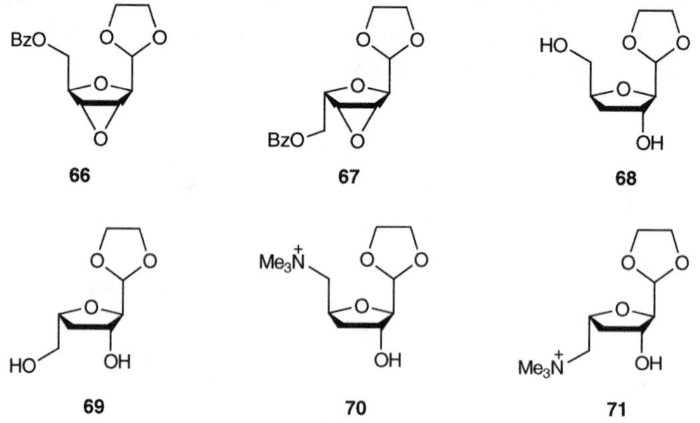

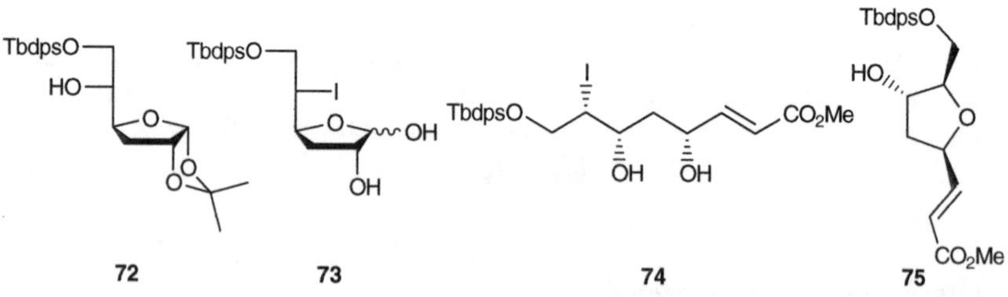

The 2,5-dialkyl tetrahydrofuran **77** (a component of covassolin) has been synthesized from **76** in which process all but two of the carbohydrate chiral centres are lost (Scheme 11).[24]

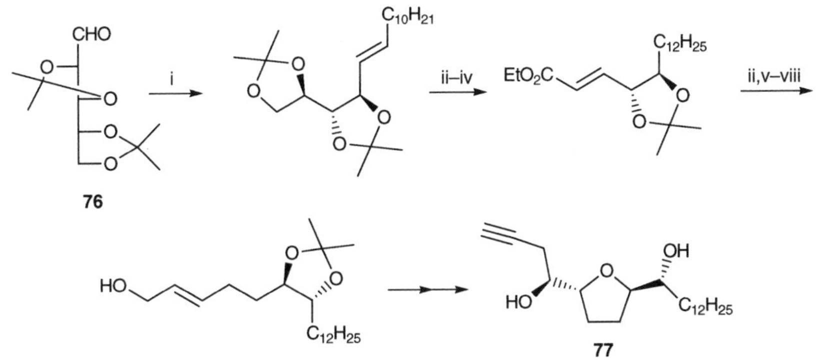

Reagents: i, $Ph_3P{=}CHC_{10}H_{21}$; ii, H_2, Pd-C; iii, IO_4^-; iv, $(EtO)_2POCH_2CO_2Et$; v, LAH, ether; vi, $(COCl)_2$, DMSO; vii, NaH, $(EtO)_2POCH_2CO_2Et$; viii, DIBAL

Scheme 11

A synthesis of (+)-furanomycin (**80**) starts with L-xylose and proceeds *via* the selenoglycoside **78**. Elaboration *via* the bis-selenide **79** allows for radical deselenation (at C5) and a radical cyclization introduced the ultimate amino acid stereocentre at this stage (Scheme 12).[25]

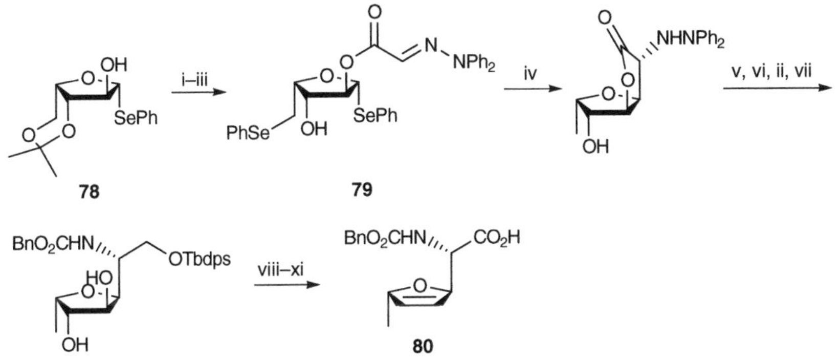

Reagents: i, $HO_2CCH{=}NNPh_2$, DCC; ii, CSA; iii, PhSeCN, PBu_3; iv, Ph_3SnH, AIBN; v, $LiAlH_4$; vi, TbdmsCl, ImH; vii, BnOCOCl, $NaHCO_3$; viii, Ph_3P, CHI_3, ImH; ix, Bu_4NF; x, Dess-Martin; xi, $NaClO_2$

Scheme 12

Treatment of D-glucal or D-galactal with catalytic samarium(III) triflate or $RuCl_2(PPh_3)_2$ in the presence of water (1 equivalent) gives the chiral furan diol **81** directly in yields of 70%.[26] But-3-enyl Grignard addition to **82** gave **83**, which underwent iodoetherification (up to 81:19 diastereoselectivity), followed

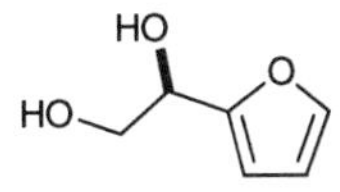

81 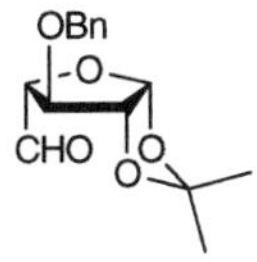 **82** 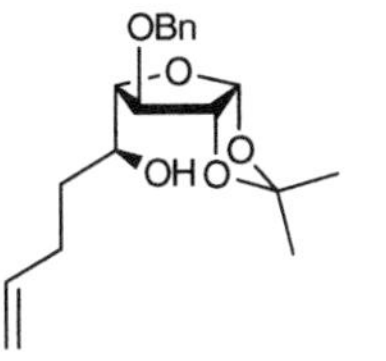 **83** 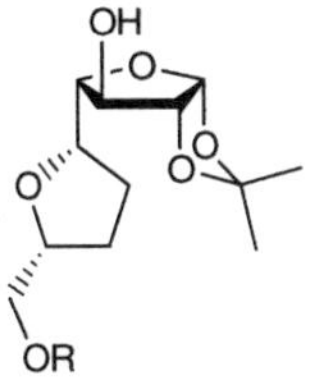 **84** R = EtCO, PhCO by iodide substitution by propionate and then debenzylation to give bis-tetrahydrofurans **84**.[27]

Another bisheterocycle synthesis from a sugar is the elaboration of ribofuranosyl chloride **85** into **86** (Scheme 13).[28] Xylose-derived **87** has been converted into the bicyclic system **88** (Scheme 14).[29]

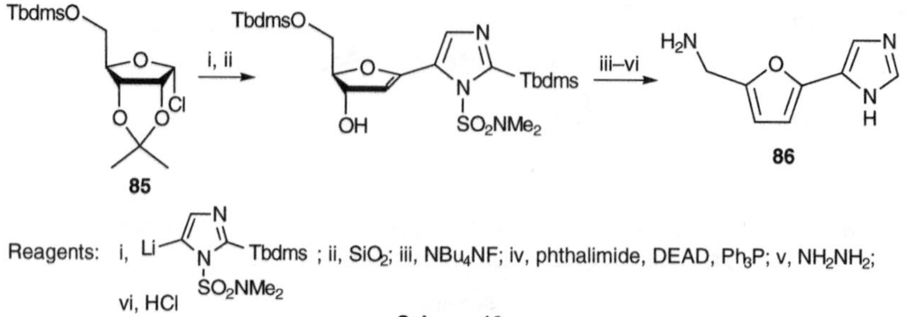

Reagents: i, Li-(imidazolyl; 2-Tbdms, 1-SO_2NMe_2); ii, SiO_2; iii, NBu_4NF; iv, phthalimide, DEAD, Ph_3P; v, NH_2NH_2; vi, HCl

Scheme 13

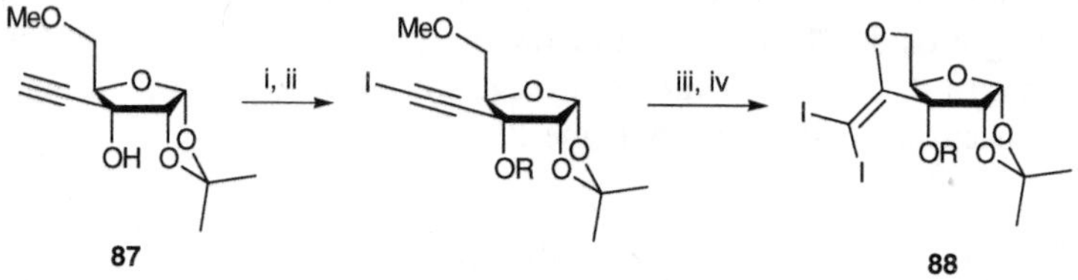

Reagents: i, NIS, $AgNO_3$; ii, NaH, MeI; iii, NBS, TsOH; iv, I_2, HTIB

Scheme 14

4.3 Six-membered *O*-Heterocycles.

– A radical domino type of process occurred on reaction of **89** with perfluorooctyl iodide to generate the bicyclic fused pyran **90** the (*R*)-isomer of which was then elaborated to the ring system **91** (Scheme 15).[30]

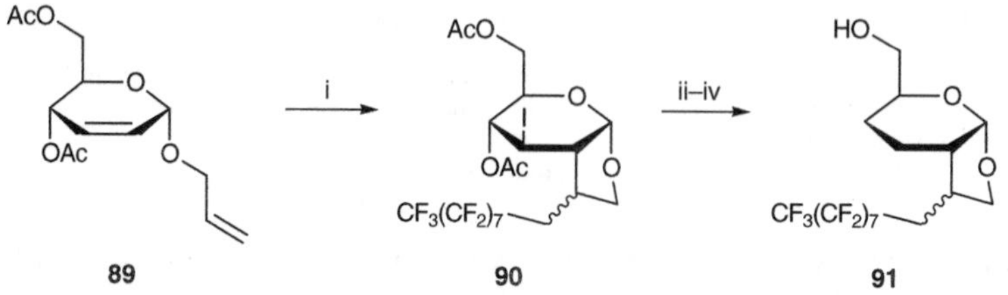

Reagents: i, $CF_3(CF_2)_7I$, NaS_2O_4, $NaHCO_3$; ii, Zn/Cu, DMF; iii, CsF on alumina; iv, H_2, Pd-C

Scheme 15

(Phenylthio)acetylenic carbohydrate derivative **92** was hydrosilylated in high yield and in short reaction times, using acetylenehexacarbonyl complexes in catalytic amounts. The complex **93** and a catalyst **94** derived from the substrate itself both catalysed the reaction in 10–30 min affording **95** in high yields (83–99%).[31]

Two approaches to the core of the zaragozic acids have appeared. The first

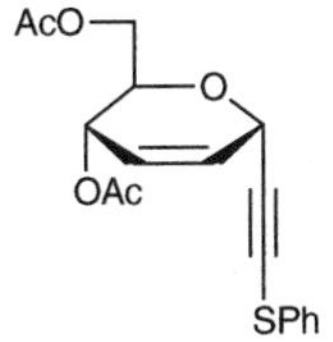

92 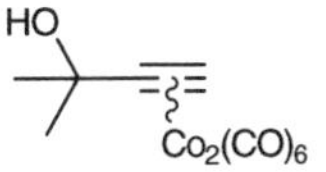 93 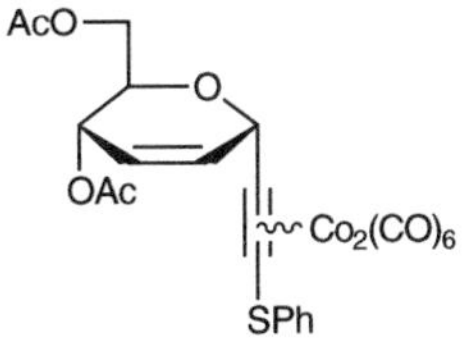 94 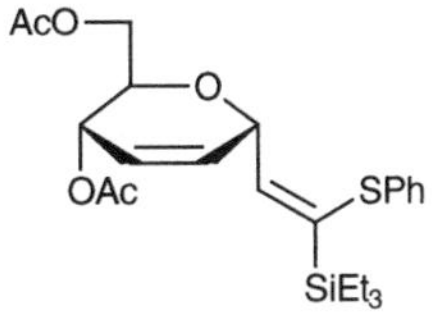 95

proceeds *via* the methylene sugar **96** (from the precursor 6-iodo derivative), fast Ru-mediated dihydroxylation of which and then silica-promoted internal glycosidation affording the anhydro sugar **97**. Treatment of the anhydride with DBU leads to **98** (a model for the core) *via* base-catalysed mesylate displacement involving an internal C–O bond migration. The more elaborate material **99** was also used to synthesize the more complex core units, **100** and **101** (Scheme 16).[32]

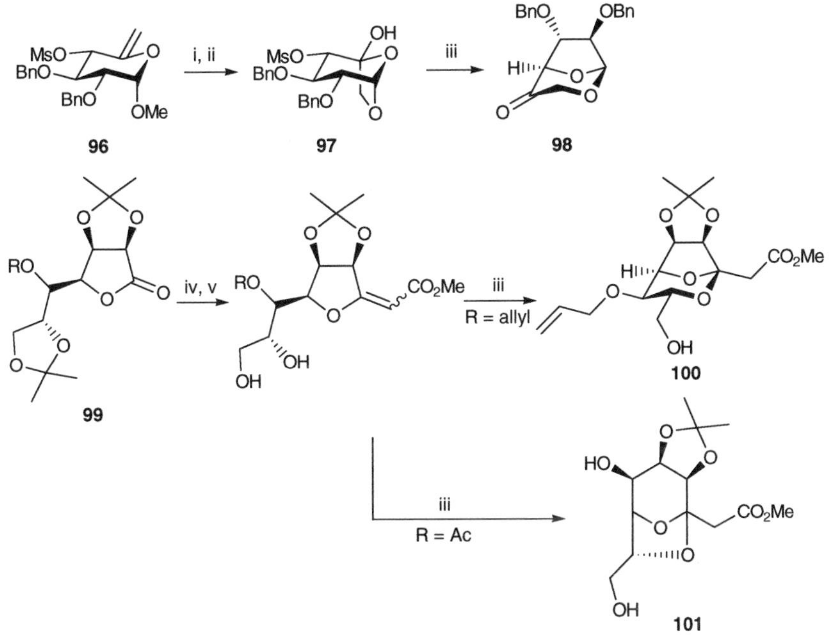

Reagents: i, $RuCl_3$, $NaIO_4$; ii, SiO_2; iii, DBU; iv, $Ph_3P{=}CHCO_2Me$; v, AcOH, H_2O

Scheme 16

The second approach utilizes the cerium(III) chloride-promoted aldol reaction between methyl (α-D-xylofuranoside)uronate **102** and D-(*R*)-glyceraldehyde acetonide, generating adduct **103**, transketolization and further modifications from this leading to different bicyclic ketals **104**, **105** and **106**.[33]

The dioxatricyclic compounds **108**, **111**, **114** have been prepared by Pd-catalysed annulation [palladium(II) acetate, triphenylphosphine, TBAHS, triethylamine] of the precursor bromoaryl ethers **107**, **110** and **113**, respec-

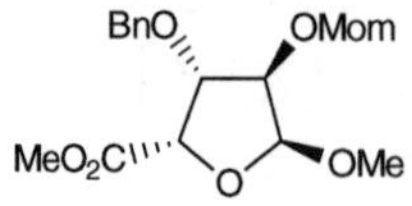

102 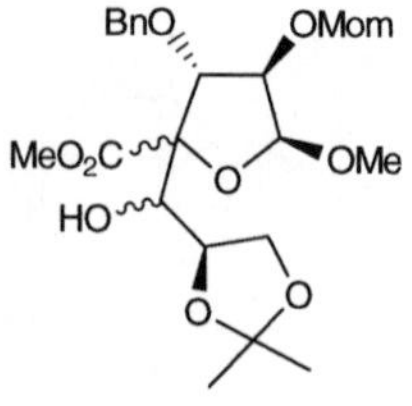 **103** 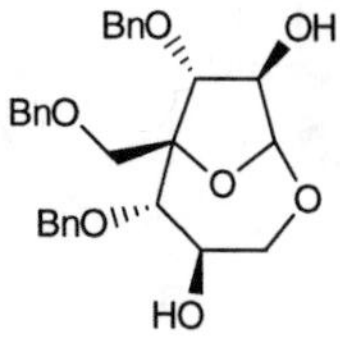 **104**

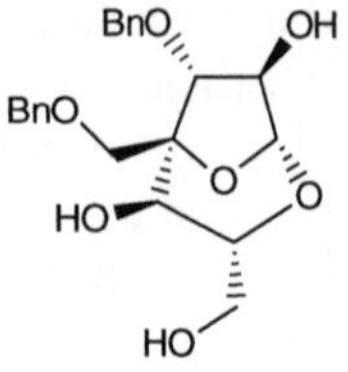

105 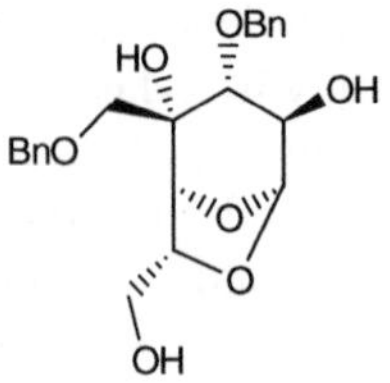 **106**

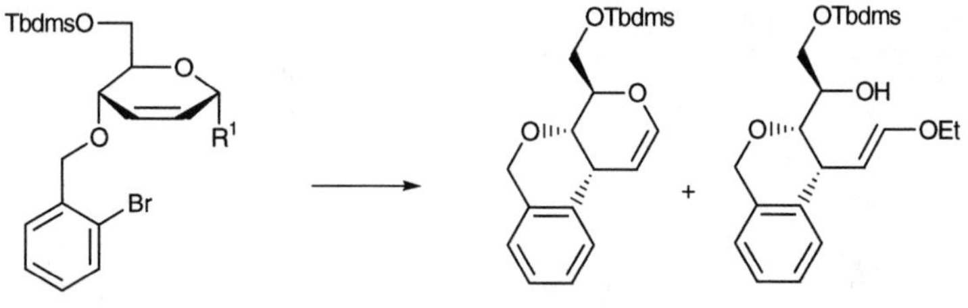

107 **108** **109**

R^1 = H, OEt, OC_6H_4p-^tBu

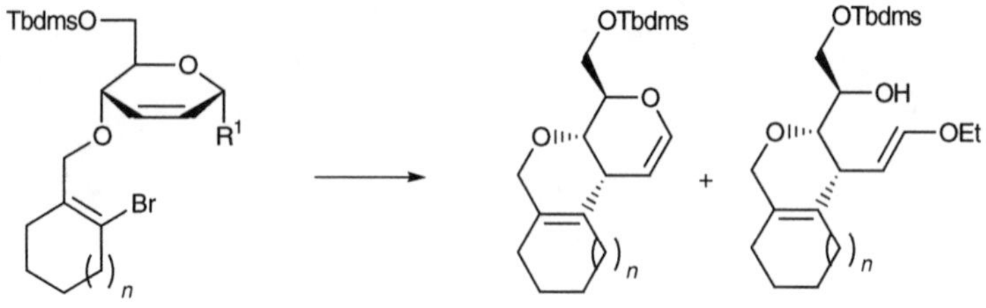

110 **111** **112**

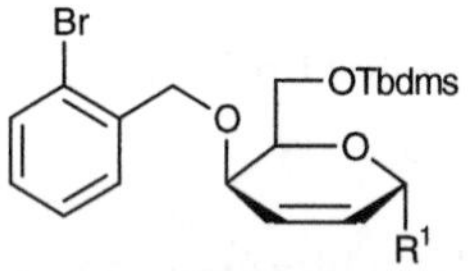

113 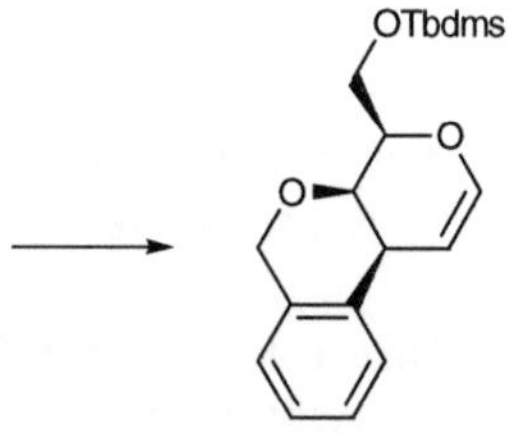 **114**

Scheme 17

tively. These reactions also lead to some formation of the bicyclic compounds, for example **107** and **110** leading to some **109** and **112**, respectively (Scheme 17).[34]

The bicyclic system **116**, the AB fragment of gambiertoxin, has been prepared from carbohydrate derivative **115**, with a radical cyclization as a key step (Scheme 18). The same paper reported the synthesis of the ABC fragment of ciguatoxin from **117**, which incorporated application of the same type of chemistry as used in the gambiertoxin synthesis in the preparation of **118**.[35]

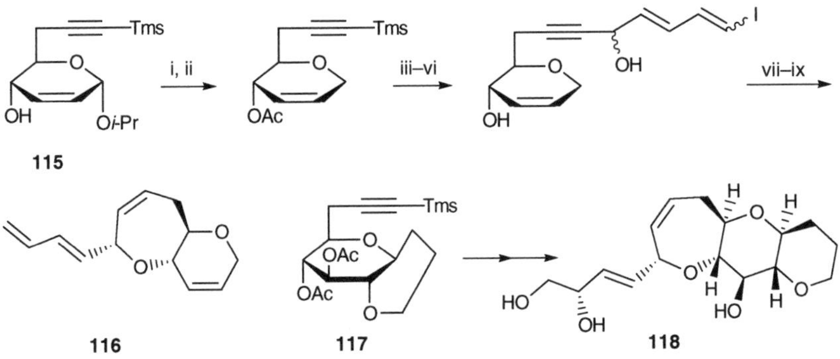

Reagents: i, Ac_2O, Py; ii, $HSiEt_3$, BF_3OEt_2; iii, K_2CO_3, MeOH; iv, $EtOCH{=}CH_2$, H^+; v, BuLi, , BF_3OEt_2, ICH=CH-CH=CH-C=O; vi, H^+, MeOH; vii, $CO_2(CO)_8$, viii, H+, ix, Bu_3SnH

Scheme 18

As similar sugar-derived bicyclic system **120**, obtained from *C*-allyl glucoside **119** by O-2 allylation and ring closing alkene metathesis, was converted to diol **121** by diastereoselective epoxidation and then ring opening.[36]

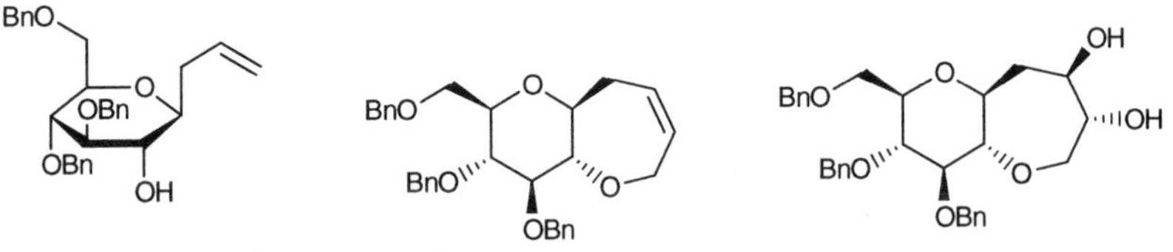

The polycyclic ethers, **123**, **125** and **126** have been synthesized by intramolecular cycloadditions of carbohydrate nitrile oxides **122** and **124** (Scheme 19).[37]

The JKLM fragment (**128**) of ciguatoxin has been synthesized from **127** (Scheme 20),[38] and the same authors reported the synthesis of the F–M fragment in the following paper.[39]

An approach to the AB-ring fragment of ciguatoxin has been reported which starts from tri-*O*-benzyl-D-glucal, which was converted to **129**, which then underwent intramolecular metathesis cross coupling to give **130**, which then underwent an intermolecular metathesis cross coupling with **131** to generate **132** (as a mixture with the oxepane side chain epimer).[40]

Without doubt, the highlight of this year's employment of carbohydrates as starting materials for complex polyether syntheses has been the report (in four parts) of the total synthesis of Brevetoxin A by Nicolaou's group. The component assemblies utilize glucose, mannose and 2-deoxyribose as starting

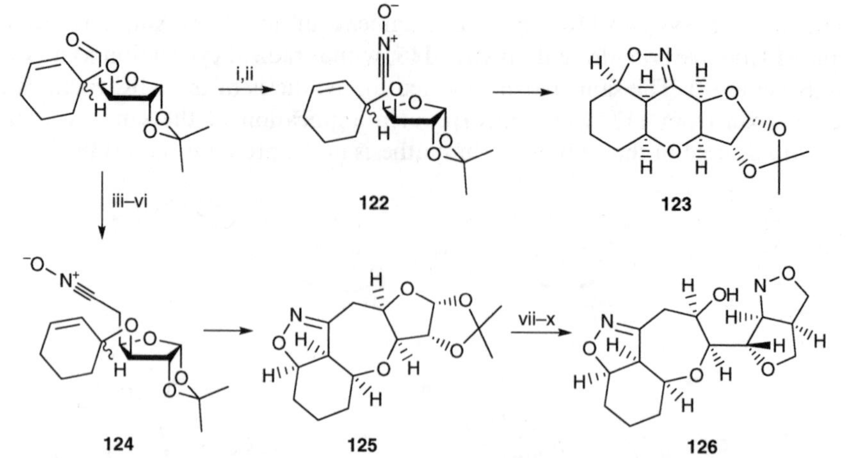

Reagents: i, NH_2OH; ii, Chloramine-T; iii, $MeNO_2$, KF; iv, Ac_2O, DMAP; v, $NaBH_4$; vi, PhNCO, Et_3NH; vii, HOTs, MeOH; viii, NaOH, DCM-H_2O, TBAB, allyl bromide; ix, TFA, H_2O; x, MeNHOH

Scheme 19

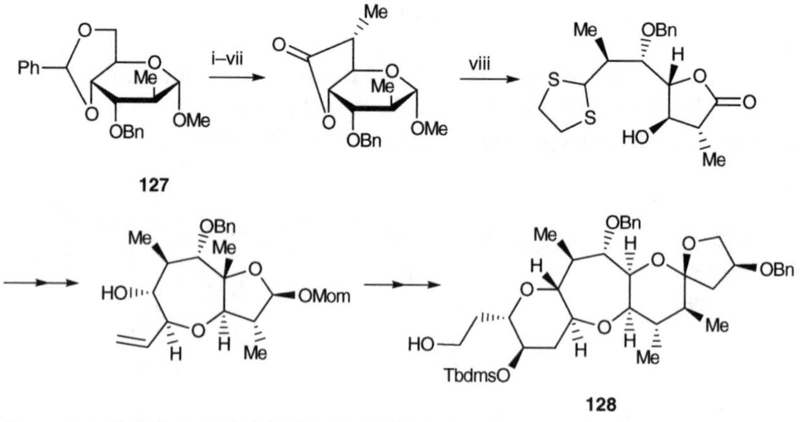

Reagents: i, H^+; ii, TsCl, Et_3N; iii, NaCN, DMSO; iv, DIBAL; v, $NaClO_2$; vi, $PhSO_2Cl$, Et_3N; vii, LHMDS; viii, 1,3-propanedithiol

Scheme 20

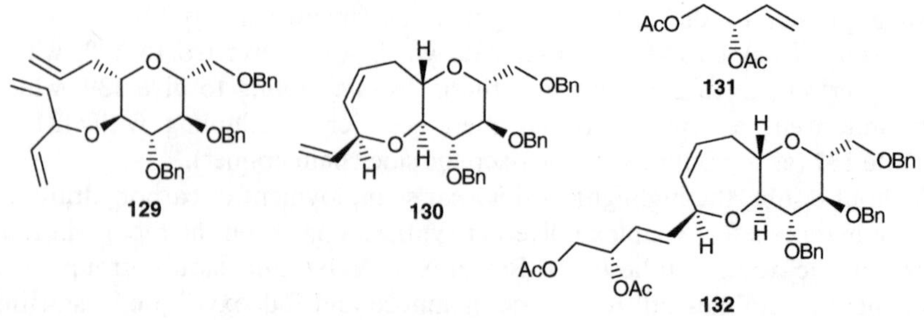

carbohydrates. Part 1 utilizes D-glucose for the synthesis of the BCD ring system **135**, for which the known bicycle **133** was elaborated to the ring C intermediate **134**, and a key double lactonization established rings B and D concomitantly (Scheme 21).[41]

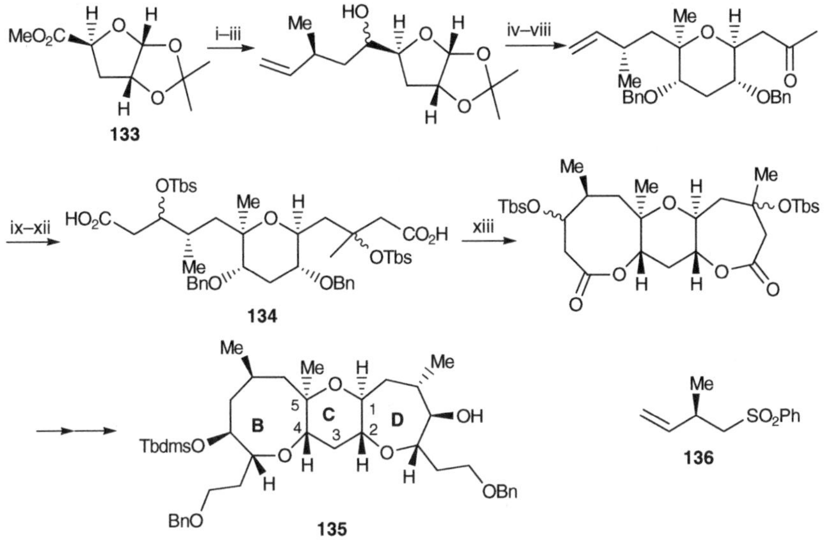

Reagents: i, n-BuLi, **136**; ii, Hg(Al), THF, H_2O; iii, MeMgCl; iv, EtSH, $ZnCl_2$; v, NaH, BnBr; vi, NCS, 2,6-lutidine, MeCN-water; vii, Ph_3P=$CHCOCH_3$; viii, NaH, THF; ix, OsO_4, NMP; x, $NaIO_4$; xi, $ZnBr_2$, CH_2=C(OBn)OTbs; xii, H_2, $Pd(OH)_2$/C; xiii, $(PyS)_2$, Ph_3P, $AgClO_4$

Schems 21

The second report describes the synthesis of an EFGH model system starting from 2-deoxy-D-ribose. The nine-membered ring E was introduced by a Z-selective Wittig reaction of **137** to give **138**, with a subsequent Yamaguchi lactonization establishing the key ring and further elaboration providing component **139**, ready for Wittig connection to the ring G precursor (Scheme 22).[42]

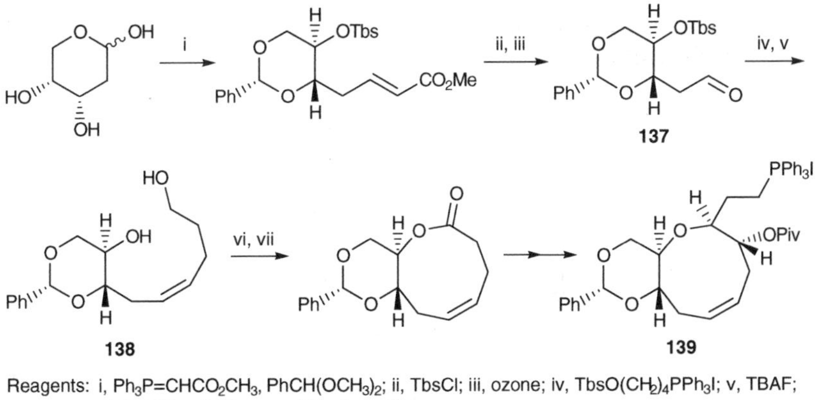

Reagents: i, Ph_3P=$CHCO_2CH_3$, $PhCH(OCH_3)_2$; ii, TbsCl; iii, ozone; iv, $TbsO(CH_2)_4PPh_3I$; v, TBAF; vi, $NaClO_2$; vii, Yamaguchi lactonization

Scheme 22

The third part of the Brevetoxin story used the mannose-derived **140**, and involves C4 radical deoxygenation and use of Sharpless asymmetric epoxidation to introduce key ring I functionality. The bicyclic intermediate **141** was then further advanced to the GHIJ ring system **142**. This paper also describes an improved synthesis of the BCDE ring system (*via* debenzylated **143**) (Scheme 23).[43]

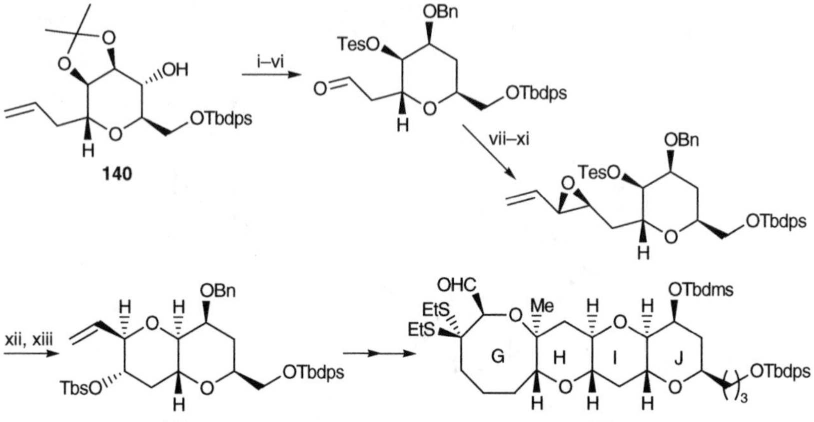

Reagents: i, NaH, CS_2, MeI; ii, Bu_3SnH; iii, CSA, TbdpsCl, MeOH; iv, Bu_2SnO, BnBr; v, TesOTf; vi, ozone, Ph_3P; vii, $Ph_3P{=}CH_2CO_2Me$; viii, DIBAL; ix, Sharpless AE; x, SO_3.pyr, DMSO; xi, $PPh_3P{=}CH_2$; xii, TBAF; xiii, TbsOTf.

Scheme 23

The completion of the total synthesis (using components from the above papers, but no new carbohydrate steps) was reported in part 4 of this series.[44] An account of a plenary lecture describing some previously unpublished work by Yamamoto's group on the synthesis of hemibrevetoxin has also appeared this year.[45]

Nicolaou's group have also reported the synthesis of everinomycin 13,384-1 A, B(A)C fragment, with the B and C rings constructed from sugars. The synthesis of the B-ring is outlined in Scheme 24.[46]

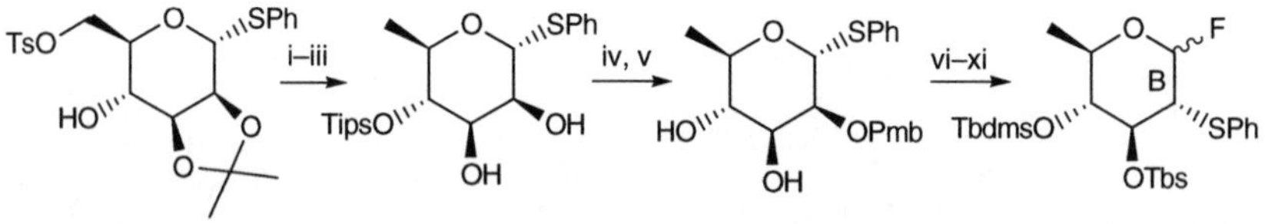

Reagents: i, TipsOTf, 2,6-lutidine; ii, $LiAlH_4$; iii, TsOH, MeOH, $(CH_2OH)_2$; iv, Bu_2SnO, PmbCl, TBAI; v, TBAF; vi, TbdmsOTf, 2,6-lutidine; vi, DDQ; viii, DAST

Scheme 24

4.4 Seven- and Larger-membered *O*-Heterocycles. – Ring closing metathesis reactions of **143** and **146** generated oxepenes **144** and **147** respectively.

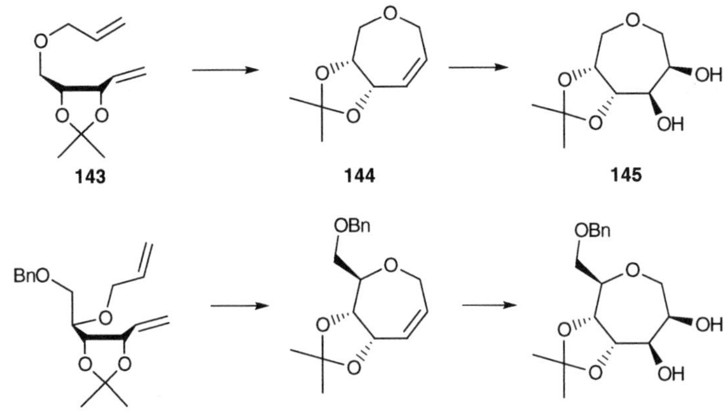

143 144 145

146 147 148

Scheme 25

Diastereoselective dihydroxylations then provides the saturated oxepanes **145** and **148** (Scheme 25).[47]

5 N- and S-Heterocycles

D-Gulonolactone-derived **149** has been converted into the protected amino acid analogue **150**, which was then elaborated to dipeptide systems (Scheme 26).[48]

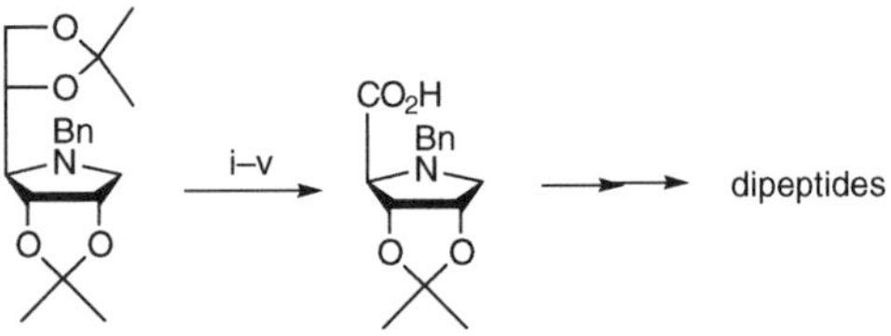

149 150 dipeptides

Reagents: i, H_2, Pd-C; ii, Fmoc-Cl, Et_3N; iii, HCl; iv, IO_4^-; v, $NaClO_2$

Scheme 26

The carbohydrate lactone **151** reacts with hydrazine to give the bicyclic system **153**, *via* the initial hydrazine adduct **152**, which cyclizes with amide bond formation, opening of the sugar ring, and the hydrazine displacement of tosylate. Intermediate **153** was then converted to pyrrolidine **154**, and a related series of examples are also reported.[49]

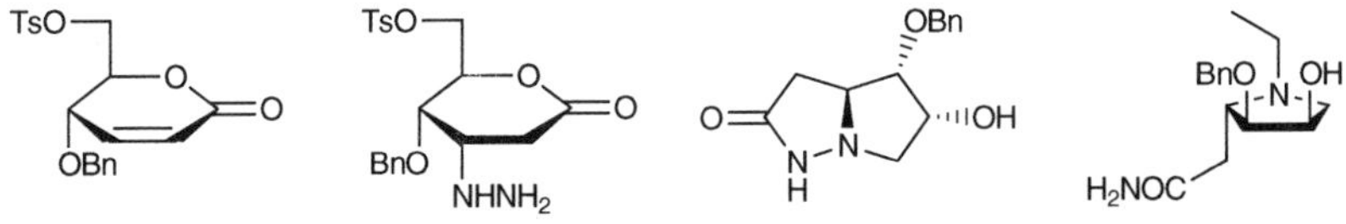

151 152 153 154

Reaction of tri-*O*-acylated glucosamine (**155**) with *O,O*-disubstituted aryliso- or isothio-cycanates gave rotationally constricted imidazolidenones or -thiones (**156**), for use as atropisomerically selective auxiliaries (applications were not reported).[50]

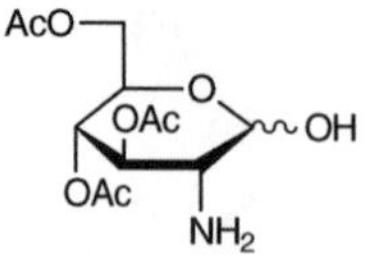

155

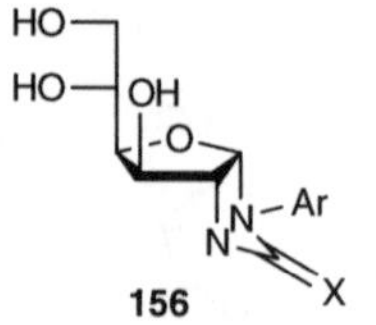

156

X = O, S
Ar = 2-Cl, 6-Me-C_6H_3
2-Et, 6-Me-C_6H_3

The lithiated dihydrofuran **157** reacts with azetidine dione **158** under boron trifluoride catalysis to give the azetidinone adduct **159**. Alternative reaction of **159** with pyridinium tosylate yields **160** and **161**. Reaction of **160** with hydroxylamine then affords **162**, whilst reaction with *m*-chloroperoxbenzoic acid and then ammonia in methanol provides spirodiketopiperazine **163**.[51]

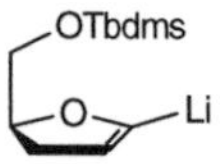

157

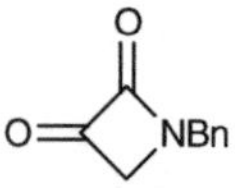

158

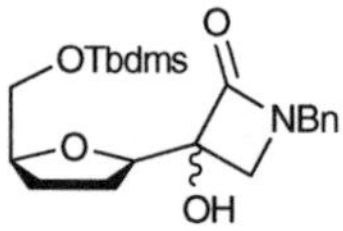

159

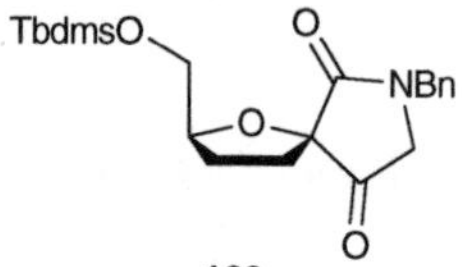

160

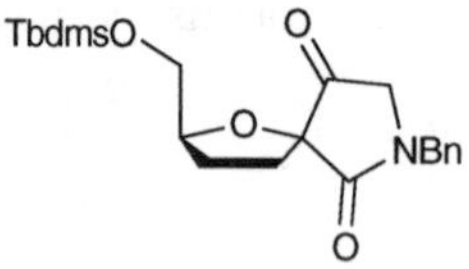

161

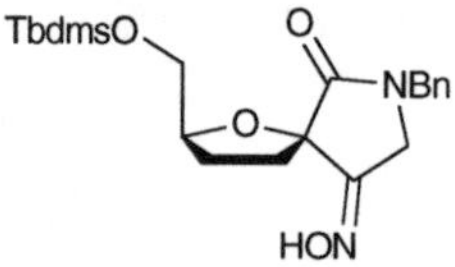

162

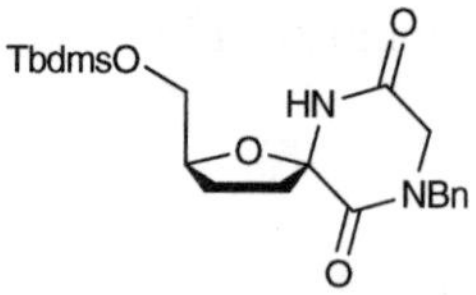

163

The (known) cyclic acetals **164** and **165** of D-erythose and D-threose react with hydroxylamines (the former with *N*-phenyl-, benzyl- and methyl-hydroxylamine, the latter with *N*-benzylhydroxylamine) yielding the novel nitrones

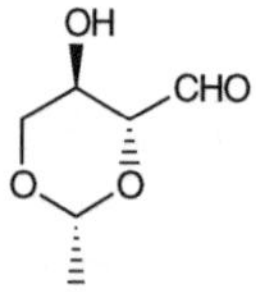

164

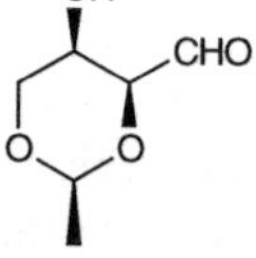

165

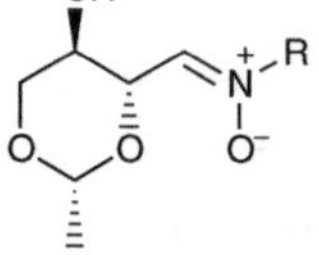

166

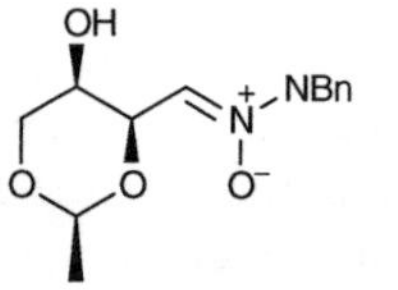

167

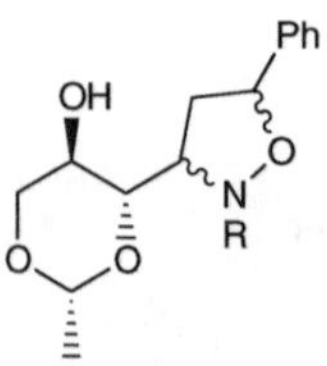

168

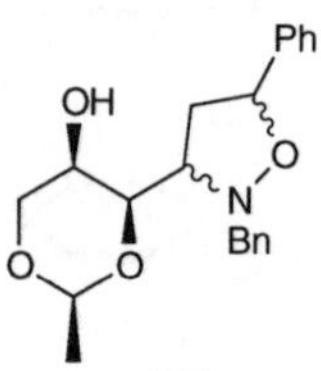

169

166 and **167**, respectively. Reaction of each with styrene yielded mixtures of four possible diastereomeric products, **168** and **169**, respectively.[52]

The diacetone-D-glucose-derived **170** undergoes Henry reaction (nitromethane, KF) and subsequent reductive deacylation to give **171** then conversion (PhNCO, triethylamine) to the nitrile oxide **172** which undergoes intramolecular dipolar addition to form the oxepanisoxazole **173**. Alternatively, **170** can be derivatized as its oxime and then converted to nitrile oxides **174** (one carbon fewer than **172**). Both the terminal acetylene and the methylated analogue undergo analogous intramolecular dipolar additions, yielding **175**. The terminal acetylene (R = H) was further elaborated into *O*-allyl bearing nitrone **176**, which undergoes a second intramolecular dipolar addition to give **177**.[53]

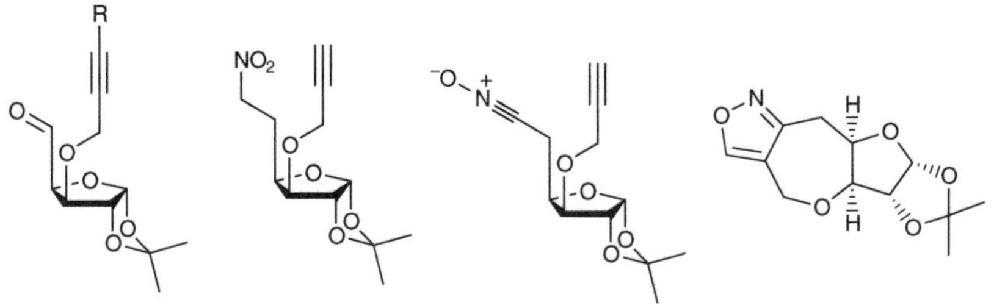

170 R = H, Me **171** **172** **173**

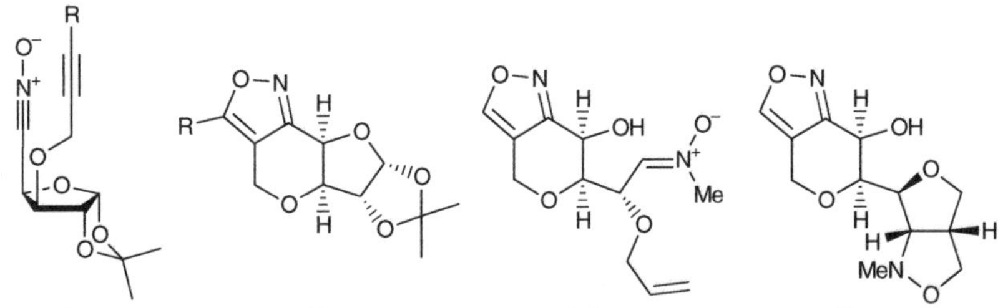

174 **175** **176** **177**

3-Amino-2-piperidones have been prepared as constrained pseudo-dipeptides (*e.g.* **181**) from carbohydrate starting materials. The sugar bromide **178** reacts with leucine *t*-butyl ester to provide **179**. Treatment with potassium carbonate causes epoxide cyclization, and subsequent reaction with sodium hyride leads to amide nitrogen ring opening of the epoxide (6-exo-tet) to give **180**, which was then elaborated to **181**.[54]

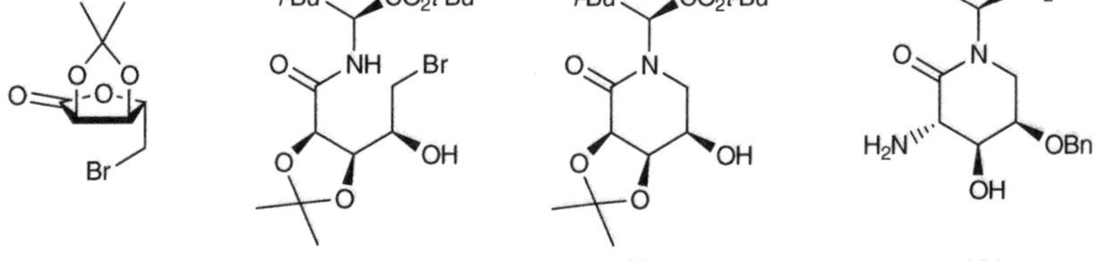

178 **179** **180** **181**

The synthese of both (+) and (−) enantiomers **183** and **184** of a plant 4-hydroxypipecolic acid has been described starting from D-*glycero*-D-*gulo*-hepton-1,4-lactone derivative **182** (Scheme 27).[55]

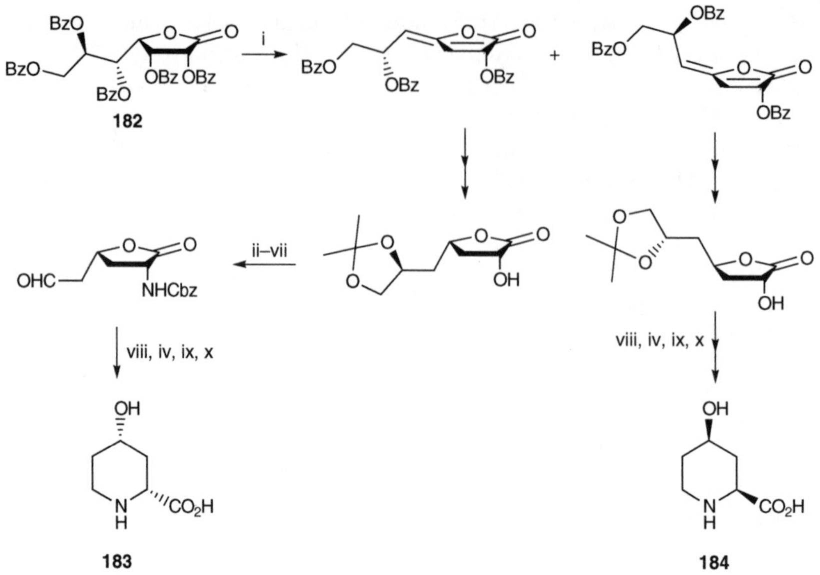

Reagents: i, Et_3N; ii, TsCl, Py; iii, NaN_3; iv, H_2, Pd-C; v, CbzCl; vi, H_3O^+; vii, IO_4^-; viii, $NaCNBH_3$; ix, MsCl; x, KOH, H_2O

Scheme 27

A diastereoselective synthesis of 1-deoxytalonojirimycin **186** involves the generation of **185** as the key stereochemical intermediate. Reductive auxiliary cleavage occurs to give the amino group required for amination-cyclization (Scheme 28).[56]

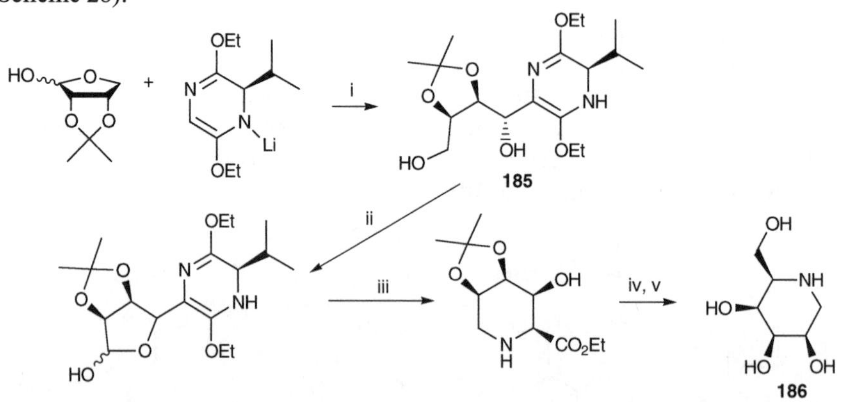

Reagents: i, −78 °C; ii, Iodosobenzoic acid, DMSO/THF; iii, 0.25M HCl, H_2, Pd-C; iv, $LiBH_4$; v, Dowex H^+

Scheme 28

D-Xylose was the starting material for a synthesis of (+)-azafagomine (**188**), *via* the known **187**. An alternative synthetic approach did not provide a route to the natural product ring system but instead yielded the seven-membered isomer of **188** (Scheme 29).[57]

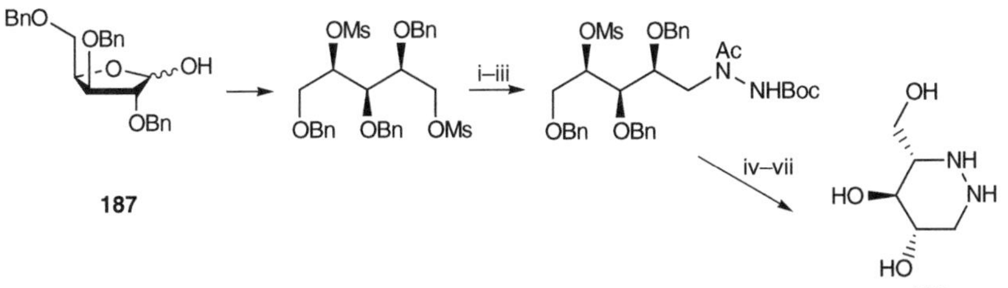

Reagents: i, $NaCNBH_3$, $BocNHNH_2$, AcOH ii, Ac_2O, MeOH; iii, MsCl; iv, TFA; v, $^iPr_2EtN$, $MeNO_2$; vi, H_2, Pd-C; vii, HCl, H_2O

Scheme 29

An asymmetric total synthesis of (−)-prosophylline **191** (an alkaloid with antibiotic and anaesthetic properties) has been achieved in 17 steps overall from D-glucal. Key intermediates are the furan **189** (*cf.* **81**) and the dihydropyridone **190** derived from it by treatment with *m*-CPBA.[58]

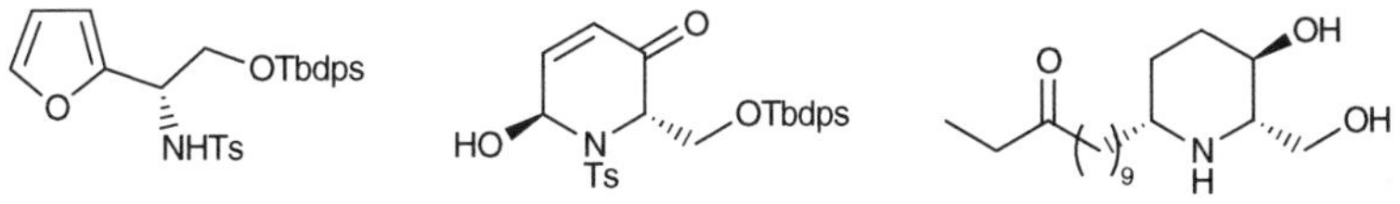

Swainsonine (**192**) has been extracted from the seeds of *Swainsona precumbens* in 0.17% yield,[59] and three new analogues of australine (**193**) have been isolated from the stalks and bulbs or unripe fruits of bluebell. They showed no biological activity.[60]

A 'microreview' on the application of ring-closing metathesis to the preparation of pyrrolizidines, indolizidines and quinolizidines included a new formal synthesis of castanospermine (**194**) involving the reaction **195**→**196**.[61] Several references to the preparation of polyhydroxylated indolizidines, *e.g.* swainsonine (**192**), can be found in a *Tetrahedron Report* on the synthesis and application of optically active α-furfurylamine derivatives.[62]

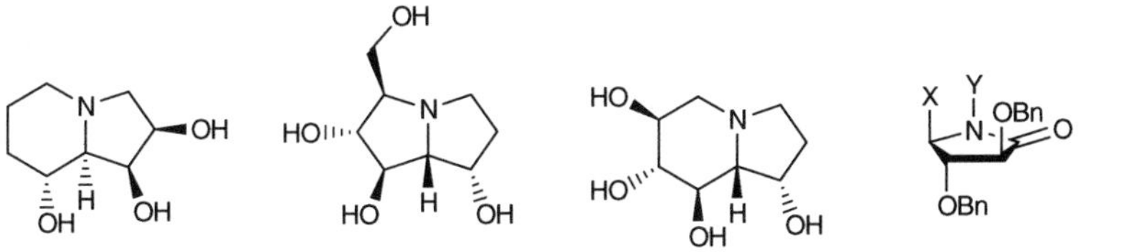

195 X = $-CH=CH_2$, Y = $-CH_2CH=CH_2$
196 X,Y = $-CH=CHCH_2-$

The enantioselective total syntheses of castanospermine (**194**), 6-*epi*-castanospermine (**197**), australine (**193**) and 3-*epi*-australine (**198**) using a chiral vinyl ether intermediate in tandem asymmetric [4+2]/[3+2] nitroalkene cycloaddition chemistry and generating nitrosoacetal **199** as key intermediate have been reported. Dihydroxylation of the double bond, then monotosylation at either the primary or secondary position, furnished the four tosylates **200**. Loss of the auxiliary and cleavage of the N–O bonds on exposure to Raney-nickel resulted in concomitant reductive amination and nucleophilic sulfonate displacement, leading to the four targets.[63] Another report details the synthesis of both 8-*epi*-castanospermine (**204**) and its (+)-1,8a-di-*epi* isomer (**205**), prepared by ester enolate addition of *t*-butylacetate enolate to **201**, giving the separable diastereomers **202** and **203** (see Chapter 2, ref. 37a for synthesis). These were converted in 13 steps to **204** and **205**, respectively, by steps involving intramolecular bicyclization by nucleophilic displacements and subsequent debenzylation.[64]

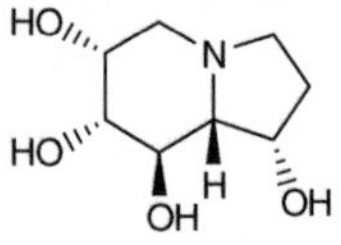

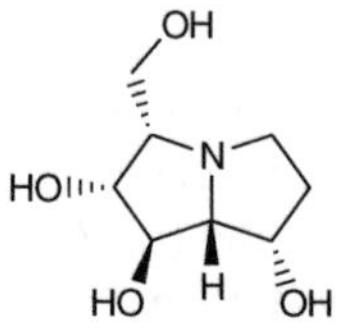

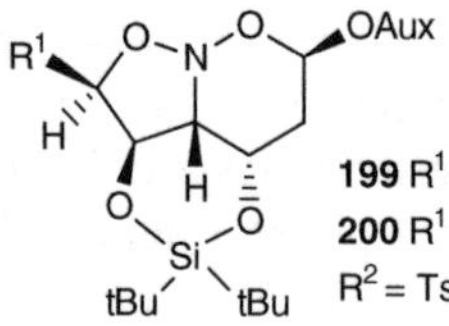

199 $R^1 = CH_2{=}CH$-
200 $R^1 = R^2OCH_2CH(OR^3)$-
$R^2 = Ts$, $R^3 = H$ or $R^2 = H$, $R^3 = Ts$

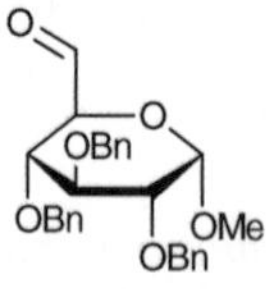

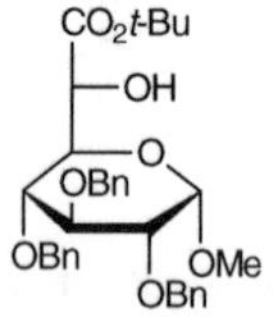

On hydrogenation, azide **206** (see Chapter 2, ref. 37 for its synthesis) furnished 1-deoxycastanospermine in a process which involved concomitant reductive amination to form the postulated intermediate **207**, attack of the ring-nitrogen atom on the nitrile group and reductive deamination,[65] and access to 1-deoxy-6-*epi*-castanospermine was provided by the 'bicyclic lactam route'.[66] Compound **211**, the product of a vinylogous Mannich reaction of aminal **210** and 2-(*tert*-butyldimethylsilyloxy)furan, has been converted to the hydroxymethyl-bearing trihydroxyindolizidine **212** by dihydroxylation ($KMnO_4$), followed by lactone → lactam rearrangement on removal of the Boc-protecting group, then reduction and *O*-deprotection.[67] The versatile cyclic nitrone **208** has also been converted to a number of bicyclic α-L-fucosidase inhibitors, such as the indolizidines **209**.[68]

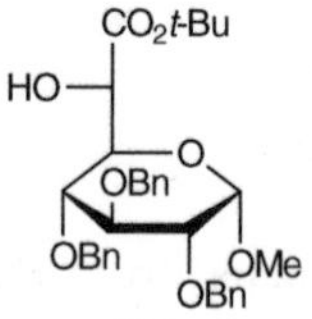

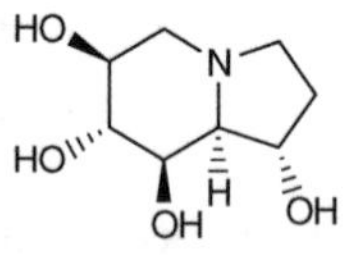

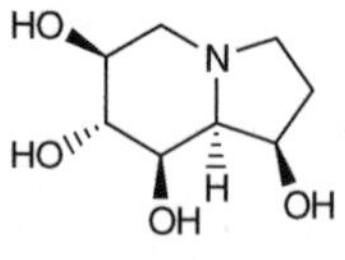

Swainsonine (**192**) has been monobutanoylated at O-2 and dibutanoylated at O-1 and O-2 by use of subtilisin in pyridine and porcine pancreatic lipase in dry THF.[59]

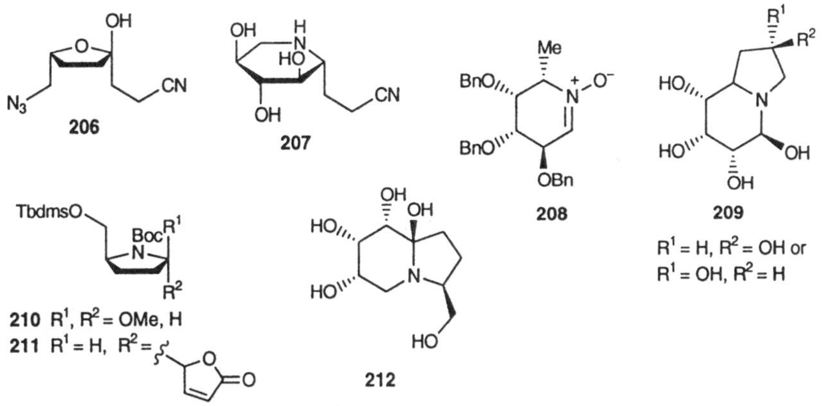

A key intermediate (**215**) towards Ecteinascidin 743 has been prepared *via* glucose-derived aziridine **214**, which was prepared in seven steps from the epoxide **213** (Scheme 30).[69]

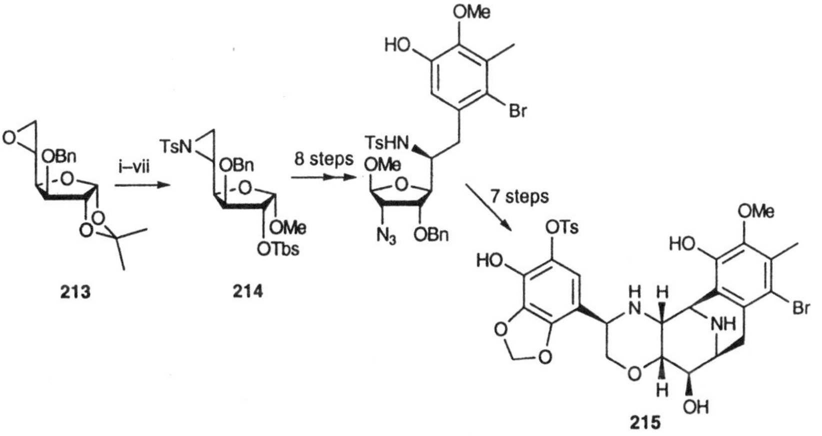

Reagents: i, $TsNH_2$, Cs_2CO_3; ii, MsCl, Et_3N; iii, HCl, MeOH; iv, DIBAL; v, $SnCl_4$; vi, NaOH, MeOH; vii, TbdmsCl, ImH

Scheme 30

Reaction of L-cysteine with **216** in the presence of pyridine affords the thizolidine lactam **217** in quantitative yield. The 5-azido-5-deoxy analogue of **216** undergoes similar reaction and was elaborated into the β-turn peptide mimetic **218**.[70] Similar heterocycles have been prepared by reaction of carbohydrates with 2-aminoethanethiol/sodium methoxide followed by tosylaltion and resulting intramolecular cyclization. In this way, sugar diastereomers **219** have been (separately) converted into diastereomers **220**, and **221** has been converted into **222**.[71] Thermolysis of azide **223** gave the tetrahydrothiazepine **224**.[72]

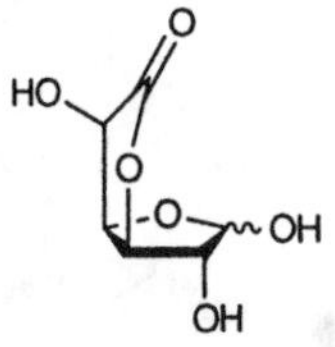

216 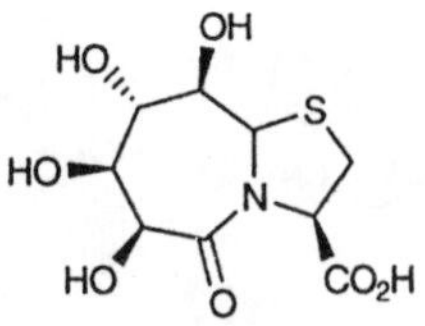 217 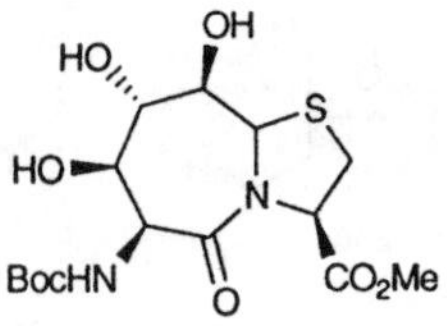 218

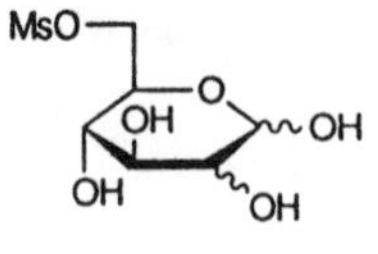

219 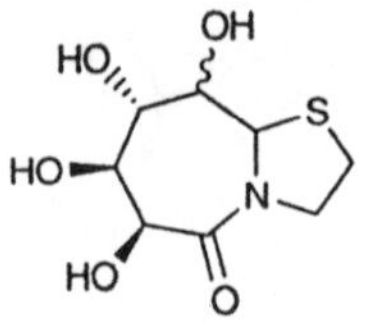 220 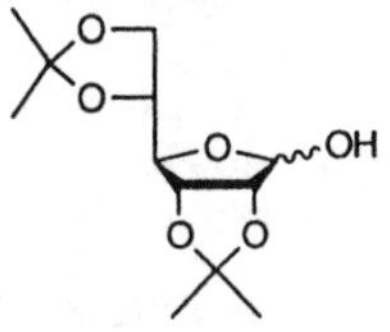 221

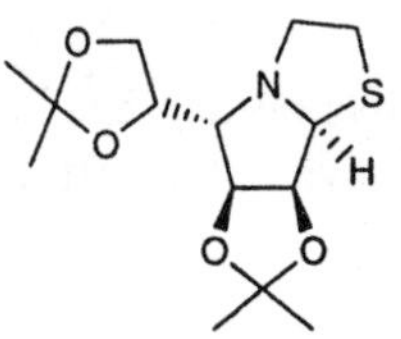

222 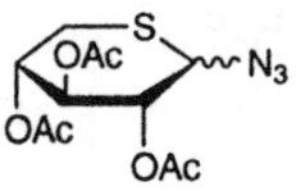 223 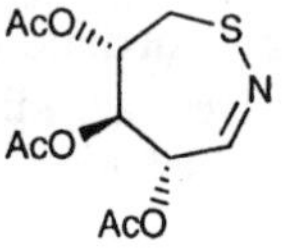 224

6 Acyclic Compounds

A review on sphingolipids, covering metabolic pathways and medicinal significance has appeared.[73] The D-ribonolactone-derived diketones **225**, potentially useful for preparation of a range of carbocyclic compounds, epimerize on standing.[74]

Pyranoid glycals **226–228** have been converted into chiral dienes **229–231**, respectively, by reaction with the corresponding aryl Grignard reagents and nickel(0) (and acetylation in the case of the products from **226** and **227**).[75]

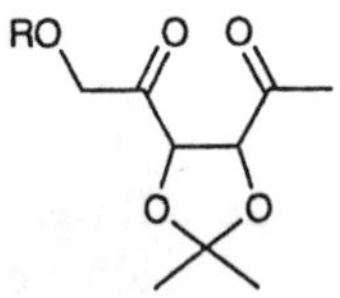

225 R = Tr, Tbdms

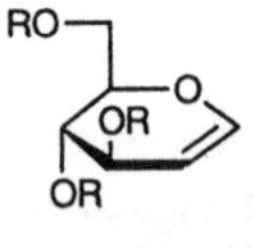

226 R = Me
227 R = Bn

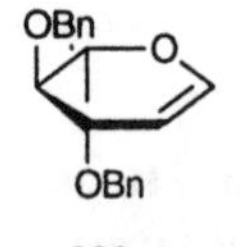

228

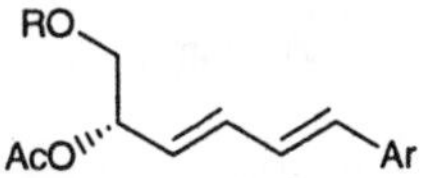

229 Ar = Ph, R = Me
230 Ar = Pmp, R = Bn

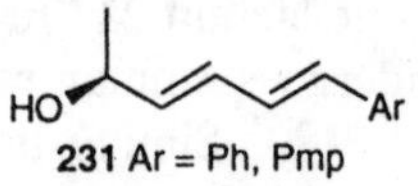

231 Ar = Ph, Pmp

The first synthesis of a β,γ-dihydroxyglutamic acid has been completed from the known D-ribose derived **232**, *via* the aziridine lactone **233** (Scheme 31).

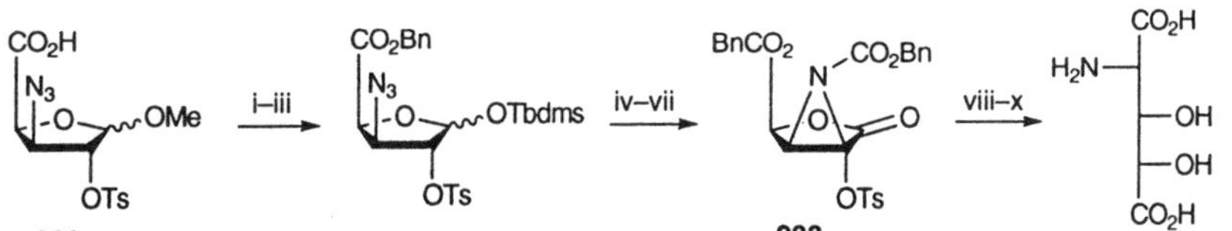

CO2H N3 OMe i–iii CO2Bn N3 OTbdms iv–vii BnCO2 N CO2Bn O O viii–x CO2H H2N OH OH CO2H OTs OTs OTs 232 233

Reagents: i, BnOH, DCC, DMAP; ii, CF_3CO_2H, H_2O; iii, TbdmsCl, Im, DMF; iv, Ph_3P, Py, H_2O, Et_3N, Δ v, BnO_2CCl, NaOH vi, Bu_4NF, HOAc, THF; vii, tetrapropylammonium perruthenate; viii, BnOH, $BF_3.OEt_2$; ix, H_2, Pd-C; x, Resin (hydroxide) and elution with H_2O-HOAc

Scheme 31

Stereoselective [2,3]-Wittig rearrangement (*n*-BuLi, −78 °C) of *E*- and *Z*-allylic stannyl methyl ethers **235** and **236**, derived from L-threitol derivative **234**, provides the homoallylic alcohols **237** and **238**. The *E* intermediate **235** affords a higher **237**:**238** ratio, the major isomer **237** being elaborated to the known polyoxamic acid derivative **239**.[76]

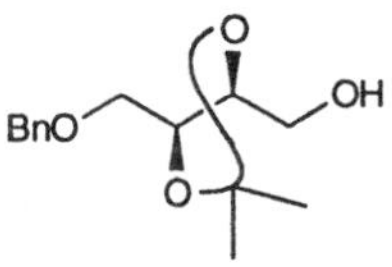

BnO O OH O 234 BnO O O SnBu3 O 235 O SnBu3 BnO O O 236

BnO O OH O 237 BnO O OH O 238 O H2N O O CO2H O 239

BnO O OH OBn BnO OBn 240 BnO O OH OBn BnO OBn 241 BnO O OH OBn BnO OBn 242

O OBn O BnO CHO BnO 243 O OBn O BnO CHO BnO 244 OH O O 245

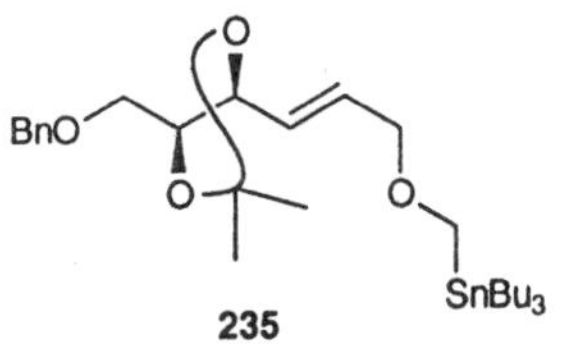

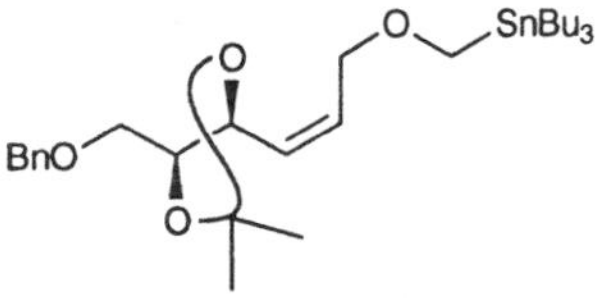

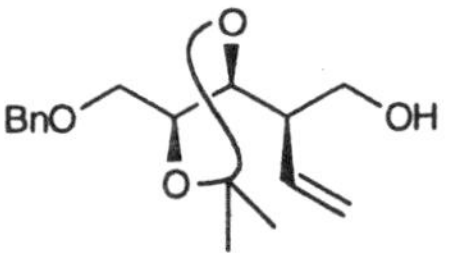

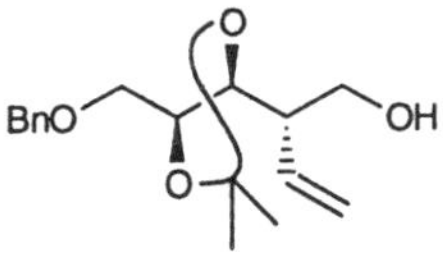

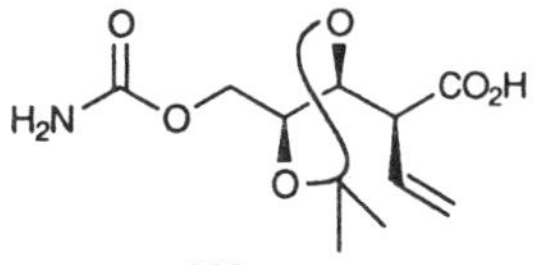

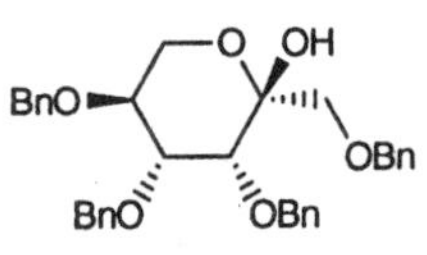

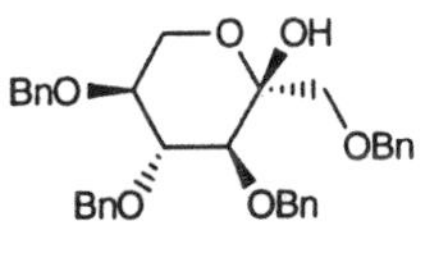

OH (CH2)14Me 246 OH O O 247 OH (CH2)14Me 248

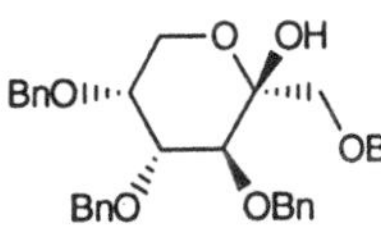

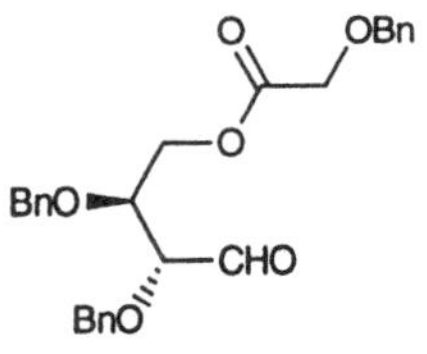

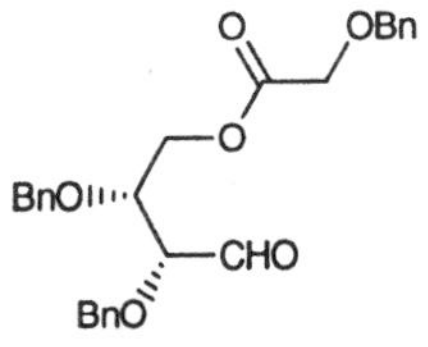

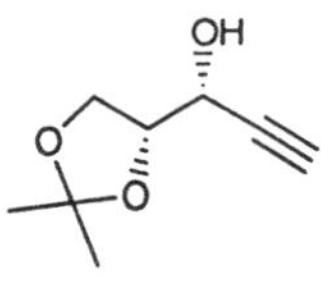

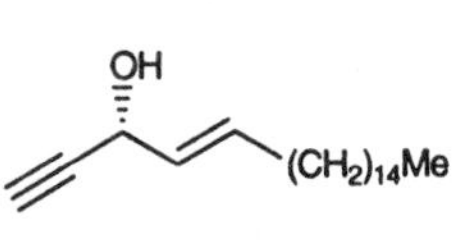

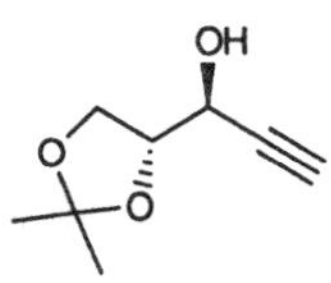

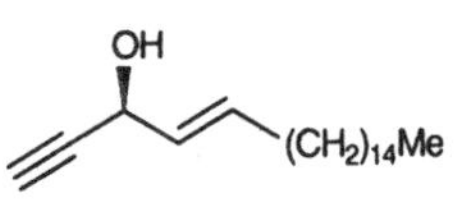

Tetra-*O*-benzyl L-tagalose and L-sorbose ketoses **240** or **241** react with iodine and iodobenzene ditrifluoroacetate to give **243**, and the D-fructose analogue **242**, an epimer of **241**, gives the epimeric L-erythrose 4-ester, **244**, these reactions proceeding by an unusual oxidative cleavage of the C2–C3 bond.[77] D-Xylose-derived **245** has been converted into **246**, and D-gulonolactone-derived **247** has been converted into **248**, the enantiomer.[78]

The natural (3*R*,9*R*,10*R*)-diastereoisomer of panaxytriol (**251**) was synthesized by coupling (CuI, hydroxylamine, triethylamine then TBAF) the sugar-derived alkynes **249** (from D-xylose) with **250** (from D-gulonolactone).[79]

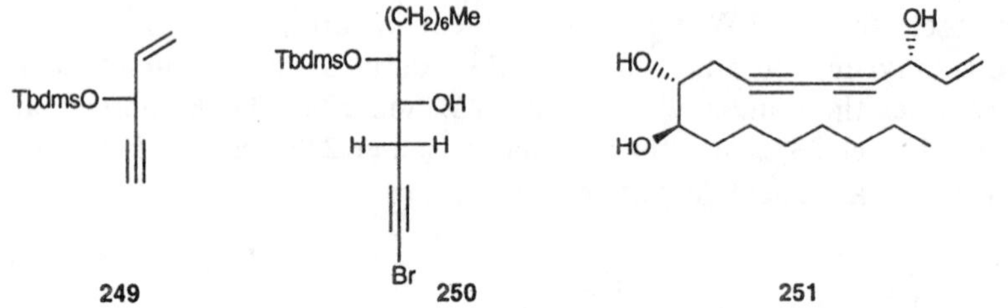

249 **250** **251**

A sythesis of AK-toxin I (**254**), which starts from D-fructose-derived **252**, has been reported. The key assembly of the *E*,*Z*,*E*-triene moiety was achieved by palladium-catalysed coupling of the *E*-vinylstannane derived from **253** with the requisite *Z*-bromodiene (Scheme 32).[80]

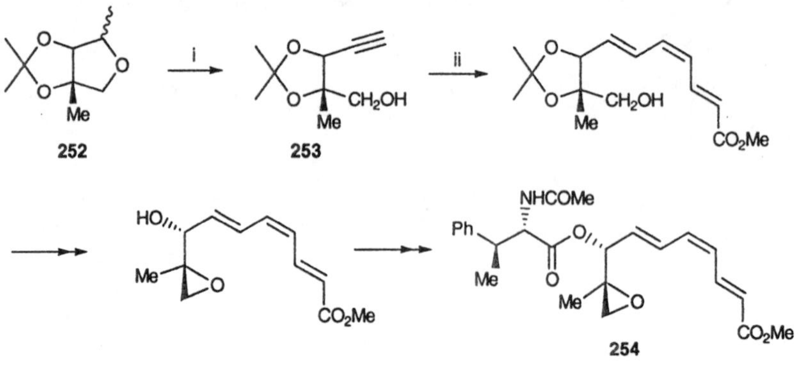

252 **253** **254**

Reagents: i, K_2CO_3, MeOH, $(MeO)_2P(O)C(=N_2)COMe$; ii, Bu_3SnH, (*Z*,*E*)-BrCH=CH-C=CHCO$_2$Me, $(PPh_3)_2PdCl_2$

Scheme 32

The first total synthesis of bengazole A **258** has been achieved by addition of 2-lithiooxazole to **255** to give **256**. A regeneration of a lithiooxazole intermediate from this product (after silylation of the free OH) allows reaction with an oxazole carboxaldehyde to provide **257** which was acylated to give the natural product (Scheme 33).[81]

Haines and Lamb have described studies on syntheses of higher sugars using aldol couplings [for example reaction of the boron enolate of **259** with C3 or

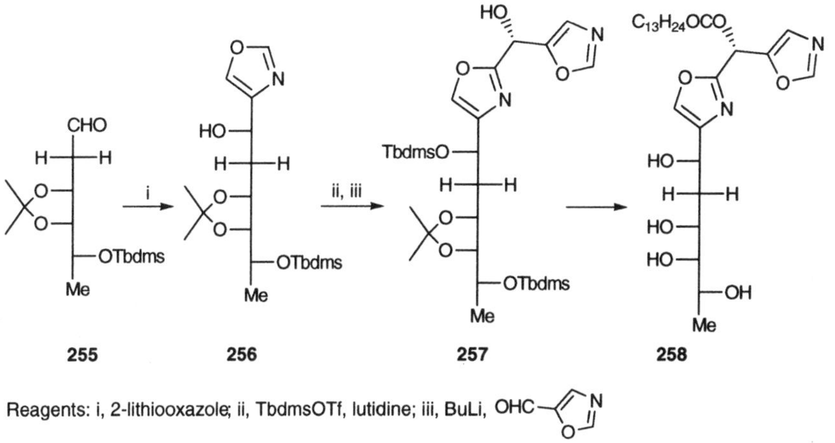

Reagents: i, 2-lithiooxazole; ii, TbdmsOTf, lutidine; iii, BuLi, OHC-(oxazolyl)

Scheme 33

C5 carbohydrate aldehydes to give aldols **260** (methodology is directed towards the herbicidins).[82] The optically active unsymmetrical phosphatidyl glyceride **263** has been prepared by periodiate cleavage and sodium borohydride reduction of **261** to give **262**, which was then elaborated to **263**.[83] A small library of phospholipids **264** has been synthesized from D-mannitol *via* the derived 2,3-*O*-isopropylidine-(*S*)-glycerol.[84]

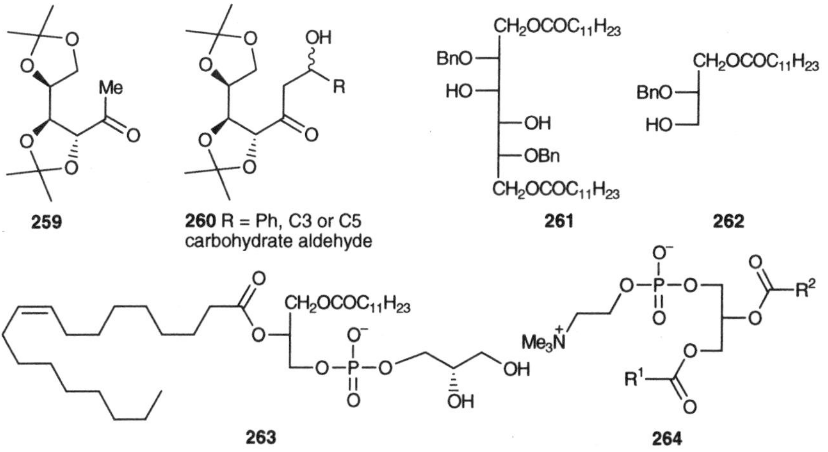

259 **260** R = Ph, C3 or C5 carbohydrate aldehyde **261** **262** **263** **264**

An improved route to D-*ribo*-phytosphingosine **268** from D-glucosamine, *via* **265**, has been reported. The intermediate **265** was also elaborated into halicylindroside analogues **266** and **267** (Scheme 34).[85]

A synthesis of sphingofungin D (**271**) from 2-acetamido-2-deoxy-mannono-1,4-lactone derivative **269** involved a key Cr-Ni coupling reaction. The diastereomeric products (**270**) were separated by derivatizing as the 3,5-*O*-isopropylidine acetals, deprotection and lactone opening to afford the isomerically pure target (Scheme 35).[86]

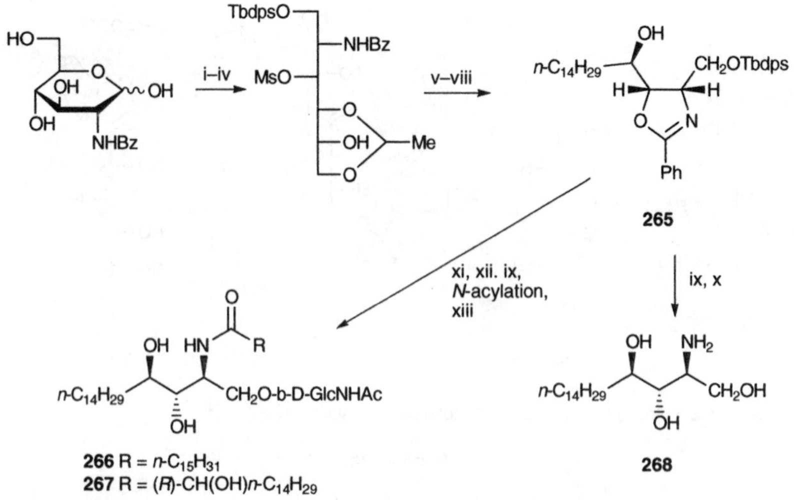

265

266 R = n-$C_{15}H_{31}$
267 R = (R)-CH(OH)n-$C_{14}H_{29}$

268

Reagents: i, MeCHO, $ZnCl_2$; ii, $NaBH_4$; iii, TbdpsCl, Py; iv, MsCl, Py; v, Py, toluene, Δ; vi, $TiCl_4$, PhSH; vii, $NaIO_4$; viii, n-$C_{14}H_{29}MgCl$; ix, H_3O^+: x, NaOH, EtOH, H_2O, xi, [structure]; xii, Bu_3SnH, AIBN; xiii, NaOMe

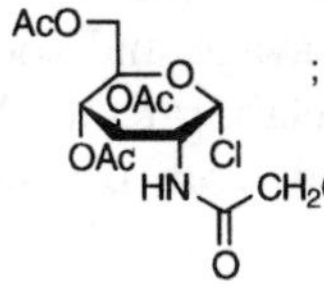

Scheme 34

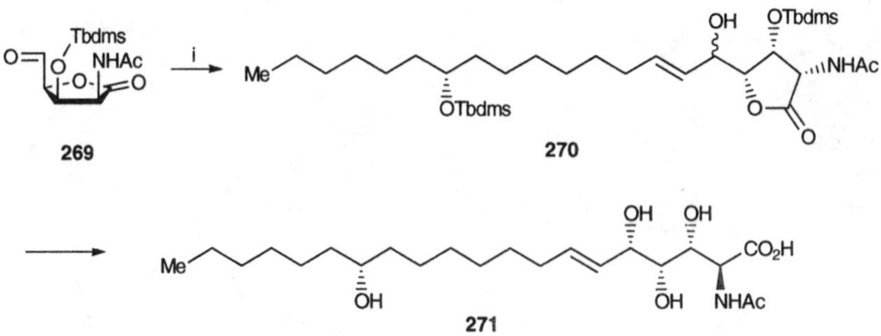

269 270 271

Reagents: i, $CrCl_2$, NiCl, DMSO, $Me(H_2C)_5$–CH(OTbdms)–$(CH_2)_6$–CH=CH–I

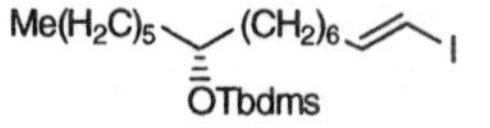

Scheme 35

7 Carbohydrates as Chiral Auxiliaries, Reagents and Catalysts

7.1 Carbohydrate-derived Reagents and Auxiliaries.

– Several reports this year have described the use of carbohydrate derivatives as auxiliaries in dipolar or Diels-Alder cycloadditions reactions. Compound **272** (which equilibrates with the acyclic oxime) reacts with ethyl glyoxylate to generate intermediate **273**, which undergoes dipolar addition to vinyl acetate to give **274** and epimers. Reaction of **274** with thymine catalysed by TMSOTf followed by acid cleavage of the sugar auxiliary affords enantiomerically pure **275**.[87]

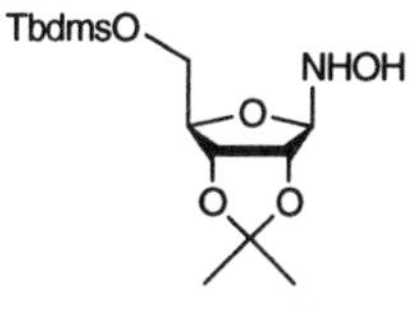

272 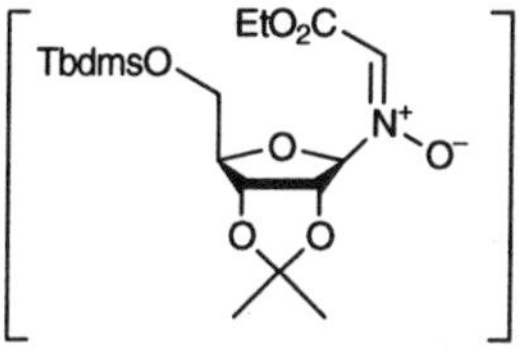 **273** 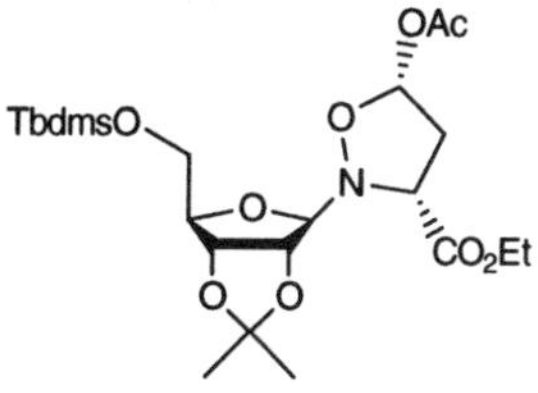 **274**

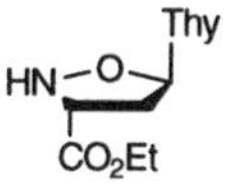

275 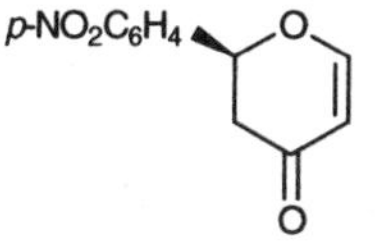 **276** 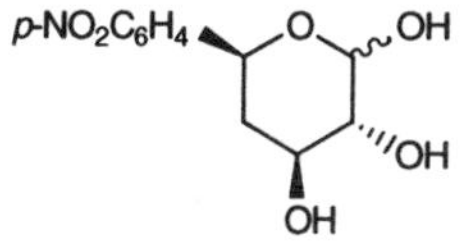 **277**

The dihydroaryl pyranone **276**, derived by cycloaddition of *p*-nitrobenzaldehyde to a glucose auxiliary-bearing diene (Vol. 28, p. 374) has been elaborated further to provide **277**, by cerium trichloride-sodium borohydride reduction, silylation, diastereoselective osmylation and final desilylation (with IR120 acid resin). In addition, the Diels-Alder reaction of glucose auxiliary-bearing diene **278** (R* = β-D-tetra-*O*-acetylglucose) with *p*-nitrobenzaldehyde and Eu(fod)$_3$ catalysis provides **279** as the major isomer. This was then elaborated to the *C*-aryl D-galactose analogue **280** (Scheme 36).[88]

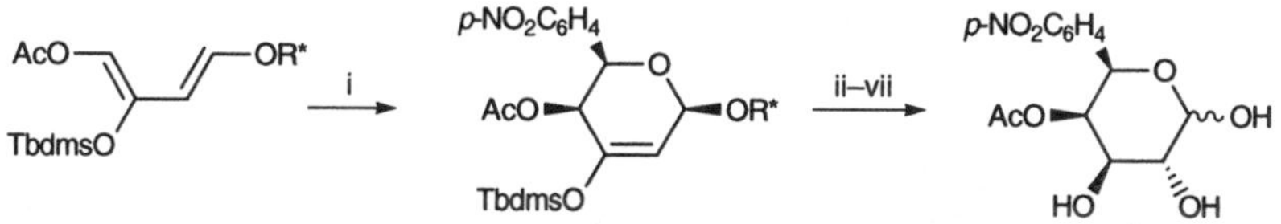

278 **279** **280**

Reagents: i, Eu(fod)$_3$, p-$NO_2C_6H_4CHO$; ii, TFA; iii, $NaBH_4$, $CeCl_3$; iv, Et_3N, CH_2Cl_2; v, OsO_4, $Ba(ClO_3)_2$; vi, Ac_2O, TfOH; vii, MeOH, Et_3N

Scheme 36

Fructose-derived auxiliary **282** has been used to control Diels-Alder reactions. It was prepared by reaction of **281** with sodium methoxide then protection with acid/2,2-dimethoxypropane, and treated with acryloyl chloride. The product underwent reaction with cyclopentadiene to give a cycloadduct which, after reductive cleavage of the ester to remove the auxiliary, yielded the bicyclic product **283**.[89] The D-mannitol derived C_2-symmetric pyrrolidine **284** was derivatized as the diene **285**, which underwent Diels-Alder reactions with e.e.s of 63–85%.[90]

Glucosylamine **286** has seen applications this year as an auxiliary. In one report, it was employed in (zinc chloride catalysed) Ugi reactions with an

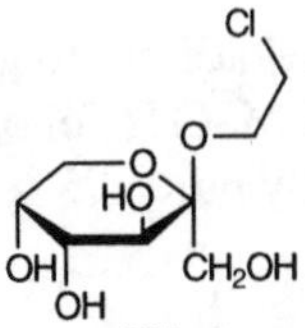

281 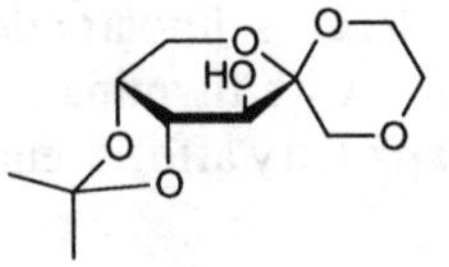 **282**  **283**

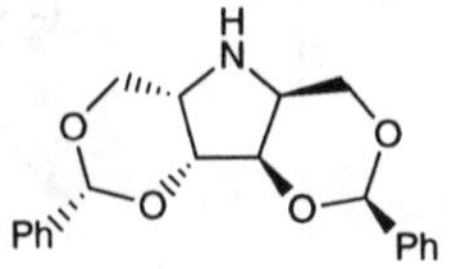

284 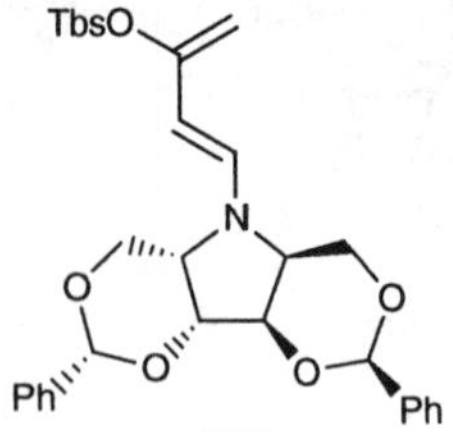 **285**

aldehyde, formaldehyde and the isonitrile **287**. The Ugi products **288** were hydrolysed with hydrochloric acid to generate a variety of amino acids.[91] Reductive amination of aldehyde **289** using **286** and sodium cyanoborohydride afforded adduct **290**, which on hydrolysis afforded (*S*)-nornicotine (**291**). This approach was also applied using **292**.[92]

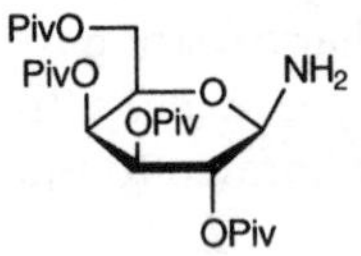

286 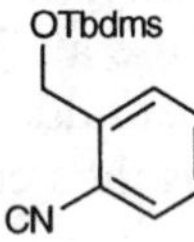 **287** 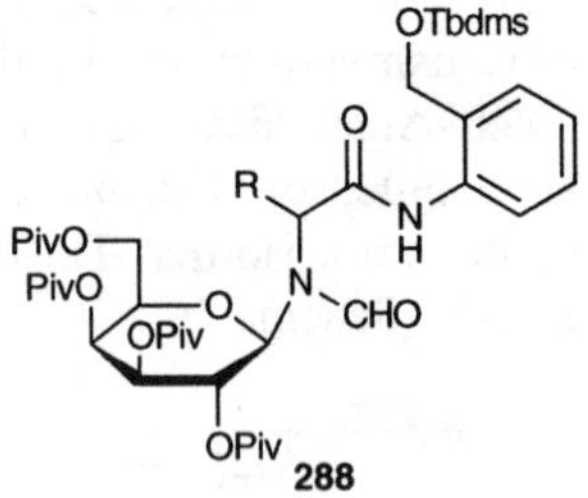 **288** 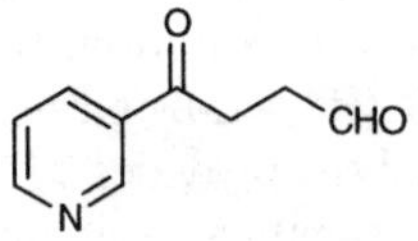 **289**

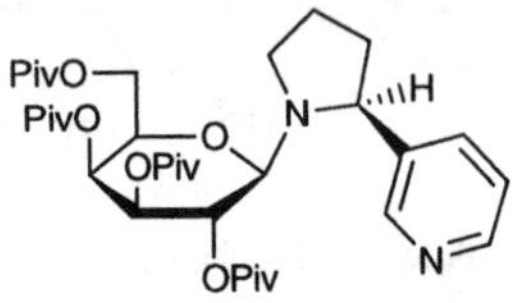

290 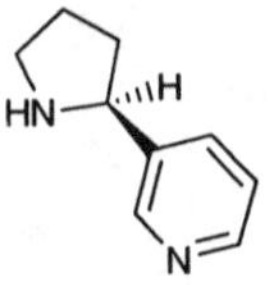 **291** 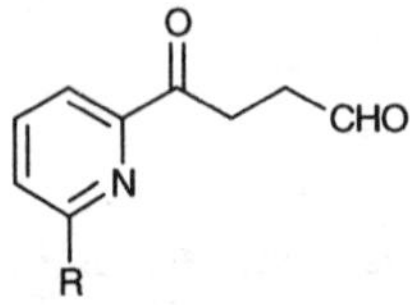 **292** R = Br or Ph

The enone **293** reacts with NBS and alcohols to generate the α-bromoketones **294**, which can then be trans-acetalated to **295** or dithioacetals **296**. The same enone **293** can be diastereoselectively epoxidized to **297**, which can then be converted into **298** and **299** (in all cases R* = tetra-*O*-acetyl-β-D-glucopyranosyl).[93]

The D-glucosamine-derived imine **300** reacts with acid chlorides to give the β-lactams **301**, which, after removal of the sugar auxiliary with BuLi, give the parent β-lactams **302**.[94]

Glucose has been used as an auxiliary to control the enantiomeric selectivity in the synthesis of butane and propane 1,2-diols **305** and **306** [R = Me, H], *via*

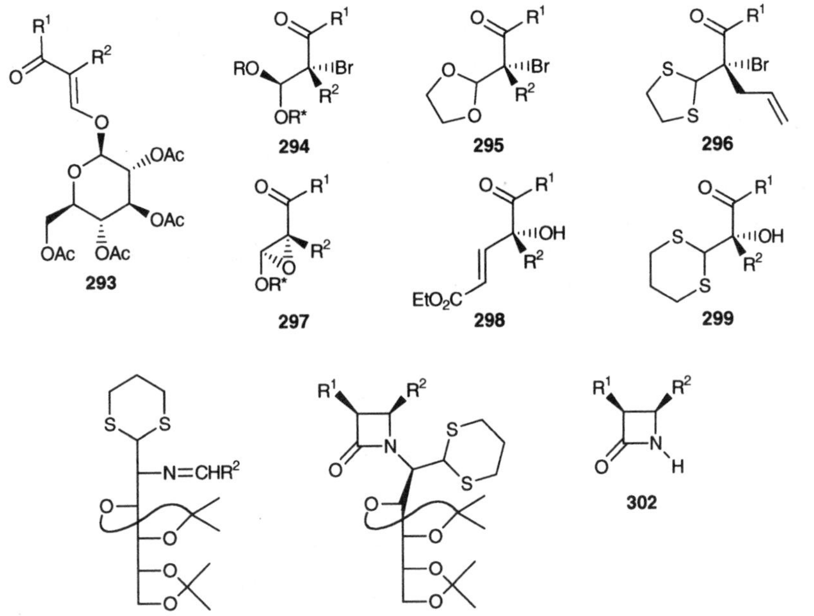

293 294 295 296 297 298 299 300 301 302

mercuration of the alkenyl glucosides **303** and the β-anomers, respectively (Scheme 37). The desired enantiomers were obtained in 81% yield and >99% e.e. The β-glucosides gave similar e.e. of the other enantiomer **306**, but with d.e. of around 58%.[95]

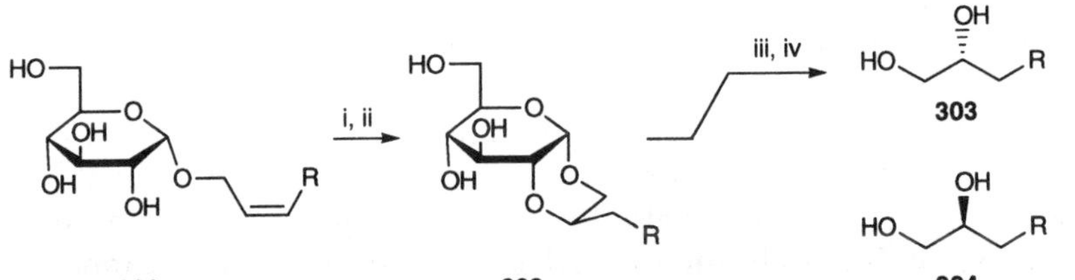

Reagents: i, $Hg(Tfa)_2$; ii, $NaBH_4$; iii, BF_3OEt_2, Ac_2O, H^+ iv, H_2O_2, OH^-

Scheme 37

The choice of base influences the outcome of sulfination and phosphinylation of diacetone-D-glucose to **307** and **308**, respectively [reaction with MeSOCl or MeP(=O)ClPh]. The R_S:S_S ratio for **307** was 7:93 using pyridine,

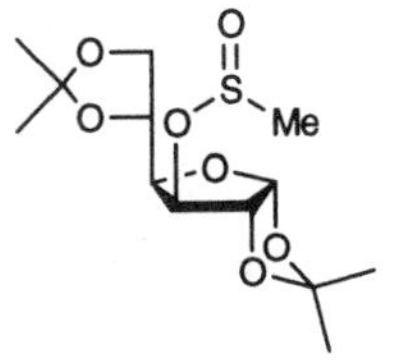

307 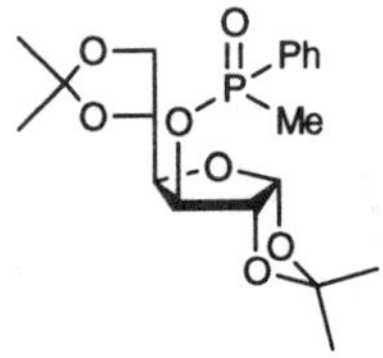 308

but reversed to >98:2 when diisopropylethylamine was employed. A similar outcome was evident in the phosphorus case, where the corresponding R_p:S_p ratios were 25:75 and 97:3.[96]

An unusual use of sugars as auxiliaries *via* their boronate complexes, in which chirality is transferred to a C_{60} derivative has been reported. Thus, **309** reacted with C_{60} (*via* the *o*-quinodimethide derived from the aryl groups – to give an adduct). Removal of the sugar and ester exchange with 2,3-dimethylpropane 1,3-diol then left the difunctionalized chiral C_{60} **310**. Analogues made from other carbohydrate boronates were also decribed.[97]

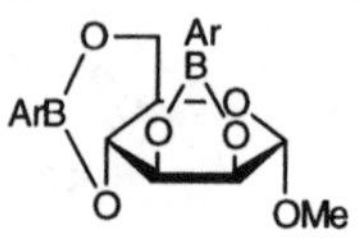

309 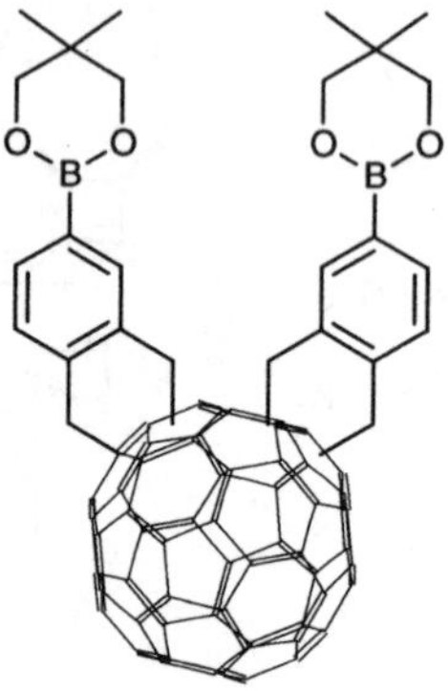 **310**

7.2 Carbohydrate-derived Catalysts. – Several further reports of the utility of the dioxiranes generated from fructose-based ketone **311** for asymmetric oxidation reactions have appeared this year. A study of the mechanism of oxidation of *vic*-diols to α-hydroxy ketones by **311** has been reported. Diol **312** gives **315** in 45% e.e. whilst diol **314** gives **315** in 65% e.e., and diol **314** gave α-hydroxy ketone **316** in 69% e.e.[98] Epoxidation of a range of eneynes with the same dioxirane (from **311**) gave epoxyalkynes of type **317** and **318** with modest to good enantioselectivity (see also Volume 32, Chapter 24).[99] The asymmetric epoxidation of 2,2-disubstituted vinyl silanes affords, after treatment with TBAF, the 1,1-disubstituted epoxides **319** in up to 94% e.e. (*e.g.* for R^1 = Me, R^2 = Ph).[100]

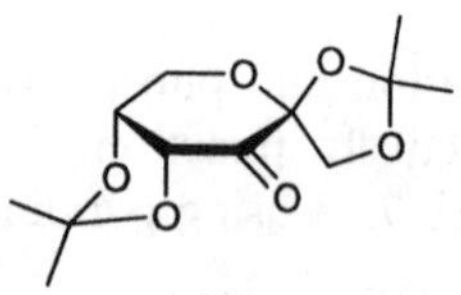

311 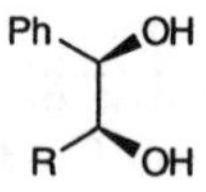 **312** R = Ph 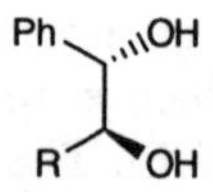 **313** R = Ph **314** R = Me **315** R = Ph **316** R = Me

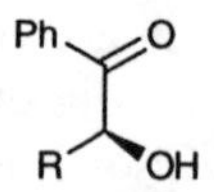

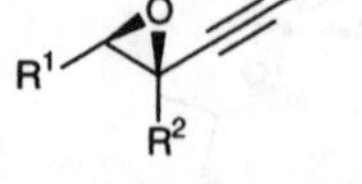

317 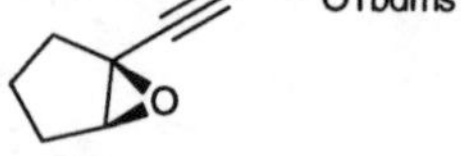 **318** 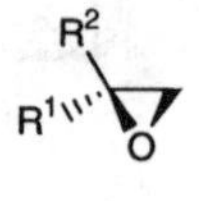 **319**

The C_2-chiral bisphosphine ligand **321** has been prepared from D-mannitol, *via* the cyclic sulfate **318** (Scheme 38). The ligand **321** catalyses asymmetric hydrogenation of substrates **322** (R = H, Ph, *p*-F-C_6H_4; R^1 = H, Me) to **323** with e.e. of ⩾98%.[101]

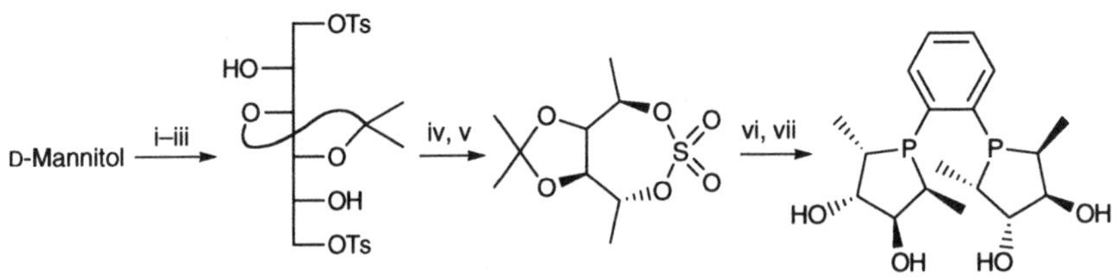

320 321

Reagents: i, Acetone, H_2SO_4; ii, AcOH, H_2O; iii, TsCl, Py; iv, LAH; v; (a) $SOCl_2$, Et_3N, (b) $NaIO_4$, $RuCl_3$; vi, (a) 1,2-$H_2PC_6H_4PH_2$, *n*-BuLi, (b) *n*-BuLi; vii, MeOH, H_2O, $MeSO_3H$

Scheme 38

1,2:5,6-Di-*O*-isopropylidine-3,4-bis(diphenylphosphino)-D-mannitol rhodium complexes catalyse asymmetric hydrogenation of **322** (R = aryl; R^1 = H), giving the *R* configuration product. An increase in reaction pressure increases the e.e. (*e.g.* with R = Ph, e.e. could be increased from 90 to 97%).[102] The glucose-derived bisphosphinites **324** and **325** have also been used for asymmetric Rh-catalysed alkene hydrogenations.[103] The glucose-derived ligand **326** catalysed the π-allyl palladium mediated conversion of **327** into **328**.[104] Several examples of disaccharide-derived bisphospho catalysts have also been reported. The Rh(cod) catalyst derived from **329** (made from α,α-trehalose) or from the β,β-trehalose-derived analogue catalyse the asymmetric hydrogenation of **322** (R = Ph, R^1 = Me) with up to >99% e.e., and in at least one case 99.9% e.e.[105,106]

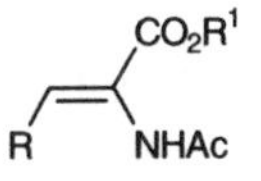

322

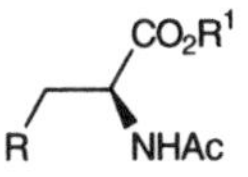

323

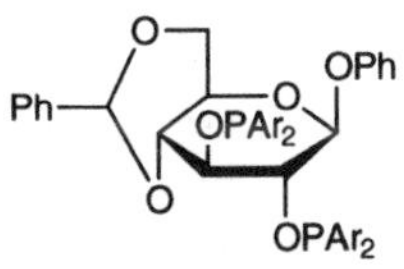

324

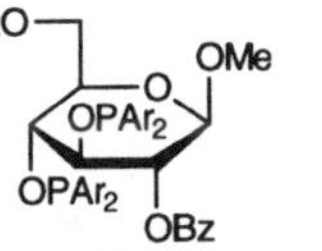

325

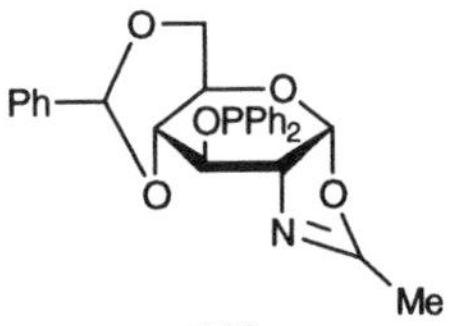

326

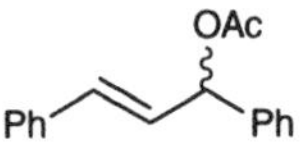

327

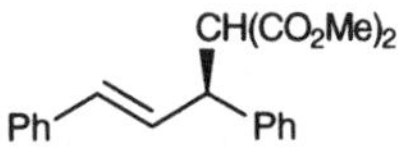

328

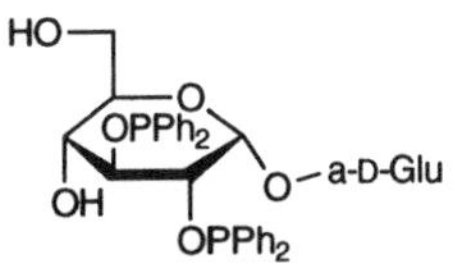

329

References

1 P. I. Dalko and P. Sinaÿ, *Angew. Chem., Int. Ed. Engl.*, 1999, **38**, 773.

2 S. Amano, N. Takemura, K. Sugihara, N. Ogawa, M. Ohtsuka, S. Ogawa and N. Chida, *Tennen Yuki Kagobutsu toronkai Koen Yoshishu*, 1998, **39**, 661 (*Chem. Abstr.*, 1999, **130**, 325 293).

3 D. J. Holt, W. D. Barker, P. R. Jenkins, S. Ghosh and D. R. Russell, *Synlett*, 1999, 1003.
4 A. Blaser and J.-L. Reymond, *Helv. Chim. Acta*, 1999, **82**, 760.
5 H. Hrebabecky, M. Masojidkova and A. Holy, *Collect. Czech. Chem. Commun.*, 1998, **130**, 2044 (*Chem. Abstr.*, 1999, **130**, 139 567).
6 J. K. Gallos, T. V. Koftis, A. E. Koumbis and V. I. Moustos, *Synlett*, 1999, 1289.
7 D. L. J. Clive, X. He, M. H. D. Postema and M. J. Mashimbye, *J. Org. Chem.*, 1999, **64**, 4397.
8 S. Amano, N. Ogawa, M. Ohtsuka and N. Chida, *Tetrahedron*, 1999, **55**, 2205.
9 A. G. Fallis, *Acc. Chem. Res.*, 1999, **32**, 464.
10 A. Roy, K. Chakrabary, P. K. Dutta, N. C. Bar, N. Basu, B. Achari and S. B. Mandal, *J. Org. Chem.*, 1999, **64**, 2304.
11 J. S. Clark, A. G. Dossetter, A. J. Blake, W.-S. Li and W. Whittingham, *Chem. Commun.*, 1999, 749.
12 H. B. Mereyla and R. R. Gadikota, *Tetrahedron: Asymmetry*, 1999, **10**, 2305.
13 P. A. Zunszain and O. Varela, *An. Assoc. Quim. Argent.*, 1998, **86**, 151 (*Chem. Abstr.*, 1999, **130**, 153 873).
14 R. I. Hollingsworth, *J. Org. Chem.*, 1999, **64**, 7633.
15 H. B. Mereyla and R. R. Gadikota, *Chem. Lett.*, 1999, 273.
16 Y. Masaki, H. Arasaki and A. Itoh, *Tetrahedron Lett.*, 1999, **40**, 4829.
17 L. Lemée, A. Jégou and A. Veyrières, *Tetrahedron Lett.*, 1999, **40**, 2761.
18 H. Toshima, H. Sato and A. Ichihara, *Tetrahedron*, 1999, **55**, 2581.
19 T. K. M. Shing and V. W.-F. Tai, *J. Org. Chem.*, 1999, **64**, 2140.
20 S. Abel, D. Faber, O. Hütter and B. Giese, *Synthesis*, 1999, 188.
21 L. A. Paquette, T.-L. Shih, Q. Zeng and J. E. Hofferberth, *Tetrahedron Lett.*, 1999, **40**, 3519.
22 V. Popsavin, M. Popsavin, L. Radic, O. Beric and V. Cirin-Novta, *Tetrahedron Lett.*, 1999, **40**, 9305.
23 D. E. Gron, T. Durand, A. Roland, J.-P. Vidal and J.-C. Rossi, *Synlett*, 1999, 435.
24 Q. Yu, Z.-Y. Yao, X.-G. Chen and Y.-L. Wu, *J. Org. Chem.*, 1999, **64**, 2440.
25 J. Zhang and D. L. J. Clive, *J. Org. Chem.*, 1999, **64**, 1754.
26 M. Hayashi, H. Kawabata and K. Yamada, *Chem. Commun.*, 1999, 965.
27 P. Bertrand, H. El Sukkari, J.-P. Gesson and B. Renoux, *Synthesis*, 1999, 330.
28 S. Harasawa, T. Imazu, S. Takashima, L. Araki, H. Ohishi, T. Kurihara, Y. Sakamoto, Y. Yamamoto and A. Yamatodani, *J. Org. Chem.*, 1999, **64**, 8608.
29 E. Djuardi and E. McNelis, *Tetrahedron Lett.*, 1999, **40**, 7193.
30 M. Hein and R. Mietchen, *Eur. J. Org. Chem.*, 1999, 2429.
31 M. Isobe, R. Nishizawa, T. Nishikawa and K. Yoza, *Tetrahedron Lett.*, 1999, **40**, 6927.
32 C. Taillefumier, M. Lakhrissi and Y. Chapleur, *Synlett*, 1999, 697.
33 P. Fraise, I. Hanna, J.-Y. Lallemand, T. Prange and L. Ricard, *Tetrahedron*, 1999, **55**, 11819.
34 K. Bedjeguelal, V. Bolitt and D. Sinou, *Synlett*, 1999, 762.
35 S. Hosokawa and M. Isobe, *J. Org. Chem.*, 1999, **64**, 37.
36 L. Eriksson, S. Guy, P. Perlmutter and R. Lewis, *J. Org. Chem.*, 1999, **64**, 8396.
37 A. Pal, A. Bhattacharjee, A. Bhattacharjya and A. Patra, *Tetrahedron*, 1999, **55**, 4123.
38 M. Sasaki, M. Inoue, K. Takamatsu and K. Tachibana, *J. Org. Chem.*, 1999, **64**, 9399.

39 M. Inoue, M. Sasaki and K. Tachibana, *J. Org. Chem.*, 1999, **64**, 9416.
40 H. Oguri, S. Sasaki, T. Oishi and M. Hirama, *Tetrahedron Lett.*, 1999, **40**, 5405.
41 K. C. Nicolaou, M. E. Bunnage, D. G. McGarry, S. Shi, P. K. Somers, P. A. Wallace, X.-J. Chu, K. A. Agrios, J. L. Gunzer and Z. Yang, *Chem. Eur. J.*, 1999, **5**, 599.
42 K. C. Nicolaou, P. A. Wallace, S. Shi, M. A. Ouellette, M. E. Bunnage, J. L. Gunzer, K. A. Agrios, G. Shi and Z. Yang, *Chem. Eur. J.*, 1999, **5**, 618.
43 K. C. Nicolaou, G. Shi, J. L. Gunzer, P. Gärtner, P. A. Wallace, M. A. Ouellette, S. Shi, M. E. Bunnage, K. A. Agrios, C. A. Veale, C.-K. Kwang, J. Hutchinson, C. V. C. Prasad, W. W. Ogilvie and Z. Yang, *Chem. Eur. J.*, 1999, **5**, 628.
44 K. C. Nicolaou, J. L. Gunzer, Q. Shi, K. A. Agrios, P. Gärtner and Z. Yang, *Chem. Eur. J.*, 1999, **5**, 646.
45 Y. Yamamoto, J. Heterocycl. *Chem.*, 1999, **36**, 1523.
46 K. C. Nicolaou, H. J. Mitchell, H. Suzuki, R. M. Rodriquez, O. Baudoin and K. C. Fylaktakidou, *Angew. Chem., Int. Ed. Engl.*, 1999, **38**, 3334.
47 J. C. Y. Wong, P. Lacombe and C. F. Dturino, *Tetrahedron Lett.*, 1999, **40**, 8751.
48 C. A. Weir and C. M. Taylor, *J. Org. Chem.*, 1999, **64**, 1554.
49 J. Rabiczko and M. Chmielewski, *J. Org. Chem.*, 1999, **64**, 1347.
50 M. Avalos, R. Babiano, P. Cintas, F. J. Higes, J. L. Jiménez and J. C. Palacios, *Tetrahedron: Asymm.*, 1999, **10**, 4071.
51 L. A. Paquette, S. Brand and C. Behrens, *J. Org. Chem.*, 1999, **64**, 2010.
52 J. Kuban, I. Blanarikova, L. Fuisera, L. Jaroskova, M. Fengler-Veith, V. Jager, J. Kozisek, O. Humpa, N. Pronayova and V. Langer, *Tetrahedron*, 1999, **55**, 9501.
53 A. Pal, A. Bhattacharjee and A. Bhattacharjya, *Synthesis*, 1999, 1569.
54 J. Piró, M. Rubiralta, E. Giralt and A. Diez, *Tetrahedron Lett.*, 1999, **40**, 4865.
55 C. Di Nardo and O. Varela, *J. Org. Chem.*, 1999, **64**, 6119.
56 M. Ruiz, V. Ojea and J. M. Quintela, *Synlett*, 1999, 204.
57 B. V. Ernholt, I. B. Thomsen, K. B. Jensen and M. Bols, *Synlett*, 1999, 701.
58 S. D. Koulocheri and S. A. Haroutounian, *Tetrahedron Lett.*, 1999, **40**, 6869.
59 G. G. Perrone, K. D. Barrow and I. J. McFarlane, *Bioorg. Med. Chem*, 1999, 7, 831.
60 A. Kato, I. Adachi, M. Miyauchi, K. Ikeda, T. Komae, H. Kizu, Y. Kameda, A. A. Watson, R.J. Nash, M.R. Wormald, G.W.J. Fleet and N. Asano, *Carbohydr. Res.*, 1999, **316**, 95.
61 U. K. Pandit, H. S. Overkleft, B. C. Borer and H. Bieräugel, *Eur. J. Org. Chem.*, 1999, 959.
62 L.-X. Liao, Z.-M. Wang, H.-X. Zhang and W.-S. Zhou, *Tetrahedron: Asymmetry*, 1999, **10**, 3649.
63 S. E. Denmark and E. A. Martinborough, *J. Am. Chem. Soc.*, 1999, **121**, 3046.
64 E. Bartnicka and A. Zamojski, *Tetrahedron*, 1999, **55**, 2061.
65 I. Izquierdo, M. T. Plaza, R. Robles, C. Rodriguez, A. Ramirez and A. J. Mota, *Eur. J. Org. Chem.*, 1999, 1269.
66 A. I. Meyers, C. J. Andres, J. E. Reseke and C. C. Woodall, *Tetrahedron*, 1999, **55**, 8931.
67 G. G. Rassu, P. Carta, L. Pinna, L. Battistini, F. Zanardi, D. Acquotti and G. Casiraghi, *Eur. J. Org. Chem.*, 1999, 1395.
68 A. Peer and A. Vasella, *Helv. Chim. Acta*, 1999, **82**, 1044.
69 A. Endo, T. Kann and T. Fukuyama, *Synlett*, 1999, 1103.
70 A. Geyer, D. Bockelmann, K. Weissenbach and H. Fischer, *Tetrahedron Lett.*, 1999, **40**, 477.

71 D. Marek, A. Wadonachi and D. Beaupère, *Synthesis*, 1999, 839.
72 J.-P. Praly, G. Hetzer and M. Steng, *J. Carbohydr. Chem.*, 1999, **18**, 833.
73 T. Kolter and K. Sandhoff, *Angew. Chem., Int. Ed. Engl.*, 1999, **38**, 1533.
74 J. B. Rodriguez, *Tetrahedron*, 1999, **55**, 2157.
75 M. Tingoli, B. Manunzi and F. Santacroce, *Tetrahedron Lett.*, 1999, **40**, 9329.
76 A. K. Ghosh and Y. Wang, *Tetrahedron*, 1999, **55**, 13369.
77 M. Adinolfi, G. Baroni and A. Iadonisi, *Synlett.*, 1999, 65.
78 W. Lu, G. Zheng and J. Cai, *Tetrahedron*, 1999, **55**, 4649.
79 W. Lu, G. Zheng, D. Gao and J. Cai, *Tetrahedron*, 1999, **55**, 7157.
80 M. Okada, H. Miyagawa and T. Ueno, *Biosci. Biotech. Biochem.*, 1999, **63**, 1253.
81 R. J. Mulder, C. M. Shafer and T. F. Molinski, *J. Org. Chem.*, 1999, **64**, 4995.
82 A. H. Haines and A. J. Lamb, *Carbohydr. Res.*, 1999, **321**, 197.
83 K. Murakami, E. J. Molitor and H.-w. Liu, *J. Org. Chem.*, 1999, **64**, 648.
84 J. Xia and Y.-Z. Hui, *Chem. Pharm. Bull.*, 1999, **47**, 1659.
85 T. Murakami and K. Taguchi, *Tetrahedron*, 1999, **55**, 989.
86 K. Otako and K. Mori, *Eur. J. Org. Chem.*, 1999, 1795.
87 U. Chiacchio, A. Corsaro, G. Gumina, A. Rescifina, D. Iannazzo, A. Pipern, G. Romeo and R. Romeo, *J. Org. Chem.*, 1999, **64**, 9321.
88 M. Helliwell, I. M. Phillips, R. P. Pritchard and R. J. Stoodley, *Tetrahedron Lett.*, 1999, **40**, 8651.
89 R. Nouguier, V. Mignon and J.-L. Gras, *J. Org. Chem.*, 1999, **64**, 1412.
90 S. A. Kozmin and V. H. Rawal, *J. Am. Chem. Soc.*, 1999, **121**, 9562.
91 R. J. Linderman, S. Binet and S. R. Petrich, *J. Org. Chem.*, 1999, **64**, 336.
92 T.-P. Loh, J.-R. Zhou, X.-R. Li and K.-Y. Sim, *Tetrahedron Lett.*, 1999, **40**, 7847.
93 R. J. Stoodley, *Carbohydr. Poly.*, 1998, **37**, 249 (*Chem. Abstr.*, 1999, **130**, 168 578)
94 J. Anaya, S. D. Gero, M. Grande, J. I. M. Hernando and N. M. Laso, *Bioorg. Med. Chem.*, 1999, **7**, 837.
95 G. Huang and R. I. Hollingsworth, *Tetrahedron Lett.*, 1999, **40**, 581.
96 I. Fernández, N. Khiar, A. Roca, A. Benabra and A. Alcudia, *Tetrahedron Lett.*, 1999, **40**, 2029.
97 T. Ishi-i, K. Nakashima, S. Shinkai and A. Ikeda, *J. Org. Chem.*, 1999, **64**, 984.
98 W. Adam, C. R. Saha-Möller,and C.-G. Zhao, *J. Org. Chem.*, 1999, **64**, 7492.
99 Z.-X. Wang, G.-A. Cao and Y. Shi, *J. Org. Chem.*, 1999, **64**, 7646.
100 J. D. Warren and Y. Shi, *J. Org. Chem.*, 1999, **64**, 7675.
101 W. Li, Z. Zhang, D. Xiao and X. Zhang, *Tetrahedron Lett.*, 1999, **40**, 6701.
102 Y. Chen, X. Li, S.-k. Tong, M. C. K. Choi and A. S. C. Chan, *Tetrahedron Lett.*, 1999, **40**, 957.
103 T. V. RajanBabu, B. Radetich, K. K. You, T. A. Ayers, A. L. Casalnuovo and J. C. Calabrese, *J. Org. Chem.*, 1999, **64**, 3429.
104 K. Yonehara, T. Hashizume, K. Mori, K. Ohe and S. Uemura, *J. Org. Chem.*, 1999, **64**, 9374.
105 K. Yonehara, K. Ohe and S. Uemura, *J. Org. Chem.*, 1999, **64**, 9381.
106 K. Yonehara, T. Hashizume, K. Mori, K. Ohe and S. Uemura, *J. Org. Chem.*, 1999, **64**, 5593.

Author Index

In this index the number in parenthesis is the Chapter number of the citation and this is followed by the reference number or numbers of the relevant citations within that Chapter.

WITHDRAWN